Agricultural Mechanics and Technology Systems

by

J.P. Hancock

Don W. Edgar

Michael L. Pate

Lori A. Dyer

W. Brian Hoover

Publisher

The Goodheart-Willcox Company, Inc.

Tinley Park, IL

www.g-w.com

Copyright © 2017
by
The Goodheart-Willcox Company, Inc.

All rights reserved. No part of this work may be reproduced, stored, or transmitted in any form or by any electronic or mechanical means, including information storage and retrieval systems, without the prior written permission of The Goodheart-Willcox Company, Inc.

Manufactured in the United States of America.

ISBN 978-1-63126-255-5

3 4 5 6 7 8 9 – 17 – 21 20 19 18 17

The Goodheart-Willcox Company, Inc. Brand Disclaimer: Brand names, company names, and illustrations for products and services included in this text are provided for educational purposes only and do not represent or imply endorsement or recommendation by the author or the publisher.

The Goodheart-Willcox Company, Inc. Safety Notice: The reader is expressly advised to carefully read, understand, and apply all safety precautions and warnings described in this book or that might also be indicated in undertaking the activities and exercises described herein to minimize risk of personal injury or injury to others. Common sense and good judgment should also be exercised and applied to help avoid all potential hazards. The reader should always refer to the appropriate manufacturer's technical information, directions, and recommendations; then proceed with care to follow specific equipment operating instructions. The reader should understand these notices and cautions are not exhaustive.

The publisher makes no warranty or representation whatsoever, either expressed or implied, including but not limited to equipment, procedures, and applications described or referred to herein, their quality, performance, merchantability, or fitness for a particular purpose. The publisher assumes no responsibility for any changes, errors, or omissions in this book. The publisher specifically disclaims any liability whatsoever, including any direct, indirect, incidental, consequential, special, or exemplary damages resulting, in whole or in part, from the reader's use or reliance upon the information, instructions, procedures, warnings, cautions, applications, or other matter contained in this book. The publisher assumes no responsibility for the activities of the reader.

The Goodheart-Willcox Company, Inc. Internet Disclaimer: The Internet resources and listings in this Goodheart-Willcox Publisher product are provided solely as a convenience to you. These resources and listings were reviewed at the time of publication to provide you with accurate, safe, and appropriate information. Goodheart-Willcox Publisher has no control over the referenced websites and, due to the dynamic nature of the Internet, is not responsible or liable for the content, products, or performance of links to other websites or resources. Goodheart-Willcox Publisher makes no representation, either expressed or implied, regarding the content of these websites, and such references do not constitute an endorsement or recommendation of the information or content presented. It is your responsibility to take all protective measures to guard against inappropriate content, viruses, or other destructive elements.

Cover images: ©iStock.com/Paul Hart (barn); nfsphoto/Shutterstock.com (cattle); Grandpa/Shutterstock.com (welder); Chukov/Shutterstock.com (tractor); muratart/Shutterstock.com (irrigation); Sunny Forest/Shutterstock.com (corn field)

Back cover image: Elena Elisseeva

Preface

In creating a modern and completely original curriculum resource for courses on agricultural technology systems, the authors of *Agricultural Mechanics and Technology Systems* set several goals for themselves. First, the text's coverage had to be basic enough for learners who have little or no prior experience with mechanical systems and yet detailed enough to challenge those with an aptitude for mechanics and technology. Second, coverage had to be broad enough to cover all the major skilled trades—including building construction, electricity, welding, and power systems—in true-to-life agricultural contexts. Third, it had to provide meaningful presentations of theory and enough depth to promote inquiry-based learning in the classroom and the lab. And fourth, the text had to be concisely written in order to balance limited classroom and lab hours with the demands of introducing multiple disciplines.

An additional imperative was to place the study of agricultural technology within the framework of the AFNR Career Cluster Content Standards, and not only attract students to the Power, Structural, and Technical Systems Career Pathway, but to also take subsequent courses.

In creating the text, the author team drew on their experience as both educators and subject matter experts, which has allowed for pedagogical soundness and a high degree of technical accuracy. They were guided by a diverse and geographically balanced panel of expert reviewers, who critiqued every chapter and made many important contributions to the text.

Working safely is an overarching theme of the text, with a dedicated chapter on safety (Chapter 5), technology-specific safety coverage in nearly every chapter, and extensive safety notes throughout the text. The repeated exposure to safety issues is intended to develop a safety-first mindset and drive home the point that safe work habits are required in every area of agricultural technology systems.

Developing an appreciation for the rich career options available in agricultural technology systems is another theme of the text. Along with an early chapter on careers (Chapter 2), the text offers practical information and advice to those participating in career development events (CDEs).

In support of the modern agricultural education model, the text promotes a deep understanding of supervised agricultural experiences (SAEs) and the many ways they can be designed to enrich the study of agricultural technology systems. Chapter 3 provides an overview of experiential learning, and numerous *SAE Connections* and *AgEd Connections* throughout the text extend and apply it. Active learning is further developed by *Hands-On Ag* features and by the chapter-ending *STEM and Academic Activities*.

In recognition of the importance of cross-curricular tie-ins, reading and writing skills are developed by the careful definition and use of technical terms in *Words to Know*, by *Before You Read* guidance provided at the start of every chapter, and by vocabulary-building exercises at the end of every chapter. The myriad applications of agricultural technology systems to science, technology, engineering, and mathematics are explored in *STEM Connections* throughout the text and reinforced in *STEM and Academic Activities*.

Special care has been taken to address the needs of visual learners. To highlight safe, correct, and current practices, four-color photographs, many of which were created specifically for this text, and line art appear on nearly every page of the text.

About the Authors

J. P. Hancock

Mr. Hancock is an Assistant Lecturer in the College of Agriculture and Life Sciences at Texas A&M University, where he is responsible for teaching classes on how to teach agricultural mechanics and safety in high school agriculture programs. In addition to teaching, Mr. Hancock is responsible for supervising student teachers in the field. Prior to joining Texas A&M, Mr. Hancock taught high school agriculture science for nearly 30 years. He holds bachelor of science and master of science degrees in Agricultural Education from Texas A&M University.

Don W. Edgar, PhD

Dr. Edgar is an Associate Professor of teacher education at University of Arkansas, where he teaches methods of teaching in agricultural education, methods of laboratory instruction, social data analysis, small gas engines, and turf machines. Dr. Edgar also teaches introduction to surveying and has taught history and philosophy of agricultural and extension education. Dr. Edgar holds bachelor of science and master of science degrees in Agricultural Education from Tarleton State University, and received his PhD from Texas A&M University, also focusing on Agricultural Education. His professional interests include methods of teaching, learner-centered instructional design, and delivery strategies. As a former secondary instructor in Texas for 14 years, Dr. Edgar's interest in young adult learning focuses his research interests to understand how students learn most effectively, especially in the laboratory and classroom settings. These interests allow his research focus to analyze the teaching and learning processes involved in secondary educational settings.

Lori A. Dyer

Ms. Dyer taught high school agriculture career technology for 30 years and continues as a Master Teacher with CASE Institutes. She is certified in three Curriculum for Agriculture Science Education (CASE) courses, and assisted with developing and field testing the CASE Agriculture Power and Technology curriculum. Ms. Dyer holds a bachelor of science degree from The Ohio State University in Agriculture, and a Masters of Instruction in Economics from the University of Delaware. She has contributed to the Agricultural Education Magazine and has developed two state-adopted Career Education curricula. Ms. Dyer actively promotes workforce development education in her local schools and is a regular contributor to online teacher resource websites. She operates a beef cow-calf herd and row crop farms with her family.

Michael L. Pate, PhD

Dr. Pate is an Assistant Professor of Agricultural Systems Technology at Utah State University, where he coordinates the agricultural systems technology undergraduate program. His agricultural work experience includes beef cattle and poultry production. Dr. Pate received his undergraduate and master's degrees in Agricultural Education from the University of Arkansas, and completed his PhD program in Agricultural Education at Iowa State University. Prior to pursuing graduate studies, Dr. Pate taught agriculture and life sciences at the high school level. His teaching philosophy is grounded in the use of extensive laboratory practicums to emphasize problem-based application of agricultural engineering principles, the function and use of agricultural technology, and the integration of technology management concepts for the food, fiber, and natural resources industry. Dr. Pate serves as a skills event coordinator for the Utah FFA Agricultural Technology and Mechanical Systems Career Development Event. His research focuses on the safety and health of individuals working in production agriculture. Dr. Pate has authored over 15 refereed journal articles. He is the current chair for the NCERA-197 Agricultural Safety and Health Research and Extension Committee and a board member for the International Society for Agricultural Safety and Health (ISASH).

W. Brian Hoover, PhD

Dr. Hoover is an Associate Professor at Murray State University in Murray, Kentucky, where he teaches undergraduate and graduate classes in the agricultural systems technology program. Dr. Hoover earned his doctoral degree from Purdue University and his bachelor of science and master of science degrees from Murray State University in the area of Agricultural Systems Technology. He has professional contributions to the area of agriculture systems over the last 14 years through teaching several courses and research in agricultural systems technology and agricultural safety. A wide variety of learning experiences on his family's farm prepared Dr. Hoover for a career in agriculture and continues to provide him a connection to crop production, livestock production, and agricultural machinery.

Acknowledgments

The authors and publisher wish to acknowledge the special contributions of Dr. Timothy L. Andera of South Dakota State University, who wrote Chapter 15: *Masonry*; Christopher Hart of Chatham Central High School, Bear Creek, NC, who wrote Chapter 25: *Farm and Landscape Irrigation*; Kevin Jump, Ag Mechanics Area Teacher for Georgia Agriculture Education, who contributed to Chapter 4: *Trends and Emerging Technology*, and wrote many of the SAE and AgEd features; and Rick Miles, freelance writer and instructor at Columbia Montour Area Vocational Technical School, who contributed to the development of the woodworking chapters. In addition, the authors and publisher would like to thank the following companies, organizations, and individuals for their contribution of resource material, images, or other support in the development of *Agricultural Mechanics and Technology Systems*.

Appleton Electric Co.

Brick Industry Association

David Kosling/USDA

Don Blazek

Doug Lautenschlager

Marshalltown Company

OSHA

Ritchie Engineering Co., Inc.—YELLOW JACKET Products Division

Roger Hanagriff, The AET Record Book

SkillsUSA

Stephen Ausmus, USDA Natural Resources Conservation Service

Tim McCabe, USDA Natural Resources Conservation Service

US Department of Agriculture, National Agricultural Statistics Service

Wes Hancock, Caldwell High School

Reviewers

The authors and publisher wish to thank the following industry and teaching professionals for their valuable input into the development of *Agricultural Mechanics and Technology Systems*.

James Corbett
Lowndes High School
Valdosta, Georgia

Mitchell Davis
North Hall High School
Gainesville, Georgia

William Doss
Van High School
Van, Texas

Jay Gibson
St. Clair County High School
Odenville, Alabama

Kevin Jump
Central Region Agricultural Education
Fort Valley, Georgia

Carole Lindsey
Norco High School
Norco, California

Andrew McCubbins
Iowa State University
Ames, Iowa

Tina Miner-James
Walton Central School
Walton, New York

Mary Ellen Pautler
Greenback School
Greenback, Tennessee

Jeff Reed
Hillsboro High School
Hillsboro, Texas

Mike Rogers
Siloam Springs High School
Siloam Springs, Arkansas

Kasee Smith
Texas A&M University
College Station, Texas

Owen Thomason
Franklin County High School
Carnesville, Georgia

Trent Wells
Fayette County High School
Fayette, Alabama

G-W Integrated Learning Solution

Together, We Build Careers

At Goodheart-Willcox, we take our mission seriously. Since 1921, G-W has been serving the career and technical education (CTE) community. Our employee-owners are driven to deliver exceptional learning solutions to CTE students to help prepare them for careers. Our authors and subject matter experts have years of experience in the classroom and industry. We combine their wisdom with our expertise to create content and tools to help students achieve success. Our products start with theory and applied content based upon a strong foundation of accepted standards and curriculum. To that base, we add student-focused learning features and tools designed to help students make connections between knowledge and skills. G-W recognizes the crucial role instructors play in preparing students for careers. We support educators' efforts by providing time-saving tools that help them plan, present, assess, and engage students with traditional and digital activities and assets. We provide an entire program of learning in a variety of print, digital, and online formats, including economic bundles, allowing educators to select the right mix for their classrooms.

Student-Focused Curated Content

Goodheart-Willcox believes that student-focused content should be built from standards and/or accepted curriculum coverage. AFNR Career Cluster Content Standards, Precision Exams Standards, and selected state standards were used as a foundation in this text. *Agricultural Mechanics and Technology Systems* also uses a building-block approach with attention devoted to a logical teaching progression that helps students build upon their learning. Agricultural examples and applications are consistently used throughout the text. We call on industry experts and teachers from across the country to review and comment on our content, presentation, and pedagogy. Finally, in our refinement of curated content, our editors are immersed in content checking, securing and sometimes creating figures that convey key information, and revising language and pedagogy.

Precision Exams Certification

Goodheart-Willcox is pleased to partner with Precision Exams by correlating *Agricultural Mechanics and Technology Systems* to their Agricultural Systems Technology I Exam Standards. Precision Exams Standards and Career Skill Exams™ were created in concert with industry and subject matter experts to match real-world job skills and marketplace demands. Students that pass the exam and performance portion of the exam can earn a Career Skills Certification™. Precision Exams provides:

- Access to over 150 Career Skills Exams™ with pretest exams and posttest exams for all 16 Career Clusters.
- Instant reporting suite access to measure student academic growth.
- Easy-to-use, 100% online exam delivery system.

To see how *Agricultural Mechanics and Technology Systems* correlates to the Precision Exams Standards, please visit http://www.g-w.com/agricultural-mechanics-technology-systems-2017 and click on the Correlations tab. For more information on Precision Exams, including a complete listing of their 150+ Career Skills Exams™ and Certificates, please visit www.precisionexams.com.

I earned a CAREER SKILLS™ Certificate in AGRICULTURAL SYSTEMS TECHNOLOGY. You can earn one, too!

Ask your instructor how you can earn a CAREER SKILLS™ Certificate for your résumé.

800.470.1215 PRECISION EXAMS precisionexams.com

Sean Lock Photography/Shutterstock.com

Features of the Textbook

Features are student-focused learning tools designed to help you get the most out of your studies. This visual guide highlights the features designed for the textbook.

Chapter Outcomes clearly identify the knowledge and skills to be obtained when the chapter is completed.

Words to Know lists the key terms to be learned in the chapter.

Before You Read literacy integration activities at the beginning of each lesson encourage development of confidence and skill in literacy and learning.

G-W Learning Companion Website Activity Icon identifies related content available on the G-W Learning companion website.

Hands-On Agriculture features are highlighted throughout the textbook to provide clear instructions for hands-on service activities. You can refer back to these procedures easily.

Thinking Green notes highlight key items related to sustainability, energy efficiency, and environmental issues.

Career Connections introduce careers in different agricultural areas and can help provide a path for career success.

SAE Connections make real-life connections to a variety of interesting SAE opportunities.

AgEd Connection features introduce you to the exciting world of leadership and personal development opportunities including career development events in a variety of areas.

Safety Notes alert you to potentially dangerous materials and practices.

STEM Connections integrate all four components of STEM education as well as the social sciences and language arts.

Illustrations have been designed to clearly and simply communicate the specific topic.

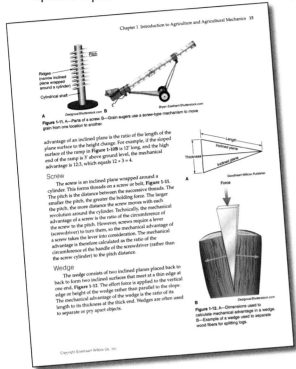

Summary feature provides an additional review tool for you and reinforces key learning objectives.

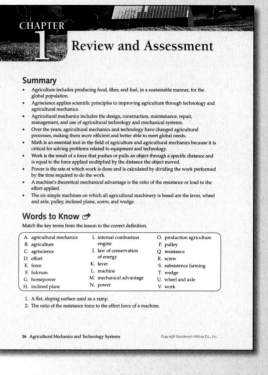

Words to Know matching activities reinforce vocabulary development and retention. All key terms are included in the text glossary and are connected to numerous online review activities.

STEM and Academic Activities extend your learning and help you analyze and apply knowledge.

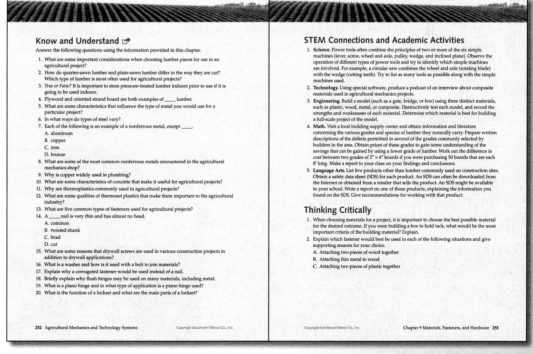

Know and Understand questions allow you to demonstrate knowledge, identification, and comprehension of chapter material.

Thinking Critically questions develop higher-order thinking, problem solving, personal, and workplace skills.

xi

Student Resources

Textbook

The *Agricultural Mechanics and Technology Systems* textbook provides an exciting, full-color, and highly illustrated learning resource. The textbook is available in print or online versions.

G-W Learning Companion Website

The G-W Learning companion website is a study reference that contains identification and matching activities, animations and videos, review questions, vocabulary exercises, and more! Accessible from any digital device, the G-W Learning companion website complements the textbook and is available to the student at no charge.

Lab Workbook

The student lab workbook provides minds-on practice with questions and activities. Each chapter corresponds to the text and reinforces key concepts and applied knowledge. It also provides hands-on practice to be completed in the school lab setting under the guidance of an instructor.

Online Learning Suite

Available as a classroom subscription, the Online Learning Suite provides the foundation of instruction and learning for digital and blended classrooms. An easy-to-manage shared classroom subscription makes it a hassle-free solution for both students and instructors. An online student text and workbook, along with rich supplemental content, brings digital learning to the classroom. All instructional materials are found on a convenient online bookshelf and are accessible at home, at school, or on the go.

Online Learning Suite/ Student Textbook Bundle

Looking for a blended solution? Goodheart-Willcox offers the Online Learning Suite bundled with the printed text in one easy-to-access package. Students have the flexibility to use the print version, the Online Learning Suite, or a combination of both components to meet their individual learning style. The convenient packaging makes managing and accessing content easy and efficient.

Instructor's Resources

Instructor resources provide information and tools to support teaching, grading, and planning; class presentations; and assessment.

Instructor's Presentations for PowerPoint®

Help teach and visually reinforce key concepts with prepared lectures. These presentations are designed to allow for customization to meet daily teaching needs. They include objectives, outlines, and images from the textbook.

ExamView® Assessment Suite

Quickly and easily prepare, print, and administer tests with the ExamView® Assessment Suite. With hundreds of questions in the test bank corresponding to each chapter, you can choose which questions to include in each test, create multiple versions of a single test, and automatically generate answer keys. Existing questions may be modified and new questions may be added. You can prepare pretests, formative assessments, and summative assessments easily with the ExamView® Assessment Suite.

Instructor's Resource CD

One resource provides instructors with time-saving preparation tools such as answer keys, lesson plans, correlations to Precision Exams Standards, and other teaching aids.

Online Instructor Resources

Online Instructor Resources provide all the support needed to make preparation and classroom instruction easier than ever. Available in one accessible location, support materials include Answer Keys, Lesson Plans, Instructor Presentations for PowerPoint®, ExamView® Assessment Suite, and more! Online Instructor Resources are available as a subscription and can be accessed at school or at home.

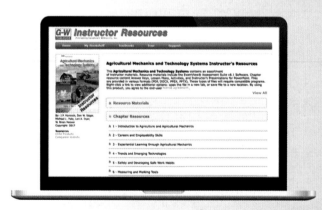

Brief Contents

1. Introduction to Agriculture and Agricultural Mechanics 2
2. Careers and Employability Skills 20
3. Experiential Learning through Agricultural Mechanics 42
4. Trends and Emerging Technologies 64
5. Safety and Developing Safe Work Habits 86
6. Measuring and Marking Tools ... 118
7. Hand Tools 140
8. Power Tools 178
9. Materials, Fasteners, and Hardware 224
10. Project Planning and Design ... 254
11. Fundamentals of Woodworking 278
12. Designing, Planning, and Constructing Woodworking Projects 306
13. Surveying 324
14. Concrete Foundations and Flatwork 348
15. Masonry 372
16. Framing Structures 404
17. Finishing Structures 434
18. Specialized Agricultural Structures 468
19. Fencing and Livestock Structures 492
20. Electric Theory 518
21. Wiring AC Circuits 542
22. Wiring DC Circuits 568
23. Electric Motors and Controls ... 598
24. Plumbing Design and Installation 624
25. Farm and Landscape Irrigation 646
26. Environmental Controls 676
27. Fundamentals of Metalworking 704
28. Oxyfuel Welding, Cutting, and Brazing 742
29. Shielded Metal Arc Welding ... 776
30. Gas Metal and Flux Cored Arc Welding 818
31. Gas Tungsten Arc Welding 856
32. Plasma Cutting 878
33. Internal Combustion Engines 896
34. Small Engine Performance, Maintenance, and Troubleshooting 918
35. Drive Trains and Power Systems 938
36. Machinery Maintenance and Management 954
37. Hydraulic and Pneumatic Power 976

Reference Section 997
Glossary 1023
Index 1050

Contents

Chapter 1
Introduction to Agriculture and Agricultural Mechanics...2
Agriscience 4
Revolutionizing Agriculture 5
Mathematics for Agricultural Mechanics................. 10
Physical Science and Physics....... 11

Chapter 2
Careers and Employability Skills 20
Careers in the Agricultural Mechanics................. 22
Job Skills..................... 23
Securing Employment............ 27
Finding a Job in Agricultural Mechanics.................. 34
Maintaining Employment.......... 37

Chapter 3
Experiential Learning through Agricultural Mechanics 42
Experiential Learning 44
Classroom and Laboratory Instruction................. 45
National FFA Organization 45
Supervised Agricultural Experiences (SAEs)..................... 48
Developing an SAE Plan........... 54
SAE Recordkeeping 58

Chapter 4
Trends and Emerging Technologies 64
Sustainable Energy............... 66
Environmental Systems........... 69
Precision Agriculture 73
Biotechnology 79

Chapter 5
Safety and Developing Safe Work Habits 86
Hazards in Agricultural Mechanics and Technology.............. 88
Safety Regulations 90
Personal Protective Equipment 93
General Safety 100
Chemical Safety 106
Electrical Safety 108
Weather Hazards 109
Insect-Borne Illness 110
Outdoor Equipment and Machinery................. 110
First Aid 113

xv

Chapter 8
Power Tools 178
Classification of Power Tools. 180
Power Tool Safety Rules 184
Drilling and Driving Tools. 186
Hydraulic Jacks and Presses 194
Cutting Tools 196
Grinders and Sanders. 209
Selecting and Purchasing
    Power Tools. 217
Care and Storage of
    Power Tools. 219

Chapter 6
Measuring and Marking Tools 118
Measuring Systems. 120
Measuring Tools 121
Marking Tools 131
Purchasing Measuring Tools 134
Care and Storage of Measuring
    Tools 135

Chapter 9
Materials, Fasteners, and Hardware. 224
Wood. 226
Metals 230
Composite Materials 235
Fasteners 237
Adhesives 244
Hardware. 246

Chapter 7
Hand Tools 140
Tool Evolution 142
General Tool Safety. 143
Naming and Classification of
    Hand Tools 143
Cutting Tools 144
Gripping Tools. 152
Impact Tools 158
Leverage Tools 163
Tools for Applying Torque 164
Extractors 169
Cleaning Tools. 170
Purchasing Hand Tools 171
Care and Storage of Hand Tools 172

Chapter 10
Project Planning and Design 254
Planning Process 256
Budgeting 258
Design Considerations 261
Sketching 263
Technical Drawings. 265
Professional Planners 272
Other Sources of Plans. 272
Construction and Material
    Standards 273
Preparing a Bid Package. 273

Chapter 11
Fundamentals of Woodworking 278
- Wood Types and Characteristics 280
- Wood Selection 280
- Cutting Wood 285
- Shaping and Smoothing Wood 295
- Drilling Wood 301

Chapter 12
Designing, Planning, and Constructing Woodworking Projects 306
- Project Planning 308
- Selecting Wood 310
- Layout 311
- Cutting the Project Parts 313
- Assembling the Project 313
- Finishing the Project 316

Chapter 13
Surveying 324
- What is Surveying? 326
- Selecting Appropriate Building Sites . . . 326
- Site Layout and Design 332
- Basic Surveying Principles 334
- Surveying Application 342

Chapter 14
Concrete Foundations and Flatwork 348
- Concrete Applications and Characteristics 350
- Forming Concrete 350
- Preparation and Materials 353
- Compression Strength and Tensile Strength 357
- Testing and Finishing Concrete 359
- Foundation Systems 362
- Concrete Forms 364
- Flatwork Placement and Curing 366

Chapter 15
Masonry 372
- Masonry Tools and Equipment 374
- Masonry Safety 378
- Masonry Building Products 379
- Bonds and Patterns 381
- Mortar 385
- Planning the Project 389
- Fundamentals of Laying Masonry Units 390

Chapter 16
Framing Structures 404
- Building Design 406
- Structure Types 409
- Framing Types 411
- Floor Framing 414
- Wall Framing 419
- Roof and Ceiling Framing 426

Chapter 17
Finishing Structures 434
- Covering and Enclosing Structures... 436
- Roofs...................... 439
- Walls...................... 445
- Windows................... 448
- Doors...................... 451
- Trim....................... 452
- Energy Efficiency............. 452

Chapter 18
Specialized Agricultural Structures 468
- Specialized Agricultural Structures... 470
- Greenhouse Structures.......... 470
- Hydroponic Systems............ 481
- Aquaculture Systems........... 482
- Aquaponics................... 485
- Prefabricated Structures........ 485
- Earthen Structures............. 486

Chapter 19
Fencing and Livestock Structures 492
- Fencing Applications............ 494
- Types of Fencing............... 495
- Planning Fence Construction...... 497
- Materials..................... 500
- Installation................... 507
- Tools and Equipment........... 511
- Inspection and Maintenance..... 512

Chapter 20
Electric Theory 518
- Conductors and Insulators....... 520
- Electric Circuits............... 521
- Protecting the Circuit.......... 527
- Alternating Current............ 530
- Direct Current................ 532
- Sources of Electric Power....... 533
- Distribution of Electric Power.... 534
- Safety around Electricity........ 535

Chapter 21
Wiring AC Circuits 542
- National Electric Code and Other Regulations............ 544
- Service Entrance............... 544
- Types of Circuits.............. 548
- Planning AC Circuits........... 549
- Electrical Devices.............. 555
- Connecting Wires.............. 559

Chapter 22
Wiring DC Circuits 568
- DC Circuits................... 570
- Sources of DC Power........... 570
- Electrical Loads................ 573
- Switches..................... 576
- Making Wire Connections....... 580
- Wiring an Agricultural Trailer.... 586
- Troubleshooting DC Circuits..... 593

Chapter 23
Electric Motors and Controls............598
- Advantages of Electric Motors......600
- Electric Motors and Magnetism.....601
- Electric Motor Components........603
- Types of Electric Motors..........604
- Motor Nameplates.................609
- Selecting Electric Motors..........610
- Installing Electric Motors..........611
- Electric Motor Maintenance........617
- Safely Using Electric Motors.......620

Chapter 24
Plumbing Design and Installation..............624
- Plumbing Codes..................626
- Plumbing Systems................626
- Plumbing Tools..................630
- Plumbing Pipe....................633
- Pipe Fittings and Valves..........640

Chapter 25
Farm and Landscape Irrigation................646
- Irrigation History................648
- Water Sources....................650
- Water Conservation...............652
- Irrigation Equipment..............654
- Irrigation Systems................656
- Irrigation System Design..........661
- Irrigation System Management.....665
- Irrigation System Troubleshooting.............668

Chapter 26
Environmental Controls...676
- Thermodynamics..................679
- Housing Livestock................684
- Grain Storage....................690
- Perishables and Other Crop Storage......................693
- Refrigeration....................694
- Evaporative Cooling..............696
- Ventilation and Exhaust Systems...697

Chapter 27
Fundamentals of Metalworking............704
- Metals Commonly Used in Agriculture....................706
- Designing and Planning Metal Projects...............709
- Marking Metal...................712
- Cutting Metal....................713
- Shaping Metal...................721
- Drilling Metal...................727
- Assembling Metal................730
- Heat-Treating Metal.............737

Chapter 28
Oxyfuel Welding, Cutting, and Brazing 742
- Oxyfuel Welding and Cutting 744
- Chemistry of Burning Gases 745
- Oxyfuel Equipment 746
- Safety 756
- Setting Up the Equipment 759
- Lighting the Torch 760
- Types of Flame 761
- Torch Shut-Down Procedure 762
- Using a Cutting Torch 763
- Welding with Oxyfuel 766
- Braze Welding and Brazing 768
- Using a Multiflame Heating Tip 770

Chapter 29
Shielded Metal Arc Welding 776
- Development of Electric Arc Welding . . 778
- Equipment 779
- Electrodes 782
- Joint Types 785
- Welding Positions 787
- Surface Preparation 789
- Controlling Welding Parameters 790
- Anatomy of a Weld—Weld Quality . . . 795
- Arc Welding Safety 800
- Practice Welding 802
- Controlling Distortion 811
- Welding Cast Iron 812
- Welding Certification 813

Chapter 30
Gas Metal and Flux Cored Arc Welding 818
- Development of Gas Metal Arc Welding 820
- Advantages and Disadvantages of GMAW 821
- Equipment 822
- Consumables 827
- Methods of Metal Transfer 833
- Flux Cored Arc Welding (FCAW) 835
- Welding Using GMAW 837
- Safety 843
- Practice Welding 846

Chapter 31
Gas Tungsten Arc Welding 856
- Development of GTAW 858
- Equipment for GTAW 859
- Consumables for GTAW 862
- GTAW Safety 865
- Welding with GTAW 866
- GTAW Practice 870

Chapter 32
Plasma Cutting 878
- Plasma Cutting 880
- Plasma Cutting Safety 883
- Using a Plasma Torch 885
- CNC Plasma Cutting 887
- CNC Plasma Cutting Practice 892

Chapter 33
Internal Combustion Engines 896
- Internal Combustion Engines 898
- Engine Classifications 900
- Engine Components 906
- Related Systems 910
- Safety 913

Chapter 34
Small Engine Performance, Maintenance, and Troubleshooting 918
- Measuring Engine Performance 920
- Engine Selection 924
- Engine Maintenance 924
- Service Information 929
- Engine Troubleshooting 930

Chapter 35
Drive Trains and Power Systems 938
- Power Trains 940
- Power Systems 947

Chapter 36
Machinery Maintenance and Management 954
- Manuals, Schedules, and Service Records 956
- Systems Inspection and Maintenance 957
- Machinery Management 966

Chapter 37
Hydraulic and Pneumatic Power 976
- Fluid Power Advantages and Disadvantages 978
- Fluid Power Fundamentals 978
- Hydraulic Components 981
- Pneumatic Power Fundamentals 989
- Pneumatic Components 991
- Pneumatic System Maintenance 993

Reference Section 997
Glossary 1023
Index 1050

CHAPTER 1
Introduction to Agriculture and Agricultural Mechanics

Chapter Outcomes

After studying this chapter, you will be able to:
- Explain the roles of agriscience and agricultural mechanics in production agriculture.
- Discuss the history and development of agricultural mechanics.
- Explain why a knowledge of mathematics is essential for agricultural professionals.
- Discuss and calculate theoretical mechanical advantage.
- Name and describe simple machines used in agricultural technology.
- Apply basic physics and physical science formulas to agricultural mechanics problems.

Words to Know

agricultural mechanics	horsepower	lever	resistance
agriculture	inclined plane	machine	screw
agriscience	internal combustion engine	mechanical advantage	subsistence farming
effort		power	wedge
force	law of conservation of energy	production agriculture	wheel and axle
fulcrum		pulley	work

Before You Read

Skim the chapter by reading the first sentence of each paragraph. Use this information to create an outline for the chapter before you read it.

While studying this chapter, look for the activity icon to:

- **Practice** vocabulary terms with e-flash cards and matching activities.
- **Expand** learning with interactive activities.
- **Reinforce** what you learn by completing the end-of-chapter questions.

www.g-wlearning.com/agriculture/

The science and practice of sustainably producing food, fiber, and fuel for a global population is known as *agriculture*. Advances in technology have changed the way we produce these things. Tools and equipment increase output while reducing effort. Science and technology are central to the support sectors within agriculture, including agribusiness, agriscience, agricultural communications, food processing, natural resources management, and production agriculture. All of these support sectors of agriculture contribute to the success and sustainability we see in agricultural production.

Agriscience

Agriculture remains an essential industry because it helps meet the basic needs of people all over the world, particularly for food and the fiber used to make clothing. Historically, production agriculture was often limited to *subsistence farming*. This means that only enough food was produced to feed the farmer and the immediate family. Subsistence remains the focus of the industry, but it is now possible to produce the raw materials needed for everyday living at such a large scale that excess products can be sold for profit.

Over the last few decades, agricultural production has increased in the United States even though the number of farms has decreased. Current estimates indicate that less than 2% of the US population is engaged in *production agriculture*—agriculture for the specific purpose of producing a marketable product. See **Figure 1-1**. Farms and ranches have become so efficient because of advances in technology and machinery that one farmer can supply food for an estimated 144 people. This was made possible partly by the development of agriscience. *Agriscience* is the field of science devoted to improving agriculture. It involves the application of scientific discoveries to improve the process of growing and reproducing plants and animals.

A — View Apart/Shutterstock.com

B — Stockr/Shutterstock.com

Figure 1-1. A—Subsistence farming produces enough food for a few people. B—Production farming is a large-scale procedure that produces enough to be sold for profit.

Agricultural Mechanics

The field of study called *agricultural mechanics* is the design, construction, maintenance, repair, management, and use of agricultural technology and mechanical systems. Mechanical applications are integrated across all disciplines of agriculture:

- Plant breeding and genetics—the use of row crop planters and tractors to prepare field plots to evaluate crop variety productivity.

- Animal breeding and genetics—the construction and maintenance of barns and corrals for livestock.
- Food science—managing refrigeration units for storage and preservation of foods or crops.
- Natural resources management—chainsaws and other machines for maintenance of forests and other natural resources.
- Agribusiness—preparing reports using computer software.

Why Study Agricultural Mechanics?

Agricultural mechanics is a fundamental component of agriculture. If you are using a tool to accomplish a task or solve a problem in agriculture, chances are you are using agricultural mechanics skills. Learning the science behind agricultural mechanics and technology will help you understand and complete various agricultural jobs.

Revolutionizing Agriculture

Everyone needs agriculture. After all, agriculture feeds and helps clothe the ever-growing world population. Estimates of world population are projected to reach 9.6 billion by 2050. Agriculture has met those needs throughout history in various ways, from farming for personal subsistence to farming for cash crop production. Today, most people engaged in agriculture do not raise products solely for their personal consumption. Instead, they are involved in production agriculture.

In 2012, US farms sold nearly $395 billion in agricultural products to both local and global commercial and retail markets. According to the USDA National Agricultural Statistics Service, less than half of one percent of farms have agriculture sales of more than $5 million, but these farms produced 32%

Thinking Green

Applying Agricultural Mechanics to Everyday Living

Studying agricultural mechanics is not just beneficial to improving production agriculture practices. It includes numerous skills that can be applied to everyday life. For example, have you ever had a leaky faucet in your home? A single faucet dripping at a rate of 5 drips per minute can waste more than a liter of water every day. Studying agricultural mechanics provides you with experience in plumbing to troubleshoot the faucet and determine the cause of the leak.

Will Rodrigues/Shutterstock.com

You can develop technical skills and knowledge to repair the faucet. This can save time and money, as well as conserve a large amount of an important resource: water.

of the total value of all agricultural products sold in the United States. Farms with sales of $1 million or more, representing just 4% of all farms, produced 66% of the total value of agricultural products sold in the United States. The agricultural industry has reached this level of efficient production through sustainable practices and the application of new inventions and technology over the course of time.

What would you say if someone asked you which discoveries made today's successful level of production agriculture possible? Did machines increase production levels? Did scientific discoveries influence plant and animal production? Did changes in animal housing revolutionize animal production? Has biotechnology helped fight animal and plant illness and disease? The answers to all of these questions would be yes. Some might argue that scientific discoveries such as pasteurization (1863), Gregor Mendel's plant-breeding discoveries (1866), refrigeration (1850s), and nitrogen fixation (1918) were the most influential. Others would argue that inventions such as cast-iron plows and mechanical reapers were the catalyst. However, most would agree that the combination of innovations is what sparked, and continues to feed, the growth of the agriculture industry. This chapter focuses on the early mechanical inventions and innovations, as well as the technologies to which they are tied and the scientific principles on which they are based.

Early Mechanical Inventions

Although some of the following inventions may seem primitive compared to today's computerized machinery, they form the basis and inspiration for today's modern agricultural machinery. These inventions allowed access to lands that were previously difficult to cultivate, increased crop production levels, and allowed for more efficient harvesting.

Steel Moldboard Plow

As US settlers progressed west, they found that their cast-iron plows were not suitable for efficiently turning the fertile, heavy soils of the prairies. Several inventors made alterations and improvements to the basic moldboard plow in an attempt to improve its efficiency. However, it was John Deere who finally designed and developed a plow that was suitable for working the prairie soil. In 1837, John Deere developed a moldboard plow with a polished steel blade. The polished steel allowed the sticky prairie soil to slide easily off the blade, reducing the amount of time and effort farmers spent plowing. Alterations and future designs, including the use of multiple blades, also decreased the amount of time needed to prepare fields for planting, **Figure 1-2**.

Cotton Gin

In 1793, Eli Whitney patented the cotton gin, **Figure 1-3**. This invention made it possible to remove the seeds from cotton fibers

Arogant/Shutterstock.com

Arogant/Shutterstock.com

©iStock.com/STRANNIK9211

Figure 1-2. Examples of the first horse-drawn steel plows used to cultivate the rich soils of the Midwest.

quickly and efficiently. The efficiency of the cotton gin allowed the expansion of cotton planting throughout the South. At the time, the most common variety of cotton (long-staple) would only grow successfully in the coastal areas of Georgia and the Carolinas. Upland cotton would grow in other areas, but it was extremely difficult to remove the seeds from the fibers of upland cotton by hand. Whitney's cotton gin reduced the need for labor to remove the seeds, making it more practical to grow upland cotton.

Mechanical Reaper and Combine

Another labor-intensive task was the harvesting and threshing of grains such as wheat. Cereal grains were traditionally reaped, or cut, by hand using scythes and sickles and then stacked in bundles, **Figure 1-4A**. The cut plants would then be threshed to separate the grain kernel from the seed head. Threshing is the process of separating grain from other plant material by a flail or revolving mechanism. The mechanical reaper, invented by Cyrus McCormick in the early 1800s, was the first mechanical aid for harvesting grain crops, **Figure 1-4B**. This invention was pulled through the fields using animal power, such as draft horses.

Everett Historical/Shutterstock.com

Figure 1-3. Eli Whitney is credited with the invention of the cotton gin.

The mechanical reaper replaced the process of cutting by hand, but people still needed to collect the grain for threshing. Later, a machine called a *combine* was developed. The grain combine both cut the grain and threshed it in the field, greatly reducing the labor needed to harvest a crop. A grain combine is a self-propelled machine that cuts, collects, and threshes grain crops such as corn, soybeans, wheat, and oats.

A   *Roman023/Shutterstock.com*

B   *Hein Nouwens/Shutterstock.com*

Figure 1-4. People reaped grains by hand until the mechanical reaper was invented. A—People still reap by hand in some areas of the world. B—An early mechanical reaper.

Steam-Powered Tractor

Until the late 1870s, most agricultural machines were powered by horses, mules, or oxen pulling the machine through the field. This involved animal husbandry skills and training, as well as the feeding and maintenance of the animals. The invention of the steam tractor decreased the need for draft animals on the farm.

The first steam engines were stationary. They were transported by teams of horses to work sites. In 1781, James Watt received a patent for an improved steam engine that used an external condenser to improve the engine's efficiency. Watt also developed a gear system to produce continuous rotary motion out of the reciprocating motion of the engine's piston. Later, skids and wheels were added to make the engine more portable.

Eventually, more complex power transmission systems were developed, allowing steam engines to be self-propelled. Later developments allowed power transmissions to reverse direction. The steam-powered tractor gained popularity for threshing and later for plowing, **Figure 1-5**. The high horsepower generated by these machines allowed farmers to use multiple plow bottoms and cover more area in less time. It also reduced the need for land and crops to support the draft animals. This land could then be cultivated for crops that could be sold at market.

The disadvantage of the steam-powered tractor was that the steam engine was large, bulky, and dangerous. Boiler failures were a serious problem. It took a great amount of skill and fuel to operate these engines.

digitalreflections/Shutterstock.com

Figure 1-5. One of the first steam-powered tractors.

Internal Combustion Engine

As commercial drilling and production of petroleum-based fuels increased during the mid-1850s, the internal combustion engine quickly replaced the steam-powered engine. An *internal combustion engine* is an engine that burns fuel within a cylinder, creating enough force to drive a piston. Internal combustion engines gained popularity due to their smaller size and reduced fuel consumption. In 1892, John H. Froelich installed a gasoline-fueled internal combustion engine to operate a threshing machine. Tractors later became available with internal combustion engines; they were called *internal combustion traction engines*. This term was shortened by the Hart-Parr company in 1907 to *tractor* in their advertisements, **Figure 1-6**. The term has been in use ever since.

Other Improvements

Over the years, many developments occurred that improved the efficiency and safety of the agricultural inventions discussed in this chapter. While some of the basic principles of their operation and design remain constant, advancements have been implemented to improve them. Advancements have included components such as construction materials, electrical power sources, hydraulics, and computer controllers. The overall size of the equipment has also changed.

Mechanical cotton harvesters are now used instead of hand-picking cotton. Cotton gins are still used for separating seeds from the lint, but now they are operated by electricity and can process up to 33,000 pounds of cotton in an hour. Compare that rate to the earlier cotton gin, which had a capacity of 50 pounds of cotton in a day.

SF photo/Shutterstock.com

Figure 1-6. An example of one of the first tractors powered by an internal combustion engine.

SAE Connection

Supervised Agricultural Experience (SAE) in Agricultural Mechanics

Are you creative? Do you enjoy building things? Would "working with your hands" be an item listed on your resume? Do you like to figure things out? If you answered yes to these questions, a career in agricultural technology and mechanics may be right for you. From John Deere's beginnings with the steel plow to the massive tractors used today, agricultural mechanics is a foundational part of agriculture.

As an agriculture student, you will likely be introduced to the concept of the supervised agricultural experience (SAE) as a part of the agricultural education curriculum. There are countless opportunities for you to engage in agricultural mechanics SAE projects. You may repair and maintain an engine or install and troubleshoot precision components for tractors, sprayers, spreaders, harvesters, and other equipment. The most difficult part is deciding where to start! Good luck, and remember to talk with your agriculture teacher about the SAE possibilities in your area.

Goodheart-Willcox Publisher

Mathematics for Agricultural Mechanics

Mathematics, or math, involves the study of quantities, patterns, shapes, and changes. Math is an essential tool in the field of agriculture and agricultural mechanics because it is critical for solving problems related to equipment and technology. Agricultural scientists, engineers, and many other agricultural professionals use math daily. Mathematical problem solving requires the ability to use information and facts to arrive at a solution. Diagrams and sketches, patterns, equations or formulas, and unit cancellation are all ways to approach solving math problems.

Common math terms used in solving problems for agricultural mechanics and technology include:

- Distance—the amount of space between two items; used to measure length or displacement.
- Perimeter—a continuous line forming the boundary of a closed geometric shape, such as a crop field.
- Area—the size of a surface, such as the total square footage of a piece of land.
- Temperature—a measure of an object's heat energy.
- Volume—the amount of space occupied in three dimensions.
- Weight—the force of gravity acting on a body.
- Force—an action that causes motion or a change of motion.
- Pressure—an amount of force exerted over a given area.
- Velocity—the rate of movement of an object expressed as distance over time.

- Power—the rate of doing work or the time it takes to perform work, usually measured in horsepower or watts.
- Work—the result of a force acting through a distance; measured in foot pounds.
- Torque—application of a force through a lever arm causing a twisting or rotational movement.
- Energy—the ability to do work.
- Mass—the amount of material in an object.

Physical Science and Physics

The laws of physical science govern much of the field of agricultural mechanics. Engineers, technicians, and other professionals in agriculture face challenges dealing with motion, force, work, electricity, and light. The *law of conservation of energy* is an example. It states that energy cannot be created or destroyed. This tells us that the energy output of a machine cannot exceed the energy put into the machine, which is helpful in understanding the efficiency of power transmissions on tractors. The amount of work being done by rotating shafts can be measured.

Work

Developments such as the internal combustion engine drastically changed agricultural production by paving the way for machinery and tools to increase the efficiency and capacity of human labor. Examples include the use of tractors for transporting seed to the field, planting the seeds in rows, and hauling hay to feed livestock.

To accomplish any agricultural production activity, work must be performed. *Work* is the result of a force that pushes or pulls an object through a specific distance.

Distance (ft) × Force (lb) = Work (ft lb)

Example:

A skid loader lifts a 2000-lb bulk bag of soybeans 7.5′ for transfer into a planter. How much work has been completed by the skid loader?

Answer:

Distance (7.5′) × Force (2000 lb) = Work (ft lb)

7.5′ × 2000 lb = 15,000 ft lb

The work done by the skid loader was 15,000 ft lb.

Compare this with a person lifting a 50-lb bag of soybeans for loading a planter a height of 4′.

Distance (4.0′) × Force (50 lb) = Work (ft lb)

4.0′ × 50 lb = 200 ft lb

The work done by the person was 200 ft lb. Compare this to the skid-steer loader example, and we see that machines allow for more work to be done with less effort.

Power

Power is defined as the rate of doing work, or the time it takes to perform work. Horsepower is often used to describe the maximum power an internal-combustion engine is capable of producing. *Horsepower* is the force needed to lift 550 lb a distance of 1 foot in 1 second. This means that working continuously, one horsepower is equal to the force needed to lift 33,000 lb a distance of 1 foot in 1 minute.

In the International System of Units, the unit of power is the watt, which equals 1 joule per second. This corresponds to the power in an electric circuit in which the potential difference is 1 volt and the current is 1 ampere.

Work (force) / Time = Power

Example:

How much power is needed to lift an 1100-lb weight a distance of 30′ in 25 seconds?

Answer:

First calculate work (force).

Distance (30′) × Force (1100 lb) = Work (ft lb)

30′ × 1100 lb = 33,000 ft lb

The work done to lift the weight was 33,000 ft lb.

Now calculate the power.

Work (33,000 ft lb) / Time (25 seconds) = Power

33,000 ft lb / 25 seconds = 1320 ft lb/sec

Now translate the power into horsepower (550 ft lb/sec = 1 horsepower)

Power = 1320 ft lb/sec / 550 ft lb/sec = 2.4 horsepower

The power expended to lift the weight was 2.4 horsepower.

Mechanical Advantage and Simple Machines

Most agricultural machines are a combination or modification of simple machines. A *machine* is any device that either increases or regulates the effect of a *force* or produces motion. Forces occur in opposite directions. Every machine is connected to an *effort* force and a *resistance* force, or load. An effort is the magnitude and direction applied to a machine. Resistance is the magnitude and direction that resists the effort force. For example, if you lift a tire with your hand, your lifting is the effort and the weight of the tire is the resistance. Your muscles and bones transfer force to move the weight of the tire.

A machine transfers energy from one place to another and allows work to be done more easily and efficiently. The use of a machine produces an advantage. The *mechanical advantage* of a simple machine is the ratio of the resistance to the effort. Mechanical advantage can be written as a ratio, as in 2:1, but the ratio is usually calculated by dividing the number before the colon by the number after the colon. In this case, the mechanical advantage would be 2 ÷ 1 = 2.

Note: In reality, mechanical advantage calculations must consider the effect of friction as well as resistance and effort. This chapter explains the *theoretical*, or ideal, mechanical advantage of various machines—the mechanical advantage that would exist if friction were not present.

Six mechanical devices, often called the *simple machines*, can be used to apply force and create a mechanical advantage. A combination of two or more simple machines forms the basis for all of the complex machines used in agriculture.

Lever

The *lever* consists of a rigid bar that is free to turn on a pivot called a *fulcrum*, **Figure 1-7**. There are three classes of levers.

- First class—the fulcrum is located between the resistance and the effort. An example is a crowbar.
- Second class—the resistance is located between the fulcrum and the effort. An example is a wheelbarrow.
- Third class—the effort is located between the fulcrum and the resistance. An example is the three-point hitch on a tractor.

The mechanical advantage of levers can be calculated by dividing the length of the input (effort) arm by the length of the output (resistance or load) arm. For example, if the length of the effort arm is 4 and the length of the resistance arm is 2, the mechanical advantage is 4:2, which equals 4 ÷ 2 = 2.

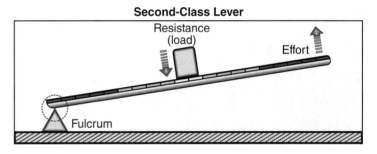

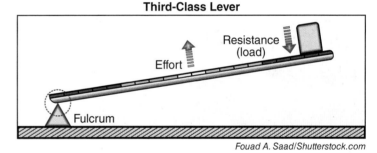

Fouad A. Saad/Shutterstock.com

Figure 1-7. The three classes of levers.

Wheel and Axle

The *wheel and axle* is a large wheel or circle attached to an axle so that they turn together, **Figure 1-8**. The mechanical advantage of a wheel and axle is similar to that of the lever. The wheel and axle can be further visualized by thinking of a manual pencil sharpener. The knob serves as the wheel, and the axle is the shaft that connects to the shaving gears. The mechanical advantage of a wheel and axle equals the ratio of the radius of the wheel to the radius of the axis.

Pulley

Pulleys are wheels that freely spin on an axle supported in a frame, **Figure 1-9**. They have grooves machined or cast in them to receive a rope or belt. Pulleys can be either fixed (attached to a support) or movable. A single fixed pulley provides a change in the direction of the applied force, but has no mechanical advantage because the ratio of effort to resistance is 1:1 = 1.

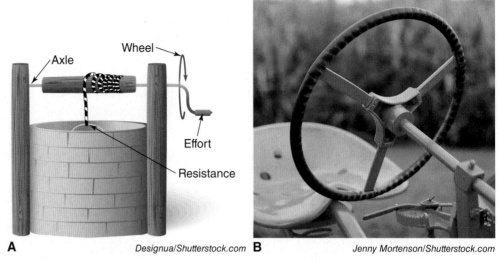

Figure 1-8. Examples of applications of the wheel and axle. A—The winch on an old-fashioned well. B—The steering wheel on a tractor.

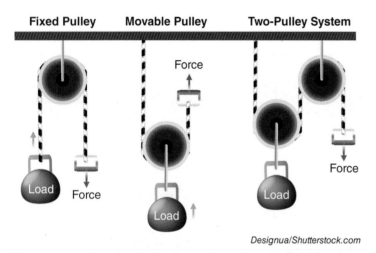

Figure 1-9. Pulleys provide a mechanical advantage to lift objects.

A single movable pulley is attached to the object to be moved instead of a stationary support. It provides a mechanical advantage of 2. If combinations of fixed and movable pulleys are used, then the mechanical advantage is the number of ropes that support the movable pulley. The fixed pulley is not counted in the calculation because its mechanical advantage remains 1.

Inclined Plane

An *inclined plane* is a flat, sloping surface. In other words, an inclined plane is a ramp. This machine is commonly used to raise an object that is too heavy to be lifted vertically, **Figure 1-10**. The mechanical

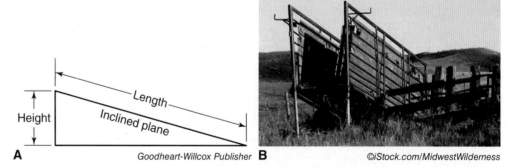

Figure 1-10. A—Dimensions used to calculate mechanical advantage in an inclined plane. B—A livestock loading ramp is an excellent example of using an inclined plane to lift a heavy object to a higher elevation without picking it up vertically.

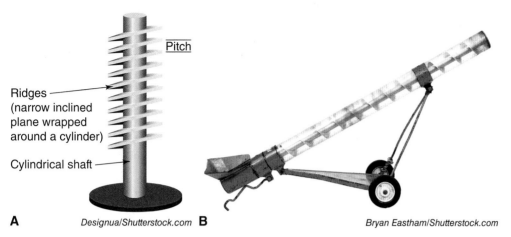

Figure 1-11. A—Parts of a screw. B—Grain augers use a screw-type mechanism to move grain from one location to another.

advantage of an inclined plane is the ratio of the length of the plane surface to the height change. For example, if the sloped surface of the ramp in **Figure 1-10B** is 12′ long, and the high end of the ramp is 3′ above ground level, the mechanical advantage is 12:3, which equals 12 ÷ 3 = 4.

Screw

The *screw* is an inclined plane wrapped around a cylinder. This forms threads on a screw or bolt, **Figure 1-11**. The pitch is the distance between the successive threads. The smaller the pitch, the greater the holding force. The larger the pitch, the more distance the screw moves with each revolution around the cylinder. Technically, the mechanical advantage of a screw is the ratio of the circumference of the screw to the pitch. However, screws require a lever (screwdriver) to turn them, so the mechanical advantage of a screw takes the lever into consideration. The mechanical advantage is therefore calculated as the ratio of the circumference of the handle of the screwdriver (rather than the screw cylinder) to the pitch distance.

Wedge

The *wedge* consists of two inclined planes placed back to back to form two inclined surfaces that meet at a thin edge at one end, **Figure 1-12**. The effort force is applied to the vertical edge or height of the wedge rather than parallel to the slope. The mechanical advantage of the wedge is the ratio of its length to its thickness at the thick end. Wedges are often used to separate or pry apart objects.

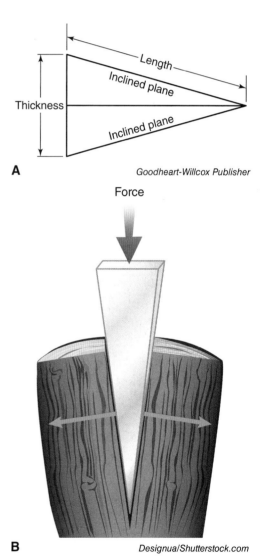

Figure 1-12. A—Dimensions used to calculate mechanical advantage in a wedge. B—Example of a wedge used to separate wood fibers for splitting logs.

Chapter 1

Review and Assessment

Summary

- Agriculture includes producing food, fiber, and fuel, in a sustainable manner, for the global population.
- Agriscience applies scientific principles to improving agriculture through technology and agricultural mechanics.
- Agricultural mechanics includes the design, construction, maintenance, repair, management, and use of agricultural technology and mechanical systems.
- Over the years, agricultural mechanics and technology have changed agricultural processes, making them more efficient and better able to meet global needs.
- Math is an essential tool in the field of agriculture and agricultural mechanics because it is critical for solving problems related to equipment and technology.
- Work is the result of a force that pushes or pulls an object through a specific distance and is equal to the force applied multiplied by the distance the object moved.
- Power is the rate at which work is done and is calculated by dividing the work performed by the time required to do the work.
- A machine's theoretical mechanical advantage is the ratio of the resistance or load to the effort applied.
- The six simple machines on which all agricultural machinery is based are the lever, wheel and axle, pulley, inclined plane, screw, and wedge.

Words to Know

Match the key terms from the lesson to the correct definition.

A. agricultural mechanics
B. agriculture
C. agriscience
D. effort
E. force
F. fulcrum
G. horsepower
H. inclined plane
I. internal combustion engine
J. law of conservation of energy
K. lever
L. machine
M. mechanical advantage
N. power
O. production agriculture
P. pulley
Q. resistance
R. screw
S. subsistence farming
T. wedge
U. wheel and axle
V. work

1. A flat, sloping surface used as a ramp.
2. The ratio of the resistance force to the effort force of a machine.

3. A large wheel attached to an axle so that both turn together to form a continuous lever.
4. The result of a force that pushes or pulls an object through a specific distance.
5. A mechanical device that is used to transfer energy from one place to another and allows work to be done more easily.
6. A pivot point on which a lever is free to turn.
7. A grooved wheel that turns readily on an axle and is either supported in a frame or attached to an object to be moved.
8. Two inclined planes joined together to form a machine with two inclined surfaces that meet at a thin edge at one end.
9. An inclined plane wrapped around a cylinder.
10. A common unit of power equal to the force needed to lift 550 lb a distance of 1 foot in 1 second.
11. The rate of doing work.
12. Agriculture for the purpose of producing a marketable product.
13. Any action that causes motion or a change in the direction of motion.
14. A machine that consists of a rigid bar that pivots on a fulcrum.
15. An engine that burns fuel within a cylinder to create a force that drives a piston.
16. The principle that energy can be neither created nor destroyed.
17. The science and practice of sustainably producing food, fiber, and fuel to meet the needs of a global population.
18. Growing only enough food or other products for the use of the farmer and immediate family.
19. The design, construction, maintenance, repair, management, and use of agricultural technology and mechanical systems.
20. The magnitude and direction of a force being applied to a machine.
21. The field of science devoted to improving agriculture.
22. The force that resists the effort force in a machine.

Know and Understand

Answer the following questions using the information provided in the chapter.

1. Explain the difference between agriculture and agricultural mechanics.
2. Agriscience is the _____.
   A. field of science involved with improving agriculture
   B. process of marketing agriculture
   C. sale of agricultural products
   D. design, construction, maintenance, repair, management, and use of agricultural technology and mechanical systems

3. *True or False?* Over the last 200 years, farm sizes have increased, but the number of farms has decreased.
4. *True or False?* Mechanization and technology advancements have resulted in decreased food production.
5. What is the law of conservation of energy, and what is the application of this principle to agricultural mechanics?
6. A tractor pulls a 650-lb skid a distance of 50′ in 40 seconds. How much horsepower was expended? Round your answer to the nearest hundredth.
7. If a force of 200.0 lb moves an object 12.0′, how much work has been performed?
    A. 1200 ft lb.
    B. 2400 ft lb.
    C. 3200 ft lb.
    D. None of the above.
8. Who is credited with the invention of the steel plow?
    A. John Deere.
    B. Cyrus McCormick.
    C. James Watt.
    D. None of the above.
9. *True or False?* Agricultural technology will need to continue to advance to meet the demands to feed a growing world population.

STEM and Academic Activities

1. **Science.** Design an experiment to evaluate fuel efficiency of internal combustion engines.
2. **Technology.** Explore phone apps that can be used to help develop farm production budgets.
3. **Engineering.** Evaluate the design of a livestock corral.
4. **Math.** Apply mathematical problem solving to each of the following problems involving agricultural mechanics.
    a. You have a roll of woven wire fencing that is 32′ in length. You need to use this fencing to create a rectangular [(length × 2) + (width × 2)] perimeter for a livestock pen. A square pen (8′ × 8′) would give you the biggest area. Which dimensions would result in the next largest area? Show all possible configurations.
    b. A team of five people pulls on a rope to lift an object using a pulley. Each person has a pulling force of 75 lb. What is the total pulling force of the team?
5. **Social Science.** Conduct research on the impact of the use of robotics on human labor demands. Write a report about your findings.
6. **Language Arts.** Read John Steinbeck's novel *The Grapes of Wrath* and compare the agricultural technology used in the book with today's technology. Write a summary of the advancements in agriculture and agricultural mechanics that resulted from the problems faced during the "Dust Bowl" era.

Thinking Critically

1. Rotating components on agricultural machinery and equipment operate with enough speed and force to cause severe injury in less than a second. What steps can you take to keep yourself safe around rotating machinery?
2. Design an experiment to test and evaluate the different classes of levers by removing nails from a piece of lumber. Which type of lever did you find most effective and why?

CHAPTER 2: Careers and Employability Skills

Chapter Outcomes

After studying this chapter, you will be able to:
- Describe at least five jobs available for all skill levels of agricultural mechanics.
- Explain ways of learning the skills and knowledge needed to obtain a job in agricultural mechanics.
- Produce well-organized documents to secure a job in agricultural mechanics.
- List sources of job leads for a job in agricultural mechanics.
- Demonstrate soft skills in agricultural mechanics.
- Maintain a job in agricultural mechanics.

Words to Know

- apprenticeship
- career development event (CDE)
- certificate
- continuing education
- cover letter
- 4-H
- job application
- National FFA Organization
- networking
- *Occupational Outlook Handbook* (*OOH*)
- portfolio
- reference
- résumé
- skilled trade
- SkillsUSA
- soft skills
- template

Before You Read

The summary at the end of the chapter highlights the most important concepts. Read the chapter and write a summary of it in your own words. Then, compare your summary to the summary in the text.

While studying this chapter, look for the activity icon to:

- **Practice** vocabulary terms with e-flash cards and matching activities.
- **Expand** learning with interactive activities.
- **Reinforce** what you learn by completing the end-of-chapter questions.

www.g-wlearning.com/agriculture

©iStock.com/DACowley

In this chapter, you will learn about some of the many careers available in the different sectors of the agriculture industry. We will also learn about the traits employers are looking for in employees, and what you can do to stand out from the competition when applying for a job. Securing a job is a process that requires knowledge and practice. Maintaining a job requires skills that can be learned and practiced on the job every day.

Careers in Agricultural Mechanics

Throughout the world, more people are employed in agriculture than in any other industry. In the United States alone, more than 22 million people are employed in agriculture-related jobs. These jobs vary greatly and include employment in everything from production agriculture to retail sales. As the amount of mechanization used in the agriculture industry has steadily increased over the years, the need for knowledgeable, skilled mechanics and technicians has also increased.

Farm equipment has evolved dramatically from the team of horses used in the early 1900s. Today's four-wheel-drive tractors have the power of 40–300 horses. As the amount of mechanization and horsepower in farm machinery has increased, the time needed to complete tasks has decreased. *Combines*, huge machines used to harvest grains such as corn, soybeans, and wheat, have dramatically changed agriculture, **Figure 2-1**. In the 1930s, before the machines were available, a farmer could harvest an average of 100 bushels of corn by hand in a nine-hour day. Today's combines can harvest 900 bushels of corn per hour, or 100 bushels of corn in less than seven minutes!

Most careers available to agricultural mechanics involve specialty skills and training. These are high-skill, high-paying jobs open to people willing to perfect their abilities in one or more skilled trades. *Skilled trades* are jobs that require a specific set of skills, such as electrical wiring, welding, construction, or masonry proficiency. Many jobs also require skill certification, which can be obtained even while students are in high school through specialty training programs or classes. Skilled trades often involve perfecting a skill over years of on-the-job practice, which produces higher skill levels that command higher pay.

Photo by Tim McCabe, USDA Natural Resources Conservation Service

Figure 2-1. The introduction and development of farm machinery has enabled farmers to greatly increase planting and harvesting efficiency.

Many jobs in this area start with an apprenticeship. An *apprenticeship* is a combination of on-the-job training and formal classroom education in the skilled trades, generally sponsored by a trade group. Some jobs may involve more advanced training, including technical college or even advanced college degrees. Most of the agricultural mechanic's skills can be applied to other industries as well, such as the home building or manufacturing sector, so career possibilities for the agricultural mechanics student are not limited to this industry alone, **Figure 2-2**.

Job Skills

Once you have decided that you have an interest in an agricultural mechanics job, you need to investigate available jobs to determine which jobs are a good fit for you. How do your existing skills and other skills you can learn match up with the available jobs? There are several ways to find this information. One excellent resource is the *Occupational Outlook Handbook (OOH)*, which details information and predicts demand for most careers in the United States. This resource is available in both hardcopy and online. The *OOH* is published by the US Department of Labor's Bureau of Labor Statistics. This resource is a one-stop location for information about a wide selection of careers available in the United States. Information includes job categories determined by the federal government for statistical purposes of employment monitoring. Categories include:

- What They Do—the basic job description.
- Work Environment—working conditions and general location (indoor, outdoor, clean, dirty, dangerous surroundings, etc.).
- How to Become One—education and/or training needed, including special licenses or certifications required and special personal or physical qualities needed.

Agricultural Mechanics-Related Career Opportunities	
Agriculture Career Sector	**Careers**
Agribusiness	Equipment repair shop manager, equipment dealer, manager of local chemical supply cooperative
Agriscience	Inventor, researcher, laboratory technician
Agricultural communication	Technical writer, editor, advertising account manager
Agricultural processing	Design or systems engineer, repair mechanic, maintenance technician, machine operator
Agriculture services	GPS field technician, software application designer, food safety inspector
Agricultural systems	Design engineer, repair technician, inventor
Natural resource management	Engineer, technician, program managers (often specialized in soil, water, forestry, air, energy production or pollution control)
Agricultural production	Farmer, rancher, mechanic, veterinarian, manager
Agriculture education	School teacher, specialty skill instructor, professor

Goodheart-Willcox Publisher

Figure 2-2. The agricultural industry is extensive, and many careers overlap into the different areas. The above chart lists only a few of the many areas to which training in agricultural mechanics and technology will apply.

- Pay—average salary, as well as the low and high ends of the wage scale. May include information on benefits available as an industry standard and often includes information on length of workweek, overtime, or unusual working hours.
- Job Outlook—which jobs are currently experiencing growth and which are expected to grow or decline in the future. This helps students gauge their job prospects when they are finished with their education, not just what is available today.
- Similar Occupations—jobs that share characteristics with the titled occupation. This helps guide students into job areas they were unaware of that may suit their skills.
- More Information—additional bits of information that might be interesting to students determining if this is a career for them to pursue. Generally, this also includes professional, trade, or union contact information that is related to that career.

Competitions and Learning Activities

The *National FFA Organization*®, an organization for students interested in agriculture and leadership, provides *career development events (CDEs)* that help students learn, practice, and demonstrate skill attainment in several competitive activities, **Figure 2-3**. You can improve your performance in many of these events by developing your agricultural mechanics skills.

Other organizations also provide skills competitions and learning experiences. *SkillsUSA*, a partnership of students, teachers, and industry working together to ensure that the United States has a skilled workforce, has dozens of skills assessments and competitive activities that test and certify students, **Figure 2-4**. *4-H* is a youth development program of the US Department of Agriculture's Cooperative Extension Service that is available in every county and parish in the country. 4-H offers many agricultural mechanics learning opportunities for skills including welding, electricity, and woodworking, but also offers projects on soils, GPS, and a variety of problem-solving activities. Students can compete in many state and local fairs and stock shows with their mechanics projects through these organizations.

David Kosling/USDA

Figure 2-3. Students who qualify compete at the National FFA Convention, where over 50,000 students are in attendance. The National FFA Organization also presents opportunities for students to attend government meetings and contribute to local and national projects.

Certification

Students enrolled in workforce training programs are often granted *certificates*. These certificates are evidence that a student has specialized

SkillsUSA

Figure 2-4. Students studying agricultural mechanics and technology may enter competitions such as those offered by *SkillsUSA*.

agricultural mechanics skills such as welding, electricity, or plumbing that are industry approved. To obtain a certificate, students are often required to complete a course of study and then pass a certification test. If students meet the minimum score, they are certified and are given a certificate that says they have met industry standards. Occasionally, certification tests require a practical demonstration of the actual skill, such as in the case of welding. Students who complete the suggested course of study and pass the test enter the job market with a marketable skill that is recognized by everyone in the chosen industry.

Apprenticeships

Apprenticeships are a great way for an agricultural mechanic to secure the job training and skills necessary to enter many of the related professions. Apprenticeships are formal training programs that include two basic components: on-the-job learning from a master or mentor, and a formal classroom education training program provided by industry experts or skilled trade unions, **Figure 2-5**. An apprenticeship allows the new worker to learn the skills needed by an industry while earning a good wage. This benefits both the apprentice, who needs training, and the industry, which needs trained workers. Apprenticeships are available through labor groups or trade unions, specific employers, or a group of employers, depending on the specific program. The US Department of Labor or your state's employment department has information on most available apprenticeship programs.

Continuing Education and Training

Many students can enter the job market with a basic set of skills, then improve their skills with *continuing education*. Generally, this training is available from local technical or trade schools, industry associations, colleges or universities, or online schools. Some employers offer tuition reimbursement to their employees for continuing education. It is wise to take advantage of the opportunity to improve your skills through continuing education, especially if the employer will pay for the training.

goodluz/Shutterstock.com

Figure 2-5. The compensation you receive from an apprenticeship may be wages, or it could be the knowledge and experience you gain from the work.

Education Needed

Because of the wide variety of careers available to skilled agricultural mechanics, the educational requirements for each specific job vary as well. Students can prepare directly for many agricultural mechanics jobs in a high school training program, certification program, or apprenticeship. However, more opportunities are available if the student has received an associate degree from a trade or technical school, or from a community college. As the job responsibilities increase, students may determine that the next logical step in their education is a bachelor's degree, which opens many doors at the engineering and technically skilled

AgEd Connection — Job Interview CDE

We spend much of our lives preparing for and working in our careers. There are many career opportunities in agriculture, and one of the best ways to prepare for your job search is to participate in a job interview CDE.

A job interview CDE will allow you to develop, practice, and demonstrate the skills needed when seeking employment in agricultural industry. In this event, you may be required to create a cover letter and résumé, complete an electronic job application, practice telephone skills, and participate in a personal interview. Awards may be available for winners, and everyone who participates will gain valuable experience that will help prepare them for actual job applications. Ask your advisor and agriculture teacher about participating in a job interview CDE today.

racorn/Shutterstock.com

levels. Engineers and managers may need a master's degree or a doctorate of philosophy (PhD). A master's degree requires two or more years of additional study beyond a bachelor's degree, and a doctorate requires another one to three years of educational work beyond a master's degree. Many scientists and educators find that a doctorate is required to reach the top of these professions.

Securing Employment

Getting a job is not always as easy as visiting the employment office of the company at which you want to work. Most companies require proof that you have the skills and training needed to perform the job. To improve your chance of being hired for the job you want, first gather the documentation you will need, and be prepared to interview for the job.

Skill Documentation

In many cases, getting a job in a skilled trade involves proving that you have the necessary skills needed. As mentioned earlier, this is sometimes accomplished with a certification from a training program. However, some training does not lead to certification, but merely helps you develop skills that will transfer to a job. One way to prove that you have these skills is to document your progress with a portfolio. A *portfolio* is a record of skills, activities, training, and education. A great way to document your skills is to include photographs of your projects before, during, and after completion.

A portfolio can be stored online using a record-keeping system such as the National FFA AgCN (agriculture career network) through your local FFA program. If you are not a member of FFA, you can use other online portfolio storage systems. Your portfolio, whether it is online or kept in a physical location, should include:

- Photos of you performing the skill (welding, finishing concrete, building, etc.). Make sure all photos show you wearing proper safety gear and working in a safe manner, **Figure 2-6**.
- Testimonials from teachers, employers, and other supervisors. Generally, these are written in letter format.
- Community service hour records, especially if they are earned using agricultural mechanic skills such as building fences for the local dog shelter or setting irrigation equipment at the local park.

Goodheart-Willcox Publisher

Figure 2-6. A portfolio is a great way to supplement your résumé by showing examples of your work, including illustrations and photographs. Ask your teacher or fellow students to help take photos of you working and of your completed projects for your portfolio.

Career Connection

Agricultural Mechanics Teacher

auremar/Shutterstock.com

Job description: Agricultural mechanics teachers teach introductory and advanced courses in agricultural mechanics and technology. Course topics include welding, power equipment, construction, and computer technology. Agricultural mechanics teachers may also advise the FFA chapter and actively participate in curriculum development.

Education required: Bachelor of Science in Agricultural Education or closely related field with the expectation of acquiring a graduate degree, such as a Master of Science or Master of Arts in Teaching.

Job fit: This job may be a fit for you if you enjoy helping people and working with technology. Individuals who are problem solvers and have natural curiosity about how things work will enjoy this career.

- Detailed records of hours worked and skills performed each day.
- Samples of work products. For example, you might include a report on the safety review you did for the local equipment dealer or a copy of the presentation you gave to the school board on repairing the athletic fields at your school.

It is also helpful to have teachers or employers write letters explaining what you learned in specific situations. These are somewhat different than letters of recommendation in that they address only the specific skill(s) you learned during a particular project or span of time. These, too, should become part of your portfolio.

Documents

Other, more traditional documents that are needed for securing a job include a résumé, cover letter, and list of references. You will often need these when filling out a job application, or sometimes an employer will ask for them after you have submitted an application.

A *résumé* is a summary of your experience, education, and skills. A *cover letter* is a letter, usually no more than one page, that introduces a job candidate and refers the reader to the accompanying résumé. Employers may also want to hear information from people who know the job applicant, either personally or professionally. These people are known as *references*. The applicant must provide their names and contact information. A reference can be a friend or family member, but employers usually prefer former teachers, co-workers, or supervisors.

These documents should be prepared as if they were your only chance of making a good impression, because they often are. Spelling and grammar errors cause many applicants to be rejected before their documents even make it to the hiring personnel. Many students say, "I won't be writing on my job; they really don't care if I have a résumé." These students are wrong.

Businesses *do* want employees who can read and write in a capable manner. The care applicants take with their initial contact information for a company indicates immediately how much they care about the job.

Résumé

When writing a résumé, remember that spelling errors can prevent you from securing a job. Generally, résumés have a grammar style that is different from other written works. The résumé needs to convey a great deal of data about an applicant in a short amount of space. For inexperienced workers, a one-page résumé is acceptable and encouraged, but it may be two pages to include references.

The résumé is also sometimes read by an electronic data scanner instead of a human. Therefore, it should follow a basic pattern so the information is displayed in an easy-to-access manner. The best way to write a résumé is to use a template from a standard word processing program. A *template* is a preformatted document for a specific form, such as a résumé. A template is set up in an acceptable, professional style and can be customized with your important facts. **Figure 2-7** is an example of a résumé template.

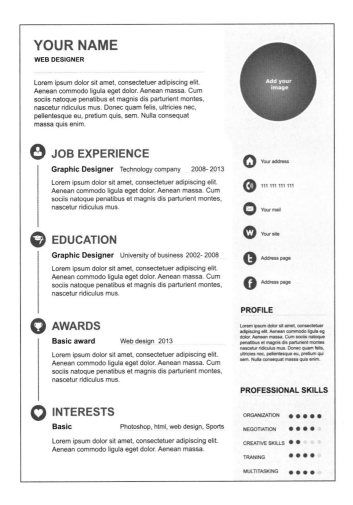

Ramcreativ/Shutterstock.com

Figure 2-7. An effective résumé is concise, highlights your skills, and presented in an easy-to-read format. Using a template is a good way to organize your information in a logical format.

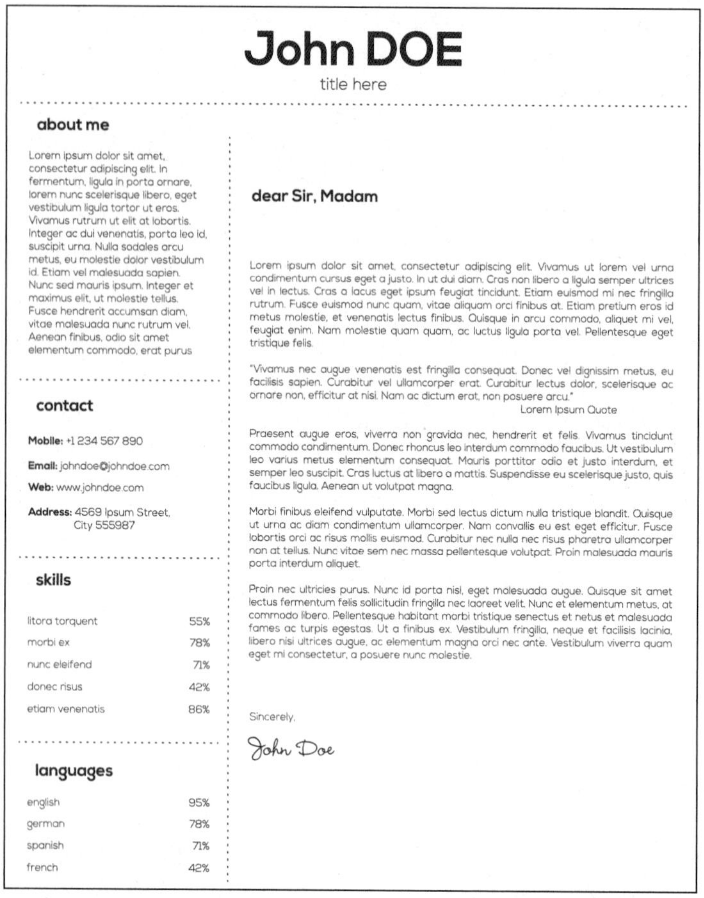

John DOE

title here

about me

Lorem ipsum dolor sit amet, consectetur adipiscing elit. In fermentum, ligula in porta ornare, lorem nunc scelerisque libero, eget vestibulum ligula tortor ut eros. Vivamus rutrum ut elit at lobortis. Integer ac dui venenatis, porta leo id, suscipit urna. Nulla sodales arcu metus, eu molestie dolor vestibulum id. Etiam vel malesuada sapien. Nunc sed mauris ipsum. Integer et maximus elit, ut molestie tellus. Fusce hendrerit accumsan diam, vitae malesuada nunc rutrum vel. Aenean finibus, odio sit amet elementum commodo, erat purus

contact

Mobile: +1 234 567 890

Email: johndoe@johndoe.com

Web: www.johndoe.com

Address: 4569 Ipsum Street, City 555987

skills

litora torquent	55%
morbi ex	78%
nunc eleifend	71%
donec risus	42%
etiam venenatis	86%

languages

english	95%
german	78%
spanish	71%
french	42%

dear Sir, Madam

Lorem ipsum dolor sit amet, consectetur adipiscing elit. Vivamus ut lorem vel urna condimentum cursus eget a justo. In ut dui diam. Cras non libero a ligula semper ultrices vel in lectus. Cras a lacus eget ipsum feugiat tincidunt. Etiam euismod mi nec fringilla rutrum. Fusce euismod nunc quam, vitae aliquam orci finibus at. Etiam pretium eros id metus molestie, et venenatis lectus finibus. Quisque in arcu commodo, aliquet mi vel, feugiat enim. Nam molestie quam quam, ac luctus ligula porta vel. Pellentesque eget tristique felis.

"Vivamus nec augue venenatis est fringilla consequat. Donec vel dignissim metus, eu facilisis sapien. Curabitur vel ullamcorper erat. Curabitur lectus dolor, scelerisque ac ornare non, efficitur at nisi. Nam ac dictum erat, non posuere arcu."

Lorem Ipsum Quote

Praesent augue eros, viverra non gravida nec, hendrerit et felis. Vivamus tincidunt commodo condimentum. Donec rhoncus leo interdum commodo faucibus. Ut vestibulum leo varius metus elementum consequat. Mauris porttitor odio et justo interdum, et semper leo suscipit. Cras luctus at libero a mattis. Suspendisse eu scelerisque justo, quis faucibus ligula. Aenean ut volutpat magna.

Morbi finibus eleifend vulputate. Morbi sed lectus dictum nulla tristique blandit. Quisque ut urna ac diam condimentum ullamcorper. Nam convallis eu est eget efficitur. Fusce lobortis orci ac risus mollis euismod. Curabitur nec nulla nec risus pharetra ullamcorper non at tellus. Nunc vitae sem nec massa pellentesque volutpat. Proin malesuada mauris porta interdum aliquet.

Proin nec ultricies purus. Nunc id porta nisl, eget malesuada augue. Quisque sit amet lectus fermentum felis sollicitudin fringilla nec laoreet velit. Nunc et elementum metus, at commodo libero. Pellentesque habitant morbi tristique senectus et netus et malesuada fames ac turpis egestas. Ut a finibus ex. Vestibulum fringilla, neque et facilisis lacinia, libero nisi ultrices augue, ac elementum magna orci nec ante. Vestibulum viverra quam eget mi consectetur, a posuere nunc molestie.

Sincerely,

John Doe

Fazakas Mihaly/Shutterstock.com

Figure 2-8. A cover letter is an opportunity to introduce yourself to the prospective employer, explain the job you are seeking, and why you would be a good candidate. Cover letters are usually read before the résumé, so a poor cover letter can often mean that the résumé will never be seen.

Cover Letter

The cover letter is another form of communication that must be completely error-free and grammatically correct. Many word processing programs have templates for this type of writing, as well. The purpose of a cover letter is to make the reader want to read the attached résumé. Remember to highlight one or two facts that will make the reader turn the page to continue on to the résumé. One good way to pique someone's interest is to refer to a person the reader knows well who recommended the applicant. Also mention the job opening and highlight a specialized skill you know the company needs. **Figure 2-8** is a cover letter template.

References

Who makes a good reference? The answer to this question is often vital for getting the job. The following are basic qualities of people who make good references:

- The person is not a relative. Even if you have worked for a relative in an ongoing family business, a relative should not be offered as a work reference.
- The person knows your work habits and is willing to sing your praises about your ability and work ethic.
- The person has a good reputation and is well respected in the industry where you are seeking employment. Generally, but not always, other students or young people do not make the best references.
- The person has known you long enough to answer questions about your work, character, and other important qualities. Be certain this person will give you a *good* reference. Some people will report that they know you but may not give a good recommendation. This may hurt your chances for employment.
- The person is willing to offer a reference in whatever format is required. Ask your references if they are willing to write a letter, send an email, or speak directly on the telephone.

Job Application

Generally, getting a job will require filling out an application. A ***job application*** is a form supplied by an employer to collect information about a prospective employee, **Figure 2-9**. If the application is the first point of contact between you and the business, make sure it is free of errors and omissions. Treat it as if it were your résumé and you want it to make a great first impression. Here are some things to consider before submitting an application:

- Is it available online? If so, should you file it online or print it out and bring it with you to an interview?
- Should it be done in your own handwriting? Do you need to sign it in person?
- Do you understand exactly what each question is asking?
- Are there any special instructions?

In some cases, the application is completed as a formality *after* the hiring process. In these cases, the application serves as a legal document. It requires that the applicant be honest and accurate, with the understanding that misrepresentations here can lead to future dismissal. By signing the form, you are agreeing that you have not lied about any felony convictions and that all statements are true.

Rawpixel.com/Shutterstock.com

Figure 2-9. Many job applications are now available online, and often you can attach and send a cover letter and your résumé electronically as well.

Job Interview

If the potential employer is impressed by the information in the documents you have submitted, you may be asked to participate in a job interview, **Figure 2-10**. Generally, a job interview is a two-part process. The employer learns more about the potential employee, and the potential employee learns more about the business.

Before the interview, make sure you have the date, time, and location exactly right. Try to arrive a few minutes early; this will make a good impression by showing that you are punctual. Arriving late to an interview is almost always the end of potential employment. Also make sure you know what you are expected to bring to the interview. For example, does the employer require a completed application form? Even if it is not required, it is often a good idea to bring a copy of your résumé along so that you have all the information (including references, contact information, and dates of training or employment) readily available.

Africa Studio/Shutterstock.com

Figure 2-10. Maintain eye contact with the interviewer during the job interview. This shows that you are attentive and interested in the available job.

Copyright Goodheart-Willcox Co., Inc.

Before any interview, you should prepare yourself both physically and mentally. You should plan to wear clothing that is clean and well representative of your work style, **Figure 2-11**. A standard for interview attire is to wear clothing that is one step up from the job for which you are interviewing. So, if the job's dress code is jeans and a T-shirt, you should wear dress slacks and a collared shirt. If the usual dress code requires wearing a shirt and tie, then adding a suit jacket to a shirt and tie is in order. This more formal attire, however, is often not necessary for a casual job in agricultural mechanics.

Wearing your hair, including any facial hair, in a neat and conservative manner is a good idea. If you wear a scent, make sure it is understated and not overpowering. Remember, employees are representatives of the company. Any employee who looks, acts, dresses, or even smells inappropriately may potentially offend customers or perhaps coworkers, which is bad for business, and employers will not tolerate it.

After you have prepared yourself physically for the interview, it is important to prepare mentally as well. You should do some basic research on the employer, including:

- How many employees does the company have?
- What business products or services does the company supply?
- What is the main part of the business?
- What role would you play in the business?
- Is the business growing?
- Are there local competitors, and how stable is the industry as a whole?

Iakov Filimonov/Shutterstock.com

Figure 2-11. A sloppy or neglected appearance can negatively affect your employment chances. Particularly for agricultural interviews, which may take place partly outdoors, a neat, job-appropriate appearance shows that you understand the work and the clothing requirements.

You should also prepare and practice your answers to basic questions that may be asked at the interview. Keep in mind the interviewer may have little experience in interviewing job applicants, and each situation is different, **Figure 2-12**. Typical interview topics include:

- Tell us about yourself. Generally, this is not an invitation to tell them where you were born or went to grade school, but rather an opportunity to tell them how you can fit into their business and what skills you have. (Good thing you did your research!)
- Why should we hire you? Again, this is a great place to explain how you can fit in with the employer's current and future business and to explain your important qualities.
- Where do you see yourself in the future? The employer wants to know if the company can count on you long term. If so, be sure to tell the interviewer that, but if you intend to work there for less than five years, be honest about that, too.

Khakimulin Aleksandr/Shutterstock.com

Figure 2-12. Be prepared to answer questions about yourself and your experience. Practicing answers to potential interview questions will make it easier for you to respond when they are asked during the interview. This will reduce the possibility that you will hesitate during your response and will show the interviewer that you are prepared and confident.

There are hundreds of ways to answer important questions. Practiced answers are generally better to help you make your point. Sometimes a question will be unique and catch the interviewee off guard. Make sure you take time to think through your answer and reply honestly. Generally, your sincerity and thoughtfulness are important qualities the interviewer is attempting to determine.

Making a Good Impression

Once you arrive at the interview, dressed appropriately and prepared to answer questions, you should attempt to make the best impression possible. Making a good impression is generally a combination of common courtesy and preparedness. The following suggestions will help you make a good first impression:

- Be early. As mentioned earlier, arriving late to an interview almost never ends well. If you have a problem and cannot avoid being late, call ahead and explain your situation. Ask to reschedule, if possible.
- Greet everyone you meet in a friendly but professional manner. Administrative assistants are often asked their opinion of the interviewees, and a sour note here can cost you the opportunity to work for this company.
- In the United States, a handshake starts many business dealings and concludes most dealings as well. Offer your hand in a firm but not crushing grip if the interviewer does not first. Maintain eye contact and a friendly look.

- Be prepared. If you have a copy of your résumé or other documents with you, make sure you have them organized and available at a moment's notice. Searching through a disorganized briefcase or folder leaves a bad impression.

Answering the Right Questions

Once you are in the interview, take a deep breath and focus on answering the questions. Carefully listen to each question, and give it a moment of thoughtful consideration before answering. Answer honestly and in such a way the interviewer will see you as someone with insight.

Often, the final interview question is, "Do you have any questions about the position or our company?" This is the time to ask questions that were not answered by your background research. These might include questions such as, "What is the possibility of advancement from this position?" or "Do you promote from within when there is an opening?" The interviewer will see that you took the initiative and showed interest in the position by doing some research before your interview. This is not the time, however, to ask about what your specific pay would be, how long your lunch break would last, or when you qualify for sick leave or vacation days. That will occur during a follow-up interview or if an offer is extended.

Follow-Up Letter

If you have made it through the job interview, there is one more document that is important for getting a job, and that is the follow-up letter. As soon as you return from the interview, sit down and compose a letter. Thank the interviewer for taking the time to speak with you, highlight the main point of your interview (restate why you are the best candidate for this job), and convey any other information that was not mentioned in the interview. For example, perhaps you were not asked about your specialized safety training; now would be a good time to mention that you are OSHA-certified for personal safety. Many templates for follow-up and thank-you letters are available within most word processing programs.

Send your follow-up letter both as electronic mail and regular mail. If possible, try to send the electronic version to the interviewer's company email address. The extra step of sending a follow-up letter shows the interviewer how serious you are about acquiring this position and could be the deciding factor in getting the job.

Finding a Job in Agricultural Mechanics

Finding a job is often the end result of education and training. Finding a job in agricultural mechanics is no different than in any other industry. Knowing where to look is the best starting point. For agricultural mechanics positions, it is usually not necessary to pay someone to assist you in your job search when there are so many free resources available.

There are several common free or low-cost places to begin a job search, including:
- Your personal network (friends and family).
- Trade unions and professional organizations.
- Government agencies.
- School employment offices.
- Areas with job opportunities.

Networking is the act of getting to know people in various professions through self-introduction or through introductions made by friends or associates, **Figure 2-13**. Networking is also a means of informing others you are looking for employment. When you are using your network, be sure to use all your resources effectively. Start by telling family and friends exactly the type of work you are looking for and exactly what your skills are. But do not limit yourself to jobs you think are well matched to your skills. Many times the not-quite-right job leads to something better. Searching for a job is a skill that takes practice, so applying for and even interviewing for jobs that seem imperfect may lead to better searches for the perfect job. Often a job is known by your network long before it is published. So be sure to tell everyone you know, and ask them to share with everyone they know. Before long, the word will be out to scores of people that you are in the job market, and your network is now in action.

Di Studio/Shutterstock.com

Figure 2-13. Your network may include friends, family, coworkers, or others in your profession. Networking with these people will help you in your search for employment.

Trade Unions and Professional Organizations

Many business associations, unions, trade organizations, and other business groups have employment posting services either online or at their physical locations. Becoming a member of these important associations is a great way to network with people in your chosen field and know about jobs when they are first announced. Often, these groups have student memberships available for a reduced cost.

Government Agencies

Every state has a government agency that assists job seekers. Generally, these agencies do not require state residency or much more than registering on their website to search large databases of job openings in the state. No government agency requires a fee to search for a job or assist you in applying for work. It is helpful to know keywords related to your area of interest. For example, if you are interested in a job as an electrician, searching for "electrical," "journeyman electrician," and "construction trades" might also deliver job openings in your area.

Be aware that the Internet is also available to unscrupulous individuals who prey on job seekers. If a job sounds too good to be true, it probably is. To ensure your safety, be wary of offers to meet in remote locations or at odd hours. Research potential employers thoroughly before responding to offers.

School Employment Offices

Whether you have been in a specialized training program or a general education program, your teachers, mentors, or others who have assisted you may know of jobs in areas in which you are interested. It is important to tell these professionals that you are seeking a job and to ask for leads. This is another example of networking, although it is a bit more complex than simply telling everyone you know.

Areas with Job Opportunities

When job searching, do not discount researching where the jobs are. Look for locations that have job openings, keeping in mind they may not be local. This seems reasonable and simplistic, but often people miss out on opportunities because they search in areas with high unemployment and few jobs, when ample jobs are available just a short distance away. Looking outside your immediate area can improve your chances. The Internet can be a valuable source of information on employment trends and recruitment. Again, beware of Internet job offers that seem too good to be true. Research any job offerings carefully to determine if they are legitimate jobs and not a scam.

Maintaining Employment

What is interesting about the skills needed for agricultural mechanics is that many on-the-job skills are not trade skills, but soft skills. *Soft skills* are skills defined by working ability, communication ability, and the ability to get along with others. These skills include, but are not limited to, the following:

- Getting along with other employees and customers. No one wants to work with a difficult coworker, and customers certainly do not want to do business with them either. Putting your best foot forward and going the extra mile to be nice is often all that is needed, **Figure 2-14**.
- Reading information and following directions. Sometimes this is as simple as being able to read the directions for new equipment or training yourself to execute a new skill.
- Making good decisions. This means making decisions that advance your employer's interest, increase profit, help customers, or improve some aspect of the operation. Try to imagine how different decisions would affect the outcome.
- Continuing to upgrade and improve your job-specific skills. Ask employers about specific training, seek out greater-skilled coworkers, and practice these new skills when you have a chance after your work is complete. Learning about new advancements or techniques in your business not only helps your employer, but it helps you stay motivated. It may also lead to a pay increase or promotion.

Monkey Business Images/Shutterstock.com

Figure 2-14. As an employee, it is important to get along with your coworkers and be pleasant to customers.

- Being punctual. Most employers will tell you that on-time attendance is one of the top employee problems, **Figure 2-15**. Arriving to work on time every day ensures that the employer can rely on you, you will not miss important information or directions, and others will not have to do your job for you. If tardiness is a problem for you, set a goal of being at work 10 minutes early every day and offer yourself a reward when you have made your goal for a period of time. Your real reward will be in retaining your job and not being docked pay.
- Being a self-starter. Employers want employees who can see what needs to be done and complete the task without being asked. This soft skill frees managers to manage bigger projects and leads to employees who can be trusted (and often rewarded) to work unsupervised. Most employees like to control their work life in some aspect, and this is one way to earn that right.
- Having a strong work ethic. This means being positive, enthusiastic, and willing to work at your day-to-day duties. Most employers believe that if their employees have a good work ethic, they will accomplish most of the other soft skills on this list. This skill can be practiced and refined by generally checking your attitude each day: Are you arriving to work on time and with a positive attitude? While you are a student, practice being positive about school and what you are required to learn. Attitude is something each person controls. If you truly do not like your situation, only *you* have the power to change it. Remember, sometimes changing how you see the situation is the only change you need.

Syda Productions/Shutterstock.com

Figure 2-15. Being on time for work is a critical soft skill. Chronic tardiness, or absence, may be grounds for termination.

CHAPTER 2 Review and Assessment

Summary

- Careers available to agricultural mechanics involve specialty skills and training.
- The *Occupational Outlook Handbook* is an excellent source of job information in the United States.
- In addition to formal schooling, education and training opportunities include the competitions and events held by student organizations, certificate courses, and apprenticeships.
- Securing a job in agricultural mechanics requires gathering documents that prove your skills to employers and hiring managers.
- Prospective employees must be well prepared physically and mentally before a job interview.
- Methods of searching for a job include networking, talking with school employment offices, and checking government agencies, trade unions, and professional organizations.
- Maintaining a job in agricultural mechanics involves continuing education, training, and the use of soft skills.

Words to Know

Match the key terms from the chapter to the correct definition.

A. career development event (CDE)
B. certificate
C. continuing education
D. cover letter
E. 4-H
F. job application
G. National FFA Organization
H. networking
I. *Occupational Outlook Handbook (OOH)*
J. portfolio
K. reference
L. résumé
M. skilled trade
N. SkillsUSA
O. soft skills
P. template

1. A record of skills, activities, training and education.
2. The act of getting to know people in various professions through self-introduction, or introductions through friends and associates; informing others you are looking for employment.
3. Evidence that a student has specialized agricultural mechanics skills; evaluated through testing of some type.
4. A form supplied by an employer to collect information about a prospective employee.
5. A job that requires a specific set of skills, such as electrical wiring, welding, construction, and masonry.

6. A competitive event sponsored by the National FFA Organization in which students learn, practice, and demonstrate skill attainment.
7. Training in which employees continue to study and gain new skills in their chosen profession.
8. A partnership of students, teachers, and industry working together to ensure that the United States has a skilled workforce.
9. A summary of your experience, education, and skills.
10. A letter that introduces a job candidate and refers the reader to the accompanying résumé.
11. A person who is willing to speak or write well of job applicants because of their past work or other association with the candidate.
12. A youth development program of the nation's US Department of Agriculture's Cooperative Extension Service that is available in every county and parish in the country.
13. A pre-formatted document for a specific form, such as a résumé.
14. An organization for students interested in agriculture and leadership.
15. A book detailing information and predicted demand for most careers in the United States.
16. A person's ability to function with other people; includes areas such as communication, language, personal habits, people management, and friendliness.

Know and Understand

1. What are jobs that require a specific set of skills called?
2. A(n) _____ is a combination of on-the-job training with some formal classroom education in the skilled trades.
3. _____ is a partnership of students, teachers, and industry working together to assure America has a skilled workforce.
4. A(n) _____ is a record of skills, activities, training, and education.
5. Name three things that a potential employer may require with a job application.
6. *True or False?* It is a not a good idea to practice answering questions before the interview because it may make the interviewee's answers seem rehearsed and unnatural.
7. A(n) _____ letter should be written and sent after the applicant returns home from the interview.
8. Name five places where one can begin a job search.
9. Describe *networking*.
10. What type of skills are defined as skills defined by working ability, communication ability, and the ability to get along with others?
11. *True or False?* Your chances of finding a job will improve if you look in an area close to your home.

STEM and Academic Activities

1. **Science.** Create a flowchart of a career related to agricultural mechanics.
2. **Technology.** Using the *Occupational Outlook Handbook*, as well as other resources, research a career related to agricultural mechanics. Make a presentation to the class about the career.
3. **Engineering.** Build a board game (or a video game if you have the capability and your teacher allows you access) that demonstrates the different paths to becoming an agricultural, or any other, engineer.
4. **Math.** Look up the average salaries for welders, welding inspectors, and welding engineers. Explain why there are differences in the expected pay.
5. **Language Arts.** In 140 characters or less, write an answer to the question, "What is an example of a soft skill?"
6. **Social Science.** Attend a local meeting of an industry organization or trade group and notice how the people at the meeting interact. Write a summary of your experience, concentrating on the networking possibilities you did or did not see at the meeting.

Thinking Critically

1. Write a one- or two-paragraph explanation of what you would do if you were the boss and an employee was tardy at least one day a week for a month.
2. Complete a one-page résumé and a one-page cover letter. If the résumé or cover letter is longer than one page, practice editing to make the document shorter but still informational. Have a classmate check both documents for spelling or grammatical errors.
3. Choose two agricultural careers that merit licensing, certification, or credentialing to become a professional within the industry. Within your state or county, examine the required licensing procedure. Investigate what types of certification or credentialing programs are also necessary/available. Write a short report based on your findings.

CHAPTER 3
Experiential Learning through Agricultural Mechanics

Chapter Outcomes
After studying this chapter, you will be able to:
- Explain the importance of learning through experience.
- Compare and contrast the four main types of supervised agricultural experiences (SAEs).
- Understand the structure and purpose of the National FFA Organization.
- Develop a plan for a personal SAE that includes an evaluation of interests and resources.
- Set both short-term and long-term goals for an SAE.
- Keep appropriate records pertaining to your SAE activities.

Words to Know
- asset
- audit
- budget
- business plan
- enterprise
- entrepreneurship SAE
- experiential learning
- exploratory SAE
- financial risk
- hands-on learning
- improvement activities
- internship
- journaling
- long-term goal
- personal interest inventory
- placement SAE
- proficiency award
- research SAE
- short-term goal
- supervised agricultural experience (SAE)

Before You Read
Arrange a study session to read the chapter with a classmate. After you read each section independently, stop and tell each other what you think the main points are in the section. Continue with each section until you finish the chapter.

While studying this chapter, look for the activity icon to:

- **Practice** vocabulary terms with e-flash cards and matching activities.
- **Expand** learning with interactive activities.
- **Reinforce** what you learn by completing the end-of-chapter questions.

www.g-wlearning.com/agriculture

There are tremendous career opportunities in the fields of agricultural mechanization and engineering. In 2014, an average of 25,340 people were hired per month. Job growth in this field is projected to increase sustainably in the future. Additional employment opportunities will become available through turnover and retirements. The demand for experienced and knowledgeable individuals will continue to grow in this industry. One of the best ways to start developing your career goals and gain valuable work experience is to complete an approved school-based agricultural education program.

Experiential Learning

Experiential learning is the process of creating understanding by doing (experiencing) something. Learning to ride a bicycle is an everyday example of this type of learning. Someone can explain how to ride a bicycle or even provide a demonstration, but in order to learn to ride, you must ride on your own. You might fall down a few times, but eventually you master the skill of balancing and pedaling to keep the bicycle from falling. Like learning to ride a bicycle, learning the skills needed for a career in agricultural mechanics takes practice and application of skills through learning experiences.

Supervised agricultural experiences allow you to begin "taking the training wheels off" as you learn essential skills. For example, you can read step-by-step instructions on how to rebuild a carburetor or install an electric fence, observe someone else doing the task, and listen to an explanation, but you do not really learn how to perform the task until you do it yourself.

Each component of school-based agricultural education is integrated with experiential learning. The three components are classroom/laboratory instruction, National FFA Organization activities, and *supervised agricultural experience (SAE)* programs. The three-circle model visually represents these components, **Figure 3-1**.

This integrated approach supports active, engaged, *hands-on learning*. Hands-on learning involves performing experiments or building projects as well as applying technical skills in FFA and SAE activities. The activities performed in each component of agricultural education help build reading, speaking, and math skills. Performing experiments and building machinery or structures require the understanding and application of scientific principles.

In the classroom and laboratory, students participate in guided practice and have ample opportunities to perform experiments. SAE programs complement classroom instruction by providing experiences. Students apply classroom instruction to an activity outside of regularly scheduled classroom or laboratory hours.

The SAE program and classroom instruction components are complemented by active involvement in FFA. Leadership development activities are an integral component of the FFA program. Classroom instruction teaches skills such as public speaking and how to conduct a meeting. The SAE and FFA programs then provide opportunities to apply the leadership skills needed in an agribusiness.

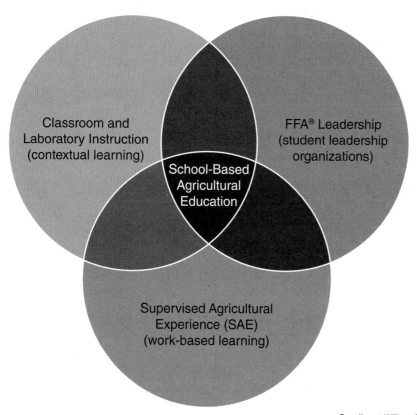

Goodheart-Willcox Publisher

Figure 3-1. Three-circle model of school-based agricultural education.

Classroom and Laboratory Instruction

The classroom serves as a formal setting to study the fundamentals of and theory behind agricultural mechanics and technology. The classroom is also where students learn how to plan and conduct a supervised agricultural experience program. Classroom instruction helps identify possible SAE options.

Laboratory settings provide controlled learning experiences. Experiments and research allow students to identify plausible solutions to mechanical or technological problems. Guided practice helps develop technical skills, such as welding or engine repair.

School greenhouses, land demonstration plots or farms, and animal production facilities simulate (mimic) real-world experiences. Practice takes place in a controlled environment in order to provide instruction and a positive outcome.

National FFA Organization

Although simulation provides an excellent environment in which to learn how to solve agricultural mechanics problems, it cannot replace real-world experiences. For example, simulation does not allow for personal interaction with employees or customers. Real-world experiences are an important part of agricultural education because agriculture is a complex and diverse industry involving people.

Leadership and communication skills are important, and it is critical that you practice them as part of your educational program. As part of the agricultural education program, the National FFA Organization serves to develop leadership, personal growth, and career success of its members. Only agricultural education students may become members of the FFA.

USDA/David Kosling

Figure 3-2. The FFA emblem is proudly displayed on this student's jacket.

FFA Levels

The FFA is organized by levels. The local chapter is the most critical level because it provides members with access to leadership development activities that meet local community needs. State associations and the national organization depend on high-quality, active local chapters.

FFA Emblem

The FFA emblem contains several symbols that represent the values of the organization. See **Figure 3-2**. These include:

- Eagle—perched atop the FFA emblem, the eagle represents patriotism and the national scope of the organization.
- Corn—grown in every state of the United States, this symbol serves to signify a unified interest in agriculture, regardless of other differences.
- Owl—representing knowledge and wisdom, this symbol acknowledges the need for experience as part of education.
- Plow—without labor, work, and effort, neither wisdom nor knowledge would accomplish much. The plow represents labor and tillage of the soil. This symbol signifies the need for action in order to get results in agriculture.
- Rising Sun—as the rising sun signals each new day, this symbol represents the progressive and forward thinking of agriculturalists working together toward common goals.

FFA Motto

The FFA motto "Learning to Do, Doing to Learn, Earning to Live, Living to Serve" describes the philosophy of learning and development provided in school-based agricultural education.

- Learning to Do—focuses on productivity using your hands as well as your mind.
- Doing to Learn—refers to the learning activities provided through school-based agricultural education. Experiencing results provides for more enduring learning.

- Earning to Live—emphasizes that FFA members should develop their skills and knowledge to support themselves in life.
- Living to Serve—emphasizes the need for cooperation, citizenship, and a willingness to help others.

Career Development Events

Challenging FFA program activities are offered for individuals or chapters. As described in Chapter 2, *Careers and Employability Skills*, career development events (CDEs) are competitions designed to polish technical and leadership skills. CDE topics include creed speaking (presenting the FFA creed from memory and explaining its meaning), prepared public speaking, extemporaneous public speaking, and job interviewing. Service opportunities range widely depending on the local chapter and community involvement.

An agricultural technology and mechanical systems CDE helps students develop technical content knowledge through practical hands-on skills. A systems approach reflects the complex interaction of processes common to agricultural production. The five systems include machinery/equipment systems, electrical systems, energy systems, structural systems, and environmental/natural resource systems. This approach emphasizes the interactive relationships among the areas of agricultural technology and mechanical systems.

Leadership

Leadership abilities are developed through participation in FFA programs. In agriculture, many organizations rely on multiple individuals with specific roles to accomplish goals. A typical agribusiness team may include a manager, sales representatives, and technicians. On a farm, a team may consist of family members and hired employees.

AgEd Connection | Awards

Award programs for proficiency and active involvement include SAE awards and FFA degrees. FFA degrees signify the progress a member is making in agricultural education. Each degree signifies professional growth from one level to the next within the FFA organization. Proficiency awards encourage students to improve and enhance their SAE program. A **proficiency award** is an honor given to FFA members who have demonstrated excellence in their SAEs. Some of the top awards for student excellence in SAE programs include Star Farmer, Star in Agribusiness, Star in Agricultural Placement, and Star in Agriscience.

pajtica/Shutterstock.com

Leadership requires the ability to guide or direct others in accomplishing a task or solving a problem. Leaders plan and organize in order to accomplish a goal. They must be able to communicate with, motivate, and engage people to accomplish complex and difficult tasks. A leader recognizes and uses the knowledge and skills of his or her team members.

Leaders should also be good citizens and model characteristics such as courage, integrity, loyalty, and selflessness. Courage is the willingness to proceed under difficult circumstances. Integrity means honesty. Loyalty is being a reliable support for another individual, group, or principle. Selflessness is placing the desires and welfare of others above one's own. Learning citizenship is a part of personal growth as you become a contributing member of society.

Supervised Agricultural Experiences (SAEs)

Participating in a supervised agricultural experience (SAE) allows you to apply knowledge gained in the school-based agricultural education program. The SAE is an important component of all courses in agricultural programs. In SAEs, students actively develop their own knowledge and skills through learning activities. See **Figure 3-3**.

All SAE programs consist of three core components: *supervision*, *agriculture*, and *experience*. Supervision is provided by a teacher, parent/guardian, or employer. The supervising individual is an experienced mentor who provides guidance and support as the student develops agricultural knowledge and skills. The mentor may also provide additional instruction or training. The SAE involves planned, sequential agricultural activities

Monkey Business Images/Shutterstock.com

Figure 3-3. Students may receive hands-on training as part of the SAE.

of education value that are conducted outside of class and laboratory instruction. SAE programs may vary for each student depending on learning needs and career goals.

Types of SAEs

Your SAE program should be a collection of experiences relevant to your educational goal. SAE programs can be a single type of activity or a variety of different experiences. SAE activities must be completed outside of regular classroom hours.

The type of experience is related to the different activities you perform in an SAE program, including how you obtain resources, how you spend your time, and how you manage SAE-related finances. The four types of experiences are exploratory, research-based, placement, or entrepreneurship experiences.

Exploratory

An *exploratory SAE* is a foundational project that allows students to explore agriculture course topics and career pathways. The purposes of this type of SAE are agricultural literacy (understanding of and knowledge about agriculture) and career exploration (learning about occupations and jobs relevant to career interests).

An exploratory SAE helps students develop a broad understanding of a particular agricultural topic, **Figure 3-4**. Ideas for exploratory projects include the following:

Monkey Business Images/Shutterstock.com

Figure 3-4. Exploratory SAEs involve recording activities and preparing reports.

- Job shadowing at an agricultural equipment dealer to explore employability skills.
- Researching the application requirements for an agricultural mechanics certification program.
- Investigating agricultural equipment emissions and air quality.
- Developing a list of high school classes that would help secure a position in the agricultural systems and mechanics field.

Students typically complete an exploratory SAE before other types in order to learn more about other agricultural mechanics–related SAEs. A final project of this experience may include a report on agricultural mechanic skills needed by agricultural professionals.

An exploratory SAE requires a considerable time investment, because exploratory activities must be initiated and managed by the student. This requires extra time outside of class. The teacher provides guidance and supervision, but it is the student's responsibility to complete the work. This work includes recording activities such as travel to worksites and developing reports. An exploratory SAE typically does not require a student to invest money in a project or take a financial risk. *Financial risk* is the uncertainty of receiving a return on money invested in a project.

Research

A *research SAE* is an original project that involves agriscience—the application of scientific principles to an agricultural problem. In research SAEs, students experiment or conduct an analysis in order to expand knowledge about a certain topic. This type of SAE requires the student to understand and apply the scientific method.

Research SAEs can be conducted using experimental or nonexperimental methods. An experiment involves orderly procedures to test hypotheses about how physical processes work under particular conditions, **Figure 3-5**. Nonexperimental research is descriptive research performed by collecting and comparing information. Nonexperimental research does not control confounding variables and does not allow the research to assume relationships between variables.

STEM Connection

The Scientific Method

The scientific method is a series of steps used to answer questions and draw conclusions. The scientific method can be used to answer scientific questions as well as simple, everyday questions.

The following are the common steps of the scientific method:
1. Ask a question.
2. Research to form a hypothesis.
3. Experiment.
4. Gather and analyze data.
5. Form a conclusion.

The first step in the scientific method is typically to ask a question based on observation. The next step is to do research to form a hypothesis. A *hypothesis* is your prediction of the answer to your question. Once you have formed a hypothesis, you design an experiment to test whether the hypothesis is true.

Experiments are conducted by controlling certain variables to determine the effect on others. *Variables* are conditions or processes that affect the results of the experiment. An *independent variable* is a condition or process that is changed by the researcher. The *dependent variable* is the variable being tested in the

Science photo/Shutterstock.com

experiment. It is not changed by the researcher, but is observed to determine the effect of changes made to the independent variable. An experiment must also provide for adequate control of confounding (external) variables that could affect the interpretation of results.

If the hypothesis is not verified by the results of the experiment, it may be modified and again be tested by experimentation. After experiments have been carried out, the data is gathered and analyzed. A conclusion is formed based on the analysis of the data.

Vlad Teodor/Shutterstock.com

Figure 3-5. Experimental research SAEs require careful planning, strict control of variables, and accurate recording of results.

Examples of experimental research SAEs in agricultural technology and mechanical systems include:
- Comparing the effects of various engine lubricants on friction reduction.
- Determining the effectiveness of fasteners on withdrawal resistance of wooden or metal construction materials.
- Comparing welding processes on filler metal consumption efficiency.

The following are examples of nonexperimental research SAEs:
- Conducting a biofuels usage survey among farmers and ranchers.
- Identifying community needs for precision agricultural technologies.
- Describing information sources that farmers and ranchers use to learn about agricultural technologies.
- Analyzing agricultural equipment sales data.

Quality research SAEs include specific research objectives that focus on an important agricultural issue, question, or principle. These SAEs provide experience in applying STEM (science, technology, engineering, and math) principles to help you prepare for an agriscience career. An exploratory or research SAE may also provide opportunities to investigate the types of resources required for placement or entrepreneurial SAEs.

Plan to invest time, and potentially money, to perform research. Research SAEs do not have financial risk, but they may require investment in supplies. Mentors provide students with guidance on the materials and equipment needed for conducting research.

Also plan to journal and keep records to verify the methods used in your research. After completing a research project, it is important to communicate your methods and conclusions. Publishing results promotes communication between scientists and encourages new scientific developments. You can publicly share your research results by participating in an FFA Agriscience Fair or by submitting a report to a scientific journal.

Placement

In a *placement SAE*, a student is placed as an employee or intern in a production agriculture enterprise, agribusiness, or agency. See **Figure 3-6**. *Enterprise* is another name for a business. Placement sites may include:

- Agricultural tire sales and repair business.
- Welding and fabrication business.
- Farm or ranch.
- Greenhouse or nursery.
- USDA Natural Resources Conservation Service (NRCS).
- Hardware or construction supply business.
- School laboratory.
- Community facility.
- Commodity agency.

Dmitry Kalinovsky/Shutterstock.com

Figure 3-6. An example of a placement SAE is working in a hardware store.

A placement SAE centers on working for someone else, either with or without pay. An *internship* is a placement experience that may last from 3 to 12 months. It focuses on providing training similar to that offered by a technical school. The US Department of Labor outlines six criteria that determine if a private sector internship can be unpaid:

1. The internship, even though it includes actual operation of the facilities of the employer, is similar to training that would be given in an educational environment.
2. The internship experience is for the benefit of the intern.
3. The intern does not displace regular employees but works under close supervision of existing staff.
4. The employer providing the training derives no immediate advantage from the activities of the intern, and on occasion its operations may actually be impeded.
5. The intern is not necessarily entitled to a job at the end of the internship.
6. Both the employer and the intern understand that the intern is not entitled to wages for the time spent in the internship.

Employment with a company is a form of placement SAE in which the student focuses on developing knowledge and skills while earning wages. Both internship and employment placement SAEs involve activities conducted outside normal classroom hours and include a training agreement that states the responsibilities of each party involved. These parties include the student as well as the teachers, parents, or employers who serve as supervisors or mentors.

Participating in a placement SAE requires a considerable investment of time devoted to working and training. The time required for travel must also be considered. Other investment considerations include supplies, such as tools required to complete work tasks at the job site. There is no financial risk associated with placement SAEs.

Entrepreneurship

An *entrepreneurship SAE* involves supervised activities conducted by students who are owners or managers for profit. An entrepreneurship SAE allows students to plan and operate either a production agricultural business or an agriculturally related business, **Figure 3-7**. Both production and agribusiness projects are considered entrepreneurial as long as is the student makes a financial investment. These projects provide excellent opportunities to make management decisions and apply classroom or laboratory instruction. The goal of an entrepreneurship SAE is to develop a profitable and productive business enterprise.

Fotokostic/Shutterstock.com

Figure 3-7. Students in entrepreneurship SAEs operate an agriculturally related business.

Production-based SAE projects may require the use of agricultural technology and mechanical systems to raise livestock or produce crops. Agribusiness SAE projects relevant to agricultural mechanics and technology may include:

- Purchasing or leasing agricultural equipment to perform custom harvest or processing operations.
- Landscaping and greenhouse ownership involving equipment operation and maintenance.
- Ownership of an equipment repair business.

Participating in an entrepreneurship SAE requires you to invest both time and money. Because entrepreneurship involves financial risk, it is important to take advantage of the mentor's assistance and supervision. The mentor can help you make decisions that increase the likelihood of a good return on your investment.

When buying or leasing equipment for a business, factors such as depreciation (reduction in value), insurance, and fees associated with ownership must be considered. Plan to maintain your enterprise through at least one production cycle. As with all SAE types, focus on developing projects that expand in scope and provide challenges each year.

Additional Components of SAE Programs

SAE programs also include other kinds of activities that allow you to practice a specific agricultural skill outside of normal class time. For example, you could include an improvement activity in your program. *Improvement activities* are learning activities that improve the value or appearance of a place of employment, home, school, or community; the efficiency of an enterprise or business, or the living conditions of a family. These activities are excellent service opportunities that allow you to contribute to your community. They generally involve a series of steps and often require a number of days for completion. Some may also require a financial investment.

Developing an SAE Plan

Choosing an SAE and identifying the most appropriate opportunities for learning and earning can be overwhelming. The most important factors to consider when planning an SAE are your interests, goals, and available resources. Consult your teachers to help identify interests and resources available to develop a successful SAE program.

Interests and Goals

When considering an SAE, first identify your agricultural career interests. A *personal interest inventory* helps you narrow down your career interests. It is a list of questions to assess your natural interests and provide guidance in developing a career. The National FFA Organization provides a personal interest survey that can be downloaded from the FFA website. You can print the form, fill it out, and share it with your agriculture teacher to help identify supervised agricultural experiences that are connected to your career interests.

Once you have decided on a career you might find rewarding or fulfilling, the next step is to determine attainable, yet challenging goals to prepare you for that career. Seek the advice of your teacher and observe what other successful students have done in their SAE programs. Your teacher can help you develop a written training plan.

An important element of developing a training plan is defining your short-term and long-term educational goals. *Short-term goals* are goals that can be accomplished in a day, next week, or even later in the year. *Long-term goals* typically take several years to accomplish. Achieving a series of short-term goals can lead to the achievement of long-term goals, such as being accepted by a college or saving for a down payment to purchase a piece of equipment for a new business.

As you write down your goals, consider evaluating them by the S.M.A.R.T. goal standards:
- S: Are your goals specific?
  - How and why will the goal be accomplished?
- M: Is your goal measurable?
  - What indicators will let you know that you have reached your goal?
- A: How attainable or achievable is your goal?
  - Have you or someone else accomplished a similar goal? Do you have the ability to achieve it?
- R: Is your goal result focused?
  - What is the reason and purpose for accomplishing the goal? What will the result will be once you have accomplished the goal?

- T: What time frame is required to accomplish your goal?
  - Does the goal have an established completion date that creates a practical sense of urgency to accomplish the goal?

Resources

A resource inventory is a list of items or sources available to help you conduct your SAE program activities. See **Figure 3-8**. Completing a resource inventory can help you identify available assets. An *asset* is the property, equipment, and other items owned by a person or company.

Assets can be current or noncurrent. Current assets are resources a company expects to convert to cash and use within one year. Examples of current assets are cash and inventory. Noncurrent assets are resources a company expects to hold for more than a year. Examples of noncurrent assets are property, equipment, and buildings.

Make a list of your assets to help your teacher identify SAE opportunities. If you are considering an entrepreneurship SAE, discuss resources such as tools, equipment, and storage that you will need to complete the activities.

Resource Inventory

1. Student Information
    a. Name: _____ Age: ____ Grade: _____ Class: _____
    b. Address: _____
    c. Phone (home): _____
    d. Phone (mobile): _____
    e. E-mail address: _____
2. Parent/Guardian Information
    a. Parent/Guardian Name: _____
    b. Occupation: _____
    c. Phone (home): _____ Phone (mobile): _____
    d. E-mail address: _____
3. Land Available
    a. Address/location: _____
    b. Leased or owned: _____ Owner: _____
    c. Phone (home): _____ Phone (mobile): _____
    d. E-mail address:
4. Building
    a. Shop: _____
    b. Storage: _____
    c. Other buildings: _____
5. Tools
    a. Power: _____
    b. Hand: _____

Goodheart-Willcox Publisher

Figure 3-8. Example of a resource inventory.

Selecting an SAE

When selecting an SAE, your objective is to seek planned, practical activities that develop necessary career skills. Hold a conference with your parents/guardians and agriculture teacher to review your plans. This planning should be based on your interests, but it is important to be flexible and open-minded about your experiences. A particular experience may trigger a new interest. See **Figure 3-9**.

Review any child-labor laws before committing to an SAE to ensure that your planned SAE is lawful. Review information about your home, work setting, community, or farm that could be useful when considering the SAE.

Agricultural Technology and Mechanical Systems SAE Ideas
Entrepreneur
Operate a lawn mower service and repair business.
Map fields, weeds, etc., for producers using GPS and GIS.
Run a custom fence building and repair business.
Provide maintenance for school laboratory equipment.
Construct compost bins to sell.
Start a chain saw basic maintenance & service business.
Start a custom vehicle refurbishing or painting business.
Start a detailing business for cleaning farm equipment on the farm (wash, wax, clean, maintain).
Start a farm equipment tire disposal business. (Turn old tires into livestock feeders.)
Start a farm fence maintenance business (cleaning fencerows, repairing).
Provide a poultry house maintenance preparation business.
Paint agricultural buildings and farm houses.
Contract with school system to maintain and service landscaping equipment.
Construct prefabricated wooden fence panels for sale to local hardware, building supply stores.
Construct spray rigs for four wheelers.
Constructing and marketing woodworking projects (birdhouses, dog houses, etc.).
Create a teaching model to show how a small gas engine works and sell to FFA chapters.
Constructing metal projects.
Placement
Work as a diesel mechanic assistant.
Work for an irrigation service company.
Assist with GPS mapping for an agronomic services company.
Work in a metal fabrication business.
Work as an assistant for an auto, truck or tractor mechanic.
Work for a local electrician.
Wire buildings for lights and receptacles.
Manage the steel inventory in the agricultural education program.
Work for a small engine repair shop overhaul and repair.
Work as a surveyor assistant.
Work for a plumbing business.
Work as an agricultural mechanics aide.
Working at a building supply business.
Working with a farm equipment dealer.
Work in a parts store.

(Continued)
Goodheart-Willcox Publisher

Figure 3-9. Examples of SAEs by category.

Agricultural Technology and Mechanical Systems SAE Ideas *(Continued)*

Exploratory

Map fields, weeds, etc., for producers using GPS and GIS.
Work as a diesel mechanic assistant.
Assist with GPS mapping for an agronomic services company.
Work in a metal fabrication business.
Work as an assistant for an auto, truck or tractor mechanic.
Work for a local electrician.
Pour concrete forms for machine sheds or other buildings.
Manage the steel inventory in the agricultural education program.
Machine and rebuild engine parts.
Work with county soil and water engineers, assisting in measuring and engineer design.
Work for a small engine repair shop overhaul and repair.
Work as a surveyor assistant.
Work for a plumbing business.
Work as an agricultural mechanics aide.
Working at a building supply business.
Working with a farm equipment dealer.

Research

Research the energy use of different types of lighting systems in farm buildings.
Research the water savings accomplished by new irrigation technologies.
Research the biggest challenges producers face when adopting new technology.
Research the differences in various styles of tillage equipment.
Test the strength of different types of welds.
Construct a hydro ram pump and calculate the efficiency and water delivery rate.
Construct prototypes of hydraulic systems.

Improvement

Restore a tractor for someone else.
Build garden sheds for homeowners.
Install electrical circuits or wiring system at a barn.
Pour concrete forms for machine sheds or other buildings.
Provide maintenance for school laboratory equipment.
Manage the steel inventory in the agricultural education program.
Complete home or farmstead improvement construction activities.
Paint agricultural buildings and farm houses.
Install plumbing fixtures or plumbing system in your own building.
Construct or recondition a welding project (such as a trailer, cooker, etc.) at home or in school provided facilities.
Build handicap ramps in the local community.
Construct concrete projects for the home or farm.
Construct metal projects.

If you are considering a placement SAE, review your schedule and consider the number of hours you will be able to work. Do not forget to include travel time and distance to the work site in your calculations.

Once you have decided on an SAE, seek advice and guidance from those most involved with your project, **Figure 3-10**. Review your progress toward achieving your goals with your parents, teachers, or employer regularly. Your experiences should logically build on each other to help you accomplish your long-term goals.

Monkey Business Images/Shutterstock.com

Figure 3-10. Planning with a mentor helps you identify potential SAE opportunities.

Planning

It is important to plan before starting an SAE. Establish goals and identify the steps needed to accomplish those goals. Planning for exploratory and research SAE includes identifying learning goals or potential areas for inquiry. Planning for placement SAEs involves identifying career goals or the steps needed to develop specific technical skills related to a career. An entrepreneurship experience may require a business plan.

A *business plan* is a formal, written statement of business goals, the reasons the goals are attainable, and the plans for reaching them. A business plan is similar to a road map—it provides directions for the future of a business. Generally, a business plan includes a sales or marketing strategy, financial information, and projected goals, **Figure 3-11**. Taking the time to make a thorough and accurate business plan is an investment in the success of the business. For example, a business plan is almost always needed to apply for a business start-up loan. The plan serves as a decision-making tool for banks, investors, and management personnel.

Business plans should follow form and content guidelines specified by the people who may read the plan. Many free business plan templates are available online. It is your responsibility, however, to choose one that reflects your proposed business. The template provides the shell, but you must write effectively to communicate your needs and explain why your business is worth investing in.

You should also develop accurate estimates of anticipated expenses and income. You may need a budget to help plan your cash flow. A *budget* is an estimate of income and expenditures for a set period of time. In addition, you should prepare a résumé, which will help a prospective investor to determine your qualifications and experiences.

SAE Recordkeeping

Agricultural production operations and agribusiness enterprises use recordkeeping to make data-based decisions in order to improve productivity and be successful. Recordkeeping helps keep your SAE documentation organized and ensures that you are using resources efficiently. Efficient

arka38/Shutterstock.com

Figure 3-11. A business plan is a road map to the future of a business.

use of resources maximizes your ability to make a profit or save money to purchase new equipment. Keeping good records of your activity also helps you organize information needed for loans or tax returns. In addition, when you apply for an FFA degree or proficiency award, thorough and organized records will save you time and frustration.

Computer programs are the most effective means of keeping records of your experiences in agricultural education, which include SAE finances, time in FFA, and SAE and other community and personal development events. Spreadsheets and written records are sometimes used, but these can be tedious and may lack connections to current industry trends in management information. Check with your teachers to determine if a specific recordkeeping program or paper-based format is required by your state FFA association.

The Agricultural Experience Tracker (AET) is the primary recordkeeping system that simplifies the process and improves financial management information, **Figure 3-12**. The AET is an online system that keeps track of your finances and your educational accomplishments. It not only allows you to record receipts and expenses, but also tracks your progress in your SAE and FFA program.

Review your SAE progress regularly and log the data. Being organized can save you significant time in the event of an *audit* (official inspection of an individual's or organization's accounts). Good recordkeeping is also useful for preparing the financial records needed for a business loan.

Income	Monthly Total
Salary (after taxes/benefits)	$
Sales (after taxes)	$
Other income (after taxes)	$
Total Monthly Income	$

Category	Expenses	Monthly Total
Overhead		
Salaries		
	Total Expenses	$

(Total monthly income) − (Total monthly expenses) =

The AET Record Book

Figure 3-12. *The AET Record Book* is a valuable tool for keeping accurate records of all your experiences in agricultural education.

Financial Records

Keeping accurate financial records is critical. Financial records include documents such as balance statements, budgets, and invoices. *Accounting*—the process of keeping or preparing the financial records of a business—is used to summarize, analyze, and report financial activity. Accounting tracks *cash flow*, which is the total amount of money being transferred into and out of a business. Cash flow depends on income and liabilities. *Liabilities* are items for which someone is responsible, such as a debt or financial obligation. *Debts* are money owed or due to a second party, such as a lender or creditor. Liabilities include loans and accounts payable.

Budget and Inventory

A budget is not just a tool to help you apply for financing. It is an active tool that you should evaluate and update frequently. Review your budget when making financial entries to monitor your expenses and income. Maintaining a budget improves the financial fitness of your business or personal bank account. The budget is important because it helps you identify limitations in time, finances, and equipment/facilities.

In many businesses, good inventory practices are one of the keys to success. Too large an inventory on hand can increase storage and overhead costs. Too little inventory may cause you to run out of an item and miss sales opportunities. Computer programs can help you keep track of inventory so that you know how much you have on hand at all times.

Journaling

In addition to keeping financial records, you should log your learning experiences. This can be accomplished through *journaling*, or writing about your experiences. Journaling can serve as a reflection time to review your experiences, allowing you to identify areas for improvement and areas in which you have been successful. In addition, regular journaling makes it easier to identify those accomplishments that make an award application stand out. See **Figure 3-13**.

Weekly Journal Worksheet	
Week of:	Instructor:
Name:	Class:
Using proper grammar and punctuation, summarize your SAE activities each day in the space provided. Include the date for each entry.	
Sunday	
Monday	
Tuesday	
Wednesday	
Thursday	
Friday	
Saturday	

Goodheart-Willcox Publisher

Figure 3-13. Example of a student journal worksheet.

CHAPTER 3
Review and Assessment

Summary

- School-based agricultural education incorporates three components: classroom/laboratory instruction, supervised agricultural experiences (SAEs), and FFA participation.
- SAEs emphasize experiential learning, which is creating understanding by doing something.
- SAE experience types include exploratory, research, placement, and entrepreneurship.
- The most important factors to consider when planning an SAE are your interests, goals, and available resources.
- When selecting an SAE, seek planned, practical activities that develop necessary career skills.
- SAE recordkeeping is necessary to keep your SAE documentation organized and ensure that you are using resources efficiently.

Words to Know

Match the key terms from the chapter to the correct definition.

A. asset
B. audit
C. budget
D. business plan
E. enterprise
F. entrepreneurship SAE
G. experiential learning
H. exploratory SAE
I. hands-on learning
J. long-term goals
K. personal interest inventory
L. placement SAE
M. proficiency award
N. research SAE
O. improvement activities
P. short-term goals

1. An estimate of income and expenditures for a set period of time.
2. Another name for a business.
3. Supervised activities conducted by students who are owners or managers for profit.
4. Goals expected to be accomplished in more than a year.
5. Property, equipment, and other items owned by a person or company.
6. Working as an employee or intern for a production agriculture enterprise, agribusiness, or agency to gain experience in a career.
7. Goals that can be accomplished within a year.
8. The process of creating understanding by doing something.
9. Official inspection of an individual's or organization's accounts.
10. An honor given to an FFA member who has demonstrated excellence in his or her SAE.

11. A list of questions to assess your natural interests and provide guidance in developing a career.
12. An original project that applies scientific principles to an agriscience problem.
13. Foundational project that allows students to explore agriculture course topics and career pathways.
14. Formal statements of business goals, reasons they are attainable, and plans for reaching them.
15. SAE learning activities that improve the value or appearance of the place of employment, home, school or community; the efficiency of an enterprise or business, or the living conditions of the family.
16. Involves performing experiments or building projects as well as applying technical skills in FFA and SAE activities.

Know and Understand

Answer the following questions using the information provided in this chapter.

1. Experiential learning involves learning by _____.
   A. observing
   B. doing
   C. reading
   D. listening
2. List the four short phrases that make up the FFA motto.
3. Career development events are competitions designed to improve technical and _____ skills.
4. The three core components of all SAE programs are supervision, agriculture, and _____.
5. A(n) _____ SAE involves the application of scientific principles to an agricultural problem.
6. *True or False?* Research SAEs can be conducted using nonexperimental methods.
7. A(n) _____ SAE involves working for someone else to gain career experience.
8. A(n) _____ SAE involves owning a business for profit.
9. *True or False?* Entrepreneurship involves taking financial risk.
10. List three important factors to consider when planning an SAE.
11. *True or False?* A building is an example of a current asset.
12. A business plan generally includes a sales or marketing strategy, _____ information, and projected goals.
13. Accounting tracks _____, which is the total amount of money being transferred into and out of a business.

STEM and Academic Activities

1. **Science.** Choose an agriscience topic of interest to you. Compile a list of potential research questions to use in developing a research SAE activity.
2. **Technology.** Compare and contrast two different software programs used for keeping financial records. Write a report that details your findings. Explain which program you would prefer to use and why.
3. **Engineering.** Interview an engineering firm director to learn the education and skills required to design equipment for production agriculture.
4. **Math.** Balance a budget for an entrepreneurship SAE. Ask your teacher for a list of fictional expenditures and revenues if you do not have an entrepreneurship SAE. Determine whether the revenue exceeds the expenditures, and provide recommendations to increase revenues or reduce expenditures.
5. **Language Arts.** After completing any type of supervised agricultural experience, review your journal and other recordkeeping documents. If possible, exchange journals with a classmate and review his/her material. What new skills have you gained or sharpened during the SAE? What did you enjoy about the project? What aspects did you find unfavorable? Did the project open new opportunities or expand your interest in the specific field of agricultural study? Present your evaluation of the completed SAE to your instructor and to your class.
6. **Social Science.** Research the career demands for agricultural mechanics and technology professionals in the agriculture industry between 1920 and the present. Create a timeline that illustrates the advances in technology and the various career fields that have developed over the years. Present your timeline to the class.

Thinking Critically

1. Analyze agricultural production sales within your state and county using statistics provided by the National Agricultural Statistics Service. Develop a market analysis that describes the industry and outlook. Describe the industry, including its current size and historic growth rate, projected growth rate, and major customer groups within the industry. Prepare a report that details your findings. State your conclusions regarding potential areas in which to begin an agricultural business.
2. Compare and contrast two different agricultural mechanics and technology careers. Identify differences and similarities in the technical skills needed for each career. Assess the skills you possess that meet the career description that you are most interested in. Prepare a written skill development plan to prepare you for one of the careers you have evaluated. Discuss your plan with your teacher or a parent.
3. A large part of your FFA chapter's efforts is establishing a program of activities that meets the FFA's National Chapter Awards Chapter Quality Standards. Review these standards and evaluate your chapter's program of activities (POA). Determine if your chapter addresses the quality standards in each of the three divisions (student development, chapter development, and community development). Use a strategic planning process to establish or modify your chapter's goals to ensure each of the standards is addressed.

CHAPTER 4
Trends and Emerging Technologies

Chapter Outcomes

After studying this chapter, you will be able to:
- Explain the role of technology in agriculture.
- Describe sustainable methods of energy production.
- Identify alternative fuels.
- Explain ways to manage agricultural technology to reduce harmful impacts on the environment.
- Describe the equipment and systems used in precision agriculture.
- Describe the role of information technology in agriculture.
- Explain the contribution of biotechnology to agricultural practices.

Words to Know 🔗

anaerobic digesters
automation
autonomous
biofuels
biologging
biosecurity
biotechnology
cloud server
diesel exhaust
  fluid (DEF)
drone
engine control
  unit (ECU)
exhaust gas
  recirculation (EGR)
forced induction
genetic engineering
geographic
  information
  system (GIS)
global positioning
  system (GPS)
hydrogen fuel cell
photovoltaic system
precision agriculture
selective catalytic
  reduction (SCR)
site-specific
  management
technology
telematics
unmanned aerial
  vehicle (UAV)
variable-rate
  technology (VRT)
yield monitoring

Before You Read

Scan this chapter and look for information presented as fact. As you read this chapter, try to determine which topics are fact and which are the author's opinion. After reading the chapter, research the topics and verify which are facts and which are opinions.

While studying this chapter, look for the activity icon to:

- **Practice** vocabulary terms with e-flash cards and matching activities.
- **Expand** learning with interactive activities.
- **Reinforce** what you learn by completing the end-of-chapter questions.

www.g-wlearning.com/agriculture

Technology has impacted nearly every facet of agriculture. *Technology* can be defined as the application of science to solve a problem. Machinery, tools, and equipment are developed to meet specific needs and solve human problems. Technological advances occur through interactions among multiple disciplines, such as engineering and physics. Technology is being increasingly applied to agricultural practices. For example, a new tractor cab design may be developed to make the cab safer and more comfortable for operators.

Production agriculture has benefited from the incorporation of technological advances achieved in other industries. The industrial age brought mechanization and synthesized fertilizers to agriculture. The technology age introduced genetic engineering and automation. The information age brings the potential for integrating the industrial and technological advances into a sustainable agriculture production system to make the best use of valuable resources.

It has been estimated that by 2050, the global population will approach nine billion, resulting in a near-doubling of global food and fiber demand. High-quality irrigation water and land are becoming increasingly scarce. Increasing production in ways that do not compromise the environment or public health is a major challenge. Intensification of agriculture through the use of high-yield crop varieties, fertilization, efficient irrigation, and protection of crops from pests and diseases remain the most likely ways to meet the challenges that lie ahead.

Sustainable Energy

Environmental technology is the use of technological methods to manage and conserve natural resources and to protect the natural environment. The production of energy using renewable resources is an example of environmental technology. Alternative means of electricity generation are growing to meet the energy demands of farming operations and to reduce costs. Several technologies, including alternative fuels, methane digesters, small-scale solar panels, wind power turbine generators, and biomass generation, are sources of renewable power for agricultural buildings, **Figure 4-1**.

ABBYDOG/Shutterstock.com

Figure 4-1. In areas that receive a dependable amount of sunshine or wind, electricity can be generated using solar panels or wind turbines, respectively.

Alternative Fuels

Most fuel produced today comes from fossil fuels found in the earth, such as petroleum and coal. Fossil fuels are nonrenewable resources, meaning they are in limited supply. The need to identify alternative and renewable fuel sources is increasing to meet the demand of a growing population. Biofuels and hydrogen fuel cells are two alternatives being developed to lessen dependence on fossil fuels. It is important to recognize that energy is needed to produce fuels. When evaluating the viability of an alternative fuel, the energy needed to produce the fuel must be compared to the energy released when the fuel is burned.

Biofuels

Biofuels are renewable resources—they are produced from natural and agricultural products that can be replaced naturally. Ethanol, a common biofuel, is an alcohol made by fermenting plant matter that is high in carbohydrates. Scientists are developing ethanol fuel from the cellulose material that makes up the volume of most plant materials. Ethanol is currently blended with gasoline in an effort to reduce emissions and decrease the use of nonrenewable fuels.

Biodiesel is a common renewable biofuel that is used in diesel engines. It is created using a complex process called transesterification, which involves a chemical reaction between methanol (a type of alcohol) and oils from vegetables, animal fat, or recycled cooking oils. Biodiesel can be used as an additive to traditional diesel fuel or as a standalone alternative fuel for diesel engines.

Hydrogen Fuel Cells

Hydrogen fuel cells run on hydrogen gas to produce electricity to power an electric motor. Hydrogen fuel cells contain an anode, a cathode, and a membrane located between the anode and cathode. Hydrogen fuel enters the anode side of the fuel cell, where the hydrogen molecule is split into a proton and an electron. The proton moves through the cell and the electron follows a circuit to power the electric motor. The proton and neutron join again on the cathode side and combine with oxygen to produce water. The water is the sole emission product and is harmless to the environment.

At least one agricultural equipment manufacturer is experimenting with a fuel cell–powered tractor. Some of the challenges include the lack of availability of hydrogen filling stations and the lack of widespread hydrogen fuel production. In addition, the presence of high-voltage electricity and compressed hydrogen gas presents safety concerns that must be resolved.

Hydrogen can be produced from natural gas and renewable resources such as water. Natural gas is currently the most common source, but scientists are working to increase the use of renewable sources of hydrogen.

Anaerobic Digesters

Anaerobic digesters, also called *methane digesters*, are growing in popularity. These digesters break down crop residue and animal manure to generate electricity or heat for farm operations. See **Figure 4-2**.

hans engbers/Shutterstock.com

Figure 4-2. Anaerobic digesters produce energy for farm operations and help reduce greenhouse gases.

Anaerobic digestion is a biological process that produces a biogas, which is composed of different gases created by the breakdown of organic matter in the absence of oxygen.

Methane digesters turn manure into a renewable energy source and can reduce odors and greenhouse gas emissions. However, the initial cost to install these systems is high. Farm size and access to electrical transmission lines can also limit the adoption of this technology.

Photovoltaic Systems

Photovoltaic systems, or solar electric energy systems, convert light energy (photons) into electricity (voltage) to provide power. These systems supply power for a number of remote agricultural applications, including pumping and electric fencing. One recent development in solar panel technology is the incorporation of photovoltaic equipment into roofing products for building construction, **Figure 4-3**.

Photovoltaic systems convert the photons into direct current (DC) electricity. An inverter is needed to convert the DC power into the alternating current (AC) that powers devices such as tools and equipment that plug into an electrical outlet.

The amount of electricity generated by a photovoltaic system depends on how much of the sun's energy reaches it. Areas in the southwestern United States can use these systems efficiently because of the high availability of sunlight or peak-sun hours. The term *peak-sun hours* refers to the amount of solar insolation (the sun's radiation that reaches the earth's surface) for a particular location if the sun were shining at its maximum value for a certain number of hours. Peak-sun hours vary by season—highest in summer and lowest in winter. At night or on cloudy days, the system may not produce enough electricity to meet electrical needs. The average solar cell of a system is about 15% efficient in converting light energy to usable electricity. An additional concern is the limited ability to store electricity generated during the daylight hours for use at night.

Federico Rostagno/Shutterstock.com

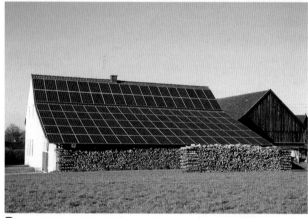
manfredxy/Shutterstock.com

Figure 4-3. Photovoltaic panels. A—Solar shingles being installed. B—Solar panels on a farm building.

Wind Energy

Turbines in fields or on farmsteads can harness energy from the wind to generate power, **Figure 4-4**. The size of a wind turbine affects the power rating. Turbines with blades that extend the length of a football field can

Stockr/Shutterstock.com

Figure 4-4. Wind turbines located on farm fields generating electricity.

produce enough power for 1,400 homes. Smaller turbines can generate from 50 to 70 kilowatts of power.

Sustainable wind speeds are critical for consistent and reliable electrical energy production. Wind resources are classified based on wind-power density classes. The classes range from 1 to 7. Class 3 and above are considered to be the most suitable for wind turbines. These areas have an average annual wind speed of at least 13 miles per hour. The disadvantages of wind energy production include cost and the variability of wind speeds when power is at the highest demand.

Environmental Systems

Management of our environment and environmental systems includes managing facilities, land, food, and commodities, along with the safety and distribution of food and commodities locally and on a global scale. Structure and planning of system components focuses on continual improvement while meeting the legislative requirements of a location.

Energy Management

Mechanical and electrical systems furnish ventilation, heating, cooling, and lighting functions. As competition for limited resources increases, agriculture production costs continue to rise. Some of the largest annual expenses for producers include cost crop irrigation, postharvest crop conditioning, and managing energy use of livestock housing systems and greenhouse production facilities.

Thinking Green

National Center for Appropriate Technology (NCAT)

The mission of the NCAT, a national nonprofit agency, is to promote sustainable living. Their work in the agricultural area includes providing assistance to small farmers and beginning farmers and supplying technical assistance and information for sustainable agriculture. NCAT-recommended guidelines for choosing technologies to improve efficiency and sustainability include:
- Start with a design that is suitable for local conditions. The design should take advantage of natural ventilation, heating, and lighting.
- Use equipment that minimizes operating energy needs.
- When possible, incorporate renewable energy supplies.

Green building technology innovations can reduce both energy costs and environmental impacts. A green building is a structure that is designed, built, and operates using environmentally sustainable methods to minimize total environmental impacts. Choosing efficient systems for heating, cooling, ventilation, and lighting reduces energy costs, minimizes pollution, and helps support the use of renewable energy on-site by cutting loads. Economic evaluation of green building technologies is highly site-specific. Their feasibility depends on variables such as electric demand (kW) and consumption rates(kWh), labor costs, and other options available for the building site.

Common agricultural structures, such as barns, grain bins, machine storage, and silos may need support systems to maintain animal health, crop quality, and human health. Ventilation, heating, cooling, and lighting systems maintain the environment of the structure. Some of the largest annual expenses for producers include crop irrigation, postharvest conditioning of crops, and energy use in livestock housing systems and greenhouse production facilities.

ssguy/Shutterstock.com

Figure 4-5. LED lighting used in indoor growing conditions to support vegetable production.

Energy-Efficient Lighting

The properties of light that enable plants to grow well and flower include light quantity/intensity, day length, spectrum, and light uniformity. Horticulturalists develop lighting that supplements or replaces natural light in areas with limited space. The use of lighting to supplement natural light is expensive. LEDs require 75% less energy and last 25 times longer than traditional incandescent bulbs without using harmful mercury found in fluorescent bulbs. LED (light-emitting diode) technologies are helping create indoor growing environments capable of commercial vegetable production, **Figure 4-5**. LED lighting technology can supply the spectrum and intensity that fruit and vegetables need without adding extra heat.

Automation

Automation means that the operation or function of equipment, processes, or systems is automatically controlled by mechanical or electronic devices instead of manually. Automation technologies are being used to maintain tighter control over heating, ventilation, and lighting of buildings. This is accomplished through sensors and controllers that monitor and improve the efficiency of heating and cooling systems. Sensors provide the controller with data on temperatures, humidity, pressure, current flow, airflow, and other essential factors. The controller is programmed to monitor the inputs and send control commands to devices of the system.

Food Distribution and Safety

Food distribution requires safe and efficient transportation of food. Effective transportation relies on well-maintained roads, railroads, airports, and seaports, as well as the vehicles that operate on them. To avoid spoilage and wasting, food must be properly handled and stored, and it must be transported in a timely manner. Perishable food may need to be refrigerated during transport and storage. Technology provides the means to safely transport, handle, and store food.

Food-borne sickness is a challenge that needs to be continually addressed as technologies and means of distribution adapt to a growing population. Food production methods vary from country to country. As foods are imported from and transported to other locations, steps must be taken to reduce sickness from food contamination and spoilage.

The overuse of antibiotics has increased the resistance of certain bacteria to antibiotics, causing an increasing number of food-related illnesses. Safe handling and packaging methods need to be adapted to reduce contamination of foods as production and processing methods change to meet increasing demands.

Some of the new technologies discussed in this chapter affect the safety of food production and processing in the future. The reduction of contaminated water and air from emissions and certain agricultural practices can increase the safety of our future food supply.

Biosecurity is the use of procedures specifically to protect humans and animals against disease or harmful biological agents, both accidental and intentional. *Agro-terrorism* refers to potential terrorist attacks using the food supply. There are many steps associated with food production, from the farm to the table. Many of these steps provide an opportunity for the introduction of toxins, pathogens, either by accident or as a deliberately harmful act. It is, therefore, critical to identify and control access points for disease and other biological agents along the chain of food distribution.

Figure 4-6. An older tractor emitting exhaust during field operations.

Pollution

Pollution is the introduction of harmful contaminants into the land, air, and water. Two major forms of pollution are air pollution and water pollution.

Air Pollution

Pollution results from manufacturing, agriculture, and other human practices. Burning petroleum products for fuel contributes to air pollution. In addition, some scientists blame methane from livestock production and chlorofluorocarbons (CFCs) from aerosol products for their contribution to air pollution.

A major cause of air pollution is vehicle emissions. Problems such as poor air quality have generated the need to address internal combustion engine emissions, particularly in diesel engines, **Figure 4-6**. Vehicle emission reduction and control technologies include engine control units (ECUs), exhaust gas recirculation (EGR), selective catalytic reduction (SCR), and performance improvements such as forced induction.

Computers are now being used to help make internal combustion engines more efficient. These computers, commonly called *engine control units (ECUs)*, are used in combination with a network of sensors and actuators to monitor and control various engine functions, fine-tuning the engine for greater fuel efficiency and increasing overall power. A *sensor* is a device that detects or responds to physical stimulus and transmits an impulse to provide data or operate a control, such as a fuel injector. See **Figure 4-7**.

Exhaust gas recirculation (EGR) allows cooled exhaust gases to be drawn into the air-fuel mixture entering an engine's cylinders. EGR reduces combustion temperatures, resulting in lower nitrous oxide emissions. Nitrous oxide is a major component of smog. EGR also improves fuel economy. The lower temperatures also help to avoid heat transfer energy losses, meaning that more of a tractor's energy is directed to providing power for its wheels. Combined with electronic fuel injection, these technologies have created more efficient and cleaner burning engines.

Figure 4-7. Engine control units (ECUs), such as the one shown above, are used in combination with a network of sensors and actuators to monitor and control various engine functions, increasing efficiency and reducing emissions.

Selective catalytic reduction (SCR) is a recent technological development that reduces diesel engine emissions. A liquid reducing agent is injected through a special catalyst into the exhaust stream of a diesel engine to break down dangerous NO_x emissions into harmless nitrogen and water. Automotive-grade urea (made of synthetic ammonia and CO_2) is commonly used as the reducing agent and is sometimes referred to as *diesel exhaust fluid (DEF)*. See **Figure 4-8**.

Forced induction is the delivery of compressed air to the intake of an internal combustion engine.

Turbochargers and superchargers increase the air pressure entering the engine. Forced induction makes higher power possible with smaller engines, improving fuel economy. See **Figure 4-9**.

Water Pollution

Safe, clean drinking water is essential for life. Water pollution may be obvious, such as debris and contaminants in lakes and streams. However, pollution of groundwater is usually less obvious. Water that looks clear and clean can contain toxins.

Sources of water pollution from human activity include sewage, agricultural pesticides, fertilizers, factory wastewater, plastics, oil pollution from petroleum spills, sediments from construction activity, and thermal pollution from factories. Efforts are underway to reduce the impact of manmade pollutants to the water supply. Alternative sources of energy can reduce water contaminants. Continued development of technologies to treat contaminated water is necessary for future clean water.

Sewage treatment is subject to local, state, and federal regulations and standards. Modern sewage treatment facilities have advanced well beyond the historical practice of dumping raw sewage into rivers and oceans. Using multistep processes involving primary, secondary, and additional treatments, modern facilities disinfect sewage-contaminated water so it can be introduced directly and safely into the environment.

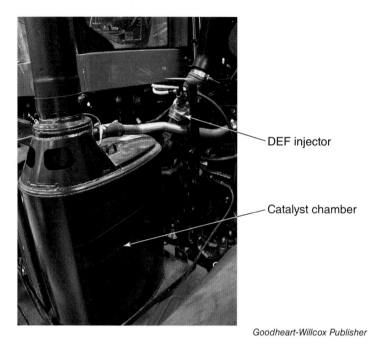

Goodheart-Willcox Publisher

Figure 4-8. Selective catalytic reduction (SCR) systems are used to help reduce diesel engine emissions. Diesel exhaust fluid (DEF) is injected through a special catalyst into the exhaust stream of a diesel engine to break down NO_x emissions into nitrogen and water.

Precision Agriculture

Precision agriculture is a systematic approach to site-specific agricultural management. Variability in crops is observed and a strategic response to improve productivity is prescribed. This approach involves *variable-rate technology (VRT)*, which enables farmers to vary the rate of crop inputs. Technology is used to minimize wasted application of seed, fertilizer, and pesticides by allowing more precise application for a given geographic location. Recent technological developments

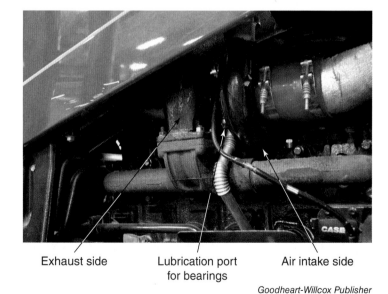

Goodheart-Willcox Publisher

Figure 4-9. Forced induction systems, such as this turbocharger on a tractor engine, increase the air pressure entering the engine, increasing power and boosting fuel economy.

Rosamund Parkinson/Shutterstock.com

Figure 4-10. Mobile technology being used in the field.

have increased yields with lower inputs. The idea is to farm sustainably and limit the use of natural resources.

Computer and Smartphone Applications

Increased computer and Internet applications continue to change the way people communicate and share information. Mobile Internet and cell phone technologies enable nearly instantaneous access to necessary information. A smartphone or other mobile device provides the ability to map, calculate, measure, and instantaneously access a variety of information on location. See **Figure 4-10**.

In agriculture, information technology aids in calculating costs and analyzing data in order to make sound farm production decisions. Software applications, often called *apps* when used on mobile devices, are continually being created, adapted, and modified for agricultural use.

There are land mapping apps that show topography, soils, bodies of water, and location. Other apps assist in the following tasks:
- Calculate tank mix for pesticides.
- Estimate crop yields.

STEM Connection

Communication Tools

There are many kinds of information technologies, including radars, mobile/telephones, FAX machines, computers, and satellites. These technologies make up information systems, such as systems for information retrieval and systems that help us solve problems or make decisions. Tablets and smartphones make mobile computing part of everyday life. The mobile apps used on these devices facilitate information retrieval and improve productivity.

Wireless networks connect various computers and communication tools. Integration of sensors with wireless capabilities allows real-time data collection in the field, such as determining soil temperature and moisture content. The information gathered and collected has traditionally been stored on physical servers—computer processers that allow people to share and store data. Servers also provide a means of hosting software resources.

A **cloud server** is a network of remote servers hosted on the Internet to store, manage, and process data. Cloud servers are rented as virtual server space rather than renting or purchasing physical servers. Individuals using a cloud server have the benefits of flexibility, scalability, ease of setup, and reliability.

Budimir Jevtic/Shutterstock.com

- Manage livestock.
- Maintain equipment.
- View product labels.
- View safety data sheets.
- Manage pest control.
- Price commodities.
- Determine weather.

A wide variety of additional features are continually being created and adapted for agricultural use.

Geographic Information System (GIS)

Geographic information system (GIS) software is a data management tool. It is used to map, model, query, and analyze large quantities of data within a single database according to a reference location. Google Earth, Arc GIS, and other software applications support this activity. Some programs download data onto removable storage media, such as a flash drive. Others upload the data to a cloud server.

Data is referenced with GPS coordinates and then mapped to create management zones or grids. These zones or grids are used to make production decisions. Information is stored as a collection of layers that can be linked together by a common locational component, such as latitude and longitude, ZIP code, census tract name, or road name. Mapping allows the farmer or rancher to visualize the data gathered from the field, such as yield, water usage, plant populations, and soil fertility, to make site-specific management decisions. See **Figure 4-11**.

Site-specific management involves observing and measuring the differences within fields, noting the specific locations of the differences, and using the information as a basis to make improvements in management or inputs. In site-specific farming, different areas within fields are managed in different ways according to their needs.

Yield monitoring is the use of sensors, such as moisture sensors, to measure and record data during harvest. Yield data can then be linked to GPS coordinates to map out high and low productive sections of a field. These are referred to as *yield maps*. Additional maps can be generated to visualize yields and profits generated from various sites. This information can be used in deciding to purchase or sell a field, as well as predict productivity for the future market.

Photo by Stephen Ausmus, USDA Natural Resources Conservation Service

Figure 4-11. GIS applications can help farmers determine how best to use available land. Here, agricultural extension agents are assessing nitrogen management at both the farm and regional levels.

76 Agricultural Mechanics and Technology Systems

The Web Soil Survey (WSS), operated by the USDA Natural Resources Conservation Service, is a web application that provides natural resource information. Producers can access soil and related information to aid in making decisions about land use. Users highlight areas on the map to collect relevant information.

Global Positioning System (GPS)

For precision agriculture and site-specific management to work effectively, a farmer or rancher must have a geographic reference point to associate with any variability within or between fields. Global positioning systems are used to provide geographic reference data. A *global positioning system (GPS)* is a navigation system made up of a network of satellites placed in orbit around the earth. These satellites transmit a signal referencing their location as they orbit the earth. A GPS receiver compares the time a signal was transmitted by a satellite and the time it was received in order to determine the user's position and display it on an electronic map. See **Figure 4-12**.

To accurately place seed and fertilizer in a field where it will be most effective for increasing yields, farm machinery must be precisely guided to a geographic reference point in the field. Satellite-based guidance for agricultural vehicles provides benefits such as the reduction of skips and overlaps, increased ability to work in poor visibility, and lower operator fatigue.

Guidance is performed with GPS receivers in combination with either a navigation aid or an integrated steering device. Navigation aids, commonly referred to as *lightbars*, help operators visualize their travel position in relation to their previous travel. These aids indicate deviations from their desired travel so the operator can make steering adjustments. Auto-guidance options use either an integrated electrohydraulic control system or a mechanical steering device installed inside the tractor cab.

Autonomous vehicles do not require a human operator to provide direction or control. Operation is controlled remotely or robotically, **Figure 4-13**. Autonomous vehicles rely on computers that collect and analyze data and then perform a function based on programmed

A
Andrey Aryagov/Shutterstock.com

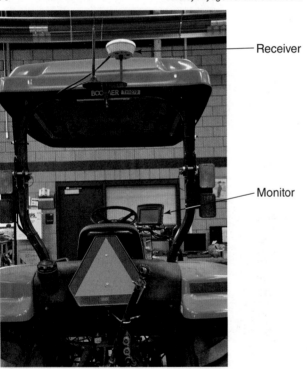

B
Goodheart-Willcox Publisher

Figure 4-12. GPS system. A—Satellite orbiting earth used for guidance and capturing images of the earth's surface. B—A GPS monitor and receiver on a tractor.

algorithms. Most auto-guidance technology is semi-autonomous, however. The operator starts and stops the machine, programs a desired travel path, and then takes control only when needed during turns and other maneuvers.

Telematics

Farmers can remotely collect and manage information from their field equipment with telematics technology. *Telematics* integrates wireless telecommunications with information technology to enable long-distance transmission of information in real time. In agriculture, electronic communication networks are embedded in farm machinery with GPS. These networks monitor and report information such as machine location, troubleshooting fault codes, hours of operation, and precision farming information such as yield during harvest. Data is captured and transferred to the Internet in real time via cellular or satellite systems to be analyzed for decision making. Ethics and security concerns include market speculation as real-time harvest data is collected and loss of privacy when this information is made available to competitors.

©iStock.com/meadowmouse

Figure 4-13. A demonstration of a GPS hands-free driving system. The tractor was driven between the spectators' chairs using only the GPS system—the driver's hands were in the air the whole time.

Unmanned Aerial Vehicles (Drones)

An *unmanned aerial vehicle (UAV)*, commonly known as a *drone*, is an aircraft that does not have a human pilot on board. A drone is controlled remotely by a human on the ground or located in another vehicle. UAVs have been used for both civil and military operations. Civil applications include surveillance, police work, and firefighting. More recently, agricultural drones have been developed to aid in crop and livestock management.

UAVs can be used to monitor livestock as well as crops, saving time and labor costs. See **Figure 4-14**. These programmable aircraft are used with video and photo surveillance equipment to monitor the health and status of livestock in minutes. This is a task that can take days when extensive amounts of land area must be traveled by horseback or on foot. The cost of UAVs is falling as technology improves, but cost may still limit use on smaller production sites.

Drones are operated by a pilot using a remote control at a ground station or are programmed to fly autonomously using a preprogrammed flight plan. They may be equipped with thermal imaging devices or high-definition cameras to monitor plant or animal health. Land used for grazing of livestock can be surveyed quickly and effectively to make counts or check the location of the herd. Plant health can be observed using images captured while the

Alexander Kolomietz/Shutterstock.com

Figure 4-14. UAVs equipped with high-definition cameras can be used to monitor plant and livestock health.

device is in flight. Images are analyzed by the farmer or operator to diagnose areas that may need intensive management for pest or disease control. Some prototypes have been developed to allow for spot-spraying of troubled areas within fields.

The use of UAVs for crop surveillance is a developing technology. UAVs linked to GPS satellites can monitor crops to identify problems that limit potential crop yield. The traditional practice of scouting crops for pests and other problems involved a labor force on foot to cover an area and take samples from various locations. Using drones, a farmer can take images of an entire crop and use color-contrast imaging to examine the health of the crop. This technology determines the amount of sunlight absorbed by the crop canopy. Images captured by UAVs can be linked with GIS software to map field borders and identify locations within a field that need specific attention.

The Federal Aviation Administration is proposing rules to govern the use of agricultural drones. Ethical issues have been identified, such as how these machines could be used to collect information from competing farms. The safety of other aircraft and their pilots is another concern. Government regulations are continually being developed to protect the privacy and safety of individuals.

AgEd Connection
Research Proficiency

New technology can be exciting. If this chapter piques your interest, consider investigating conducting a research proficiency. The Agriscience Integrated Systems Research proficiency has a category for Power, Structural, and Technical Systems research. This category would allow you to conduct research in a variety of areas that are mentioned in this chapter, such as precision technology or alternative energy sources.

Have a conversation with your agriculture teacher today about getting started with a research proficiency. Investigate information sources locally available or available online and begin gathering information about a topic in this chapter. It could lead to an exciting career in agriscience and technology!

Chaivaporn Baokaew/Shutterstock.com

Wireless Sensor Networks

Innovations in livestock management technologies include the use of remote sensing, or *biologging*, to record multiple animal health variables. Wireless sensor networks convey and process data measurements pertaining to behavior and physiology. Limitations of this technology include battery life and costly maintenance. Reliable transmission of data over long ranges is also a concern. Animals wear collars or ankle monitors to collect and transmit data to a central hub for analysis by the farmer. The present role of biologging is to aid agriculture researchers in gathering and processing genomic data to study protein function.

Biotechnology

Biotechnology is the use of living organisms to make or modify products that serve a useful purpose. Recent advances in plant and soil systems have been focused on improving crop yield and health. New and valuable traits are being incorporated into crops, and new crop varieties are being developed. An example is golden rice, **Figure 4-15**, which is biofortified with beta

ArtisticPhoto/Shutterstock.com

zcw/Shutterstock.com

Figure 4-15. Golden rice, a variety of rice enriched with beta carotene, is produced through genetic engineering.

carotene. Additional technologies have been developed to improve plant health monitoring as well as the management of limited resources while protecting the environment.

Genetic engineering is the scientific process of making changes to the genetic code of plants or animals to produce a desired trait, such as resistance to parasites. Scientists perform genetic recombination either by introducing foreign DNA to encode a desired trait in the plant (recombination) or by manipulating the genetic signals or material within the plant.

Gene Mapping

Gene mapping (determining the location of genes on chromosomes) gives agricultural scientists an increased understanding of the structure and function of crop plants and livestock. This knowledge allows the development of breeding practices to enhance crop and livestock species, resulting in breeds and varieties that are more nutritious, produce greater yields, and can better resist the negative effects of pests or other adverse conditions.

Like any research, gene mapping is a tool that can be abused. As research and development continues, legal and ethical issues continue to surface. Questions arise as to how the technology will be used. Other concerns focus on the research techniques used on animals and potential long-term adverse effects on humans.

Genetically Modified Crops

Genetically modified crops are agricultural plants in which the DNA has been altered to introduce a trait that does not occur naturally in the species. See **Figure 4-16**. These traits may include resistance to pests, disease, drought, spoilage, or chemicals. An example of crops modified for chemical resistance is Roundup Ready® crops. These crops are resistant to the active ingredient in Roundup® brand herbicides, allowing the chemical herbicide to be sprayed over the crop to kill undesirable plants without harming the intended crop. Roundup Ready® crops include alfalfa, canola, corn, cotton, sorghum, and soy. Other crops, such as wheat, are being developed.

Some crops are modified to increase their resistance to pests. An example is the modification of certain plants to include *Bacillus thuringiensis (Bt)*, a soil-dwelling bacterium that is commonly used as a biological pesticide. Research has been performed in nonfood crops (tobacco and cotton) and food crops (potatoes and corn) with success. The use of Bt toxins in plants destined for human consumption requires extensive evaluation of their safety. Potential impacts on the environment must be evaluated as well. The advantage of Bt-modified plants is a decreased need for pesticides.

The main food source for approximately half of the world's population is rice. Modifying rice to increase production, increase nutrient content, or resist pests has the potential to combat hunger, poverty, and malnutrition on a global level. While genetically modified rice has been developed, controversy

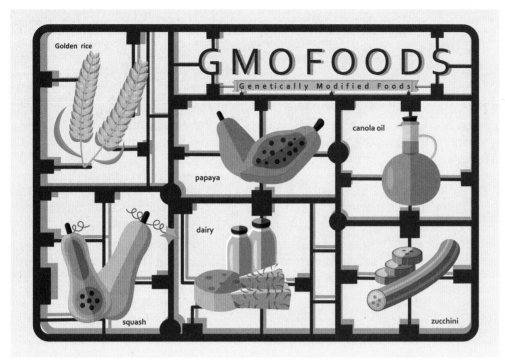

Wanwisspaul/Shutterstock.com

Figure 4-16. Many foods on the market today are either genetically modified or contain genetically modified ingredients.

surrounds its use. A recent development is a non-GMO "green" super-rice that can resist adverse conditions and achieve above-average yields without the use of fertilizer or pesticides.

Robotics

Robotic technologies are now applied to various agricultural labor tasks once performed by humans. Robots are machines guided by a computer program. They may function autonomously or semi-autonomously. Dairy farmers are increasing their use of robotic milking systems due to fluctuating costs and availability of laborers to milk, **Figure 4-17**. Robotic technologies offer efficient labor management and a reduction of strain on human health by eliminating repetitive movements that cause physical problems. Some researchers have documented a decrease in mastitis (inflammation of udder tissue) and improved milk production because of more frequent milking sessions as cows learn to use the system. The initial investment and learning curve may be high, and initial costs may limit the application of robotic technologies to large dairies.

Research is being conducted to find other agricultural applications for robots. Researchers at Harvard University are working to develop a robotic pollinator to counteract the honeybee crisis caused by colony collapse disorder. Other uses for robotics include the manufacture of agricultural equipment, such as tractors and combines. See **Figure 4-18**.

Figure 4-17. A robotic milking system can help increase production while reducing labor costs.

Figure 4-18. Robotic welding machines used in fabrication of farm equipment.

CHAPTER 4

Review and Assessment

Summary

- Several alternative electricity generation technologies, such as wind power, biological energy sources, and anaerobic digesters can meet the energy demands of farming operations and reduce costs.
- Alternative fuels such as biofuels and hydrogen fuel cells are being developed to lessen dependence on fossil fuels.
- Safe handling and packaging methods must be adapted to reduce food contamination as methods of production and processing change to meet increasing demands.
- Vehicle emissions are a major cause of air pollution. Vehicle emission reduction and control technologies include engine control units (ECUs), exhaust gas recirculation (EGR), selective catalytic reduction (SCR), and forced induction.
- Precision agriculture technologies can be used to manage the application of nutrients and water within a field based on various information sources.
- In geographic information systems, data is referenced with GPS coordinates and then mapped to create management zones or grids that are used for making production decisions.
- Agricultural drones are being developed to aid in crop and livestock management.
- Genetic engineering introduces traits that do not naturally occur in a species. Traits introduced in crops include pest resistance, disease resistance, drought resistance, spoilage resistance, and chemical resistance.

Words to Know

Match the key terms from the chapter to the correct definition.

A. automation
B. autonomous
C. biotechnology
D. cloud server
E. drone
F. geographic information system (GIS)
G. global positioning system (GPS)
H. precision agriculture
I. selective catalytic reduction (SCR)
J. site-specific management
K. telematics
L. variable-rate technology (VRT)
M. yield monitoring

1. A systematic approach to site-specific agricultural management.
2. A system designed to capture, store, manipulate, analyze, manage, and present all types of spatial or geographical data.
3. Injection of a liquid reducing agent into the exhaust stream of a diesel engine to break down nitrous oxide.

4. A system that uses electronic communication networks embedded in farm machinery with GPS to monitor and report on information including machine location, hours, troubleshooting fault codes and precision farming information such as yield during harvest.
5. A navigation system made up of a network of satellites placed in orbit around the Earth.
6. Not requiring human intervention for direction or control.
7. Control of the operation or function of equipment, processes, or systems by mechanical or electronic devices, without requiring an operator.
8. Machine that is either flown using a remote control operated by a pilot at a ground station or programmed to fly autonomously using a pre-programmed flight plan and GPS coordinates.
9. Describes any technology that allows the machine operator to change the application of science to solve a problem or make the best use of resources.
10. Managing crop needs based on specific data collected from a field site.
11. Obtaining information during crop harvesting by using sensors to measure and record data.
12. A network of remote servers hosted on the Internet to store, manage, and process data rather than using a local or physical server.
13. Encompasses those techniques and tools that use living organisms to make a product or run a process for commercial use.

Know and Understand

Answer the following questions using the information provided in this chapter.

1. The industrial age, the technology age, and the _____ age have brought technological advances to the agriculture industry.
2. List two common types of biofuels.
3. The source of power for a photovoltaic system is _____.
   A. wind
   B. crop residue
   C. the sun
   D. a generator
4. What is a green building?
5. A(n) _____ is needed to convert DC power generated by solar panels into alternating current.
   A. inverter
   B. wind turbine
   C. transformer
   D. light-emitting diode
6. List two benefits of exhaust gas recirculation (EGR).

7. _____ is a software system used to map, model, query, and analyze large quantities of data within a single database according to a reference location.
    A. A global positioning system (GPS)
    B. A geographic information system (GIS)
    C. Variable-rate technology (VRT)
    D. Site-specific management
8. *Drone* is another name for a(n) _____.
    A. unmanned aerial vehicle
    B. robot
    C. GPS
    D. hydrogen fuel cell
9. What is *biologging*?
10. List three traits that can be introduced to agricultural crops through genetic engineering.

STEM and Academic Activities

1. **Science.** Develop an experiment to determine the effect of variables on the accuracy of handheld GPS receivers. For example, trees and buildings can interfere with the signal reception. Graph your data and write a report that details your findings. Explain the importance of an accurate GPS signal.
2. **Technology.** Compare and contrast two different software programs used for GIS mapping. Write a report that details your findings. Explain which program you would prefer to use and why.
3. **Math.** Develop a spreadsheet that compares the costs of using three different GPS guidance systems for tractors. Your spreadsheet should include software subscription fees, installation costs, and hardware costs.
4. **Language Arts.** Read an article in a farming or ranching magazine that discusses a new trend in agricultural technology. Write a summary that identifies the applications of this new technology, as well as the pros and cons of adopting the technology.
5. **Social Science.** Research several types of agricultural robots that have been developed in the past few years. What jobs will these robots displace? What types of jobs will the increasing use of agricultural robots create?

Thinking Critically

1. Research information on genetic modification of crops for human consumption. Find sources that support GMO use and sources that oppose it. Write a report detailing your findings and include your personal opinion.
2. What forms of pollution are generated by the agricultural industry? What greenhouse gases are emitted? In what ways can agricultural producers limit pollution and the emission of greenhouse gases? Write a report that answers these questions.
3. What is your vision of the future of farming? What technological innovations do you think will be most important in meeting agricultural challenges in the next 50 years? Be prepared to answer these questions in class.

CHAPTER 5
Safety and Developing Safe Work Habits

Chapter Outcomes

After studying this chapter, you will be able to:
- Describe hazards that affect agricultural mechanics and technology workplaces.
- Discuss safety regulations and restrictions affecting agricultural work.
- Identify and select personal protective equipment for use in the workplace and in the agricultural mechanics laboratory.
- Discuss general safety practices and rules.
- Explain the role of safety data sheets in ensuring chemical safety.
- List steps that can be taken to improve electrical safety.
- Evaluate outdoor working conditions and take steps to increase outdoor safety.
- Describe safety precautions for working with outdoor equipment and machinery.
- Explain why workers in agricultural mechanics and technology should know basic first aid procedures.

Words to Know

- all-terrain vehicle (ATV)
- American National Standards Institute (ANSI)
- American Society of Agricultural and Biological Engineers (ASABE)
- combustible liquid
- corrosive
- decibel (dB)
- Department of Labor (DOL)
- electrocution
- explosive
- Fair Labor Standards Act (FLSA)
- first aid
- flammable liquid
- flash point
- Globally Harmonized System of Classification and Labeling of Chemicals (GHS)
- hot work
- National Institute for Occupational Safety and Health (NIOSH)
- Occupational Safety and Health Administration (OSHA)
- oxidizer
- personal protective equipment (PPE)
- power take-off shaft (PTO)
- reactivity
- rollover protection systems (ROPS)
- safety data sheets (SDS)
- scaffolding
- signal word
- standards
- Underwriters Laboratories (UL)

Before You Read

Arrange a study session to read the chapter aloud with a classmate. Take turns reading each section. Stop at the end of each section to discuss what you think its main points are. Take notes of your study session to share with the class.

While studying this chapter, look for the activity icon to:
- **Practice** vocabulary terms with e-flash cards and matching activities.
- **Expand** learning with interactive activities.
- **Reinforce** what you learn by completing the end-of-chapter questions.

www.g-wlearning.com/agriculture

Agriculture is a unique industry, one in which families work as well as live on the job site. This exposes family members, including children as well as hired workers, to the potential for injury on a daily basis. The people working in agriculture are faced with various hazards such as machinery entanglement, vehicle collisions and rollovers, grain engulfment, falls, and chemical exposures. Respiratory diseases, musculoskeletal disorders, and hearing loss are prevalent among people working in agriculture. Working with livestock also presents the potential for injuries, such as being stepped on or crushed by large animals.

Agriculture continues to rank as one of the most dangerous industries in the United States. It has been estimated that every day 167 agricultural workers suffer a work-related injury in the United States. The agriculture, forestry, fishing, and hunting industries have a nonfatal occupational injury rate of 5.5 per 100 workers. The best safety practices involve removing or reducing the hazards, **Figure 5-1**.

Hazards in Agricultural Mechanics and Technology

Hazards come in many forms and exist in every workplace. This chapter focuses specifically on those that apply to agricultural mechanics and technology. When hazards are not addressed, injuries may result. The *National Institute for Occupational Safety and Health (NIOSH)* collects data on agricultural hazards and injuries. This information gives an indication of the magnitude of the dangers associated with working in agriculture. NIOSH provides the following statistics:

- In 2012, 374 farmers and farm workers died from work-related injuries, resulting in a fatality rate of 20.2 deaths per 100,000 workers. Tractor

Everett Historical/Shutterstock.com Balefire/Shutterstock.com

Figure 5-1. Work conditions for agricultural workers have changed with advances in technology to keep them safe and comfortable. Sprayers instead of hoes are used to control weeds. A lot of manual laborers are still needed to harvest sensitive and perishable crops such as strawberries and apples.

overturns were the leading cause of death for these farmers and farm workers.

- In 2012, 59% of tractors used on farms in the United States were equipped with *rollover protection systems (ROPS)* designed to protect operators from injuries caused by vehicle overturns or rollovers, **Figure 5-2**. If ROPS were placed on all tractors manufactured since the mid-1960s that are used on US farms, the prevalence of ROPS-equipped tractors could be increased to over 80%.

Some jobs in agricultural mechanics and technology are considered more hazardous than others. According to the United States Department of Labor, hazardous agricultural occupations are those that involve performing the following tasks:

Figure 5-2. This tractor is equipped with rollover protection and a shade canopy.

- Operating a tractor over 20 PTO horsepower or connecting or disconnecting its implements.
- Operating or assisting to operate a corn picker, cotton picker, grain combine, hay mower, forage harvester, hay baler, potato digger, mobile pea viner, feed grinder, crop dryer, forage blower, auger conveyor, unloading mechanism of a nongravity-type self-unloading wagon or trailer, power posthole digger, power post driver, or nonwalking-type rotary tiller.
- Operating or working with a trencher or earthmoving equipment, forklift, potato combine, or power-driven circular, band, or chain saw.
- Working in a yard, pen, or stall occupied by a bull, boar, or stud horse maintained for breeding purposes; a sow with suckling pigs; or a cow with a newborn calf (with umbilical cord present).
- Felling, buckling, skidding, loading, or unloading timber with a butt diameter of more than six inches.
- Working from a ladder or scaffold at a height of over 20 feet.
- Driving a bus, truck, or automobile to transport passengers, or riding on a tractor as a passenger or helper.
- Working inside a fruit, forage, or grain storage designed to retain an oxygen-deficient or toxic atmosphere; an upright silo within two weeks after silage has been added or when a top unloading device is in operating position; a manure pit; or a horizontal silo while operating a tractor for packing purposes.
- Handling or applying toxic agricultural chemicals identified by the words *danger*, *poison*, or *warning* or a skull and crossbones on the label.
- Handling or using explosives.
- Transporting, transferring, or applying anhydrous ammonia.

Safety Regulations

During the Industrial Revolution, machines were introduced in most industries, including agriculture, to increase production and improve efficiency. The designers of these machines did not always have safety in mind. Working conditions were very unsafe at times, and there were few rights to protect workers.

The US *Department of Labor (DOL)* was created in 1913 as part of an effort to ensure the welfare of all individuals employed or seeking employment. Part of the DOL's mission is to improve working conditions. The DOL administers and enforces more than 180 federal rules and regulations. Over the years, as new technology and new worker issues developed, several other legislative acts were passed. Several of these apply to people working in agriculture.

The Fair Labor Standards Act (FLSA)

The *Fair Labor Standards Act (FLSA)* was passed in 1938. The FLSA is administered by the Wage and Hour Division of the DOL. Part of the FLSA created so-called "child labor laws," which are provisions to ensure safe working conditions for youths and to ensure that their employment does not jeopardize their health, safety, or education.

The largest impact the FLSA has on agricultural mechanics and technology is the limits it places on youths working in hazardous agricultural occupations. These provisions were designed to keep young workers safe and to ensure that their educational opportunities are not affected. Many of the young people who work on farms and ranches also live on those farms and ranches. As family members, they are often expected to work. However, agricultural jobs are not necessarily safe for young workers. According to NIOSH:

- From 1995 to 2002, an average of 113 youths under the age of 20 died from farm-related injuries, with most of these deaths occurring to those 16–19 years of age (34%).
- Of the leading sources of fatal injuries to youths, 23% involved machinery (including tractors), 19% involved motor vehicles (including ATVs), and 16% were due to drowning.
- In 2012, an estimated 14,000 youths were injured on farms; 2,700 of these injuries were due to farm work.

Under the FSLA, if you are not yet 16 years of age, there are certain restrictions on your employment in agriculture, particularly in occupations declared hazardous by the US Secretary of Labor. Once you turn 16, these restrictions no longer apply. Note that youths of any age may work at any time in any job on a farm owned or operated by their parent or guardian. The following list includes those occupations deemed hazardous by the US Secretary of Labor.

The Occupational Health and Safety Administration indicates that if you are:

- Under 12: You may work outside of school hours in any nonhazardous job on a small farm that is exempt from the federal minimum wage provisions as long as you have parental consent.

- 12 or 13: You may work outside of school hours in nonhazardous jobs on a farm where your parent(s) work or with written parental consent.
- 14 or 15: You can work outside of school hours in any nonhazardous agricultural job.
- 16 or older: You can work in any farm job at any time.

Occupational Safety and Health Act

In 1970, the Occupational Safety and Health Act was passed, which created the *Occupational Safety and Health Administration (OSHA)* as a division of the DOL. This gave the federal government the authority to set and enforce safety and health standards for most workers in private industries. Several states have their own safety and health administrations in addition to the federal administration. Besides establishing standards, OSHA enforces these rules through inspection of worksites. OSHA provides training and outreach to employers to ensure compliance. Federal OSHA inspectors report directly to the Secretary of Labor.

OSHA has established general safety standards as well as specific standards for certain industries. *Standards* are rules that describe methods for ensuring safety. Some general standards include proper hazard communication, handling of materials, and exposure limits. Standards specific to agriculture include testing to ensure the safety of rollover protective structures for tractors, protective frames, and enclosures for wheel-type agricultural tractors. Employee training is also a critical component of this standard. Minimum requirements for safety guards on farm field

Grandpa/Shutterstock.com

SAE Connection
Job Safety

Supervised agricultural experience (SAE) projects vary widely. In agricultural mechanics, you may be operating field equipment or making repairs using an arc welder. Personal protective equipment (PPE) required for many of these activities includes safety eyewear, gloves, and other protective gear. Your teacher, employer, and family all want you to work safely so you can continue those activities and return home in good health.

Read this chapter carefully and remember to follow safe practices when working in your SAE. If you have safety questions, ask your instructor or your employer for the best method of proceeding. Learn to recognize and avoid potential hazards. Safety is the responsibility of everyone, and no safety shortcut is worth the risk of potential damage to people or property.

equipment are specified as well. Additionally, field sanitation requirements for workers are specified. The goals of these standards are to protect employees from hazards found in the agricultural workplace.

Federal Insecticide, Fungicide, and Rodenticide Act (FIFRA)

In the United States, the Federal Insecticide, Fungicide, and Rodenticide Act (FIFRA) protects workers from potential pesticide exposure and ensures that workers are trained about pesticide safety. This act gives the Environmental Protection Agency (EPA) the responsibility of determining which pesticides are safe enough to be used in the United States. The EPA also works with other federal and state agencies to enforce the pesticide regulations, to test new pesticides, and to update the regulations as necessary.

American National Standards Institute (ANSI) Safety Standards

Safety equipment that has met or exceeded the standards set by the *American National Standards Institute (ANSI)* should be your final line of defense when exposed to a hazard. Safety equipment intended for eye protection or foot protection should be tested and certified to meet the standards set by ANSI.

In addition, ANSI has established safety standards for many areas that apply to agricultural mechanics and technology. These include clothing and equipment standards, ladder safety, and construction safety, among others.

National Institute for Occupational Safety and Health (NIOSH)

The National Institute for Occupational Safety and Health (NIOSH) was first created under the Occupational Safety and Health Act of 1970. NIOSH is part of the US Centers for Disease Control and Prevention. NIOSH performs critical safety research and provides recommendations to prevent worker injury and illness. NIOSH certifies respirators and provides recommendations for personal protective equipment. The National Personal Protective Technology Laboratory is a component of NIOSH that helps identify equipment designed to keep workers safe.

Underwriters Laboratories (UL)

Underwriters Laboratories (UL) certifies and validates a variety of products to ensure safety. The UL listing mark means the product meets safety requirements established by the UL. The mark is commonly found on appliances, electrical panel boards, fuses, and other products. Using products that have the UL mark greatly decreases the chance of an accident due to faulty equipment.

American Society of Agricultural and Biological Engineers (ASABE)

The *American Society of Agricultural and Biological Engineers (ASABE)* is an international professional organization that establishes engineering standards for agriculture. These standards define materials, products, processes, tests, testing procedures, and performance criteria. Several standards have been established to improve personal safety during equipment operation. Using equipment that meets the ASABE standards is a valuable safety technique.

Personal Protective Equipment

Personal protective equipment (PPE) is equipment worn to minimize exposure to a variety of hazards and is typically designed to protect a specific area of the body, **Figure 5-3**. It is important to be able to identify the hazards, then select the right type of PPE and use it appropriately. A hazard assessment should begin with a walk-through survey of the agricultural mechanics facility to develop a list of potential hazards in the following basic hazard categories:

- Impact
- Penetration
- Compression (rollover)
- Chemical
- Heat/cold
- Harmful dust
- Light (optical) radiation
- Biologic

THPStock/Shutterstock.com

Figure 5-3. Head-to-toe protection is available for performing work safely.

Eye and Face Protection

Using the appropriate eye or face protection reduces exposure to hazards to the eyes and face, such as flying particles, molten metal, liquid chemicals, acids or caustic liquids, chemical gases or vapors, or potentially harmful light radiation. Some of the most common types of eye and face protection include safety glasses, goggles, welding helmets/shield, and face shields.

Many occupational eye injuries occur because workers are not wearing any eye protection; others result from wearing improper or poorly fitting eye protection.

Potential eye or face injuries include but are not limited to:
- Dust, dirt, metal, or wood chips entering the eye during activities such as chipping, brushing, grinding, sawing, hammering, or drilling.
- Chemical splashes from corrosive substances, hot liquids, solvents, or other hazardous solutions.

Copyright Goodheart-Willcox Co., Inc.

- Objects swinging into the eye or face, such as tools, springs, chains, or ropes.
- Radiant energy from welding, harmful rays from the use of lasers, or other radiant light (as well as heat, glare, sparks, splash and flying particles).

Common prescription corrective lenses do not provide adequate protection against most eye and face hazards. Individuals who need corrective lenses must wear either eye protection that incorporates the prescription into the design or wear additional eye protection over their prescription lenses. It is important to ensure that the protective eyewear is positioned so that it does not interfere with the prescription lenses, which would limit the wearer's vision. Individuals who wear contact lenses also must wear eye or face PPE when working in hazardous conditions.

Safety Glasses

These protective eyeglasses have impact-resistant lenses and safety frames constructed of metal or plastic. All safety glasses should have side protection as well, **Figure 5-4**. Eye protection must conform to the American National Standards Institute (ANSI) Z87.1-1989. Look for this stamp on the inside of the safety glass frame. Safety glasses with suitable filter lenses may be appropriate for light gas torch operations such as brazing and soldering.

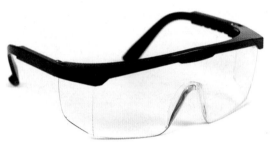

Ronnachai Palas/Shutterstock.com

Figure 5-4. Safety glasses offer protection from the front and the side.

Goggles

Goggles are tight-fitting and provide eye protection over the eyes, the eye sockets, and the facial area immediately surrounding the eyes, **Figure 5-5**. Two different types of goggles are intended for two specific protective uses. Impact-protection goggles protect the eyes from flying debris but contain perforations on the sides of the goggles. These goggles are not intended to be used for chemical splash protection. Splash goggles, which contain shielded vents at the top of the goggles, are appropriate for chemical splash protection and also provide limited eye impact protection. Goggles do not offer protection for the face and neck. Goggles with filter plates or tinted glass are available for work with intense light sources such as gas welding or oxygen cutting operations.

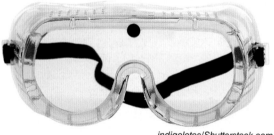

indigolotos/Shutterstock.com

Figure 5-5. Safety goggles cover more space around the facial area than safety glasses. Safety goggles are designed to protect against liquids that may splash into the face or against impact or flying debris.

Welding Helmet or Shield

A welding helmet has a shield constructed of heat-resistant material. This material is fitted with a filter lens to protect eyes from burns caused by infrared light or intense radiant light produced from arc welding or arc cutting operations, **Figure 5-6**. Welding helmets offer protection to the eyes and face from flying sparks, metal spatter, and slag chips produced during hot work operations. OSHA requires filter lenses to have a shade number appropriate to protect against harmful light radiation from the work being performed. Types of lenses vary tremendously. Options include passive or auto-darkening lenses. Auto-darkening lenses may be either fixed or variable shade with anywhere from two to four sensors. Standard viewing sizes vary from 2″ × 4″ to 4″ × 5″.

Face Shield

Constructed with transparent sheets of impact-resistant material, face shields must extend from the eyebrows to below the chin and across the entire width of the user's head, **Figure 5-7**. Some face shields may be polarized for glare protection or tinted for use during light gas welding. Some face shield visors come in steel or nylon mesh material. Mesh visors provide good airflow for worker comfort and are typically used in the logging and landscaping industries to protect the face from flying debris when cutting wood or shrubbery. The ANSI Z87.1-2003 standard states that face shields are considered secondary eye protection and must be used in conjunction with safety glasses or goggles. Face shields used in combination with goggles or safety spectacles provide additional protection against impact hazards. Face shields should be contoured to protect the sides of the neck along with frontal protection.

Netfalls-Remy Musser/Shutterstock.com

Figure 5-6. This welder is fully protected against intense heat and flying sparks from this welding torch.

Bruce Au/Shutterstock.com

Figure 5-7. Face shields offer good visibility and airflow while not being constricting.

Head Protection

Protection from potential head injuries is a key element of any safety program. A head injury can create a permanent impairment or even be fatal. Wearing a safety helmet or hard hat is one of the easiest ways to protect against a head injury, **Figure 5-8**. Hard hats protect against impact and penetration hazards in addition to electrical shock and burn hazards. Head protection should be worn if any of the following risks apply:

- Objects falling from above.
- Heads bumping against fixed objects, such as exposed pipes or beams.
- Accidental head contact with electrical hazards.

Sukpaiboonwat/Shutterstock.com

Figure 5-8. It is important that hard hats fit properly to ensure comfort while working. A hard hat that is too big may slip off the worker's head, and one that is too small may cause discomfort or not provide sufficient protection.

Hard hats must be worn properly and manufacturer recommendations must be followed in order to provide the proper protection. Protective helmets or hard hats should:
- Resist penetration by objects.
- Absorb the shock of a blow.
- Be water-resistant and slow-burning.
- Have clear instructions explaining proper adjustment and replacement of the suspension and headband.

Protective headgear must meet the ANSI Z89.1-1986 standard (Protective Headgear for Industrial Workers) or provide an equivalent level of protection. Helmets purchased before July 5, 1994 must comply with the earlier ANSI standard (Z89.1-1969) or provide equivalent protection.

Hard hats are divided into three industrial classes:
- **Class A** hard hats provide impact and penetration resistance along with limited voltage protection (up to 2,200 volts).
- **Class B** hard hats provide the highest level of protection against electrical hazards, with high-voltage shock and burn protection (up to 20,000 volts). They also provide protection from impact and penetration hazards by flying/falling objects.
- **Class C** hard hats provide lightweight comfort and impact protection but offer no protection from electrical hazards.

"Bump hats" are designed for use in areas with low head clearance. They are recommended for areas where protection is needed from head bumps and lacerations. These are not designed to protect against falling or flying objects and are not ANSI-approved.

It is essential to check the type of hard hat used to ensure that the equipment provides appropriate protection. Each hat should bear a label inside the shell that lists the manufacturer, the ANSI designation, and the class of the hat.

Hearing Protection

Some jobs in the agricultural mechanics field may produce a number of conditions where the noise level and duration may cause irreversible damage. Tinnitus is a ringing in the ears that can be caused by damaged auditory nerves.

Determining the need to provide hearing protection can be challenging. Exposure to excessive noise depends upon a number of factors, including:
- The loudness of the noise as measured in *decibels (dB)*.
- The duration of each worker's exposure to the noise.
- Whether individuals move between work areas with different noise levels.
- Whether noise is generated from one or multiple sources.

Typically, the louder the noise, the shorter the exposure time before hearing protection is required, **Figure 5-9**. For instance, someone may be exposed to a noise level of 90 dB for eight hours per day before hearing protection is required. On the other hand, if the noise level reaches 115 dB, hearing protection is required if the anticipated exposure exceeds 15 minutes. Impact or impulse noise (loud momentary explosions of sound occurring at an interval rate of more than one second) exposures must not exceed 140 dB.

Manufacturers of hearing protection devices must display the device's noise reduction rating (NRR) on the product packaging. The following are the three most common types of hearing protection:

- Single-use earplugs are disposable earplugs made of expandable foam. The foam forms to your ear canal.
- Reusable earplugs may be made from silicone, plastic, or rubber. Although manufactured as "one-size-fits-most," they are usually available in several sizes. Reusable plugs should be cleaned after each use. Reusable earplugs may also be referred to as preformed earplugs.
- Over-the-ear earplugs, also known as earmuffs, require a good seal around the ear. Earmuffs come in various sizes and profiles to fit most users. Glasses, facial hair, long hair, or facial movements such as chewing may reduce the protective value of earmuffs. See **Figure 5-10**.

Noise Exposure Limits	
Sound level (decibels)	Duration (hours per day)
90	8
92	6
95	4
97	3
100	2
102	1½
105	1
110	½
115	¼ or less

OSHA

Figure 5-9. The need for hearing protection depends on the noise level and the duration of exposure to that noise.

Breathing Protection

Ventilation and safe work practices are the primary means to prevent exposure to respiratory hazards. When these are not feasible or available, however, then wearing a respirator may be necessary. Individuals who wear a respirator must conduct a workplace assessment to identify airborne hazards.

Respirators fall into one of two categories based on how they protect the user. The first category of respirators removes contaminants from the air. Respirators of this type include particulate respirators, which filter out airborne particles, and "gas masks," which filter out chemicals and gases. An N95 respirator filters particulates—solid particles in the air such as

A Steve Collender/Shutterstock.com B Yotsatorn Laonalonglit/Shutterstock.com

Figure 5-10. Hearing loss may occur without protection against exposure to excessive noise. A—Earplugs are inexpensive, easy to use, and effective against lower-decibel noises. B—Earmuffs cover the entire circumference of the ear and should be used for protection against exposure to louder sounds.

sawdust and animal dander. Certain pesticides require the use of elastomeric respirators equipped with specific gas cartridges to absorb toxic fumes.

The second category of respirators protects by supplying clean, breathable air from another source. These respirators are used when the work environment air is deficient in oxygen. An oxygen-deficient atmosphere means the oxygen content is below 19.5% by volume. Respirators can be tight-fitting or loose-fitting. OSHA has established several standards that dictate the proper use and selection of a respirator, **Figure 5-11**. Contact your local extension agent or safety professional to help provide recommendations on how to best protect your respiratory system.

Protective Clothing

When hazards cannot be removed through engineering or work practices, protective clothing and shoes may be necessary. Selecting protective clothing, like other personal protective equipment, should begin by assessing the hazards it will be used to protect against.

Body Protection

Examples of body protection include laboratory coats, coveralls, vests, jackets, and aprons. When chemical or physical hazards are present, check with the clothing manufacturer to ensure that the material selected will provide protection against the specific hazard.

Protective clothing comes in a variety of materials, each effective against particular hazards, such as:

- Paper-like fiber—used for disposable suits to provide protection against dust and splashes.
- Treated wool and cotton—adapts well to changing temperatures. It is also comfortable, fire-resistant, and protects against dust, abrasions, and rough and irritating surfaces.
- Duck cloth—closely woven cotton fabric that protects against cuts and bruises when handling heavy, sharp, or rough materials.
- Leather—often used to protect against dry heat and flames.
- Rubber, rubberized fabrics, neoprene, and plastics—protect against certain chemicals and physical hazards.

A   Goodheart-Willcox Publisher   B   Goodheart-Willcox Publisher

Figure 5-11. Respirators come in several varieties, but all are designed to prevent inhalation of hazardous material. A—Full-face elastomeric respirator. B—Half-face elastomeric respirator.

Gloves

A variety of gloves are available to protect against various hazards. Hazards include skin absorption of harmful substances, chemical or thermal burns, electrical dangers, bruises, abrasions, cuts, punctures, fractures, and even amputations. The nature of the hazard and the work task involved determine which type of gloves should be used.

Selecting the right pair of gloves can be challenging. It may be necessary to consider the area needing protection (hand only or forearm included), duration of contact with the potential hazard, comfort, and size. Some gloves can reduce your grip on certain materials, so be sure to take into account your grip requirements.

Most gloves can be classified into one of four groups based on their construction materials and intended purpose:

- Leather, canvas, or metal mesh—provide sufficient protection for use with rough, sharp, or heavy materials. This type should be considered when protection is needed against cuts or burns, or when sustained contact with heat may be encountered.
- Fabric and coated fabric gloves—protect against dirt, slivers, chafing, and abrasions. Adding a plastic coating will strengthen some fabric gloves while improving slip resistance, **Figure 5-12**.
- Chemical- and liquid-resistant gloves—different kinds of rubber or various types of plastic. As a general rule, the thicker the material, the more chemical resistance. Thick gloves, however, may impair grip and mobility.

Kzenon/Shutterstock.com

Figure 5-12. Hand protection is necessary to prevent damage from cuts, burns, chemicals, and other hazards.

- Insulating rubber gloves—typically reserved for working with electrical hazards. Special considerations must be followed for the selection, use, and care of insulating rubber gloves.

Foot and Leg Protection

Do not overlook the need for foot and/or leg protection. Personal protective equipment should be used to protect all parts of the body that may be subject to hazards, **Figure 5-13**. Instances where protective footwear may be needed include:

- When heavy objects might roll or fall onto feet.
- Working with sharp objects that could pierce ordinary shoes.
- Exposure to molten metal that might splash on feet or legs.
- Working on or around hot, wet, or slippery surfaces.
- Working when electrical hazards are present.

Safety footwear must meet or exceed standards set in ANSI Z41-1991. Footwear that meets ANSI approval will have a protective toe and offer impact and compression protection. Different footwear may protect in different ways. It is always important to check the product labeling or consult the manufacturer to make sure the footwear will protect the users from the hazards they may face.

Zerbor/Shutterstock.com

Figure 5-13. Protective footwear should be worn to protect your feet from hazards in the workplace, including slippery surfaces, sparks, and sharp objects.

PPE Size and Care Considerations

Appropriate use of PPE also includes knowing how to properly care for and maintain the PPE. This includes proper cleaning and storage. Periodic cleaning and inspection will extend the useful life of protective equipment. PPE must be selected to fit appropriately on each individual. Protection that is either too large or too small is inappropriate for use, even if it meets all other requirements. A daily inspection of the PPE is essential. If safety eye wear is shared between individuals, it should be disinfected after each use. Do not store protective equipment in direct sunlight, such as on the rear window of a car, because sunlight and extreme heat can damage it.

General Safety

It is extremely important to be trained in the proper use of machinery and safety procedures, even while you are in school. Working in the agricultural mechanics laboratory presents hazards similar to those on a farm or ranch. The machinery in the laboratory is the same that is used in the agricultural workplace, and it presents the same potential for injury.

Hand Tools and Power Tools

Both hand tools and power tools expose users to hazards such as thrown objects, cuts, punctures, entanglements, and harmful dusts. Because hand tools are not powered, improper use and poor maintenance pose the greatest hazards associated with hand tools.

Some examples include:
- If a screwdriver is used as a chisel, the tip of the screwdriver may break off and hit the user or bystanders.
- The head of a hammer with a handle that is loose, splintered, or cracked may fly off, striking the user or other people working.
- Chisels, wedges, or punches that have mushroomed heads might shatter on impact, sending sharp fragments flying.
- Saws or other cutting tools should be directed away from aisle areas and away from other people working in close proximity. Cutting tools should be kept sharp, because a dull tool can slip, moving the cutting part of the tool toward the body. Cracked saw blades must be removed from service.

Notice that all of the hazards in the preceding list can be prevented by using hand tools only for their intended purposes and paying close attention to inspection and maintenance. Screwdrivers should not be used as chisels. Hammers, chisels, wedges, and punches should be inspected frequently and either repaired or removed from service if necessary. Cutting tools should be sharpened or have the blades replaced as needed.

Power tools have additional hazards depending on whether their power source is electric, pneumatic, or hydraulic. Electric shock or electrocution is a serious danger. Most electric tools are double-insulated or have grounding to keep the user safe. Pneumatic and hydraulic power sources present high-pressure hazards that can cause an embolism, a potentially fatal obstruction in blood vessels that can get stuck while traveling through the bloodstream.

STEM Connection

Reaction Time

Accidents often occur because workers are fatigued, dehydrated, or impaired due to other common factors. Farm workers put in long hours outdoors and are highly susceptible to fatigue. Scientific and medical experiments have shown clearly that fatigue can greatly reduce a person's reaction time. When possible, farm workers should schedule hazardous activities early in the day, while they are well rested and have the energy, concentration, and reaction time to perform the activities safely.

Tyler Olson/Shutterstock.com

As little as 12 pounds of compressed air pressure can blow an eye out of its socket. Nail guns are one of the most common causes of punctures associated with construction tasks.

To prevent hazards associated with the use of power tools, follow these general precautions:

- Do not carry a tool by the cord or hose.
- Do not yank on the cord or the hose to disconnect it.
- Keep cords and hoses from rubbing on sharp edges.
- Disconnect tools when you are not using them, before servicing and cleaning them, and when changing accessories such as blades, bits, and cutters.
- Keep all people not involved with the work at a safe distance from the work area.
- Secure work with clamps or a vise, freeing both hands to operate the tool.
- Avoid accidental starting. Do not hold fingers on the switch button while carrying the tool.
- For best performance, properly maintain tools by following the instructions in the user's manual.
- Be sure to keep good footing and maintain good balance when operating power tools.
- Wear proper apparel for the task. Loose clothing, ties, or jewelry can become caught in moving parts.
- Remove all damaged portable electric tools from use and tag them: "Do Not Use."
- Power tools must be fitted with guards and safety switches. Extremely hazardous situations can occur when they are used improperly. Any exposed moving parts of power tools must be safeguarded, **Figure 5-14**.

A — Benoit Daoust/Shutterstock.com
B — Robert Wydro Studio/Shutterstock.com

Figure 5-14. Hand and power tools may be easy to use, but care must be taken when using them to prevent injury. A—Compound miter saw blade guard. B—Table saw guard.

Shop and Lab Practices That Improve Safety

Cleaning and keeping the workplace organized are excellent practices to reduce or eliminate safety hazards. Keeping a workspace clean and organized can prevent most slips and trips. In general, slips and trips occur due to a loss of traction between the shoe and the walking surface, or an accidental contact with an object. Falls may result in injuries serious enough to require medical treatment. The causes of slips, trips, and falls can vary considerably but may include:

- Spills of oil, water, or other liquids on walkway surfaces.
- Floors covered with wood dust or grinder dust.
- Mats that have become unanchored or loose.
- Cluttered walkway surfaces.
- Electrical cord extensions that cross walkways.
- Heavily polished or freshly waxed floors.

An additional benefit of having a clean, organized workspace is being able to locate tools and equipment when you need them. OSHA rules for floor safety recommendations for slips, trips, and fall prevention include:

- Keep floor surfaces clean and dry.
- Ensure that wet-floor warning signs are posted in and around wet floor locations.
- Provide and maintain adequate drainage for floors.
- Maintain clear walkways and prevent obstructions.
- Report and clean up spills immediately.
- Use non-slip coatings or surfaces in slippery locations.
- Minimize carpet and matting trip hazards.
- Use good housekeeping procedures.

Using designated storage for construction materials such as lumber or metal can help reduce floor clutter. Marking walkways and designated work zones helps ensure that work areas are safe, uncongested areas. Observe the following rules while working in safety zones for stationary machines:

- Only one person is allowed within the limits of the safety zone associated with the machine.
- No one other than the operator should be allowed to touch control switches and adjustment levels.
- Only one person should use the machine at a time unless approved by the instructor or unless it is a two-person operation.

Color Codes and Signal Words

Tape, signs, or paint are often used to mark work zones, hazards, and safety areas. The colors used have specific meanings that you should recognize and understand. ANSI has established a color code for safety signs that includes the colors red, orange, yellow, green, blue, purple, gray, black, and white. Assigning meanings to these colors helps workers know at a glance where hazards exist, as well as where to find firefighting and first aid equipment.

The ANSI standard also includes *signal words* that call attention to a specific hazard or safety message. Examples of signal words are **CAUTION**, **WARNING**, and **DANGER**. See **Figure 5-15**.

Fire Extinguishers

Arc welding and other hot work should have proper fire safety procedures. *Hot work* is any process that can be a source of ignition when flammable material is present or can be a fire hazard regardless of the presence of flammable material in the workplace. A hot engine muffler or chemical fumes can also create a fire hazard. Understanding the three elements of fire can help you stay safe. In order for fire to start, there must be:

- A fuel source.
- Oxygen.
- Heat.

These three elements are sometimes called the fire triangle. See **Figure 5-16**. Eliminating any one of these elements will prevent or extinguish a fire.

The potential hazards and damage caused by fires can be eliminated if the response is quick and effective. Most fires start small, and using fire extinguishers is the best way to put them out. Fire extinguishers are specific to the class of fire they are designed to extinguish. Fires are classified based

ANSI-Specified Colors and Signal Words		
Color	**Signal Word**	**Application**
Red	Danger	Used to identify a dangerous activity or condition that will result in serious injury or death if not avoided. Also used to designate fire protection equipment and flammable-materials containers.
Orange	Warning	Used to identify a dangerous activity or condition that may result in death or serious injury if not avoided.
Yellow	Caution	Used to identify a dangerous activity or condition, such as falling or tripping hazards, that may result in minor or moderate injury if not avoided.
Green	Safety	Used to indicate location of safety/first-aid equipment or provide pertinent safety instructions.
Blue	Notice	Identifies activity that may lead to property damage or other issues not related to personal safety.
Purple		Not assigned to specific signal words or hazards. Meaning can be assigned by end user.
Gray		
Black		
White		
Combinations of Black, White, and/or Yellow		

Goodheart-Willcox Publisher

Figure 5-15. It is important to know the meanings associated with the different ANSI-specified colors and signal words.

on the type of fuel source, **Figure 5-17**. Classes of fire are:

- A—Ordinary combustibles, such as wood, paper, some plastics and textiles.
- B—Flammable liquid and gas fires, such as oil and gasoline.
- C—Fires that involve live electrical equipment, requiring the use of electrically nonconductive extinguishing agents. (Once the electrical equipment is de-energized, extinguishers for Class A or B fires may be used.)
- D—Combustible metals, such as magnesium, titanium, sodium, etc., which require an extinguishing medium that does not react with the burning metal.

Most extinguishers used in agricultural mechanics and technology shops and labs are ABC-rated; class D fires are very rare. Fire extinguishers should be inspected annually and properly maintained.

Understanding how to use a fire extinguisher properly is critical. When using a fire extinguisher, remember P-A-S-S:

Pull the safety pin.
Aim at the base of the fire.
Squeeze the handle, keeping a distance of eight feet from the fire.
Sweep the nozzle back and forth over the base of the fire.

Luciano Cosmo/Shutterstock.com

Figure 5-16. Fire is not possible if all three of these elements are not present.

Jovanovic Dejan/Shutterstock.com

Figure 5-17. Not all fires are alike. If one should occur, know the type of fire and the class it falls under before attempting to extinguish it.

Chemical Safety

Chemical hazards and toxic substances pose a wide range of health hazards such as skin, eye, and nasal irritation; sensitization to allergies; and cancer. Chemicals also pose physical hazards such as flammability, corrosion, and *reactivity*. Several factors influence how a chemical will behave and the hazards the chemical presents, including the severity of the response:

- Concentration of the chemical.
- Physical state of the chemical (solid, liquid, gas).
- Physical processes involved in using the chemical (cutting, grinding, heating, cooling, etc.).
- Chemical processes involved in using the chemical (mixing with other chemicals, purification, distillation, etc.).
- Other processes (improper storage, storage in sunlight, refrigeration, etc.).

Safety Data Sheets (SDS)

To help ensure that chemicals are used safely in the workplace, OSHA established the Hazard Communication Standard. This standard is used to ensure that all individuals receive the information they need in order to use and handle the hazardous substances they work with safely. It required chemical manufacturers and distributors to supply a formal document for each chemical, the physical and chemical properties of the chemical, as well as toxicity, flash point, procedures for spills and leaks, storage guidelines, and exposure control. This document was called a material safety data sheet (MSDS).

On March 26, 2012, the Department of Labor adopted the United Nations' *Globally Harmonized System of Classification and Labeling of Chemicals (GHS)* and OSHA incorporated it into the Hazard Communication Standard. Under this change, material safety data sheets are now referred to as *safety data sheets (SDS)*. The new SDSs also have new signal words and pictographs to help communicate chemical hazards, **Figure 5-18**.

Types of Chemicals

Chemicals are often categorized according to the specific hazard they present, although a chemical can present

Rainer Lesniewski/Shutterstock.com

Figure 5-18. These pictographs used on safety data sheets alert workers to dangers from chemicals being used.

more than one hazard or a combination of several hazards. Types of chemicals include:

- An *explosive* is a chemical that causes a sudden, almost instantaneous release of pressure, gas, and heat when subjected to sudden shock, pressure, or high temperature.
- A *flammable liquid* is any liquid having a flash point below 100°F (37.8°C). The exception is any mixture having components with *flash points* of 100°F (37.8°C) or higher that make up 99% or more of the total volume of the mixture.
- A *combustible liquid* is any liquid having a flash point at or above 100°F (37.8°C), but below 200°F (93.3°C). The exception is any mixture having components with flash points of 200°F (93.3°C) or higher, that make up 99% or more of the total volume of the mixture.
- An *oxidizer* is a chemical other than a blasting agent or explosive that initiates or promotes combustion in other materials, thereby causing fire either of itself or through the release of oxygen or other gases.
- A *corrosive* is a chemical that causes visible destruction of, or irreversible alterations in, living tissue by chemical action at the site of contact. They are subdivided as acids and bases.

Compressed Gases

Compressed gases present a potential for simultaneous exposure to both mechanical and chemical hazards. Chemical hazards vary depending on the type of gas in storage. Some gases may be flammable or combustible, while other gases are considered inert, or chemically inactive. Gases can displace oxygen, creating a hazardous atmosphere in which you could suffocate. Compressed gases are often stored in highly pressurized metal cylinders, which creates a "sleeping giant." If the soft brass valve is broken, the cylinder can violently erupt and propel the cylinder like a rocket, creating vast destruction. If a cylinder is exposed to excessive heat, the gas inside expands, potentially causing the cylinder to explode and send metal fragments into the air like shrapnel.

Gas cylinders are often used in many different industrial and laboratory sites. Therefore, it is important to learn how to handle and store them safely. When a cylinder is in use, secure it in a suitable cylinder truck, chain, or other steadying device to keep it from being knocked over. When a cylinder is not being used with an approved regulator, always store the cylinder with the cap on straight and snug. Never store cylinders without securing them in a rack or using restraints to secure them to a fixed object such as a wall. This is to prevent them from falling and creating an explosion. The storage area should be free from open flame or any ignition source. If the storage area is outside, it should be protected from weather conditions. Oxygen should be stored separate from fuel-gas cylinders or combustible materials by a minimum distance of 20 feet (6.1 m) or by a noncombustible barrier at least 5 feet (1.5 m) high with a fire-resistance rating of at least one-half hour.

Safety Note

Wear the appropriate personal protective equipment for the hazards presented. This includes closed-toe shoes and safety glasses. Read the gas label. Make sure that you identify the gas and know the response plan in case of an emergency. This can be found in the SDS for the gas. Also adhere to the following safety rules for handling gas cylinders:
- Never use oxygen to remove dust or debris. This can saturate clothing, which creates a highly flammable condition.
- Make sure the regulator is approved for the gas you are using.
- Stand to the side of the regulator when opening and open the valve slowly.
- Open oxygen cylinder valves all the way to prevent the high-pressure gas from leaking out through the valve.
- Tighten connections appropriately and regularly check for leaks using soapy water or an approved leak detection solution.
- Never use petroleum or flammable solutions on regulators, hoses, or valves.
- Do not attempt to repair a regulator.
- Inspect hoses and equipment daily. Replace faulty hoses and equipment immediately.

Electrical Safety

Electricity is a versatile and important source of power for production agriculture. Electricity, however, can be hazardous if not handled with respect and approached with the appropriate safety practices. On average for the last five years, 161 fatalities have occurred from electrocution. *Electrocution* is death from electrical shock. Electrical hazards can create personal and property damage. Electrical hazards can be created by worn electrical wire insulation, improper wiring, improper wire size or type, or corrosion of electrical connections. Agricultural work environments also present other electrical hazards. For example, tall equipment, such as grain augers, machinery, and irrigation pipes may come into contact with overhead power lines. Electricity can expose workers to electrical shock, fires, burns, and explosions. Excessive and prolonged exposure to electricity can have harmful effects on the body. See **Figure 5-19**.

Agricultural buildings are subject to dusty, moist, and corrosive environments, making proper wiring and grounding a priority. The National Fire Protection Association's National Electrical Code (NEC) provides guidelines for the sizing and installation of electrical circuits. Be sure that equipment is grounded properly. Protect people from shock by installing a ground fault circuit interrupter for electric circuits that are found in wet or damp locations. Never work on an electrical circuit that is energized.

Effect of Electric Current on the Human Body

Current Level (Milliamps)	Effect
0.5–3	Tingling sensations to slight shock felt; not generally painful but disturbing.
3–10	Muscle contractions and pain that may lead to strong involuntary reactions that can cause other injuries.
10–40	"Let-go" threshold—current at this level causes involuntary muscle contractions that prevent individuals from letting go of the power source. Person may be thrown away from power source.
30–75	Respiratory paralysis; death is likely.
100–200	Ventricular fibrillation; extremely painful with probability of death.
200–500	Heart ceases to pump and nerve damage may occur; death is highly probable.
1,500+	Tissue and organs start to burn, likely resulting in death.

Goodheart-Willcox Publisher

Figure 5-19. Electric current is measured in amperes (amps). This table shows the damage to the human body when exposed to different levels of electricity.

Weather Hazards

Exposure to extreme weather can be dangerous. Dehydration, heat exhaustion, and heat stroke are real dangers during hot, humid weather, **Figure 5-20**. Drink plenty of cool, clean, potable water and take frequent breaks to reduce your risk of heat-related illness. Wearing appropriate clothing that is lightweight, light-colored, and breathable will help keep you cool during hot weather. Sunscreen, sunglasses, and wide-brim hats can also protect you from sun exposure.

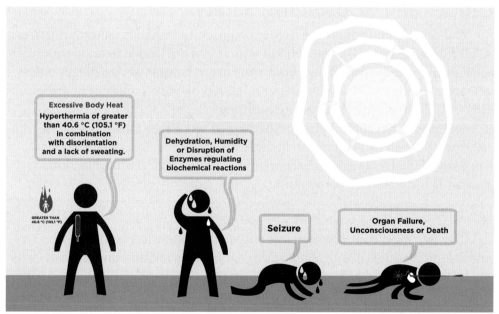

Crystal Eye Studio/Shutterstock.com

Figure 5-20. Staying hydrated is important when working in hot conditions.

When water hazards such as irrigation canals or ponds are present, care must be taken to prevent drowning. Working with a partner and having the proper rescue equipment and a proper emergency protocol can improve your safety.

Cool, wet, windy weather can cause hypothermia, abnormally low body temperature. Rain gear, gloves, waterproof work boots, and headgear are essential. Be aware of the potential for rain or snow while working outdoors, and be sure you have the appropriate gear. Knowing the signs of a heat-related illness or hypothermia can greatly improve your chances of survival if the condition is treated quickly and appropriately.

Insect-Borne Illnesses

Exposure to insects such as ticks, fleas, and mosquitos can increase your chance of contracting diseases such as Lyme disease or West Nile virus. Insect-borne illness is a serious reality. In fact, a case of bubonic plague, passed on by fleas, was identified and treated in one western state as recently as 2015. Using an insect repellent containing 20%–30% diethyltoluamide, or DEET, and appropriate work clothing can reduce your risks of exposure. Appropriate workplace controls, such as removing standing water, can also reduce insect populations.

Outdoor Equipment and Machinery

Agricultural machinery helps improve work efficiency and makes jobs such as tilling and harvesting much easier. However, using machinery and farm equipment to gain advantages in speed and power also produces safety hazards.

Ladder and Scaffolding Safety

Fall-related injuries are far higher in agriculture than in other industries. A fall from a height of 12 feet can kill you. If you are under 16 and working in agriculture, you are prohibited from working at heights above 20 feet unless you are working on a farm owned or operated by your parent or guardian. This includes painting or repairing buildings.

Fall protection should be provided when working at heights during construction. Fall protection can include guard railing, safety netting, or a personal fall arrest system. Each must meet the OSHA Fall Protection standards for materials, setup, and use.

Ladders

The two most common types of ladders are extension ladders and foldable stepladders. Common hazards for ladders include uneven and slippery surfaces, improper use of equipment, and damaged equipment. Be sure to place portable ladders three feet above the landing surface. Always face the ladder when ascending or descending, while maintaining three points of contact at all times. Keep your body centered on the ladder. Do not let the center of your body pass either side rail.

Before stepping onto a ladder, consider the:
- Duty rating of the ladder—what is the maximum weight it can hold?
- Height of the ladder—is it too short or too tall?
- Condition of the ladder.
- Instructions unique to the selected ladder.

When using ladders, follow these guidelines:
- Consider work height when selecting a stepladder.
- Use a ladder only in the fully open position on firm, level ground.
- Do not use a ladder that is folded or in a leaning position.
- Never sit or stand on the top two rungs.

Scaffolding

Scaffolding is an elevated, temporary work platform. The three basic types of scaffolds include:
- Supported scaffolds—one or more platforms supported by rigid, load-bearing members such as poles, legs, frames, or outriggers.
- Suspended scaffolds—one or more platforms suspended by ropes or other non-rigid, overhead supports.
- Other scaffolds—scissor lifts, man lifts, and personnel hoists. These are sometimes thought of as vehicles or machinery, but can be regarded as another type of supported scaffold.

Scaffolds must be able to support their own weight and at least four times the expected load, which can include workers, equipment, tools, and materials. Scaffold hazards include falls from elevation due to lack of fall protection; collapse of the scaffold caused by instability or overloading; being struck by falling tools, work materials, or debris; and electrocution, principally due to proximity of the scaffold to overhead power lines.

All-Terrain Vehicles (ATVs)

An *all-terrain vehicle (ATV)*, also known as a quad or four-wheeler, is defined by ANSI as a small motor vehicle that travels on low-pressure tires, with a seat that is straddled by the operator and is steered using handlebars. Each year, hundreds of injuries and fatalities are caused by improper use of ATVs. Most ATV-related injuries and deaths occur during recreational use.

However, ATV use in the workplace is increasing, especially in the agricultural industry. ATVs are specifically designed for off-road tasks such as carrying loads to the field, checking fences, or spraying weeds. Overturning or rolling over an ATV is the most common hazard. Rollovers are generally caused by excessive speed, unstable load, rough terrain, or excessive inclines. With ATVs weighing several hundred pounds, the forces of a rollover can be devastating. It is critical to not allow extra riders on ATVs designed for single riders. Using proper safety gear such as riding helmets and safety goggles is strongly recommended, **Figure 5-21**.

Figure 5-21. ATVs can be fun, but also dangerous. A—ATVs are often ridden on uneven ground, and are therefore susceptible to rollovers. B—Be sure to wear an ATV helmet approved by the US Department of Transportation.

Tractors

Although tractors are essential to the agricultural industry, they account for the majority of deaths, **Figure 5-22**. The *power take-off shaft (PTO)* of a tractor is an example of a revolutionary invention that helped modernize agriculture, but also poses a substantial hazard to workers. The PTO is one of the oldest and most persistent safety hazards associated with farm machinery, **Figure 5-23**. The PTO, or implement input driveline, is basically a metal rod attached to a power source (tractor) at one end, and an attachment at the other. When the engine is running, power flows along the shaft and transfers energy from the engine to the attachment. The shaft can rotate at two common speeds of 540 or 1,000 revolutions per minute (rpm). That means the shaft is rotating approximately nine times a second. Many accidents occur as a result of entanglement with the PTO.

Figure 5-22. Tractors may be slow-moving, but they are still susceptible to rollovers and other dangers associated with moving vehicles.

Other specific tractor hazards include:
- Rollovers.
- Being crushed by a tractor that is improperly hitched.
- Clothing and hair entanglement in improperly guarded moving belts or chains.
- Compressed hydraulic fluid released through a leak penetrating the skin or eyes, causing severe injury, such as gangrene, a condition that occurs when body tissue dies as a result of blood loss.
- Noise levels. Rules that should be followed when operating tractors are shown in **Figure 5-24**.

vallefrias/Shutterstock.com

Figure 5-23. The power take-off shaft made agricultural mechanics both more productive and more dangerous.

First Aid

Any injury that occurs during class time should be reported immediately to your instructor. *First aid* is emergency care provided for injury or sudden illness before professional medical treatment is available. Knowing basic first aid procedures for treating shock, bleeding, wounds, burns, choking, electric shock, eye injury, fainting, heat stroke, hypothermia, and unconsciousness are important for reducing the severity of an injury. Infectious diseases can be passed from one person to another during first aid treatment, so it is important to follow bloodborne or airborne pathogen safety procedures. First aid kits, safety showers, and eyewash stations should be located close to the work area and kept in working order in case of an emergency. There should be enough of these items for the number of people working in the area. Proper first aid response training is necessary.

There are several ways to be properly prepared for first aid administration:

- Have onsite personnel who are trained in first aid and cardiopulmonary resuscitation (CPR). Contact the American Red Cross, National Safety Council, or local emergency medical service or hospital to locate training in your area.
- Make first-aid kits for various areas of the farm or ranch. Remember to restock the kit after use and to replace expired items annually.

Know how to safely operate the tractor.
Make sure you are familiar enough with the tractor to drive it on a public road. If you have questions about how to operate a tractor, ask a knowledgeable person to assist you.
Obey the traffic laws.
Make sure you follow all traffic laws, including speed limits and traffic signs and signals. Tractor drivers must follow the rules of the road at all times. Display the slow moving vehicle (SMV) symbol in its proper location on the tractor, usually at the rear, facing traffic coming up behind you. Make sure your SMV sign is clean and visible for an appropriate distance behind your tractor.  Mark Herreid/Shutterstock.com
Drive with your lights on.
If your tractor is equipped with headlights and brake lights, make sure they are in proper working order. Use your lights to improve your visibility on the road. Some states may require that you operate yellow or red flashing hazard lights while operating a tractor on a public road. Check the local laws in your state to determine the right course of action to follow. Make certain that your lights are visible from the rear when hauling implements behind the tractor. Most collisions occur when the tractor is rear-ended or when making a left turn. Make sure your brake lights, turn signals, and SMV sign are clearly visible to traffic approaching from the rear.
Pick the right time to drive on a public road.
Choose a time to move farm equipment when traffic is light. Plan the route so you avoid major highways where traffic may be heavy with cars traveling at high speeds. Drive when highway visibility is good. Avoid driving when visibility on the road is 500′ or less.  David Scheuber/Shutterstock.com
Using tractor brakes safely on the road.
Before taking your tractor onto a public road, make sure you have locked the brake pedals together. If you have to hit the brakes in a hurry, this will assure that both sets of rear wheel brakes will engage simultaneously.
Use an escort vehicle.
Use an escort vehicle with a flashing warning light if the equipment you are hauling extends into the opposite lane of travel.
Wear the seat belt.
If the tractor is equipped with rollover protection, wear the seat belt. Without the seat belt holding you in place, it would be easy to be thrown off the tractor in the event of a crash.

Goodheart-Willcox Publisher

Figure 5-24. Following these rules will help you avoid accidents and injury when operating a tractor.

CHAPTER 5 Review and Assessment

Summary

- Many agricultural mechanics and technology occupations are considered hazardous by the US Department of Labor.
- Safety regulations have been issued by the DOL to help ensure the welfare of people who are employed or are seeking employment.
- Personal protection equipment helps protect workers from various dangers and hazards.
- Following general safety practices and understanding the meaning of color codes, signs, and signals can significantly reduce hazards in the agricultural mechanics workplace or laboratory.
- Types of dangerous chemicals include explosives, flammable liquids, combustible liquids, oxidizers, and corrosives.
- Compressed gas presents both a mechanical and chemical hazard.
- Safety must be practiced around high amounts of electricity to prevent injury, including electrocution.
- Dehydration, heat exhaustion, and heat stroke are real dangers from working in extremely hot weather, and hypothermia is a risk in very cold weather.
- Tractors account for the majority of deaths in the agricultural industry, and an increasing number of accidents are occurring with all-terrain vehicles (ATVs).
- Knowing first aid or having someone in the workplace trained in first aid can help minimize the extent of injury when an accident occurs.

Words to Know ↗

Match the key terms from the chapter to the correct definition.

A. combustible liquid
B. decibel (dB)
C. explosive
D. Fair Labor Standards Act (FLSA)
E. first aid
F. flammable liquid
G. hot work
H. OSHA
I. oxidizer
J. personal protective equipment (PPE)
K. rollover protection system (ROPS)
L. Safety data sheets (SDS)
M. Standards

1. A division of the DOL that has the authority to set and enforce health and safety standards for workers in most private industries.
2. Any liquid having a flash point below 100°F (37.8°C).
3. Rules that describe methods for ensuring safety.

4. Device designed to protect operators from injuries caused by vehicle overturns.
5. Any liquid having a flash point at or above 100°F (37.8°C), but below 200°F (93.3°C).
6. Equipment or clothing worn to protect a specific area of the body.
7. The unit of measure for noise.
8. Provisions to ensure safe working conditions for youths.
9. Documents used to communicate the physical and chemical properties of chemicals.
10. A chemical that causes a sudden release of pressure, gas, and heat when subjected to sudden shock, pressure, or high temperature.
11. A chemical other than a blasting agent or explosive that initiates or promotes combustion in other materials.
12. Any process that can be a source of ignition when flammable material is present.
13. Emergency care provided for injury before professional medical treatment is available.

Know and Understand

Answer the following questions using the information provided in this chapter.

1. What is another name for the implement input driveline?
2. What federal body has the authority to set and enforce safety and health standards for most workers in private industries?
3. Name three general safety standards specific to agriculture.
4. Name five types of personal protective equipment (PPE).
5. _____ are designed to protect operators from injuries caused by vehicle overturns or rollovers.
6. Pneumatic and hydraulic power sources present high-pressure hazards that can cause _____.
7. The most common hazard using an all-terrain vehicle (ATV) is _____.
8. Name three potential hazards from tractor use.
9. What is the difference between flammable liquid and combustible liquid?
10. Which type of fire extinguisher is used on fires from combustible metals such as magnesium, titanium, and sodium?
    A. Class A.
    B. Class B.
    C. Class C.
    D. Class D.

STEM and Academic Activities

1. **Science.** Conduct an experiment to test various rollover protection systems that will prevent a raw egg from cracking when it is dropped from a distance of 3' above the ground. Use construction paper, egg cartons, tape, and straws to build your rollover protection structure.
2. **Technology.** Conduct a research report on gas monitors used to detect potentially hazardous atmospheres. Explain how gas monitors operate and what problems they solve.
3. **Engineering.** Design and test an evacuation procedure to determine the most effective route to remove students from the agricultural mechanics laboratory in case of an emergency.
4. **Math.** Calculate the volume of the agricultural mechanics laboratory and determine the ventilation rate to complete a full exchange of air in one hour.
5. **Language Arts.** Read and report on the book "The Jungle." Discuss the connection between the events described in the book and regulations to prevent youths from working in hazardous occupations.
6. **Social Science.** Conduct a research report on child labor conditions in the United States during the Industrial Revolution. Discuss the key leaders and organizations instrumental in changing those conditions.

Thinking Critically

1. Perform a safety audit of your agricultural mechanics laboratory and develop a plan to improve the working conditions and student safety. Review this with your teacher to discuss ways to improve laboratory safety.
2. Conduct a worksite visit to a local farm or ranch. Interview farm managers or owners to identify areas on their worksite in which they would like to reduce hazards. Identify safety alternatives to improve workplace conditions.

CHAPTER 6
Measuring and Marking Tools

Chapter Outcomes

After studying this chapter, you will be able to:
- Understand and use units of measurement in the US Customary and metric systems of measurement.
- Distinguish between accuracy and precision and understand the importance of each in measuring tools.
- Select appropriate measuring tools to fit various tasks in agricultural mechanics.
- Select appropriate marking tools that are designed to be used with a variety of project materials.
- Explain considerations for purchasing measuring and marking tools for use in agriculture mechanics.
- Describe how to purchase, care for, and properly store precision measuring tools.

Words to Know

accuracy	gauge	scribe	thickness gauge
argent pencil	laser level	sliding T-bevel	tolerance
caliper	level	soapstone	torpedo level
carpenter's level	mason's level	Speed® Square	torque
center punch	micrometer	spirit level	torque wrench
combination square	plumb	square	try square
depth gauge	plumb bob	string level	unit of measurement
dumpy level	precision	tape measure	Vernier caliper
feeler gauge	prick punch	theodolite	water level
framing square	scratch awl		

Before You Read

Before you read the chapter, read all of the figure captions. What do you know about the material covered in this chapter just from reading the captions?

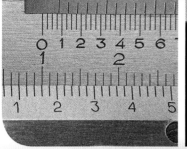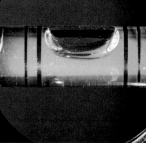

While studying this chapter, look for the activity icon to:

- **Practice** vocabulary terms with e-flash cards and matching activities.
- **Expand** learning with interactive activities.
- **Reinforce** what you learn by completing the end-of-chapter questions.

www.g-wlearning.com/agriculture

kwest/Shutterstock.com

Have you ever heard someone say, "If the job is worth doing, it is worth doing well"? What is meant by that phrase? One element of doing a good job is ensuring that parts fit correctly, function together as intended, and are crafted with consistency.

Measuring tools enable you to be consistent and precise in your work. Whether measuring to see if a machine can fit through a doorway, machining precise assembly pieces, cutting a hundred 2 × 4s to a specific length to frame a building, or weighing agricultural produce for trade, you need accurate measuring tools to do the job well.

Measuring Systems

Agriculture in the United States is built on local economies that are heavily influenced by a world economy. Most countries of the world use the metric system of measurement. The United States, however, uses the US Customary system of measurement. As you study, familiarize yourself with both systems of measurement. Memorizing many of the unit conversions within each system and between the two systems is helpful when preparing to work in any market.

Product manufacturers and material suppliers may give measurements in the appropriate standardized unit of measurement for either or both systems. A *unit of measurement* is a specific quantity accepted as a standard of measurement. One foot is equal to 12 inches, 1 pound is equal to 16 ounces, and 1 kilometer is equal to 100,000 centimeters. These values were determined

AgEd Connection: Agricultural Technology and Mechanical Systems

An agricultural technology and mechanical systems (ATMS) career development event (CDE) helps students develop technical knowledge and skills in electrical systems, energy systems, environmental and natural resource systems, machinery and equipment systems, and structural systems. Students participating in this CDE will be required to use many of the measuring tools discussed in this chapter. For example, students may have to use a tape measure to determine the length of a board used for a rafter in a structural system, or they may use a micrometer to measure an engine part to determine if it meets the required tolerance.

Each tool listed in this chapter could potentially be used during an agricultural technology and mechanical systems CDE. Ask your teacher today how you can be a part of your local ATMS CDE team.

Blend Images/Shutterstock.com

for the purpose of standardization. Standard unit sizes serve as the reference base of each measuring system.

Measuring Tools

The usefulness of a measuring tool depends on its accuracy. In other words, to verify the *accuracy* of a tool, multiple measurements of a specific quantity, distance, or weight taken with the tool need to be the same each time, and the measurement must conform to an accepted standard.

Many jobs demand that measuring tools be accurate and also very precise. *Precision* is the ability to be exact within fractions of a unit when taking measurements. A tool that will measure to thousandths of an inch is more precise than a ruler that is only marked to sixteenths of an inch.

The degree of measuring instrument precision required for a project is often determined by the work material. Soft, rough-cut lumber used to build a fence does not need to be measured as precisely as steel parts assembled in a grain harvesting machine. As a general rule, tolerances for wood projects are not as strict as those for metal projects. *Tolerance* is the maximum allowable variation between measurement specifications and the actual dimensions of the final part or product. Maintaining strict tolerances is an indication of quality workmanship and requires more skill, better tools, and more time.

Rulers

The first measuring tool most students learn to use is the ruler. Rulers are commonly graduated in feet, inches, and fractions of an inch on one edge and in centimeters and millimeters on the opposite edge. See **Figure 6-1**. The ruler forms the foundation of several important measuring and layout tools.

Different lengths and styles of rulers are available for specialized work, **Figure 6-2**. When the rigid structure of the ruler is needed or when the objective is to check for straightness rather than measuring a distance, a ruler or yardstick is used. Some carpenters and cabinet makers prefer to use a folding ruler, **Figure 6-3**, for its sturdiness, pivoting sections, and ability to be folded into a small package.

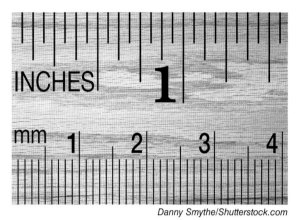

Danny Smythe/Shutterstock.com

Figure 6-1. Understanding the fractions that make up an inch and the unit divisions of a meter is essential knowledge in the field.

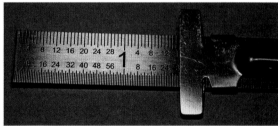

Goodheart-Willcox Publisher

Figure 6-2. This steel ruler is equipped with a sliding depth gauge that is used for taking accurate inside measurements. The measurement scale is divided into 32nds and 64ths of an inch.

Goodheart-Willcox Publisher

Figure 6-3. A six-foot folding ruler is handy for repeatedly measuring the same distance. For example, while digging the post holes for the construction of a fence, you must measure each hole to confirm that all are the same specified depth.

Tape Measures

One of the most essential measuring tools is the *tape measure*. See **Figure 6-4**. Tape measures are extendable, flexible rulers that retract into a durable case for storage. Tapes range in length from three feet to one hundred feet. The blade of the tape measure is usually made of steel but can also be made of strong cloth-like materials. It is marked in US Customary or metric units. As with rulers, some tapes include both measurement systems on the same blade. Retraction of the blade is usually accomplished by a spring within the case. Some long tapes have to be rewound by turning a hand crank on the side of the case.

The metal hook on the end of a tape measure blade is designed to catch the edge of objects for outside measurements and push flush against surfaces for inside measurements. It slides backward and forward on the blade the exact distance as the thickness of the hook, **Figure 6-5**. This slight movement allows inside and outside measurements to be accurate without accounting for the hook's width. The hook also stops the blade from rewinding too deeply into the case.

Professionals employed in many different fields use tape measures to determine various linear measurements. This includes measuring the outside dimensions of an object, the inside dimensions of a structure, and the distance between two surfaces.

When using a tape measure, it is important to keep the blade straight and properly aligned. If you allow the blade to sag, measurements will be longer than the actual length being measured. Extra precision is possible by holding the one inch mark even with your starting measuring point and subtracting one inch from the result to get the final measurement. Always remember to "measure twice and cut once." Discarding materials due to mistakes from incorrect measuring can be very expensive.

When retracting a tape measure, use one finger to apply slight pressure to the bottom of the blade. The resulting resistance slows the rewind, which extends the life of the blade and the retracting spring. If the sharp edges of the blade are in contact with your fingers when the blade retracts rapidly, serious cuts can result. Occasionally wipe the blade with a clean cloth to remove dust and other contaminants.

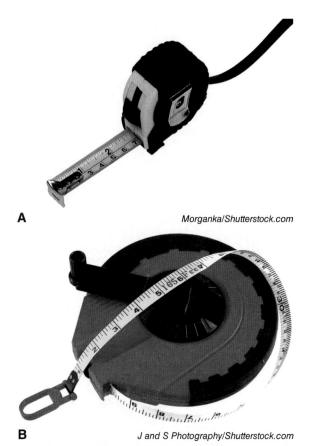

A — Morganka/Shutterstock.com
B — J and S Photography/Shutterstock.com

Figure 6-4. Tape measures, available in many lengths and styles, are the standard tool for measuring lengths in agricultural mechanics. A—A typical self-retracting tape measure. B—A tape measure that uses a manual crank on the side of the case to retract the blade.

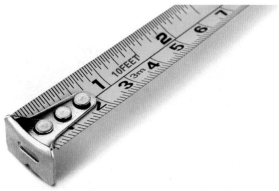

Fotonic/Shutterstock.com

Figure 6-5. Close examination of the hook on the end of a tape measure reveals that the rivets securing the hook to the blade are placed through slotted rather than round holes. This allows the hook to slide back and forth the exact thickness of the hook material. When the blade is pressed against or hooked on an object, the accuracy of the tape measure will not be affected.

Calipers

A *caliper* is an adjustable measuring tool used to measure inside and outside dimensions of small parts. The simplest type of caliper measures just one dimension and does not record a numeric measurement. It must be used in conjunction with a ruler to determine the measurement, **Figure 6-6**.

Modern calipers have sets of jaws that align either with the inside diameter or width of an opening or with the outside edges of a part. This allows them to measure both inside and outside dimensions. Some calipers have a sliding tail, known as a *depth gauge*, that extends into a recess to measure depth. These tools are capable of a high level of precision and accuracy.

Calipers are classified according to the means by which they take and display measurements. A *Vernier caliper* uses a graduated main scale and graduated sliding bar, known as a Vernier scale, to indicate the precise width between the open jaws. A dial caliper, **Figure 6-7**, has a round dial for precision measurement, and a digital caliper, **Figure 6-8**, provides a digital readout to make taking and reading precise measurements easier. Depending on the type, calipers may have graduations for measuring US Customary or metric distances.

Reading a Vernier Caliper

Reading a Vernier caliper is not difficult, but it does require good vision and some practice. Using the main and sliding graduated scales in combination makes it possible to measure differences as small as one thousandth of an inch (0.001″).

The Vernier top scale has divisions which are exactly 0.025″ apart. The main lower scale has twenty-five 0.001″ graduations, which are used to divide any of the Vernier scale divisions of 0.025″ into 25 parts (some may be divided into 50 parts). When the zero mark on the Vernier scale aligns with a graduation on the main scale, the last Vernier scale mark also aligns with a main scale graduation. In this position, the small graduations are not significant to the measurement.

When the zero mark does not align with a graduation on the main scale, counting the misaligned graduations on the Vernier scale until a mark on the scale does line up directly with the main scale determines by how many thousandths the zero missed the 0.025 inch graduation. Read the number on the

Wittaya Budda/Shutterstock.com

Figure 6-6. Some calipers must be used with other measuring tools because they have no markings for measurements.

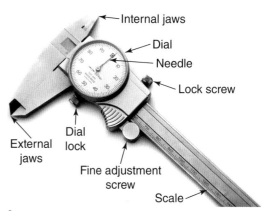

A Theerapol Pongkangsananan/Shutterstock.com

B Paul Tobeck/Shutterstock.com

Figure 6-7. A—To correctly calibrate a dial caliper before use, clean the jaws and close them completely. Loosen the dial lock and turn the dial until the needle is aligned exactly with the zero mark. Retighten the dial lock. B—A dial caliper can be used to measure the outside diameter of a steel pin. The reading on the main scale is between the 0.3″ and the 0.4″ marks. The reading on the dial is 0.075″. Add these two values together to get the final measurement, 0.375″.

124 Agricultural Mechanics and Technology Systems

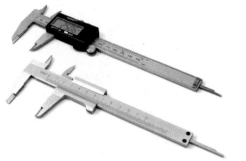

Denis Dryashkin/Shutterstock.com

Figure 6-8. Digital calipers are similar to Vernier calipers. However, digital calipers allow the user to read the measurements directly from the digital display instead of having to add the measurements on the main and Vernier scales.

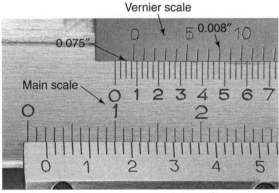

Laborant/Shutterstock.com

Figure 6-9. The top set of scales on this caliper is in US Customary units and the bottom graduations are metric. To read the US Customary measurement, note that the zero mark on the Vernier scale is just past the third mark (0.075″) on the main scale. The graduation on the top that lines up exactly with a mark on bottom is the eighth mark (0.008″). Add the two readings, 0.075″ + 0.008″, to get a final reading of 0.083″.

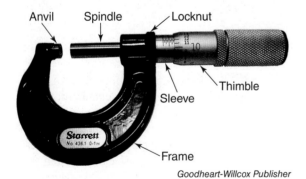

Goodheart-Willcox Publisher

Figure 6-10. In order to read measurements correctly, it is important to be able to name the parts of a micrometer.

main scale that does line up directly. Add this number in thousandths to the reading on the Vernier scale that is just before the zero mark. An example is shown in **Figure 6-9**.

Micrometers

Micrometers are used to take very precise measurements. They may be US Customary or metric and some can measure differences as small as one thousandth of an inch (0.001″). Using and reading a micrometer accurately requires knowledge of the micrometer parts, an understanding of the measurement markings, and the ability to add a column of three decimal numbers, **Figure 6-10**. Outside micrometers are used to measure the outside dimensions of small parts, inside micrometers measure the distance between surfaces, and depth micrometers measure the depth of holes.

The most common sizes of outside micrometer are 0–1″ and 1″–2″. A 0–1″ micrometer measures objects up to 1″, and a 1″–2″ micrometer measures objects that range in size from 1″ to 2″. On engines and similar equipment, very small tolerances are extremely important for the machine to function properly.

Before using a micrometer, clean the anvil and spindle faces. Follow the manufacturer's instructions to determine if the tool is correctly calibrated, and if not, read how to recalibrate it. Depending on the part to be measured, hold the micrometer and turn the thimble with one hand while holding the part with the other, or use two hands on the micrometer, one turning the thimble and the other holding the frame. Turning the thimble causes the spindle to move. When the spindle lightly touches the part, stop turning the thimble. Turn the locknut to hold the spindle in place, remove the part, and take the measurement from the micrometer.

Look at **Figure 6-11**. The first step in reading a US Customary micrometer is to identify the largest numbered graduation completely visible on the sleeve scale. The "1" on the sleeve above the index line represents 0.100″. Next, count the scale graduations after the fully visible numbered graduation on the sleeve scale. It is followed by three marks below the index line that represent 0.025″ each, or 0.075″. Then, locate the closest mark to the index line on the thimble. It is "12," representing 0.012″. Add these together for the final measurement: 0.100″ + 0.075″ + 0.012″ = 0.187″.

Gauges

Gauges are measuring tools that are used to determine precise part clearances and dimensions, material thickness, or enclosed system pressures. They have preset measurements built into their structure or design. In the case of clearance, thickness, and part size, measurements are determined by the fit of the tool to the part being measured.

Feeler gauges are designed to fit into spaces to determine the clearance between two surfaces. **Figure 6-12** is an example of a feeler gauge with a number of blades. Each blade is etched with its exact thickness.

Other gauges measure outside dimensions. *Thickness gauges*, sometimes called "go or no go" gauges, are made with graduated slots into which the material fits (go) or does not fit (no go) for measurement. See **Figure 6-13**. Different types of pressure gauges can measure enclosed system pressures ranging from a vacuum to thousands of pounds of force.

Levels

An object is considered *level* when it is exactly horizontal (perpendicular) to the gravitational pull of Earth. When an object is aligned perfectly parallel to the pull of gravity, it is considered *plumb*.

The most common method of checking an object for levelness is to use a tool that includes a liquid-filled glass or plastic vial called a *spirit level*. These clear vials are slightly curved in the center and marked with lines equidistant from the center. Enough colored alcohol is added to the vial to leave a small bubble that fits between the center lines. See **Figure 6-14**. When the tool is positioned on a level surface, the bubble aligns exactly in the middle of the two center marks.

Manufacturers use alcohol in spirit levels due to its inherent low surface tension and viscosity, which lets the bubble travel smoothly within the sharp-edged tube. Alcohol is also much less likely than water-based liquids to freeze and break the glass.

Carpenter's and Mason's Levels

A *carpenter's level* consists of a rectangular frame of stabilized wood, aluminum, or plastic with one or more vials embedded along its length. See **Figure 6-15**. The center vial is aligned parallel with the long edge of the

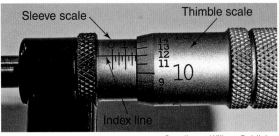

Goodheart-Willcox Publisher

Figure 6-11. Each graduation on this micrometer sleeve represents 0.025″. Each numbered graduation on the sleeve is 0.100″ (four 0.025″ divisions between every number). Each thimble graduation represents 0.001.

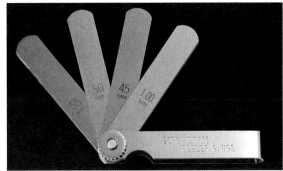

Ehrman Photographic/Shutterstock.com

Figure 6-12. Engine mechanics use a feeler gauge to check for clearances between parts when assembling engines. If blades are not kept clean and flat, measurements will be inaccurate.

Goodheart-Willcox Publisher

Figure 6-13. These thickness gauges are used to determine the diameter of metal wires (left) and the thickness of metal plate (right). Do not force a material to fit into this style of gauge. Such an action can permanently damage the gauge's measurement slots.

Deyan Georgiev/Shutterstock.com

Figure 6-14. When the bubble is between the two center marks as shown, the tool is sitting on a level surface. Tilting the level until the bubble touches either of the outer marks indicates a slope of about 2 to 5 degrees, depending on the curvature of the vial.

Pushish Images/Shutterstock.com

Figure 6-15. A carpenter's level is equipped to check horizontal surfaces for level and vertical surfaces for plumb. Some designs can also confirm 45° inclines.

Tyler Olson/Shutterstock.com

Figure 6-16. Checking the levelness of boards that support the floor of a structure is critical during construction. A carpenter's level is commonly used for this and numerous other building tasks.

tool for checking levelness, **Figure 6-16**. Near one end—or both ends if a 45° vial is not installed—a vial is aligned perpendicular to the long edge for checking vertical objects for plumb. If the tool includes a vial set at a 45° angle, the level can be used to determine the proper location of a structural brace. Carpenter's levels are usually 2 to 4 feet long. A *mason's level* is built like a carpenter's level but is usually four feet or longer in length for leveling and plumbing longer surfaces, **Figure 6-17**.

Torpedo Levels

A *torpedo level*, **Figure 6-18**, is typically 9″-10″ long and is often tapered in thickness on each end. Torpedo levels are handy for taking readings in confined spaces. Many torpedo level designs have magnets built into the frame to free an extra hand when working with metals. Styles designed for use by plumbers or electricians have a V-groove machined into the edge that assists in aligning the tool with pipes or conduits. These levels often have multiple sets of indicator lines surrounding the bubble. The additional lines indicate slopes required for drain lines. A laser may be incorporated into a torpedo level for establishing a level line over an extended distance.

To check a level for accuracy, hold it in place to get a level reading on a stationary object and note the position of the bubble. Next, reverse the direction of the tool to take a second reading in the same location and check the position of the bubble again. The bubble position should indicate level and be situated in the same spot in the vial for both measurements.

String Levels

String levels, also known as line levels, are small tubes containing a single vial that are suspended from a string stretched tautly between two points. See **Figure 6-19**. It is useful for jobs such as building fences or laying bricks that require that points be level even when set long distances apart. When using a string level, you must keep the string very taut; allowing it to sag will corrupt the reading on the level.

Safety Note

When using a laser level, take care never to look directly into the laser beam or direct the beam toward anyone's face. Be aware that the laser beam can be reflected in unintentional directions by hard, slick surfaces.

Plumb Bobs and Lines

A *plumb bob* is another measuring tool that uses a string. It consists of a pointed weight that is tied to the end of a string. When the string is hung from the opposite end, gravity stretches the string in a vertical, plumb line. See **Figure 6-20**.

Ozgur Coskun/Shutterstock.com

Figure 6-17. A mason's level is used during the installation of paving stones for a patio. While evenly laid pavers make for a level patio, some projects require the use of the level to lay the stones at a slope that provides adequate water drainage away from a foundation, steps, or other areas.

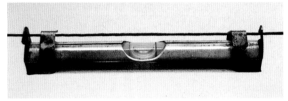

Goodheart-Willcox Publisher

Figure 6-19. When using a string level, you must choose a nonelastic string and pull it very tightly between points to ensure the trueness of the level.

Soraluk Chonvanich/Shutterstock.com

Figure 6-18. Torpedo levels are often used when installing building components where lack of space prohibits the use of longer leveling tools.

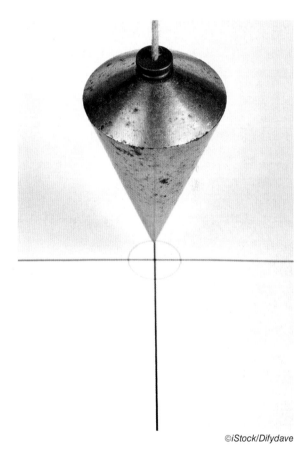

©iStock/Difydave

Figure 6-20. A plumb bob may be used to establish a vertical line or to accurately position another measuring tool directly over a specific point.

Surveying Levels

A surveyor's optical instrument called a *dumpy level* is often used in agriculture for planning new construction sites, laying out fence lines, and determining drainage slopes. Dumpy levels resemble telescopic sights mounted on a tripod, **Figure 6-21**. A similar instrument used for the same purposes is called a *theodolite*. Both tools rely on the principle that the human line of sight is always a straight line. A theodolite can also be tilted to sight at a specific vertical angle, while a dumpy level can only sight at horizontal angles. Horizontal angles are changes in direction from north to south, east, west, or any direction in between along a level horizontal plane. Vertical angles indicate the slope, or rise and fall of the land.

A modern version of these tools, the *laser level*, incorporates a laser beam instead of the scope for increased speed, improved accuracy, and reduced labor requirements. See **Figure 6-22A**.

Red-colored glasses, **Figure 6-22B**, are often worn when working with laser measuring tools to enhance the contrast of the laser beam, making it easier to see. Although protective eyewear is required for more powerful lasers that are used for other purposes, it is not generally needed for lasers designed for measuring purposes.

Squares

Many different types of squares are used to measure and mark common building angles. The most commonly used angle by far is the 90° angle, or a *square* angle. In constructing or repairing stationary structures, certain key components must be confirmed level or plumb before a square is used to correctly position all the other components. When working on mobile equipment such as livestock feeders or trailers, it is often impractical to set the equipment on a level surface while working. In these instances, squares are used to position parts relative to each other without regard to whether the equipment is level.

Framing Squares

The *framing square*, or carpenter's square, is the first tool used when trying to position lumber squarely for construction. Standard framing squares are made of steel or aluminum. They have a longer leg, the *blade*, which is 2″ wide by 24″ long, and a shorter leg,

krugloff/Shutterstock.com

Figure 6-21. This surveyor is using a dumpy level to plan the location of a new forest road. The dumpy level focuses the naturally straight human line of sight into a level line.

A *Dmitry Kalinovsky/Shutterstock.com*

B *chrisbrignell/Shutterstock.com*

Figure 6-22. A—This laser level can be set up directly on a flat, level surface or mounted on a tripod. Once in a level position, the tool projects a laser beam in a straight, level line or in a level plane. B—These red-shaded lenses should be worn to enhance the visibility of the laser beam and to protect your eyes from over-exposure to the laser.

the tongue, which is 1 1/2″ wide by 16″ long. Both the blade and the tongue are marked in inches. A framing square can be used to position parts or for marking materials. Angles for cutting rafters and stair stringers are commonly laid out using the markings on a framing square. Half-scale versions of carpenter's squares are available for working in confined spaces.

Try Squares

A *try square* consists of a steel blade and a stock attached at a right angle to the blade. The blade may or may not be marked as a ruler, and the stock is much thicker than the blade. The thicker stock catches on the edge of the material as the blade lies across the surface, as shown in **Figure 6-23**. Try squares are used in woodworking and metalworking for "trying," or checking, parts for square and for marking lines perpendicular to an edge. Try squares, like framing squares, can also be used to determine whether surfaces are flat.

mihalec/Shutterstock.com

Figure 6-23. The handle of this try square is resting against the edge of a board to position the blade of the tool squarely across, or perpendicular to, the edge of the board. A try square may be used to check, or "try," the cut end of a board for square or as a guide for marking a board prior to cutting.

Speed Squares

In 1925, Albert Swanson invented the versatile *Speed® Square*. An aluminum tool in the shape of a right triangle, the Speed Square combines many of the functions of a framing square and a try square, **Figure 6-24**. One leg of the triangle is a blade marked in inches and fractions of an inch. The other leg is a fence that catches on the edge of the material, similar to the stock of a try square. The hypotenuse of the Speed Square is marked in degrees for measuring and marking angles. The tool's thick, strong construction allows it to be used as a saw guide for cutting 90° or 45° angles.

Authentic Swanson® Speed® Squares are sold with a pocket-sized reference book that includes directions and charts for laying out common construction components such as various types of rafters and stair stringers.

Goodheart-Willcox Publisher

Figure 6-24. Use the Speed® Square to instantly identify 90° and 45° angles and easily find other angles when cutting rafters or other structural parts.

Combination Squares

Another common measuring tool is the *combination square*, **Figure 6-25**. Combination squares are available in several varieties. The standard form incorporates a rigid steel ruler and a standard head—a sliding frame that includes a 90° and a 45° fence. A spirit level and a hidden scribe for marking materials are included as part of the standard head. Other common

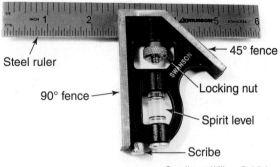

Figure 6-25. This standard combination square can perform many different measuring and layout tasks. The ability to change heads on the tool generates even more possibilities.

attachments include a protractor and a centering head. A combination square is useful for a number of measuring needs:

- The US Customary or metric ruler takes accurate measurements.
- The ruler functions as a straightedge for drawing lines or checking surfaces for adequate flatness.
- The tool functions as a try square when using the standard head.
- The standard head works like a Speed Square when using the 45° angle.
- The spirit level checks structures for level and/or plumb.
- The standard head slides on the blade, allowing the blade to be used as a depth gauge.
- The protractor head adjusts to measure or mark specific angles.
- The centering head can determine the exact center of round materials.
- The hardened steel point, or *scribe*, hidden in the body of the standard head of many combination squares makes scratch marks for cutting or laying out materials.

Sliding T-Bevels

A *sliding T-bevel* is a measuring tool that resembles a try square with an adjustable blade. See **Figure 6-26**. Sliding T-bevels are used to duplicate angles. The blade has an elongated slot on one end but is not marked with any measurements. It may be locked in any position by tightening the knob or wing nut at the joint. After aligning and pressing the stock up against the edge of the material, the blade is moved into the correct angle outline and locked in place. The tool is then transferred to the new material, which is marked or positioned at the copied angle. Precisely measured angles are possible by pairing the sliding T-bevel with a protractor.

Figure 6-26. A sliding T-bevel is used to duplicate angles for repeated transfer to other materials.

Torque Wrenches

Torque is a twisting force that, when applied to an object, causes the object to rotate. It is expressed in foot-pounds, inch-pounds, or Newton-meters. For some applications, such as assembling machines or mounting wheels, it is important to know exactly how much torque must be applied to fasteners. A *torque wrench* is used to apply a specific amount of torque to a nut or bolt.

In agricultural mechanics, two commonly used types of torque wrenches are the beam-type wrench and the click-type wrench. See **Figure 6-27**. Both types function like socket wrenches and use standard sockets.

Beam-type wrenches are inexpensive and require no adjustment. However, they can be difficult to read when working in and around equipment. After the wrench is placed on the fastener, force is applied to the handle until the indicator beam points toward the desired torque value on a scale attached to the handle.

Click-type torque wrenches are more complex and expensive. They are adjusted to the desired setting with a micrometer-like mechanism on the handle prior to use. The tool emits an audible "click" sound when the desired amount of torque is achieved on the fastener. Unlike ordinary wrenches, torque wrenches should be treated as precision measuring tools and handled with care.

Marking Tools

Marking tools are used to place marks that guide the cutting, bending, or connection of materials. Marking tools must be suited to the materials to be effective. One type will not work for all materials. They may be as simple as a common pencil or as sophisticated as a diamond-tipped scribe.

Along with the composition of the marking tool, the width of the mark can also be very important. If parts have to be cut within a tolerance of 1/32", a marker that leaves a line that is over 1/8 thick does not offer enough precision for accurate cutting. The width of the mark made by the marking tool should match the precision of the measuring tool.

Pencils

Common pencil leads are composed of graphite or a graphite and clay mixture, not lead. Regulating the ratio of graphite to clay allows control over the hardness of the core. The common pencil uses an abrasion process in which residue from the graphite and clay core is rubbed onto the material to leave a mark that should remain through the work process.

A   ©iStock.com/DanielmAnderson

B   ©iStock.com/WichienTep

Figure 6-27. A—The scale and indicator of a beam-type torque wrench must be read as each fastener is tightened. B—A click-type torque wrench is preset and locked to a torque setting using the twisting scales on the handle. In addition to hearing the noise, the user feels the handle stop when the correct torque setting is reached on a fastener. Torque wrenches are used with standard socket wrench attachments.

132 Agricultural Mechanics and Technology Systems

Anthoy DiChello/Shutterstock.com

Figure 6-28. The shape of a carpenter's pencil prevents sharpening in a standard pencil sharpener, but the design lets the pencil make marks of different widths.

Goodheart-Willcox Publisher

Figure 6-29. Soapstone is a naturally occurring soft stone that can be formed into different shapes to be used as a marking tool for metal or stone. Holders are available to fit the most common marker shapes.

Regular pencils work well on most wood products. Carpenter's pencils are available for marking on lumber products that have rough-textured surfaces. A carpenter's pencil is about the same size as a child's "jumbo" pencil except that it has a flat, rectangular-shaped body and core. The width of the line can be varied by the direction in which the pencil is used, **Figure 6-28**. Drawing with the narrow edge held perpendicular to the line creates a narrower mark than when pulling across the flat edge.

Carpenter's pencils cannot be sharpened in a conventional sharpener. They are typically sharpened by whittling with a pocketknife or by some type of powered abrasive tool. Avoid sharpening a pencil on a sanding tool that will later be used on fine woodwork. This practice can transfer graphite to the finished surface of the wood.

Soapstone

The marking device of choice for many metalworkers and masons is *soapstone*. It is a white, naturally occurring soft stone composed mostly of talc. Soapstone looks like chalk, but the two should not be confused. Soapstone is used for marking ferrous metals for cutting and welding. It leaves a temporary mark that will withstand the heat of welding. The temporary nature of the white marks is the reason soapstone is popular with masons and other stoneworkers.

Soapstone is available in 5″ × 1/2″ × 3/16″ rectangular bars or 5″ × 1/4″ round rods. Holders for each style are also available, but are generally not necessary. See **Figure 6-29**. Soapstone can be ground or sharpened to a fine point using a power grinder for more precise marks. It is not a good marker for aluminum and stainless steels.

Colored Markers

A variety of wax crayons and oil-based pencils are available for marking bright metals and other slick surfaces. For marking wood and other materials, crayons use a waxy medium for marking and are available in assorted bright colors. Grease pencils have an oil-based core and will mark on many surfaces, but, like crayons, they are limited to relatively wide marks and are not well suited for precision work. Paint pens use liquid paint to mark objects with easy-to-see and relatively permanent identifying marks.

Silver pencils, known as *argent pencils*, use a heat-resistant substance for marking many metals for welding. They are sold as pencils or as flat or round refills that require special holders. Argent pencils are capable of very high precision and, although still temporary, the mark quality is often superior to soapstone in its ability to resist accidental erasure or smudging.

Felt-tipped markers that use permanent ink are useful for marking on stainless steel, aluminum, and other smooth materials. Common in black or blue, these markers are adequate marking tools, except in high-heat applications. Silver-inked versions are better, but visibility on bare metals is often inadequate.

Scribes and Scratch Awls

Sharp-pointed scribes and *scratch awls* leave a precise, permanent mark that is unaffected by heat. Scratch awls look similar to an ice pick, **Figure 6-30**, and are usually constructed from hardened steel. They will mark any materials that are softer than themselves. To mark harder materials, scribes with carbide, tungsten, or diamond tips are needed.

Choose a scribe or awl for marking materials when high precision is needed. The scratches made by these instruments are thin and may be difficult to see on rough work. In certain cases, a utility knife may be substituted as a less precise marking tool.

Teodora D/Shutterstock.com

Figure 6-30. The sharp point of a scratch awl is used to score marks in many materials including wood, plastics, and metals.

Care should be taken to place the mark correctly when using scratch methods due to the permanence of the marks on the material. Missing the measurement and making a second mark can lead to confusion, mistakes, and wasted materials.

Punches

Punches are commonly used in combination with a hammer or other impact tool to mark a variety of materials. Punch marks are used to mark the location of holes and the intersections of lines. The punch deforms the surface of the material to form a relatively permanent mark. Punch marks are more durable than most other methods of marking materials and are precise enough for detailed work.

A *center punch* is used for marking the locations for drilling holes. The indention left by the center punch helps center the drill bit and hold it in place when starting the hole. This helps ensure that the hole is drilled in the proper location. A *prick punch* is used primarily to mark metals. The tip of a prick punch is sharpened at a 30° to 40° angle and is smaller in diameter than the tip of a center punch, **Figure 6-31**. The smaller size is useful when marking precise points. Sometimes a series of prick punch marks is used to designate a cut line or a bend line. To mark the center of a hole that must

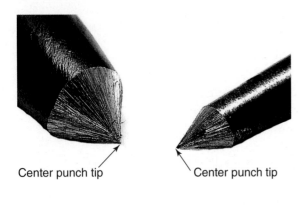

Goodheart-Willcox Publisher

Figure 6-31. Because the prick punch has a smaller tip, it can be used to position marks more precisely than the center punch.

be precisely located, the location can first be marked with a prick punch. A center punch can then be used to enlarge the mark. A prick punch can also be used to back out broken or damaged threaded connectors out of a hole.

Automatic punches are available that do not require the use of a hammer. Often equipped with extremely hard points, automatic punches use an internal spring-loaded mechanism to strike a blow when downward pressure is applied to the tool's shaft. See **Figure 6-32**.

Purchasing Measuring Tools

The quality of the measuring instruments a worker uses has a strong influence on the quality of the work he or she is capable of producing. When purchasing measuring tools, it is important to remember that the precision and accuracy of measurements determines the fit, appearance, and functionality of the final product. It is not necessary to purchase the most expensive tools, but the tools need to be of adequate quality to match the job requirements. The Internet and modern shipping methods have made the acquisition of quality tools much easier than in the past.

Most tool manufacturers offer some type of guarantee on the quality of their products. A published guarantee can be an indicator of quality. Companies that have developed a reputation for quality usually produce accurate tools in order to uphold their good name.

Select tools that are made from durable materials and produced with higher degrees of finish. Moving parts should operate freely, but too much slack leads to a lack of precision and accuracy.

Safety Note
Store scratch awls with the sharp point covered and protected to prevent damage or accidental finger or hand injuries.

Goodheart-Willcox Publisher

Figure 6-32. The automatic punches shown here do not require the use of a hammer. Instead, applying downward pressure to the shaft of an automatic punch activates an internal spring-loaded impact mechanism. This propels the point downward to leave a punch mark in the material.

Hands-On Agriculture

Making a Water Level

A *water level* consists of two plastic graduated cylinders connected by a clear, flexible plastic hose. A 1/4″ diameter hose is common, but any diameter or length can be used. Instead of using a bubble suspended in alcohol to determine level, a water level is based on Pascal's law that pressure is uniformly transmitted throughout a confined liquid. The water in the two cylinders seeks the same level because the cylinders are connected to a common reservoir. Water levels work well for setting up or correcting building foundations. They can also be used to level objects that have obstructions or a great distance between them.

To make your own water level:
- Connect the ends of two plastic syringe barrels to each end of a length of aquarium air tubing.
- Use a permanent marker to make a prominent ring around the middle of each syringe as shown in **Figure A**. These marks will be the "level" indicator marks.
- Hold the two ends of the syringes together and upright while filling the system with water until the level of the water in both syringe barrels is exactly at the level mark. See **Figure B**.

Make sure there are no air bubbles in the system. (Using a few drops of food coloring in the water will make it easier to see air bubbles.)
- Practice using the level by stacking books on two desks sitting side by side. Have one person hold one end of the level next to the first stack while a second person builds the second stack using the opposite end of the water level to determine the height of the stack.
- Check your work by using a carpenter's or mason's level set across the two stacks.

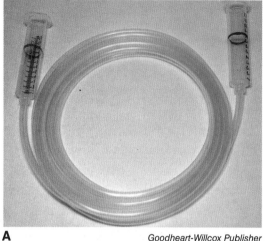

A *Goodheart-Willcox Publisher*

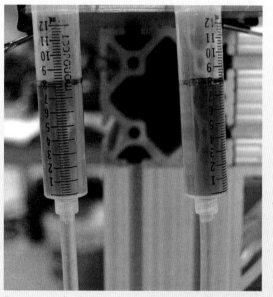

B *Goodheart-Willcox Publisher*

Care and Storage of Measuring Tools

Quality measuring tools should be properly maintained and carefully stored. Remember to treat measuring tools as precision instruments that require a different level of care than other hand tools. Do not store precision tools in a typical toolbox where they might bang against other tools. Many come with protective storage cases that reduce the risk of damage during storage and transport. Placing certain types of tools in lined drawers or hanging others on wall pegs are also good storage methods. Regularly wiping measuring tools with a clean rag during use and prior to storage should be a habit. In humid climates, it is advisable to leave a film of light oil or wax on the steel surfaces of measuring tools to prevent rusting.

CHAPTER 6
Review and Assessment

Summary

- The two most common systems of measurement in agriculture in the United States are the US Customary system and the metric system.
- Measuring tools require precision and accuracy to contribute to quality workmanship.
- Calipers and micrometers allow you to measure small items with higher precision than other common measuring tools.
- Many tools are available to take linear measurements and to determine degrees of level, plumb, or square in agricultural mechanics applications.
- Torque wrenches are considered measuring tools because they can be turned or adjusted to apply a specific torque value to a fastener.
- Marking tools should be selected according to the type of material on which they will be used and the degree of precision required by the job and the measuring tool.
- When purchasing measuring tools, remember that the tools should be of adequate quality to match the job requirements for accuracy and precision.
- Measuring tools are instruments of precision that should be handled, used, and stored gently and with respect.

Words to Know

Match the terms from the chapter to the correct definition.

A. accuracy	H. level	O. spirit level
B. argent pencil	I. plumb	P. square
C. caliper	J. precision	Q. thickness gauge
D. torque wrench	K. scribe	R. tolerance
E. center punch	L. sliding T-bevel	S. try square
F. depth gauge	M. soapstone	T. water level
G. feeler gauge	N. Speed® Square	

1. Descriptor for an object aligned parallel to the Earth's gravitational pull.
2. A "go or no go" device with graduated slots for measuring outside dimensions.
3. A tool that leaves a heat-resistant silver mark that is ideal for marking many metals for welding.

4. An adjustable tool used to reproduce angles.
5. An L-shaped tool, consisting of a steel blade and stock, which is used for checking perpendicular parts.
6. A tool that marks the location of holes to be drilled.
7. An instrument that leaves a mark consisting mostly of talc.
8. An instrument that has a bubble floating in a sealed vial of liquid to indicate horizontal or vertical alignment.
9. An adjustable measuring tool for measuring inside and outside dimensions of small parts.
10. Term for two objects set perpendicular, or 90°, to one another.
11. The ability to record the same measurement every time for a specific quantity, distance, or weight.
12. Tool used to make scratch marks for cutting or laying out materials.
13. Tool used to measure clearances between close-fitting parts.
14. The maximum allowable variation between specifications and actual dimensions.
15. An instrument used to apply an exact amount of twisting force to a nut or bolt.
16. The ability to measure exact measurements down to fractions of a unit.
17. Descriptor of an object aligned horizontal (perpendicular) to the Earth's gravitational pull.
18. Versatile, triangular-shaped measuring tool that could be used as a saw guide for 45° and 90° cuts.
19. A basic measuring tool that functions based on Pascal's law and atmospheric pressure.
20. A device used for probing into and measuring holes.

Know and Understand

Answer the following questions using the information provided in this chapter.

1. When the same quantity is measured repeatedly with the same tool and the same measurement in standardized units is reached, the tool is said to have _____.
2. A tool with _____ is capable of measuring objects down to very small fractional divisions of a unit of measure.
3. Why is the metal hook on the end of a tape measure blade made to slide back and forth a short distance?
4. The difference between a Vernier caliper, a dial caliper, and a digital caliper is the ease or simplicity of the way the _____ are read.
5. Spirit levels use the natural force of _____ to locate level or plumb lines or planes.
6. *True or False?* To check a level for accuracy, take a reading on a level surface and reverse the direction of the tool for a second reading in the same location; both measurements should match.
7. The stock on a try square or the fence of a Speed® Square can be set against the edge of a board so that a square or _____ line may be drawn across the width to mark a cut location.

8. If you determine during construction of a six-sided gazebo that a large number of boards will need to be marked and cut at a 60° angle, a(n) _____ can be used to duplicate all of the angles after the first board is measured and cut.
9. _____ pencils leave a silver mark on metal that is more precise and often has better staying potential than soapstone.
10. Precision measuring tools commonly come with a(n) _____ that reduces the risk of damage during storage and transport.
    A. owner's manual
    B. reference booklet
    C. protective case
    D. None of the above.

STEM and Academic Activities

1. **Science.** Investigate gravity and Pascal's law further. Prepare a short presentation explaining how Pascal's law and the pull of gravity help us determine level and plumb.
2. **Technology.** Study the booklet that accompanies a Swanson® Speed® Square. Work with a partner to demonstrate how to use a Speed Square to mark a 2" × 4" to be cut for use as a rafter.
3. **Engineering.** Gather a collection of different marking tools. Use each marking tool to draw one line in a series of side-by-side parallel marks on a scrap of lumber and on a length of flat steel bar. Which marks are the easiest to see? Which ones offer the greatest precision? Rub the marks lightly with a shop rag. Do some of the marks resist wear better than others?
4. **Math.** Practice measuring different nuts, bolts, and other hardware using a caliper and/or a micrometer.
5. **Social Science.** In the early days of measurement, kings established their own units of measure. For example, a foot was the length of the king's foot. Confusing the issue, each new king could change the standard. Thus over time, a unified standard was needed and established. People still use body parts to estimate length. Find your own inch and foot measurements. Is an inch the width of a thumb or a joint on a finger? Practice estimating lengths of objects using these newly developed measurements.
6. **Language Arts.** Communicating with others is an important step in many agricultural mechanics jobs. Practice written communication by describing the measuring tool of your choice in a five-sentence paragraph. Do not use the name of the tool in your description. Pay close attention to spelling and grammar. Check your work by allowing others to read your description. Do they know the name of the tool you are describing?

Thinking Critically

1. Work with a partner and select several different combinations of measuring tools and marking devices. Mark sections of scrap lumber or other material in lengths of exactly 5 1/2″. Which combination offers the greatest precision? Is there a difference in accuracy? How did you check your work?
2. Use the sliding T-bevel and a protractor to mark a 45° angle across a board. What tool can you use to quickly double-check your work?

CHAPTER 7
Hand Tools

Chapter Outcomes

After studying this chapter, you will be able to:
- Explain how tools have been developed and relate to how new tools originate.
- Recall safety knowledge and establish safe practices when using all hand tools.
- Classify and identify hand tools based on their design and function.
- Select the appropriate hand tools for performing different tasks in agriculture mechanics.
- Demonstrate safe and proper use of hand tools for common tasks.
- Summarize the factors to consider when purchasing hand tools.
- Describe how to properly organize, maintain, and store hand tools.

Words to Know

adjustable wrench	dead-blow hammer	locking pliers	slip-joint pliers
anvil	diagonal pliers	mallet	slotted-tip screwdriver
aviation snips	die	miter box	socket wrench
backsaw	extractor	needle-nose pliers	soft-faced hammer
ball-peen hammer	file	nut driver	tap
brace and bit	gouge	Phillips-head screwdriver	tapered punch
chamfered	groove-joint pliers		Torx®
claw hammer	hacksaw	pipe wrench	vise
combination wrench	hex key	rasp	wood chisel
coping saw	kerf	ratchet handle	
cross-peen hammer	keyhole saw	ripsaw	
crosscut saw	lineman's pliers	sledge hammer	

Before You Read

Read the chapter title and tell a classmate what you have experienced or already know about this topic. Write a paragraph describing what you would like to learn about the topic. After reading the chapter, share two things you have learned with the classmate.

While studying this chapter, look for the activity icon to:

- **Practice** vocabulary terms with e-flash cards and matching activities.
- **Expand** learning with interactive activities.
- **Reinforce** what you learn by completing the end-of-chapter questions.

www.g-wlearning.com/agriculture

kwest/Shutterstock.com

A tool is a device manufactured to aid in the completion of a certain task. Hand tools help you perform tasks that are difficult or impossible to do strictly with your bare hands. This chapter covers the hand tools commonly used to construct, maintain, service, and repair agricultural systems and equipment.

Tool Evolution

Satisfying the basic needs of food, shelter, and security was the motivation for the development of early hand tools. The very first tools were natural materials with unique shapes that were probably discovered by chance and used based on need. Evidence of manufactured stone tools dates back to 2.6 million years ago on the African continent, **Figure 7-1**.

Early tools were used for pounding or as a cutting edge, or both. Basic handheld tools were needed for securing shelter, hunting, and processing dead animals for food and fur. Once early humans began to use tools, the need for better tools developed with time.

One of the earliest improvements in the manufacture of tools was hafting. Hafting is the act of fitting a handle to a stone, bone, or metal tool. This addition gave the user better control of the tool and allowed the tool to be swung or leveraged with greater force. Care in selecting the best materials was, and still is, important to improving the quality of the tool. The development of more sophisticated tools also created the need for special implements designed just for tool production.

jps/Shutterstock.com

Figure 7-1. A stone axe estimated to be from the early Paleolithic era, approximately 2 million years ago.

The introduction of copper and bronze into toolmaking greatly influenced the complexity and capability of hand tools. Iron tools significantly changed the development of tool technology and increased the rate at which civilizations could grow and develop. The introduction of steel tools led to the creation of tools that were not possible with weaker materials. Centuries later, the development of the Bessemer process in 1856 and subsequent improvements to the steel-making process led to the wide availability and use of steel. Today, a greater variety of hand tools provides a wider scope of capabilities and continues to increase the efficiency of work. Steel and steel alloys are still used today for producing some of the finest tools available.

A quick look at most tools reveals that they consist of one or more simple machines. There are six classical simple machines: the inclined plane, screw, pulley, wedge, lever, and wheel and axle. Tools that have no moving parts generally have a single simple machine featured within their design. For example, a lever used for prying or a wedge used as a cutting edge are practical applications of simple machines. See **Figure 7-2**. Two or more simple machines combined as a unit to perform a task form a mechanism.

Mechanisms allow more sophisticated tools to perform a greater number of operations and more difficult tasks.

General Tool Safety

Each shop has its own safety and operational rules, but there are general hand tool safety rules you should *always* follow:

- Always read and follow the safety and maintenance procedures provided by the tool manufacturer for a particular tool.
- Use only tools that are in good condition. Report damaged or broken tools immediately and remove them from service.
- Wear safety glasses at all times. Proper eye protection should be worn when using any tool. Use ANSI Z87.1 approved eyewear. Use safety goggles or a full face shield to fit over prescription eyeglasses.
- Use each tool only for its designed purpose.
- Use only tools that you have been trained to use.
- Properly clean and store hand tools to keep them in safe working order.
- Keep work areas clean and clutter-free at all times. Confirm that adequate lighting and ventilation are installed throughout the shop.

Naming and Classification of Hand Tools

Hand tool names usually reflect their design or the job that they do. Some tools are named after the inventor or the company that introduced the tool to market. These facts usually make figuring out and remembering common tool names a simple process. Sometimes tool naming is complicated by regional differences. Different regional names, however, usually still refer to the way the tool is used.

In agriculture, hand tools can be classified by how they work or by the specific part of the industry in which they are used. Common classifications include woodworking tools, metalworking tools, mechanic's tools, plumbing tools, electrician's tools, and concrete and masonry tools. Other tools in agriculture are used in fields such as horticulture and livestock husbandry.

A
Zbigniew Guzowski/Shutterstock.com

B
Lee Torrens/Shutterstock.com

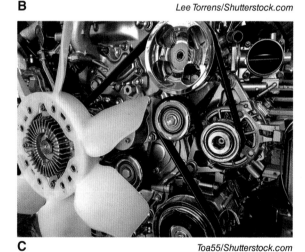
C
Toa55/Shutterstock.com

Figure 7-2. Simple machines, including levers, inclined planes, wheel and axles, pulleys, wedges, and screws, are the basic building blocks of both hand tools and complex machines. Determine which simple machines are used in these photographs.

SAE Connection

Proper Tool Selection and Use

What tools do you use in your SAE? What tools have you been trained to use in your job or agriculture class? A clear understanding of the purpose and proper use of each hand tool presented in this chapter will help you to perform your job or SAE activities safely and efficiently.

Mike Flippo/Shutterstock.com

In this chapter, tools are classified according to the way they function. Note, however, that some important tools work in ways that combine two or more classifications.

Cutting Tools

Cutting tools are some of the oldest hand tools. Knives and axes have been used as cutting tools for thousands of years. The principal action of most cutting tools is to guide a serrated or sharpened blade through a material to cut the material to the desired dimensions. The cutting edge of the tool must be constructed of, or tipped with, a harder material than the stock being cut. Holes can also be bored with some hand tools. During the cutting process, many cutting tools remove small bits of waste, while others just separate the material.

The most basic of cutting tools is the knife. Different kinds of utility knives are made to cut various materials. One of the most common utility knife designs uses a reversible, replaceable blade that is made in the form of a trapezoid. See **Figure 7-3**.

Handsaws

The business end of any saw is its teeth. Saw teeth are sized and shaped according to the type of material the saw is intended to cut. The size of the saw tooth correlates to the number of teeth per inch (or centimeter) on the blade. Softer materials require a blade with larger saw teeth (fewer teeth per inch) set at greater angles. Harder materials require a blade with smaller saw teeth (more teeth per inch) set at steeper angles.

Most saws are designed to cut in only one direction—usually on the push stroke—but some cut on the pull stroke or in both directions. Teeth are angled

nuwatphoto/Shutterstock.com
Figure 7-3. Removing the screw in the handle of this utility knife separates the tool for access to the extra blades stored inside.

slightly forward to act as tiny points or chisels that dig their way into the material. To reduce friction as the saw blade cuts through the material, the angled teeth are set alternately on either side of the blade. The result is a *kerf*, or cut slot, that is wider than the bulk of the saw blade.

Crosscut Saws and Ripsaws

Crosscut saws have pointed knifelike teeth designed for cutting lumber across the grain. *Ripsaws* are similar to crosscut saws, but they have chisel-shaped teeth and are used for cutting lumber parallel to the grain. Both crosscut saws and ripsaws commonly have a D-shaped wood or composite handle and a wide blade that can range from 16″ to 36″ in length. See **Figure 7-4**. The longer blades are mainly used for crosscut saws that are intended for cutting timber. Four to six feet long, two-man crosscut saws with handles on both ends have been used for many years, and are still available for cutting logs.

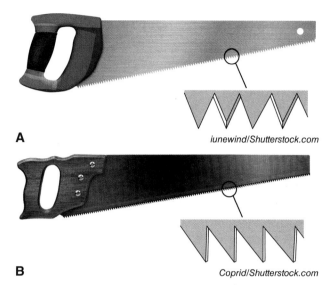

Figure 7-4. A—Crosscut saws have knifelike teeth designed to cut wood perpendicular to its grain. B—Ripsaws have chisel-shaped teeth designed to cut wood parallel to its grain.

Hacksaws

A *hacksaw* cuts material using replaceable blades held under tension across an adjustable metal frame. See **Figure 7-5**. Hacksaws were originally designed for cutting metal, but they can also be used for cutting plastic, wood, or other materials.

Most hacksaw blades are 10-12 inches long, but shorter blades exist for special purposes. Blades are thin and usually about 1/2″ wide, with between 14 and 32 teeth per inch. Instead of each small tooth being alternately set on either side of the blade, hacksaws usually have a wavy set blade, which consists of alternating groups of teeth offset in both directions to widen the kerf. Under most conditions, hacksaws should be assembled and used with the teeth angled forward to cut on the push stroke.

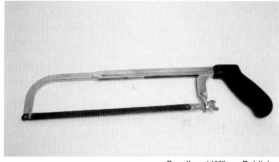

Figure 7-5. This hacksaw frame can hold blades of different lengths. The blades are kept under tension by tightening the wing nut located at the back of the frame.

Coping Saws

A *coping saw*, **Figure 7-6**, has a lightweight, U-shaped metal frame and uses thin, narrow blades for cutting shapes in sheet materials. Replaceable blades are available to cut wood, metals, and plastics. The short, straight handle is designed to be rotated in your hand to turn the saw blade and frame when cutting intricate curves. Tightening the handle tensions the blade. Although it is preferable to set up the saw blade

Figure 7-6. Coping saws are used to cut intricate patterns in wood and other materials. The thin blade can be turned easily to change the direction of the cutting action.

to cut on the push stroke, sometimes better control is possible by mounting the blade to cut on the pull stroke.

Keyhole Saws

A *keyhole saw* was developed to make inside cuts on covering materials such as paneling and plywood. Keyhole saws have wooden, plastic, or metal handles and thin, tapering steel blades, **Figure 7-7**. The teeth are usually coarse for a small saw, ranging from eight to twelve teeth per inch, but blades used to cut materials such as sheet metal or plastic have a higher teeth-per-inch count. Typical replaceable or permanent blades are 12" or 15" long, but some are as short as five inches. The narrower the blade, the tighter the curve it can cut; finer blades are preferable for cutting plywood. A version of the keyhole saw is often used when installing and repairing drywall.

Stocksnapper/Shutterstock.com

Figure 7-7. Keyhole saws are often the hand tool of choice for cutting through interior wall panels to install electric switches and power outlets.

Safety Note
When using a keyhole saw to cut into an existing wall, be careful not to accidentally cut into electrical or plumbing systems. Serious injuries and damage can result.

Backsaws

A *backsaw* is a precision tool for cutting very straight lines. Backsaws are crosscut saws with small teeth and a stiffening rib on the top of the rectangular blade. The teeth have a minimum offset to produce very narrow kerfs. They are mostly used in trim work and in the construction of fine furniture.

When cutting precise angles in trim work, a worker often uses a *miter box* in conjunction with the backsaw. See **Figure 7-8**. Miter boxes guide the saw blade through a straight cut at the desired angle. They may be standard for common angles, custom-made for specific angles, or adjustable to any angle. Like keyhole saws, backsaws have been replaced in many situations by more efficient power saws.

Using a Saw

Before using a saw, begin by marking the material with an easy-to-see mark. Clamp the workpiece securely in a vise or against a workbench so that it will not move unexpectedly during the cutting process. Use your free hand to guide the saw, but do not let your hand get into the path of the blade.

Remember that the width of the kerf must be considered when cutting pieces to an exact length. Place the saw edge on your mark so the mark remains visible and the kerf will be formed on the waste side of the line. Begin by dragging the blade backward with light pressure for a couple of strokes to establish a shallow cut line. Keeping your arm close to your body, push the saw forward and allow the blade to do the work.

Use just enough downward pressure to produce cutting action. Do not apply pressure on the backstroke. Most saws should be worked at an angle

a_v_d/Shutterstock.com

Figure 7-8. The reinforcement on the blade of the backsaw keeps it straight while the slots of the miter box guide the saw blade through precise angles during cutting.

to the surface of the material for greatest efficiency. This reduces the effort required to push the saw through the material and the number of teeth in contact at any moment with the material. Keep a minimum of three teeth engaged at all times during the cut.

Hand-Powered Drills

Drills are cutting tools that use sharpened bits (inclined planes wrapped around a shaft) and leverage from the user to bore holes in all types of construction material. Hand-powered drills consist of a handle or manual drive mechanism that provides leverage and a chuck that holds the interchangeable drill bits, which serve as the cutting edges of the tool. Drill bits for hand-powered drills are usually one of two common types: auger bits or twist drill bits.

Auger bits for boring wood have a coarse spiral wrapped around a central shaft. At the tip, a small screw bites into the wood to establish and maintain the center of the hole. As the bit is turned, one, two, or three cutting edges, positioned at a slight angle to the surface, dig into the material. Additional cutting edges perpendicular to the surface slice through the grain to produce a clean, round perimeter for the hole. Auger bits commonly have tapered, square shanks.

Twist drill bits are similar to auger bits: both have multiple cutting edges, a centering mechanism, and a way to remove waste material from the hole. Instead of a square shank, however, a twist drill bit has a round shank.

The development of the cordless electric drill led to the drastic decline in the use of hand-powered drills. Nevertheless, hand-powered drills still offer a few advantages. They are better for drilling deep, large bore holes in large lumber and posts, and they can be carefully controlled when precise bore depth is needed. They also need no form of electricity, which is ideal when working in remote locations.

Remember these guidelines when using any drill:
- Mark hole locations with a center punch.
- Make sure the bits are sharp and the correct type for the material being drilled.
- Clamp or otherwise secure material so it will not move during the drilling process.
- Always apply enough pressure to the back of the drill to keep the cutting edges of the bit engaged and removing material.
- Use a backing board to keep lumber from splintering as the bit passes through the opposite side.
- Be prepared for the sudden loss of resistance as the bit completes the hole.

Brace and Bit

When building agricultural fences and buildings, a ***brace and bit*** is a valuable tool to have in your toolbox. A brace consists of an offset frame, two handles, and a chuck for holding a bit. See **Figure 7-9**. The top handle

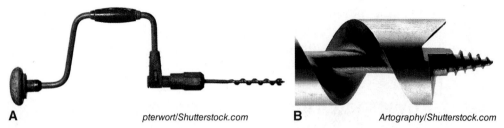

A *pterwort/Shutterstock.com* B *Artography/Shutterstock.com*

Figure 7-9. A—This brace is equipped with an auger bit for boring holes in wood. Always apply enough pressure to any drill to ensure that it is continually removing material. B—A close-up of an auger bit showing the cutting edges and the small screw attached at the tip of the bit.

is round and flat, which allows the user to apply adequate pressure on the drill bit. The other handle spins around the offset portion of the handle as the tool is rotated during use. Typically, an auger bit is used with a brace, but the chuck will accept the round shank of a twist drill bit as well.

Other Hand-Powered Drills

Three other types of hand-powered drills include the hand drill, pin vise, and push drill. Hand drills, often referred to as eggbeater drills, use a pair of beveled gears to drive the drill bit, as shown in **Figure 7-10**. Hand drills have mostly been replaced by cordless power drills. However, they are still used in the construction of fine furniture and for other precision work.

Thatsaphon Saengnarongrat/Shutterstock.com

Figure 7-10. Although electric drills have replaced most hand-powered drills, hand drills are still useful in remote locations where power is unavailable.

A pin vise is simply a handle with a collet-type holder designed to grip a small drill bit, file, or reamer. It is turned like a screwdriver to drill a hole. As you can imagine, a pin vise drills very slowly, but it is capable of great precision, **Figure 7-11A**.

At first glance, a push drill looks like a screwdriver, but on closer examination it is much more complex. Push drills use a ratcheting spiral mechanism to convert some of the user's pushing force on the handle to rotating force to drive the drill bit. See **Figure 7-11B**. It uses a different style of bit from standard twist drill bits. The bit drills when the handle is pushed down and clears material when the handle is released.

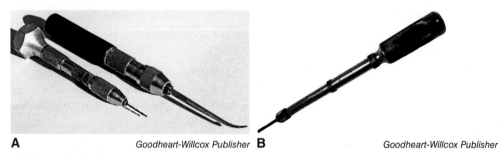

A *Goodheart-Willcox Publisher* B *Goodheart-Willcox Publisher*

Figure 7-11. A—Pin vises are used by machinists and others who need to drill very small holes in precise locations. B—Push drills are sometimes preferred by fine cabinetmakers for drilling pilot holes for screws. Note the design of the drill bit.

Figure 7-12. A—Tapered taps are general-purpose taps used to start threading operations. Bottom taps are square-ended, without a taper, and are used to thread holes that do not pass all of the way through a part. B—A die is fitted into a handle that usually accompanies the set. The entire tool must be held square as the user carefully turns the bit to bore into the material. C—A typical tap and die tool set.

Taps and Dies

When work materials require threaded connections, taps and dies are used. A *tap* is used to cut threads along the inside of a hole. See **Figure 7-12A**. A *die* is used to cut threads on the outside of a round stock to form a bolt or screw, **Figure 7-12B**. Taps and dies are marketed in sets that include handles, **Figure 7-12C**.

The sizes of taps and dies are based on standard thread sizes. Unified National Coarse Thread (UNC) and Unified National Fine Thread (UNF) are measurement systems with fractional inch divisions for thread size. Metric threads are different and cannot be interchanged with UNC or UNF threads. Steel pipes are connected using the National Pipe Thread Taper (NPT). Special taps and dies are available for this purpose.

One of the most common uses for a set of taps and dies is cleaning old threads when repairing machinery. This process is called chasing the thread. See **Figure 7-13**. Taps designed specifically for chasing threads in electric components or tire valve stems are also available. Basically, the tap or die is used to cut or scrape away rust and other deposits that contaminate threaded components.

When using taps or dies, follow these rules:
- Always use the proper handle with the tap or die.
- Keep taps and dies sharp by properly cleaning and storing them in a secure compartment away from other tools.
- When cutting threads in metal, use cutting oil as a lubricant.

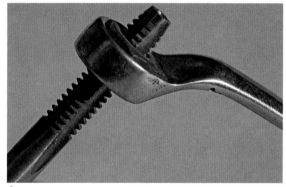

Figure 7-13. Taps and dies may be used to cut new threads or to clean and reshape existing threads. A—A tap is being used to "chase the threads" in the nut held in this wrench. B—This valve stem tool includes a tap, a die, an extractor, and a wrench for removing the valve stem cores used on pneumatic tires.

- Use a "two steps forward, one step back" motion to clear the cutting surfaces of metal chips while cutting. This keeps the cutting tool from binding and prevents damaging newly cut threads.
- Use a soft brush to clean up waste chips. Do not wipe away metal chips with your bare hands or blow them away with compressed air.

Chisels and Planes

A chisel is a cutting tool that uses a sharpened inclined plane to remove small pieces of material from a larger stock. Chisels may be forced into the work by hand alone, or they may be driven with a mallet or hammer.

Similar to chisels, planes use a very sharp inclined plane mounted in a stabilizing frame to shave thin slivers from the surfaces of wood. Planes are powered by the user's strength alone. They are not designed to be driven with a mallet or hammer. The shape and construction of chisels and planes vary greatly depending on their intended use.

Wood Chisels

A *wood chisel*, **Figure 7-14A**, is sharpened at a 25° to 30° angle on one side only, and its effectiveness depends on its sharpness. The shape of the tip determines the shape of the cut in the material. Common wood chisels are 3/4" or 1" wide and perfectly flat. The edge is perpendicular to the edge of the blade. *Gouges* are chisels with U-shaped cutting edges that produce a concave cut into the material. Many other types of wood chisels have been developed for making cuts of specific shapes in lumber.

The single inclined plane of a wood chisel should be kept very sharp with regular sharpening on one side only. When using a power sharpener, you must cool the blade edge frequently to maintain the temper, or hardness, of the blade.

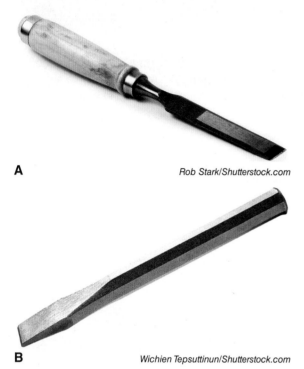

Rob Stark/Shutterstock.com
Wichien Tepsuttinun/Shutterstock.com

Figure 7-14. A—The single sharpened side of a wood chisel's blade is ground to a narrow angle to create a very sharp edge. B—The two sides at the tip of a cold chisel are ground to a steep edge. It must be sharp and very tough to withstand repeated impact with metal.

Cold Chisels

Metal-cutting chisels are much more rugged than wood chisels. The most common cold chisel has a straight edge that is sharpened on both sides near the tip. See **Figure 7-14B**. They are used to cut and clean the surface of metal in areas where other tools are inefficient or cannot reach.

Cold chisels are formed from a solid piece of tool-grade steel. The term "cold" is used because the metal is worked during manufacturing without being heated. Hammer blows on the flat end are used to drive the chisel.

Safety Note

Always wear approved safety glasses and leather gloves when using a cold chisel. Use a hammer that is engineered for striking other tools. Hammers designed for driving nails can actually be broken by hammering on a cold chisel. The head of a cold chisel is beveled to reduce the mushrooming effect of hammering, **Figure 7-15**.

Hand Planes

Hand planes are capable of shaving thin layers from the surface of wood. See **Figure 7-16**. Planes are used for smoothing and straightening the surfaces of lumber. The blade, or iron, of a plane is sharpened on one side only, just like a wood chisel. It is held in place at a 45° angle with the sharpened side down.

Frames of common bench planes are usually 9"–22" long. The longer the plane, the better job it will do for straightening lumber. A 9"–10" smooth plane is the most common and useful for general work. Jack planes are 12"–17" long and are good for straightening longer boards. Block planes are small in size, which makes them ideal for confined spaces.

Files and Rasps

Files and *rasps* are cutting tools used to remove small, controlled amounts of material from a workpiece. They are made from hardened tool steel and come in a variety of profiles or shapes, **Figure 7-17**. Files can be single cut or double cut. On the single cut, the teeth all go in one direction. On double cuts, the teeth are ground to cut both on the push and pull strokes. Files may be used for shaping or smoothing most materials.

Rows of teeth cover the full width of a file blade. Smaller, more densely grouped file teeth create a finer, smoother cut than larger, coarse teeth. Rasps have rows of multiple triangular teeth that remove material more aggressively than files. For work on softer materials, a rasp may be more effective than a file.

Common rasps and files are made with a tapered rectangular tang that is intended to be inserted into a handle. Do not use a file or rasp without a handle attached because the steel tang may injure your hands. The teeth of rasps and single-cut files cut during the push stroke only. Slightly lift a file and release pressure during the return stoke to minimize damage to the teeth.

As a file is used, especially on softer materials, the spaces between the teeth can become contaminated and clogged with waste particles. A short, stiff-bristled brush called a file card can be used to clean the file as needed. Use nonmetallic bristles rather than a steel brush, because steel bristles can dull the file. Soapstone can be

Goodheart-Willcox Publisher

Figure 7-15. Mushrooming is the dangerous flattening and distortion of the head of tools that are repeatedly struck with a hammer. Small shards of steel can break off and be propelled long distances. Use the correct type of hammer and keep chisel and punch heads correctly ground to reduce the chances of injury.

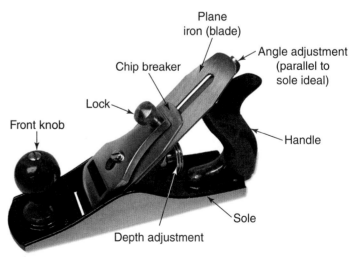

Alex Hubenov/Shutterstock.com

Figure 7-16. Power tools have replaced hand planes for most jobs, but hand planes are still an effective tool for shaving the edge off a door sticking in its frame or for straightening a twist or warp from a board.

Marafona/Shutterstock.com

Figure 7-17. Files and rasps are available in many different shapes, sizes, and cuts. Each is designed for specific tasks. Always use a file with the handle securely in place.

Goodheart-Willcox Publisher

Figure 7-18. Working a little soapstone between the teeth of a file helps to prevent small chips of filed metal from clogging up the file.

rubbed into the teeth before use to reduce clogging. The talc in the soapstone serves as a dry lubricant. See **Figure 7-18**.

Gripping Tools

Gripping tools hold parts and materials in a secure manner for work operations. Their designs provide a stronger, safer holding method than using just your hands or feet. Gripping tools pull and hold parts together, thus freeing your hands to perform other functions.

Maximizing grip surface area is the key to using any gripping tool. Always choose a gripping tool that will provide the most surface area of grip and will hold the work material as tightly as possible without damaging the surface.

Vises

A *vise* is a clamping device that is mounted to a stationary workbench or stable structure. A worker primarily uses a vise to hold a part or parts in place while performing work on the material. Vises use two jaws (one stationary and one movable) and a leverage and screw system to adjust the clamping force. Larger, more heavily constructed vises are capable of applying greater force than smaller types. When operated by hand, the movable jaw of a moderate-size vise can easily generate in excess of 7,000 lb of force against the stationary jaw. Applying the appropriate amount of force ensures that the work material is held solidly enough for a variety of shaping and cutting activities. A vise can also be used to press parts together for fastening.

Mechanic's and Machinist's Vises

Both a mechanic's vise and a machinist's vise have two horizontal jaws with an adjustment screw passing through a protective tunnel under both jaws. See **Figure 7-19**. The configuration of mechanic's and machinist's vises is generally similar. However, the machinist's vise is commonly constructed under stricter tolerances with higher quality materials. It usually has steel jaws that can be covered or replaced with softer materials for holding more delicate parts.

Mechanic's vises often incorporate a small anvil, adequate for light hammering chores, as part of the frame. Most are also constructed to swivel on a base plate that is mounted to the top of the workbench.

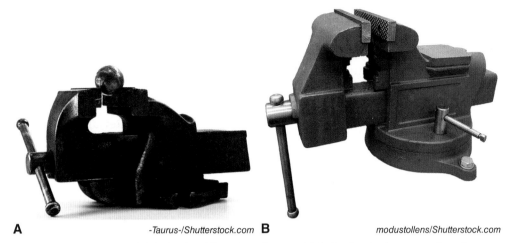

Figure 7-19. A—A machinist's vise. Note that the machinist's vise lacks an anvil surface and a pipe clamp. B—A typical mechanic's vise. Mechanic's vises are constructed a bit more coarsely than machinist's vises, but both have jaws that remain parallel and are capable of exerting tremendous clamping force.

Woodworking Bench Vises

Vises intended for woodworking are often mounted to the underside of a workbench at the end or front edge of the bench. They are mounted with the jaws flush or just below level with the working surface of the bench, as shown in **Figure 7-20**. Most have relatively wide jaws that offer a large surface area for clamping lumber. Soft wooden inserts may be used to reduce damage to lumber from the pressure of the vise's clamping force. Many woodworking bench vises also have a retractable block that serves as a bench stop, or dog, which prevents large pieces of lumber from sliding while chiseling or other work is performed.

Pipe Vises

Plumbers use several different types of pipe vises to hold pipe during cutting, threading, or attachment fitting work. One of the most common pipe vises uses a screw threaded through a yoke to push the movable V-shaped jaw down toward the stationary set of jaws mounted to the base. Yoke-type vises come in a variety of sizes to accept different sizes of pipes. Plumbers also commonly work with a pipe vise that uses a screw-driven chain to pull the pipe into the stationary jaw.

Vise Maintenance

Vises are sturdy pieces of equipment that can be expected to deliver years of trouble-free service with just a little care

Figure 7-20. Woodworking bench vises are an indispensable woodworking tool. Modern versions of the vise are made entirely of metal, although traditional styles were constructed mostly of wood. A wood jaw liner like the one shown must be installed on metal vises to prevent marring workpieces made of wood.

and maintenance. Light lubrication of their moving parts will help to keep them working properly. Too much lubricant can attract and trap dirt, leading to difficult operation and premature wear. Keeping the jaws in good condition allows vises to grip snugly without causing excessive damage to parts.

Clamps

Similar to vises, clamps are used to hold parts and materials together and stationary while work is performed. Unlike vises, however, clamps are portable and are moved into different positions based on the task. Many different types of clamps are made for holding all types of materials for a multitude of operations. In general, the more sturdily a clamp is constructed, the more force it can apply.

C-Clamps

C-clamps consist of a C-shaped metal frame, a threaded rod with a handle on one end, and a swiveling pad on the clamping end. See **Figure 7-21**. C-clamps are strong, versatile tools that can be used with most materials. When using a C-clamp to hold wood, it is advisable to use scrap pieces of lumber to protect the work from being damaged by the jaws. The handles of C-clamps are intentionally smaller and lighter duty compared to the rest of the tool. This is done to limit the user's tendency to apply excessive pressure. Maintain C-clamps by keeping them clean and lightly lubricating the threaded rod.

Sashkin/Shutterstock.com

Figure 7-21. The leverage of the sliding handle combined with the long screw enables the C-clamp to grip material very securely.

Parallel Clamps

For woodworking projects, parallel clamps are often the best choice for holding lumber together for fastening. Parallel clamps, sometimes called hand screws, are usually two parallel hardwood jaws pulled together with a pair of threaded handles. The wooden jaws offer a large contact area and are less likely to damage wooden projects than clamps that have metal jaws. A typical set is shown in **Figure 7-22**.

To get maximum compression from a parallel clamp, line up the jaws to a close fit with the material, using both screws, until the workpiece is snugly clamped. Then, tighten the back handle, the one farthest from the work, a little more. This takes advantage of the leverage within the design of the clamp and can exert a surprising amount of clamping force. Although either side of the clamp can be angled to hold a tapered material, keeping the jaws as close to parallel as possible during use protects the integrity of the clamping surfaces and reduces damage to the jaws. Since the clamp face is wood, it cannot be used for hot metalworking.

Charles Neal/Shutterstock.com

Figure 7-22. Parallel clamps are an excellent choice for securing easily marred material. The wide surface area of the parallel jaws evenly distributes the clamping force to reduce the risk of material damage while still providing a secure hold, even on tapered materials.

Bar Clamps

A bar clamp uses parallel jaws to clamp across an extended distance. The typical design has a fixed jaw and screw fastened to one end of a steel bar and a movable jaw that slides along the rest of the bar. The movable jaw can be locked in place where needed. For holding furniture, gates, and other large projects square during assembly, no other tool works as well.

Bar clamps come in several varieties. Some use an attached, solid, rectangular metal bar 18″ to 6′ long, **Figure 7-23**. Instead of the metal bar, some types are made to be attached and used with lengths of 1/2″ or 3/4″ diameter steel pipe. These are referred to as pipe clamps but are limited in range by the length of pipe selected. A newer class of bar clamp has a lever and cam rather than the traditional screw adjustment to adjust clamping tightness. Lever clamping systems function faster, but provide a weaker clamping force.

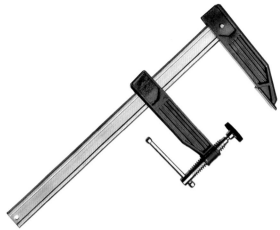

Natan86/Shutterstock.com

Figure 7-23. Bar clamps can extend a clamping force over a longer distance than other clamps.

Spring Clamps

Spring clamps use stored energy from a torsion or compression spring to temporarily hold small parts in place. As a rule, spring clamps do not apply a strong clamping force relative to other types, but they are quick to install and easy to remove.

There are many varieties of spring-operated clamps, **Figure 7-24**. Since they do not squeeze with great force, they can be used in many applications where material damage is a concern. When secure positioning is required, a spring clamp may be used to hold parts in place while stronger clamps are applied.

Magnetic Clamps

Strong permanent magnets are used by magnetic clamps to hold smooth-surfaced ferrous parts in place. Some are light duty and easily employed by

Yingko/Shutterstock.com   *TRINACRIA PHOTO/Shutterstock.com*

Figure 7-24. Spring clamps use an internal spring or the natural flexibility of their construction material to provide the force for holding parts in place. Their clamping action is usually not as strong as other types of clamps, but they are easy to put into place. Spring clamps are often used as a ground connection for welding machines.

Goodheart-Willcox Publisher

Figure 7-25. Magnetic clamps are handy for holding metal parts, but they should not be exposed to extreme heat. The clamp on the left is used to hold steel parts at a 90° angle during assembly. The clamp on the right may be switched off and on and is designed with threaded holes for temporary mounting to parts or templates.

> **Safety Note**
> A relatively small magnetic clamp can generate several hundred pounds of clamping force. Industrial versions of magnetic clamps use an electromagnet to move the permanent magnets into position, creating a tool capable of thousands of pounds of holding force. Extra care should be taken when working with these tools to avoid injury.

hand, but stronger versions use multiple magnets held inside a steel carrier to hold pieces together. A switch or lever on the surface of the clamp aligns the magnetic poles of the interior magnets for clamping. Moving the switch to *Off* takes the magnetic poles out of alignment and releases the part. See **Figure 7-25**.

Pliers

Pliers work by gripping objects to turn, twist, hold, or possibly cut them. A pair of pliers consists of two levers connected at a fulcrum. The jaws of most pliers are much shorter than the handles to provide a mechanical advantage in gripping objects.

Pliers are designed with various jaw configurations; each type of jaw is used for a specific purpose. Each pair of pliers should be used only for its designed purpose. A single style of pliers should not be used as an all-purpose pair. Blacksmiths use long-handled pliers known as tongs to handle hot pieces of metal safely. Other task-specific designs are made for mechanics, fence builders, electricians, and other professionals.

Slip-Joint Pliers

Slip-joint pliers, also known as combination pliers, have an elongated, figure-eight-shaped hole at the fulcrum that allows the pliers to be shifted to two positions. One is for gripping small objects, and the other is for gripping larger parts. See **Figure 7-26**. The jaws of a pair of slip-joint pliers include a flat, parallel section to grip flat materials, a rounded section for gripping odd shapes, and a bypass notch for cutting wires.

Lineman's Pliers

Lineman's pliers combine a large, flat jaw surface and a heavy-duty, pinch-type wire cutter for use in electrical work, **Figure 7-27**. A single rivet provides a sturdy non-adjustable fulcrum. Sometimes called side-cutting pliers, lineman's pliers are also useful for work on thin sheet metal.

Groove-Joint Pliers

Another common type of pliers, named Channellock in 1933 by Howard Manning of the Champion-DeArment Tool Company, has a long slot at the fulcrum with multiple half-moon-shaped bosses that establish several open position sizes. See **Figure 7-28**. The tool was an immediate success, and in 1963, the Champion-DeArment Tool Company changed its name to Channellock, Inc®. in an effort to preserve their patent rights on the distinct tool.

Generically known as *groove-joint pliers* or tongue-and-groove pliers, they are useful for gripping

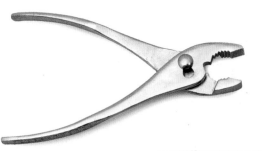

warat42/Shutterstock.com

Figure 7-26. Common slip-joint pliers have a slotted hinge joint that allows the tool to be shifted into two positions, one for gripping smaller parts and another for larger objects.

large, odd-shaped objects. The pliers range in size from 4 1/2″ to 20 1/2″ long. Each size has relatively long handles for extra leverage.

Needle-Nose Pliers

When you need a tool to work in tight places and grip small objects, *needle-nose pliers* may be the best choice. Needle-nose, or long-nose, pliers have long, narrow, tapering jaws capable of reaching deep into tight places. **Figure 7-29** shows a typical pair. They may be used for gripping very small parts or for twisting wires into small loops. Needle-nose pliers have a single-position joint at the fulcrum. Variations of needle-nose pliers include stainless steel models for fishermen and long reach models with extra length handles.

Cutting Pliers

Pliers may be designed to cut in several different ways: two wedge-shaped blades may come together to pinch through material, a single blade may pinch through material as it approaches a backing plate or anvil, or blades may bypass each other in a scissor-like action. The cutting design depends on the intended purpose of the pliers.

Diagonal pliers are designed to cut light-gauge wire. They have two cutting edges that pinch together to cut the wire, **Figure 7-30A**. A single rivet positioned at the fulcrum is the pivot point. Using diagonal pliers for gripping will degrade the cutting edge, but they are often an effective tool for removing unwanted finishing nails or staples.

Bolt cutters, cable cutters, and tin snips are other types of cutting pliers. *Aviation snips*, which cut sheet metal used for covering buildings or roofing, may have a straight jaw or a jaw angled to the left or the right, **Figure 7-30B**. Due to the greater force required for cutting compared to just gripping, many cutting pliers incorporate a compound lever system in the handles to multiply the user's squeezing force.

Clamping or Locking Pliers

In 1924, blacksmith Bill Petersen was granted the patent for the first set of *locking pliers*. These were marketed as Vise-Grip® pliers, a trade name that is often used synonymously with any locking pliers. See **Figure 7-31**. They offer strong clamping strength combined with a jaw adjustment screw and quick

dcwcreations/Shutterstock.com

Figure 7-27. Lineman's pliers, named after workers who maintain electrical supply lines, are strong pliers designed for cutting and twisting wires.

Sorapop Udomsri/Shutterstock.com

Figure 7-28. The groove in groove-joint pliers allows them to be adjusted to grip objects of many different sizes. Versions of this tool are widely used in a number of professional trades and industries.

BW Folsom/Shutterstock.com

Figure 7-29. The narrow jaws of needle-nose pliers are designed for working in tight spaces and for twisting small wires. Some include pinch-type cutters for snipping wires.

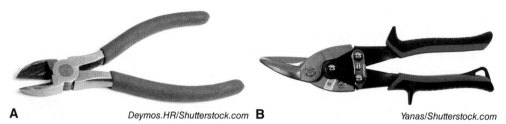

Figure 7-30. A—Diagonal wire-cutting pliers are made for cutting, not gripping, and are available in many different sizes and materials. Note the design of the jaw. B—Aviation snips are designed with a compound lever system in the handles to cut stronger metals with less user effort. The angled jaw on this model allows for angled cuts.

release feature. The mechanism in locking pliers has been used to develop clamping devices of many sorts.

Some locking pliers have specially designed jaws that are used to attach and remove hose and wire clamps. Some crimp in place and others use spring tension. Other pliers are manufactured with the tool itself serving as a temporary clamp. These tools use compound leverage and a mechanism to lock the jaws in place.

Figure 7-31. Locking pliers incorporate compound leverage and a locking mechanism to solidly clamp parts for performing a variety of jobs.

Impact Tools

Tools designed for striking channel great amounts of energy into a concentrated force to produce useful work. The most common striking tool, the hammer, has been used for thousands of years. Basically, a hammer consists of a head and a handle. The weight of a hammer refers to the head weight and excludes the handle weight. Specific heads are designed for delivering blows to particular materials or for certain purposes. In addition to serving as a comfortable hand grip, the handle increases the user's leverage, control, and power while swinging the tool.

Other tools used in conjunction with hammers are made to accept the striking force. Anvils of different types absorb the excess energy of hammer blows during blacksmith or metal work. Punches and chisels of various types are used to concentrate the blow of a hammer to a very accurate point.

Safety Note

Always wear protective eyewear when using any impact tool.

Claw Hammers

Claw hammers are essential for driving nails. The curved claw hammer is the most common and can be found in sizes ranging from 10 to 24 ounces. Straight claw hammers, sometimes called rip or framing hammers, are usually on the heavier side and are built to withstand harsh use.

Hammers designed for driving nails have slightly convex faces designed to contact the nail during the downward swing without marking the wood. The edges of the face are beveled to prevent chipping. The hardened steel is not excessively damaged from contact with nails.

Curved claws can pull out nails with minimal damage to wooden surfaces, although a piece of scrap wood is commonly used for extra leverage and greater protection of delicate surfaces. The straight claws of rip hammers are useful for prying boards apart. A pry bar designed for pulling nails should be used instead of a claw hammer when a large number of nails needs to be removed. Hammer handles are designed for swinging, not for prying.

The handle of a hammer is made to form a natural grip at the base. Gripping in the correct position increases power and accuracy. This technique is demonstrated in **Figure 7-32A**. Novice users often "choke up" or hold the handle too close to the head. This poor technique is inefficient and unsafe, and should be avoided. See **Figure 7-32B**. Concentrating on the head of the nail, or target, will improve your accuracy and speed.

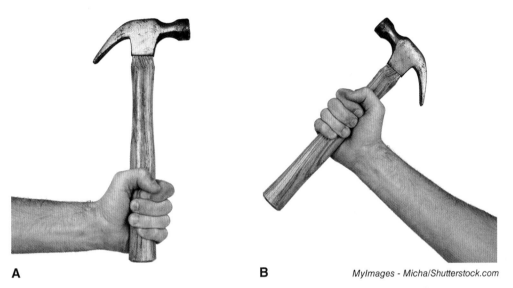

A   B   *MyImages - Micha/Shutterstock.com*

Figure 7-32. A—Examine a hammer and you will notice a slight hour glass shape near the bottom of the handle. Grip the handle in this area for the best control and the most leverage. B—Avoid the temptation to "choke up" close to the head. This inappropriate grip leads to poor aim and lack of power.

Metalworking Hammers

A *ball-peen hammer* is used for shaping metal and driving punches and chisels. The head ranges in weight from 4 to 32 ounces. The head of a ball-peen hammer has a face that is cylindrical and a face that is hemispherical, or ball shaped, **Figure 7-33**. The cylindrical face, like the face of a claw hammer, is beveled on the edges, slightly convex across the face, and used for driving tools and striking metal. The ball end of the hammer is used to precisely direct the force of the hammer for applications such as rounding the head of a rivet or in peening. Peening is the process of repeatedly hammering a metal surface. Cold metal may be peened to improve surface characteristics, while peening hot metal as it cools relieves stresses developed due to welding.

RedDaxLuma/Shutterstock.com

Figure 7-33. Ball peen-hammers are constructed from strong steel that is a bit softer than the steel used for claw hammers. The softer material reduces the hazard of small pieces shattering off the head when hammering heavy metal objects.

Other metalworking hammers include *cross-peen hammers*, which have a rounded horizontal wedge opposing the flat hammer face, and the body hammer, which is used for repairing dents in auto bodies and other sheet metal work. The typical blacksmith's hammer is a two- to four-pound cross-peen or sledge hammer.

Sledge Hammers

Sledge hammers are large hammers, usually weighing from 3 to 10 pounds. Heavier versions weigh as much as 20 pounds. The head is usually octagonal with two identical, opposing flat faces, **Figure 7-34**. Sledge hammers are used for heavy duty work such as driving posts, breaking concrete, or dislodging stuck machinery. Handles may be wood or a composite material. Heavier sledges with longer handles should be swung with two hands.

When using a two-handed hammer, grip the base of the handle with your less dominant hand. Place your dominant hand close to the head at the start of the swing and allow it to slide down the handle toward the base as the hammer moves through the swing. Concentrate on the target and not the hammer. Although powerful and designed for tough jobs, the sledge hammer is not the best hammer for all heavy-duty work.

TFoxFoto/Shutterstock.com

Figure 7-34. This builder is using a sledge hammer to shift sheets of floor decking into a tight fit on top of the floor joists for a new barn.

Soft-Faced Hammers and Mallets

When working on machinery and equipment, you often need to tap parts into or out of position without damaging them. Sometimes tools are designed to be driven, but not with a hard steel hammer. For these situations, soft-faced hammers and mallets are used. *Soft-faced hammers* are designed to shift parts into position without causing material damage or producing sparks. The hammer faces are usually replaceable inserts made of a soft metal, such as brass or lead, or synthetic plastic materials. See **Figure 7-35A**.

Mallets, as a general rule, have larger heads than soft-faced hammers of a similar weight. Natural materials such as wood, rawhide, and rubber are commonly used to form mallet heads, although synthetic materials are also common. Mallets are often the best choice for driving wood chisels or leatherworking tools.

Dead-blow hammers are designed to deliver a single blow while minimizing the rebound of the hammer head. Their heads are hollow and filled with sand or steel shot, **Figure 7-35B**. Dead-blow hammers are often used in working with sheet metal.

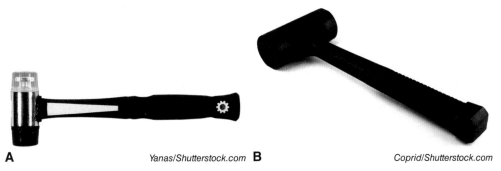

A *Yanas/Shutterstock.com*   B   *Coprid/Shutterstock.com*

Figure 7-35. A—Soft-faced hammers are engineered to line up delicate metal parts without causing damage. Using a soft-faced hammer for heavy hammering will permanently damage the head. B—The construction of a dead-blow hammer minimizes the rebound of the head after a strike.

Anvils

Anvils are used to back up or support the work when shaping metal parts. Early anvils were natural stone, but later anvils were formed from bronze or iron. Modern anvils are constructed of steel or cast iron. The shape and size of an anvil depends on where it was made and its intended purpose. Farriers use anvils every day for shaping and fitting shoes for horses.

Career Connection

Farrier

Sus Pons/Shutterstock.com

Job description: Farriers make and fit shoes for horses and mules. Trimming horse feet and general horse foot care are also common duties. Farriers are skilled in working and shaping metal. A comprehensive knowledge and understanding of equine anatomy and physiology—especially the legs and feet—is required when working with a veterinarian to diagnose and treat a horse's foot ailment. The typical farrier travels to the horses rather than having horses brought to a central location. Many farriers specialize in specific types of work, such as shoes for racing horses or corrective shoes for horses with leg or hoof problems. Farriers may be self-employed or work for larger businesses.

Education required: To become a farrier, students can complete a training program at a school that typically requires a six-week commitment. Although not required, apprenticeship opportunities are available and recommended to provide first hand experience. Farriers are not licensed, but successful completion of a training program leads to accreditation. The American Association of Professional Farriers promotes continuing education and support for its membership.

Job fit: This job may be a fit for you if you enjoy working in a variety of environments, you like working with horses and metals, and you like mostly independent work. You must be able to handle the tough physical demands of the job and patiently care for animals on a daily basis. A high degree of self-motivation, a willingness to always keep learning through your work, and a passion for horses can help you succeed in this industry.

A few steps must be taken to safely secure an anvil before use. Firmly mount the anvil to a solid surface to reduce the rebound effect during striking. Large blocks of wood can be used for this purpose. The face of a properly mounted anvil should be about wrist high. The face of the anvil is hardened and intended for shaping hot and cold metal. The table section is not hardened and holds parts for chiseling. Use the horn for shaping curved parts.

A variety of tools, called hardies, fit into the square hardy hole and are used for cutting or forming detailed shapes. Traditionally, hardies were made by the blacksmith for specific purposes. The pritchel hole is a round hole—smaller than the hardy hole—used for bending rod or as a backup for punching holes or removing fasteners. See **Figure 7-36**.

Body dollies are similar to anvils and are used to back up sheet metal while the metal is being shaped with a hammer. Unlike anvils, dollies are handheld and are small enough to be used on sheet metal that is still attached to large equipment. Dollies come in a large assortment of sizes and shapes. In some instances, a hardy built for use with an anvil can also be used as a dolly.

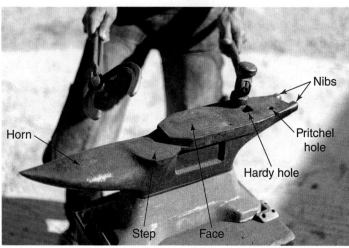

Sus Pons/Shutterstock.com

Figure 7-36. Each area of an anvil is designed to perform different tasks. This anvil is designed to be used by a farrier. The two protruding nibs on the back of the face are specifically engineered for bending horseshoes to the correct size.

Punches

Any complete tool set should include several different types of punches. The center punch not only marks the location, but also creates a depression that guides and stabilizes a drill bit as it starts the hole. The point of a standard center punch is ground to a 60° to 90° angle, and the head is beveled, or *chamfered*, to reduce the mushrooming effect caused by repeated blows from a hammer. See **Figure 7-37**.

A prick punch is ground to a sharp point, similar to a center punch, but the prick punch is narrower and leaves a smaller mark. In addition to marking metals, it can be used to back broken or damaged threaded fasteners out of a hole. The difference between these two can be seen in **Figure 7-38**. The prick punch is used to dig far enough into the remaining portion of a damaged fastener to transfer energy from a striking hammer in tiny bursts to drive the fastener in a rotation out of the hole.

Pin punches, **Figure 7-39**, are used to drive roll pins and other types of pins from their holes. The tip of a pin punch is flat with a long, narrow shaft that

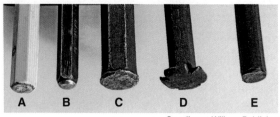

Goodheart-Willcox Publisher

Figure 7-37. Punches are sold with beveled edges around the head to counteract the mushrooming effect of repeated hammering. They must be periodically re-ground to maintain a safe working condition. The punches shown are in various stages of wear. A—New. B—Properly re-ground. C—Upset and in the early stages of mushrooming. D—Severely mushroomed (dangerous). E—Re-ground, but beginning to upset.

is sized by diameter. The correct diameter punch is chosen to match the pin. Similar to a pin punch, a nail set has a flat or concave tip with a narrow, tapered shaft. A nail set is positioned over the heads of small nails and struck to drive the small nails flush with or below the surface of the work material.

When tapping delicate parts into place or working around flammable substances, use bronze punches. They are softer than steel and will not produce a spark when hit with a hammer.

Leverage Tools

One of the most common misuses of hand tools is prying objects with the wrong type of tool. Tools properly designed to impart leverage allow you to move or separate components that would be difficult to move or may be damaged with other methods. They must be constructed of strong material and reinforced correctly to handle the force.

The act of prying is potentially harmful to the workpiece, so it is important to select the tool best suited for the job. A tool designed for pulling nails, for example, has sharp edges and is not intended to replace a pry bar designed for mounting tires.

Tapered Punch

Although its appearance and name would indicate otherwise, a *tapered punch* is used for aligning holes in two or more parts so the parts can be bolted or screwed together. Like other punches, tapered punches, sometimes called alignment or drift punches, are formed from a single piece of tool steel. They have a long, tapered shaft ending in a flat point. Some tapered punches called heel bars include a short, stout hook on the head for light prying. See **Figure 7-40**.

Pry Bars

Pry bars are sized according to the intended purpose; the more leverage required, the longer and heavier the bar. Many pry bars are used for pulling nails. Their reinforced construction and configuration makes them better suited for this purpose than claw hammers. Nail-pulling bars such as crowbars, flat bars, and cat's paws have paired claws similar to those found on a claw hammer.

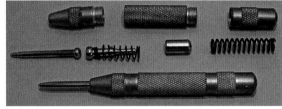

Goodheart-Willcox Publisher

Figure 7-38. A—The prick punch is narrower, the point is ground to a more acute angle, and it leaves a smaller mark than the center punch. The larger impression created by a center punch is better for stabilizing a drill bit. B—The automatic punch is spring-loaded and operated with hand pressure only. Note the parts in the disassembled view.

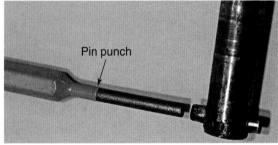

Goodheart-Willcox Publisher

Figure 7-39. Pin punches have narrow, straight tips that are ground flat on the end for removing pins from machine shafts.

Goodheart-Willcox Publisher

Figure 7-40. Tapered punches are stronger than pin punches because their tips are designed with long tapers. The tapered shaft can be used to draft parts into alignment and line up holes for the insertion of fasteners. Large tapered punches with hooked ends are designed for prying machine parts for assembly and disassembly.

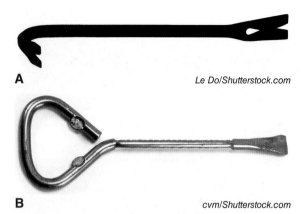

Figure 7-41. Like many hand tools, pry bars are available in a variety of shapes and sizes to meet different types of prying needs. A—A common style of crowbar. B—A paint key is a simple pry tool used for opening paint cans. Using a screwdriver for this task can damage the screwdriver tip.

Crowbars, or wrecking bars, are usually made from a hexagonal shaft of tool steel hooked and forked on one end for pulling nails and flattened and slightly bent on the other end for prying lumber apart. Flat bars are formed from a flat bar of steel and function as a light-duty crowbar. Cat's paws and similar tools have short, sharp-rounded claws designed for digging and slightly elevating the head of a nail for pulling with minimal damage to the lumber. **Figure 7-41** shows other types of pry tools.

Pry bars are often required when servicing and repairing equipment. Pry bars are available for tasks such as mounting tires, adjusting brakes, and separating mechanical parts.

Tools for Applying Torque

Tools that apply torque use leverage to impart twisting forces on an object. Screwdrivers and wrenches are tools that increase and concentrate the amount of torque applied by hand. Drills are cutting tools that use torque to cut and remove material from a hole.

Wrenches

Wrenches use leverage for turning nuts, bolts, and other objects. Most are constructed from hardened tool steel to increase both the strength and the useful life of the tool. Some wrenches are sized to fit one or two specific measurements, while others are adjustable.

Open-End, Box-End, and Combination Wrenches

One of the most iconic tool designs is the combination wrench. *Combination wrenches* consist of an open-end wrench on one end and a box-end, or completely closed, wrench on the other. Both ends of a combination wrench usually fit the same size hexagonal-shaped nuts and bolts. The open end also fits older square fasteners of the same size.

The two designs on a combination wrench have opposing characteristics. The U-shaped mouth on a standard open-end wrench contacts the fastener on only two corners, which makes the tool more likely to slip than a box-end wrench. Standard open-end wrenches are also offset from the handle about 15° to increase usability in confined spaces. The angle allows the user to flip the tool over to change the angle of approach.

The mouth of a box-end wrench may be 6-point or 12-point, but both types contact the fastener on all six corners. A 6-point wrench is the strongest and safest to use because it places a greater amount of surface area in contact with the fastener. A 12-point wrench has the ability to fit in twice as many positions, which is an advantage when working in a limited space. See **Figure 7-42**.

Safety Note

Remember that the right tool for the job—whether it be a pry bar, hammer, or any other type of hand tool—is the tool that was designed to safely perform the task.

All three wrench configurations are available in Society of Automotive Engineers (SAE) fractional inch sizes and metric sizes. They are most commonly sold in sets of graduating size. Variations of open-end and box-end wrenches are designed to meet a variety of specific needs.

Socket Wrenches

Socket wrenches consist of short steel tubes (sockets) configured much like box-end wrenches that fit into *ratchet handles* for loosening or tightening bolts and nuts. A handle and collection of sockets are commonly sold in a set, **Figure 7-43**. One end of the tube is recessed and has six or twelve angular notches to fit the hexagonal head of nuts and bolts. The other end has a square hole that fits assorted ratchet handles.

Standard sockets fully engage the bolt head or nut, usually better than a box-end wrench. Deep sockets have longer bodies that are capable of reaching bolts in deeper recesses. They also provide more clearance for bolts that protrude a long way through the nut. An 8-point socket, although not common, is available for use on fasteners that have square heads.

The ratcheting mechanism in the handle allows the wrench to turn in one direction only. Moving a lever on the back of the head reverses the direction. The square drives that fit into the sockets are commonly 1/4″, 3/8″, 1/2″, or 3/4″, depending on the size and strength of the set. Sockets have small devices on the inside of the square drive recess designed to lock the socket in place when a small, spring-loaded ball on the driver engages the hole.

Socket sets often include a variety of other attachments that help to increase efficiency or accessibility. Some common examples are extension bars and breaker bars. Extension bars fit between the handle and the socket to reach fasteners that are inaccessible by any other means. A breaker bar is a long, heavy-duty handle used with sockets when extra leverage is needed to free stuck fasteners. It has a pivoting mechanism instead of a ratchet to increase the amount of force it can apply. See **Figure 7-44**.

Adjustable Wrenches

An *adjustable wrench*, or Crescent® wrench, has a stationary jaw and a movable jaw that allows the mouth to be adjusted to any size within its opening

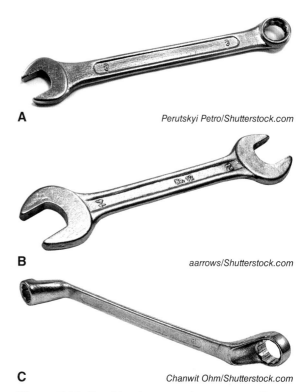

A

Perutskyi Petro/Shutterstock.com

B

aarrows/Shutterstock.com

C

Chanwit Ohm/Shutterstock.com

Figure 7-42. Combination wrenches are usually the same size on both ends, while open-end and box-end wrenches often have different sizes on each end. A—Combination wrench. B—Open-end wrench. C—Box-end wrench.

design56/Shutterstock.com

Figure 7-43. The ratcheting handle of a socket wrench allows for working in confined spaces and eliminates the need for re-setting the wrench multiple times on a fastener to complete the tightening or loosening process. A variety of different size sockets and attachments, such as the extension shown here, greatly increase the versatility of a single ratchet handle.

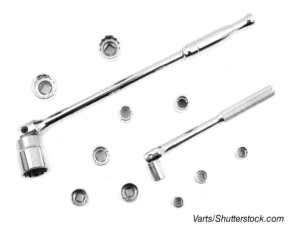

Varts/Shutterstock.com

Figure 7-44. Breaker bars are used with socket attachments for jobs requiring a stronger application of torque. They do not have a ratcheting action, but their rugged design can withstand more force than ratcheting mechanisms.

koosen/Shutterstock.com

Figure 7-45. Designed for convenience, the adjustable wrench can be set to fit any size fastener at or below its maximum opening. The tradeoff for this versatility is that an adjustable wrench is usually a looser fit than a single-size wrench. Always pull the handle of an adjustable wrench in the direction of the movable jaw to reduce the risk of fastener damage or personal injury.

range, **Figure 7-45**. It can be used on SAE or metric size fasteners. The size of an adjustable wrench is taken from its overall length, which commonly ranges from 4″ to 24″. Larger adjustable wrenches are handy for bending small sheet metal parts held in a vise.

To use an adjustable wrench correctly:

- Always apply pressure toward the movable jaw. This applies more pressure to the sturdy, fixed jaw and decreases the chance of adjustable jaw movement and slippage.
- Make sure that both jaws fully engage the flat surfaces of the fastener. Adjust the fit by turning the knurled wheel below the jaws.
- Always pull on the wrench rather than pushing on it. Injuries are less likely when pulling if the wrench slips.
- In most situations, use a properly fitted open-end or box-end wrench instead of an adjustable wrench on hexagonal nuts and bolts. The extra few steps to the tool box are steps to improve safety.
- Do not use an adjustable wrench as a hammer or strike it with a hammer.

Pipe Wrenches

Another common adjustable wrench is the *pipe wrench*. Pipe wrenches, also called Stillson wrenches, were originally patented in 1869 by Daniel Stillson. They are designed to grip and hold or turn pipes of various diameters. The hardened steel jaws have jagged, sharp teeth that are replaceable. A knurled nut is used to adjust the extension of the movable jaw that reaches up and over the stationary jaw riveted to the end of the frame. See **Figure 7-46**. When the handle is pulled, the movable jaw rocks back slightly, causing the teeth of both jaws to dig aggressively into the surface of the pipe. Releasing pressure on the handle allows a small spring to slightly open the jaws. At this point, without removing it from the pipe, the wrench can be moved back for another bite and pulled again to turn the pipe or fitting.

Do not use a pipe wrench on surfaces on which damage from the teeth will be a problem. On these surfaces, or for holding hardened steel shafts, use a strap wrench. This style of wrench tightens a cloth or synthetic belt around the outside or circumference of the material with enough gripping force for most purposes. A chain wrench applies torque like a strap wrench but uses a roller chain instead of the strap.

modustollens/Shutterstock.com

Figure 7-46. Pipe wrenches have sharp teeth on the jaws and are designed to tighten and bite into the surface of steel pipe and fittings. Using a pipe wrench on other fasteners will likely damage the fasteners to the point that the proper tools will no longer fit.

Screwdrivers and Hex Keys

A screwdriver has only one purpose—turning screws. All screwdrivers have a handle, a shank, and a tip. Handles may be wood or plastic. Handles typically have a hexagonal or square shape with rounded corners, a rubberized covering, or preformed finger grooves to improve grip. Heavy-duty screwdrivers may have handles long enough to accommodate two-handed use. Variations of screwdriver handles may be hollow for storing extra bits, or they may include a ratcheting mechanism.

There is no standard length for a screwdriver shank. It can be as short as 1 1/2" in the case of "stubby" screwdrivers to as long as 18" for an extended reach. The usual range is 4" to 8" long with a round, square, or hexagonal shape. Square and hexagonal shanks may be gripped with a wrench to provide extra leverage to loosen particularly stubborn screws. Many designs incorporate a built-in nut at the top of the shank specifically for this purpose.

Tips are hardened steel to extend their useful life. They may be formed from the end of the shank or they may be interchangeable. Interchangeable tips are known as *bits*. Often screwdriver tips are magnetic to help hold the screw when starting and removing it. However, the use of magnetized tools should be avoided when working around delicate electronic components.

Screwdriver Tips

The tip of a screwdriver defines its use. There are many different types of tips in use, but only a few are commonly used. The most recognizable is the **slotted-tip screwdriver**, also known as "standard" or "flat-head," which fits screws with a single straight-walled slot that extends through the diameter of the head. See **Figure 7-47A**. When using a slotted-tip screwdriver, be sure the tip closely fits the slot of the screw. If it does not, you risk damaging the screw and the tool, **Figure 7-47B**. Using a screwdriver improperly as a scraper, a pry bar, or to open paint cans can result in damage to the tip. Slotted tips may be

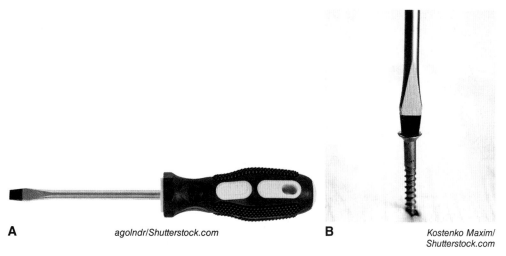

A   agolndr/Shutterstock.com   B   Kostenko Maxim/Shutterstock.com

Figure 7-47. A—Standard screwdrivers are designed only for turning slotted screws. B—The tip of the tool should completely fill the slot of the screw to prevent damage to the fastener and the screwdriver.

168 Agricultural Mechanics and Technology Systems

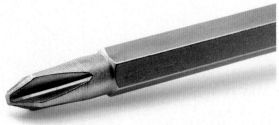

poligrafistka/Shutterstock.com

Figure 7-48. The Phillips screwdriver was designed as an improvement to the standard screwdriver, offering a self-centering tip with more surface area engaged in the fastener.

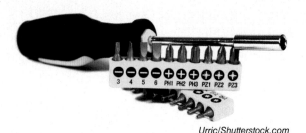

Urric/Shutterstock.com

Figure 7-49. Interchangeable screwdriver tips in different styles and sizes can be used in a single screwdriver handle designed to accept the bits.

LifetimeStock/Shutterstock.com

Figure 7-50. Nut drivers are much like socket wrenches attached to screwdriver handles. They are designed for turning fasteners with hexagonal heads. Many nut drivers have hollow shanks to provide clearance for the end of screws when attaching nuts.

reground. Depending on the hardening process used in its manufacture, however, regrinding may ruin the tip.

Worldwide, the *Phillips-head screwdriver* is the most commonly used screwdriver type. The driver has a 57° point and four tapered flutes that are rounded by design. See **Figure 7-48**. The taper of a Phillips tip causes it to be self-centering and the rounded corners are designed to cause the tip to cam out, or slip, when excessive torque is applied. This is intended to protect the screw, but repeated camming out damages both the screw and the screwdriver. Phillips tips are available in several sizes, but #1 and #2 tips are by far the most common.

Square-drive and star-drive screwdriver tips were designed for more positive contact with screw heads than standard slotted or Phillips-head screwdrivers. This feature was meant to reduce cam-out tendency also. Square-drive, or Robertson, screws have a square recess into which the square tip of the driver fits. *Torx*®, or star-drive, screws have a six-pointed, star-shaped recess with rounded corners. Square-drive and Torx® screws and drivers both come in multiple sizes. **Figure 7-49** shows an economical way to build a screwdriver collection equipped to deal with the most commonly used screw head styles.

Nut Drivers

Many screws have hexagonal heads that are similar to those of nuts and bolts, or in the case of machine screws, accept and are used with nuts. A *nut driver* resembles a screwdriver with a socket wrench attached to the tip. See **Figure 7-50**. Nut drivers may be used instead of wrenches for driving these types of screws or nuts. Most are equipped with a hollow shank to provide clearance for the threaded portion of the screw or bolt extending through a nut.

Hex Keys

The adoption of headless set screws for securing pulleys to drive shafts arose in the early twentieth century as part of a movement to improve the safety of working conditions for industrial workers. Previously the heads of bolts on spinning equipment would snag workers' clothes and cause injuries. The "Allen Safety Set Screw" was the answer. These screws are turned by inserting a *hex key*, or Allen wrench, into the hexagonal recess in the top of the screw. Recessed hex drives are also commonly found on screws and bolts with heads.

The typical hex key is an L-shaped tool formed from a single piece of hexagonal bar stock, **Figure 7-51**. Using the short end of the tool to engage the fastener provides for more torque, and using the long end provides greater reach.

Controlling Torque

The handle length and grip style of tools designed for delivering torque is a determining factor in the amount of force that any given tool may apply. When the handle resists bending and slippage in the hand, greater leverage can be obtained. Tool handles must be comfortable enough for the user to maintain sufficient grip during use.

Some tool handles are intentionally designed to be short to limit leverage or are not comfortable to use under high torque. Occasionally the temptation arises to use a cheater bar, or short length of pipe, to extend the length of a tool handle for more leverage. This is an unsafe practice that often leads to damaging the fastener or tool and injuring the user.

Many complex machines require securing parts by evenly tightening a group of fasteners to a specific torque value. Using a torque wrench and following the correct tightening pattern guarantees that an accurate and even amount of force is applied to all of the part fasteners. Tightening fasteners in this manner reduces the chance of part distortion from the incredible amount of holding force that is possible with some screws and bolts.

Extractors

A screw or bolt breaking off below the surface of a piece of equipment is one of the most frustrating things that can occur on a job. When it does occur, however, a screw *extractor* can often be used to solve the problem. Several types of screw extractors are available, including the tapered, straight-fluted

BaLL LunLa/Shutterstock.com   mikeledray/Shutterstock.com

Figure 7-51. Hex keys, or Allen wrenches, are usually short sections of hexagonal tool steel bent at a right angle. Other versions are equipped with T-handles or packaged in folding sets. Hex key bits are also available as socket wrench attachments and for interchangeable bit screwdrivers. All styles of the tool are designed to turn fasteners with a six-sided hole.

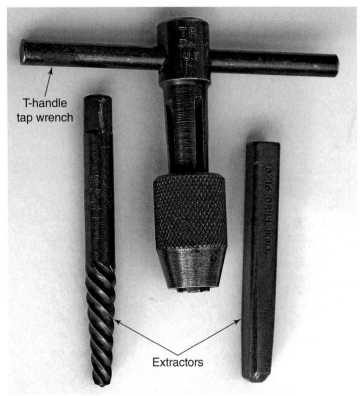

Figure 7-52. Fasteners occasionally break off in their holes. Different types of screw extractors are available to help. All require using a drill to bore a correctly sized pilot hole in the damaged fastener. The extractor is inserted into this pilot hole and turned to reverse the fastener from its threaded hole.

extractor and the spiral-fluted extractor. See **Figure 7-52**.

Both types require the remnant of the fastener to be drilled out with a pilot hole approximately two-thirds the diameter of the connector. Many extractors have the correct pilot hole size stamped on the shank. Next, the extractor is tapped into place and turned counterclockwise using a T-handle tap wrench. The cutting edges of the extractor engage the sides of the pilot hole and back the fastener out. Use caution, as too much torque can break the extractor off in the hole.

As the name suggests, the T-handle tap wrench is also used to apply torque to taps for cutting and chasing threads. It has an adjustable collet-type connection that grips the square end of taps and extractors.

Cleaning Tools

Cleaning tools are used prior to other operations to provide clean, uniform working surfaces. They are also used after most operations to prepare surfaces for finishing. Cleaning tools can be considered safety equipment, because their use eliminates or reduces possible contact with hazardous materials.

Scrapers

Scrapers come in many shapes and sizes and are intended for a variety of different uses. See **Figure 7-53**. They may be used to remove paint, dirt,

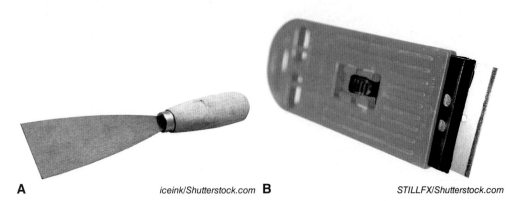

Figure 7-53. Scrapers with fixed blades or replaceable blades that are much like single-edged razors are useful for removing heavy accumulations of contaminants from flat, hard surfaces.

or other residue or to remove stuck gaskets during equipment repairs. Some have fixed blades that may be stiff or flexible with sharp or blunt edges, **Figure 7-53A**. Most blades are steel, although softer materials are also used. Many scrapers use replaceable blades that resemble razor blades, **Figure 7-53B**.

Always remember to use a scraper when necessary. Never use a tool designed for other tasks as a substitute.

Brushes

Brushes with many different bristle configurations are used for jobs ranging from cleaning parts to applying finishing materials. Soft paintbrushes with natural or synthetic bristles are the most common. These brushes are used to remove dirt and dust from materials, including contaminants from between parts with limited clearance. Stiffer versions, often called part-washing brushes, are used with liquid solvents to clean greasy, oily residue from machine parts.

Several types and styles of wire brushes are available for various jobs, **Figure 7-54**. Stiff wire hand brushes are used to remove oxidation and other contaminants in preparation for metal work. A file card is a short-bristled, flat wire brush designed to clear the teeth of files.

A table brush is a soft-bristled hand broom for cleaning equipment and workbench tops. They have natural or synthetic bristles attached to a short handle.

Robert Wolkaniec/Shutterstock.com

Figure 7-54. Wire brushes are available in assorted styles for aggressive cleaning and removal of oxides and other heavy surface contaminants.

Brooms

Brooms and dust pans should be used to clean work areas before and after many tasks in the agricultural workshop. The typical household broom is made from a bundle of natural broomcorn bristles bound to a single wooden handle. The bristles are flattened by horizontal stitching and cut square across the end. Heavy-duty versions of these are useful for general shop cleaning. Another common type of broom is the push broom, or floor brush. Push brooms are long wood or plastic boards with multiple rows of bristles attached to a single wooden, metal, or fiberglass handle. Softer bristles work best for sweeping fine material from smooth floors. Stiffer bristles are needed if floor surfaces are rough or if the waste material is heavy or wet.

Purchasing Hand Tools

When purchasing hand tools, it is wise to remember the saying, "You get what you pay for." It is not necessary always to purchase the most expensive option, but an inexpensive tool may not be a bargain. Cost and quality are not the only factors to think about when making the purchase. Other important

Safety Note

Keeping floors and other work surfaces clean and free of clutter makes the working environment safe and more comfortable. Store brooms and other cleaning brushes conveniently out of the way to ensure that they are not a hazard. This is especially true in environments where brooms could become a fire hazard. Natural bristles contaminated with cleaning oil, solvent, or fuel spills should be stored away from the work area.

factors to consider are availability, compatibility with other tools, and duration of need.

Major tool manufacturers offer replacement guarantees on the quality of their products. A good guarantee is an indication of quality, but this should not be your only deciding factor. Name brands that have developed a reputation for quality strive to produce reliable tools in order to maintain their reputation. Look for tools that are made from better materials and sold with a higher degree of finish. Hand tools that have moving parts should operate freely without excessive looseness or play between parts.

The Internet and direct shipping have multiplied buying options exponentially. Tool vendors and major chain stores also offer plenty of tool choices. However, if tools require consumables, buying from a local supplier may be the wisest choice. Tools that will be used on a single job may not need to be of the same quality as tools that are used frequently or on a daily basis. Jobs involving expensive raw materials may require better quality tools than tasks using ordinary consumables.

Care and Storage of Hand Tools

Quality hand tools are an important investment and should be maintained and stored with care. People who work with tools are often judged by the condition and organization of their tools. Keeping tools clean and neatly stored is a sign that you take pride in your work.

Tools, especially those with steel and wood components, can be damaged by storing them in dirty conditions. Simply wiping the tool with a clean rag is often enough. Tools with teeth for cutting or gripping may need periodic cleaning with a brush. You should use a brush that has bristles stiff enough to be effective, but softer than the tool. Tools with moving parts should be

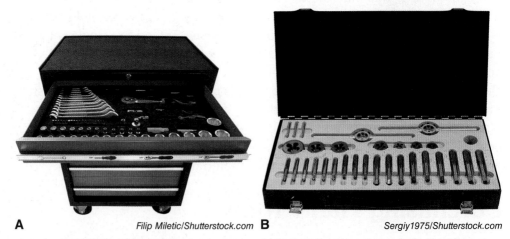

A *Filip Miletic/Shutterstock.com* B *Sergiy1975/Shutterstock.com*

Figure 7-55. A—A good tool box with plenty of space is an ideal storage solution for many hand tools. Larger, roll-around tool boxes work best when only a small number of workers will use the same tools within a shop. B—Some tools, like this tap and die set, need to be stored in dedicated containers to reduce the chance of parts being lost or damaged.

lightly lubricated at the joints. Organizing tools by type for storage makes them easy to find when needed. Tools with sharp edges should be stored in a manner that prevents the edges from colliding with each other or other hard surfaces.

A simple toolbox may be sufficient for small tool collections. Larger toolboxes with casters and several lined drawers are good for organizing tools when only a few workers will use the tools. See **Figure 7-55**. When many workers will share the same set of tools, wall-mounted pegboards or portable racks are a good idea. Mounting tools over silhouettes (cutouts in the shape of each tool) provides an organized system of storage that is easily managed. All tools can be quickly checked for cleanliness and returned to their proper places. **Figure 7-56** shows one type of setup.

donatas 1205/Shutterstock.com

Figure 7-56. When many workers use a group of tools, an open storage solution may be the best choice. A tool wall in a secured tool room such as the one pictured has a place for every tool and makes it is easy and quick to locate a specific tool.

CHAPTER 7 Review and Assessment

Summary

- The quality of hand tools has steadily improved over time as better materials and manufacturing methods have been developed.
- It is important to understand and follow relevant safety precautions when using hand tools.
- Hand tools can be classified in many ways, including how they work, for what material they are designed, and for what industry they were developed. Classifications of tools often overlap.
- Cutting tools separate or shape raw materials by removing waste material or by a shearing action.
- Gripping tools include vises, clamping tools, and many different types of pliers.
- Impact tools include hammers and mallets, some of which are used to strike punches or to form metal against an anvil.
- Pry bars use leverage to pull nails or force apart certain materials. Wrenches, screwdrivers, nut drivers, and hex keys all use leverage to rotate fasteners.
- An extractor is used with a T-handle tap wrench to remove screws and bolts that break off below the surface of a piece of equipment.
- Cleaning tools are used to collect waste materials and keep the workplace safe and usable.
- Quality is one of the most important factors to consider when selecting and purchasing tools.
- Quality hand tools are an investment. They must be organized and stored properly as well as regularly cleaned and lubricated to keep them in safe working condition.

Words to Know

Match the terms from the chapter to the correct definition.

A. chamfered
B. combination wrench
C. diagonal pliers
D. die
E. extractor
F. groove-joint pliers
G. kerf
H. lineman's pliers
I. mallet
J. miter box
K. mushrooming
L. needle-nose pliers
M. rasp
N. ratchet handle
O. ripsaw
P. slotted-tip screwdriver
Q. socket wrench
R. tap
S. vise
T. wood chisel

1. A tool comprised of a short, steel tube that looks like a box-end wrench that fits into a ratchet handle to tighten nuts and bolts.
2. A tool used to cut boards in a direction parallel to the grain of the wood.

3. A tool with a blade sharpened to a 25° to 30° angle on only one side to create a sharp edge.
4. A piece of equipment used to guide a backsaw when cutting precise angles in trim work.
5. A heavy-duty tool combining flat jaws and a side-cutting, pinch-type wire cutter.
6. A tool used to cut threads on the inside of a hole.
7. The physical slot made by a saw or other tool that cuts by removing a thin section of waste material.
8. An instrument that can be used to remove a damaged fastener.
9. A tool that typically mounts to a bench and secures the workpiece in a tight grip while other tools are used.
10. A tool with a long slot at the fulcrum, or joint, which allows for multiple working positions and for gripping objects within a large range of sizes.
11. A tool that has an open-end wrench on one end and a box-end wrench on the other.
12. Deformation damage to a metal tool from repeated impacts during use.
13. A tool that provides leverage to rotate a socket to loosen or tighten fasteners.
14. A cutting tool with rows of multiple triangular teeth that remove controlled amounts of material aggressively from a workpiece.
15. A tool with long, narrow jaws that are engineered to grip very small objects and work in tight places.
16. A beveled edge, which is included on some tools to reduce the damaging effects from repeated impact.
17. A tool used to cut threads on the outside of round stock.
18. An impact tool with a head made of wood, rubber, or rawhide that is often used to drive wood chisels or leatherworking tools.
19. A tool with two cutting edges designed specifically for cutting wire but not for gripping materials.
20. A common tool with a flat-head tip, a shaft, and a body.

Know and Understand

Answer the following questions using the information provided in this chapter.

1. Tools made from _____ significantly changed the development of tool technology and increased the rate at which civilizations could grow and develop.
2. When a saw blade cuts, material is removed from a slot the width of its offset teeth, called the _____.
3. A back saw is often used with a(n) _____ for cutting precise angles in trim work.
4. Drill bits for hand-powered drills are usually either _____ bits or twist drill bits.
5. Always apply enough _____ to the back of a hand drill to keep the cutting edges of the bit engaged and removing material.
6. _____ and _____ are used in a process called "chasing the thread" to clean contaminants and corrosion from the threads of fasteners.
7. The effectiveness of a wood chisel depends on its _____.

8. *True or False?* A hammer designed for driving nails can be safely used to strike a cold chisel.
9. *True or False?* Files have rows of multiple triangular-shaped teeth that remove material more aggressively than rasps.
10. When operated by hand, a moderate-size vise can easily generate force in excess of _____ lb.
11. Woodworking bench vises are commonly mounted _____ the working surface of the table.
    A. ten to twelve inches below
    B. flush with
    C. six to eight inches above
    D. in the middle of
12. Parallel clamps are not suitable for hot metal work because they are made of _____.
13. The jaws of most pliers are much shorter than the _____ to provide a greater mechanical advantage.
14. When using any hammer, concentrate your focus on the _____ to maximize your accuracy and power.
    A. swing of the hammer
    B. surface of the material
    C. hammer's head
    D. target
15. A(n) _____ is a tool that fits into the square hole in an anvil and is used to cut or form a detailed metal shape.
16. What is a tapered punch designed to do?
17. A(n) _____ wrench consists of an open-end wrench on one end and a box-end wrench on the other.
18. Under most conditions, using a properly fitted box-end wrench is preferable to using a(n) _____ wrench on hexagonal nuts and bolts due to a lower risk of slippage.
19. An Allen wrench fits into what shape of recess in the top of a screw?
    A. Hexagonal.
    B. Square.
    C. Six-pointed star.
    D. Slotted.
20. What are five factors to consider when purchasing a hand tool?
21. Why is it important to keep hand tools clean and properly organized?

STEM and Academic Activities

1. **Science.** Choose your favorite hand tool and prepare a short presentation explaining how it incorporates one or more simple machines into its design.
2. **Technology.** Study different styles of wrenches. Sketch a handle design for a wrench to limit the amount of torque that the wrench can apply.
3. **Engineering.** Gather a collection of different types of pliers. Use each to grip into a scrap piece of soft lumber. Which pliers hold the wood with the most force? Why?

4. **Math.** Choose a pair of pliers and measure the length of the handles and the jaws. Calculate the mechanical advantage offered by the leverage of the handles. Look up prices for that type of pliers from different tool manufacturers. What is the cost difference in purchasing quality, brand-name pliers compared to generic brands?
5. **Social Science.** Choose a hand tool and research its development throughout history. Put together a short presentation outlining the history of the tool and explain the benefits it provided compared to the old tools/methods.
6. **Language Arts.** Communicating with others is an important step in many agricultural mechanics jobs. Practice written communication by describing the hand tool of your choice in a five-sentence paragraph. Do not use the name of the tool in your description. Pay close attention to spelling and grammar. Check your work by allowing others to read your description. Do they know the name of the tool you are describing?

Thinking Critically

1. Identify and locate the separate simple machines that are used in the design of the locking pliers shown in the illustration.

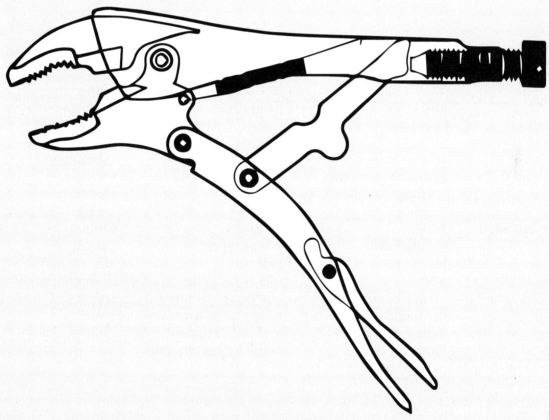

2. It is common to see hand tools used incorrectly for different tasks. Think of two or three situations in which you have witnessed a hand tool being used for something other than its designed purpose. What were the potential dangers or problems associated with using this hand tool in an improper manner?

CHAPTER 8: Power Tools

Chapter Outcomes

After studying this chapter, you will be able to:
- Briefly describe the development of power tools.
- Explain the different methods of classifying power tools.
- Identify and follow safety instructions and safe procedures when using power tools.
- Explain how power tools are designed to perform different tasks.
- Select and use appropriate power tools for drilling, driving, cutting, shaping, and removing material.
- Summarize the factors to consider when purchasing power tools.
- Properly maintain and store power tools commonly used in agriculture.

Words to Know

abrasives	double-insulated	jack stand	planer
band saw	drill press	jig saw	portable power drills
battery charger	floor jack	jointer	portable power tools
bimetal blade	flutes	kickback	radial arm saw
bit	grinder	lithium-ion (Li-ion) battery	rake angle
blade set	ground fault circuit interpreter (GFCI)	lodging	reciprocating saw
bottle jack			ripping
brad nailer	grounded	machine guard	router
charge memory effect	gullet	miter saw	sander
chuck	hammer drill	Morse taper	shields
chuck key	hollow ground blades	nail gun	skip tooth configuration
circular saw	hook tooth configuration	nickel-cadmium (NiCad) battery	
cordless power tool			stationary power tools
cribbing	hydraulic press	nickel-metal-hydride (NiMH) battery	swarf
crosscutting	impact driver		table saw
die grinder	impact socket	oscillating multi-tools	

Before You Read

Before reading this chapter, flip through the pages and make notes of the major headings. Analyze the structure of the relationships of the headings with the concepts in the chapters.

While studying this chapter, look for the activity icon to:

- **Practice** vocabulary terms with e-flash cards and matching activities.
- **Expand** learning with interactive activities.
- **Reinforce** what you learn by completing the end-of-chapter questions.

www.g-wlearning.com/agriculture

kwest/Shutterstock.com

Early power tools were developed to perform the tasks of hand tools with increased efficiency and less manual effort. The first power sources were wheels driven by running water, wind, or harnessed livestock. Power tools driven by natural sources were large tools permanently installed and restricted to the geographic location of the power supply. For example, sawmills were located on creeks with reliable flowing water. The advent of steam power allowed shops that used power tools to be located closer to sources of raw materials or to methods of transportation used for commerce.

History credits the invention of the first handheld power tool to the German engineering firm C&E Fein when they introduced their electric drill in 1895. It was a cumbersome and inefficient tool, but it paved the way for the development of modern power tools.

Classification of Power Tools

Like hand tools, the name of a power tool usually reflects its intended use or its method of operation. Sometimes the tool inventor's name or developing company becomes the common name of the tool. Often, tools used in different areas of the country have different names. It is important that you attempt to use the proper or official names of tools when communicating with others to avoid confusion.

Power tools can be classified by power source, size or portability, or the type of work they are built to do. Professionals in different occupations use different collections of tools. For example, carpenters use tools designed for woodworking, and machinists use tools built for metalworking. Mechanics use an entirely different set of tools for their jobs. Many power tools are versatile enough to perform multiple tasks and may be better classified by the type of job they do. Drills, for example, may be used by woodworkers, metalworkers, and mechanics, as well as by plumbers, electricians, and many others. Knowing how tools are classified can help you select the best tools for performing different tasks.

Bigger stationary tools are often more powerful than their smaller, portable counterparts. Mounting a stationary tool to the floor or to a secure workbench strengthens its stability. This reduces safety risks and lets the user perform repeated tasks with precision and consistency.

All early power tools were stationary and were driven by natural sources or, later, by bulky steam or internal combustion engines. See **Figure 8-1**. The cost of early engines and electric motors was so high that it was much more economical to use a single power source to operate many tools.

Delmas Lehman/Shutterstock.com

Figure 8-1. Steam power was used in the past to generate power for agricultural equipment.

It once was commonplace to have all of the stationary power tools in a shop linked by a series of overhead drive belts and pulleys, all driven by a single power source. This is similar to today's farm practice of relying on a single tractor outfitted with a variety of agricultural equipment to perform many different functions.

Stationary Electric Power Tools

The development of increasingly more efficient electric motors led to the constant improvement of electrically powered tools. *Stationary power tools* are constructed with relatively heavy bases intended to stabilize the machine during operation. Most include mounting systems to be secured to the floor or workbench tops.

Today, most stationary tools are powered by a dedicated electric motor. These machines use the same standard alternating current (AC) electric power that is used in most American homes. Agricultural applications commonly use 120-volt or 240-volt power. However, larger industrial machines may require the use of higher voltages.

Stationary electric equipment may be connected to the power source by a wall plug or by hard wiring. Hard wiring is directly connecting the electric wiring of a tool to the wiring of the electric supply system—no wall plug is necessary. Power tools that are usually operated in stationary positions but are occasionally moved require plugging a power cord into a wall receptacle.

Portable Electric Power Tools

Portable power tools are traditionally powered through an electric power cord plugged into standard 120-volt AC current. Larger tools may require 240-volt AC current. Portable grinders, drills, and saws may also be powered by direct current (DC) power sources. Many portable welding machines have convenience outlets connected to the same DC generator that produces the welding current. Some portable electric power tools are configured to use either AC or DC current; these are machines intended for use in field applications.

Safety Note

Extension cords are used as temporary wiring additions to the permanent electrical system of a building. Although not ideal, extension cord use is inevitable on many job sites. Grounded three-conductor extension cords made of strong, durable material should be used for powering portable tools and equipment. A few safety rules will help you prevent tool and cord damage and avoid injury:
- When choosing the proper extension cord to use on a job, check that the cord's amperage rating is high enough to handle the amperage of the power tools or equipment that will be connected to it.
- Never use a frayed or damaged extension cord.
- Always be mindful of the location of extension and power cords when using portable power tools.

Safety Note

Although many of today's portable power tools are capable of safely using AC or DC current, check for an AC/DC rating marked on the tool or in the manufacturer's instructions—especially with older tools—before attempting to connect to a direct current power source. Injury or tool damage could result from connecting to an unsuitable power supply.

Battery-Operated Power Tools

Another common DC power source for a portable power tool is a rechargeable battery. Battery-operated tools are also known as *cordless power tools* and commonly use between 9.6 volts–36 volts DC power. Tools equipped with similar motors deliver higher speeds proportional to higher voltage inputs. Conversely, as more torque is applied to a power tool, it uses more amperage or current.

Three main types of rechargeable batteries are used in cordless power tools. *Nickel-cadmium (NiCad) batteries* are the oldest example. NiCad batteries are the least expensive and last for about 1000 charge cycles, but they can easily deteriorate when not charged and discharged properly.

Nickel-metal-hydride (NiMH) batteries can run two to three times longer on a single charge than NiCad batteries. NiMH batteries, however, are very sensitive to improper charging methods and storage conditions. Deep discharging will shorten their life span, and exposure to extreme high or low temperatures degrades the battery.

Both NiMH and NiCad batteries can develop a *charge memory effect*. If they are improperly charged, a charge memory develops in these batteries, allowing only a portion of the battery to accept a charge.

Lithium-ion (Li-ion) batteries, although more expensive, outperform the other battery types used in power tools, **Figure 8-2**. Due to their high power density, Li-ion batteries offer a more uniform delivery of power. The performance of NiCad and NiMH batteries starts slowing down at about 70% charge. Li-ion-powered tools deliver uniform speed and torque down to about 20% of a full charge. Li-ion batteries last only for 350 to 500 charges, but they do not suffer from a charge memory effect.

The key to preserving the life of power tool batteries is using a quality charging system. *Battery chargers* use standard 120-volt AC current to fully charge a battery in 1 to 16 hours. Fast charging is not always best. NiCad and NiMH batteries last longer using chargers that deliver a full charge in about an hour, but Li-ion batteries can be damaged by chargers that offer less than a 3-hour complete charge time.

Jeffrey B. Banke/Shutterstock.com

Figure 8-2. Manufacturers of cordless power tools may use a single lithium-ion battery designed to fit multiple cordless tools within their product line.

The run time on battery-operated tools varies greatly depending on the energy demands placed on the tool. Tools for drilling or driving fasteners may have a 30-minute battery life, while tools that are used for grinding or sawing are often limited to a 15-minute run time or less before a fully charged battery is needed. Contractors who rely heavily on battery-operated tools address the limited run time by having multiple batteries and charging stations set up on a job site.

Safety Note

The cadmium in NiCad batteries is considered a heavy metal that is highly toxic to the environment when disposed of improperly.

Pneumatic Power Tools

Another way to supply power to portable tools is to use pneumatic power, or compressed air or gas. Pneumatic, or air, tools are driven by compressed air supplied by an air compressor or by compressed CO_2 delivered from a high-pressure container called a cylinder.

Air compressors are pumps that collect atmospheric air into a high-pressure tank. The compressed air can then be released in controlled amounts through flexible hoses to operate power tools. See **Figure 8-3**. Compressors may be powered by electricity or by internal combustion engines. Pneumatic tools usually operate at 80 to 100 pounds per square inch (psi) of air pressure. Supply hoses must be reinforced to withstand this range of pressure and are usually connected to tools using locking connectors. A few examples of locking connectors are shown in **Figure 8-4**.

Visionsi/Shutterstock.com

Figure 8-3. Portable and stationary air compressors are used to provide pneumatic power to operate tools.

Pneumatic tools have some advantages over other power tool types. They typically weigh less than other types and pose less of a spark hazard than electric tools. This characteristic reduces the fire risk in environments where flammable gases may be present. Using compressed air as a power source can also be advantageous when compressed air is also necessary for other jobs, such as inflating tires or spray painting.

The size and number of tools that may be operated at one time by a single air compressor is limited by the cubic feet per minute (cfm) air output of the compressor. For example, an air compressor capable of a sustained output of 12 cfm cannot be expected to operate a paint sprayer that requires 7 cfm and a sander that requires 9 cfm at the same time, although it would be adequate for either one individually.

Compressing atmospheric air concentrates its moisture content. To protect pneumatic tools from damage, a water filter or separator should be installed in air supply lines. Filters are usually combination units that include a water separator, a cartridge-type air filter, a pressure regulator, and a gauge to monitor line working pressure. See **Figure 8-5**. Filters remove large particulates from the air supply that could cause premature wear within the tool.

Goodheart-Willcox Publisher

Figure 8-4. Male and female quick-connect couplers are used to facilitate the connection of pneumatic tools to compressed air supply lines.

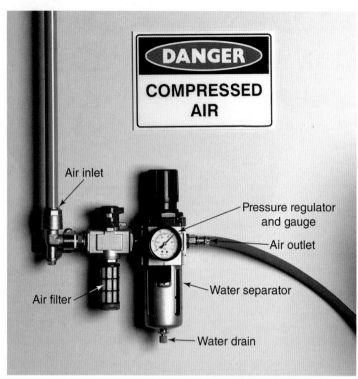

Figure 8-5. A combination unit housing a pressure regulator, water separator, and pressure gauge. Note the separate air filter. For efficient operation, pneumatic power tools require that the air supply be kept clean, free from moisture, and regulated to a stable pressure.

In addition to requiring dry, clean air, pneumatic tools need to be lubricated to reduce wear. Lubricants reduce friction and cool the tool while it is in use. Lubricants can be introduced into the air supply through automatic oilers, or the tools may be lubricated manually.

Gasoline- and Diesel-Powered Tools

Sometimes electricity is not readily available at a job site. In these situations, professionals complete their work using tools and equipment powered by gasoline or diesel engines.

Internal combustion engines convert the energy from a burning fuel source to usable motion. Fuels can be petroleum- or ethanol-based or blends of the two. Gasoline, diesel, and liquid petroleum gas (LPG) are the most common fuels.

Within this category, a distinction can be made between the terms *tool* and *equipment*. Equipment is larger and is not designed to be moved easily by one person. Tools are smaller and are meant to be operated by one person, or at most, by one person and a helper.

Gasoline- and diesel-powered tools are usually heavier and designed for tougher work than their electric counterparts. They are meant for big jobs where the size and power they offer speeds up job completion. Some common examples of handheld, engine-powered tools are chainsaws, concrete cutters, posthole diggers, and leaf blowers. See **Figure 8-6**.

Power Tool Safety Rules

Safe operation of power tools means following appropriate rules and procedures. Get yourself into the habit of recognizing, addressing, and following safety protocol whenever you use power tools. Follow these general power tool safety rules at all times:

- Always read and follow the safety and maintenance procedures provided by the power tool manufacturer for a particular tool.

Figure 8-6. Internal combustion engines are used to power heavy equipment and tools, like this posthole digger, in remote areas.

AgEd Connection: Use of Power Tools

Agricultural mechanics curriculum includes training in the safe operation of a variety of power tools. Most agricultural mechanics labs contain an assortment of power woodworking and metal fabrication tools, from table saws to portable welders. The safe and effective use of power tools allows work to take place in a more efficient manner than by hand or with hand tools. Take advantage of the training offered by your high school agriculture teacher and become proficient in the use of power tools. That training will benefit you in your agricultural mechanics projects and increase your employability skills when looking for a job that utilizes power tools. Prior training and experience is valuable to an employer and can be obtained while in class!

auremar/Shutterstock.com

- Check that an electric power tool is either *grounded* in such a way that any stray electrical current is directed to the ground without passing through the user or that it is double insulated. *Double-insulated* tools usually have plastic housing parts wherever a user normally touches the machine. The electric current is isolated from any exposed metal parts. Double-insulated tools are only required to have a two-conductor power cord. Grounded tools have a third ground conductor in the power cord.
- Ensure that electrical outlets that are located in damp locations or near sources of water are protected by a *ground fault circuit interrupter (GFCI)*. **Figure 8-7** shows the distinct look of a GFCI electrical outlet. GFCIs are designed to monitor the electric current flow along the hot and neutral wires of the circuit, and if there is a difference (as little as 0.005 amp) between these current values, electricity to the circuit is interrupted.
- Use all *machine guards* and *shields* installed by the power tool manufacturer. Both parts serve to protect the user from moving parts and flying particles and should never be removed.

Brandon Blinkenberg/Shutterstock.com

Figure 8-7. When operating electric power tools outdoors or in wet environments, you must connect to an electric circuit protected with a ground fault circuit interrupter. Periodic testing of the outlet is advisable and involves pressing the "Test" button and confirming that the "Reset" button pops out and trips the circuit.

- Secure the workpiece properly and firmly before performing any operations.
- A *kickback* when using certain power cutting tools can be very dangerous. This interruption can cause the tool to suddenly jump backward out of the work and toward the operator. Prevent kickbacks from happening by removing any nails, screws, or staples from previously used lumber, and avoid cutting boards through naturally occurring knots.
- Use only power tools that are in good condition. Report damaged or broken tools immediately and remove them from service.
- Wear OSHA-approved (ANSI Z87.1) eyewear and/or face protection at all times. Respiratory (if needed) and hearing protection should also be worn.
- Tie back and secure long hair, and do not wear jewelry. Clothing should fit snugly with no hanging pieces. Wear proper foot protection when working with heavy materials or heavy tools.
- Properly clean and store power tools to keep them in safe working order.

Additional pertinent safety information can be found in Chapter 5, *Safety and Developing Safe Work Habits*.

Drilling and Driving Tools

Power drills and driving tools are machines that use *bits* or other attachments to create holes, shape metals, or attach materials using assorted fasteners. Most of the tools in this broad category are portable, although there are several prominent examples of stationary machines as well.

Drill Bits

Drill bits are designed to drill holes in many different materials. The design and construction among bit types varies greatly, depending on the material on which they are designed to be used. All drill bits, however, have three things in common: at least one cutting edge, a shank that must be attached to the drill by means of the chuck, and a method of moving swarf from the hole. *Swarf* is the shavings created and displaced by the drill bit as a result of drilling into the material. See **Figure 8-8**.

David Orcea/Shutterstock.com

Figure 8-8. Swarf created by a working drill bit is an indication that the moving bit is being applied with sufficient pressure to cut the material. It is also a hazard for machine operators. Always wear eye protection when using a drill, and use a table brush or similar tool to clean up sharp debris left by the drilling process.

Some common types of drill bits are shown in **Figure 8-9**. Most common drill bits are sharpened to a 118° angle at the point because this provides optimum strength and durability for drilling in most materials. Drill bits intended for harder steels, such as stainless steels, are often ground to a 135° angle to begin the drilling process faster, thus reducing the amount of time the bit is creating friction without cutting.

Types of Drill Bits

Auger
Designed for boring relatively large holes in soft materials. Threaded tip helps to draw bit into the work without a pilot hole. May have single, double, or triple flutes.

Artography/Shutterstock.com

High-Speed Steel (HSS) Twist
Multipurpose bit used on wood, metals, and plastic materials. Always use a center punch to initially guide the bit, and drill pilot holes for larger bits.

Giuliano Del Moretto/Shutterstock.com

Spade or Paddle
Commonly used for boring holes in lumber. Relies on operator pressure to penetrate the wood. Versions with threaded points aggressively pull themselves into the work.

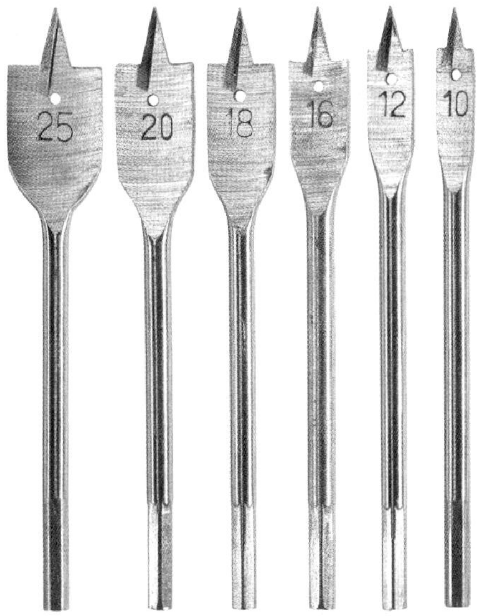

mihalec/Shutterstock.com

Masonry
Designed with a hardened, chisel-like tip that is used to chip through hard, brittle materials such as stone, tile, and concrete. Some are heavily reinforced and/or equipped with SDS shanks for use in hammer drills.

OlegSam/Shutterstock.com

Forstner
Complex bits capable of boring nearly flat bottomed holes of often large diameter. Used frequently by fine carpenters and furniture makers. Best suited for use in drill press to maintain stability.

Biehler Michael/Shutterstock.com

(Continued)

Figure 8-9. Common types of drill bits.

Types of Drill Bits (Continued)

Step Drill Designed with multiple bit sizes on the same self-piloting bit. Limited to use on thin materials. The diameter of the hole is determined by the depth to which the bit is allowed to penetrate.	*Eimantas Buzas/Shutterstock.com*
Countersink Used to shape the top of the hole to fit the fastener when screws designed to fit flush with the surface of the work are used. May employ a short piloting point or the entire length of the pilot bit.	*Christopher Elwell/Shutterstock.com* *Goodheart-Willcox Publisher*
Tile and Glass Uses hardened tips that may be imbedded with abrasive materials to cut glass and ceramic tile. Lubricating and cooling the bit with a water bath during drilling is often recommended.	*Coprid/Shutterstock.com* *Goodheart-Willcox Publisher*
Hole Saw A saw blade formed in a circle for cutting large holes. Available for cutting a variety of materials. Incorporates a pilot bit that keeps the saw aligned.	*Dan Kosmayer/Shutterstock.com*

Drill Bit Materials

Material used to manufacture drill bits must be harder and stronger than the work material. Bits designed for drilling steel have to be significantly harder than bits designed for boring holes in wood.

High-speed steel (HSS) is the material most commonly used to make drill bits. HSS bits are good for drilling wood, most plastics, aluminum, and soft steels. Cobalt bits include 5%–8% of cobalt blended into the steel. This creates a more durable bit that is sufficiently hard for drilling into hard materials such as stainless steel. Carbide bits are the hardest bits readily available, but they are also the most brittle and require extremely stable drilling conditions.

Using carbide bits in handheld drills should be avoided because they are likely to break if excessive side stress is applied to the bit.

Drill bit surfaces are finished differently to improve performance in a variety of materials. Most common drill bits have a polished surface to reduce friction and speed the flow of chips or swarf. As the quality and price of drill bits increase, the bits are coated with black oxide, titanium nitride (gold color), or titanium carbonitride (bluish-gray color) to further reduce friction and to extend the useful life of the tool.

A sharp drill bit cuts continuously and creates less friction than a dull drill bit, which tends to rub into the material. Allowing a drill bit to turn against the material without cutting generates excess friction, which creates heat buildup. The excess heat dulls the drill bit faster and may damage the work material.

Drill Bit Sizes

The size of a drill bit determines the diameter of the hole that it can create. There are three standard measuring systems for drill bits. Drill bit size is usually expressed in inches and fractions of an inch. Size may also be indicated in millimeters and decimal fractions of a millimeter. Machinists, professionals who fabricate and repair machinery and machine parts, often use a system of sizing drill bits that uses a series of letters, A to Z, and whole numbers, 1 to 80. A drill bit chart is essential when drilling precision holes.

Drill bits may be purchased separately, but usually they are sold in sets that include a drill index. A drill index is a storage box with holes or pockets that arrange each drill bit by diameter from smallest to largest. Drill bits should always be stored in the drill index to preserve their sharpness and to make finding the correct size easier.

Portable Drills

Portable power drills are handheld tools intended primarily for boring holes in wood, metals, and other common construction materials. See **Figure 8-10**. They work by applying torque to a drill bit to cut a circular hole through the material. Many drills have a variable speed setting that allows the operator to match drilling speed to the type of work material. Power drills do not follow the general rule of relying on the weight of the tool to apply pressure to the work. Instead, it is necessary to manually apply enough force to a drill to keep the bit cutting at all times.

Portable drills are usually powered by electricity, either 120-volt AC or rechargeable batteries. Pneumatic portable drills are used in environments where flammable gases are present and a flash fire hazard exists. Powerful drills powered by an internal combustion engine are used for tasks such as digging fence postholes or boring blast holes in quarries.

Naruedom Yaempongsa/Shutterstock.com

Figure 8-10. A typical cordless electric power drill.

190 Agricultural Mechanics and Technology Systems

ALEXANDER LEONOV/Shutterstock.com

Figure 8-11. Secure drill bits in a Jacobs chuck by using the correct size chuck key to tighten the locking ring at each of the three holes around the chuck. This process must be done each time the drill bit is changed. Do not forget to remove the chuck key before drilling.

Drill bits are secured in the drill using a *chuck*, or bit-holding device. The most common type of drill chuck is the three-jawed Jacobs chuck, **Figure 8-11**. This device requires the use of a geared tool known as a *chuck key* to tighten and loosen the hold on the bit.

A common variation of this setup is the keyless chuck. The user grips and turns the barrel of the keyless chuck to tighten or loosen the chuck. Another type of chuck retains the drill bit using a ball-and-socket mechanism in a hole shaped to receive the specially modified bit shank.

Hammer Drills

A *hammer drill* is a tool that resembles a conventional power drill but features a stronger driving force. Hammer drills combine the torque of a drill with a mechanism that works like someone hitting the back of the drill with a hammer as you drill the hole. They are useful for drilling holes in hard materials such as concrete, masonry, and rock, **Figure 8-12**. The impact action is unnecessary when drilling in wood, and it will damage many bits if used on metal. Some hammer drills use Special Direct System (SDS) bits, which have heavily reinforced shanks with a series of longitudinal slots. Spring-loaded keepers fit into the slots to eliminate the need for hand tightening.

The precautions for operating a hammer drill are much the same as those for other power drills. However, the additional protection of a full-face shield, hearing protection, and respiratory protection is recommended when drilling concrete and masonry.

©iStock/Dan70

Figure 8-12. Hammer drills are heavy-duty tools that are capable of delivering greater force and power than standard power drills.

Impact Drivers

Impact drivers combine the torque of a power drill with an impact mechanism much like a hammer drill. In a hammer drill, however, the impact is applied in a straight line parallel to the drill bit. The impact in an impact driver applies additional torque, much like hitting the back of a wrench with a hammer as the bolt is being turned. Impact drivers have a two-stage mechanism. At first they work like a drill, applying torque to a fastener until enough resistance is built up to cause the hammering action to kick in. This mechanism extends the life of the driver and reduces damage to fasteners from excessive hammering.

Impact drivers are divided into two basic types based on the intended use and the method for holding attachments. The oldest type has a square driver designed to receive reinforced impact sockets for driving nuts and bolts. *Impact sockets* look similar to those that attach to ratchet handles used by mechanics, but they are made of much stronger steel. In fact, impact sockets may be used with standard socket wrench sets, but standard sockets are not engineered to withstand the hammering effect of an impact driver. When impact sockets are paired with an impact driver, the combination is often referred to as an *impact wrench*. See **Figure 8-13**.

The second type of impact driver looks like a stubby-bodied power drill equipped with a quick-connect hexagonal socket to accept an assortment of screwdriver bits, nut drivers, drills, and other attachments. See **Figure 8-14**. For general construction and most do-it-yourself work, this type of impact driver is indispensable. Driving screws with an impact driver and a screwdriver bit has replaced the use of a hammer and nails for many applications. The development of long-lasting, powerful lithium-ion batteries has also greatly increased the effectiveness and popularity of these tools.

Drill Presses

Most well-equipped shops include a stationary *drill press*, **Figure 8-15**. Drill presses are large, variable-speed power drills mounted to a vertical stabilizing shaft. The shaft and drill assembly are attached to a base with an adjustable table for securing workpieces. Using a drill press instead of a portable power drill helps to increase the precision of the alignment and angle of holes. It can also safely apply much more force because the workpiece can be clamped to the table, and the motion of the spindle and the bit is stabilized through its attachment to the column, or frame, of the drill press.

Floor-mounted and benchtop drill press models are available. The size of a drill press is determined by the maximum diameter of the work material that can be drilled exactly in the center. The size may be checked by measuring from the center of a chuck-mounted bit to the closest location on the column and doubling that measurement. For example, a drill press that is 7″ from center to column is a 14″ drill press. This press is capable of drilling a hole in the center of material with a maximum 14″ diameter.

A Jacobs chuck is the common way of attaching bits to smaller drill presses. Some larger drill presses accept Morse taper shank bits. A *Morse taper* is a tapered shank that holds the drill bit entirely by the friction

Yotsatorn Laonalonglit/Shutterstock.com

Figure 8-13. Pneumatic impact wrenches use heavy-duty sockets to loosen and tighten fasteners.

OlegSam/Shutterstock.com

Figure 8-14. Cordless impact drivers can be very powerful and are usually more compact than cordless drills. Note the quick-connect hexagonal socket to accept different bits.

192 Agricultural Mechanics and Technology Systems

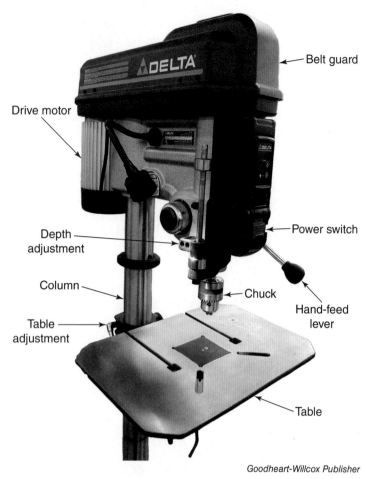

Goodheart-Willcox Publisher

Figure 8-15. Using a drill press instead of a handheld power drill provides added precision for the location and placement of drilled holes.

between the shank and the tapered socket. The spindle of many drill presses is designed with a Morse taper to accept the chuck assembly or a Morse taper drill bit.

The drilling, or rotating, speed of a drill press is adjusted by changing the arrangement of a belt or combination of belts within the drive motor housing. The belts are positioned between stacks of pulleys of different sizes attached to the drive motor and the spindle. Moving the belts changes the drive ratio, thus changing the speed and torque delivered at the drill bit. The physical properties of the work material and the size of the drill bit determine the optimum speed for drilling. See **Figure 8-16**.

A common drill press accessory is the drill press vise. These vises are machined flat on the base to be clamped securely to the drill press table. The parallel sides of the vise are capable of holding flat or round objects. Using a drill press vise is the safest and most stable method to hold materials while drilling.

Drilling Tool Rules

When operating a drilling tool, follow all the power tool safety rules outlined earlier, along with these specific rules:

- Tighten, if necessary, a Jacobs chuck using all three holes. Remove the chuck key before drilling.

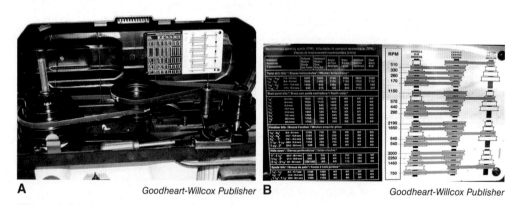

Figure 8-16. A—Most drill presses operate at variable speeds to accommodate the characteristics of different raw materials. Note the numerous possible belt configurations on the pulleys of this drive motor. B—Refer to the drill speed chart located either directly on the drill press or in the operator's manual to set the drilling speed prior to drilling. The chart on the right displays belt positions for the different speeds.

Copyright Goodheart-Willcox Co., Inc.

- Select the proper speed and the correct bit for the work material.
- Start all holes with a center punch.
- Apply enough force to the drill to ensure that the bit is cutting continuously. Power tools rely on momentum to deliver force to the work and should be operated without loading the motor to a point that it slows. To achieve this effect with a drill, adjust the speed to ensure that enough torque is applied to the bit that it will cut under pressure.
- Be careful not to apply too much side pressure when using a portable power drill, because this can break the drill bit.
- Use a pilot bit (smaller diameter bit) followed by a larger bit to drill larger holes.
- Use a piece of scrap lumber to back up the material being drilled and to protect the surface of the drill press table.
- Use a brush or broom to remove swarf; do not handle swarf with bare hands.

Nail Guns and Brad Nailers

Nail guns are tools that drive nails into wood and similar materials with a single power stroke. Most are operated using a pneumatic power supply, **Figure 8-17**. However, some models use gunpowder-charged cartridges or electromagnetic forces to drive fasteners. Some electric nail guns are actually self-contained pneumatic tools with the air supply built into the body of the tool. Another type of self-contained nail gun uses the combustion of liquefied petroleum gas to drive nails.

Nail guns are generally offered in two basic sizes. Large nail guns are used for framing buildings, and smaller models are used for finish and trim work inside buildings. A 3 1/2" 16d nail is typically used for framing. The length and diameter of finish nails varies greatly with the type of material being fastened. Nails for nail guns are fastened together into sticks or coils, which are fed into clips or channels.

©iStock/tinabelle

Figure 8-17. This style of pneumatic nail gun is called a coil nail gun and is primarily used for roofing projects. Note the drum that is used to hold the nails.

Brad nailers are even smaller versions of nail guns that drive very small fasteners for fine trim work. Fasteners for brad nailers are usually 18 gauge nails, 1/2" to 1 1/4" long with very small heads, but some tools also accept staples.

Nails driven with nail guns or brad nailers are less likely to split wood than nails driven by a hammer because the fastener is driven with one smooth motion. Friction from the tool's action of driving the nail melts the adhesive coating holding the nails together. The melted adhesive lubricates the nail as it is being driven and sets up to enhance the nail's holding ability once in place.

Using a nail gun requires adherence to specific safety precautions and special training prior to use. Not all nail guns have the same safety features. Nail guns that have a sequential trigger have a sliding safety tip that must be depressed before the gun will fire. Sequential guns are much safer than semiautomatic-firing, or bump-fire, nail guns that fire every time the trigger is pulled without cycling a safety tip.

Safety Note

Nails driven by a nail gun have sufficient force to fully penetrate a board in a single stroke. If the nail comes in contact with severely warped grain or a knot in the lumber, the nail could be deflected as a dangerous projectile. Metal structures near wood framing can produce the same result. Care must be taken to maintain appropriate air pressure and to work around potential deflections. Large nail guns must be held securely as they have the potential to kick back slightly when operating.

Hydraulic Jacks and Presses

A tool that uses hydraulic pressure is capable of exerting a tremendous amount of pressure with a limited amount of input energy. The principle of hydraulics uses an incompressible liquid to transfer and multiply force within a closed connected system. **Figure 8-18** illustrates this basic principle.

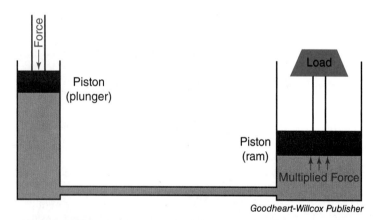

Figure 8-18. Pascal's law states that an increase in pressure at any point in a confined fluid results in an equal pressure increase at every other point in the connected system. When a downward force is applied to the smaller piston in this illustration, the force is multiplied due to the increased surface area of the larger piston.

Hydraulic applications are used for tasks such as lifting heavy objects, spreading objects apart, and applying a squeezing force to an object, as well as for other tasks.

Hydraulic jacks are low-profile tools capable of lifting heavy objects such as vehicles and equipment. *Bottle jacks* are vertical cylinders with a piston or ram that is driven upward by the application of hydraulic force, **Figure 8-19A**. Bottle jacks may be operated by a manual or pneumatic-driven pump. Jacks that lift from 2 to 20 tons are common. The stroke, or distance the piston travels, determines the amount of movement a bottle jack can produce. Increasing the diameter of the output piston multiplies the output force a jack can produce. Bottle jacks are small and portable.

Floor jacks use a horizontally mounted bottle jack that pushes against a lever to lift the saddle arm and platform vertically, **Figure 8-19B**. This design loses about three-quarters of the force produced by the standard bottle jack, but results in a compact tool that can lift objects proportionally higher. The wheels on the jack make it convenient to move on the shop floor, but most floor jacks are heavy and lack portability.

When using a jack or jacks to lift a vehicle, you must physically block the wheels that remain in contact with the ground to keep them from rolling. Commercially made wheel chocks are available for this purpose, but wood blocks of adequate size left over from other projects can also work.

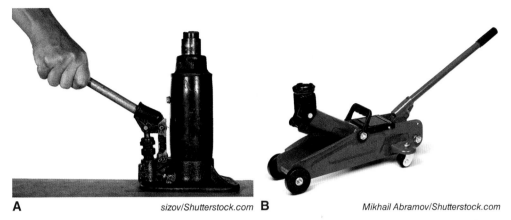

A sizov/Shutterstock.com B Mikhail Abramov/Shutterstock.com

Figure 8-19. Simple hydraulic jacks put Pascal's law into practical use. A—The portable design of a bottle jack makes it ideal for agricultural field work. B—The low profile design and durable wheels of a typical floor jack make it easy to slide under vehicles and equipment.

Remember that the purpose of a jack is to lift heavy objects, not to hold them in place. Never trust a jack alone to hold up a vehicle that must be serviced from the underside. Use a *jack stand*, which is designed for stability, or cribbing constructed from short timbers to hold vehicles up while working underneath them. See **Figure 8-20**. The height of jack stands can be adjusted using a mechanical device such as a screw or pin.

Cribbing is a method of stacking timbers in a crisscross pattern to form a structure resembling an old-time corn crib, **Figure 8-21**. Using cribbing is a stable way to temporarily support heavy equipment while it is under construction or in for repairs.

Hydraulic presses are sturdily framed machines that use a hydraulic cylinder to apply pressure to objects for a variety of purposes. The most basic types use a bottle jack mounted in an H-shaped frame. These are designed for pressing bushings and bearings and for moderate bending of metals. Basic hydraulic presses may be operated with a manual pump or with an electric or pneumatic pump. Many include a pressure gauge to monitor the force exerted by the press. One common use of a hydraulic press is to perform testing on the strength of welded objects.

Sophisticated versions of hydraulic presses are used in the metalworking industry to shear, punch, bend, and cold-form metal. **Figure 8-22** shows one type of machine. A small version of these heavy-duty machines can exert forces in the range of 40 to 60 tons.

Another type of hydraulic press is designed for bending pipe and tubing. Pipe benders may be stationary, or they may be portable devices that can be taken to the job site. Both types use rollers to press pipe or tubing around a mandrel—a curved piece or form that determines the radius of the bend. Rollers and mandrels are matched sets

ericlefrancais/Shutterstock.com

Figure 8-20. After agricultural equipment is lifted with a mechanical or hydraulic jack, jack stands are moved into place and used to support the heavy load. Never rely on mechanical or hydraulic jacks alone to support a load for an extended period of time.

Figure 8-21. Cribbing, a method of using stacked timbers to support a load, is a very secure method of supporting heavy objects during service or construction.

based on the diameter of the pipe. Modified sets of rollers and mandrels are available for bending square tubing and angle iron.

Cutting Tools

Power saws and other power cutting tools are machines that use blades (or other attachments) to separate or shape materials. Some blades and attachments can be readily sharpened, while others must be replaced when dulled or damaged from use. Power cutting tools may be stationary or portable. The most common power source for this family of tools is an electric motor, but other methods are also used.

Cutting Tool Rules

When using any type of power cutting tool, observe the power tool safety rules described earlier and these general safety precautions:

- Unplug or disconnect the power supply before changing blades or bits.
- Allow the cutting tool or machine to reach full operating speed before making contact with the work material.
- Allow the saw to determine the speed of the cut. Do not force the blade into the work or the work into the blade.
- Let the cutting tool or machine come to a complete stop before retrieving the cut parts or pieces.

Circular Saws

Circular saws are portable power saws originally designed for cutting lumber. As the name implies, the blade of a circular saw is round and ranges in size from 3 1/2" to over 10 1/2". The diameter of the blade determines the capacity of the saw, or the thickest lumber it is capable of cutting through. The most common saw size drives a 7 1/2" blade and is well-suited to cutting 2" lumber and 3/4" plywood.

Figure 8-23 shows the two basic circular saw types that are available, the sidewinder and the worm drive. The sidewinder—the most common type—uses an in-line, spur-gear motor that sits at a right angle to the face of the blade, and the handle is mounted on the top of the saw. Worm-gear-driven saws use a gearbox to transfer motor motion 90° to the blade and have the handle mounted at the rear of the saw.

Figure 8-22. This hydraulically powered ironworker is capable of many functions needed for forming steel parts. It will shear, notch, punch, and bend many different types of metal.

A Photoexpert/Shutterstock.com   B   TFoxFoto/Shutterstock.com

Figure 8-23. Two saw types are commonly used by building carpenters. A—This sidewinder saw drives the blade directly using a motor oriented perpendicular to the blade. B—The motor of this worm-drive saw is parallel to the blade and transfers energy through a gearbox.

Circular saws typically are powered using standard 120-volt AC current. However, the development of more reliable rechargeable batteries has increased the popularity of cordless models. At one time, a pneumatic-powered circular saw was produced for the military and was capable of cutting timbers while submerged in the water. Another type of circular saw uses gear reduction to slow the rotational speed of its blade for cutting structural steel and metal roofing materials.

Circular Saw Blades

Blades for circular saws vary based on the materials they are intended to cut. Most blades are classified by the size, number of teeth, and material used to construct the blade. See **Figure 8-24**.

Blades for *ripping*, or cutting with the grain, have fewer and larger teeth that remove wood quickly. *Crosscutting* blades are designed to cut across the grain of lumber, as in cutting a board to length. Crosscutting blades have more teeth and shallower gullets. A *gullet* is the space between saw teeth that allows for waste removal. Using a blade with more teeth results in a smoother cut and a slower cutting speed.

Combination blades combine groupings of smaller teeth with periodically spaced deep gullets to achieve acceptable efficiency for both crosscutting and ripping. Plywood blades are set up with a large number of very small teeth to reduce splintering and produce a very smooth cut. Attempting to use a plywood blade on framing lumber usually results in a slow cutting speed, burned lumber at the cut from friction, and possible damage to the blade from excessive heat.

Goodheart-Willcox Publisher

Figure 8-24. Circular saw blades are configured to accommodate the variety of materials they are designed to cut. Note the size, shape, and number of teeth on each of these blades.

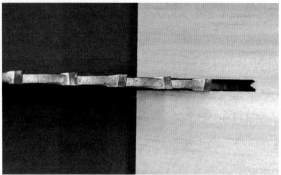

Figure 8-25. The set of the teeth on a saw blade creates a kerf that is wider than the thickness of the blade in order to provide clearance. This reduces friction and allows waste material to be removed as the cutting action progresses.

Saws cut by chiseling away tiny bits of material in the path of the blade. It is important to understand that these tiny bits add up to waste material. The path cut away by a saw is called the kerf, shown in **Figure 8-25**. The width of the kerf is the same as the width of the widest part of the blade. The teeth of a saw are alternately turned slightly to right and left of center to create *blade set*. Blade set ensures that the teeth remove enough material so that the body of the blade has clearance as it passes through the kerf. Too little clearance for the blade allows friction to build up heat in the blade. *Hollow ground blades*, or thin kerf blades, do not use blade set to create clearance. Instead, the tips of the teeth are thicker than the body of the blade, resulting in the removal of less material and reduced waste.

Most circular saw blades are made from steel. High-speed steel is used in higher quality blades to increase durability and maintain blade sharpness longer. Blade teeth can be tipped with sharp inserts made from tungsten

STEM Connection

Saw Blade Designs

Use a caliper to record an outside measurement of the thickness of a saw blade, **Figure A**, and an inside measurement of the kerf it produces, **Figure B**. Subtract the width of the kerf from the thickness of the blade. The difference is the clearance. Take these measurements on a crosscutting blade, a ripping blade, and a combination blade. If other types of saws are available, measure them as well. Determine whether the clearance is different for different types of saw blades.

Figure A

Figure B

carbide or titanium carbide. Carbide tips greatly extend the useful life of a blade, in part by increasing the length of time a blade will retain its sharpness.

Two types of blades, abrasive and diamond impregnated, do not have teeth. They rely instead on the coarseness of the abrasive materials infused on the outer ring of the blade to wear through the work material. Abrasive blades are used for cutting masonry tile and steel, while diamond blades are available for cutting glass, ceramic tile, and concrete.

Table Saws

Table saws are stationary versions of circular saws that are ideal for ripping lumber into narrower boards and for making cuts in plywood. They are sometimes called *tilting arbor saws* because the motor and blade can be moved to change the depth and/or angle of the cut. Based on their intended purpose, most table saws are equipped with larger, more powerful motors than handheld circular saws. However, many modern table saws could be considered to be semi-portable. They are constructed to be lightweight enough to be transported to the job site, yet sturdy enough to be set up as a stationary saw.

The capability to make long, accurate cuts and complex angled cuts has much to do with the table saw's design and operation. The guides and adjustments on most table saws offer more possibilities and much greater precision than those on most handheld circular saws. **Figure 8-26** shows the parts of a typical table saw. Special table saw blades are available for cutting dados (straight grooves) or rabbets (recessed edges) to create strong woodworking joints.

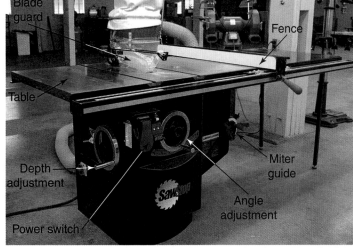

Goodheart-Willcox Publisher

Figure 8-26. A table saw can be used to make a variety of intricate cuts on small or large pieces of work material.

Radial Arm Saws

Radial arm saws are always stationary tools. Invented by Raymond DeWalt, radial arm saws basically consist of a heavy-duty circular saw mounted to a rotating, tilting, yoke-like frame that slides below a horizontal support arm. Radial arm saws are ideally suited to be used as a cut-off saw. They also are capable of rip, miter, and bevel cuts. If equipped with the proper attachments, radial arm saws can also cut dados or form moldings. Due to its versatility, the radial arm saw has been used in the past in place of a table saw for many operations. It also requires clearance on only three sides because it rips lumber from side to side rather than from front to back, as a table saw does.

Radial arm saws are particularly susceptible to kickback and binding in the work. Teeth in blades designed for radial arm saws are arranged at a different angle than the teeth in most circular saw blades; the blades are therefore not interchangeable.

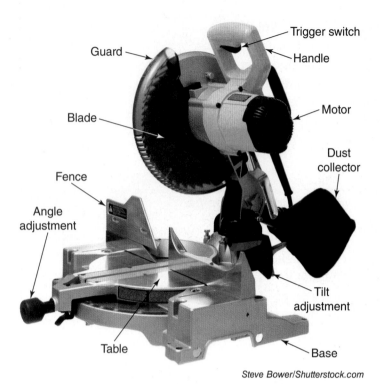

Steve Bower/Shutterstock.com

Figure 8-27. The blade and motor of a miter saw pivot down into the lumber to make the cut. The angle may be adjusted to most common miters needed for framing and finish woodwork. On this compound miter saw, the tilt of the blade can also be changed to make bevel cuts.

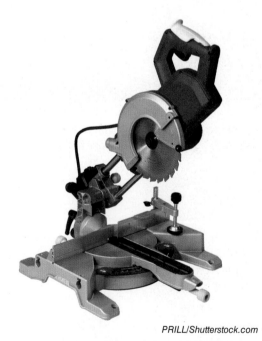

PRILL/Shutterstock.com

Figure 8-28. The sliding mechanism of this sliding compound miter saw enables the tool to make cuts in wider lumber than a standard miter saw.

Today, few options exist to purchase a radial arm saw because safer, more efficient methods of cutting lumber are available. Most of the radial arm saws that are still in use are found in professional woodworking shops.

Miter Saws

Miter saws have taken the place of radial arm saws and portable circular saws for crosscutting lumber on many job sites. Miter saws consist of a circular saw mounted on a pivot that is attached to a stabilizing frame or table. If taken to the job site, the saw may be used on a raised surface such as a portable miter saw stand, a workbench, or a trailer bed. It can also be operated directly on the ground if needed.

Miter saws are available in three common configurations. The simplest version crosscuts lumber at 90° or down to 45° in either direction. A compound miter saw, **Figure 8-27**, cuts all the same angles as the basic version, and its motor and blade can be rotated to cut a bevel in a secondary plane as well. Sliding compound miter saws operate on a short horizontal slide much like the radial arm saw. See **Figure 8-28**. This increases the capacity, and allows compound sliding miter saws to cut wider boards than saws without a slide.

A portable miter saw can be permanently mounted to a stationary stand if necessary. Miter saw stands can be purchased or constructed. Either option should include extensions of the saw base and fence to the left and the right. When set up correctly, the extensions should safely support long lumber as it is being cut and reduce the need for a second person to catch long pieces. Miter stands also include a method for securing a stop, or built-in jig, for cutting multiple pieces to the same length. The combination of a miter saw and jig shown in **Figure 8-29** eliminates the need to measure and mark every piece, potentially saving huge amounts of time.

Band Saws

A *band saw* uses a continuous belt-like metal blade that is driven around at least two pulleys to cut material. The pulleys are wrapped with soft, rubber-like bands called *tires* that grip the blade without damaging the teeth. Stationary band saws may be vertical, horizontal, or convertible between the two positions.

Horizontal band saws that use a liquid cooling system, often called *wet saws*, are commonly used as cut-off saws for cutting long pieces of metal into shorter pieces. See **Figure 8-30**. Handheld, portable versions are useful for fieldwork.

Vertical band saws can be used for cutting many different types of material. Many commercial lumber mills use large band saws to reduce timber to boards. The blade on a band saw can be thinner than a traditional circular mill saw blade capable of cutting materials of the same thickness. The result is a thinner kerf and less waste. Butchers use a specialized type of vertical band saw for cutting frozen meat.

The bigger a band saw is, the longer the blade it requires. However, a band saw's size is not determined by the length of its blade, but rather by the size of the material it will cut. Size is measured at the throat, or the distance from the blade to the machine frame. **Figure 8-31** shows the parts of a band saw.

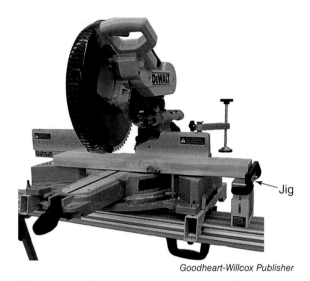

Goodheart-Willcox Publisher

Figure 8-29. This miter saw stand is equipped with an adjustable stop, or jig, that makes it easy to cut multiple pieces to the same length without measuring each piece.

Garsya/Shutterstock.com

Figure 8-30. A horizontal band saw can be adjusted to automatically control the pressure and feed rate applied to the blade. This allows the same saw to be used for cutting a variety of metal objects.

Band Saw Blades

Band saw blades vary in composition, length, thickness, width, tooth style, and teeth per inch (TPI). It is important not only to select a blade that will fit a particular saw but to also select the blade with the proper characteristics for cutting the selected material.

The regular tooth configuration, which is useful for thin cuts in most materials, has each tooth beginning at the base of the last and a very slight 0° to 1° positive or forward rake angle. The *rake angle* is the number of degrees off perpendicular to the back of the blade each tooth is angled. For wood, plastics, and soft metals, a *skip tooth configuration* may work better because the extended gullet allows more room for chips to clear from the kerf. A *hook tooth configuration* may be used with thicker wood, plastics, and some metals because it uses an aggressive 10° positive rake angle to make faster cuts. **Figure 8-32A** shows the three blade configurations and blade terms.

The number of teeth per inch (TPI) on a blade not only helps to determine the size and type of material a band saw cuts, it also influences the speed at which the blade operates and the feed rate, or how fast the blade cuts. Common TPI counts for band saws range from 2 to 32 TPI. Blades with coarse teeth (2 to 6 TPI) are suited for cutting thick wood, and blades with fine teeth (18 to 32 TPI) should be selected for cutting thinner materials, plastics, and metals. The coarser the teeth, the faster a blade can cut. Finer teeth require a reduced feed rate and produce a smoother finish than a blade with a low TPI.

When deciding how many teeth per inch are required for a given job, the accepted rule is that at least three teeth must be engaged in the work while the band saw is cutting. This practice is done for maximum efficiency and to prevent damage to the blade. In general, harder materials require a higher TPI to distribute the cutting pressure to a greater number of teeth in order to reduce blade wear.

There are generally three types of blade set found on band saw blades: alternate set, raker set, and wavy set. See **Figure 8-32B**. On blades with alternate set, each successive tooth is alternately

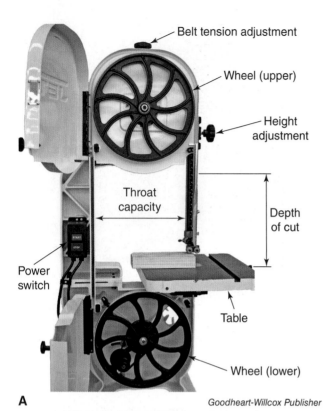

A

Goodheart-Willcox Publisher

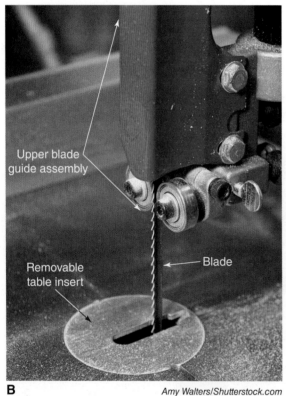

B

Amy Walters/Shutterstock.com

Figure 8-31. Vertical band saws are found in many different applications. Different blade styles and speeds are used based on the material being cut. A—Study the parts of a band saw. B—A close-up of the band saw blade and upper blade guide assemly.

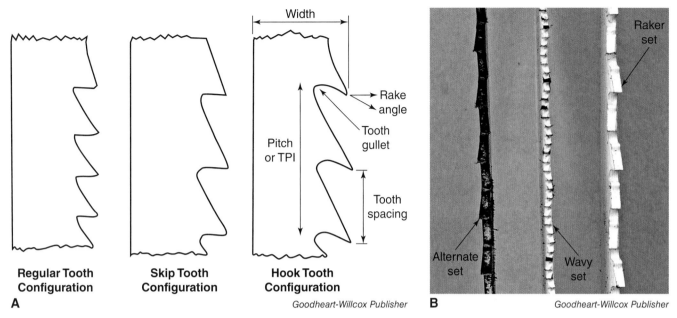

Figure 8-32. A—Study the different parts and configurations of band saw blades. Most of the blade terms apply to all saw blades. B—A decision on which of the three basic types of blade set to use to cut a material requires knowledge of the material characteristics and how each set cuts.

angled to the left and right of the center of the blade. These blades produce fast, smooth cuts in wood. Instead of angling each tooth, blades with wavy set have alternating groups of two or three teeth angled to either side. Wavy set is restricted to fine-tooth blades and is used for cutting thin pipes, tubing, and sheet material. Blades with raker set have groups consisting of one set of teeth to the right, one set to the left, and a raker tooth (no set). Raker set is used for cutting thick metal and dense woods.

Reciprocating Saws

Several types of saws use a reciprocating motion to cut through a variety of materials. Handheld *reciprocating saws* operate by mechanically moving a blade in a fast back-and-forth motion similar to the motion of a handsaw. See **Figure 8-33**. Reciprocating saws can be powered by standard AC electric power, a DC generator, or rechargeable batteries. These saws are commonly used during demolition and in other situations where a less precise or rougher cut is acceptable.

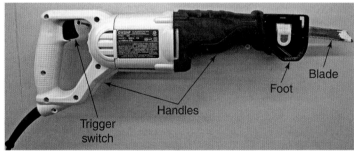

Figure 8-33. Reciprocating saws are demolition tools that provide convenient, aggressive cutting power. They are not designed for finesse operations.

Special setups are available for jobs such as pruning trees, cutting pipe, and emergency rescue operations.

Just like the blades of hand-operated saws, reciprocating saw blades are designed to cut in only one direction. Most cut on the pulling stroke. Only one end of the blade is supported by the tool, so blade construction must be relatively stronger, thicker, and more rigid than band saw blades.

Figure 8-34. When using any reciprocating saw, you should always ensure that the blade is long enough to extend fully through the stock in both the extended and the retracted position. A blade that is too short can bind and violently force the saw backward.

Blades are available for cutting wood, metals, and other building materials. Reciprocating saw blades can be purchased in lengths from 3"-12". When selecting a blade, you need to choose one that will easily stay fully engaged beyond the depth of the work material, as shown in **Figure 8-34**. A blade that is too short can snag and cause the saw to push away from the cut. Excessive blade length can make contact with undesired objects on the opposite side of the cut. The unsupported end of the blade will also have a tendency to flex or wobble.

Blades for reciprocating saws can be made out of high-carbon steel (the least expensive and most common), high-speed steel (required for harder materials), or a bimetal bond of carbon steel and HSS. *Bimetal blades* combine the toughness of carbon steel with the hardness of HSS, resulting in a blade that, although more expensive, outperforms plain carbon steel blades and lasts approximately 10 times longer.

Tool manufacturers use different mechanisms to convert the rotary motion output of an electric motor to the linear motion of the reciprocating blade. Regardless of the mechanism used, the result in a handheld tool is the potential for strong vibration that must be kept under control. Some reciprocating saws are equipped with a variable-speed option. The higher speeds are for cutting soft wood, and slower speeds should be used when cutting harder, denser materials. To maintain control when using a reciprocating saw, keep the foot of the saw in firm contact with the surface of the work. Failure to maintain foot contact with the material results in the blade catching in the kerf, which causes the reciprocating motion to be transferred back to the user.

Jig Saws

Jig saws are moderately powered handheld saws that use a reciprocating blade action similar to reciprocating saws. See **Figure 8-35**. The base of a jig saw is bigger than the foot of a reciprocating saw, and jig saws use smaller, more delicate blades. Jig saws are used to make precise, often curved cuts in wood, metal, plastics, and other building materials. Many jig saws are equipped with adjustments that allow the base to be tilted for bevel cuts. Some have variable-speed motors to allow for better quality cutting in a large variety of materials. Jig saws may be powered by standard 120-volt AC current, or they may be cordless versions powered by rechargeable batteries.

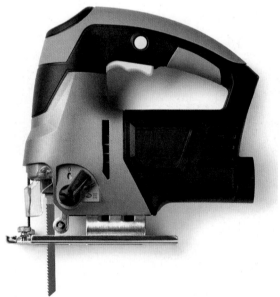

Figure 8-35. The size and design of a jig saw allow it to be used in many situations and for multiple types of cuts.

Jig Saw Blades

Most jig saw blades are only 2″ to 3 1/2″ long. This is adequate because the stroke of the blade is only 1/2″ to 1″ long. Blades are marketed according to the material they are designed to cut. However, when choosing which blade to use for a particular job, it is helpful to consider several factors. Carbon-steel blades are the most widely available and are adequate for soft woods and plastics. Stepping up to high-speed steel blades is necessary for cutting soft metals such as aluminum or brass; also, HSS blades last nearly five times longer than carbon-steel blades. For ferrous metals or hard abrasive materials, a bimetal blade cuts more efficiently and has a longer cutting life than softer blades.

High TPI counts can cut harder materials and produce smoother but slower cuts than rapid-cutting blades with low TPI counts. The way the teeth of jig saw blades are sharpened and set also affects their cutting ability.

Teeth that are formed sharp in the initial blade-making process (milled) are designed for quick but rough cuts. Milled teeth usually outlast ground teeth when used in similar materials. See **Figure 8-36**. Ground teeth go through a secondary sharpening process during blade making, which produces sharper teeth that are capable of fast, smooth cuts in wood. Some blades combine ground teeth with a tapered back to give a clean cut with a reduced kerf width.

Most jig saw blades are designed with the teeth cutting toward the base of the saw on the pulling stroke. Blades designed to cut on the downward stroke are used to cut some plywood and laminates. This design ensures that the cleanest side of the cut is on the top of the material. Special blades with abrasives bonded to the cutting edge are designed for cutting materials such as concrete board, glass, and tile.

Figure 8-36. Most jig saw blades are designed to cut on the pulling stroke like blades A and B. Blade C cuts on the pushing stroke and is used for cutting veneers used in cabinetmaking. Blades A and B have milled teeth, while blade C has teeth that were formed by grinding.

Scroll Saws

Scroll saws use a reciprocating blade action similar to reciprocating saws and jig saws. The difference is that scroll saws are stationary tools, and their blades are secured at both ends, **Figure 8-37**. Extremely intricate cuts—much finer than possible with a band saw or jig saw—are possible in wood, plastics, and soft metals.

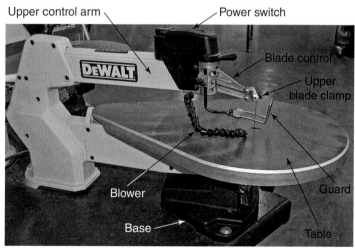

Figure 8-37. A typical scroll saw can make delicate and intricate cuts.

For a stationary power tool, scroll saws are not very powerful and are fairly safe to operate. The blade's stroke length is usually less than 1″ and the teeth, although sharp, are small. They can still cut fingers, but the potential damage is far less than with other, more powerful saws.

Scroll saws, unlike band saws, are capable of center cuts that do not open to the edge of the work. By piercing a hole with a drill bit first, then threading the blade through the workpiece before mounting the blade to the saw, the user can make a cut from the center of an object.

Scroll Saw Blades

Although a few light-duty saws use shorter ones, most scroll saws use a 5″ blade. Blade widths vary from almost 1/4″ to less than 1/16″. Blades have either plain ends that are clamped into the blade holders or pin ends with tiny round cross pins that fit into slots on the blade holders. The plain-ended type of blade is the most common.

Tooth arrangements vary greatly to accommodate use in many applications. The general rules for selecting saw blades also apply to scroll saws. More teeth per inch create a smoother cut and cut through denser materials than coarse blades. Coarse blades remove waste material better as the saw cuts. Small, fine blades remove less material, leave a narrower kerf, and are better for tight turns and intricate curves. Twisted, or spiral blades, are designed with cutting edges on all sides to enable easy, sharp turns with the blade.

Oscillating Multi-Tools

Oscillating multi-tools are portable power tools designed with a mechanism that rapidly vibrates a blade or other attachment back and forth in a narrow arc. The entire motion is only a shift of 3° to 4°, but it vibrates at up to 21,000 strokes per minute.

Attachments are mounted to the front of the tool with the cutting edge or working surface extended out to the front. The result is a tool with the ability to reach into tight places or make cuts flush with an existing flat surface. Attachments are available to cut wood, steel, tile, and other materials. Additional attachments allow completion of other jobs, such as scraping or smoothing surfaces. The tool and some of its various accessories are displayed in **Figure 8-38**. It can be powered through a traditional electric cord or it may be cordless. When used in the proper situations, oscillating multi-tools are both versatile and efficient.

Oleksandr Chub/Shutterstock.com

Figure 8-38. The unique configuration of oscillating multi-tools allows them to reach into places where other power tools cannot. Attachments for the tool provide flexibility to cut, scrape, dig, and sand.

Routers

The basic *router* is a portable woodworking tool consisting of a powerful electric motor, a rotary cutting tool, and an adjustable base plate to control cutting depth. See **Figure 8-39**. Portable routers are versatile tools, capable of performing jobs traditionally completed by a variety of other, less efficient hand and power tools. Routers are used for shaping the edges of lumber and other materials, cutting grooves or slots, creating rabbets or dados, and carving designs into the surface of wood.

The three main types of routers are fixed base, plunge, and trim routers. "Fixed base" is a slight misnomer because the base can be adjusted, but it remains fixed in position while in use. Plunge routers allow the bit to be smoothly lowered into the work surface for cutting that does not extend to the edge of the board. Trim routers are smaller machines well suited to trimming the edges of laminate countertops or working in areas with limited clearance.

Router Bits

Bits for routers range from large bits 3 1/2" in diameter to tiny bits less than 1/16" in diameter. Routers operate at speeds from 12,000 rpm to 24,000 rpm. Higher rpm speeds limit the size of bit that may be used safely.

Unlike drill bits, which are designed to cut down and bore a hole, router bits are designed to cut sideways. The *flutes* of a router bit are its cutting edges. Bits are available in 1/4" and 1/2" shanks and are attached to routers using a collet-type chuck. Collets are interchangeable on some routers, while others only accept a single size shank, usually 1/4". Router bits with 1/2" shanks are more stable and produce less vibration. Bits with larger shanks, or otherwise increased mass, dissipate heat better than bits of lighter construction. Reducing heat buildup prolongs the sharpness and longevity of the bit.

The profile, or shape, of a router bit determines the shape of the wood remaining at the edge of the cut. A bit with straight flutes will leave a straight-walled cut surface. A convex shape created on the edge of a cut is done using a bit with concave flutes. See **Figure 8-40**. Some bits are equipped with pilot bearings that guide the bit evenly along the edge of a surface to ensure that the cut is uniform in depth for the entire length of the work.

One of the most useful bits for agricultural purposes is a simple roundover bit. It is used to produce safe, rounded corners on the edges of lumber used in the construction of feeders and other livestock equipment and facilities.

Robert Hale/Shutterstock.com

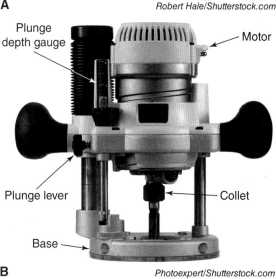
Photoexpert/Shutterstock.com

Figure 8-39. A—A fixed base router. B—A router equipped with a plunge attachment.

Jeffrey B. Banke/Shutterstock.com

Figure 8-40. A few of the most common edge-forming router bits. The shape of a router bit is the reverse of the material it will remove. Note the pilot bearing used on each bit.

Like other bits and cutting attachments, router bits are available in several choices of material. The least expensive bits are constructed of high-speed steel. HSS bits are adequate for softer woods, but they dull quickly, especially when used on harder woods or on knots in soft woods. Carbide-tipped bits stay sharper and last longer than HSS bits. Solid carbide bits are the most expensive, but are engineered to last the longest.

Router bits need to be extremely sharp for clean, efficient cutting. When manufactured, router bits are balanced and sharp. Only some router bits may be re-sharpened; sharpening a router bit is a job for a professional because it requires specialized equipment and skill to maintain the balance.

Router bits should be stored in a manner that keeps them clean, dry, and positioned so the cutting edges are not damaged. Today, many bits are sold in their own individual plastic box or within a boxed set. A traditional alternative storing method is to bore holes through the top of a block of wood to hold the shanks of the bits. The holes in the holder need to be spaced adequately to keep the bits from touching.

Router Tables

The torque a portable router produces can easily cause the user to momentarily lose control of the tool. When performing edge work, using a router table may be a good choice. See **Figure 8-41**.

A router table is essentially a large accessory that turns a portable router into a stationary machine. In this design, a router is secured to the bottom of the table with the bit protruding through the top. The workpiece is brought to the machine instead of manipulating the router around the workpiece. Commercial tables are equipped with an adjustable fence and a groove for a miter guide to provide extra stability. Using a router table is an effective way to shape the edges of numerous project materials, such as top rails for a board fence or custom trim work in a home.

Steven Belanger/Shutterstock.com
Figure 8-41. When mounted to a router table, the router bit is the only part that extends through the tabletop. Keeping lumber pressed flat to the top of the table controls the depth of the cut.

Planers and Jointers

Planers are used to shave layers of wood from lumber to smooth surfaces, remove saw marks, or adjust the thickness. See **Figure 8-42**. Planers are used to surface rough-cut lumber and to trim it to the desired thickness. Stationary models shave the entire width and length of a board, removing no more than 1/16" at a pass. Handheld portable planers are limited to cutting widths of about 3" to 6 1/2".

Planers work by passing the board under a rotary cutter with two or three very sharp blades that uniformly slice away a layer of the lumber. The result is finished lumber with the same thickness throughout

Dan70/Shutterstock.com
Figure 8-42. Power planers use a very sharp rotating blade to shave away thin layers of wood from the surface of lumber.

and a uniform surface finish. A planer will keep both sides of a board parallel, but it will not remove a bend or correct a warp.

Jointers use rotating blades to straighten and finish the edges of lumber for applications in which boards are fitted side by side. See **Figure 8-43**. Jointers produce very flat surfaces, with bends and warps removed. A jointer can be used to do most of what a planer does, except that it is not as efficient at planing to uniform thickness. Using the fence on a jointer can produce an edge at a perfect 90° angle or produce a precise mitered angle along the edge of a board.

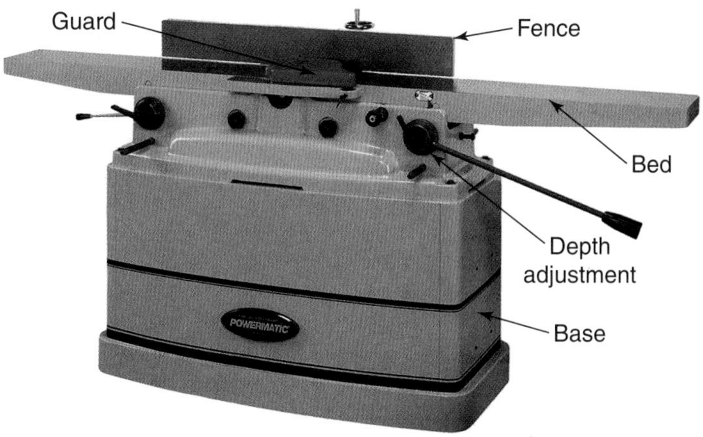

Goodheart-Willcox Publisher

Figure 8-43. Jointers are similar to planers, but they are configured to shape and straighten the edges of lumber for tighter fits during construction.

Planer and Jointer Rules

Follow the power tool safety rules and observe these rules to safely operate a planer or jointer:
- Do not plane boards with serious splits or loose knots. Planing through lumber coated with a finish can damage the blades.
- Feed lumber so that the blades cut with the grain. Cutting against the grain causes tearing and splintering of the wood.
- Remove small amounts of material with multiple passes instead of a large amount with one pass. This better preserves the condition of the equipment and produces smoother finishes.
- Keep your hands and fingers away from the rotating blades. Use a push stick to manipulate smaller pieces.
- Handle boards from the side. Do not stand directly in the line of feed.
- Keep the board in constant contact with the table. If you are using the fence, the board must also remain in constant contact with the fence.

Grinders and Sanders

Power grinders and sanders are machines that use consumable abrasives to shape or condition the surfaces of stock materials. *Abrasives* are hard, sharp materials used to shape, clean, or finish a workpiece. They are termed *consumable* because they wear away during use and must be replaced regularly. Grinders are used primarily with metals, whereas sanders are typically used with wood materials. These machines may be handheld or stationary.

Grinder and Sander Rules

When using any type of grinder or sander, observe the power tool safety rules described earlier and these general safety precautions:
- Always operate grinders and sanders in well-ventilated areas.
- Allow the tool to reach operating speed before contacting the work material.

- For portable tools, apply only light pressure and let the tool do the work. Keep the tool in constant motion to reduce digging in/oversanding and heat buildup.
- For stationary tools, ensure that the machine is mounted in a secure position. Apply even pressure and move the work material continuously across the face of the stone/sander.

Grinders

Grinders are machines that use high-speed rotation and an abrasive to smooth and shape metal. They are also used to remove small amounts of metal from surfaces and edges and to polish metal surfaces. Portable grinders include angle grinders and die grinders. Stationary models include bench grinders, pedestal grinders, and surface grinders. The type of grinder chosen depends on the materials and the type of job.

Grinder Abrasives

Grinders use abrasive materials that are formed by bonding adhesive particles together to create solid, usually nonflexible, synthetic stones. These materials are sometimes reinforced with fibrous fillers that strengthen the stone and extend the working life of the abrasive consumables. New layers of abrasives are exposed as the wheel or stone is worn away from use.

Angle Grinders

Angle grinders are essential handheld tools for grinding and shaping metal parts to be welded. They are also used for rough sharpening operations and for cleaning parts. Angle grinders, also called disc grinders, can be fitted with an assortment of grinding stones, wire brushes, sanding attachments, and buffing wheels to meet any number of cleaning and polishing needs, **Figure 8-44**. The disc is typically held at an angle of 15° to 30° to the surface of the material.

KKulikov/Shutterstock.com

Figure 8-44. A typical pneumatic angle grinder. Many different sizes and types of attachments are available for disc grinders. Always match the correct diameter, arbor size, and rated rpm to the power tool.

In addition to the power tool safety rules and the rules specific to grinders and sanders, observe the rules listed here when working with angle grinders:
- Inspect the disc or attachment for defects before starting the machine.
- Always use both hands to hold the angle grinder by the provided handles.
- Ensure that sparks and debris are directed in a safe direction and away from others.
- Always engage the material by gently touching the grinder to the work.
- Avoid free spinning unnecessarily.
- Be cautious of kickback if the grinder is allowed to bind. Binding may occur when attempting to grind inside of a hole or depression or when excessive pressure is applied.

- Keep the tool in constant motion to reduce digging in and heat buildup.
- The outside perimeter of the attachment spins at a faster rpm than areas toward the center. The result is that the most effective grinding, or fastest removal of material, occurs at the edge. Holding the disc flat on a surface results in less material removal and a flatter surface texture, but it typically leaves swirl marks on the surface.
- Wait for the disc to stop spinning when finished with the grinder, and place the machine on the work table with the disc facing up.

Bench Grinders

Bench grinders are stationary tools mounted to a workbench or table, **Figure 8-45**. Pedestal grinders are usually mounted to a stand called a pedestal that is securely affixed to the floor. Bench grinders are useful for shaping, cleaning, and polishing small metal parts. They may also be used to sharpen and recondition tools.

Bench and pedestal grinders are often set up with a coarse abrasive wheel on one side (for rapid removal of material), and a finer abrasive wheel (for finishing) on the other side. They may also be equipped with wire brushes for cleaning parts, or with soft buffing wheels for polishing. In most applications, the grinding stones are enclosed, with only the actual working area exposed. Many bench and pedestal grinders have built-in containers for water that is used to keep parts cool as they are being shaped. This is particularly important when sharpening tools to maintain the edge-holding characteristics of tool steel. Allowing tool steel to change color due to heat generated by the grinding process causes a change in the molecular structure of the steel and results in a loss of hardness.

The following rules and the power tool safety precautions apply to working with bench grinders:

- Adjust eye shields for a clear view of the grinding area.

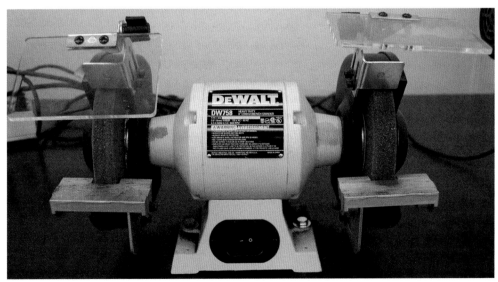

Goodheart-Willcox Publisher

Figure 8-45. Some stationary grinders are designed to be mounted directly to a workbench. Other models are mounted on a dedicated pedestal.

212 Agricultural Mechanics and Technology Systems

Goodheart-Willcox Publisher

Figure 8-46. Keep the tool rest adjusted to within 1/8″ of the grinding stone.

- Check the tool rests for correct spacing (≤1/8″ clearance between stone and rest) before starting the grinder, **Figure 8-46**.
- Turn the grinder on while standing to one side.
- Use the tool rest to support the part during grinding. It may be adjusted for grinding angles as long as the standard ≤1/8″ clearance between stone and tool rest is maintained.
- Keep the part as close to alignment with the radius of the stone as possible.
- Apply even pressure and move the piece continuously across the face of the stone; never grind on the side of the stone.
- Keep the part cool by dipping it into water while grinding.
- Pliers may be used to hold pieces that are too small to be held by hand.

Die Grinders

Die grinders are pneumatic tools used for finish work on small parts and for shaping surfaces with intricate contours that fit closely with other parts. Grinding attachments, which consist of small grinding stones, sanding drums, or tungsten carbide or high-speed steel burrs, fit into the tool and typically operate at high speeds. Often the tips are shaped to reach into tight spaces. See **Figure 8-47**. Veterinarians use a type of die grinder to repair teeth of domesticated animals.

The extreme speed at which die grinders often operate restricts the diameter of attachments that should be used. Attempting to use attachments intended for slower-speed machines can create a hazardous situation or result in failure of the tool. Unlike using a grinding stone, using a carbide burr on steel creates very sharp shards of waste material, so extra eye protection and protective clothing are needed.

Stationary Surface Grinders

Stationary surface grinders are expensive stationary tools mounted to a bench or table. They use high-grade grinding stones to remove material from large surfaces that must be ground to near perfect flatness. Surface grinders are most often found

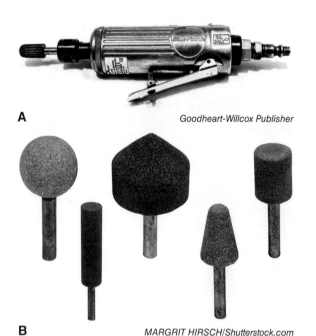

MARGRIT HIRSCH/Shutterstock.com

Figure 8-47. A—A typical pneumatic die grinder. B—The size and shape of many die grinder bits allows them to reach into places that other grinders cannot access.

Copyright Goodheart-Willcox Co., Inc.

in machine shops or shops that do a great deal of precision automotive or engine work. Engines undergoing comprehensive repair and rebuilding have parts that require this type of grinding.

Sanders

Sanders are machines that use abrasives to smooth rough wood surfaces and to remove small amounts of a wood surface. Sanders may also be used in metalworking applications. In fact, many grinders can be set up with sanding attachments. Like grinders, sanders may be portable or stationary.

Portable sanders include disk sanders, belt sanders, sheet sanders, random orbital sanders, detail sanders, and oscillating multi-tools. Stationary sanding machines include several types of belt sanders, disk sanders, oscillating spindle sanders and various combinations of these examples. Sanders operate by many different methods. Some spin around a single axis like grinders, while others vibrate, oscillate, orbit, or rotate around multiple axes.

Sander Abrasives

Sanders use abrasive particles that are bonded to flexible backings of reinforced paper, cloth, or plastics, **Figure 8-48**. These backing materials are held in place on the sander by a variety of methods. Belts and drums are held by friction, while some sanders use clips to hold sanding pads in place. Others rely on adhesive or a hook-and-loop system to fasten pads to the sander. As a sanding consumable is used, the effectiveness of the abrasive degrades as the abrasive is worn away from the backing material.

Goodheart-Willcox Publisher

Figure 8-48. Sanding belts are commonly a cloth-based belt with abrasive material bonded to one side.

Disk Sanders

Although similar to angle grinders, disk sanders usually operate at slower speeds than angle grinders and do not necessarily come equipped with the same type of guards. As stated earlier, angle grinders are typically used with the disk at an angle of 15° to 30° to the surface of the material, whereas disk sanders are often held flat to sand or polish large surface areas. Disk sanders are aggressive and capable of removing large amounts of material quickly. They have a tendency, however, to leave behind distinctive circular marks or striations that may require additional sanding. A disk sander might be used to rough-shape wood or as a finishing tool on metal surfaces, such as refinishing the hood of a tractor.

Portable Belt Sanders

Belt sanders use a paper or cloth belt coated with abrasive that wraps around a pair of rotating drums. The resulting motion is a sanding action that moves in a straight line as opposed to the circular motion associated with other types of sanders. A belt sander is very efficient for quickly removing considerable amounts of stock from the face of a piece of wood. Often a belt sander is the first step in smoothing a surface. A random orbital sander can then be used to smooth out the surface (and remove the striations, or grooves, created by the belt sander) before finishing.

When using a belt sander to smooth a surface, begin by turning on the sander and allowing the motor to achieve full speed. Next, using two hands (one on the trigger handle and the other on the forward handle for stability), ease the sander to the surface with a slight forward motion. The rear roller should contact the wood first, followed by the rest of the sanding face pressing onto the wood with a slight rocking motion.

Belt sanders have a substantial amount of torque, which causes the belt to tend to grab the wood and drive the sander forward. Work the sander with the grain of the wood as much as possible, and do not allow it to stay in one place. Remember that a belt sander removes material quickly, so stop and check the progress frequently. It is much better to have patience and allow the weight of the sander to do the work rather than pushing down too hard and risking damage to the work material that needs to be repaired later.

Before changing the belt on a belt sander, read the manufacturer's instructions. Typically, a belt sander has a lever on one side that releases the tension between the front and back rollers. Once the tension is released, slide the sanding belt off of the rollers. To install a new belt, center a new belt onto the rollers, close the lever, and plug in the sander. With the sander running, use small adjustments to the tracking control knob to align the belt correctly on both rollers. As a new belt wears in, it will stretch and require tracking adjustments to keep it aligned on the roller drums.

Sheet Sanders and Random Orbital Sanders

Sheet sanders are designed to work with rectangular, partial sheets of standard sandpaper. Random orbital sanders, on the other hand, require round disks, typically attached to the sander base with either adhesive or hook-and-loop fasteners. See **Figure 8-49A and B**. The ability to use standard sandpaper is a major advantage for sheet sanders, since sheets of sandpaper are readily available at most hardware stores or home improvement centers. The sheets are cut into halves or quarters and are then securely clamped

A   VladaKela/Shutterstock.com   B   Alexander Ishchenko/Shutterstock.com   C   Levent Konuk/Shutterstock.com

Figure 8-49. Select the appropriate type of hand sander for the task and the material being finished. A—Random orbital sander. B—Sheet sander. C—Detail sander.

onto the flat base of the sander. Sheet sanders have a motor that moves the sanding pad in a circular motion at high speed.

The motion of a random orbital sander is similar, except that the random orbit mechanism has an offset drive bearing that also moves the pad in an elliptical pattern. When these two motions are combined, the pad of the random orbital sander moves in a randomized manner. This reduces the tiny circular swirl marks that a standard sheet sander tends to leave behind.

To combat the swirls left behind by a sheet sander (particularly when moving against the grain), the operator must keep the sander moving, working as much with the grain as possible. Using progressively finer grits of sandpaper with each step of the sanding process will remove the swirls left by the previous sanding. Final sanding by hand may be required for really fine finishes.

Detail Sanders

Some surfaces are too small or are located behind obstructions that prevent the use of sheet or random orbital sanders. Detail sanders that are designed for use on wood have a triangular sanding pad that is smaller than the standard 1/4 sheet sanders, **Figure 8-49C**. These sanding pads have an orbital drive system like sheet sanders, and the sandpaper is also attached using a hook-and-loop system or adhesive. Another type of detail sander, often used in auto body work, is actually a small disk sander that uses a 2" diameter or less sanding attachment. When using these tiny disk sanders, keep the tool moving; allowing it to remain in one position too long can burn the surface from the concentrated friction.

Stationary Sanders

Stationary sanders work like portable sanders except the workpiece is brought to the larger, more powerful machine. Most stationary sanders include a support table that functions much the same as the tool rest on the bench grinder, **Figure 8-50**. It can be adjusted for producing angles as long as the clearance between the sanding surface and the table is kept to a minimum of 1/8". Care must be taken to hold the work in a manner in which the action of the sander draws the work toward the support table and not toward the operator.

Abrasives

Many different materials are used as abrasives, including naturally occurring materials such as flint and garnet, and other substances such as aluminum oxide, silicon carbide, ceramic, diamond, and cubic boron nitride. It is interesting to note that there is no sand in sandpaper. The synthetic materials used are more consistent in grain size and last longer than naturally occurring sand.

Andresr/Shutterstock.com

Figure 8-50. This stationary disc sander is equipped with a table that supports the workpiece while it is being sanded.

Material	Mohs Hardness
Diamond	10
Boron carbide	9–10
Silicon carbide	9–10
Aluminum oxide	9
Tungsten carbide	9
Zirconia	8
Emery	7–9
Steel	4–8
Glass	7
Silica sand	6–7
Pumice	6
Titanium dioxide	5.5
Aluminum	2.5–3
Gold	2.5–3
Steatite (Soapstone)	1–2.5

Goodheart-Willcox Publisher

Figure 8-51. The Mohs hardness scale is a measurement of a material's relative hardness and resistance to scratching. It was developed to compare different minerals and is the typical comparison method for categorizing abrasives.

The most important factor in selecting an abrasive type is its hardness. Hardness is the resistance of a material to scratching or deformation. Abrasives must be harder than the material to be shaped or finished. If not, the work material quickly wears away the abrasive. See **Figure 8-51**.

The second important factor to consider is the size of particles in the abrasives. The grain size indicates the coarseness of the abrasive side of the paper. The grain size ranges from very coarse (#8) to extremely fine (#1200). A higher grit number indicates a smaller abrasive grain and a finer abrasive product, **Figure 8-52**. Abrasives of lower grades have larger particles that produce a coarse cut and remove material at a faster rate than those of a finer grade. Coarse abrasives leave more noticeable marks or lines on the finished surface that may have to be removed, but they remove more material while creating less friction and less heat buildup in the stock.

Friction and Heat Buildup

Grinders and sanders create friction as they work. This friction results in a buildup of heat that can damage the finished material being shaped. Friction can be reduced by selecting the largest grit size that can be used effectively for the job and by using slower sanding speeds. Following the general rule of letting the weight of the machine or part do the work can also greatly reduce the amount of friction generated.

When it is an option, spreading the abrasive work over a larger surface allows the heat generated by friction to dissipate faster. For example, if a stationary belt sander with a 48″ × 6″ belt is used instead of a 21″ × 3″ portable belt sander, the heat is spread out over four and a half times the surface area. When using a bench grinder to shape small metal parts, excess heat can be removed from the work by periodically dipping the part in a container of cool water.

Lodging

As a grinder or sander is used, the abrasive particles are broken down with particles of the material being worked and may become lodged between the grains of the abrasive. The result, called *lodging*, reduces the effective cutting action of the abrasive and increases the heat buildup from friction.

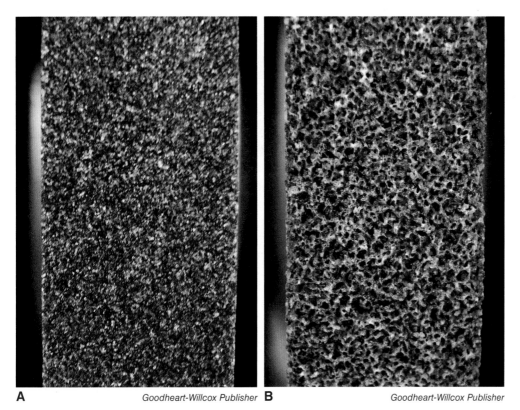

A *Goodheart-Willcox Publisher* **B** *Goodheart-Willcox Publisher*

Figure 8-52. The size of the grit particles that make up an abrasive affect the speed at which it is capable of removing material. The fine stone (A) removes material much more slowly than the coarse stone (B).

The solution is to use a tool called a *wheel dresser* to remove the outer layer of worn, contaminated grit and restore the abrasive cutting surface, **Figure 8-53**.

Wheel dressers are available in two basic types. The first uses a bonded diamond or similar very hard abrasive on the end of a stick or rod. The second type of dresser has a row of free-spinning, star-shaped hardened wheels. Both types are run on the surface of a grinding stone to dislodge contaminants and worn abrasives. The use of a dressing wheel or dressing stick also helps true the stone, keeping the stone surface perfectly round and balanced.

Selecting and Purchasing Power Tools

The job at hand, power supply, versatility, quality, investment, and availability of parts and service are all factors that must be considered when selecting the proper power tools. Is the job a one-time occurrence, or is it an ongoing or repeated task? Are there different tools that may perform the same job? Some jobs require very specific tools. For other jobs, multipurpose tools may be an option. However, as a rule, multipurpose tools are usually not as efficient as their more specialized counterparts. Tools designed for a single task can also be expected to have a longer useful life than tools designed as a compromise among several tasks.

Goodheart-Willcox Publisher

Figure 8-53. A grinding wheel dresser can be used to correct both lodging and grooves in the face of the grinding stone.

The available source of power at a job site may influence the selection of power tools. For jobs where AC power is readily available, corded power tools usually offer the most power for the investment. If an air compressor is already available or if you will be working in an area where flammable gases may be present, pneumatic tools could be a prudent choice. Many jobs are suited to the use of cordless power tools. The convenient utility of cordless tools is undeniable, and this convenience can drive the decision about selecting a power source. If cordless tools are selected, it is wise to select tools that use the same batteries.

Availability of parts is important when using power tools that have attachments that require frequent replacement. Brand-name tools are more likely to accept attachments that are standard or universal. Tools that use only attachments that fit a single brand or model should be avoided unless a ready supply is available.

There is more pressure on companies who manufacture and sell name brand tools to offer uniform quality tools and service because they have a national reputation to maintain. Bargain tools may not be the best deal. On the other hand, the quality difference between very expensive power tools and their moderately priced competition may not be worth the added investment if the tool is to have only limited use. Carefully weigh the economic factors when purchasing tools.

Safety must always be a major consideration. All power tools sold in the United States are required by law to be manufactured according to safety standards set forth by the Occupational Safety and Health Administration (OSHA). The American National Standards Institute (ANSI) and the Power Tool Institute also help to develop and maintain standards of safety in the manufacture of power tools. Selecting the correct tool—one that was designed for the job—goes a long way toward maintaining safety. Again, purchasing name brand tools ensures that safety concerns were addressed by the manufacturer.

When purchasing previously used tools, you as the buyer ultimately have the responsibility of safety on your shoulders. In these situations, do your best research and exercise good judgment.

Care and Storage of Power Tools

Power tools are an important investment. Tools in good condition are safer and more efficient to use. It just makes good sense to follow a few simple rules to ensure that they remain in optimum condition:

- Keep power tools clean. Wipe away dust with brush or a rag. Some tools are better cleaned with compressed air or by using a shop vacuum. Do not use compressed air without adequate eye protection or in situations in which debris may be blown into electric motors or into electric or electronic components.
- Lubricate power tools that require lubrication on a regular schedule.
- Neatly wrap or roll up power cords for storage. This is an ideal time to inspect for damage. Avoid storing power cords wrapped around sharp edges or bent sharply at the attachment point to the body of the tool.
- Store power tools where they will be protected from the elements or theft. Power tools should not be allowed to bounce against other tools in storage or transport. Many portable power tools are sold in durable molded plastic or metal boxes. Such boxes are excellent storage solutions.
- Store attachments, batteries, bits, blades, and other accessories with the tools they fit.

CHAPTER 8 Review and Assessment

Chapter Summary

- Power tools were developed as faster, more efficient options for performing tasks traditionally performed using hand tools.
- Power tools may be classified in several ways, such as according to size, source of power, type of work performed, or method of operation.
- Safely using power tools requires a person to wear the correct PPE and to be aware of things such as moving parts, flying debris, sharp edges, pinch points, electric power supplies, and general conditions of the work environment.
- Drilling tools have a large selection of different drill bit designs that are used for boring round holes in a variety of materials.
- Hydraulic tools are used to lift, press, bend, and form materials.
- The correct cutting tool for a job depends on the type and complexity of the cut(s) required for the job.
- Material can be cut by removal of waste material that creates a kerf or by a shearing motion.
- Grinding and sanding tools use abrasives to wear away material and shape, clean, and sharpen the surface.
- Several factors influence the choices made when selecting and purchasing power tools.
- Proper maintenance and storage extend the safe, useful life of power tools.

Words to Know

Match the terms from the chapter to the correct definition.

A. abrasives	H. flutes	N. lodging
B. blade set	I. grounded	O. nail gun
C. charge memory effect	J. hook tooth configuration	P. rake angle
D. chuck	K. jointer	Q. reciprocating saw
E. cribbing	L. kickback	R. router
F. die grinder	M. lithium-ion (Li-ion) battery	S. swarf
G. drill press		T. table saw

1. A power source for cordless power tools that delivers uniform speed and torque down to about 20% of a full charge.
2. A method of using stacked timbers to support a heavy load.
3. Pneumatic tool that drives a sharp spike into wood and similar materials with a single power stroke.

4. The part of a drill that holds the bit in place.
5. Handheld demolition tool that operates by mechanically moving a blade in a fast back-and-forth motion.
6. Sharp-edged waste material created by the action of drill bits boring into metal.
7. A stationary tool that uses rotating blades to straighten and finish the edges of lumber for applications where boards are fitted side by side.
8. The difference between the overall width of a saw blade's teeth and the thickness of the blade material.
9. Small, sharp-edged particles bound to certain consumable power tool attachments that are used for the removal of material.
10. A stationary or semi-portable version of a circular saw that is ideal for ripping lumber into narrower boards and for making cuts in plywood.
11. A saw blade type that uses an aggressive 10° positive rake angle to make fast cuts.
12. Term that refers to a power tool that uses 120-volt AC power and has three separate conductors included in the power cord.
13. A large, variable-speed power drill mounted to a vertical stabilizing shaft with an adjustable table for securing workpieces.
14. The number of degrees off perpendicular to the back of the blade each tooth in a band saw blade is angled.
15. A portable woodworking tool consisting of a powerful electric motor, a rotary cutting tool, and an adjustable base plate to control cutting depth.
16. The result of soft metal particles building up as deposits in the surface of an abrasive attachment.
17. Adverse result from charging certain battery types before they are completely discharged, which causes only a portion of the battery to accept charge.
18. The spiral hollows formed in drill bits to assist in the removal of waste material or the cutting edges of router bits.
19. Pneumatic tools used for finish work on small parts and for shaping surfaces with intricate contours that fit closely with other parts.
20. Event that can occur when a power tool sticks or binds momentarily in the work.

Know and Understand

Answer the following questions using the information provided in this chapter.

1. List the three ways that power tools are classified into groups.
2. For jobs that last long periods of time, the limited run time of _____ power tools can be dealt with by using a system of multiple units and charging stations.
3. *True or False?* Pneumatic power tools usually operate at between 350 and 430 psi.
4. Explain why the third ground prong on grounded electric power tools and the machine guards and shields on any power tool should never be removed.
5. The waste material created by a bit as it drills into metals and other material is known as _____.

6. A drill _____ is used to arrange drill bits by size and to store them in a convenient and safe manner.
   A. jig
   B. index
   C. chuck
   D. punch
7. Explain why many drills are capable of operating at variable speeds.
8. The development and use of cordless _____ drivers to drive screws has replaced the use of a hammer and nails for many construction applications.
9. A nail gun with a(n) _____ trigger incorporates a sliding safety tip that must be cycled and depressed each time a nail is delivered.
10. Pipe benders use _____ power to shape pipes between rollers and a mandrel.
11. Saw blades that, instead of blade set, use teeth tips that are thicker than the body of the blade to create clearance during a cut are called _____.
12. Circular saw blade teeth tipped with sharp inserts made from _____ greatly extend the useful life of the blade, in part by increasing the length of time a blade will retain its sharpness.
    A. copper
    B. bronze
    C. aluminum
    D. tungsten carbide or titanium carbide
13. Miter saws are used to make straight and angled cuts _____ the grain of the lumber.
14. *True or False?* In general, saw blades designed to cut harder materials have more teeth per inch than blades designed for softer materials.
15. When using a(n) _____ router, you can lower the bit into the center of the board surface to create a cut that does not extend to the edge.
16. How are the abrasive materials for grinders formed?
17. The _____ of a bench grinder should always be kept adjusted to within 1/8″ from the working surface the grinding stone.
18. All power tools sold in the United States are required by law to meet safety standards set forth by the _____.
19. List three rules for keeping power tools in optimum condition.
20. *True or False?* The plastic or metal boxes in which portable power tools are sold should never be used for storing the tools.

STEM and Academic Activities

1. **Science.** Use a microscope or strong magnifying lens to examine the surface of a new sanding disc. Then examine a sanding disc that is worn. What differences do you see? How will this affect their usefulness?

2. **Technology.** Use a CAD drawing program to illustrate the configuration of a carbide-tipped saw tooth. Add text to your drawing explaining how this configuration alters the kerf formed by the blade.
3. **Engineering.** Compare a Jacobs chuck to the collet of a router. How are they the same? What are the differences? What are the advantages of each system for holding a bit suited to the two different applications?
4. **Math.** Measure the throat size and depth of cut size on a stationary or bench-top horizontal band saw. Based on your findings, determine if you could use the saw to safely crosscut a 6" × 6" timber or a 2" × 4" × 2' board. Why could a 12" horizontal band saw not be used to cut to the center of a 30"-wide board?
5. **Language Arts.** Develop an illustrated safety poster for a selected power tool. Present your work to the class.

Thinking Critically

1. In the 1960s, the Hurst Performance company developed the iconic Jaws of Life®, which was first used to free race car drivers from wrecked vehicles. Soon, rescue responders around the globe were also using the tool. Now, this broad class of hydraulic rescue tools includes tools for spreading parts, cutting mangled structures, and rams for lifting. Use the Internet or your library to research the function of the hydraulic system within these power tools and explain why this makes the tool a very important rescue instrument.
2. Imagine that you are building a pole barn. Explain six required job tasks that can be completed using a circular saw, table saw, and miter saw (two tasks for each type of saw). Why would you choose that style of power saw for the task?

CHAPTER 9
Materials, Fasteners, and Hardware

Chapter Outcomes

After studying this chapter, you will be able to:
- Describe dimensional lumber and engineered lumber used for agricultural projects.
- Describe types of metals used for agricultural projects.
- Describe composite materials used for agricultural projects.
- Choose appropriate fasteners for agricultural projects.
- Choose appropriate hardware for agricultural projects.

Words to Know

adhesive	grain	nail	self-piercing screw
alloy	hardwood	nonferrous metal	softwood
aluminum	hinge	nut	soldering
bolt	I-beam	nylon	span
brad nail	joist	oriented strand board (OSB)	spring steel
brass	laminated dimensional lumber (LDL)		stainless steel
bronze		oxidize	steel
cast iron	laminated lumber	piano hinge	strap hinge
composite material	laminated structural member	plain-sawn lumber	thermoplastic
concrete		plate	threads
continuous hinge	laminated veneer lumber (LVL)	plywood	tool steel
copper		pressure-treated lumber	twisted shank nail
corrugated fastener	lead		veneer
dimensional lumber	machine screw	quarter-sawn lumber	wall anchor
ductility	malleability	rivet	washer
engineered lumber	masonry nail	roofing nail	
ferrous metal	mild steel	screw	
finish nail	mortar	self-drilling screw	

Before You Read

Before reading this chapter, review the objectives. Based on this information, write down two or three items that you think are important to note while you are reading.

While studying this chapter, look for the activity icon to:

- **Practice** vocabulary terms with e-flash cards and matching activities.
- **Expand** learning with interactive activities.
- **Reinforce** what you learn by completing the end-of-chapter questions.

www.g-wlearning.com/agriculture

kwest/Shutterstock.com

The basic materials used in agricultural mechanics projects and processes are wood, metal, and composites. The wood may be natural or engineered. The metals may be those that contain iron, those without iron, or those that are mixtures of metals. The composites may be made of wood, wood by-products, plastic, or other materials. The materials used for a project determine both the method of construction and the outcome of the process.

Wood

Wood used in construction and other projects is either softwood or hardwood. *Softwood* is wood from the trunk of a gymnosperm (a tree that is generally evergreen and bears its seed without a fruit). *Hardwood* is lumber cut from an angiosperm (a tree whose seeds are inside an ovule or fruit). See **Figure 9-1**.

Softwoods are usually less dense than hardwoods and are generally used for structural members, such as wall framing. Spruce, pine, and fir are the most commonly used species of softwoods. Although they are different species, they have similar *grain* (the pattern the cells of a tree make because of the growth rings) and similar structural strength. They may be used interchangeably. Another commonly used softwood lumber is southern yellow pine (SYP). SYP is a relatively dense, harder softwood cut from longleaf, shortleaf, loblolly, and slash pines.

Wood is strong, flexible, and relatively lightweight, and it can be machined into many forms. It can be used as a structural member in framing for a building or as a decorative element on a piece of furniture. As a natural fiber, wood can have natural defects as well. Some important considerations when choosing lumber pieces include where these defects are in relationship to where the wood will be cut for the project at hand and what the defect does to the natural strength of the wood.

Dimensional Lumber

Wood is available as dimensional lumber, which can be further categorized as quarter-sawn, plain-cut, or engineered. When a tree is cut to make *dimensional lumber* (lumber that is cut to a standardized width and depth specified in inches), it is either quarter-sawn or plain-sawn.

Quarter-Sawn Lumber

Quarter-sawn lumber is cut from large-diameter hardwoods. This style of cutting exposes the grain in such a way that the annual growth rings are perpendicular to the wide face of the board. This cutting method gives the wood more integrity and strength, and makes it less prone to cupping, warping, or twisting. Quarter-sawn lumber is used in fine furniture projects

A Valentina Razumova/Shutterstock.com

B Dionisvera/Shutterstock.com
Figure 9-1. Types of trees. A—A gymnosperm typically has evergreen needles and produces softwood. B—An angiosperm has broad leafs and produces hardwood.

and projects where strength and aesthetics are important. Quarter-sawn lumber is more costly to produce than plain-sawn lumber because more waste results.

Plain-Sawn Lumber

Plain-sawn lumber is cut in large slabs across one face of the log. This produces some boards with nearly perpendicular grain, but most boards have a variety of grain patterns. This cutting method creates less waste, but it leads to more cupping, twisting, and warping of individual boards. Most lumber used in agricultural projects will be of the plain-sawn variety.

Pressure-Treated Lumber

Pressure-treated lumber is wood that has been processed with chemical preservatives to be highly resistant to rot and insect damage, **Figure 9-2**. Therefore, it is used for applications where the lumber will be in constant contact with soil, animal waste, salt water, and other highly corrosive environments. It is also used where the general humidity is high or other destructive elements, including insect attack, are prevalent. In agricultural environments, pressure-treated lumber is used for fencing, livestock buildings, and most greenhouse applications.

Pressure-treated lumber is processed using a water-based, oil-based, or creosote treatment. In recent years, pressure treatments have been modified to meet new standards that require the wood to be nontoxic. This is important in situations where the lumber may come into contact with food production systems, in children's play structures, and in other potentially harmful situations. The intended use of the wood will determine which type of pressure-treated lumber should be used.

Pressure-treated lumber contains residual moisture, much like newly harvested wood, and must be properly stored before use to minimize warping, twisting, and cupping. For example, if pressure-treated lumber is to be used indoors, it should be stored indoors prior to use. Storing the lumber inside will allow it to reach a stable moisture level.

A   Chris Hill/Shutterstock.com

B   Joe Gough/Shutterstock.com

Figure 9-2. Pressure-treated lumber is commonly used in outdoor projects. A—Pressure-treated lumber was used to construct this fence. B—These pressure-treated posts are stored so that air can flow freely between them.

Storing pressure-treated lumber is much like storing green, or newly harvested, wood. Each piece must be supported at several points to allow airflow in and around each board. It must also be raised off the ground to prevent it from drawing moisture from the ground. This attention to storage detail minimizes warping, twisting and cupping. Special, coated fasteners and hardware must be used with pressure-treated lumber.

Engineered Lumber

Engineered lumber includes a wide variety of wood products made from the scrap from other wood manufacturing processes. Engineered lumber can be made into a stronger, more flexible, or better suited material than natural lumber by taking advantage of the natural strength of wood fiber in its design. Once it is determined what strength and other characteristics are needed in the engineered lumber, the engineer can manipulate the wood's natural structure to achieve those design goals.

Laminated lumber is an example of engineered lumber. *Laminated lumber* is a wood product formed by adding adhesives and then compressing layers of wood in specific orientations. Engineered lumber can be made to specifications for a particular job, such as spanning a specific distance or bearing a specific load. Some types of laminated lumber include plywood, laminated veneer lumber (LVL), and oriented strand board (OSB).

SAE Connection
Hardware Store Placement

Almost every town in America has a hardware store. From chain stores like Lowes and Home Depot to the local hardware stores owned by the gentleman who knows everyone in town, hardware stores blanket our country. That being said, these stores offer wonderful opportunities for you to gain work experience and have a placement SAE in agricultural sales.

Work hard to familiarize yourself with the items in this chapter, and you are well on the way to gainful employment. A solid understanding of the materials, fasteners, and hardware covered in this chapter will give you an advantage in a hardware store job interview. Knowing the difference between plywood and OSB or the difference between a machine screw and a drywall screw is an everyday requirement for your local hardware store employee.

Tyler Olson/Shutterstock.com

Plywood

Plywood is a manufactured wood product made by cutting lumber into very thin pieces called veneers. Each *veneer* (thin layer of wood) is laid in the opposite direction of the preceding layer and then laminated (glued) under high pressure, **Figure 9-3**. For plywood, the veneer layers are laid in alternating patterns of grain with the outside layers laid so that the orientation is with the length of a standard 4′ × 8′ sheet of plywood. Both outside layers are oriented this way because the natural strength of wood is with the long edge of the veneer.

Plywood is engineered to take advantage of the natural tendencies of wood fiber. Each layer of veneer contains the grain of the natural fiber. The more plies (layers) that are in the sheet, the stronger the material. Three to seven layers are typical. This gives plywood vastly increased strength over dimensional lumber of the same thickness. When veneer sheets are laid side by side, widths much wider than natural tree dimensions can be created. Plywood is ideal for flooring, sheathing, and forming everything from concrete forms to flooring to manufactured projects, such as furniture and storage cabinets.

Laminated Veneer Lumber (LVL)

Laminated veneer lumber (LVL) is similar to plywood and is formed using the same basic process. LVL may also be called *laminated dimensional lumber (LDL)*. LVL is made with a series of veneers laid into dimensional lumber shape. Common dimensions of LVL range from 1 3/4″ to 3 1/2″ wide with depths from 7 3/4″ to 24″. It is generally available in lengths ranging from 8′ to 24′. Unlike plywood, LVL veneers are generally all laid in the same direction. Any defect present in a single veneer is overcome by its neighboring veneers when they are glued and pressed together. Their design makes these members stronger than natural dimensional lumber of the same dimension (or size).

LVL members can be ganged together to carry more load over great distances, such as the length or width of a large machine shop or storage barn. LVL members are generally useful only indoors in dry conditions. However, this leaves a broad range of agricultural uses for this environmentally friendly engineered lumber. LVL lumber looks very similar to natural lumber on the face, but the edge has the look of plywood, **Figure 9-4**. Both surfaces take common finishes well.

A ©iStock.com/wabeno

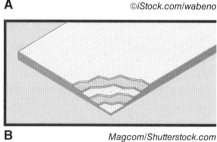
B Magcom/Shutterstock.com

C TFoxFoto/Shutterstock.com

Figure 9-3. Plywood is a common building material. A—Plywood is available is various thickness and with different types of veneer. B—Construction of plywood. C—Plywood sheathing being used in a roof.

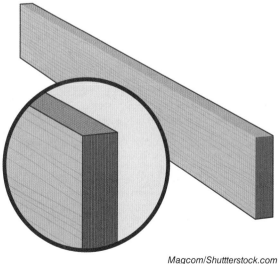
Magcom/Shutterstock.com

Figure 9-4. Laminated veneer lumber is constructed from numerous thin veneers laminated together.

Oriented Strand Board (OSB)

Oriented strand board (OSB) is commonly used in agricultural applications where strength and cost are important but appearance is not. This would include sheathing for buildings, subflooring, and forming, **Figure 9-5**. *Oriented strand board (OSB)* is a laminated wood product made by chipping wood fibers into very thin layers and orienting each chip into a common direction. Then resin (a type of glue) is added, and the mixture is pressed into a thin sheet. The sheets are combined in an alternating pattern, as in plywood, to increase the strength of the final material. The orientation of the strands makes OSB strong across a wide span. (In construction, *span* refers to the distance between structural supports.) OSB is generally interchangeable with plywood in sheathing, flooring, and roofing applications. OSB should not be confused with *chipboard* or *particleboard*, which are manufactured with chips of wood or coarse sawdust mixed with resin and pressed into 4′ × 8′ sheets.

A *stockphoto-graf/Shutterstock.com*

B *Steven Belanger/Shutterstock.com*

Figure 9-5. Oriented strand board. A—A sheet of oriented strand board (OSB) is made up of many smaller pieces of wood bonded together for strength. B—OSB is commonly used for sheathing.

Laminated Structural Members

Laminated structural members are engineered pieces used in floor joist and ceiling joist applications. A *joist* is a structural member laid horizontally from rim to rim on a floor or ceiling that carries the load. Laminated structural members are *I-beams* made with OSB as the center component and dimensional lumber as the top and bottom pieces, **Figure 9-6**. Another advantage of these laminated members is that they are more resistant to twisting, cupping, and warping than natural wood and cause fewer problems when used in dry interior applications.

TFoxFoto/Shutterstock.com

Figure 9-6. Laminated structural members are commonly used as floor joists.

Metals

Whether you are building projects in the school lab or making products in industrial applications, it is important to know and understand which metal, or combination of metals, is the best choice for the job. Some of the characteristics that influence what type of metal you use for a particular job include strength, weight, ability to hold an edge, corrosion resistance, and finish.

Ferrous Metals

Ferrous metals are iron-based metals, such as cast iron and steel. Iron is seldom used in its pure state. Ferrous metals have several beneficial characteristics:

- *Malleability*—the property of metal that allows it to be hammered or pressed (compressive force) into shape without breaking.
- *Ductility*—the property of metal that allows it to be permanently deformed (reshaped) by stretching (tensile force) without breaking.
- Luster—the ability of a metal to reflect light.

Ferrous metals include the various types of iron and steel as well as iron and steel alloys. Iron-based *alloys* (a base metal mixed with another element) are prized for their durability and versatility in machined parts, **Figure 9-7**. The characteristics of iron-based alloys vary according to their composition. Therefore, their composition determines the applications for which they are best suited.

Dmitry Kalinovsky/Shutterstock.com

Figure 9-7. This steel rod is being machined into a worm screw shaft.

Cast Iron

Cast iron is an iron alloy high in carbon, **Figure 9-8**. It is a hard, brittle metal that resists wear and is difficult to crush. Depending on the elements used in its composition and the production process, cast iron may be classified as white cast iron, malleable iron, or gray iron. White cast iron is used primarily in applications requiring wear and abrasion resistance. Malleable iron is white cast iron that has been treated with a special heating and cooling process. The process makes the metal malleable. Gray iron is not as strong as malleable iron, but it is less expensive and is easily formed into intricate and complex shapes.

Although cast iron alloys are not greatly affected by extremely hot or cold conditions (because they expand and contract less than other metals), a cast iron piece is relatively brittle. This means it may crack freely if the temperature is changed quickly or it is dropped onto a harder substance. Cast iron is difficult to weld or repair.

Omegaphoto/Shutterstock.com

Figure 9-8. Cast iron is a high-carbon iron alloy. It is typically cast to the desired shape rather than machined.

Steel

The most common ferrous metal is steel. *Steel* is a mixture of mainly iron and carbon. With the addition of carbon, iron becomes harder and more workable. There are hundreds of kinds of steel manufactured, and each is designed with a specific application. Steel types vary in many ways, including the following:
- Carbon content.
- Alloying elements.
- Forming and hardening processes.
- Strength and durability.

- The ability to hold a cutting edge.
- Corrosion resistance.

While working in an agricultural mechanics shop, you will encounter various types of steel. Four of the most common types are mild steel, stainless steel, tool steel, and spring steel.

Mild Steel

Mild steel is a general purpose steel made with a small quantity of carbon added in the production process. There are many variations of mild steel that improve or change characteristics of the steel. However, as a group, these steels are hard, easily welded, and are often formed into structural components because they hold their shape well. Most agricultural mechanics will learn to weld on mild steel because it is a common metal and can easily be worked with the addition of heat.

Stainless Steel

There are many types of *stainless steel* with highly different compositions, but they all share one common ingredient: chromium. Enough chromium must be added to make the steel corrosion resistant. Stainless steels may or may not be magnetic, and they may be lustrous when polished. Stainless steel is used for machined parts in many agricultural applications, **Figure 9-9**. It is also used in medical applications (such as surgical tools and hypodermic needles), for corrosion-resistant tubs or vats in food production systems, and in other areas where corrosion resistance is critical. While stainless steel is useful for many applications, it can be costly and difficult to repair with conventional welding and cutting methods. It is often replaced rather than repaired.

mattomedia Werbeagentur/Shutterstock.com

Figure 9-9. Stainless steel's lustrous appearance, high strength, and corrosion resistance make it a good material for parts that are decorative, subjected to extreme wear, or are exposed to the elements.

Tool Steel

Tool steel is a tough steel with a relatively high carbon content. Tool steel is machined into dense, tough, hard tools that can be used to work other steels. These tools are hard and tough enough to turn steel bolts or nuts and other parts without breaking.

Spring Steel

Spring steel is a relatively mild steel that can be formed into parts that have a lot of "give" before breaking and will return to their original shape after bending repeatedly. Spring steels are used for parts such as tension springs, compression springs, clock springs, and various types of wire. See **Figure 9-10**.

©iStock.com/ ishmeriev

Figure 9-10. Spring steel has unique ability to return to its original shape after being deformed. It is used to make springs and a wide variety of tools, such as rake tines and piston ring compressors.

Nonferrous Metals

Nonferrous metals are metals that do not contain iron. Nonferrous metals are not magnetic, often conduct electricity well, and have other ductile and malleable qualities. In comparison, ferrous metals are magnetic, heavier, and more prone to corrosion than nonferrous metals. Some of the most common nonferrous metals encountered in the agricultural mechanics shop include copper, brass, bronze, aluminum, and lead.

Copper

Copper is a ductile metal that has excellent heat and electrical conductivity. It is the metal of choice for wiring in a structure. Copper conducts electricity well, maintains a relatively cool temperature, and is easily drawn into wire. It is also much less expensive than other good metal conductors, such as gold and silver. However, copper does *oxidize* (combine with oxygen to form corrosion) rapidly when exposed to air. This is one of the main reasons wiring is covered by insulation made of plastic or rubber. The covering also prevents electricity from arcing across wires.

Copper is widely used for plumbing pipe because of its relatively inert quality when exposed to water, ease of assembly, and durability. Copper pipe can be an expensive plumbing solution and takes some skill to solder. *Soldering* is a process used to join metal pieces by melting another alloy and allowing it to flow between the two metals to be joined. In the case of copper plumbing, the pieces will be joined with a waterproof seal when the melted alloy cools. Copper is also a base metal in a variety of nonferrous alloys, such as *brass* (a metal made of copper and zinc) and *bronze* (a metal made of copper and tin), **Figure 9-11**. Due to its conductivity, relatively low cost, antibacterial properties, and ease of working, copper is also a common choice in making everything from cookware to computer chips.

Constantine Pankin/Shutterstock.com

Figure 9-11. Water pipes are typically made of copper, and valve bodies are usually made of brass.

Brass

Brass is an alloy of copper mixed with zinc. Brass is used in plumbing fixtures and fittings because of its hardness and endurance even in the presence of water. It is also a very lustrous metal, which makes it ideal for decorative hardware, such as latches and hinges.

Bronze

Bronze is a metal alloy consisting mainly of copper. Tin is usually the main additive. This alloy of copper is often used in malleable forms, which are highly prized in the landscape industry as garden statuary or ornaments, **Figure 9-12**. Bronze is especially resistant to seawater corrosion and commonly used for boat and ship fittings. Bronze is also used for specialty bearings and bushings in engines.

Aluminum

Aluminum is a material commonly used in agricultural applications due to its relatively low cost, light weight, and strength. Aluminum is prized for being lightweight and maintaining its shiny appearance when used in equipment, such as cargo/stock trailers, engine blocks, and trim components, **Figure 9-13**. These same qualities make it a good material for structural members in buildings and for components in electronic devices.

Lead

Lead is a toxic, very dense, and heavy metal used in anchoring materials, in batteries, and as shielding material for radiation. When ingested, lead collects in muscle tissue. With long exposure, it can cause death or mental problems. Lead was once commonly used for pipes and solder in plumbing systems. However, its toxicity led to its replacement in most plumbing pipe systems and soldering in general.

Lead was also used in ceramic coatings and in oil-based paint, but both of these uses have been banned. Lead must be disposed of in an environmentally responsible manner.

Ball00025/Shutterstock.com

Figure 9-12. Bronze is highly resistant to the elements and is a good material choice for statuary and decorative hardware.

Safety Note

Exercise caution around old, peeling paint or when sanding old paint. It may contain lead. Lead-based paint should be removed only by an expert. If the paint is not peeling or chipped, it can be contained by covering it with new paint as part of general maintenance.

Baloncici/Shutterstock.com

Figure 9-13. Aluminum is corrosion resistant, strong, and lightweight. It is often used to make lower-weight versions of parts or components traditionally made from steel or cast iron.

Composite Materials

Many composite materials are important to the agriculture mechanic. *Composite materials* are materials that are made of a variety of substances in a way that makes the finished product different than the original members of the mixture.

Concrete

Concrete is a composite material made of Portland cement, aggregate (coarse gravel or fine sand), and water. Portland cement is a manufactured powder made of heated and crushed limestone, ash, and other ingredients. When mixed, these ingredients form into a plastic state, which then cures to a very hard state.

Concrete is used in agriculture in many building operations from barn floors to silo walls. Concrete is prized for its low cost, durability, compressive strength, and ability to be formed into any shape, **Figure 9-14**.

pryzmat/Shutterstock.com

Figure 9-14. Concrete is commonly used in the foundations, footings, and floors of buildings.

Mortar

Mortar is cement paste consisting of Portland cement, sand, and water mixed in a ratio that is plastic when wet and rock hard when cured (dry), **Figure 9-15**. It is used to bind masonry bricks, blocks, tile, or other units to each other or to other concrete structures. Mortar holds concrete block walls together and keeps bricks in a solid mass on the side of a structure.

Thermoplastics

Thermoplastics are a group of plastics that can be repeatedly melted and reformed. They are used extensively in agriculture equipment. Thermoplastic parts are lightweight, heat and chemical resistant, noncorrosive, and inexpensive. Thermoplastics are also easily formed into many shapes, including gears, bushings and spacers, pipe, and conduit. Because of their versatility, thermoplastics may also be used as substitutes for sheet metal, cast metal, wood, and glass parts.

John Leung/Shutterstock.com

Figure 9-15. When mortar dries and cures, it forms a tight bond with bricks, masonry blocks, and tile.

Thermoplastics include the following types:
- Polyethylene (PE)—a light, flexible synthetic resin made by polymerizing ethylene. The term *polymerization* refers to chemically forming a long-chain organic compound from short-chain compounds. Polyethylene is used in construction, generally for plumbing supply pipe. It is also used for plastic bags, food containers, and other packaging.
- Polystyrene—a synthetic resin that is a polymer of styrene. It is chiefly used for packing material, insulation, and food service containers.

Figure 9-16. PVC pipe comes in various sizes. It can be easily cut to length and permanently assembled and sealed using an solvent cement.

Tao55/Shutterstock.com

- Polyvinyl chloride (PVC)—a tough, chemically resistant synthetic resin made by polymerizing vinyl chloride. It is commonly used in waste and vent systems in plumbing, **Figure 9-16**.
- Acrylonitrile butadiene styrene (ABS)— a hard, tough plastic that costs more to produce than polystyrene. However, it has a desirable gloss and electrical insulating properties that make it useful for a wide variety of applications, including 3-D printing, plumbing pipe, and molded plastic panels for vehicles and agriculture equipment.

All these plastics are formed by mixing long-chain organic hydrocarbons with other chemicals or elements to form specific types of plastic. These plastics are recyclable, depending on the length of the molecular chain. Adding corn starch or other plant starches to the mixture can form a biodegradable form of thermoplastic that will degrade in a natural environment in a few months. This is one feature of plastics that will be important in the future of agriculture. There will be a market for plant-based starches and more machine components will be made from these plastics.

Nylon is a thermoplastic that is light in weight and heat tolerant. It is easy to make into a variety of cast or machined parts. Nylon is a relatively old form of thermoplastic that has found a new use in agricultural machines. It is likely to be the material of choice for creating items with 3-D (three dimensional) printers, **Figure 9-17**. Nylon can be combined with other materials, such as glass and carbon, creating a composite that has great heat resistance. This makes nylon ideal for use in engines. While most agricultural machine shops are not set up to work with nylon in its raw form, many parts are available in this material. An agricultural mechanic should recognize the material and understand its properties when dealing with it in repair situations.

STEM Connection

Monomers

In chemistry, sometimes small molecules can easily join together with other molecules of their type because of the molecular design. These are called monomers. Often monomers can form very long chains together, and these are called polymers. The prefix *poly* means "many." Polymers are formed along a backbone of elements, often carbon, as in the hydrocarbon plastics. Polymers also occur in nature such as DNA and RNA and cellulose, each being made of small pieces of repeating molecules.

Tyler Olson/Shutterstock.com

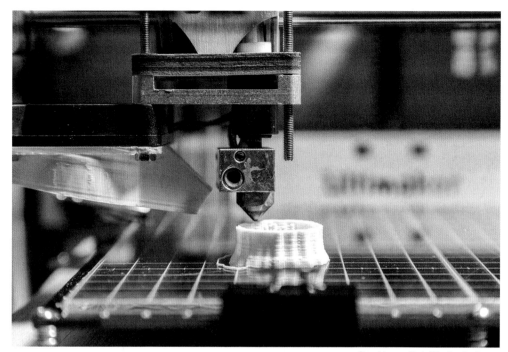

Sergi Lopez Roig/Shutterstock.com

Figure 9-17. Nylon is one of the most popular types of plastic used for 3-D printing. Parts made from nylon are strong, durable, and heat-resistant.

Thermoset Plastics

Thermoset plastics are a group of plastics that, unlike thermoplastics, cannot be melted and reformed once they have been cured. They are not easy to recycle. When thermoset plastics are produced, the component parts are chemically bonded together forming a new chemical which cannot be changed back into component chains, unlike thermoplastics. Thermoset plastics are very easy to mold into a shape, have very stable natures, and resist heat well. These qualities make them important to the agricultural mechanics industry as components of machines. An example of a thermoset plastic is a synthetic rubber tire. While these materials cannot usually be refabricated by the agriculture mechanic, it is important for an agricultural mechanic to be familiar with thermoset plastics because they are commonly used in the parts of many machines. Their repair is specialized work.

Fasteners

There are many types of fasteners, each with its own characteristics and intended use. Often times, the type of fastener used will be determined by building codes and their specific requirements for different applications. If you are working on a project for which hardware is supplied, use those fasteners. The fasteners were carefully chosen by engineers to work best in the application. Common types of fasteners include:
- Nails.
- Screws.

- Bolts.
- Hurricane and earthquake fasteners.
- Plates.
- Wall anchors.
- Corrugated fasteners.
- Rivets.

Nails

A *nail* is a small metal spike that can be driven into wood or other materials to hold materials together. Nails are used in most framing and structural applications. Always use the proper type and size for a specific job. See **Figure 9-18**.

Safety Note
Always wear eye protection when driving in nails or screws. The sparks given off by a driven nail or screw are tiny shards of metal. These shards could reach your eyes if you do not wear safety glasses.

Common Nails

Common nails were first cut from long iron wires in the 1800s. The standardized sizes and naming conventions used today are derived from how wire nails were sized and sold. The letter *d* stands for *denarius*, a Roman coin similar to the English penny. The sizes of nails were related to their cost in pennies. Refer to **Figure 9-19** for a list of common nail sizes.

Today, common nails can be purchased loose. However, they are more often purchased in a coil or strip with a D-shaped head so that they can be driven with a pneumatic nail gun. Most repetitive nailing is done with a nail gun to make the nailing job more uniform. Besides common framing nails, most roofing applications and brad/finish nailing applications in large jobs are done using pneumatic nail guns and proprietary nails (of specific styles or brands).

Cut Nails

Cut nails are cut from sheets of iron. These nails have a rectangular, somewhat tapered shank and a rectangular head with a somewhat blunted point. Because these nails are thicker, they have more holding power than common nails and are often used in wood or masonry applications. Today, cut nails are largely decorative due to the additional cost and availability of more efficient fasteners.

Box Nails

Box nails look similar to common nails, but they have a much smaller diameter shank and less holding power than a similarly sized common nail. These nails should not be used for framing as they are

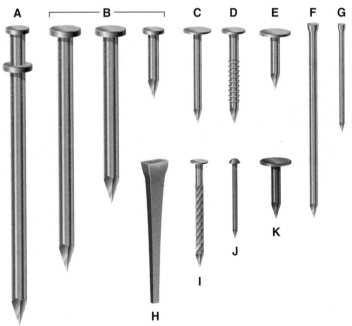

Bill Fehr/Shutterstock.com

Figure 9-18. Types of nails A—Duplex nail. B—Common nails. C—Box nail. D—Ring shank nail. E—Roofing nail. F—Finishing nail. G—Brad nail. H—Cut nail. I—Twisted shank nail. J—Brass round head nail. K—Copper roofing nail.

weaker than common nails and do not form as sturdy of a joint. These nails are ideal for joining thinner materials, such as those used for wooden boxes.

Brad Nails and Finish Nails

Brad nails are very thin wire nails that have almost no head. These nails are used to attach thin trim materials. Brad nails are often small enough that the hole they leave at the surface of the wood is not filled with wood filler but is left to be filled by the paint or finish. *Finish nails* are thin wire nails that have shaped heads that can be driven below the surface of wood. They are similar in design to brad nails except the wire is somewhat thicker and the head is bigger and often concave so that a nail set can easily drive it below the surface of the wood. Because of the finish nail's larger size, the remaining hole should always be filled.

Roofing Nails

Roofing nails are short shanked, large-headed nails used to attach roofing paper, asphalt shingles, and roofing sheet metal. A roofing nail has a much larger head than a common nail. It is easily driven the short distance through the roofing material into the under sheeting. It often has a ringed shank to increase its holding power and prevent it from being pulled out of the roof due to changes in weather. Because of the short, thickened shank of this nail, it is not used in joining wood components of a structure.

Masonry Nails

Masonry, because of its hardness and brittleness, presents a special problem when it comes to attaching wood at a joint. This special requirement led to the design of a masonry nail. A *masonry nail* is a nail that has a hardened, extra thick shank that is often grooved to give it increased holding power. This type of nail is used for joining wood to masonry or concrete. There are alternative fasteners for masonry work that will be discussed later in this chapter.

Twisted Shank Nails

Twisted shank nails are specialty nails that are about the size of a box nail, but have a twisted shank. The twisted shank makes the nail very difficult to remove once it has been driven into a joint. The nail turns as it is driven and, when it is removed, it must be removed by turning it in the opposite direction. These nails are useful in decking applications or where wind and weather cause great structural stresses to joints. Often times, these nails are coated with zinc or other corrosion-resistant material.

Common Nail Sizes	
Penny Size (d)	Length
3 penny (3d)	1¼"
4 penny (4d)	1½"
6 penny (6d)	2"
8 penny (8d)	2½"
10 penny (10d)	3"
12 penny (12d)	3¼"
16 penny (16d)	3½"
20 penny (20d)	4"
40 penny (40d)	5"
60 penny (60d)	6"

Goodheart-Willcox Publisher

Figure 9-19. This table shows the lengths of common nails of different penny sizes.

Figure 9-20. Although there are a variety of screw types, they all have a threaded shank and a head that provides a means of driving the screw. The head of the screw shown here is slotted for a Phillips screwdriver and is tapered for countersinking.

Paul Wishart/Shutterstock.com

Figure 9-21. Wood screws are designed to bite well into wood.

Barna Tanko/Shutterstock.com

Figure 9-22. Deck screws are similar to wood screws, but they have a heavier shank and are typically coated to prevent corrosion.

Screws

A *screw* is a fastener that has three basic parts: a head, a threaded shank, and a tip, **Figure 9-20**. The *threads* on a screw are a ridge of metal that spirals from the tip to the head of the screw. This ridge helps secure the screw and pieces of material being joined. Screws may also have a partially threaded shank where the top part is smooth, allowing it to seat securely in the top piece of material (generally wood or plastic) while the tip part of the screw is "biting" into the attaching member. Generally, screws are tapered in the shank from the head to the tip. Screws are generally put into material with screwdrivers (either hand or power drivers) or specialty wrenches. Just as nails are designed for specific applications, so are screws.

Wood Screws

Wood screws are used for joining wood and for other applications as well, **Figure 9-21**. Wood screws have a tip that can be pointed or flat and a shank that can be threaded completely to the bottom. The threads are very sharp and help form a tight connection with the wood. Wood screws with a flat tip are designed for being driven into predrilled holes. The topmost part of the shank (near the head) of a wood screw is unthreaded, allowing this part to pass through the hole. This allows the head to be pulled tight into the top piece of wood. When combined with glue or adhesive, this forms a very tight, stable joint.

Deck Screws

Deck screws are designed specifically for outdoor wood decking and wood stairs. They have a heavier gauge shank than regular wood screws, **Figure 9-22**. Because they are designed for outdoor use, most deck screws have a water-resistant coating. The head is usually bugle-shaped so it can be countersunk into the wood with little chance of causing the wood to split.

Sheet Metal Screws

The difference between a sheet metal screw and a basic wood screw is the shape of the head. The head of a sheet metal screw is slightly rounded (convex) but flattened on the outside rim, **Figure 9-23**. Sheet metal screws are used to attach sheet metal to wood or metal components.

Machine Screws

Machine screws are small-diameter, straight-shanked, threaded screws, **Figure 9-24**. Unlike the screws previously mentioned, machine screws have UNF, UNC, or ISO metric threads, like a bolt. Machine screws usually have convex or rounded heads with slots for driving. They are useful for attaching metal to metal in holes that are predrilled. Machine screws are also available with a head designed for use with a screwdriver or socket wrench. They can be used with a nut or in a threaded bore.

Drywall Screws

Drywall screws are designed with a bugle-shaped head that compresses into drywall without cracking it, **Figure 9-25**. They are threaded tip to top with a wide or coarse thread. Drywall screws are used in many construction projects where thinner material is attached to framing members, not just in drywall applications. These screws are commonly found in agricultural mechanic shops because they:

- Are easy to power drive with a screw gun or drill.
- Have good holding power.
- Are inexpensive.
- Work with a variety of materials, especially woods.

Self-Drilling and Self-Piercing Screws

Many types of screws are self-drilling or self-piercing. *Self-drilling screws* are fasteners configured with the tip end of the threads more open, allowing the threads to start a hole by cutting through the material. These are generally not reusable. See **Figure 9-26**. *Self-piercing screws* are screws that have a long taper and an extremely sharp point. The threads extend right to the very tip of the screw, so that it grabs immediately.

Bolts

A *bolt* is a threaded fastener, usually with a flattened tip and straight (not tapered) shank, used with a nut to secure materials. A *nut* is a tubular, threaded piece of metal whose threads must match the threads of the bolt with which it is used so they will fit together securely. Bolts are typically inserted through a predrilled bore in material and secured by a nut on the other side of the material. Often, machine screws are also referred to as bolts.

anmbph/Shutterstock.com

Figure 9-23. Sheet metal screws have a large head with a flattened rim. The large surface area of the head helps prevent it from tearing through the sheet metal.

Skiving/Shutterstock.com

Figure 9-24. Machine screws vary widely in size, length, and head design. They are identifiable by their straight shank and bolt-like threads, which allow them to be used with nuts.

vseb/Shutterstock.com

Figure 9-25. Drywall screws have bugle-shaped heads, sharp points, and threads that cut aggressively.

MARGRIT HIRSCH/Shutterstock.com

Figure 9-26. The tip of a self-drilling screw is angled like the tip of a drill bit. The edge of groove in the screw tip cuts the material as the screw is turned. In this way, the screw drills its own pilot hole as it is screwed into the material.

BLUR LIFE 1974/Shutterstock.com

Figure 9-27. As a nut is tightened, its internal threads increase tension on the threads of the bolt, holding the nut and bolt in place and applying pressure to the material between them.

Kondrachov Vladimir/Shutterstock.com

Figure 9-28. A lock washer (left) compresses as the bolt and nut are tightened together. This increases tension on the bolt's threads. A flat washer (right) increases the surface area through which the nut and bolt's holding force is applied.

Nuts

In order to keep bolts from working out of the material, nuts are attached to the tip end of the bolt. Nuts can also be used to tighten the bolt head firmly against the material being held together, **Figure 9-27**.

Washers

A *washer* is a flat steel disk that has a hole in the center. The outer diameter of a washer is larger than the head of the bolt with which it is used, and the hole in the washer is nearly the same diameter as the bolt's shank. Flat washers are used to keep bolts from pulling through a material by providing a solid resting place for the bolt to tighten against. Lock washers lock against the nut so that the bolt and nut threads do not turn out of each other, which would cause the bolt to become loose in the material, **Figure 9-28**.

Hex Bolts

Hex bolts are bolts with a hexagonally shaped head, **Figure 9-29**. They are generally used with a nut and tightened using a socket wrench or box end wrench. Hex bolts are used with both metal and wood materials.

Carriage Bolts

Carriage bolts are often used in heavy framing. Their shanks may be completed threaded from the end to a squared part of the shank immediately below the domed top, or there may be a portion of the cylindrical shank left unthreaded. A nut is used to draw a carriage bolt tight into wood. This type of bolt can be tightened from one side only, **Figure 9-30**.

Lag Bolts

Lag bolts are very large, screw-like bolts (often having a diameter of up to 1/4″) that are used in heavy wood framing. These bolts are turned into the wood in a location where heavy members are joined, but there is no possibility of adding a nut to the tip end. For example, from wood to block walls or joints where the mechanic only has access to one side of the members. Lag bolts are often started in a predrilled hole in the face member and expected to cut into the adjoining member, pulling it tight to the face, **Figure 9-31**.

Chapter 9 Materials, Fasteners, and Hardware   243

Hurricane and Earthquake Fasteners

In parts of the country where hurricanes and earthquakes are prevalent, additional structural fasteners are often required by local or state codes. You should always check with the local authorities before starting on a structural project to obtain permits and acquire the latest building codes for the area.

Hurricane and earthquake fasteners are designed to provide added strength to joints where the stress is likely to be greatest during a hurricane or earthquake. For example, where rafters meet roof plates or exterior walls meet sill plates. Check local codes for the needed product for this type of weatherproofing. See **Figure 9-32**.

Plates

Plates are relatively thin metal pieces designed for general strengthening of a project. They are made in Tee, ELL, or straight shapes with predrilled holes, **Figure 9-33**. Screws are used to attach the plates. Plates are often used when additional strength is needed in a common joint or to repair an existing joint that has given way with age.

Wall Anchors

A *wall anchor* is a fastener that provides support for a screw, allowing that screw to support more weight. The screw and wall anchor may be supplied as a set, or the screw may be separate. The wall anchor is usually inserted into a predrilled hole in a wall. The screw is driven into the anchor. The anchor spreads out and presses against the sides and back of the wall. The design allows the fastener to provide strength without being attached to a structural member.

This fastener is used in instances when the screw cannot be driven directly into a structural member. For example, when putting a screw or nail into a masonry, sheetrock, or a hollow wall. See **Figure 9-34** for the common types of wall anchors.

Corrugated Fasteners

Corrugated fasteners are wavy, thin pieces of metal, about 1/2″ to 3″ long, that are driven into a wood joint (usually a miter or butt joint). Because the

Sergioato23/Shutterstock.com

Figure 9-29. Hex bolts are so named because of their hexagonal heads.

AlexLMX/Shutterstock.com

Figure 9-30. The head of a carriage bolt has no provision for tightening the bolt. A carriage bolt is tightened by turning the nut.

Coprid/Shutterstock.com

Figure 9-31. Lag bolts are very similar to screws, but they are much heavier and are designed to be turned with a wrench.

Steven Frame/Shutterstock.com

Figure 9-32. Hurricane fasteners, also called hurricane tie downs, are metal strips used to tie one structural member to another. This provides an additional level of protection from wind damage in areas that are prone to hurricanes.

fastener pulls each piece toward the joint, it is better suited for securing miter joints than a nail, which might push against the opposite member as it is driven in, **Figure 9-35**.

Rivets

Riveting is a relatively old method of assembling metal on metal parts where little movement of the joint is desired. A *rivet* is a metal pin with a head and a tail piece. It is made from a soft piece of steel or other alloy. A rivet must be applied with a specialty tool called a rivet gun or pneumatic riveter. The tail piece is put through a predrilled hole on the back side (in both pieces of material), and the head is placed on the tail from the front side. The rivet gun is clamped down over both pieces until sufficient pressure causes the head to flatten and the tail to clench over, holding the rivet and the material in place. Rivets are usually spaced closely together, **Figure 9-36**.

Adhesives

An *adhesive* is a substance, generally liquid, that forms a physical and/or chemical bond with materials and causes them to adhere. Wood glue, a common adhesive, is used to fasten some joints together and to

Johnny Habell/Shutterstock.com

Figure 9-33. Plates are often added at joints in framing. The plates provide reinforcement and strengthen the joints.

Style of Wall Anchors	Name	Uses	Benefits or Drawbacks
 A *Robin Lund/Shutterstock.com*	Expansion-type molly bolt	Used for heavier hanging in drywall or plaster situations	Only as strong as drywall or plaster it is seated into; causes damage if removed improperly
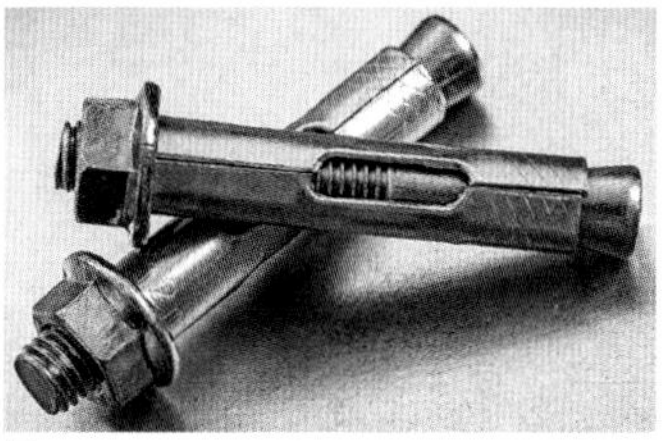 B *mihalec/Shutterstock.com*	Sleeve-type anchor bolts	Can be used for heavier hanging in solid surfaces.	Can be driven into less dense material, but works best with predrilled hole; can be reused
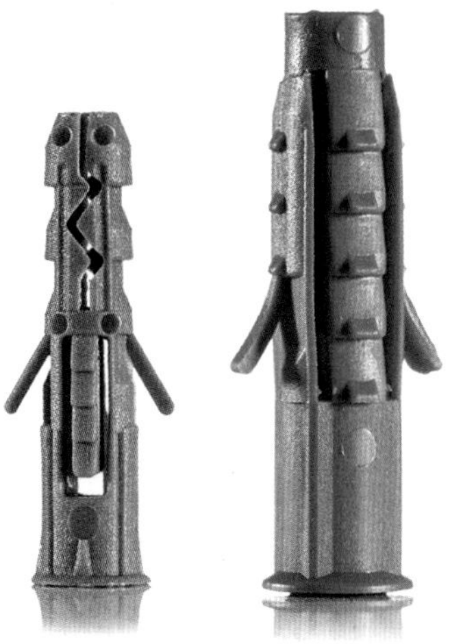 C *mihalec/Shutterstock.com*	Plastic expansion anchors	Very lightweight hanging; can be used in concrete, drywall, or other solid surfaces	Easy to install; inexpensive; no strength when pulled down, such as hanging from a ceiling; use only in wall applications with light loads (10 pounds or less)
 D *Cico/Shutterstock.com*	Threaded dry wall anchors	Used for lightweight hanging directly into dry wall	Easy to install, little damage to dry wall. Do not hold heavy weights

Goodheart-Willcox Publisher

Figure 9-34. Common types of wall anchors. A—Molly bolt anchor for drywall. B—Hex nut sleeve anchor for concrete. C—Light-duty plastic expansion anchors. D—Self-drilling drywall anchor.

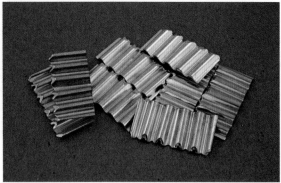

BW Folsom/Shutterstock.com

Figure 9-35. Corrugated fasteners are used to secure miter joints, such as those at the corners of a picture or window frame.

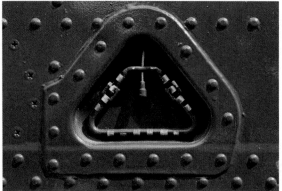

edography/Shutterstock.com

Figure 9-36. Riveting has been a common method of joining metal parts for many years.

increase the holding power of both nails and screws in woodworking applications. In certain applications, glue forms a tighter bond than other fasteners because it bonds with the materials chemically and/or physically, depending on the material and the type of glue, **Figure 9-37**.

Hardware

In terms of construction projects, hardware may be defined as the decorative fasteners and other functional pieces, such as hinges, handles, locksets, and latches.

Hinges

When a project calls for a part to be movable yet remain attached, a hinge is usually needed, **Figure 9-38**. A basic *hinge* consists of two flat plates connected by a centered pin. Each plate is free to rotate against this pin, allowing the piece to open and close. As is the case with most hardware, there are various types of hinges designed for specific applications. Hinges can add a decorative element to many projects and are, therefore, available in many different metal finishes and styles. Basic hinges that are commonly used by agricultural mechanics include:

- Butt hinges.
- Flush hinges.
- Strap hinges.
- Piano or continuous hinges.

Butt Hinges

The most common use for butt hinges is for attaching doors. Butt hinges are recessed into the edge of the door and the door frame, **Figure 9-39**. Butt hinges are generally used in a system of two or three on a piece the size of a door because the door's weight hangs against the hinge pins. Specialized butt hinges include rising butt hinges and security butt hinges. Rising butt hinges pivot on the center pin and pick the door up as it moves it away from the jam. This allows for a tighter seal at the bottom of the door. Security butt hinges have a hinge point that is interconnected. Even if the pin is removed, the hinge stays intact and does not allow the door to be released.

Flush Hinges

Flush hinges work in a fashion very similar to butt hinges. However, they do not need to be recessed to work. Since they do not require a recess to be cut into the parts being connected, they can be used with a wide variety of materials, including metal, **Figure 9-40**.

Adhesive	Color and Application Method	Materials
Acrylic resin glue	Clear Liquid (powder/water mix)	Waterproof, multipurpose (wood, metal, glass, concrete)
Aliphatic resin glue	Cream Squeeze bottle	Multipurpose
Casein glue	Light beige Liquid (powder/water mix)	Wood
Contact cement	Tan, green, milky White Liquid (tube or brush)	Multipurpose (wood, cloth, leather, plastic, rubber, metal, ceramic)
Epoxy	Clear Liquid (resin/hardener) Stick putty	Porous and nonporous
Hide glue (liquid)	Clear amber Squeeze bottle	Multipurpose
Hide glue (solid)	Clear amber Heated, brushed on	Multipurpose
Plastic resin glue (urea formaldehyde)	Light tan Two-part (resin/water)	Multipurpose
Polyurethane glue	Tan Spray, liquid	Waterproof, multipurpose (wood, metal, plastic, ceramics, stone)
Polyvinyl acetate (PVA) glue	Clear white Squeeze bottle	Multipurpose, primarily exterior applications (windows, doors)
Solvent-based contact cement	Light amber, clear Spray, brush	Plastic laminate to particleboard

Goodheart-Willcox Publisher

Figure 9-37. This table lists the characteristics and suitable applications for common types of adhesives.

Strap Hinges

A *strap hinge* is a hinge that allows a door or lid to lay flat back against a wall or side surface. Strip hinges are pinned in the center like other hinges, but they are completely exposed rather than being partially concealed. One strap is attached to the outside of the door (or lid). The other strap is attached to the surface of the wall or box.

Strap hinges are also available in *T* styles. In this type of hinge, the door part of the mechanism is strap-shaped (often a triangular plate), and the wall or box side plate is rectangular. T-strap hinges hold less weight than double strap hinges because the rectangular plate is smaller than a strap plate. T-strap hinges can be made in a secure style by using specialty screws with security heads and folding the hinge back on itself. See **Figure 9-41.**

Guru 3D/Shutterstock.com

Figure 9-38. A typical hinge consists of two plates that are connected by, and pivot around, a connecting pin.

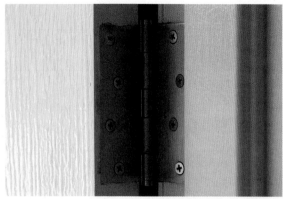

Thissatan Kotirat/Shutterstock.com

Figure 9-39. Butt hinges must be recessed into the door and frame to allow the door to close completely.

Kalabi Yau/Shutterstock.com

Figure 9-40. Flush hinges do not need to be recessed because one plate has a cutout that is the same size and shape as the other plate, allowing the door to close completely.

Glen Jones/Shutterstock.com

Figure 9-41. Strap hinges are commonly used on shed doors, gates, and chests.

Piano Hinges

A *piano hinge* (also called a *continuous hinge*) is a fastener with two plates with alternating pins that allow one plate to fold back upon the other. Piano hinges are generally used in box applications where the lid needs to be supported along the entire length. The piano hinge is named for its original use on a piano lid, **Figure 9-42**. These hinges are often decorative as well. Piano hinges come in lengths up to 8′ but can be cut to fit the project. Often the plates are predrilled with small screw holes.

Handles

Handles are parts on an object that are designed to be grasped by a person's hands. Handles generally are useful in opening a hinged part of a project, such as a door, or moving a part of a project, such as a drawer. Not only are handles useful; some are used for decorative purposes as well, **Figure 9-43**. Handles that are designed to be operated without putting the entire hand in or around the handle are also called pulls. Pulls are handles that are used to open a door or drawer by pulling only. Generally, the moving part returns to the starting position on its own or with a gentle nudge on the pull.

Latches

A latch is a mechanism consisting of two parts, such as a bar that fits into a notch or holder, that is used to hold items together in a non permanent way. A latch is useful when an item, such as a door or a lid, is hinged. The pieces can be held together with a latch. When access is needed, the parts can be unlatched and the door or lid swung open on the hinge. See **Figure 9-44**. Latches can be used to secure things like windows, gates, or case lids. Latches are often incorporated into locksets.

Locksets

A lockset is the mechanism that, in conjunction with the handle, works to keep a door closed. A lockset consists of the handle, a strike plate and a latch bolt and/or a dead bolt. The strike plate is the section of the lockset on the door jamb part of the door frame. The strike plate receives the latch bolt, **Figure 9-45**. If the latch bolt is not spring-loaded but is fixed in the lockset, it is called a dead bolt, **Figure 9-46**. A lockset is often required on a hinged door or even a box lid to keep the door or lid from swinging freely.

Chapter 9 Materials, Fasteners, and Hardware 249

Sorapop Udomsri/Shutterstock.com

Figure 9-42. Piano hinges are a good option for a long joint, such as a keyboard cover, because they provide support along the entire joint, preventing sag between hinges and potential binding.

alessandro0770/Shutterstock.com

Figure 9-43. Handles are available in a wide variety of styles. Handles can be both functional and decorative.

Peter Baxter/Shutterstock.com

Figure 9-44. Simple latches are used on doors, gates, and lids.

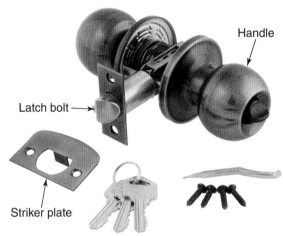

nikitabuida/Shutterstock.com

Figure 9-45. Locksets include the parts that are to be installed in the door and the doorjamb. The handle and latch bolt are installed in the door, and the striker plate is installed in the doorjamb.

You Touch Pix of EuToch/Shutterstock.com

Figure 9-46. A dead bolt is similar to a latch bolt, but it is not spring loaded. The key or knob is turned one way to extend the bolt or turned the other way to retract it.

CHAPTER 9 Review and Assessment

Summary
- The basic materials used in agricultural projects are wood, metal, and composites.
- Softwoods are usually less dense than hardwoods and are generally used for structural members, such as wall framing.
- Wood is available as dimensional lumber, which can be further categorized as quarter-sawn, plain-sawn, or engineered.
- Pressure-treated lumber is wood that has been processed with chemical preservatives to be highly resistant to rot and insect damage.
- Engineered lumber includes a wide variety of wood products, such as plywood, laminated veneer lumber (LVL), and oriented strand board (OSB).
- Some of the characteristics that influence the type of metal to use for a particular job include strength, weight, ability to hold an edge, corrosion resistance, and finish.
- Ferrous metals are iron-based metals, such as cast iron and steel.
- Nonferrous metals are metals that do not contain iron. Examples include copper, brass, bronze, aluminum, and lead.
- Composite materials are made of a variety of substances in a way that makes the finished product different than the original members of the mixture. Concrete, mortar, and thermoplastics are examples of composite materials.
- Common types of fasteners include nails, screws, bolts, hurricane and earthquake fasteners, plates, wall anchors, corrugated fasteners, and rivets.
- Adhesives are a group of substances that form a physical or chemical bond between materials, causing those materials to adhere to each other.
- For construction projects, hardware includes decorative fasteners and other functional pieces, such as hinges, handles, latches, and locksets.

Words to Know
Match the key terms from the chapter to the correct definition.

A. adhesive
B. alloy
C. cast iron
D. composite material
E. concrete
F. copper
G. corrugated fastener
H. dimensional lumber
I. ductility
J. engineered lumber
K. grain
L. hardwood
M. hinge
N. I-beam
O. joist
P. malleability
Q. oriented strand board (OSB)
R. plate
S. plywood
T. pressure-treated lumber
U. rivet
V. screw
W. softwood
X. threads
Y. tool steel
Z. wall anchor

1. A ductile metal with excellent heat and electrical conductivity.
2. The pattern the cells of a tree make because of the growth rings.
3. A base metal mixed with another element.
4. Lumber cut from an angiosperm (a tree whose seeds are inside a fruit).
5. Dimensional lumber processed using chemical preservatives to make it highly resistant to rot and insect damage.
6. A substance (generally a liquid) that forms a physical and/or chemical bond with materials and causes them to adhere.
7. A composite material made of Portland cement, aggregate, and water.
8. Wood that is cut to a standardized width and depth, specified in inches.
9. A fastener that has a head, a threaded shank, and a tip, and is twisted into wood or other material to secure it.
10. An iron alloy high in carbon that resists wear, is brittle, and is difficult to crush.
11. A manufactured wood product made of very thin wood layers (veneers) and laminated under high pressure.
12. A laminated structural member made with oriented strand board as the center component and dimensional lumber as the top and bottom pieces.
13. Lumber cut from a gymnosperm (a tree that is generally evergreen and bears its seed without a fruit).
14. A material made of a variety of substances in a way that makes the finished product different than the original members of the mixture.
15. A metal pin with a head and a tail piece made from a soft piece of steel or other alloy.
16. A laminated wood product made from very thin rectangular wood strands and resin that is compressed into cross-oriented layers (similar to plywood).
17. A wide variety of wood products made from scrap from other wood manufacturing processes.
18. The property of metal that allows it to be hammered or pressed (compressive force) into shape without breaking.
19. The property of metal that allows it to be permanently deformed (reshaped) by stretching (tensile force) without breaking.
20. Two flat plates held together by a pin that allows a lid or door to be attached but not stationary.
21. A relatively thin metal piece designed for general strengthening of a project; made in Tee, ELL, or straight shapes with predrilled holes.
22. A ridge of metal that spirals on an incline from the tip to the head of a screw.
23. A structural member laid horizontally from rim to rim on a floor or ceiling that carries the load.
24. A wavy, thin piece of metal, about 1/2" to 3" long, that is driven into a wood joint to pull each piece toward the joint.
25. A fastener that provides support for a screw, allowing that screw to support more weight.
26. A tough metal with a relatively high carbon content that is often machined into hard implements that can work other steels.

Know and Understand

Answer the following questions using the information provided in this chapter.

1. What are some important considerations when choosing lumber pieces for use in an agricultural project?
2. How do quarter-sawn lumber and plain-sawn lumber differ in the way they are cut? Which type of lumber is most often used for agricultural projects?
3. *True or False?* It is important to store pressure-treated lumber indoors prior to use if it is going to be used indoors.
4. Plywood and oriented strand board are both examples of _____ lumber.
5. What are some characteristics that influence the type of metal you would use for a particular project?
6. In what ways do types of steel vary?
7. Each of the following is an example of a nonferrous metal, except _____.
    A. aluminum
    B. copper
    C. iron
    D. bronze
8. What are some of the most common nonferrous metals encountered in the agricultural mechanics shop?
9. Why is copper widely used in plumbing?
10. What are some characteristics of concrete that make it useful for agricultural projects?
11. Why are thermoplastics commonly used in agricultural projects?
12. What are some qualities of thermoset plastics that make them important to the agricultural industry?
13. What are five common types of fasteners used for agricultural projects?
14. A _____ nail is very thin and has almost no head.
    A. common
    B. twisted shank
    C. brad
    D. cut
15. What are some reasons that drywall screws are used in various construction projects in addition to drywall applications?
16. What is a washer and how is it used with a bolt to join materials?
17. Explain why a corrugated fastener would be used instead of a nail.
18. Briefly explain why flush hinges may be used on many materials, including metal.
19. What is a piano hinge and in what type of application is a piano hinge used?
20. What is the function of a lockset and what are the main parts of a lockset?

STEM Connections and Academic Activities

1. **Science.** Power tools often combine the principles of two or more of the six simple machines (lever, screw, wheel and axle, pulley, wedge, and inclined plane). Observe the operation of different types of power tools and try to identify which simple machines are involved. For example, a circular saw combines the wheel and axle (rotating blade) with the wedge (cutting teeth). Try to list as many tools as possible along with the simple machines used.
2. **Technology.** Using special software, produce a podcast of an interview about composite materials used in agricultural mechanics projects.
3. **Engineering.** Build a model (such as a gate, bridge, or box) using three distinct materials, such as plastic, wood, metal, or composite. Destructively test each model, and record the strengths and weaknesses of each material. Determine which material is best for building a full-scale project of the model.
4. **Math.** Visit a local building supply center and obtain information and literature concerning the various grades and species of lumber they normally carry. Prepare written descriptions of the defects permitted in several of the grades commonly selected by builders in the area. Obtain prices of these grades to gain some understanding of the savings that can be gained by using a lower grade of lumber. Work out the difference in cost between two grades of 2″ × 4″ boards if you were purchasing 50 boards that are each 8′ long. Make a report to your class on your findings and conclusions.
5. **Language Arts.** List five products other than lumber commonly used on construction sites. Obtain a safety data sheet (SDS) for each product. An SDS can often be downloaded from the Internet or obtained from a retailer that sells the product. An SDS might be available in your school. Write a report on one of those products, explaining the information you found on the SDS. Give recommendations for working with that product.

Thinking Critically

1. When choosing materials for a project, it is important to choose the best possible material for the desired outcome. If you were building a box to hold tack, what would be the most important criteria of the building material? Explain.
2. Explain which fastener would best be used in each of the following situations and give supporting reasons for your choice.
   A. Attaching two pieces of wood together
   B. Attaching thin metal to wood
   C. Attaching two pieces of plastic together

CHAPTER 10
Project Planning and Design

Chapter Outcomes
After studying this chapter, you will be able to:
- Describe factors to consider when planning an agricultural mechanics project.
- List factors to consider when creating a project budget and getting price quotes.
- Describe design considerations for an agricultural mechanics project.
- Explain how sketches are used to demonstrate the overall look of a project.
- Explain how technical drawings, including computer-aided design drawings, are used to show an accurate representation of an object.
- Discuss the role of professional planners, such as architects and engineers, in a project.
- List sources for plans for agricultural mechanics projects.
- Explain how the quality of design, materials, and construction processes affect a project.

Words to Know 📲

aesthetics	board foot	isometric drawing	scale drawing
American Society of Mechanical Engineers (ASME)	compression strength	National Electrical Code (NEC)	style
	computer-aided design (CAD)		tensile strength
architect	cutting list	orthographic drawing	zoning code
bill of materials	engineer	quote	zoning law
		rendering	

Before You Read
Take two-column notes as you read the chapter. Fold a piece of notebook paper in half lengthwise. On the left side of the column, write main ideas. On the right side, write subtopics and detailed information. After reading the chapter, use the notes as a study guide. Fold the paper in half so you only see the main ideas. Quiz yourself on the details and subtopics.

While studying this chapter, look for the activity icon to:

- **Practice** vocabulary terms with e-flash cards and matching activities.
- **Expand** learning with interactive activities.
- **Reinforce** what you learn by completing the end-of-chapter questions.

www.g-wlearning.com/agriculture

*E*ssential to any successful project are well-laid-out plans. Planning involves design, code specifications, material lists, cost research, and steps for project development. This chapter examines project design principles and the use of technology, such as computer-aided design (CAD) programs, to facilitate planning and design.

Planning Process

When planning an agricultural mechanics project, there are several factors to be considered. These factors should be considered as part of the project planning:

- The need for the project.
- The codes or legal restrictions of the area.
- Local customs or adaptations.
- Materials available.
- Time and money available.

The planning process begins when you identify a need or problem that can be addressed with an agricultural mechanics project. A project could address an issue with storage, animal containment or housing, **Figure 10-1**, or a method to make another process easier. The need for a project generally starts with a problem to be solved. A lack of tool and equipment storage space or problems with escaping livestock, for example, present opportunities to implement a problem-solving strategy.

©iStock/Hoover Tung

Figure 10-1. This chicken coop was constructed as an agricultural mechanics project to shelter a small group of chickens for a family.

Safety and Zoning Codes

It is important to consider safety during project planning. Published safety rules are established for job sites. If a project will contain an electrical component, electrical safety must be considered. If large power equipment will be needed for a project, the proper protective gear must be worn and the appropriate safety rules must be followed.

Some agricultural lands and improvements are exempt from local codes. All codes are specific to the area, however, and must be checked locally. Depending on the purpose of the project, all applicable laws, regulations, and codes must be checked and followed. In many areas, environmental restrictions are part of the laws. The safety of the user, builder, and livestock should also be considered in the design and planning of the project.

Electrical safety rules are published by the National Fire Protection Association® (NFPA). These codes are specifications for the safe installation of electrical wiring and appliances. The *National Electrical Code® (NEC)* covers safety related to electrical installation in residential and commercial properties and also in barns and agricultural structures. The NEC is published by the NFPA every three years. The NEC is often adopted by local governments as the standard for new buildings or renovation of existing facilities. Before designing and building any project with electrical components, you should consult the local electrical codes to make sure that all applicable laws are followed.

The *American Society of Mechanical Engineers (ASME)* is an organization that provides educational programs, training, codes and standards, and other information related to engineering. An *engineer* is a professional who designs solutions to problems with creativity and specialized knowledge of science and math. The ASME codes cover a wide range of product and process safety standards that apply to individual materials and to systems. Following these codes is entirely voluntary. However, the codes detail the best safety practices of experienced engineers who know the material or system.

Zoning laws, or *zoning codes*, are local rules for private property use that generally cover safety, land and building uses, and aesthetics. *Aesthetics* means the look or style of something, such as a building or landscape. Zoning codes are local (or regional) laws that control how property can be used. While zoning codes are as varied as the local areas they represent, they often have common characteristics:

- Codes restrict where certain businesses can be established. For example, zoning laws might prohibit an agriculturalist from constructing a greenhouse in a residential area.
- Codes restrict certain practices, such as how close to dense populations livestock can be raised.
- Codes cannot be arbitrary or written just to restrict one property owner's use of her or his property.

Zoning codes are often administered by a board of elected or appointed officials. There are rules for seeking the zoning board's approval or permission for proceeding with a planned project. Being aware of and following all local codes should help avoid restrictions and interruptions to the project.

Local Adaptations

Although not usually directly considered, local preferences can influence a project. The local availability of materials, labor, and general public acceptance of the project may affect the plan. An example of local adaptations that influence design can be seen in fencing types or the style of barns characteristic to a given area. Many styles of fences and barns will adequately maintain livestock. Over time, however, specific styles have come to be associated with certain areas. For example, a black three- or four-board oak

258 Agricultural Mechanics and Technology Systems

A
©iStock.com/thoth11

B
jessicakirsh/Shutterstock.com

Figure 10-2. A—The gambrel-style roofline of this Wisconsin dairy farm barn has become an iconic feature of barns in the American Midwest. B—Many horse farms in Kentucky use a three- or four-board oak fence to enclose pastures.

fence is a traditional fence style used on Kentucky horse farms. The distinct look of a Wisconsin dairy farm barn is a style prevalent throughout the Midwest. See **Figure 10-2**.

Materials

Available materials for a project often depend on the project location and should be considered in the design. Readily available materials generally cost less than materials shipped long distances. Many materials may match the needs of the design, but the durability, cost, and environmental suitability of the materials should be considered in the design process.

Using the most corrosion-resistant material in a highly corrosive environment or a material that is easy to clean and sanitize in a food manufacturing project are examples of material suitability. Other qualities to consider in this regard include how a material is finished, insulation properties, and the weight verses strength of the material in comparison to other materials.

Material suitability and durability are closely related. In general, agricultural projects need to last for an extended period of time with minimal deterioration to be cost effective and reliable. Durable materials may need to resist chemicals, livestock waste, insects, soil contact, and changes in temperature, humidity, and sunlight. Before you chose a material, consider the exposure hazards and the material's capacity to withstand these hazards. An engineer or ASME standard can help with these considerations.

Cost of materials is often related to suitability. Local materials generally have lower cost and have immediate availability compared to materials collected and delivered from other areas. Cost can also be related to durability. Pressure-treated lumber, for example, is more costly than similarly sized lumber that is not treated. However, pressure-treated lumber is the most suitable for projects where lumber contacts the ground.

Budgeting

A budget is a plan that details amounts of money to be spent for a certain purpose, such as containing livestock or building a barn. When determining the budget, you will find it useful to consider the project cost versus the utility of the project or substitutes available. If cattle can be contained with $1000 worth of electric fencing, is spending $10,000 on a beautiful white board fence really necessary? There are other considerations besides money, but understanding how a project will be used is a major part of making an informed decision. Often, there are many possible substitutes for a given project, and a second look may find a superior project or option to solve the problem.

After determining that a project offers the best solution, you can plan a budget based on money and time available. Costs for materials should be compared by checking the same material quality and quantity at multiple suppliers. When budgeting for labor cost, compare the cost of the labor to the skill of the worker. Highly skilled and licensed workers may be more expensive per hour. However, their skill may allow them to complete work faster, thus saving money in the long run, **Figure 10-3**. Investigate a contractor's previous work and customer satisfaction to determine the best labor option.

pryzmat/Shutterstock.com

Figure 10-3. Skilled professional labor, like this plumber, can complete tasks quickly and efficiently. Budgeting for professional labor upfront has the potential to save a lot of time and money by the end of a project.

Often an entire project cannot be accomplished at one time due to budgetary or time constraints. A project plan may include several phases or stages to accomplish the project. The first stage might include completion of all preparation work and construction of basic parts of the design to establish the foundation of the project. Later stages may include more involved or intricate parts with aesthetic improvements or additional features. The budget is written to detail the phases of the project as well as the total project cost.

Total Project Cost

The total cost of a project may include materials, labor, permits, equipment rental, and design experts. Incidental costs might also include the cost of having the electric or utility company provide site service, worker comfort and safety items, access roads, and signage to aid in deliveries. All costs should be investigated and accounted for prior to starting the project.

Some costs may be part of the labor cost or general contractor's fees. It is important to know about these costs. If a general contractor or hired labor is required, having a contract for these services is recommended. Among other things, the contract should cover each of the related cost items and state who is responsible for paying the cost.

Bill of Materials

Once you decide on a project and the plans are complete, a listing of all materials used in the project should be made. **Figure 10-4** shows a bill of materials. A *bill of materials* is a complete list of all needed parts and supplies used in a project. It covers all fasteners, hardware, adhesives, and

Bird House Bill of Materials			
Quantity	Item	Unit Cost	Total Cost
2	1" × 8" × 8' Cedar board (untreated)	$8.39	$16.78
24	#6 Wood screws	$0.18	$4.32
1	Wood glue (8 oz bottle)	$3.23	$3.23
1	Dowel pins (1 pkg)	$1.65	$1.65
1	Can of paint	$6.48	$6.48
		TOTAL COST	$32.46

Goodheart-Willcox Publisher

Figure 10-4. A bill of materials for a basic bird house shows all the items needed for the project.

finish materials, as well as wood and metal. A specialty tool may also be included in the bill of materials. A complete bill of materials and a *cutting list* (exact dimensions and cutting instructions for each piece of material in a project) are valuable aids in budgeting for a project.

Calculating Material Cost

The cost of materials may be the deciding factor on whether or not a project gets built. Therefore, it is very important to calculate it correctly. Once the bill of materials is produced, it can be given to a supplier. The supplier will prepare a *quote*, which is a document that shows the price the supplier

AgEd Connection — Computer-Aided Design Software

Supervised agricultural experience (SAE) programs require planning and strategy. Teachers work with students and parents to evaluate potential SAE programs based on available resources, student interest, and other factors. An agricultural mechanics project also requires planning. You must often determine material cost and to do so you need a materials list. A materials list is usually based on a drawing of the project that one intends to build.

These drawings can be created with pencil, paper, and a ruler or using computer software. Computer aided design (CAD) software can be very expensive. However, you can download free alternatives to get you started. There are tutorials available for most CAD software, and with a little practice, you can begin drawing simple project plans right away. A computer drawing of your agricultural mechanics project will enhance your SAE and also allow you to recreate your designs more easily. Take the opportunity to research available CAD software to determine which one will best suit your project design needs.

RAGMA IMAGES/Shutterstock.com

will charge for each item on the bill of materials. Gathering quotes from multiple suppliers allows you to compare and determine the best source from which to purchase materials. This information can also be helpful in preparing the project budget.

When buying a large quantity of lumber, the price may be quoted for the total board feet of lumber. A *board foot* is a piece of lumber that measures one square foot by one inch thick (1″ thick × 12″ wide × 12″ long). Using the board foot quote rather than pricing each individual size of material separately makes it easier to price large loads of lumber when several dimensions of boards are needed.

Steel or other metal is often priced based on the gauge (thickness of the metal) or the schedule (thickness of the pipe walls) of pipe. Sheet steel comes in standard gauges from 3 to 38, which represents thicknesses from 1/4″ to 1/128″. The larger the gauge number, the thinner the material.

When buying steel pipe, you should provide the outside diameter (OD), the inside diameter (ID), and the schedule numbers to reference a certain size pipe. For the novice pipefitter, all three dimensions should be given to the manufacturer to make clear the exact size needed. Most steel pipe comes standard in 10′ lengths.

When preparing a bill of materials, you must incorporate all the materials including fasteners (nails, screws, bolts), hinges, concrete, and other components, such as gates, fencing, and appearance items for the finished project. When all components and materials are considered during the planning and design process, there are limited surprises about the project cost and outcome.

Design Considerations

After deciding that a particular project is the best option for solving a problem or filling a need, it is time to work on the project design. Begin by determining if the project is needed solely for functionality without concern for appearance. If appearance is a concern, the project might need to meet certain criteria to be pleasing to the eye. The two are not necessarily mutually exclusive. Purely functional pieces can be beautiful. However, beauty may be an afterthought rather than a planned part of the design. *Architects* (professionals who design buildings or landscapes) often create designs that are ascetically pleasing as well functional. Factors to think about when designing a project include the function of the project, style choices, engineering considerations, time requirements, and construction order.

Function

Function is the basis of the design. What is the project's function or purpose? Does the project need to contain a specific component? Does it make a process easier? Often, when a problem in the agricultural field needs to be solved, it is because a suitable item does not currently exist or a modification is needed. This requires attention to the functionality of the design.

Style

Many times, several styles are available to complete a project. *Styles* are distinctive manners of expression and can vary greatly. Thought should be given to the style of the final project before completing the design. Style considerations are especially important when you are building a project that needs to fit in with other components. For example, new cabinets in an office should match or go well with existing cabinets. A building addition to a farmstead may have more aesthetic appeal if it matches the style of existing buildings.

Engineering Considerations

Engineering structures so that they can withstand stresses and strains is an important part of the design process. Considerations include *compression strength* (the ability to withstand heavy weights or forces that reduce size) and *tensile strength* (the ability to withstand force that pulls or stretches). It is always a good idea to have an engineer design components that take precise technical knowledge. Sometimes, however, it is understood that common elements of a structure (framing or rafters, for example) will be strong enough to accomplish their task. Many local building codes require an engineer's approval on building plans to be granted final approval.

Timetable

When planning a project, you must determine a schedule or timetable. Different aspects of the job require a precise completion order. In an attempt to avoid various conflicts, you should try to follow the schedule set at the beginning of a project. When constructing a project that will contain livestock, for example, you have very little time to waste. The temporary relocation of livestock can cause additional problems. A project that is designed to contain livestock should be completed before the need is dire.

Following a time line is also important when ordering materials and scheduling work. Having the required labor without the needed materials can cause costly delays. However, ordering materials too far in advance of their use may mean that you have to pay expensive storage charges.

Your ability to accurately estimate the time needed for each step or the entire length of a project will improve as you gain experience. After working on a few projects, you will learn the time needed for common tasks. Time should be allotted for:

- Designing the project.
- Securing permits (if needed).
- Obtaining the materials.
- Acquiring the needed labor.
- Completing the necessary steps of building the project.

Construction Order

During the design phase of the project, the designer thinks through the order of completing steps in a project. Generally, after a project is designed and the materials are secured, the first step is to complete any preparatory work on the site or materials.

Assembly may need to be completed in stages. Determine which subassemblies should be started first. After all subassemblies are completed, the remainder of the project can be completed by building up from the component pieces to the final project, **Figure 10-5**. Time spent planning the construction order saves time during the building process. If one step cannot be completed until an earlier (overlooked) step is done, valuable time and money may be wasted.

Sue Smith/Shutterstock.com

Figure 10-5. Steps for the construction of a project must often be performed in a specific order. For example, the concrete foundation pad must be poured and allowed to cure before the wall framing can be started.

Sketching

When preparing a plan or design for presentation to others, you must create a graphic representation of the project. In basic form, it can be in a rough sketch done free hand on paper, **Figure 10-6**. Generally, a project design is sketched in perspective or by using isometric projection to demonstrate the overall look of the object, part, or building. In a perspective drawing, objects are made to appear with depth and distance as they do in real life.

hanayama/Shutterstock.com

Figure 10-6. Note the series of squares and rectangles that make up this hand drawn sketch of a house. The lighter lines may be erased when the drawing is finalized.

Isometric Sketches

Isometric drawing is a method for displaying three-dimensional objects in two dimensions by rotating the object 30° from the front view and tilting it forward a specific number of degrees. The drawing is created with 120° angles between each of the main axes (X, Y, Z). The view shows all three dimensions of an object in a single drawing. This style of drawing is a very common type of rendering used for technical purposes, **Figure 10-7A**. A *rendering* is a precise drawing or the act of creating a drawing. However, it is not an ideal method of creating working drawings/sketches. While the parts and pieces are depicted in scale to the other parts of the object, they cannot be measured from the drawing and translated to a scale measurement.

Isometric graph paper, **Figure 10-7B**, is available in a variety of figure sizes (triangle sizes). To use this graph paper, it is best to place the graph paper under a drawing paper to serve as a guide for edge lines. Although the graph paper can be drawn on directly, the graph lines will show and may be distracting in the final drawing.

Drawing Principles

Isometric drawing is similar to hand sketching. Follow these basic principles to create high-quality drawings:

- Look for the basic shape of the object when you are sketching. Build complex objects from simple shapes. For example, if you are drawing a box, start with a square or rectangular shape. If you are drawing a rounded object, start with a basic circle.
- Sketch with light lines until the basic shape is established. Then add darker/heavier lines to finish the drawing.
- Determine and draw details that define the idea to finish a sketch. Pay particular attention to special parts, assemblies, or other specifics.

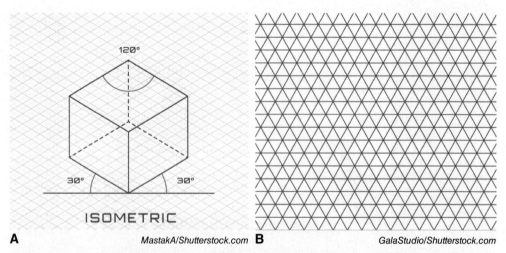

MastakA/Shutterstock.com   GalaStudio/Shutterstock.com

Figure 10-7. A—In the isometric drawing of a cube, all lines and faces are equal and the view of the cube is angled 120° between each of the projected X, Y, or Z axes. Horizontal lines are drawn at 30° from the normal horizontal axis. B—Isometric graph paper can be used in creating drawings.

- Choose one original viewpoint and maintain it as the drawing point. Generally, it is easy to start with the top left point or bottom left where the front and top or side views adjoin. If the object is available, actually view if from this angle.
- Use isometric graph paper to draw accurate measurements at the correct angle. Isometric graph paper templates can be accessed on the Internet.
- Remember that in perspective drawing, items farther away appear smaller. However, in isometric drawing, the lines are mathematically constant and skewed at an angle from a front view.

Technical Drawings

Technical drawing or drafting is done to scale, which differs from hand sketching or drawing. A *scale drawing* is an image that uses a ratio (the scale) to reduce or enlarge the dimensions of an object for an accurate representation of the object. Very tiny objects, such as computer components, or very large objects, such as buildings, can be precisely rendered on paper. Technical drawings also use multiple views to communicate all aspects of the object.

Drawing to Scale

When you draw to scale, you must understand ratios. The scale indicates the ratio of the drawing length to the actual length of the real object. For example, a house plan drawn to 1":36" scale would show a house that is 24' long as 8" long on the scale drawing. Any ratio can be chosen for a drawing as long as the scale allows the illustration to fit on the available drawing space. It is also important that the drawing be scaled correctly to show enough detail to be useful.

Imagine a drawing where the scale is 1":3' (or 1":36".) The ratio is always written as scale size to real size.

If your real object is 144"tall, how long does the scale line need to be drawn for the height of the object?

1":36" = X":144" **OR** 1"/36" = X"/144"

(1 × 144) ÷ (X × 36)

144" ÷ 36" = X

X = 4"

If your scale drawing is 2.5" long, what is the real length of the object?

1":36" = 2.5":X" **OR** 1"/36" = 2.5"/X"

(2.5 × 36) ÷ (X × 1)

90" ÷ 1 = X

X = 90"

Illustrations are created with a scale measurement tool that uses common measurement ratios on its edges. The tool was designed as an aid for drafters to eliminate manually calculating each line measurement. The instrument

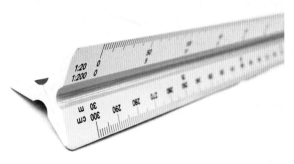

Bokic Bojan/Shutterstock.com

Figure 10-8. On this engineer's scale, note how each edge uses a different scale.

is available as a flat ruler or a ruler built similar to an equilateral triangle with three points and six edges. See **Figure 10-8**.

The engineer's scale is typically divided into ratios of decimal inches; 1:10, 1:20, 1:30, 1:40, 1:50, and 1:60, with each decimal inch scale on a different edge. The architect's scale is not as standardized as an engineer's scale and can contain a wider variety of scales on each edge. The most common scales consist of a 1:1" scale (a standard ruler with 1/16" marked lines). The five remaining edges have two scales each (3/4", 1/2", and 3/8") that are read one from each end of the tool.

Specifications

When completing a technical drawing or drafting by hand, you should follow several specifications or standards used in drawing plans. The standards are set by national engineering and architectural societies or other professional organizations. The specifications cover almost every aspect of drawing, from line thickness to placement of the views on the paper. Details need to be standardized so the drawing can communicate as much information as possible to as many people as possible. The specifications and standards reduce the miscommunication from local terms that might be a problem in describing objects. Specifications that even a beginning student draftsman should know include:

- Edge or outline lines that are visible are the darkest, thickest lines on the drawing.
- A plan title box should go at the bottom. It should include the draftsman's name, the object name/part number, the date of the drawing, and the scale.
- Edge lines that are not visible in actual view should be drawn in as dotted lines. If the edge is partially visible and partially hidden, it can be started as a solid line and then changed to a dotted line.
- Dimension lines should have arrow heads and should be kept to a minimum on the drawing. They should not be redundant or repetitive. Each part should be labeled clearly with its dimensions.

Dimensions

When drafting a plan for a project, labeling the dimensions precisely is critical. Dimensions are drawn for each part and for the whole project. **Figure 10-9** shows some basic types of lines that are used on plans (engineering or architectural) to indicate measurements or directions:

- Border lines—dark, heavy lines that mark the drawing surface edges.
- Object lines—solid lines (lighter in weight than border lines) that are used to mark the outline of an object. Within this category, dashed lines that mark object edges and intersections that cannot be seen from the view are called hidden lines.

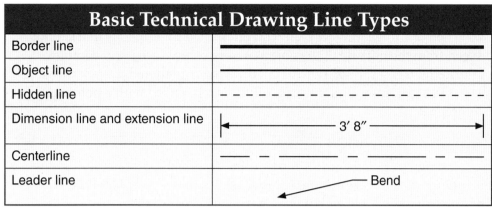

Figure 10-9. These dimension line types are used on plans to indicate measurements or directions.

- Dimension lines—fine, solid lines with arrowheads at the ends. If possible, include the dimension within the line. Arrowheads at the end of dimension lines should be three times as long as they are wide.
- Extension lines—fine, solid lines that extend from each edge of an object to which a measurement refers. The lines should have a small gap from the actual object and extend far enough to have room to show the dimension label in an uncluttered manner.
- Centerlines—fine, long lines that pass through the center of circles and symmetrical objects. They start as a solid line with a small gap, a dash, and another small gap and finish with a solid, long line. It is the generally accepted convention that the dash is positioned at the center of the circle.
- Leader lines—lines that point out informational text on a plan. They are inclined straight lines that terminate with an arrowhead touching the object to which the text refers and a small perpendicular line that connects to the text or dimension. Leader lines should never cross each other, should not cut across corners of an object, and should be radial (pass through the center if extended through the circle) when pointing to a circle.

Types of Technical Drawings

Drafting or technical drawing is a skill used by engineers, architects, designers, skilled tradesmen, and construction workers to share information. Several types of technical drawings have specific uses for different workers:
- Detail drawings.
- Sectional drawings.
- Orthographic drawings.
- Assembly drawings.
- Plan drawings.

Detail Drawings

When drafting a project plan, you often encounter details that cannot be seen from the outside of the object or very complicated and complex parts. In these cases, a separate detail drawing is rendered. Detail drawings are put in a callout or pullout plan, which is referenced with a note on the main plan.

Sectional Drawings

Sectional drawings (also simply called sections) are generally used when the interior of an object is more complicated than the outside would suggest. They are sometimes called cutaway drawings because the outside of the object is drawn with one section cut away. The cut in the cutaway drawing should be labeled with the center or cut line, and a notation on the main drawing should indicate from where the section is cut. There are many styles of cutting that can be used:

- Full section—the entire piece is cut away, usually along the centerline.
- Half sections—half the part is cut away, usually along a centerline.
- Offset section—part of the object is rendered from the outside view and the remainder is rendered in a cutaway view.
- Removed sections—interior sections shown with the detail removed from the part. The removed section and the main part are labeled to indicate where the section is removed.
- Broken out sections—used when a part can be shown "broken" away. The broken line is drawn in a jagged line.

Orthographic Drawings

An *orthographic drawing* is a two-dimensional image of a three-dimensional object. Generally, the drawing includes three views of the object, such as the front, side, and top, **Figure 10-10**. In some instances, more views are required. For objects that do not have a natural front view, the view that affords the most detail is rendered as the front. If the top and bottom have significant differences, they are both rendered. Orthographic drawings are made as if the object was placed in a glass box, viewed from each side, and then rendered with two-dimensional lines.

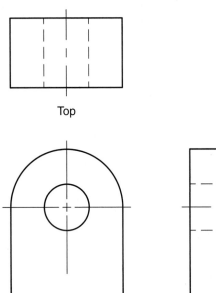

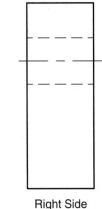

Goodheart-Willcox Publisher

Figure 10-10. A basic orthographic drawing of a part shows the front, top, and side view of the part.

Assembly Drawings

Assembly drawings show how the parts of an object are put together to make the finished piece/assembly. Often these are the "exploded" view included in parts catalogs. They are also used to assemble the pieces and subassemblies into the final parts. Each piece is typically labeled with an identification number and rendered in order with the piece it is connected to in the operational assembly.

Plans

Drawings of large scale projects, such as landscapes, buildings, public utilities, and plots, are referred to as plans. Architects render building plans, including layout and design details. Engineers include structural and strength details of the project. There are slight differences in the drawing conventions used by architects and engineers. You should pay careful attention to the type of drawing you are using. These plans might include:
- Plot plan—the common view of landscapers, architects, and civil engineers. The plans, also called site plans, include roads, driveways, utilities (both above and below ground), soil contours, and proposed projects.
- Floor plan—the plan rendered for approval before a building is constructed. Floor plans show the layout and size of buildings, individual rooms, and the layout of electrical and plumbing equipment. They commonly include windows, doors, and other structural components. They may also include layouts of furniture, textures, and colors when used by designers.
- Foundation plan—a detailed drawing of how a site is excavated, footers are poured, and foundation walls are assembled, including connections to the building to be built. These plans are rendered by a building design engineer who has special knowledge of soil and its structure.
- Structural plan—a drawing rendered by an engineer with the inclusion of mechanical systems layout and structural strength details (not shown on floor plans).
- Utility plan—the locations of public utilities, such as water, electricity, storm water drainage, and sewage removal. The plan shows how they will be affected by the proposed project and is often required by government agencies before a building or repair permit will be approved.

Hand Drawing Tools

Hand drawing tools are commonly available tools. A writing utensil, preferably a pencil, and a piece of paper are the simple tools needed for basic hand sketching. As you gain experience with technical drawing, other tools will make it easier to a create quality drawings:
- Drawing board—a board or table top with square corners and even sides, generally made of composite materials with steel or laminate edges. It sometimes includes a pencil tray and the ability to adjust to an angle for drawing or writing. A paper can be attached in the center that is parallel to the bottom and sides. The straight sides allow a T-square to be used to draw lines parallel to the edges.

- T-square—a tool with a T-shaped head that fits against the edge of the drawing board or table and an oversized blade that is used for drawing lines parallel to the bottom of the edge of the board, **Figure 10-11A**. It is made of plastic, wood, aluminum, or a combination of materials.
- Compass—a steel or plastic drafting instrument used to draw a circles with a precise radius, **Figure 10-11A**.
- A scale measurement ruler that can be used to convert and transfer measurements to a drawing quickly.
- Set square or triangles—a hard plastic tool that generally comes in two styles: a right triangle with opposing 45° angles and a right triangle with opposing 30° and 60° angles. They can be used individually to draw perpendicular lines or used in combination, **Figure 10-11B**, to draw at a variety of angles.
- French curve—a specialty drawing tool usually made of plastic, steel, or other metal, **Figure 10-11C**, used for drawing organic (nonstandard) arcs and other curved lines. The tool may also contain templates for drawing standardized parts, such as sinks, toilets, and doorways.
- Brush—a thin brush used to remove pencil and eraser dust when correcting mistakes. Removing dust keeps the drawing clean and reduces the smudges that might obscure the design lines.

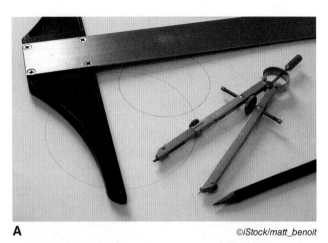

©iStock/matt_benoit

tgpooley/Shutterstock.com

sasaperic/Shutterstock.com

Figure 10-11. A—A T-Square and compass are used for drawing lines and circles. B—You can draw various angles on a plan by using set triangles. C—French curve tools are used for drawing arcs and other curved lines.

Computer-Aided Design (CAD) Programs

Computer-aided design (CAD) is the use of computers and software to design, draw, and model parts, objects, or structures. CAD began as computer-aided drafting. Even for those just entering the drafting field, CAD is much more common today than drawing by hand. Drafting by hand, however, is often taught as an introduction to CAD. This practice helps the student understand the basics of drafting and begin to think in terms of two dimensions versus three dimensions.

As personal computers have become more powerful, the CAD software has become more detailed. The designer can use this software in making decisions about more than the geometric drawing components. Originally, CAD software simply allowed the user to display geometric calculations graphically. The user generally could not make enough calculations to display an object drawn in two dimensions as a three-dimensional object.

Hands-On Agriculture

Orthographic Drawing

On graph paper, practice accurately drafting an orthographic drawing of a small rectangular box chosen by your instructor.

1. Use 11″ × 17″ white paper.
2. Draw a 10 1/2″ × 16″ box in the center of the paper. This creates a border around the drawing area.
3. Add a rectangular box 1/2″ high across the 16″ bottom side of the first box and divide it into five sections.
4. In the far left section, add the plan number. In the next section, add the date. All writing in the label boxes should be printed in capital letters.
5. In the middle section, add the scale that is used to draw the plan (for example, 1″:1.5″).
6. In the next section, write the plan name (for example, *Nail Box*).
7. In the far right section, add the creator's name.
8. Obtain a small rectangular box and measure each side. Choose a front, side, and top view of this box. (Sometimes it is helpful to label the view names.) When viewing the object, imagine that it is in a glass box. Each face of the box is one *projection* or view of the object. This allows for six projections or views of any object. Generally, however, the minimum number of views or projections is preferred in technical drawings.
9. At the top left on the remaining space inside the large rectangle, draw a top view of the item to be drawn. If done correctly, it will look rather like a two-dimensional rectangle on your paper. Label this with the correct dimensions for the box.
10. In the bottom left area, sketch a side view of the item. Label it with the correct dimensions.
11. In the remaining area, draw the top view. If the top and bottom views are radically different, draw the bottom view also. Make sure to label all dimensions for each projection.

When you have finished, have your instructor check the drawing for accuracy.

Today, however, most basic CAD programs can turn two-dimensional drawings into three-dimensional displays. They can also allow the designer to work directly on the three-dimensional model to make changes, **Figure 10-12**. Greater design capabilities mean greater freedom and more complexity for modeling and displaying floor plans in three dimensions. The programs are also used to design prototypes and projects for 3-D printing. CAD programs can be used to produce very detailed renderings of both architectural plans and engineering drawings. The software can quickly compute how a change in one part of the design affects dependent parts and alter all related measurements.

As the power of CAD software has expanded and the technology has become more available, the cost has decreased. The programs are also more user friendly. Designers learn to use various program features and commands and then practice to become proficient. Modeling objects in both two dimensions and three dimensions makes it easy for viewers to visualize the final product.

In addition to creating drawings and plans, designers can use design files and 3-D printers to make three-dimensional solid objects. Engineers, fabricators, architects, and other industry workers must master this technology as it becomes more prevalent in the industry. This 3-D printing technology allows replacement parts, such as a gear for a machine, to be printed as needed.

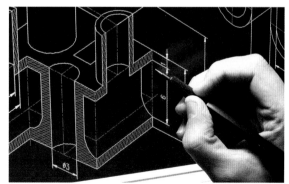

FERNANDO BLANCO CALZADA/Shutterstock.com

Figure 10-12. Using a stylus on a touch-screen monitor, an engineer makes alterations on a part scale drawing created using CAD software.

Professional Planners

Some projects can be elaborate enough to make the required designs too detailed or difficult for an inexperienced drafter. These cases call for the professional experience of an architect or an engineer, **Figure 10-13**. Local zoning codes will often require that a professional design or approve plans.

The cost of hiring architects or engineers can often be worth the specialized skill and knowledge these professionals bring to buildings or other projects. For example, a project that does not meet local building codes may need to be reworked. The cost for this rework might have been avoided by consulting an architect or engineer before starting the project.

Monkey Business Images/Shutterstock.com

Figure 10-13. A complex building project may require the expertise of a professional architect. In accordance with local code, an architect may have to examine and approve plans before building can begin.

Other Sources of Plans

Besides using a professional to design plans that will fit your criteria, you can consult other plan sources. The Internet, project books, and publications from the United States Department of Agriculture (USDA) or university cooperative extension services may have a plan that suits your needs.

Internet Sources

The Internet has a treasure trove of plans for agricultural mechanics projects. Plans available online are often very specific for solving a given problem. However, you should read the plans carefully before deciding to use them. Problems can compound quickly when you use an unverified plan. Unless it is found on a reputable, trusted website, there is no way to know if a plan posted online will work as claimed.

Project Books

There are many reliable books that contain plans for agricultural mechanics projects. You can find project books at a local hardware store, lumber yard, or library. Project books can also be purchased online. Generally, published plans offer the added security that someone constructed and tested the plan. A drawback to these project books is the cost. If the buyer is interested in one project only, the entire book must be purchased instead of just that individual project. This cost, however, could be eliminated if you can find and checkout project books at your local library.

Extension Publications

The United States Department of Agriculture (USDA) and local offices of university cooperative extension services often provide quality plans for agricultural mechanics projects. Many agriculture engineering departments at these universities make a variety of project plans available via the Internet. Often, they are also available as printed materials.

Adapting Plans

A plan that you purchase may not be exactly what is needed. You should evaluate the plan and, possibly, adapt it for your own use. When evaluating or adapting plans, you should consider several questions:
- Does this plan solve the given problem?
- Is this the best way to solve this problem?
- Does this plan or project fit with other components of the system or styles in use?
- Will measurement conversions be required?
- Are the correct materials and tools available?

Construction and Material Standards

For any project, the quality of the design, the materials, and the construction work determine the quality of the finished project. If you follow the guidelines established by national organizations, such as NFPA and ASME and other industry overseers, the craftsmanship and quality of the project should be commendable. When you adhere to local laws and codes, the project will pass inspection from local authorities. The use of quality materials that conform to recognized construction standards will allow you to create strong, reliable projects that are protected against preliminary failure. Along with compliance to these factors, safety rules on the job site should be at the forefront of worker's minds.

Preparing a Bid Package

Large projects are sometimes *contracted out*. This involves employing a contractor or company to complete the project for you. When contracting out a project, you must seek bids for completing the project and then write a legal contract that covers the entire scope of the project.

When seeking bids, you must prepare a bid package. The items that should be covered in the most basic bid package include:
- Drawings or blueprints of the project.
- Project timeline and plans, including start date, completion date, and milestone dates.
- Material specifications and quantity.
- Information on project delays, penalties, and cost overruns.
- Your contact information or that of the project manager.

There are many ways to put a bid package together. As long as all pertinent information is included, the package can be very informal or formal depending on the scope of the project.

Contractors will use the bid package to prepare a quote, or bid, for completing the project. The quote should include material charges, labor charges, timelines and schedules, and special requirements such as the need for permits and inspections. You can then compare bids from various contractors based on similar criteria. While one contractor's bid may be cheapest overall, the quality of materials may be substandard or worker qualifications may be lacking.

CHAPTER 10 Review and Assessment

Summary

- Planning for an agricultural mechanics project involves considering the need for the project, safety and zoning codes, local adaptations, and available materials, time, and money.
- A budget is a plan that details amounts of money to be spent for items needed for a project. The budget is written to detail the phases of the project as well as the total project cost.
- A bill of materials for a project can be given to a supplier who will prepare a price quote. The prices from the quote can be used in preparing a project budget.
- Factors to consider when designing a project include the function of the project, style, engineering considerations, time requirements, and construction order.
- Hand sketching allows a designer to share his or her vision with others through an accurate drawing.
- An isometric drawing shows a three-dimensional object by rotating the object and tilting it forward to show all three dimensions in a single drawing.
- A technical drawing uses a ratio (the scale) to reduce or enlarge the dimensions of an object for an accurate representation of the object.
- Different types of technical drawings (detail, sectional, orthographic, assembly, and plan) are used for specific purposes.
- Drawings of large-scale projects, such as landscapes, buildings, public utilities, and plots, are referred to as plans.
- Computer-aided design (CAD) programs can be used to design, draw, and model parts, objects, or structures.
- Design professionals, such as architects and engineers, can plan an agricultural mechanics project using their knowledge of materials and design.
- Other sources of project plans include the Internet, project books, and publications from the United States Department of Agriculture or cooperative extension services.
- Adhering to construction and material standards will help ensure that structures are properly built and safety rules are followed.

Words to Know

Match the key terms from the chapter to the correct definition.

A. aesthetics
B. American Society of Mechanical Engineers (ASME)
C. architect
D. bill of materials
E. board foot
F. compression strength
G. computer-aided design (CAD)
H. cutting list
I. engineer
J. isometric drawing
K. National Electrical Code (NEC)
L. orthographic drawing
M. quote
N. rendering
O. scale drawing
P. style
Q. tensile strength
R. zoning code

1. A distinctive manner or expression that is often related to the form, appearance, or character of a project.
2. A document that shows a price given by a dealer or seller for an item to be purchased.
3. A local rule for private property use that generally covers safety, land and building use, and aesthetics.
4. A method for depicting three-dimensional objects in two dimensions by rotating the object from the front view and tilting it forward, showing all three dimensions of an object in a single image.
5. The use of computers and software to design, draw, and model parts, objects, or structures.
6. A complete list of all parts and supplies used in a project, such as wood, metal, hardware, fasteners, and adhesive.
7. A two-dimensional image that shows a three-dimensional object from different views, usually the front, side, and top.
8. An organization that provides educational programs, training, codes and standards, and other information related to engineering.
9. The look or style of something, such as a building or landscape.
10. The ability to withstand heavy weights or forces that reduce size.
11. A professional who designs solutions to problems with creativity and specialized knowledge of science and math.
12. A set of rules that covers safety related to electrical installation in residential and commercial properties and in agricultural structures.
13. An image that uses a ratio to reduce or enlarge the dimensions of an object for an accurate representation of the object.
14. A precise drawing of an object or the act of creating a drawing.
15. A list of exact dimensions and cutting instructions for each piece of material in a project.
16. Lumber that is 12″ long by 12″ wide by 1″ thick or some combination that equals this size.
17. A professional who designs buildings and landscapes.
18. The ability to withstand forces that pull apart or stretch something.

Know and Understand

Answer the following questions using the information provided in this chapter.

1. What factors should be considered when planning an agricultural mechanics project?
2. What is the National Electrical Code® (NEC)?
3. *True or False?* Zoning codes restrict where certain businesses can be established.
4. What is a project budget?
5. What are some examples of costs that should be considered when determining the total cost of a project?
6. What are some factors to think about when designing a project?
7. How does an isometric drawing differ from a perspective drawing?
8. A rendering of a machine part has a scale of 1":7.5". If the length of the part on the drawing is 3.5", how long is the actual part?
9. On a technical drawing, a line with an arrowhead on one end that is used to point out informational text on the drawing is called a(n) _____.
    A. object line
    B. extension line
    C. leader line
    D. centerline
10. What is the purpose of an assembly drawing?
11. What information is typically shown on a floor plan?
12. Which tool can be used to hand draw organic (nonstandard) arcs and other curved lines?
    A. French curve
    B. set triangles
    C. compass
    D. T-square
13. What are some uses for CAD software?
14. When might a design professional, such as an architect or engineer, be needed for a project?
15. Besides designing your own plan or consulting a professional, what are some other sources of plans for agricultural mechanics projects?
16. What are some questions that you should ask yourself when evaluating or adapting plans for a project?
17. What factors determine the quality of a finished agricultural mechanics project?
18. Name five items that should be included in every bid package.
19. Why is it important to compare bids based on similar criteria rather than automatically selecting the lowest bid?

STEM and Academic Activities

1. **Science.** Set up a science fair project or an experiment to determine the relative mechanical properties of natural versus man-made materials. For example, test the tensile strength of a 1″ × 6″ × 8′ pine board, a 3/4″ × 5 1/2″ × 8′ plywood piece, and a piece of laminated beam the same size. Enter a local science fair with your project.
2. **Technology.** Learn to use available or free CAD software to design a box project. Include a perspective, orthographic, or 3-D drawing of the box.
3. **Engineering.** Plan a project to build a model bridge with toothpicks or paper cups. Create a project plan that includes the project goal, the scope and timetable, steps to implement the project, and final evaluation criteria for the project. Submit the written plan to your teacher.
4. **Math.** Choose an agricultural mechanics project and compile a bill of materials. Using this bill of materials, calculate the total project cost from two local and one Internet supplier of materials.
5. **Social Science.** Determine what local building codes govern the construction of a new greenhouse built on school property. Choose one code and research why this code is law in your area.
6. **Language Arts.** Without using any graphics, drawings, or pictures, write or explain the plans for building a storage box that is 10″ × 22″ × 11″. Assume that your audience is an experienced woodworker, but remember to describe the steps with enough detail to satisfactorily build the box.

Thinking Critically

1. What is the impact of local adaptations on project planning? Why have these local preferences become an important consideration in planning a project?
2. Why are local zoning codes used to determine how private property is used? Determine what officials create or oversee local codes in your area.

CHAPTER 11
Fundamentals of Woodworking

Chapter Outcomes
After studying this chapter, you will be able to:
- Identify wood types and characteristics.
- Select the type of wood best suited to a project.
- Describe types of handsaws, power saws, and stationary saws used for cutting wood.
- Describe tools used for shaping and smoothing wood.
- Describe tools used for drilling holes in wood.

Words to Know ↗

appearance lumber	kerf	push stick	stainability
blind cut	milling	rasp	stress graded
dressed dimensions	miter gauge	relief cut	utility lumber
file	nominal dimensions	rip fence	
hand plane	nonstress graded	router	
hone	planing	shaver	

Before You Read
Examine the charts in the chapter before you read it. Write down questions you have about them. Try and answer the questions as you read.

While studying this chapter, look for the activity icon to:

- **Practice** vocabulary terms with e-flash cards and matching activities.
- **Expand** learning with interactive activities.
- **Reinforce** what you learn by completing the end-of-chapter questions.

www.g-wlearning.com/agriculture

Woodworking is both a fun and functional activity. Wood can be made into products that are useful and decorative. Using the best wood and the right tools for the job can lead to better results for your projects. Using the incorrect wood type or the wrong tools may ruin the final project or may even be dangerous. For example, suppose a support railing is made from wood that is not strong enough. The railing might break or collapse, leading to injury.

Wood Types and Characteristics

Two basic types of wood are hardwoods and softwoods. Hardwoods come from deciduous trees. Softwoods come from coniferous trees as detailed in Chapter 9, *Materials, Fasteners, and Hardware*. Because wood is a natural product made from trees, there is some variation among boards. For example, the color, grain pattern, strength, and flexibility may vary among pieces of wood. In general, the characteristics of pieces of wood from one species are similar. For example, pieces of wood from pine trees will be similar even though they come from different pine trees. The characteristics of wood include the type of wood (hardwood or softwood), color, grain pattern, density (hardness), flexibility, rot resistance, and insect resistance. These factors should be considered when selecting wood for a project. The standard types of wood available at many lumberyards are listed in the chart in **Figure 11-1**.

Wood Selection

When selecting wood for a project, first consider what is being built. Then choose a type of wood to fit the design requirements. Density or hardness is often the first characteristic to be considered. The harder the wood, the less easily it cuts. Using very hard wood can make some cuts quite difficult to accomplish. On the other hand, easily cut wood often does not wear well enough to make it a durable finished project. *Stainability*, or the ability of wood to take stain or paint, is also important. If the project cannot be painted, a wood with an interesting grain pattern and a good ability to take stain evenly should be selected. After these basic criteria are considered, the cost or availability of a specific wood is often the deciding factor in selecting wood for a project. When selecting lumber, it is important to understand lumber grades, dimensions, and defects.

Grades

Lumber grades separate the hardwoods and softwoods based on their general uses. Hardwoods are graded by the National Hardwood Lumber Association (NHLA). The NHLA sets the standards or grades for hardwood lumber, helping buyers to determine the exact quality of this natural product. These grades are based upon the length of clear lumber (wood without defects) that can be cut from a piece of wood for use in the furniture or millwork trades. See **Figure 11-2**.

Wood Characteristics

Type of Wood	Tree Type	Hardness	Stainability	Relative Cost
Birch	Deciduous	Hard, often used in veneered plywood for making cabinets.	Often gets blotchy with stain but takes paint well.	Moderately priced but often a lower cost option in plywood veneer.
Cedar/Western Red Cedar	Coniferous	Relatively soft, straight grain; resistant to rot; slightly aromatic.	Reddish-colored wood, often oiled for the finish.	Moderately priced.
Exotics		Many types and densities available; consult lumberyard for availability and price.	Varies by color and grain patterns.	Generally highest-priced options.
Fir/ Douglas Fir	Coniferous	Hard, one of the most dense softwoods, almost no grain pattern.	Finish with paint due to the weakness of the grain pattern.	Moderately priced.
Hard Maple	Deciduous	Very hard, often has unique grain pattern and stable boards.	Takes stain well.	Often lower cost than other hardwoods, not always available.
Pine	Coniferous	Soft, often used in craft projects.	Stains well if sealed first but is often painted.	Least expensive and generally available in most lumberyards.
Poplar	Deciduous	Soft, often used in crafts and toys, especially if painted.	Often dark grey or greenish streaks, takes paint well.	Often least costly of hardwoods.
Red Oak	Deciduous	Hard, interesting grain; standard in furniture making and flooring.	Excellent.	Moderately priced.
Redwood	Coniferous	Soft, straight grain lumber; useful in outdoor furniture projects due to its resistance to rot.	Reddish cast; often stain is not true to color when going against the natural coloring.	Moderately priced.
Soft Maple	Deciduous	Hard, but not as hard as hard maple; great for furniture and carved items; easy to work with.	Takes stain well.	Often lower cost hardwood than other hardwoods; not always available.
White Oak	Deciduous	Hard, slightly harder than red oak, outdoor furniture and other furniture.	Takes stain well.	Moderately priced.

Goodheart-Willcox Publisher

Figure 11-1. This chart describes the different characteristics of various types of wood.

Hardwood Grades			
Grade	Industry Designation	Minimum Size	Clear Area
Firsts and Seconds	FAS	6" × 8'	83⅓%
Firsts and Seconds—One Face	F1F	6" × 8'	83⅓% (on one side only)
Select	SEL	4" × 6'	83⅓%
No. 1 Common	1C	3" × 4'	66⅔%
No. 2A Common	2C	3" × 4'	50%
No. 2B Common	3AC	3" × 4'	33⅓%
No. 3A Common	3BC	3" × 4'	25%

Goodheart-Willcox Publisher

Figure 11-2. Hardwoods are graded by minimum size and the percentage of clear lumber that can be cut from a board of this size.

At lumberyards, the Firsts and Seconds (FAS), First and Seconds—1 Face (F1F), and Select grades are often called Selects because the yard mixes these three grades of lumber in the same lot. Grades No. 2A Common (2C) and No. 1 Common (1C) are generally called No. 2 Commons. However, the No. 2 Commons lot may contain lumber from both 2A and 1C grades, although not in a standardized amount.

Softwoods can be grouped in the following categories:
- *Nonstress graded*—wood used as construction material. Nonstress graded lumber, also called common lumber, is usually lumber less than 2" thick.
- *Stress graded*—wood used in applications where strength and stability are important. Often called dimensional lumber, stress graded lumber is always 2" to 4" thick and is used for framing and other stress-bearing applications.
- *Appearance lumber*—wood used for making projects that appear free from defects. It is graded according to the overall look of the piece.

Each category has a unique lumber grading system. See **Figure 11-3**. In the stress graded category, there is also a designation of utility. *Utility lumber* is stress graded wood used for rough framing and may have distinct warping or twisting.

Dimensions

When selecting wood for a project, you need to understand the dimensions used for lumber. Standard dimensions for lumber in the United States are given in *nominal dimensions*, which are the measurements of a piece of wood before kiln drying. After kiln drying, wood is milled and planed. *Milling* is cutting a piece of lumber to a certain size, and *planing* is shaving and smoothing the edges of the cut lumber. The final measurements of a piece of lumber after it has been kiln dried, milled, and planed are called

Softwood Lumber Grades

Stress Graded (Dimensional) Lumber

Grade Name	Size	Uses
Select Structural	2″ × 4″ and larger	Post, timbers, beams, studs, structural members.
No. 1	2″ × 4″ and larger	Framing members.
No. 2	2″ × 4″ and larger	Framing members.
No. 2	2″ × 4″ and larger	Framing members.
No. 2 & BTR	Same criteria as No. 2 but may also contain some No. 1 and/or selects.	Framing members; the stronger the lumber, the fewer the defects.
No. 3 & BTR	Same criteria as No. 3 but may also contain No. 2s.	Framing members; the stronger the lumber, the fewer the defects.

Nonstress Graded (Common) Lumber

Grade Name	Defects	Use	Finish
Construction/No. 1	Tight and moderate number of knots.	Some furniture, shelving, cornices, paneling, and siding.	Accepts paints well; some can be stained.
No. 2 Standard	Larger and more numerous knots.	Some furniture, shelving, cornices, paneling, and siding.	Accepts paints well.
No. 3 Utility	Splits, knotholes.	Subflooring, sheathing, crates, and small furniture parts.	Does not accept paint well.
No. 4 Economy	Many splits and knotholes, much waste area.	Sheathing, subflooring, and concrete forms.	Does not accept paint well and is rough.
No. 5 Economy	Large waste areas, coarser defects.	Pallets and rough concrete frames.	Does not accept paint well.

Appearance (Select) Number

Grade Name	Defects	Use	Finish
Finish—May also contain custom patterns.	Rare, smooth finish.	Quality furniture, trim, siding, flooring, and paneling.	Stained or painted.
Selects A and B—Often sold together as B & BTR.	No knots, splits, checks, or visible defects.	Fine furniture, cabinetry, trim, and flooring.	Stained or painted.
Select C	Small tight knots; may be perfect on one side.	Most furniture, shelving, and some trim and flooring.	Stained or painted.
Select D	Pin knots; numerous small blemishes; may be mostly on one side.	Some furniture, shelves, and some trim and flooring.	Stained or painted.

Goodheart-Willcox Publisher

Figure 11-3. Softwoods are categorized as stress-graded, nonstress-graded, or appearance lumber. Each category has its own grade classification system.

SAE Connection

Woodworking Opportunities

Do you enjoy working with wood? There is a certain satisfaction that comes from designing and building things from wood. A basic understanding of the material covered in this chapter will open the door for numerous opportunities for future earnings. Anything from small toys and furniture to large cabinets and frames for buildings can be constructed from lumber. Knowing the difference between types of wood and the best uses for each type, along with types of fasteners available and how to use them, goes hand in hand with the use of tools and finishing products to produce a quality product that will sell. You could work for someone with a construction business or a cabinet shop. You could be self-employed and build your own furniture or even refinish old furniture.

While studying this chapter, ask your instructor for guidance in researching potential jobs. Determine if you have the interest in a woodworking career by taking the opportunity to construct a project during class. Make sure to document the hours you spend in the research, layout, design, construction, and finish of your project to help complete an accurate proficiency application when the time comes. Woodworking jobs are great for SAE projects.

guruxox/Shutterstock.com

Lumber Sizes	
Nominal size	**Dressed size**
1″ × 4″	¾″ × 3½″
1″ × 6″	¾″ × 5½″
1″ × 8″	¾″ × 7¼″
1″ × 10″	¾″ × 9¼″
1″ × 12″	¾″ × 11¼″
2″ × 4″	1½″ × 3½″
2″ × 6″	1½″ × 5½″
2″ × 8″	1½″ × 7¼″
2″ × 10″	1½″ × 9¼″
2″ × 12″	1½″ × 11¼″
4″ × 4″	3½″ × 3½″

Goodheart-Willcox Publisher

Figure 11-4. Notice the difference between the nominal sizes and dressed (actual) sizes of dimensional lumber.

the *dressed dimensions*, and the dressed dimensions are smaller than the nominal dimensions. For example, a piece of lumber with nominal dimensions of 2″ × 4″ has actual or dressed dimensions of 1 1/2″ × 3 1/2″. Lumber comes in a wide variety of standardized sizes, as listed in **Figure 11-4**, and these sizes are generally available at lumberyards without requiring a special order.

Lumber Defects

All species of hardwood and softwood lumber will contain varying degrees of imperfections called defects. Wood from high grades will have fewer defects than wood from low grades. Your ability to recognize these defects will help you select wood that will be appropriate for a project. See **Figure 11-5** for examples of the following natural defects found in lumber:

- Bow—a warp on the face of a board from end to end.
- Check—a crack along the annual growth rings.
- Crook—a warp along the edge of the board; also referred to as a crown.
- Cup—a curvature from edge to edge on a board.

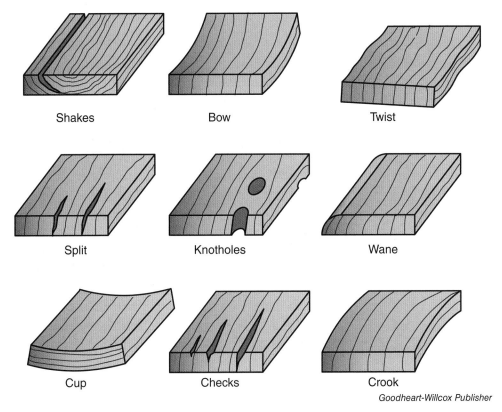

Figure 11-5. Some defects are easy to recognize. The more subtle defects are also detectable by individuals with a trained eye.

- Knot—a defect left in wood from a growth of a branch. Large knots weaken the wood.
- Knothole—an opening or hole in a piece of wood where a knot has fallen out. Loose knots are also to be avoided.
- Shake—a separation of the grain between the growth rings.
- Split—a crack that originates on the ends and extends lengthwise on the board.
- Twist—one or multiple bends in a board.

Cutting Wood

There are several useful hand tools and power tools for cutting wood. Handsaws are generally less expensive than power saws and are available in a wide variety of styles. The project being built and the material being used will generally determine the type of saw you will need.

Handsaws

Handsaws, both rip cut and crosscut models, are commonly used on wood of large dimensions where a smooth cut is not necessary. Backsaws and dovetail saws are used on smaller dimensions of wood where a fine cut is desired. Coping saws and keyhole saws are designed to make curved cuts and internal cuts on wood.

Crosscut and Ripsaws

The crosscut saw is used for cuts across or perpendicular to the grain of the wood. The saw has a wide set in its teeth, creating a wide kerf. A *kerf* is the gap created when material is removed by the saw blade. This gap allows the blade to pass through the wood without binding. Binding occurs when the wood closes against the saw blade, causing it to be stuck in place. Crosscut saws are used at an angle of about 45° between the blade and the wood. Ripsaws are used to cut wood lengthwise or parallel to the grain with the saw held at an angle of about 60° between the blade and the wood.

When starting to make a cut with either handsaw, the saw should be positioned on the waste side of the line. If the board will be sanded after the cut, leave enough room on the waste side so the board may be sanded down to the line. The saw is first drawn back twice, creating a groove for the blade to start a kerf. The saw is then gently pushed back and forth using the full length of the blade. The best woodworkers keep their elbows in close to the body and gently push the saw blade without twisting or bending the blade, letting the saw do the work. Watch the cut and keep the saw parallel with the line. See **Figure 11-6**.

Backsaws and Dovetail Saws

Backsaws and dovetail saws both have a reinforced rigid spine down the back of the saw. These saws are small, have fine teeth, and are designed for accurate finish cuts. Cuts are made with the saws in the horizontal position, and care should be taken not to splinter or break the wood on the final strokes. The backsaw may be combined with a miter box to help cut specific angles on the wood stock, **Figure 11-7**.

Coping Saws

Coping saws are designed to make intricate curved cuts in wood. See **Figure 11-8**. With the teeth facing the operator of this saw, the cutting action is on the back stroke. The blades for the saw have very fine teeth. Care must be taken to not twist the blade in the kerf, causing the saw to bind or the blade to break. The saw's design allows for detailed inside cuts not connected to the material edge. To perform such a cut, first drill a hole that is large enough to pass the blade through within the material cutout area. Next, insert the blade through the hole before connecting it

Tyler Olson/Shutterstock.com

Figure 11-6. Using a handsaw effectively requires proper technique. Be careful not to force the blade as it is cutting.

Stocksnapper/Shutterstock.com

Figure 11-7. A backsaw has a stiffened spine piece that helps in making very accurate cuts, especially when used with a miter box.

to the coping saw frame. Adjust the blade tension so that it will not flex during the cut, and proceed with the cut.

Keyhole Saws

Keyhole saws are often used to make internal cuts in sheet goods for electrical boxes. Before a cut can be made, a hole must be drilled through the stock. Then the saw is inserted through the hole. Using the saw's long narrow blade, curved cuts can be made with this saw. When using a keyhole saw to cut into an existing wall, be careful not to accidentally cut into electrical or plumbing systems. Serious injuries or damage can result.

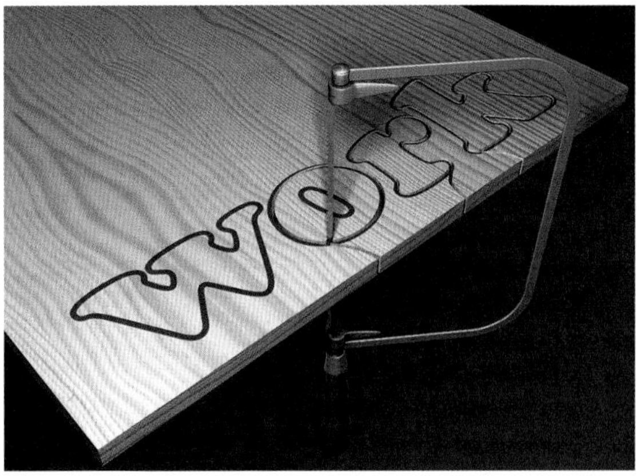

albund/Shutterstock.com

Figure 11-8. The C-shaped blade holder of a coping saw allows the very fine blade to be turned through the cut, making this saw ideal for cutting out intricate shapes.

Power Saws

Power saws are handheld saws that are electrically powered, with power coming from an electrical outlet or a battery. The most popular portable power saws are the circular saw, reciprocating saw, and jigsaw.

Circular Saws

The circular saw is a general purpose saw and can be used for cross cutting and ripping dimensional lumber or wood sheet goods, such as plywood. Ripping is cutting wood lengthwise or parallel to the grain. When marking material to be cut with a saw, keep the blade thickness in mind. In order to allow for that thickness, cuts should be made on the waste side of a mark. Cutting on the wrong side of or through the middle of the line will make the desired part too short.

Circular saws are powerful tools. When used properly, they can make quick work of many jobs. Follow all the power tool safety rules provided by the manufacturer and remember these guidelines that apply to circular saws:

- Adjust the depth of the blade to about 1/8" to 1/4" deeper than the thickness of the wood to be cut. A good rule of thumb is that no more than three teeth should be exposed through the wood when the saw is set in place for the cut.
- Operate the saw with both hands, and push away from your body as the cut progresses. See **Figure 11-9**. Avoid using a power saw from a body position that is uncomfortable or causes overreaching.
- Keep the body of the saw on the stationary side of the cut and not on the side that will drop away.

Anton Kuba/Shutterstock.com

Figure 11-9. When possible, hold the circular saw with both hands and push the saw away from your body.

Safety Note
Dull blades can easily bind and cause the saw to kick back. This may throw the wood or saw into the operator, causing injury or harm.

If the saw binds during the cut, do not pull the saw backwards while it is running. This will cause the saw to kick back (jerk strongly toward the user). To cut a board with a circular saw, first make sure the wood is safely supported by saw horses or some other stable devices. Set the supports up so the waste piece can freely fall away after the cut. Make sure the supports are properly placed. Cutting between two supports that are placed too far apart will cause the saw to bind and kick back. Clamp the board down to ensure it will not move while making the cut. Position the base of the saw on the wood and line the blade up along the waste side of the cutting line. Make sure the blade is not touching the wood when the saw is started because this will splinter the wood. Allow the saw to gain full speed and gradually make the cut across the board. Once the cut is complete, release the trigger and ensure that the retractable guard retracts to its original position before setting the saw down.

Reciprocating Saws

A reciprocating saw has a long straight blade held in a powered base that creates a back-and-forth motion with the blade. This saw is valuable in demolition because it can be used to cut through most materials, including wood, sheet rock, lathe, and even metal and plastic pipes.

Follow these safety rules, as well as the power tool safety rules, when operating a reciprocating saw:

- Check that the blade is long enough to pass completely through the material throughout the entire cut, **Figure 11-10**.
- Do not force a saw into the work.
- Always keep the foot of the saw in contact with the wood that is being cut.
- Use a drill or other suitable tool to start plunge cuts (cuts that begin away from the edge of the wood) in soft materials. Do not attempt to begin plunge cuts with the saw.
- Use caution when cutting into existing walls to avoid cutting into electrical wiring, plumbing, or gas lines.
- Do not handle blades and blade holders with your bare hands after a making a cut. These parts will be hot from friction. Do not wrap power cords around hot parts.

Robert Wydro Studio/Shutterstock.com

Figure 11-10. Reciprocating saws can be used to cut most materials, including wood, metal, and plastic pipes. Hold the tool with both hands and make sure the blade passes completely through the material being cut.

To make a cut using the reciprocating saw, hold the saw firmly with both hands. Start the saw and engage the wood, keeping the saw blade on the waste side of the line. Once the cut is complete, set the saw down and ensure nothing comes into contact with the blade until it cools down.

Jigsaws

The fact that a jigsaw is not a very powerful tool makes it a little less intimidating to the novice user than some other power tools. By understanding the saw's unique characteristics and its limitations, the user can quickly become proficient in using this tool. Follow basic power tool safety rules and these simple guidelines:

- After changing blades, ensure that new blades are fully secured.
- Check that the blade is long enough to pass completely through the wood throughout the entire cut. The jigsaw is not capable of blind cuts. A *blind cut* is one that does not pass completely through the wood to the other side.
- Always keep the base of the saw in contact with the wood that is being cut.
- Do not begin plunge cuts with a jigsaw. Although plunge cutting in soft materials is possible with a jigsaw, it is safer to use a drill or other suitable tool to start interior cuts.
- Use caution when cutting into existing walls or countertops to avoid cutting into electrical wiring, plumbing, or gas lines.
- Cutting through blue painter's tape or masking tape can be done to help reduce splintering at the kerf. The base of a jigsaw may be covered with tape to protect finished wood surfaces.

A jigsaw is useful for cutting small diameter circles and arcs in wood. Jigsaws are also useful for cutting scrolls and other irregular shapes. To make a cut in a piece of wood, make sure the wood is safely supported and the path of the blade is unobstructed. Keep the base of the saw resting on the material for the duration of the cut, **Figure 11-11**. Allow the saw to come to full speed and gradually guide the saw through the cut, ensuring the blade does not bind on a tight radius. After the cut, lay the saw on its side; do not stand it on its blade.

Stationary Saws

Stationary saws are powered with a motor and are mounted on a stand or bench. The types of stationary saws used in woodworking include miter saws, table saws, radial arm saws, vertical band saws, and scroll saws.

Miter Saws

A miter saw is used to cut boards at a wide range of angles, **Figure 11-12**. When using this saw, it is important for your safety and for accurate cuts to hold the material firmly in place. When holding the wood in position, be careful to keep your hands and arms out of the cut zone. Do

Kzenon/Shutterstock.com

Figure 11-11. Jigsaws allow for creativity when cutting wood. Note that the base of the tool is resting on the wood that is being cut.

not cross your arms when holding the wood in place and using this saw. The advantage of using this saw is that cuts can be made at an angle with great accuracy, including a 90° angle across the grain of the wood. The saw should be secured on a stable surface before use, and long pieces of wood should be supported during the cutting process. Use caution when operating a miter saw. As with all power saws, improper use can be dangerous. Miter saws are not designed to perform rip cuts (cuts made down the length of a board).

Table Saws

A table saw is power saw that is mounted with the motor underneath a worktable and the saw blade extending above the table surface. This general purpose saw can be used for cross cutting and ripping dimensional lumber or wood sheet goods, such as plywood. The procedures for using a table saw are different from those for using a circular saw because the saw remains stationary and the wood is moved across the blade.

As for any stationary power tool, choose a location carefully. Consider the position of the power supply. Extension cords on the floor can be a hazard. An overhead or floor outlet may be a better choice than a standard wall receptacle. There must be adequate clearance both to feed material to the saw and to catch or receive cut materials as they come off the saw. When cross cutting lumber and plywood, you also need adequate clearance on both sides of the saw.

PointImages/Shutterstock.com

Figure 11-12. Boards can be cut at a wide range of angles with a miter saw.

Table saws are often used with other devices to improve safety or accuracy of cuts. A dust collection system may be used to remove dust created during cutting. The system should be cleaned on a regular basis. A *rip fence* is a piece of metal that can be attached to the table for a saw and used to guide a board as it is being cut to improve accuracy. The rip fence position can be adjusted to accommodate boards of various widths. A *miter gauge* is a device that can be used to set the position of a board so that it is cut at a particular angle. The miter gauge is placed in a groove on the saw table. See **Figure 11-13**.

Observe the power tool safety rules and a few specific rules included here when operating a table saw:

- Make sure there is clearance to feed wood and receive cut wood.
- Handle the wood with both hands, and push the wood away from your body as the cut progresses. Do not stand directly in line with the blade or get your hands near the moving blade.
- Always use a rip fence to guide a long cut or a miter gauge to guide a cut across the board. To prevent boards from binding, do not use the fence and the miter gauge at the same time.

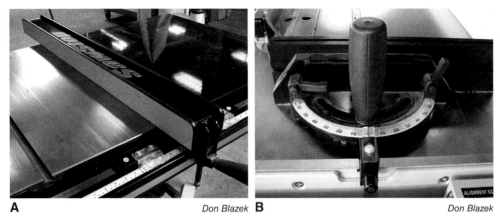

Figure 11-13. A—A rip fence guide is used for ripping stock. B—The miter gauge is used for making cross cuts. It can be adjusted to make a variety of angle cuts.

- Use a push stick when cutting short or narrow pieces to keep your fingers away from the blade. A *push stick* is a piece of wood or plastic used to guide lumber through the saw.

A table saw can be used for rip cutting, dado (slot) cutting, miter cutting, and cross cutting with some added precautions. The basic use of this saw is to rip cut lumber, see **Figure 11-14**. Note the careful placement of the wood tightly against the fence and the use of a push stick to guide lumber through the saw. Using a push stick is safer than placing your hands near the saw blade. When the table saw is set with a large open area around the machine, it can also be used to make accurate cuts on large plywood pieces, again using the fence as a guide. This type of cutting often requires one person to operate the saw and another person (or a tool) to help hold up the large piece being cut. A human helper should act only to hold the material parallel to the table, not to push or pull the material through the blade.

Radial Arm Saws

A radial arm saw has a power-driven blade mounted on a horizontal metal arm. The blade can be pulled along the arm to make crosscuts in lumber. See **Figure 11-15**. A drawback of using this machine is that if the operator pulls the saw more quickly than the cut can be made, the blade may come out of the kerf and jerk or bounce up onto the wood, leaving marks that may damage the wood. The benefit of using a radial arm saw is that it is safer than a table saw for making a crosscut, dado (slot), or rabbet (a groove cut along the edge of a board).

When cutting with the radial arm saw, the material must be firmly held by the operator against the fence and flat on the saw table. The cutting line should be

Figure 11-14. This figure shows lumber being rip cut. Note that a push stick is being used to guide the lumber.

Figure 11-15. To use a radial arm saw to make a crosscut, hold the stock firmly against the fence, turn on the saw, and allow it to reach full speed. Then slowly pull the blade across the piece to be cut. Use caution when working around the spinning blade.

lined up with the blade before the saw is turned on. When the saw is started, the operator must gradually pull the saw across the width of the board. Once the cut is made, the saw should be returned to the original position and the power turned off.

Vertical Band Saws

A vertical band saw is a machine that has a long continuous metal blade with saw teeth on one side. The blade is turned with the use of two pulleys. The saw is fixed in place on a tabletop. The cutting blade extends down through an opening in the tabletop. The wood is placed against the saw at the position where the cut is to begin. The wood is moved against the saw blade to complete the cut. A band saw can be used to make small cuts along straight lines in wood, **Figure 11-16**. This saw also can be set at a bevel and (with accessories) can be used to cut miters and compound miters. However, the miters tend to not be accurate because of the flex in the saw blade. The operator must exercise care when using a vertical band saw because the blade is exposed at the cutting point.

Figure 11-16. Using a vertical band saw requires the operator to work in close proximity to the moving blade.

To safely use a vertical band saw, adhere to the power tool safety rules and these guidelines:

- Check to see that the blade is sharp, undamaged, under proper tension, and tracking correctly on the pulleys. Always mount the blade with the teeth pointing down toward the table.
- Adjust the blade guide to be as close as possible to the material without interfering with the feed. Allowing the blade to be unsupported will produce errors due to deflection of the blade and decrease the life of the blade.
- Do not push material by placing your hands directly in line with the blade. Keep your hands and fingers at least 2″ from the moving blade.
- Keep the wood being cut in contact with the saw table. Use a block of wood cut in a V shape for holding round pieces of wood while being cut.
- Stop the motion of the saw before backing the blade out of long, curved cuts.
- Use *relief cuts*, which are cuts made to release waste material and reduce the chances that the blade will bind.
- Provide additional support for long pieces of wood.
- Stop the saw immediately if the blade hangs up or starts to make a clicking sound, which is an indication of damage to the teeth.

To use a band saw, first check the distance between the table and the blade guide assembly (which holds the cutting blade). The blade guide assembly should be located approximately 1/4″ above the surface of the wood to be cut. Once the assembly is set, the material can be cut safely following the guidelines listed above.

Scroll Saws

A scroll saw is a bench-mounted jigsaw. A scroll saw is useful for cutting small diameter circles and arcs in wood as well as scrolls and other irregular shapes. The difference between a jigsaw and a scroll saw is that the blade of a scroll saw is attached at both ends, **Figure 11-17**. Although considered a stationary power tool, many scroll saws are light enough to be moved easily by one person. Often, the scroll saw will not have a dedicated place in a shop and is stored after use. When moving a scroll saw, remember to carry the saw by lifting the base and not by lifting on the upper arm. The table on most scroll saws can be tilted, allowing the

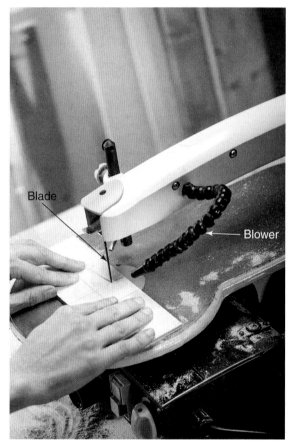

donatas 1205/Shutterstock.com

Figure 11-17. Because a scroll saw's blade is attached at both ends, it is under tension and can cut tighter curves than a jigsaw. The blower will keep guide markings visible and reduce the amount of debris in the kerf.

operator to make bevel cuts. On some expensive machines, the table is left horizontal and the head of the saw is tilted.

Scroll saws are commonly equipped with a blower to remove the sawdust from the area of the cut. **Figure 11-17** shows that proper positioning of the blower tip. Before use, securely clamp or mount the saw to a workbench or stand to reduce the vibration from the movement of the saw blade. Follow the power tool safety rules and these guidelines to make quality cuts:

- Always mount the blade with the teeth pointing down toward the table.
- Set the blade to the proper tension. If it flexes when cutting, it is too loose. If the blade breaks frequently, this is a sign that the tension is not set correctly. Setting the tension properly will take practice and a little trial and error.
- If the saw has variable speed settings, set the saw to slow speed for metals, plastics, hard woods, and very thin materials. High speed can be used for cutting soft materials.
- Keep the wood being cut in contact with the table of the saw.
- Use your thumb and forefinger on both hands to hold down and guide the wood through the cut. Do not push the wood by placing your fingers directly in line with the blade. Keep your fingers at least 2″ from the moving blade. Use a push stick if needed.
- Let the cutting action determine the rate of feed. Do not force material faster than the saw will cut. If feed rate is unsatisfactory, it may be necessary to choose a different blade.
- Stop the motion of the saw before backing the blade out of long, curved cuts.
- Use relief cuts to release waste material and to reduce the chances that the blade will bind.

Career Connection

Cabinetmaker

Job description: The cabinetmaker designs, fabricates, repairs, and installs wood cabinetry and related items for use in a variety of applications. Successful cabinetmakers must be proficient in the use of a variety of hand and power tools. They must also be creative and detail oriented, possess strong math and skills, and have the ability to read technical drawings and follow instructions.

Education required: Most cabinetmakers have, at minimum, a high school diploma. Although some cabinetmakers learn on the job, many seek additional training through trade schools, and apprenticeships.

This job may be a good fit for you if: You enjoy working with your hands and are attentive to detail.

goodluz/Shutterstock.com

Shaping and Smoothing Wood

Shaping tools for use in woodworking include both hand tools and power tools. The tools used to shape and smooth wood include files, rasps, planes, shavers, chisels, routers, rotary cutters, and sanders.

Files and Rasps

A *file* is a hand tool with raised areas or teeth used to cut or smooth material, such as wood, metal, or plastic. Files are used to remove tiny bits of material when the file is passed over the wood with slight pressure. The direction and size of the teeth determine how much material is removed.

File teeth are graded into three basic grades: bastard, second cut, and smooth. The bastard grade is the coarsest grade and is used for removing the most material. The second cut file has finer teeth than the bastard grade and is used for removing material quickly. A file with second cut teeth gives a smoother finish and cut than a bastard grade file. A smooth grade file is used for smoothing material and readying it for sanding. See **Figure 11-18**.

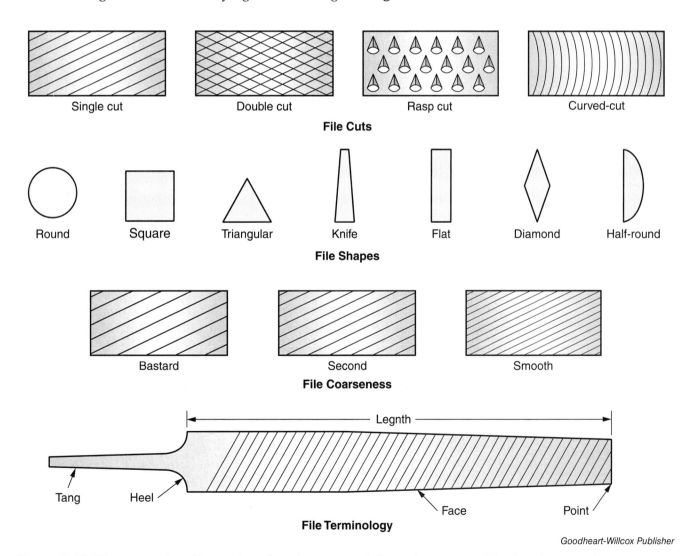

Goodheart-Willcox Publisher

Figure 11-18. Files are produced in a variety of grades, cuts, and shapes to accommodate almost any need.

A *rasp* is a hand tool that is similar to a file except that the cutting teeth are much coarser. Rasps come in three types with different sizes, or grades, of teeth: wood, for removing the most material at one time; cabinet, for removing a medium amount of material; and pattern maker's, for removing the smallest amount of material. Cabinet grade rasps can be further subdivided into bastard, second, and smooth grades. Rasps are generally flat or concave (half round) in shape and squared off at both ends. They do not require an extra handle.

Files and rasps are used to shape or smooth the edges of wood and should be used with two hands, one on the handle and the other hand on the other end of the file. The file is run across the edge of the wood on a slight angle. Care should be taken to not chip the end grain of the wood. See **Figure 11-19**.

Don Blazek

Figure 11-19. When using a file, hold it with both hands and move it across the work surface at a slight angle.

Hand Planes

A *hand plane* is a tool used for shaping wood by removing small shavings. When the plane is moved over the wood, the plane blade slices off any high spots to create a smooth surface. The two most important planes are the bench plane and block plane. The bench plane is used for smoothing flat surfaces. The block plane is used for smoothing corner edges and the end grain of boards.

The blade of a plane should be set so that the removed material is a very thin curl of shaved wood. If a plane removes too much material because it is cutting too deep or if the curl is wispy and broken, the plane blade setting is too shallow. Adjust the blade of the plane to extend just enough to slice paper thin shavings from a piece of scrap lumber. The cap iron, or chip breaker, should be adjusted as close to the cutting edge as possible to provide maximum support without interfering with the shavings as they pass through the slot.

When using a plane, keep the base, or sole, of the tool flat on the surface of the wood. Turn the plane at a slight angle to the direction of travel and push evenly to slice off thin slivers from the surface. Always lay the plane on its side when finished and retract the blade into the frame to prevent damage to the cutting edge. See **Figure 11-20**.

A sharp plane blade is the most important characteristic of this tool and honing the iron may be necessary every time it is used. *Hone* means to grind or shape to a razor-sharp edge, often using a whetstone.

rtem/Shutterstock.com

Figure 11-20. Bench planes should be adjusted so they shave off a very thin piece of wood. Note the base is held flat against the wood.

Shavers

A *shaver* (also called a draw knife) is a hand tool with a blade that is used to smooth wood by removing small shavings when drawn along the wood. A shaver works much like a plane does. However, a shaver has a slightly different holding mechanism. These tools should be used razor sharp. They are designed to remove small amounts of material from specific locations, such as edges of doors that are dragging or spindles with rough spots.

The blade of the tool is centered between two handles. The tool is operated by holding both handles and drawing the shaver toward you along the edge of the wood. The amount of wood removed is determined by the angle that the shaver is held in relation to the plane of the wood.

Chisels

Wood chisels are important tools for cutting out excess wood in tight spaces, smoothing out rough spots, and cutting mortise and tenon joints. A mortise and tenon joint has an opening or slot cut into one piece of wood and a peg or extension cut into the other piece of wood. The peg is fitted into the slot to create a tight joint. Since this tool must have sharp cutting edges to work properly, honing the chisel before use is important. Chisels can be used with the beveled edge facing the wood or facing away from the wood, depending upon where the chiseling is needed. If the chisel can be laid flat and parallel to the wood, the bevel should face up away from the wood.

sergign/Shutterstock.com

Figure 11-21. Chisels must be razor sharp to be effective.

If the handle of the chisel needs to be elevated to keep it from coming into contact with the wood, then the bevel should face the wood, and the angle of the chisel will determine the amount of wood removed. Angling the chisel too high will cause the blade to dig into the wood. See **Figure 11-21**.

The chisel can be used with a wooden mallet to remove large amounts of wood. It also can be used to lightly shave small amounts of wood from a surface. In either case, the wood should be secured in place with a vise or clamp. You should never hold the wood while chiseling, and both hands should be engaged with the chisel and/or mallet.

Safety Note

Always make cuts with the blade facing away from your body. Carry a wood chisel with the sharp edge facing down, and store the tool in a manner that protects the very sharp edges.

Routers

A *router* is a power tool that can be used for cutting a decorative edge or design on wood. A router might be used on a piece of lumber or on molding. See **Figure 11-22**. For their size, routers are very powerful machines that produce considerable torque. Brace yourself in a secure stance and clamp the wood before using the tool.

Lamarinx/Shutterstock.com

Figure 11-22. Routers are used for cutting a decorative edge or design on wood.

A router has a cutting device called a bit. Bits spin clockwise, and wood should be fed from left to right. Feeding in this manner allows the cutting edge to contact new material first and reduces the chances of the bit kicking sideways. Use a guide for stability. Many commercial guides are available, and guides can be custom made. Some guides may be as simple as a straight edge or piece of wood clamped in place. Little or no additional edge finishing will be required when the router is used properly.

Follow the power tool safety rules and these guidelines to use a portable router safely and efficiently:

- Hold the router using the provided handles. They are designed for best control of the machine. Do not let your fingers get anywhere near the moving bits.
- Always keep the base of the router in complete contact with the wood.
- Allow the machine to do the work. Do not force it into the material. Forcing the router is hard on the motor, causes burning and roughness on the edge of the cut, and damages the bit.
- Use a guide whenever possible to aid in creating quality cuts.
- When changing bits, ensure that the new bit is secured by using the proper chuck wrench.
- Adjust the bit and base to produce the desired design profile. Make a test cut on scrap wood before cutting expensive materials.
- Use multiple thin passes to reduce the possibility of splintering that may happen on deep cuts or cuts that remove a great deal of material.
- Use a plunge router for blind cuts that do not pass completely through the work. Do not attempt plunge cutting with a router that is mounted in a table.

Rotary Saws

A rotary saw, also called a spiral saw, is a tool used for making cuts without a pilot hole in plywood or other thin material. Rotary saws are useful for cutting holes or circles for electrical boxes when hanging drywall or plywood used as paneling. Wall panels are tacked in place and the saw bit is plunged through the wall material at the location of the electrical box. The bit is then rotated around to the perimeter of the box to remove the cutout.

Rotary saws may be used like a plunge router or a jigsaw. Rotary saws can be used on small projects to complete a number of activities. A range of cutting bits are available for cutting different materials. When using a rotary saw to cut a hole, make sure the bit is secure in the tool. It should project far enough from the base to cut through the material. Grip the rotary saw securely for a controlled and accurate cut.

Sanders

Sanders are tools used to remove small amounts of wood or wood finishes (paint) and to smooth wood. Sanding is easily accomplished by using sandpaper by hand or by using a power sander. Choosing the correct sandpaper for the job is important. Sandpaper is heavy backing paper with sand grit adhered to one side. Sandpaper comes in a variety of grits, which identify how rough the surface is. Sandpaper designations are coarse (40–60 grit), medium (80–120 grit), fine (150–180 grit), very fine (220–240 grit), extra fine (280–320 grit), and super fine (360 grit and above). When sanding wood, begin with a coarse or medium grit sandpaper, depending on how rough the wood is or how much finish medium must be removed. Then use progressively finer grits of sandpaper until the wood is completely smooth and clean.

When sanding by hand, sandpaper can be wrapped around a small block of wood or a purchased sanding block. This will help keep the paper in direct contact with the wood and prevent rounding over edges or digging into divots. When sanding, move the sander parallel with the grain of the wood to avoid scratching the surface. Sanding against the grain (perpendicular) or across the grain will leave scratches that will be visible when applying a stain finish to the project.

Handheld power sanders come in four basic types: belt sander, palm sander (or orbital sander), random orbital sander, and detail sander (or mouse sander). See **Figure 11-23**.

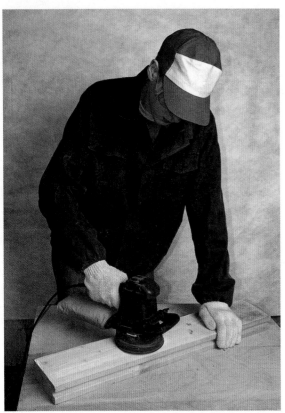

Konstantin Gushcha/Shutterstock.com

Figure 11-23. An orbital sander can be used to quickly smooth the surface of a board.

- Belt sanders are used to remove large amounts of material from wood. A belt sander uses a loop of sandpaper called a belt that travels around a set of wheels. A typical handheld belt sander will sand an area of about 4″ × 6″ as the belt turns. This sander is useful for removing paint or other finishes or preparing very rough finishes for final sanding. Make sure that the wood is secured in place before applying the belt sander to the surface.
- A palm sander fits in one hand, uses a square sandpaper pad (sheet), and vibrates back and forth. It will leave sanding marks where it vibrates against the grain of the wood.

- A random orbital sander turns a sandpaper pad (sheet) in a loop pattern while the pad also spins in a circular pattern. This type of power sander is useful for first sanding and final sanding of woods. It does not leave marks, even when sanding against the grain of the wood.
- A detail sander, also called a mouse sander, has a triangular-shaped sanding head and sheet of sandpaper. This sander is useful in tight places and corners, but it is generally used only for light or finish sanding.

Drilling Wood

Drilling is making a hole in wood or other material, typically using a tool with a rotating or cutting tip. Drilling operations can be done in wood with hand tools, such as a hand drill. However, drilling is often done with power tools, such as the power drill shown in **Figure 11-24**. It is important to note the difference between a power drill and a power driver. A power drill is used with a drill bit to bore holes into wood. A power driver is used with either a screwdriver tip or a nut driver tip to insert screws into wood. A power drill can be used to drive screws with the proper tip at a low speed, but the power driver is a single-use tool.

When drilling holes through wood, make sure the wood is supported. To prevent splintering at the exit hole when drilling through wood, back up the wood with a scrap piece of wood. This helps ensure that the bit will cut through the entire depth of the hole and will not be pushed through the last bit of material.

When using a portable power drill, follow all the power tool safety rules along with these guidelines:
- Select the proper speed and the correct bit for the work material.
- Start all holes with a center punch.
- Be careful not to apply too much side pressure, as it can break the drill bit.
- Use a pilot bit (smaller diameter bit) followed by a larger bit to drill larger holes.

Robert Wydro Studio/Shutterstock.com

Figure 11-24. A battery-powered drill is commonly used for drilling wood.

CHAPTER 11
Review and Assessment

Summary

- Some characteristics of wood include the type of wood (hardwood or softwood), color, grain pattern, density (hardness), flexibility, rot resistance, and insect resistance.
- When selecting wood for a project, considerations include density, hardness, stainability, and cost of the wood.
- Grades for hardwood lumber are based upon the length of clear lumber that can be cut from a board. Softwoods are grouped in categories according to use: stress graded, nonstress graded, and appearance lumber.
- Standard dimensions for lumber in the United States are given in nominal dimensions, which are the measurements of a piece of wood before kiln drying.
- All species of hardwood and softwood lumber will contain varying degrees of imperfections called defects. Wood from high grades will have fewer defects than wood from low grades.
- Several hand tools and power tools are useful for cutting wood. Handsaws are generally less expensive than power saws. Handsaws include crosscut saws, rip saws, backsaws, dovetail saws, coping saws, and keyhole saws.
- Power saws are handheld saws that are electrically powered. Popular power saws include circular saws, reciprocating saws, and jigsaws.
- Stationary saws are powered with a motor and are mounted on a stand, table, or bench. Examples include miter saws, table saws, radial arm saws, vertical band saws, and scroll saws.
- The hand tools used to shape wood include files, rasps, hand planes, shavers, chisels, and sanders. Routers, rotary saws, and power sanders are power tools used to shape and smooth wood.
- Sandpaper comes in a variety of grits, which identify how rough the surface is. Progressively finer grits of sandpaper can be used until the wood is completely smooth and clean.
- A power drill is typically used to bore or drill wood. Drilling holes in wood also can be done using a hand drill.

Words to Know

Match the key terms from the chapter to the correct definition.

A. appearance lumber
B. blind cut
C. dressed dimensions
D. file
E. hand plane
F. hone
G. kerf
H. milling
I. miter gauge
J. nominal dimensions
K. nonstress graded
L. planing
M. push stick
N. rasp
O. relief cut
P. rip fence
Q. router
R. shaver
S. stainability
T. stress graded
U. utility lumber

1. The final measurements of a piece of lumber after it has been kiln dried, milled, and planed.
2. The measurements of a piece of wood before kiln drying, milling, and planing.
3. Shaving and smoothing the edges of wood.
4. The gap created when material is removed by a saw blade.
5. A piece of metal that can be attached to the table for a saw and used to guide a board as it is being cut.
6. The ability of wood to take stain or paint.
7. A power tool that can be used for cutting a decorative edge or design on wood.
8. Stress graded wood used for rough framing that may have distinct warping or twisting.
9. A hand tool with a blade that is used to smooth wood by removing small shavings when drawn along the wood.
10. Wood used in applications where strength and stability are important.
11. A piece of wood or plastic used to guide lumber through a saw.
12. Wood used as construction material.
13. Cutting a piece of lumber to a certain size.
14. A cut made to release waste material and reduce the chances that the saw blade will bind.
15. A hand tool that is similar to a file except that the cutting teeth are much coarser.
16. A device that can be used to set the position of a board so that it is cut at a particular angle.
17. A tool used for shaping wood by removing small shavings when the plane is moved over the wood.
18. A hand tool with raised areas or teeth used to cut or smooth material, such as wood, metal, or plastic.
19. Wood used for making projects that appear free from defects.
20. To grind or shape to a razor-sharp edge.
21. A cut that does not pass completely through the wood to the other side.

Know and Understand

Answer the following questions using the information provided in this chapter.

1. What are some characteristics of wood?
2. What is often the first characteristic of wood to be considered when selecting wood for a project?
3. Grades for hardwoods are based on what standard?
4. Describe the three categories into which softwoods can be grouped.
5. What are some examples of natural defects that may be found in lumber?
6. For what purpose are rip cut and crosscut hand saws commonly used?
7. When using a handsaw, what causes a saw to bind?
8. What are three guidelines to follow when using a circular saw?
9. What is a reciprocating saw?
10. What are some types of stationery saws used in woodworking?
11. What is the purpose of a miter saw?
12. What types of cuts are made with a table saw?
13. What is a vertical band saw and how are cuts made with this saw?
14. What types of cuts are made with a scroll saw?
15. How is a file used to shape or smooth wood?
16. How should the blade of a hand plane be set?
17. What are some uses for a wood chisel?
18. What is an example of a typical use of a rotary saw?
19. What are four basic types of handheld power sanders?
20. What can be done to prevent splintering at the exit hole when drilling through wood?

STEM and Academic Activities

1. **Science.** Find a piece of scrap lumber with the heartwood or center wood from a branch. Determine how many "rings" you can count in this piece of lumber. How old was the tree from which this piece was harvested? Where was the tree likely to have been harvested based on its species and age? Can you see a ring that indicates ideal growing conditions? What about rings that indicate stressful conditions? What affect might growing conditions have on wood characteristics? Write a short paragraph explaining your answers to these questions.
2. **Technology.** Research Cross Laminated Timber (CLT) route to the market.
3. **Engineering.** Using three species of wood or wood products build, test, and compare the tensile strength (could also test compressive strength, flexibility, nail holding ability or other engineering standards) of each. Provide the results in a demonstration to class members.

4. **Math.** Calculate how many board feet of lumber are in a given project. Calculate the number of fasteners needed and using the Internet determine the cost per fastener and use this to determine the total cost of the fasteners for the given project.
5. **Social Science.** Determine the predominate wooden chair style for each decade of the last century. Explain what influenced each style and the changes.
6. **Language Arts.** Looking at the words to know, select all the words that are the name of tools. Write those words in red, then write all the other words in another color. Define all the red words with only a picture of the tool, and write the definition of the other words using your own description. Check the glossary for correct answers.

Thinking Critically

1. Very coarse sandpaper removes the most material in the quickest manner. Why is this not always the best way to start the finishing process? Explain.
2. Why would a stain only finish not be appropriate for a food bowl? Clarify the difference between stain, paint, and sealer finishes and explain which would be the best to use on a food-handling bowl.

CHAPTER 12
Designing, Planning, and Constructing Woodworking Projects

Chapter Outcomes

After studying this chapter, you will be able to:
- Complete project planning steps involving drawings or plans, materials lists, and a procedures plan.
- Select appropriate wood for a project.
- Accurately mark layout lines and cutting lines.
- Properly cut wood for a project.
- Complete subassemblies and final assembly for a woodworking project.
- Apply a wood finish and hardware to complete a woodworking project.

Words to Know

biscuit jointer	finish	sketch	waste side
construction procedures	materials list	subassembly	wood joint
	registration marks	technical drawings	

Before You Read

Before reading this chapter, skim the photos and their captions. As you read determine how these concepts contribute to the ideas presented in the text.

While studying this chapter, look for the activity icon to:

- **Practice** vocabulary terms with e-flash cards and matching activities.
- **Expand** learning with interactive activities.
- **Reinforce** what you learn by completing the end-of-chapter questions.

www.g-wlearning.com/agriculture

Finding or creating a good design is the first step in constructing a woodworking project. The process starts with an idea to create a beautiful woodworking piece or a functional piece to solve a problem. Many projects are easily made of wood in the agricultural mechanics shop. Boxes, shelves, tables, or benches are the basic elements of many projects. See **Figure 12-1**. Even complex projects are often just a combination of these basic elements. For example, cabinetry is essentially a box with shelves built in it.

Project Planning

Project planning begins with finding or creating a plan for the item to be built. The plans for many types of projects are available from various sources. You may find a plan to meet your needs by searching in project plan books, on the Internet, or in university cooperative extension service publications. If you cannot find an existing plan that will meet your needs, an original plan will be needed for the project.

The basic steps for completing a woodworking project are listed below. Think about each of these steps during the planning phase of the project:

- Produce a *sketch*, a hand drawing with little or no detail, numbers, or measurements.
- Create *technical drawings*, a set of plans that includes measurements and construction details needed to build the project.
- Write a *materials list*, a record showing all items needed in the construction of a project, such as wood, hardware, and paint.
- Write *construction procedures*, a step-by-step plan for making *subassemblies* (parts that are put together separately and used as a unit in a finished product) and the final product.
- Construct the project or a prototype (model or sample).
- Apply a finish to the wood. A *finish* is a product, such as paint or wood stain, used to seal the wood; protect it from damage, rot, and other hazards; or add beauty to the project.

Goodheart-Willcox Publisher

Figure 12-1. Most wood projects start as a simple box, shelf, table or bench. This nail tote is a basic box with dividers to separate nails of different types and sizes.

Drawings and Plans

A quick sketch is often the first drawing of a project. Sketches are meant only to capture ideas for a finished project. See **Figure 12-2**. A sketch might include only basic shapes without any numbers or size indications. Simple sketches can be drawn in the form of an isometric drawing. Isometric drawing is a method for depicting three-dimensional objects in two dimensions by rotating the object 30° from the front view and tilting it forward, showing all three dimensions of an object in a single image. Sketches are not drawn to scale and often do not even consist of all views. Engineers and designers can use this drawing method to depict the project and share the idea with others. Sketching is a skill that can be developed with practice.

Technical drawings are more detailed and precise than sketches. These drawings are often orthographic drawings, as discussed in Chapter 10, *Project Planning and Design*. Technical drawings may include several views of the project, most often the top, side, and front or bottom (depending on which viewpoint shows the most important details). Measurements are also included. A good computer-aided design program can make completing these drawings as easy as drawing basic shapes with a computer mouse. In **Figure 12-3,** the nail box shown in the sketch in Figure 12-2 is depicted in a orthographic drawing.

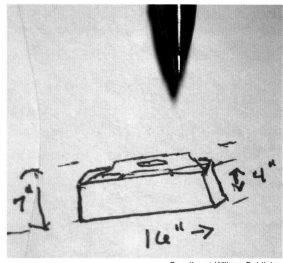

Goodheart-Willcox Publisher

Figure 12-2. This sketch of a nail box was made to get ideas down on paper.

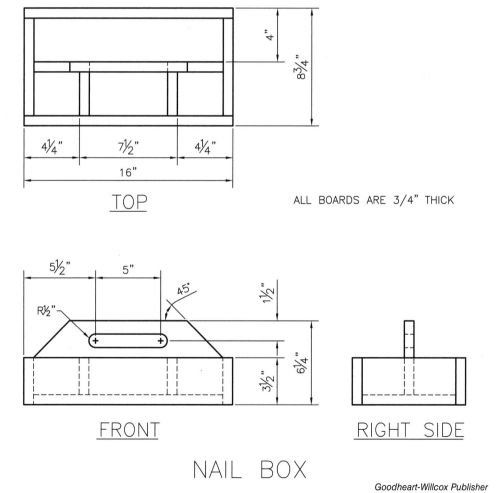

Goodheart-Willcox Publisher

Figure 12-3. An orthographic drawing of the nail box in Figure 12-2 shows the front, top, and side views of the box. This particular drawing was made with computer-aided design program.

Nail Box Material List

Quantity	Description	Size
1	Pine board	1" × 4" × 54"
1	Pine board	1" × 6" × 15"
1	Pine board	1" × 8" × 15"
1 sheet	#120 grit aluminum oxide sandpaper	9" × 11"
1 sheet	#180 grit aluminum oxide sandpaper	9" × 11"
As req'd	Wood screws	#8 × 1½"
As req'd	Finishing nails	#7 × 1½"
As req'd	Wood glue	—
As req'd	Finishing materials (wood stain, polyurethane, paint, etc.)	—

Goodheart-Willcox Publisher

Figure 12-4. The materials list for the nail box project contains all the materials needed to complete the project.

Nail Box Cutting List

Quantity	Part Name	Size
2	Long side	1" × 4" × 16"
2	Short side	1" × 4" × 7¼"
2	Partition	1" × 2¾" × 3¼"
1	Handle	1" × 6" × 14½"
1	Bottom	1" × 8" × 14½"

Goodheart-Willcox Publisher

Figure 12-5. This cut list for the nail box project has more pieces of wood than the materials list because several of the cut pieces come from the same piece of lumber.

Materials List

After the technical drawings are completed, a materials list and cut list can be created. A materials list should include lumber, fasteners, finish, hardware, and other parts or supplies that will be needed to construct the project. A cut list should include all the wood pieces needed for a project and the size of each piece specified in the plans. Remember that several wood parts may be cut from the same piece of dimensional lumber. Each piece does not need to be cut from a separate board. If only a cut list is included with the plans, take the time to construct a materials list so that the correct amount of lumber and other items can be secured. A materials list and cut list for the nail box project are shown in **Figures 12-4** and **12-5**.

Construction Procedures Plan

When you purchase a prepared plan, a construction procedures plan is usually included. This plan tells the woodworker the steps required for making the project. Following the procedures plan saves time because someone has thought through the construction sequence and identified any preliminary steps that must be taken or subassemblies that must be put together before the final assembly can begin. Subassemblies are units that are built separately and incorporated into the project during final assembly. See **Figure 12-6**. When producing your own plans, think through the construction sequence to make sure all preliminary steps are addressed and subassemblies are completed before final assembly begins. This helps ensure that everything fits well and no parts need to be taken apart later. Thinking through each step of the plan helps the woodworker visualize the final project and prepare to make it in a logical sequence.

Selecting Wood

When selecting the lumber, choose pieces that are appropriate for the final project. (Expensive hardwood for a livestock gate or treated deck boards for a fine cabinet are not good choices). Choose lumber that needs the least amount of cutting. For example, if the part requires a finished size of 5" wide × 3/4" thick, start with a 1" × 6" board rather than a 2" × 10" board. Also notice the defects in the wood and decide if they will add to the beauty of the final piece or if they will detract from the strength or look of the project. You may find it helpful to sketch the basic parts on the lumber to determine the placement of defects or to determine where each part will fit on the board. This sketching can be done in chalk or with light pencil lines.

Layout

Before wood for a project can be cut, filed, or chiseled, the pieces must be marked accurately with the proper dimensions. This process of measuring and marking lines is called layout.

For each project piece, use the dimensions in the drawing plans or the cut list. Measure each dimension of the piece and then mark cutting lines for those dimensions. Remember, when marking pieces to be cut from the same board, the saw kerf will destroy some wood with each cut. Take this into account when marking final length measurements for each piece.

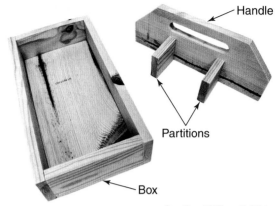

Goodheart-Willcox Publisher

Figure 12-6. Always plan the assembly sequence before beginning a project. When assembling the nail box, the partitions should be fastened to the handle before the handle and partition subassembly is installed in the box. If the handle is installed in box before the partitions, nailing or screwing the partitions to the handle would be difficult.

SAE Connection

Planning

All projects are the result of their planning. Careful planning and preparation is likely to result in a better quality product. Attention to detail is necessary to maintain efficiency and arrive at a desirable outcome. Imagine you plan to build a picnic table and start the job with only an idea in your head. You might go to the store and purchase lumber, fasteners, and paint, but chances are good that you will either purchase too much or leave something out. That will result in wasted time and money that could have been prevented with careful planning.

The same idea applies to your supervised agricultural experience (SAE) project. Careful planning allows you to document your activities and consider your financial investments. For example, taking pictures along the way gives proof of your activities and provides valuable material for completing proficiency applications. As you work

Stasique/Shutterstock.com

through this chapter, think about how important it is to carefully plan a woodworking project and how planning and preparation relate to your SAE project as well.

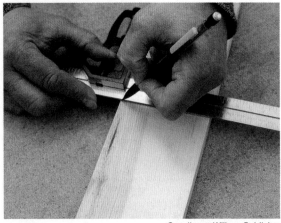

Figure 12-7. A square can be used to accurately mark the length of a board.

Figure 12-8. A student is tracing a scroll pattern onto the wood.

Figure 12-9. Registration marks can be used when the pattern must be repeatedly placed in the same location.

Laying out (marking) for length involves measuring from the end of a board along the length to the position where the cut should be made. Once the measurement is marked, a square is used to transfer the mark across the width of the board. See **Figure 12-7**. Laying out for width involves measuring from the edge of a board across the width and then extending that mark the entire length of the board. The lengthwise line can be accurately marked by using a square or bench rule as a gauge, a marking gauge, or a chalk line.

As a general rule, the grain should always run lengthwise of a piece of cut wood. Running the grain with the width of the piece will cause it to break easily. Cut the pieces to length and width, if needed. Remember that it is more efficient to cut longer pieces first. If a mistake is made, the board can be used for a shorter piece, resulting in less waste. Locate and mark other cuts that need to be made, such as for handles, holes, or scroll work.

Curves are best laid out using a pattern or a compass. Generally, using a pattern is the best option if the design is very detailed, **Figure 12-8**. When using a pattern, it is important to place the pattern in an exact location on the wood so that the shapes are aligned correctly in relation to one another or a part of the board, such as the edge. This is accomplished by using registration marks. *Registration marks* are points on a pattern that are used to align the pattern in the correct location when it is repeated or to align parts of a pattern that are printed on separate sheets of paper, **Figure 12-9**. The marks may be lines, crosses, circles or some other symbol. Using registration marks is especially important when the pattern is used repeatedly in a series to cut a design that is repeated over a distance. A pattern can be cut from a piece of paper, card stock, or thin wood. Remember, if there is an error or minor imperfection in the pattern, it will be transferred to the design. Occasionally, an object that has the correct dimensions but is not an exact copy of the design can be used as a pattern. For example, a tin can might have the correct radius for a circle. You could draw around the can instead of a using a paper pattern. See **Figure 12-10**.

Using a compass is an exact way of drawing curves on wood, as shown in **Figure 12-11**. When using a compass, it is important to hold the lumber securely so that nothing but the compass moves along the drawing line. Remember that it is the circle's radius that is measured with the compass. This method is generally too complicated to use for intricate designs, but it is the method of choice for arcs and circles that must be precise.

Cutting the Project Parts

Cutting wood is generally done with some type of saw. It is important to know the width of the saw kerf so that it can be subtracted from the total length of the lumber to be cut from. Remember that as more cuts are made, more wood will be lost to kerfs. For example, in cutting one piece from an 8′ board, only 1/8″ will be lost. But in cutting eight pieces, a full inch of lumber will be lost in saw kerf. This is an important consideration when buying lumber using only a cut list. Remember to consider the waste when calculating the total amount of lumber to purchase.

When making saw cuts, the blade should not be centered exactly on the cutting line. If the saw blade were to be centered on the cutting line, the resulting cut piece would be shorter than the desired measurement by the width of the saw blade, **Figure 12-12.** While this short piece may not be noticeable on a rough cut project, it may completely ruin a furniture or cabinetry piece. The woodworker should always cut on the waste side of the line. The *waste side* is the edge of the line that is opposite the project piece and nearest to the part of the board that will be discarded. This cutting method results in a cut piece that is the full measure needed. If it is not possible to cut on the waste side of the line, the woodworker should always cut at the same location on the line so that the error is equally distributed in the project and not concentrated in a single piece.

Goodheart-Willcox Publisher

Figure 12-10. In some cases, objects can be used in place of templates. Here, a small paint can is being used mark a corner radius on a board.

Goodheart-Willcox Publisher

Figure 12-11. A compass is commonly used to draw circles and curves on a piece of wood.

Assembling the Project

Once all the parts of a project have been cut according to the cut list, it is a good practice to sand each piece prior to creating subassemblies or doing the final assembly. The plan should then be examined to see how the project is to be assembled. Several types of fasteners are available for use in woodworking projects, but wood joints can be used to assemble a project by simply using the wood.

Basic Wood Joints

A *wood joint* is a connection between two pieces of wood where they intersect that is created using the wood itself rather than fasteners, such as nails or screws. Some types of wood joints that are commonly used for

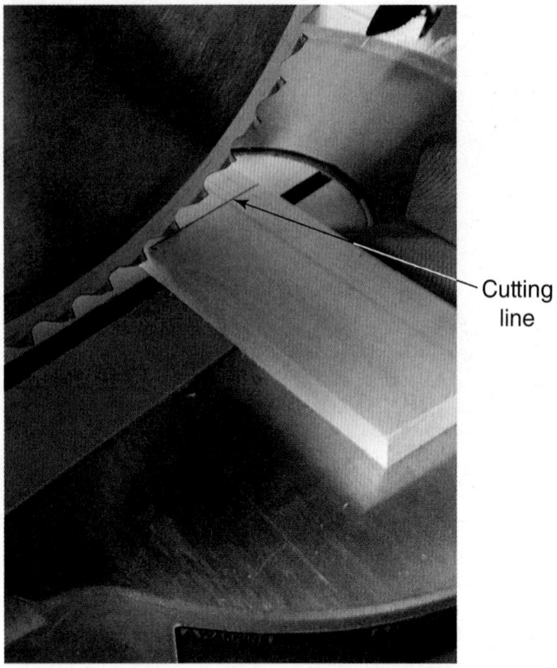

Figure 12-12. In this example, the saw blade is centered on the cutting line, making the resulting piece approximately 1/16″ shorter than the desired length.

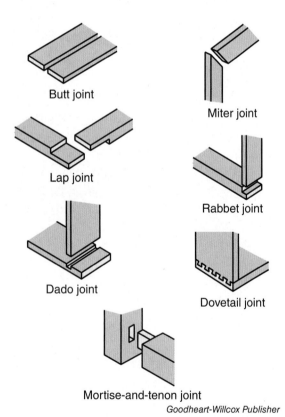

Figure 12-13. There are a variety of joints used to join wood when assembling a project. The most common joint types are shown here.

woodworking projects include the butt joint, lap joint, mortise-and-tenon joint, dovetail joint, rabbet joint, dado joint, and mitered joint. The chart in **Figure 12-13** shows several joints that are useful for the woodworker. Most joints can be cut with the use of a backsaw or dovetail saw and chisel. They also can be cut using a power miter box, table saw, router, or radial arm saw. Wood joints are frequently used in cabinets and shelving units. Wood joints create an extremely strong connection and may be used in post and beam construction for buildings.

Fasteners

Depending on the type of woodworking project you are constructing, there are a variety of fasteners available to choose from. Nails can be driven with a hammer or nail gun, and screws can be installed with a screw gun or power driver. Several types of glues, adhesives, and cements are available for use in various applications.

Nails

Nails are manufactured in a variety of sizes, finishes, and styles. Nails are selected depending on the type of project and its exposure to the weather. Refer to Chapter 9, *Materials, Fasteners, and Hardware,* for a complete description of nail types.

When using a hammer to drive nails, center the nail on the width of the board being fastened and drive the nail straight into the wood. Common nails should have the head of the nail driven tightly against the wood. Finishing nails are smaller than common nails and have a smaller head. Finishing nails are driven with a hammer only to within 1/8″ of the surface. A nail set is then used to set the nail head below the surface of the wood so filler may be applied to fill the hole. This gives the wood a surface with no visible nail heads. See **Figure 12-14**.

The general rule of thumb concerning the length of nails is to choose a size that is twice the thickness of the board you are nailing through. For example, if you are nailing a 3/4″ board to a 2″ board, you will need a 1 1/2″ nail.

Screws

Like nails, screws are also manufactured in a variety of sizes, finishes, and styles. Screws are selected depending on the type of project and its exposure to the weather. Refer to Chapter 9, *Materials, Fasteners, and Hardware,* for a complete description of the screw types.

When using screws to assemble a framing project or rough project, a screw gun or power driver will drive screws even with the surface of the wood. See **Figure 12-15**. Care should be taken to ensure that the boards do not separate as the screw bites into the second board. If this occurs, continue to turn the screw until the boards draw together. As with finish nails, screws can be placed with the heads recessed below the surface of the board. To do so, you should first predrill the location with a pilot countersink bit. When the screw is installed, it can be set into the wood below the surface and covered with a wood plug glued into the hole to conceal the screw.

Adhesives

An adhesive, commonly called glue, is a substance that bonds with materials and causes them to adhere. Because screws have threads, it is unlikely that pieces of your project that have been screwed together will pull apart. Nails are more likely to pull loose due to stress or movement of parts or pieces of a project. Using wood glue with nails creates a much stronger joint than nails used alone. See **Figure 12-16**. Adhesives may also be used with wood joints to create a stronger or tighter joint.

If the project is to be stained, wipe off any excess glue that seeps from the glued joint with a damp rag. Adhesives are also available for the assembly of sheet goods (plywood), floor coverings, and other uses.

Biscuit Jointer

Using a biscuit jointer (also called a plate jointer) and biscuits to join wood parts is a method that is often used in constructing cabinets and shelves. The *biscuit jointer* is a tool used to make crescent shaped cuts in two pieces of wood that are to be joined together. A biscuit (a thin, flat, oval piece of wood) is then glued into the joint between both pieces, creating a strong bond between the pieces. See **Figure 12-17**. Biscuits come in various sizes, such as 0 (small), 10 (medium), and 20 (large), to fit the size needed for various projects.

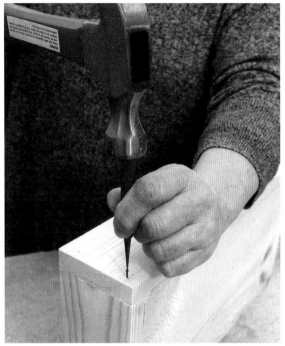

Goodheart-Willcox Publisher

Figure 12-14. When using a nail set, tap it lightly with a hammer until the head of the nail is slightly below the surface of the board. Use care to prevent the tip of the nail set from slipping off the nail head.

Goodheart-Willcox Publisher

Figure 12-15. A screw gun can be used to quickly drive screws flush with the surface of the wood.

> ## STEM Connection | Adhesion
>
> Adhesion is the chemical property of a substance to attract to an unlike substance. Glue or adhesive adheres to other substances because of a chemical and physical bond between the adhesive and the molecules on the surface of the other substance. When these bonds are formed, they are very strong and difficult to break apart.

Finishing the Project

Once a project has been completely assembled, you have reached the finishing stage of the project. The finishing stage consists of the following three steps:

1. Sanding the project a final time to eliminate rough surfaces, pencil lines, and marks.
2. Applying a wood finish, such as paint, stain, or polyurethane.
3. Installing any hardware required by the construction plan.

Woodworking projects can be finished with a variety of products that bring cosmetic appeal as well as provide a layer of protection for the wood.

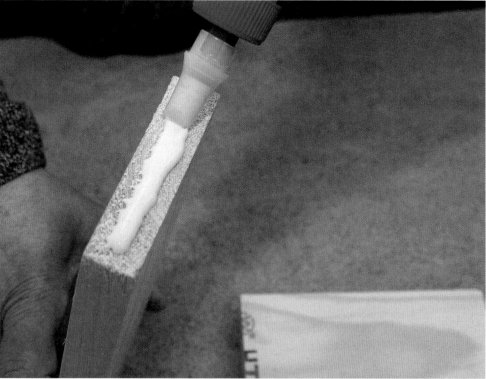

Goodheart-Willcox Publisher

Figure 12-16. Wood glue applied to joint surfaces before nailing will help create stronger joints. If a joint is to be glued but not nailed, clamps can be used to hold the joint together tightly until the glue dries.

The finish is the final touch in making the project, but it is generally the first thing viewers notice about the project. A good piece can become better with a fine finish. Conversely, a poor finish can ruin an otherwise quality piece.

Types of Finishes

Commonly used finishes for wood products include paints, stains, and sealers. Some types of oil finishes, such as tung oil or linseed oil, can also be used for wood. However, their use is limited for agricultural applications.

- Paints are designed to cover or conceal the wood with a colorful protective coating. They are classified oil-based or water-based latex. Paints are also classified by their degree of sheen, such as gloss, semi-gloss, eggshell, interior satin, or flat.
- Stains are oil-based or water-based products designed to add color to a project while still allowing the grain of the wood to be seen. A variety of stain colors is available.
- Sealers, such as shellac and polyurethane, are designed to seal the pores of wood to prevent moisture from getting into the wood. A wide variety of types are available in either glossy or satin finishes. Some sealers are clear and some can add color to the wood. Sealers may also be called varnish. Either oil-based or water-based products can be used on a wood project that will be used indoors. However, an oil-based sealer should be used on a project that will be exposed to the elements.

Preparation for Painting and Staining

Before applying paint or stain to a project, a final sanding should be done. See **Figure 12-18**. All remaining sawdust on the project surface must be brushed off or blown off with compressed air. Always make sure there is adequate ventilation when using paint or stain products. Use a drop cloth to catch any drips or splatters. Choose a brush in an appropriate size for the project, and make sure the paint or stain has been completely stirred or shaken.

Projects that are outdoors or will be exposed to the weather should be painted with an oil-based paint to protect the wood. Another option is to

A *Goodheart-Willcox Publisher*

B *Goodheart-Willcox Publisher*

C *Goodheart-Willcox Publisher*

Figure 12-17. Biscuits can be used to form a strong joint between wood parts. A—A biscuit jointer is being used to make cuts in a board. B—Typical biscuit. C—Parts being assembled with biscuits in place.

Figure 12-18. Before applying paint or stain to a project, a final sanding should be done.

stain the project, then add two layers of polyurethane. This will protect the project from the elements but still allow the wood grain to be seen. Products that are not exposed to the elements can be left natural, stained, or painted to suit your taste.

Paint

Paints should be applied to a project with even coats covering the entire surface. Brush strokes should be applied with the grain of the wood and care should be taken to ensure no runs in the paint are evident, **Figure 12-19**. If a second coat is required, the first coat must be completely dry before the next application. Brush cleanup will be determined by the type of paint used on the project. Oil-based paint will require paint thinner or mineral spirits as a cleaner. A latex-based paint will require soapy water to clean the brush. If in doubt, read the label on the paint can for cleaning instructions.

Stain

Stains are applied to a project in a similar fashion as paints, but there is no need to brush them on with the grain. Stains are applied to a section of a project and are then wiped off with a clean rag. See **Figure 12-20**. When applied, the stain immediately begins to soak into the wood. Only the excess stain is removed. Because of this, it is important to be consistent in the amount of time you allow the stain to remain on the project before wiping off the excess. The longer the stain is allowed to remain on the surface, the darker the final finish will be. For stain products that are oil-based, paint thinner or mineral spirits will be needed to clean the brush.

Sealer

Polyurethane or shellac can be applied over raw wood or as a finish coat over a stained project. Like paint, polyurethane is applied with the grain of the wood in even coats to ensure there are no runs on the project. Cleanup will be determined by the type

Figure 12-19. Paint should be applied with the grain of the wood to help ensure a smooth finish.

of polyurethane used. Paint thinner or mineral spirits is used for oil-based products, and water is used for water-based products.

When the first coat of a sealer dries, it generally leaves a rough surface on the project. If a second coat will be applied, the project needs to be lightly sanded with very fine steel wool to smooth the surface after the first coat is dry. Then it must be patted down with a tack cloth to remove any metal fibers from the steel wool. Because the pores of the wood were sealed with the first coat, the second coat will dry more smoothly. Generally, the more coats of the product that are applied, the more smooth and protected the surface will be. However, each coat should be lightly sanded before adding the next coat until the final coat.

Hardware

The hardware commonly found on many indoor and outdoor projects includes hinges, knobs, and pulls. A variety of other hardware components for multiple applications are available at local hardware stores. For example, a hasp or lock may be used on a storage bin or box.

Hinges

Hinges are used on a variety of projects, such as cabinet doors, wood fence gates, or doors for a shed. Many variations of hinges exist for cabinets, including full overlay, lipped, and flush. A variety of finishes are available

Goodheart-Willcox Publisher

Figure 12-20. Stain can be applied with a brush or a rag. After allowing the stain to penetrate the wood's surface, use a clean cloth to wipe away excess stain.

for hinges. If hinges are used with knobs and/or pulls on cabinets, they should match in color and style. Barrel hinges are typically used for doors in buildings.

Knobs and Pulls

Knobs and pulls are placed on cabinet doors and drawers. The user holds the knob or pull when opening or closing the door or drawer. Knobs and pulls serve the same purpose and are often selected to enhance the appearance of the finished piece. For some cabinets, both knobs and pulls are used. In other cases, only knobs or only pulls may be used. Care should be taken when marking locations for knobs and pulls so that they are placed in the same location for all related cabinets. See **Figure 12-21**. Doors and drawers should be predrilled for this hardware, which is installed with screws from inside the cabinet.

Goodheart-Willcox Publisher

Figure 12-21. Pulls make it easier to open the drawers and doors on this cabinet. Notice how the drawer pulls align and are equidistant from the edges of the drawers.

CHAPTER 12 Review and Assessment

Summary

- Finding or creating a good design is the first step in constructing a woodworking project. The plans for many projects are available from various sources.
- Technical drawings, a materials list, construction procedures, and finish materials should be considered during the planning phase for a woodworking project.
- Technical drawings include measurements and other details for a project. They may include several views of the project, most often the top, side, and front or bottom.
- A materials list should include lumber, fasteners, finish, hardware, and other parts or supplies that will be needed to construct the project. A cut list should include all the wood pieces needed for a project.
- A construction procedures plan gives the steps required for making the project, including steps for subassemblies.
- When selecting lumber, a woodworker should choose pieces that are appropriate for the final project. The amount of cutting needed and defects are considerations in selecting lumber.
- Before wood for a project can be cut, it must be marked with accurate dimensions. As a general rule, wood should be marked so that the grain runs lengthwise on the piece. Curves are best laid out using a pattern or a compass.
- Cutting wood is generally done with some type of saw. It is important to know the width of the saw kerf so that it can be subtracted from the total length of the lumber to be cut from.
- Once all the parts of a project have been cut, it is a good practice to sand each piece prior to creating subassemblies and doing the final assembly.
- Wood joints or fasteners, such as screws and nails, can be used to join pieces of wood in a project.
- Once a project has been assembled, the finishing stage of the project takes place. Finishing includes a final sanding or cleaning of the wood, applying a wood finish, and installing hardware.

Words to Know

Match the key terms from the chapter to the correct definition.

A. biscuit jointer
B. construction procedures
C. finish
D. materials list
E. registration marks
R. sketch
G. subassembly
H. waste side
I. wood joint
J. technical drawings

1. A connection between two pieces of wood where they intersect that is created using the wood itself rather than fasteners, such as nails or screws.
2. Points on a pattern that are used to align the pattern in the correct location when it is repeated or to align parts of a pattern that are printed on separate sheets of paper.
3. A tool used to make crescent shaped cuts in two pieces of wood that are to be joined together.
4. Parts that are put together separately and used as a unit in a finished product.
5. A record showing all items needed in the construction of a project, such as wood, hardware, and paint.
6. A set of plans that includes measurements and construction details needed to build a project.
7. The edge of a cutting line that is opposite the project piece and nearest to the part of the board that will be discarded.
8. A hand drawing with little or no detail, numbers, or measurements.
9. A product, such as paint, stain, or sealer, used to seal wood; protect it from damage, rot, and other hazards; or add beauty to the project.
10. A step-by-step plan for making subassemblies and the final product.

Know and Understand

Answer the following questions using the information provided in this chapter.

1. What are some sources where you might find plans for a woodworking project?
2. Describe the isometric drawing method that can be used to sketch woodworking projects.
3. How does a technical drawing differ from a simple sketch of a project?
4. What information should be included on a cut list for a project?
   A. The size of each piece of wood specified in the plans.
   B. The hardware needed to complete the project.
   C. The supplies needed to complete the project.
   D. All of the above.
5. Why is it important to identify all subassembly steps for a project?
6. What are two factors that should be considered when selecting wood for a project?
7. What does laying out a board for length involve?
8. What are two methods of laying out curved shapes on wood?

9. When making saw cuts, why is the blade not centered exactly on the cutting line?
10. What are some types of wood joints that are commonly used for woodworking projects?
11. How are finishing nails used in a woodworking project?
12. *True or False?* Adhesives are sometimes used with nails or wood joints when joining pieces of wood.
13. What are two examples of projects for which a biscuit jointer may be used?
14. What three steps are involved in the finishing stage of a woodworking project?
15. What are three types of products that are commonly used to finish wood projects?
16. How should paint be applied to a wood project?
17. What are three examples of projects for which hinges may be used?
18. How are knobs or pulls installed on a cabinet door?

STEM Connections and Academic Activities

1. **Science.** Create a science fair project testing the relative density of five types of wood. Display your project at a science fair.
2. **Technology.** Design a board for a game made of wood, and use a computerized numerical control (CNC) machine to cut the board. If you do not have access to a CNC machine, produce the board with available tools.
3. **Engineering.** Determine an environmentally friendly use for sawdust. If possible, collect sawdust and use it for the purpose you have identified.
4. **Math.** When cutting a board, calculate the amount of lumber that will be lost to the saw kerf when cutting six pieces from an 8′ board with a standard miter saw blade. How much wood would be lost when cutting 10 pieces 8″ long from an 8′ board? How much waste would be left over in actual measurement?

   Measure and compare the thickness of three species in a variety of widths of finished lumber.
5. **Social Science.** Explain, using charts or graphs, the economic impact of woodworking on the local economy. Include how students, professionals, and hobbyists use wood products in your area.
6. **Language Arts.** Tell the story of wood, from tree to final product. Use images, graphs, first person accounts, and other means to explain the process that goes into growing, harvesting, and processing lumber.

Thinking Critically

1. What is the difference between a cut list and a materials list? Why is it important to know the difference? If you can only get one of these lists, which would be the most useful to you and why?
2. What are three criteria that you should consider when choosing a finish for a woodworking project?

CHAPTER 13 Surveying

Chapter Outcomes

After studying this chapter, you will be able to:
- Name three general soil categories and briefly describe their characteristics.
- Explain how to conduct a basic soil quality test.
- Calculate cuts and fills volume using the end area method.
- Describe how to lay out a square building corner and check it for accuracy.
- List reasons why land surveys are performed.
- Briefly explain the various surveyor methods used to measure distance.
- Summarize the process of correctly setting up, leveling, and measuring with a dumpy level, auto level, and laser level.
- Define differential and profile leveling and explain each process.

Words to Know

backsight	electronic distance	global positioning	scope
batter boards	measuring (EDM)	system/geographic	surveying
benchmark	device	information system	swell
chaining pin	end area method	(GPS/GIS)	taping
cuts and fills	foresight	grade	target rod
dumpy level	fill dirt	overburden	turning point

Before You Read

After reading each section (separated by main headings), stop and write a three- to four-sentence summary of what you just read. Be sure to paraphrase and use your own words.

While studying this chapter, look for the activity icon **to:**

- **Practice** vocabulary terms with e-flash cards and matching activities.
- **Expand** learning with interactive activities.
- **Reinforce** what you learn by completing the end-of-chapter questions.

www.g-wlearning.com/agriculture

As mankind became less nomadic, there was a greater need to mark boundaries of land. When boundaries were established, descriptions of the property and its boundaries were also needed. Therefore, an accurate method of establishing land boundaries and ways of describing them were developed. Today, technological advancements have further enhanced these systems and methods.

What Is Surveying?

Surveying is the practice of taking essential measurements to determine accurate points or places on land. All types of construction projects make use of surveying. Engineers, foresters, geologists, and construction personnel use surveying as a means to plan and lay out projects. Buildings, highways, railroads, canals, dams, and irrigation and drainage networks are built by workers who rely on surveying data to ensure correct specifications are met for the structure.

Surveying is a very precise process. The measurement data surveyors (professionals who specialize in surveying) collect must be recorded in a usable format, as most land surveys are considered legal documents. Inaccurate surveying can cause problems and result in monetary losses. An incorrectly laid out surface for a building foundation or ill-prepared foundation pads for a bridge could result in disastrous consequences. If a surveyor is off even a few degrees when measuring property line boundaries, a landowner may lose valuable acreage. Such factors underscore the meticulous nature of a surveyor's work.

If you learn and understand basic surveying concepts, you will better understand how the process is critical to construction. Some common surveying tasks, such as soil analysis, distance measurement, and layout work, are considered necessary skills when working in an agricultural setting. A career as a professional surveyor, however, requires postsecondary education, course work, training, and first-hand experience in the surveying field.

Selecting Appropriate Building Sites

Land used for agriculture is bought and sold on a regular basis. Reliable methods of legally marking and describing the land are needed. Landowners may want to change the profile of the land for future needs and uses. Therefore, having an accurate description of land boundaries and their locations is vital for construction purposes. Understanding the makeup of the land itself is also important. Whether building a new home or a structure for animals, it is important to know about soils and how to prepare the land for building sites, **Figure 13-1**.

Soil Analysis

As a general rule, soils are characterized into three categories: sand, silt, and clay. Types of soils differ by particle size. Sand has the largest particle

Henryk Sadura/Shutterstock.com   ©iStock.com/kadmy

Figure 13-1. Surveying equipment is positioned and used during the planning and layout stages of construction. Accurately mapping the elevations of the build site and any necessary elevation changes is an early but crucial aspect of a building project. Using the instruments to establish correct boundaries is also an essential step in the process.

size. Picture the sand used in children's sandboxes or on the beach at the ocean. Sand particles are large, coarse, and rough to the touch. Sand feels gritty when rubbed between your fingers.

Clay particles are the exact opposite of sand particles. Clay particles are small and hardly felt by touch. The texture is similar to that of flour used in baking. You can hardly feel individual clay particles, and they almost feel smooth when compared to sand. When you rub clay particles between your fingers, they fill the lines in your fingertips due to their small size. They also appear almost shiny.

Silt is a combination of sand and clay. Its consistency varies slightly based on the content of both materials. Silt particle size falls between the particle sizes of clay and sand.

A simple illustration of each soil's particle arrangement can be seen by filling three buckets with three different items. Fill the first bucket with golf balls, the second bucket with baseballs, and the last bucket with soccer balls. The bucket filled with golf balls represents clay, the bucket filled with baseballs represents silt, and the bucket filled with soccer balls represents sand. See **Figure 13-2**.

When choosing a location for a building (a build site), you should conduct an analysis of the soil. Using a basic knowledge of soil particles, you can determine many things about the soil. If it is sand, there will be an abundance of air and water pockets in the soil. Water will drain through the soil quickly. Again, picture the large number and size of the spaces between the soccer balls in the bucket.

If the soil is mostly clay, less air space exists in the soil, and water will not drain well through it. Its upper surfaces will hold water. Now picture the smaller spaces between the small golf balls in the bucket.

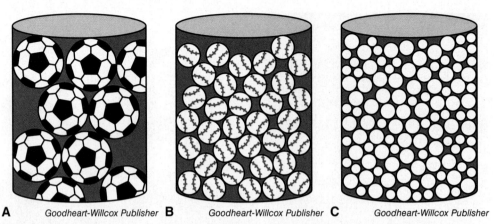

Figure 13-2. An illustration of the particle arrangement of different soil types. Note the size, shape, and amount of space between the representations for the sand, silt, and clay particles. A—Sand particles are the largest of the three and also have the most space in between particles. B—As a mixture of sand and clay, silt has characteristics of both soils. C—Clay particles are the smallest and most compactable.

The permeability of the soil (how easily water or air can pass through it) impacts how well the soil can be compacted. Soil compaction is a crucial part of the construction process. It involves applying pressure to the soil to drastically reduce the air and water pockets between the soil particles. It is much easier to compact clay soil than other soil types because the small particles will fit in a tight configuration. Sand is the least compactable soil type and does not hold its shape well.

AgEd Connection: Surveying

Surveying, site layout, construction layout, contour lines for terraces, and other activities involving levels and transits may all be part of an agricultural technology and mechanical systems CDE. Understanding these tools and their uses has a variety of practical applications in the job market as well.

People and companies are constantly buying and selling land. When land is purchased, legal documents are used to prove the boundary locations. It is a land surveyor who creates these documents. The construction of a building on that same site requires the knowledge and skill of an engineer using similar equipment to determine the slope of the land and what is necessary to make it level or what is necessary to construct a building on the site. Ask your teacher about working as an apprentice to a local surveyor or working with or shadowing an engineer on a building site.

Soil Quality and Testing

You can perform a basic soil test to determine the amount of sand, silt, and clay present in soil. This test does not give exact results, but it is one that can be beneficial if you want a basic understanding of your soil.

To conduct a basic soil test, you will need a clear glass jar with a lid and a source of clean water. You should take the soil sample below the level of grass or organic materials at the surface. Complete the following steps:

1. Remove the lid of the container.
2. Fill the container with about 2″ of soil material.
3. Add water until the level of the container is about 3/4 full. Add a small amount of detergent (about a teaspoon) to aid in soil dispersion. See **Figure 13-3A**.
4. Screw the lid on the container tightly and shake it vigorously for at least one minute, ensuring that the water and soil mix thoroughly.
5. Place the container where it will not be disturbed for at least 12 hours.
6. Measure the depth of the sand, silt, and clay layers after the 12 hours have passed. Sand will settle as the bottom layer, silt will be in the middle layer, and clay will settle as the top layer. Any organic material will rise to the top. Each layer will have a distinct appearance, **Figure 13-3B**.
7. The amount of each of the three materials present in the 2″ sample can be converted to the percent form. For example, if there is 1″ of sand, 1/2″ of silt, and 1/2″ of clay, then the soil sample contains 50% sand, 25% silt, and 25% clay.
8. After determining the percentage of each material, mark the intersection point from the three categories on **Figure 13-4A** to find the soil type of your sample. **Figure 13-4B** shows characteristics of each soil category.

Soil testing laboratories can process a sample and provide precise measurements at little to no cost. These laboratories usually provide you with a test kit, which includes a container for the soil and directions on how to get the needed materials. After the collection, you send the sample to a laboratory, and in a few weeks you receive the results.

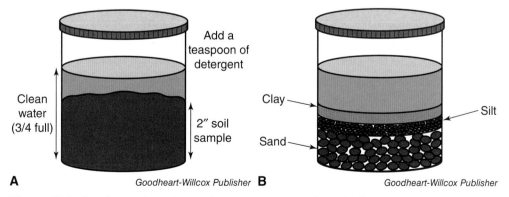

Figure 13-3. A soil sample test can show percentages of sand, silt, and clay in the soil. A—Prepare the sample correctly. B—Measure each layer of soil particles after the particles have settled in the jar.

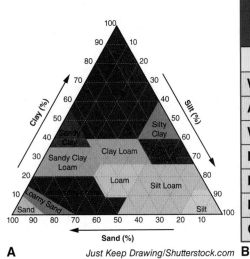

Influence of Soil Texture on Properties and Behavior in Soil			
Property/Behavior	**Sand**	**Silt**	**Clay**
Water-holding capacity	Low	Medium to high	High
Aeration	Good	Medium	Poor
Water erosion	Low	High	Low
Drainage rate	High	Medium to slow	Slow to very slow
Porosity	Low	Medium	High
Bulk density	High	Medium	Low
Compactibility	Low	Medium	High

A *Just Keep Drawing/Shutterstock.com* B Goodheart-Willcox Publisher

Figure 13-4. A—The percentage makeup of your soil sample will determine its texture. B—Different soil compositions have different characteristics that could be advantageous or detrimental to building.

Land Preparation

Soil is normally referred to as loose or compacted. When loose, soil has air in it, which expands the physical space the soil occupies. Soil increases in volume due to moisture and air content, and this expansion is called *swell*. Clay soil will swell approximately 30%, but sand will only swell about 12% due to air and moisture levels.

When compacted, the soil has the air removed. The moisture in the soil is also removed to some extent. Soil compaction ability can be determined by a soil analysis at the proposed build site. A building foundation requires a firmly compacted soil base to properly support the weight of the structure. The tighter the soil particles can be compacted together, the greater the stability of the soil. Without a strong base, the building could face problems from further soil settling, water seepage, and other environmental factors.

When describing and calculating soil quantity, you measure the values in volume. The cubic yard (yd^3) is the standard unit of measurement for large soil quantities. Consider this simple example of loose versus compacted soil. Soil that fills 1.0 yd^3 in a stockpile will take up 1.25 yd^3 of space when loaded loosely into a container and only 0.90 yd^3 when it is compacted.

At a building site, *cuts and fills* are the amounts of soil to be removed from or added to the landscape. Any soil, rock, or material that is not wanted at the site is called *overburden*. Overburden often includes the top layer of unwanted soil that should be removed for construction purposes. Any soil that is used to fill low areas or to create higher levels is called *fill dirt*. An accurate measurement of cuts and fills is needed to properly lay the foundation for the site construction.

Once the soil is tested for adequate use, the next job at many building sites is to determine the boundaries of the structure. Once boundaries are established, determinations are made for cuts or fills needed for the site.

Removal of Overburden

Building on a good foundation is very important. Determining the elevation or grade of a construction site also determines the amount of soil to be taken away (removal of overburden) or added (fill). Normally, heavy equipment and trained operators are used to remove or fill material as needed. The amount of cuts and fills should be calculated to determine costs and time associated with this necessary process.

Calculating Cuts and Fills

Several methods can be used to calculate cuts and fills. One of the easiest and quickest methods is the *end area method*. In this method you calculate the area of each end of a given space, take an average of those two areas, and then multiply by the length between the two end areas to find the volume. If there is a long distance to traverse and many grade changes along the length measurement, an accurate measure may not be gained, but the method will provide an estimate.

Look at **Figure 13-5**. For the basics of learning the method and due to the slope of land, the space in this illustration is defined by triangles on each end. Therefore, the area formula for a triangle is used. Start with one end (A_1) and make the following calculations.

Area of a triangle = 1/2 × (base × height)

The area of End A_1 equals:

A_1 = 1/2 × (16′ × 12′)

A_1 = 96 ft²

The area of the other end, End A_2 equals:

A_2 = 1/2 × (12′ × 10′)

A_2 = 60 ft²

Next, the average area of the two end areas multiplied by the length between them equals:

Volume = [(96 ft² + 60 ft²) ÷ 2] × 45′

Volume = 78 ft² × 45′

Volume = 3510 ft³

The volume in cubic feet needs to be converted to cubic yards for the final measurement. This is done by dividing the value by 27.

Volume = 3510 ft³ ÷ 27

Volume = 130 yd³

The estimated cut or fill volume of the space shown in **Figure 13-5** is 130 yd³. The building plans will show whether the volume of dirt must be cut from or filled into the area.

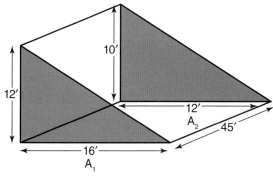

Goodheart-Willcox Publisher

Figure 13-5. The cut or fill volume of this space can be accurately estimated using the end area method.

Career Connection

Surveyor

Job description: Surveyors make measurements to determine property lines and the locations for buildings, bridges, and roads. Surveying requires work both in the field and indoors. The work done by surveyors is important for land development, construction, cartography, and topography. Great precision and attention to detail is required to perform a surveyor's duties.

Education required: A four-year bachelor's degree and licensing through state agencies is typically required. Many surveyors work for engineering firms on big projects, while others work independently doing smaller jobs.

Job fit: This job may be a fit for you if you enjoy applying algebra, geometry, and statistics to solve real-world applications. An interest in geography, and some background in engineering, technology, law, and government would also be helpful for the position. Knowledge of the building and construction trade would be helpful for many of the job duties. Finally, if you prefer to work independently and enjoy working mainly outdoors, surveyor could be a profession to consider.

pryzmat/Shutterstock.com

Site Layout and Design

Determining the first corner of a building site is extremely important. It will determine the layout of the building and affect the cuts and fills measurements that are contingent on its height. The corner is usually determined by owner preference. However, many considerations, such as prevailing winds, drainage of the soil below the site, and access to utilities, must also be taken into account. Once all the factors have been carefully considered, the corner is set.

Laying Out a Corner

Most corners are laid out in a 90° angle. Once you set the single outermost point, lay out lines radiating from it. You can then use the Pythagorean theorem ($a^2 + b^2 = c^2$) to check that the corner measures exactly 90°.

Look at **Figure 13-6**. The legs of this triangle (9′ and 12′) are designated as a and b respectively. The other unknown leg, designated as c, is the hypotenuse. After plugging the values into the Pythagorean theorem, you solve for c.

$$c^2 = (9' \times 9') + (12' \times 12')$$

$$c^2 = 225 \text{ ft}^2$$

$$c = \sqrt{225} \text{ ft}^2$$

$$c = 15'$$

The length of the hypotenuse, or c, equals 15′. This measurement is the key to keeping your corner square.

Now that you know all the lengths of the triangle, measure out 9′ from your first corner point to mark side *a*. Next, start at the first corner point again and measure out 12′, following a path perpendicular to side *a*. This is side *b*. The length between the farthest ends of sides *a* and *b* creates side *c*, or the hypotenuse. This hypotenuse must measure 15′ for the corner to be 90° and perfectly square. This example demonstrates a builder's trick commonly called the 3-4-5 method. If you have squared the corner accurately, the distance of the first leg will be three units, the distance of the other leg will be four units, and the distance of the hypotenuse will be five units. It does not matter what distance measurement is used—inches, feet, centimeters, etc. The method will work due to the Pythagorean theorem. In the example given, the first leg was 9′ (3 × 3′), the other leg was 12′ (4 × 3′), and the hypotenuse was 15′ (5 × 3′). You can check the accuracy of the 3-4-5 method for laying out square building corners by inserting the values into the Pythagorean theorem ($9^2 + 12^2 = 15^2$) and determining if the equation is true (81 + 144 = 225).

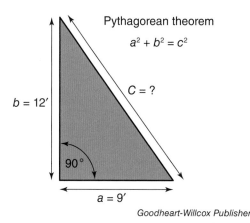

Figure 13-6. You can check that a corner is square by using the Pythagorean theorem. If you solve for the hypotenuse, you can use the measurement to ensure the corner is accurately angled at 90°.

Staking the Site

Most construction surveying contains a majority of layout work. This could be setting the boundaries of the work site or determining site conditions that would involve fills and cuts. Layout work requires the use of grade stakes to convey this information.

Grade stakes are wooden stakes driven into the ground at planned locations around the build site. Surveyors mark information on the stakes in a shorthand language that corresponds to information provided on the building plans. Workers read the information on the stakes and use them as a guide for where and exactly how much material to remove or fill. The grade stake for the first corner is one of the most important stakes because it sets the first boundary for the building site and the elevation.

Although some grade stakes are driven into the ground until the top is located at the correct elevation, land contours may not allow this practice. More often 3′ to 5′ of the stake is left above ground. The surveyor will write pertinent information necessary for the workers on this part of the stake. The information might include how much material to remove or fill and at what grade. *Grade* is the slope, pitch, or incline of a surface or line. Grade is marked on the stakes to indicate the elevation difference that is needed between the stakes. The correct grade level at different points on the site is important for a number of reasons, including establishing the correct height of the foundation footings and walls, proper water runoff away from the structure, and accurate floor elevation.

Once the building site has the first corner and the rest of the building laid out, the dirt work must be completed. After this is finished, other processes must be accomplished by the surveyor or contractor.

Staking the Foundation

Depending on the foundation of the proposed building, the level for the foundation is set. Normally, this requires all exterior building corners to be set. The areas between the corners are set up with temporary string lines. Footings on which the building will sit must be excavated. The footings provide a solid foundation for the building and can be concrete, packed stone/gravel, or other easily compressible materials. The outer edges of the building are designated by the string reference lines for excavation purposes.

Once the excavation of the foundation is complete, the surveyor or contractor will set up forms to determine the exact foundation place and elevation. There are many methods available, but one of the most common is the use of batter boards.

Batter boards are stakes and horizontal cross pieces used as a temporary framework when marking corners and lines of a foundation. Usually, two stakes are driven into the ground with a movable board spanning them. Batter boards are set up parallel with exterior reference lines and used in pairs at corner building points, **Figure 13-7**. Therefore, if you have a building foundation in the shape of a square, you would need eight sets of batter boards, or two pairs at each corner.

The batter boards are used to run a reference line between one corner to the other, and the height of the cross board is moved to set the foundation elevation. The surveyor repeatedly refers to these reference lines as he or she sets the elevation. The reference lines that run from the batter boards also give the workers constructing the foundation an exact guide for their forms to hold the foundation material.

Basic Surveying Principles

Land surveys are used to establish exact locations, boundaries, or subdivisions of land. They are often referred to as cadastral or property surveys. Land surveys are performed for the following reasons:
- Establish markers or monuments to define or preserve boundaries.
- Legally relocate markers or monuments.
- Determine lengths and direction of old land survey lines.
- Subdivide large parcels of lands.
- Calculate areas, distances, and directions for survey data.
- Provide a technical and written description of land and its boundaries.

Surveyors must use different instruments to calculate the pieces of data that are needed as part of the land survey.

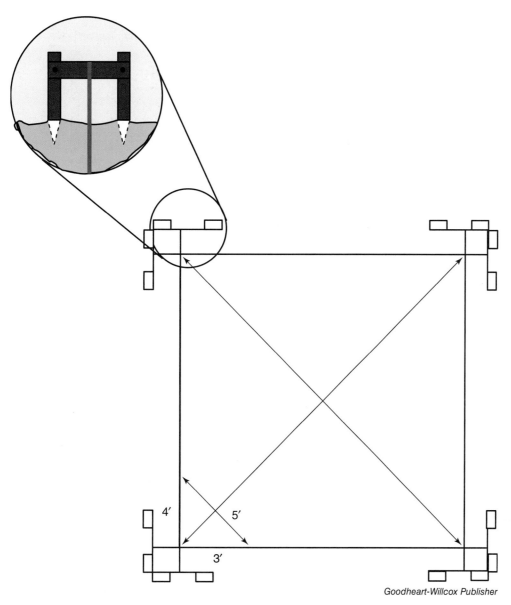

Goodheart-Willcox Publisher

Figure 13-7. Batter boards are commonly used for foundation layouts. Reference lines radiating from the boards (red line in the inset image) must be straight, level, and squared in the corners in order for the foundation to be level and square. The lines are checked again and again throughout the layout process to ensure there is no error in measurement.

Distance Measurement

Measuring distance is one of the tasks most commonly done by a surveyor. The distance between two points on a two-dimensional map is a horizontal distance. Land is measured on a flat plane due to this fact, and the measurement is not based on elevation changes. Even on sloping ground, there are methods you can use to measure horizontal distance, **Figure 13-8**. Maps are most often used to determine and mark boundary points.

In the United States, most surveying measurements are read in feet and tenths of a foot to simplify calculations. In the metric system, surveying measurements are given in meters and millimeters. There are many ways to measure distance in surveying, such as pacing, taping, and using electronic equipment.

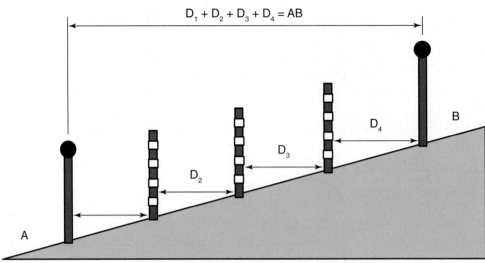

Figure 13-8. This figure shows a horizontal distance measurement method performed on sloped ground. The tape measurement process is done in steps, with *A* as the starting point and *B* as the ending point. Each distance (D_1, D_2, D_3, D_4) is measured to get the final distance.

Pacing, a practice that involves a person walking out the distance and calculating the distance based on his or her pace factor, is a basic method. After determining your pace factor by walking a known distance several times and dividing that distance by your average number of steps, calculating an unknown distance simply involves multiplying your pace factor by the number of paces taken. Odometer wheels that roll on the land and measure the distance based on wheel revolutions can also be used. These methods, however, are not always accurate.

Measuring with a metal tape has been done for many years, and a measuring tape is durable and accurate. However, it takes considerable time to measure long distances this way. An *electronic distance measuring (EDM) device* is a piece of equipment that has circuit boards that calculate distance between two points through algorithms based on the speed of light (186,000 miles per second). These devices are easy to use and are accurate, especially at long distances, but they are expensive. A *global positioning system/geographic information system (GPS/GIS)* is software and equipment that use advanced technology and linked satellites to measure distance. GPS uses satellites in orbit around the earth to determine where a point is located. GIS is the software that computers use to translate information from GPS data. This method is gaining popularity due to its pinpoint accuracy, but it is also expensive.

Taping Measurements

Taping is a process used in land surveying where a metal tape marked with length measurements is used to determine the distance between two points. The main disadvantage when taping is that all the variables associated with physically measuring distance could cause error in the final measurement. Temperature fluctuations can cause the metal tape to expand or contract. Over long distances, the tape will sag in the middle due to its weight. It may produce inaccurate readings because it does not follow the contours of the land. These and other errors can be corrected, avoided, or compensated for with proper training and practice.

There are six main steps for measuring distance when using the taping process.

1. First, "line in" the points to be measured by determining the shortest distance between the two points.
2. Apply tension to remove sag to the tape. Be cautious to not apply too much tension and break the tape.
3. Connect a plumb bob to one or both ends of the tape so it remains horizontal. This is mainly needed when obstructions or land features exist along the measurement line, requiring the tape to be elevated.
4. When measuring longer than 100′, mark the distance with chaining pins. *Chaining pins*, **Figure 13-9**, are metal stakes used during surveying to incrementally mark taping measurements on the ground. Chaining pins usually come in a set of 11 so that you can measure up to 1,000′ per set, in 100′ increments.
5. Once the distance is measured, read the tape to the nearest 1/10.
6. After an accurate measurement is read, record the distance in your field notes. Field notes allow you to continue your measurements and later describe what was measured and the recorded length.

alterfalter/Shutterstock.com

Figure 13-9. Chaining pins (far left in the photo) are used during the surveying practice known as taping. Each chaining pin is used to mark a location to the next pin placement. Therefore, each pin represents a known distance except for the last measurement.

Electronic Measurements

Electronic distance measurement (EDM) devices are becoming more available and economical. This equipment calculates distance between two points based on the speed of light (186,000 miles per second) and the time needed for the light to travel from one point to another. The calculations are recorded on electronic data loggers for retrieval either on site or away.

The ability to calculate distances effectively and with little error has revolutionized the procedures used in surveying. The increased accuracy level of electronic methods has made taping nearly obsolete. Despite terrain conditions and, to some extent, weather conditions, distances can be measured easily, quickly, and very accurately with EDM devices and electronic systems.

Surveying Equipment and Use

Surveying requires many pieces of specialized equipment. Besides a way to measure distances using tapes, EDM devices, or satellite systems, many times it is necessary to ensure levelness or heights of respective points at a location or building site. The most common instruments used to do so are the dumpy level and auto level.

Dumpy Level

The *dumpy level*, or builder's level, is one of the most commonly used surveying instruments. It is an optical tool that is used along with a target rod to measure the elevation rise and fall of a piece of land. It is also used to sight and determine locations along a horizontal line. You do not need to be a professional surveyor to understand how to use the level. It is one of the simplest leveling instruments, **Figure 13-10**, which makes it a good teaching tool.

Before survey work can begin, the dumpy level must be mounted securely on a tripod stand. The stand holds the instrument over a designated location. All three legs of the tripod must be locked in a sturdy position on a firm footing. A plumb bob may be hung directly under the center of the tripod to assist in placing the instrument precisely over a given location.

A dumpy level normally has four screws used in opposite pairs to adjust the position to level. A spirit (bubble) level on the tool functions as the leveling guide. A good rule of thumb is to turn opposing screws toward each other or away from each other to level the instrument across one axis, **Figure 13-11**. The instrument *scope* is an optical device (telescope) through which the user looks to sight a level horizontal line. Once the instrument has been leveled on one axis, the instrument scope should be turned 90° and aligned with the other axis/screws to level the tool along that axis. After the instrument is leveled, it can be rotated 360° on the tripod and used to sight horizontally in any direction.

If the level only has three screws, use the same procedure of turning two screws at a time to find the level position while viewing the bubble level on the instrument. Then, adjust one of the previous leveling screws with the remaining screw to find the level position of the instrument.

A helper is needed to hold the target rod at the desired locations on the property or build site for the measurement process. A *target rod* is a stick or pole marked with measurements and used with a leveling instrument to determine elevation. The surveyor looks through the leveling instrument scope and aligns the scope crosshairs on the target rod to determine the point's elevation, **Figure 13-12**. Measurements are logged in the field notes for each of the necessary locations.

julie deshaies/Shutterstock.com

Figure 13-10. The dumpy level, or builder's level, is used to measure the elevation rise and fall of a piece of land. Note the four screws used to adjust the instrument.

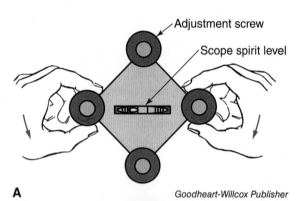

A   Goodheart-Willcox Publisher

B   thanee khansri/Shutterstock.com

Figure 13-11. Leveling adjustment screws are used to find the level position for the instrument. On a four-screw instrument, turn opposing screws in the same direction to try and find a level reading on the scope's bubble vial. After one axis is level, turn the scope along the other axis and adjust the other set of screws.

A   *CandyBox Images/Shutterstock.com*   B *Edler von Rabenstein/Shutterstock.com*

Figure 13-12. A—An engineer's target rod is used in surveying practices. B—A helper holds the target rod while the surveyor aligns the scope crosshairs on the rod at the elevation measurement mark. Note that the helper is holding a metric target rod.

The engineer's target rod is graduated in feet and tenths and hundredths of a foot or in metric measurements. See **Figure 13-13**. The large red numbers indicate feet and the smaller black numbers located between the feet graduations signify 1/10th of a foot. Between each 1/10th mark, there are four black line markings. Each marking represents a 1/100th of a foot or 1/10th of an inch. The distance between these marks and from the top to the bottom of the mark itself is equal to 1/100th of a foot.

Auto Level

Auto levels are similar to dumpy levels, but they normally have only three screws for adjustment. The instrument should be leveled using all three screws once the level is placed at the desired location on the tripod stand. The engineer's target rod is also used for measuring elevations with the auto level.

The main difference between the dumpy level and auto level is that the auto level has a compensator built into the instrument. Refractors that are part of the compensator help ensure that the tool is level by automatically eliminating slight movements from level. The dumpy level is more expensive than an auto level, but it is also more accurate.

Dumpy and auto levels are delicate tools manufactured to perform precise work. Extending the life of the instrument means that you must take proper care of the tool. Always keep the level clean, lubricated, and protected from water. Store it in a protective case while not in use or when in transit. After it has been mounted on the tripod, hold the tool in a level position when moving it around the build site. If problems are detected, send the instrument to a qualified technician or to the manufacturer for repairs.

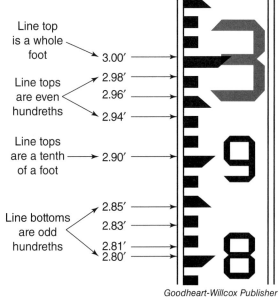

Goodheart-Willcox Publisher

Figure 13-13. Study the marks on the engineer's target rod and the corresponding value for each.

> **Safety Note**
> When using a laser level, never look directly into the laser beam or direct the beam toward someone's face. Be aware that the laser beam can be reflected in unintentional directions by hard, slick surfaces.

Laser Level

The use of laser levels is becoming common in the surveying industry due to their accurate measurements and operational efficiency and the ease with which the instrument can be set up. They are stationed and set up in a level position using methods similar to those used for dumpy or auto levels. The surveyor, however, does not have to sight a target rod through a scope. Instead, the tool projects a laser beam that is captured by the laser receiver unit. The laser beam is directed either in a straight line or in a level plane completely around the tool.

A surveyor does not need assistance when taking measurements with a laser level. After properly setting up the laser, the surveyor attaches the laser receiver unit to a target rod and carries the rod to the specific measuring points. When the surveyor stops at a measuring point, the receiver directs the surveyor—using audible signals, LCD screen icons, or both—to move the receiver up or down the target rod until it captures the level laser beam being projected 360° around the instrument. The surveyor tightens the receiver at this level point on the target rod and records the measurement in the field notes. In contrast to the process required for the dumpy level, no movement of the laser level is required for measurements, which reduces the opportunity for errors.

Leveling

The two major types of leveling used to determine elevations at specific points are profile and differential leveling. Both processes begin at a *benchmark*, a point with a known or expected elevation from which all site elevation measurements are referenced. From the benchmark point, a similar method is used for each type of leveling with the exception of a few changes.

Differential Leveling

Differential leveling is used to determine the difference in elevation between two points, often with a long distance between them. Differential leveling is unconcerned with the elevation of reference points between the two points. These points are used merely to advance the leveling process. The primary focus of differential leveling is to determine the ending point elevation based on a starting elevation.

Between the benchmark and final reading, the instrument is often moved many times, and corrections to the instrument must be made to calculate the final elevation. This allows a greater distance to be traveled between the beginning and ending points.

For example, if the benchmark elevation is 1000′ and the end point is three miles away, you might have to move the instrument 5 to 20 times because of topography to determine the final elevation. In these situations, *turning points* are established that act as temporary elevation reference points. For differential leveling, turning points (TP) are set up at intervals with the level placed an equal distance in between the points.

Look at **Figure 13-14**. A benchmark elevation is established at 236.07′. The level is set up at the first position (SU_1). You begin by reading the rod

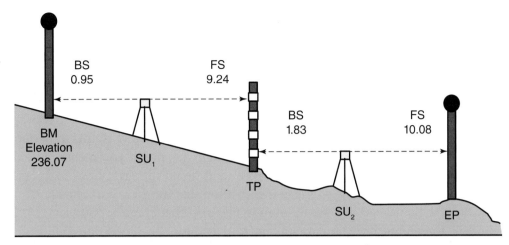

Point	BS	HI	FS	Elevation
BM	0.95	237.02		236.07
TP	1.83	229.61	9.24	227.78
EP			10.08	219.53
	+2.78		−19.32	−16.54

Goodheart-Willcox Publisher

Figure 13-14. An example of a differential leveling diagram and the field notes table is shown here. Although this example only features one turning point, the outlined steps can be followed for as many turning points as necessary until the end point is reached.

at the benchmark (BM) to find the backsight (0.95′). A *backsight* (BS) is a rod measurement taken at a known elevation point. It measures the vertical distance on the rod from the established elevation to the horizontal sight line extending from the instrument scope. Each time the instrument is set up or moved, a single backsight must be read at the previous reference point.

If you add the backsight to the benchmark elevation, the answer is the height of the instrument's horizontal sight line (0.95′ + 236.07′ = 237.02′). This elevation is listed in the HI column within the table.

Next, stay at the first setup and read the rod for the foresight (9.24′) at the turning point (TP). A *foresight* (FS) is a rod measurement taken to determine the elevation at an unknown point. If you subtract the foresight from the height of the instrument's horizontal sight line, the answer is the elevation of the turning point (237.02′ − 9.24′ = 227.78′). Now that the TP elevation is known, it becomes the reference point.

Finally, move the level to the second setup (SU_2) and take a backsight (1.83′) on the TP. Add the backsight to the TP elevation for the new instrument height (1.83′ + 227.78′ = 229.61′). Stay there and take a foresight (6.17′) at the end point (EP). Subtract the foresight from the instrument height and the answer is the elevation of the end point (229.61′ − 10.08′ = 219.53′).

You can check your work by adding the backsights (BS) column together and totaling the foresights (FS) column together (add all the negative values). The sum of these two columns should equal your change in elevation (−19.32′ + 2.78′ = −16.54′). See the field notes table in **Figure 13-14**.

Profile Leveling

Profile leveling is used to determine elevations at different points along a specific line of sight and is commonly performed for the construction of roads and fences. It creates a side view of elevation levels along a route and may feature very different elevation points along the sight line used.

The main difference between differential and profile leveling is the use of many supplemental foresights between designated turning points to better gauge the elevation of the terrain. From a benchmark point, you measure points positioned at designated intervals and at noted topographical features along a line of sight for the changes in elevation that might occur. The level may have to be moved and set up at new, assumed benchmark points. Multiple foresights, however, are taken from a single setup position at the marked points along the line of sight.

For example, if the starting elevation is at 1,000′ and a point 500′ away is the ending point along a line of sight, every marked high or low point can be evaluated for its change in elevation. A reading must be made and recorded at each chosen point to determine its elevation based on the initial or benchmark measure. These supplemental elevation measurements help create a true profile image of the route. See **Figure 13-15**.

Surveying Application

You now know that there are several types of surveying and that several instruments are used. The question that you may ask yourself is, how do I plan a building site, lay out the foundation, and ensure that everything is where I want it to be? Consider this example and think about what you might do to complete this project.

You have selected a site to build a barn that is relatively flat and free of most obstructions. You have removed trees, shrubs, grass, and other vegetation from the build site. Soil at the build site has been collected and sent to a laboratory for analysis. Lab results show soil characteristics that are adequate for proper water drainage and load-bearing capability for the proposed structure.

Initially, you will determine the location of your first building corner. Next, you will lay out the rest of the structure based on the outer dimensions. Once you have laid out these preliminary markers, you will need to determine the height of the four outside points with a surveying level. If the corner elevations are not exactly the same, you will need to determine which point will be your benchmark height. Depending on the location, you will probably choose the highest point because water could infiltrate the structure at the lowest point of elevation. Then you will determine your cuts and fills based on the topography of the building site.

Once the soil has been properly distributed and compacted, you will once again determine a corner point. You will use batter boards to lay out the foundation outer lines as well as set the elevation of the foundation. Remember, you can use any type of foundation material (concrete, stone, or dirt), but the building needs a solid foundation for the life expectancy of the structure.

Once the initial foundation lines are laid out through the batter boards, you will use the leveling instrument (auto, dumpy, or laser) to determine the exact elevation of all corner points of the structure. Remember that if your rod measurements increase, the point of elevation is higher; and conversely, if the rod readings decrease, the point of elevation is lower. Always ensure that you have set up the leveling instrument correctly and double-check your readings to ensure accuracy. Grade leveling may be necessary in order for the foundation footings to be correctly excavated. This ensures the building has a solid and secure foundation with a level floor height.

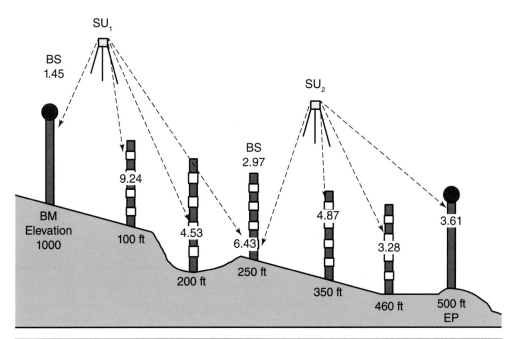

Point	BS	HI	FS	Elevation
BM	1.45	1001.45		1000
100			9.24	992.21
200			4.53	996.92
250	2.97	997.99	(6.43)	995.02
350			4.87	993.12
460			3.28	994.71
500 (EP)			(3.61)	994.38
	+4.42		−10.04	−5.62

Goodheart-Willcox Publisher

Figure 13-15. An example of a profile leveling diagram and the field notes table is shown here. Note the use of the supplemental foresights at each of the two setup points. To correctly check your work during profile leveling, only total the foresight values from the turning point(s) and end point. Do not total the supplemental foresight values. Turning point and end point values are commonly offset with parenthesis or are listed in a separate column in the field notes.

CHAPTER 13 Review and Assessment

Summary

- Descriptions of a piece of land and its boundaries are often needed. Surveying is the practice of taking essential measurements to determine accurate points or places on land.
- The soil must be considered when preparing land for building sites. Soils are grouped into three categories: sand, silt, and clay. These soils have different particle sizes and permeability characteristics.
- A simple soil test can offer basic information about soil composition.
- Soil compaction is necessary to construct a sturdy foundation on which a structure or road can be built. Determining the elevation or grade of a construction site determines what soil is needed to be taken away (removal of overburden) or added (fill).
- The end area method for calculating cuts and fills requires finding the average area of two land ends. That number is multiplied by the distance between the ends to calculate the volume.
- Determining the first corner of a building site is extremely important. A method commonly called the 3-4-5 method is often used to check that building corners are laid out perfectly square.
- Grade stakes are planted around the build site and marked by surveyors to convey cut and fill plans, grade requirements, and other information.
- Batter boards are used on many build sites to help establish the lines and level of a building foundation and its overall footprint.
- Land surveys are completed to define land boundaries and gather data on topography and land distance measurements.
- When measuring distance, surveyors can use pacing, odometer wheels, taping, or electronic distance measuring devices/systems as their preferred method.
- Dumpy levels and auto levels are set up and operated in similar ways. A correction device within the auto level automatically adjusts for slight level misalignments.
- Laser levels project level beams of light that can be captured on a laser receiver on a target rod to determine point elevations.
- Differential leveling is performed to determine the difference in elevation between two points, often with a long distance between them.
- Profile leveling is used to determine elevations at different points along a specific line of sight and is commonly performed for the construction of roads and fences.
- Preparing a building site involves removing obstructions, doing a soil analysis, determining a building corner, leveling the site, and completing cuts and fills.

Words to Know

Match the key terms from the chapter to the correct definition.

A. backsight
B. batter boards
C. benchmark
D. chaining pins
E. cuts and fills
F. dumpy level
G. electronic distance measuring (EDM) device
H. end area method
I. foresight
J. fill dirt
K. global positioning system/geographic information system (GPS/GIS)
L. grade
M. overburden
N. scope
O. surveying
P. swell
Q. taping
R. target rod
S. turning point

1. A stick or pole marked with measurements and used with a leveling instrument to determine elevation.
2. Increases in soil volume due to moisture and air content.
3. A temporary elevation reference point used in differential and profile leveling.
4. Soil, rock, or material that is not wanted on land or a building site.
5. A rod measurement taken to determine the elevation at an unknown point.
6. A point with a known or expected elevation from which all site elevation measurements are referenced.
7. Soil that is used to fill low areas or to create higher levels in an area of land.
8. A process used in land surveying where a metal tape marked with length measurements is used to determine the distance between two points.
9. A method used to calculate cuts and fills in which you calculate the area of each end of a given space, take an average of the two areas, and then multiply by the length between the ends to find the volume.
10. Stakes and horizontal cross pieces used as a temporary framework when marking corners and lines of a foundation.
11. Amounts of soil to be removed from or added to the landscape or build site.
12. An optical device on a surveying level through which the user looks to sight a level horizontal line.
13. A rod measurement taken at a known elevation point.
14. An optical tool that is used to measure the elevation rise and fall of a specific piece of land.
15. Software and equipment that use advanced technology and linked satellites to measure distance on the earth.
16. A piece of equipment that calculates the distance between two points through algorithms based on the speed of light.
17. Metal stakes used during surveying to incrementally mark taping measurements on the ground.
18. The slope, pitch, or incline of a surface or line.
19. The process of taking essential measurements to determine accurate points or places on land.

Know and Understand

Answer the following questions using the information provided in this chapter.

1. List the three general soil categories from the largest particle size to the smallest.
2. After a basic soil test is performed, you measure 1/8" of sand, 3/8" of clay, and 1 1/2" of silt in the 2" sample. What is the percentage of each soil type in the sample?
3. What is removed from soil when it is compacted?
4. What is the standard unit of measure for large soil quantities?
   A. Pounds (lb).
   B. Square feet (ft^2).
   C. Cubic yards (yd^3).
   D. Feet (ft).
5. You can check the accuracy of the 3-4-5 method for laying out square building corners by using the _____ theorem.
6. What are batter boards and how are they used in laying out foundations?
7. List three reasons why land surveys are conducted.
8. What do the circuit boards inside modern electronic distance measuring (EDM) devices use to calculate distance?
9. On an engineer's target rod, the four black line markings between each numbered tenth of a foot represent what fraction of a foot?
   A. 1/16th.
   B. 1/100th.
   C. 1/32nd.
   D. 1/1000th.
10. How does a surveyor determine the elevation at a specific point of land using a laser level?
11. Each time the level is set up or moved during the differential leveling process, how many backsight reading(s) must be taken at the previous reference point?
12. *True or False?* The main difference between differential and profile leveling is that only differential leveling uses turning points and foresights.

STEM and Academic Activities

1. **Science.** Conduct a basic analysis of soil to be used for a surveying project. Explain the soil analysis process and your results.
2. **Engineering.** Describe the characteristics of compacted soil and loose soil. Investigate the natural factors that influence the location and design of building foundations.
3. **Math.** Using the end area method, calculate the volume of fill dirt on a job site that requires an elevation change from 1000' to 1500' and is 200' wide by 1000' long. Imagine the space is defined by a triangle on each end.

4. **Social Science.** Use the Internet or your library to research different civilizations that used land surveying processes early in the formation of their societies. Explain why it was important.
5. **Language Arts.** Write a short story about the daily life of a surveyor with a focus on the requirements of his or her specific duties.

Thinking Critically

1. As populations continue to grow, how can the work of professional surveyors impact society?
2. You and a partner are given property that requires surveying. What steps are necessary to survey the property to determine boundaries and elevation changes between boundaries?
3. Describe the differences between the dumpy level and the auto level. List two examples of where each could be used and explain why. What are the advantages of using a laser surveying level?

CHAPTER 14
Concrete Foundations and Flatwork

Chapter Outcomes

After studying this chapter, you will be able to:
- List the characteristics of concrete.
- Describe how concrete is created and tested.
- Discuss preparation, materials, and tools needed for concrete projects.
- Describe how concrete is tested for readiness and finished.
- Describe foundation systems for structures that use concrete.
- Discuss types of concrete forms and their uses.
- Discuss placement and curing methods for concrete.

Words to Know

Abrams cone	edger	knee board	slump cone
admixture	exothermic	pier footing	slump test
aggregate	expansion joint	Portland cement	spread footing
battered foundation	float	power trowel	spreader
bleed water	floor scraper	rebar	straightedge
broom finish	floor squeegee	safety roller	tamper
bull float	form	screed	trowel
clinker	hydration	slab-on-grade	vapor barrier
control joint	insulated concrete form	foundation	
Darby level	(ICF)	slump	

Before You Read

Before you read the chapter, interview someone in the workforce (your supervisor, a parent, relative, or friend). Ask the person why it is important to know about the chapter topic and how this topic affects the workplace. Take notes during the interview. As you read the chapter, highlight the items from your notes that are discussed in the chapter.

While studying this chapter, look for the activity icon to:

- **Practice** vocabulary terms with e-flash cards and matching activities.
- **Expand** learning with interactive activities.
- **Reinforce** what you learn by completing the end-of-chapter questions.

www.g-wlearning.com/agriculture

Concrete is a composite material made of cement, *aggregate* (coarse gravel and fine sand), and water. Concrete is often a good choice for building agricultural structures because of its characteristics, **Figure 14-1**. It is a simple process to mix, pour, and finish concrete, and these steps will be detailed in this chapter.

Concrete Applications and Characteristics

Two common uses of concrete in agriculture are for building foundations and flatwork. A foundation is the lowest load-bearing part of a building and is often below ground level. Concrete footings are often used to create a foundation. Flatwork is any concrete element that is positioned horizontally, such as a floor, patio, or sidewalk.

Alison Hancock/Shutterstock.com

Figure 14-1. Concrete is easy to mix and pour, and it is durable and relatively inexpensive.

Concrete has several desirable characteristics, including the following:
- Relatively inexpensive.
- Durable and low maintenance.
- Fire resistant.
- Easy for the novice to work.
- Resistant to corrosive chemicals, including animal waste.
- Great compressive strength.
- Good tensile strength (with reinforcement materials).
- Useful for decorative elements.
- Easy to make with local building materials.

Although concrete is the ideal material to use in many agricultural construction projects, it has some drawbacks. Concrete is heavy, and it lacks tensile strength if used without additional materials or reinforcements. Concrete is relatively permanent, which could be an advantage because a concrete structure or component will last for many years. However, this could be a disadvantage if the structure needs to be temporary or movable. These limiting factors should be considered when selecting concrete as a building material.

Forming Concrete

The formula for concrete is simple, and all ingredients are usually available in even the most remote locations. Concrete is made in a step-by-step process that causes a chemical reaction between Portland cement and water. *Portland cement* is a substance that is made from limestone and clay and hardens as it cures, even under water. This process is *exothermic*, meaning it releases heat as part of the reaction. As a result, concrete warms as it begins hardening. The hardening of the concrete forms a rock-like mass when the cement paste (Portland cement and water) forms strong chemical bonds around the course aggregate (gravel) and the fine aggregate (sand).

In making Portland cement, a basic mixture of limestone is heated to a high temperature to form *clinkers*—stone-like, unburnable lumps left after the heating process. These clinkers are then ground with additional ingredients to a very fine powder and, when mixed with water, form cement paste. When cement paste is added to the aggregate and cured (the chemical reaction is completed), the resulting mass is rock-like concrete. See **Figure 14-2**.

Budimir Jevtic/Shutterstock.com

Figure 14-2. Once the proper ingredients are mixed together, concrete is created and ready to be used.

Hydration

When cement is mixed with water, the various chemicals in the Portland cement combine with water in an exothermic chemical reaction called *hydration.* Hydration produces the strong bonds that are characteristic of cured concrete. The qualities of concrete can be altered significantly by changing the base chemicals in Portland cement.

When making concrete, the strength of the final product depends on the complete reaction of all the cement and water in the mix. The hydration reaction will continue for months or even years in concrete if it is well mixed and placed correctly. The strength of the concrete increases as the reaction continues.

Hydration occurs in three distinct phases that make concrete construction possible. In the first phase, water is added to the Portland cement. A quick rise in temperature occurs as the reaction starts. This is followed by a dormant or resting phase. This resting phase allows concrete to be mixed at a ready-mix plant and hauled for several miles (or 1 to 2 hours) to where it is placed. The secondary hydration reaction starts the third phase. In the third phase, heat is again given off and strong chemical bonds start to form around and among the aggregates, forming the rock-like mass known as concrete. This phase can last as long as there is water available to continue the hydration process.

For proper hydration, it is important to use the correct amount of water. Using excess water will result in weak bonds being formed. Lack of water interferes with the ability to complete the hydration process, making the concrete too stiff to be worked. Although the bonds are strong, they may be unconsolidated, causing them to be porous and weak.

Aggregates

Stones or gravels are often used as the course aggregate for concrete. The cement paste forms around the stones to provide a stable and durable concrete. Aggregates for making concrete vary by the job requirements. Course aggregates should be solid and not easily fractured. Limestone is an ideal course aggregate when it is no larger than one third of the entire slab thickness. It also works well if the concrete is less than 6" in diameter or if the area is larger than 18". For example, a 6" concrete slab could have up to 2" of stone. Course aggregates should also be clean and free of dust, organic matter, or other debris.

Course aggregates can also be broken or crushed concrete from other projects or demolitions. In specialized constructions, the course aggregates can be very light-weight materials that reduce the weight of the finished product or change the porosity of the concrete.

Fine aggregates are almost always sand. Particles the size of grains of sand allow the cement paste to fill in the voids between the larger course aggregate pieces and create tight bonds. Fine aggregates or sand should be dry or damp. If water is present, the amount of water must be known so that the water-to-cement ratio will not be adversely affected. Wet sand makes for weak concrete if the water is not accounted for in the recipe. Fine aggregates must also be free of dust, organic matter, or other debris.

Slump Test

Slump is a measure of the consistency of fresh concrete as determined by performing a slump test. A *slump test* is a method of checking the consistency of fresh concrete by seeing how much a sample of the concrete subsides (slumps down) when the surrounding container is removed. A *slump cone*, also called an *Abrams cone*, is a container 12" high with an 8" diameter bottom (widest part) and 4" diameter top. The cone is filled with concrete in three layers. A rod is used after each addition to settle the concrete. After the third addition, the concrete is struck off even with the top of the cone. The cone is then removed by pulling straight up, and the resulting subsistence or slump is measured. See **Figure 14-3**.

If a mix of concrete contains only Portland cement, water, and aggregate, slump tests indicate the water-to-cement ratio. The less slump, the stronger the concrete. Today, with the modern mixing methods and many possible additives, the slump test does not always indicate the final strength of the concrete. However, the test is often performed at job sites to determine the consistency of the concrete from load to load. A slump

TFoxFoto/Shutterstock.com

Figure 14-3. Slump cones are used in slump tests to determine the quality of concrete before pouring begins.

test may also be useful when mixing concrete yourself using the three basic ingredients. A slump of 1″ to 3″ is a good general rule of thumb for standard mid-strength concrete. Adjustments can be made if needed. Generally, the lower the slump number, the stiffer the wet concrete and the stronger the final product. A slump that fails to hold its shape is of no use, and a new test should be performed.

Admixtures

When mixing concrete, the basic recipe can be enhanced by the addition of admixtures. An *admixture* is something added to concrete (such as coloring agents, retarders, or accelerators) during the mixing or pouring phase to change the characteristics of the mixture. Generally, these chemicals do not improve correctly mixed, well-placed, and finished concrete. They do, however, allow improving less-than-ideal conditions in mixing, pouring, placing, and finishing concrete. Some basic types of admixtures include the following:
- Accelerating admixtures—speed the initial curing time of concrete and aid in pouring in cold weather.
- Retarding admixtures—slow the rate of curing and are especially useful in warm weather applications.
- Plasticizers—make low-water mixes more workable (more fluid) and easier to pour or place. These admixtures are also called superplasticizers.
- Air-entraining admixtures—cause microscopic air bubbles to form in the concrete as it cures. This allows the concrete to withstand more freezing and thawing with less cracking when finished correctly. Air-entrained concrete must be finished with a special magnesium float. A *float* is a tool used to smooth the surface of concrete as part of the finishing process.
- Water-reducing admixtures—reduce the required water content for a concrete mixture, which can result in stronger concrete.

Admixtures that do a variety of jobs, including coloring concrete, improving water-holding capacity, and decreasing corrosion of reinforcement bars, may also be added. These admixtures, however, are generally specialty applications that should be discussed with the ready-mix supplier.

Preparation and Materials

Preparation for a concrete project involves evaluating the site and completing any needed groundwork. The amount of materials needed for the job must be determined and ordered. Forms must be prepared to hold the fresh concrete for a foundation or flatwork. The possible need for a moisture barrier must be considered. The tools and equipment needed for the job must be determined and acquired.

Site Preparation

When preparing for concrete flatwork, the soil is of utmost importance. The strength of the soil is determined by the underlying soil structure. The

SAE Connection

Construction Work

When it comes to the construction of a building, we know how important it is to build on a good foundation. Without a good foundation, a building would eventually fall. The weight of the heavy materials would sink unevenly in the soil and cause the structure to crack or even fall over.

Construction workers knowledgeable in the preparation and use of concrete are in high demand. Work is readily available in the construction of homes, driveways, roads, farm buildings, and other structures requiring good foundations. If you are looking for a potential supervised agricultural experience (SAE) project and enjoy construction work, pay attention to the information in this chapter.

Foto-Ruhrgebiet/Shutterstock.com

structure is determined by the arrangement of soil mineral components, organic matter, and pore space. The soil structure will have a great effect on the stability of the concrete. Soil suitability can be determined from a variety of sources, including soil maps available from the local government offices and local engineers.

Another easy-to-use source is a soil survey available on the United States Department of Agriculture (USDA) website. An array of soil descriptions, characteristics, and suitability ratings is provided. This information is a starting point for determining soil limitations. The USDA rates soils from shallow excavations and excavations for basements on a five-point scale from very limited to no limitations. After limitation ratings have been determined, it is important to make a site inspection. During the inspection, check the current depth to the high water table, any disturbance or change in the soil from the time the soil survey was made, and other site conditions that affect soil strength and stability or weight-bearing capacity. Soil that has limited use for shallow excavations or buildings with basements is also limited for concrete construction.

Wet soils should not be disturbed to dig foundations or prepare for flatwork because disturbing wet soil ruins the natural structure of the soil, causing it to be weaker. Soil in dry climates or with drought-like conditions can also have structure issues. Extremely dry soil should be wet thoroughly so that the soil does not draw water from the fresh concrete, causing the water-to-cement ratio to be incorrect and weakening the final product.

Soil Bearing Capacity

The bearing capacity of soil is the ability of the soil to hold up a load without failure or falling apart. When a foundation is put into the soil, the foundation allows the total weight of the building and contents to be spread over all the surrounding soil. An engineer can figure the total load weight and determine the bearing capacity of the soil from soil maps. These two figures are matched to ensure that neither the soil nor foundation fails.

Groundwork

Groundwork refers to changes made to the ground (soil or rock) in which a foundation is created or on which a structure is built. When starting groundwork for a foundation, it is important to excavate with care. Soil that is filled into an excavated hole is known as backfill. Backfill placed into a hole will settle and the broken soil structure of the backfill makes it weaker than undisturbed soil. A good machine excavator will dig only the trench required and leave the surrounding soil undisturbed. If this is not possible, hand digging can leave the surrounding soil in excellent condition. If soil must be disturbed when digging, particularly when it has low bearing capacity, it should be compacted when it is returned to the excavated area.

Material Estimates

Weight of the materials is an important consideration if you will mix the concrete yourself or need to store materials before creating the concrete. Portland cement powder is heavy. A standard bag weighs 94 pounds per cubic foot and makes one cubic foot of concrete. The weight of the concrete will increase with the addition of aggregates and *rebar* (steel rods used for reinforcement). Weight will also increase when the mixture is vibrated to remove air pockets and consolidate the material.

Volume

When estimating the amount of concrete that will be needed for a project, it is best to measure the volume of the concrete forms. A *form* is a mold or frame (typically made of metal or wood) used to hold fresh concrete until it sets enough to hold the desired shape. See **Figure 14-4**. Volume can be figured by measuring the thickness, length, and width of the forms to be filled. For example, a sidewalk that is 4″ thick by 100′ long and 4′ wide has a volume of 133.33 cubic feet. Remember, all measurements must be in the same standard measurement before calculating. Thus, 4″ equals 1/3′, and 1/3′ equals 0.33333′. To find the volume, multiply the three numbers: $0.33333' \times 4' \times 100'$ equals a volume of 133.33 cubic feet (ft^3). In excavated forms where the bottom is not level, the space should be compacted and measurements taken at several locations along a string tied between two diagonal points at the top of the form. The measurement is taken from the string to the compacted base. The measurements are then averaged for the depth dimension.

TFoxFoto/Shutterstock.com

Figure 14-4. Forms must be in place and secured with stakes before concrete can be poured.

STEM Connection

Calculating Volume

Finding the volume of a concrete slab requires measuring the length, width, and average depth of the slab. Each of these numbers must be in the same unit of measure. If the measurements are in different units, convert them all to the same unit before proceeding. Use this method to convert units:
- Divide inches by 12 to get feet.
- Divide inches by 36 to get yards.
- Divide feet by 3 to get yards.

Once all measurements are in the same units, multiply each measurement by the other two dimensions to get cubic volume. If the numbers are in feet, the answer will be in cubic feet. For mixing or ordering concrete, the volume needs to be in yards. So if the answer is in cubic feet, divide by 27 to get yards. Once the measurements are in units of cubic yards, the concrete can be ordered.

Safety Note

Concrete is a caustic material, which means it can burn or corrode the skin or eyes by chemical reaction. Proper personal protective equipment (PPE) should always be worn when working with plastic concrete. This personal protective equipment includes safety goggles, gloves, a long-sleeve shirt, and pants to protect the skin and eyes from chemical burns. Once concrete is cured, it is very heavy and often brittle. This means eye protection, hard hats, and steel-toed shoes are often required on job sites.

Mixing Concrete

Once a total volume needed is known, concrete can be ordered from a ready-mix company or mixed at the job site. The tensile strength of concrete can be increased by adding fiber to the mixture at the ready-mix plant. Tiny polypropylene fibers that form a matrix of fibers going in all directions make the concrete strong.

If mixing is done at the job site, the recipe should be followed exactly for the quality and quantity of concrete needed. A standard recipe for concrete is written as a ratio of cement to sand to course aggregate, in that order. While water is a critical part of the mixture, the amount is dependent on many other things and is not listed in the recipe ratio. A common mix of concrete would be a 1:2:3 mix, where one part of cement, two parts sand, and three parts of stone are combined with water to make a workable mix. The mix should not be runny but should still flow well. These ingredients are generally added by weight or shovelfuls because the weight and volume of these ingredients are relatively the same. One cubic foot of cement weighs approximately 94 pounds and comes in a single bag.

Occasionally, the recipe ratio lists the number of bags of cement required to fill a certain volume. For example, suppose a bag mix is prepared for a space that is 12 cubic feet. The recipe would call for two bags of cement (2 cubic feet), 4 cubic feet of sand, and 6 cubic feet of stone. Water weighs about 7.34 pounds per gallon and can be added by weight or volume. In this mixture, between 10 and 14 gallons would be needed. The amount of water needed will depend on the quantity of water in the sand, the relative humidity, and how fluid the concrete needs to be.

Because the quantity of water in the sand influences the quantity of water to be added to the concrete mixture, it is important to know how much water is in the sand. A quick test to determine the amount of water in the sand is a hand-squeeze test. When a handful of sand is squeezed, dry sand does not hold any shape and filters through your fingers. Damp sand makes a ball that cracks when your hand opens. Damp sand contains about 1/4 gallon of water per cubic foot of sand. Wet sand remains in a ball after it is squeezed but does not leave water on your hand. This sand has about 1/2 gallon of water per cubic foot. Very wet sand, however, remains in a ball and leaves your hand wet. Very wet sand has about 3/4 gallon of water per cubic foot. To make a workable mix, this water needs to be accounted for when adding water because excess water weakens the cement bonds.

Vapor Barriers

A *vapor barrier* (also called a moisture barrier) is a material, such as thin plastic sheeting, used to retard the movement of moisture from one area to another. Vapor barriers are an important consideration for slab concrete, especially indoor concrete. Water will migrate through finished and cured concrete. If the slab is in a heated space, the floor will constantly move water from the cool outside soil into the warm dry building, causing damp and slick floors. A vapor barrier can stop this from occurring. A vapor barrier is placed below the concrete. Traditionally, the vapor barrier has been a layer of thin plastic sheeting placed below a layer of aggregate, such as small stones. Current recommendations, however, suggest placing a barrier sheet that is 5 to 15 mils thick and is sealed well directly below the concrete layer. (A mil is 0.001".) Moisture migration is not a problem in concrete that is outside or in unheated structures.

Concrete Tools and Equipment

The tensile strength of concrete is much lower than the compressive strength. Tensile strength can be improved by placing wire mesh or rebar in the form before the concrete is poured. Rebar is a round rod that has ridges around the outside to improve bonding with the concrete. Special tools, such as bolt cutters or hand grinders, are needed to cut rebar to the correct size for placing in forms. **Figure 14-5** shows a bolt cutter. If making a few cuts for each concrete pour, a hand grinder with a fine carbide blade can be used. If many cuts need to be made, a metal chop saw, **Figure 14-6**, is often used.

kadmy/Shutterstock.com

Figure 14-5. Hand grinders are strong enough to cut through metal.

Compression Strength and Tensile Strength

All materials have physical properties. One of the most important physical properties of concrete is its strength. Concrete has very high compressive strength, meaning it can withstand heavy weights before it is distorted out of shape. This same concrete has very low tensile strength, or ability to withstand being pulled apart. For this reason, rebar or wire mesh is often used in concrete to improve the tensile strength.

Safety Note

Any time metal is cut, there is the possibility of tiny metal shards being released. (The shards may appear as tiny sparks.) Therefore, eye protection should always be worn when cutting metal.

STEM Connection

Compression Strength and Tensile Strength

All materials have physical properties. One of the most important physical properties of concrete is its strength. Concrete has very high compressive strength, meaning it can withstand heavy weights before it is distorted out of shape. This same concrete has very low tensile strength, or ability to withstand being pulled apart. For this reason, rebar or wire mesh is often used in concrete to improve the tensile strength.

Peangdao/Shutterstock.com

pongathee kluaythong/Shutterstock.com

Figure 14-6. A metal chop saw can be used to cut rebar.

Christina Richards/Shutterstock.com

Figure 14-7. This worker is using a spreader to distribute the fresh concrete evenly.

When placing concrete, many specialty tools are needed. All the needed tools should be available at the start of the process so that they are ready at the exact time they are needed. When beginning to place concrete, it is also important to have enough labor available. On large jobs, it is often necessary to have workers both to place concrete and to begin finishing concrete.

Concrete should be delivered close to the place of final placement. Concrete should not be dropped from large heights, as this causes the aggregate to segregate and come out of the mix. If at all possible, ready-mix concrete should be delivered to the farthest corner of the form first and worked back toward the truck. This saves having to move concrete excessively and walking on concrete that is already in place.

Tools that are needed for placing concrete include the following:

- *Spreaders*—Tools or devices used to distribute concrete as it is poured. Spreaders are used to pull or push fresh concrete into place, **Figure 14-7**.
- Heavy rakes—Tools used to move fresh concrete into place, **Figure 14-8**, particularly when the mixture is somewhat dry and stiff.
- *Tampers*—Tools or materials used to press down and settle substances, such as soil or concrete. Tampers are used to settle the aggregate slightly under the surface in very dry mixture concrete, **Figure 14-9**. Tampers should not be used in standard mixes, as this causes the aggregate to segregate and settle into layers, causing cracking of the surface.

Chapter 14 Concrete Foundations and Flatwork 359

Figure 14-8. Because this concrete is relatively dry and stiff, a rake is preferred over a spreader.

Nylon line can be used to mark the top edge of concrete that will be poured onto a slight slope. Once concrete is placed, it should be leveled off at the top of the form. This is accomplished by using a straightedge, **Figure 14-10**. A *straightedge* is an even piece of lumber or an aluminum or magnesium tool similar in shape to a 2″ × 4″ piece of lumber. A straightedge is used to strike off the concrete level at the top of the form. Many times a *screed* (a straight 2″ × 4″ piece of lumber or a straightedge slightly longer than the width of the form) is used, **Figure 14-11**. A screed is used in a sawing motion to level fresh concrete and settle the aggregate into the paste, leaving a surface that can be finished. A screed should be used at a slight tilt to keep a small quantity of concrete in front of the screeding motion. This will fill in any gaps in the surface and level the high spots. Immediately after screeding, a bull float (also called a Darby level) should be used. A *Darby level* or *bull float* is a tool with a long handle attached to a large, flat surface that is moved over the top of plastic concrete to completely settle aggregate, level the surface, and smooth high and low spots, **Figure 14-12**.

Testing and Finishing Concrete

Bleed water is water that rises to the surface of concrete as it cures. Bleed water will begin to appear

Figure 14-9. This gas-powered tamper is being used on this dry, gravel surface.

Figure 14-10. This worker is using a straightedge to carefully strike off concrete to create an even top surface.

Figure 14-11. This worker is moving a screed back and forth until the concrete is ready to be finished.

Figure 14-12. A worker is using a bull float to level concrete poured for a driveway.

soon after the concrete is placed and floated. This water, **Figure 14-13**, must be allowed to evaporate before final finishing can take place. Bleed water is not needed in the concrete hydration reaction and should not be disturbed until evaporated. Mixing it into the surface when finishing will make a weak cement paste at the surface. The surface will later crack and flake, causing an unsightly finish, **Figure 14-14**. In very dry mixes, the bleed water may rise and evaporate very quickly, and finishing can be started soon after bull floating. In many cases, flatwork can be effectively finished at this point, depending on the use of the slab.

Readiness Test

Concrete is ready for final finishing when the bleed water has stopped rising to the surface (the surface looks damp but not shiny wet) and when an indentation of less than 1/4″ to 1/8″ remains when a person stands on the slab. Any slab with exposed edges (edges not butted up directly to a wall or other slab) should have an *edger* used on them to settle aggregate, provide a seal in the concrete, and give a smooth finish to the slab, **Figure 14-15**. Edging also helps prepare the slab so that when forms are removed, the slab edge does not chip or crack. After edges are smoothed, *expansion joints*, also called *control joints*, are placed in the slab. These are grooves cut into the slab that allow the concrete to crack at these points so that it does not crack in a more random, unsightly pattern. These joints can be cut with a finishing tool, or a saw can be used to cut along a straight edge, **Figure 14-16**.

Finishing

A final finish is selected based on the intended use of the concrete. All final finishing must be completed after bleed water evaporates. For a sidewalk, patio, or other outdoor slab, a final float finish can be completed. A float finish leaves the surface fairly smooth, but it will not be slick and slippery when wet. When a very smooth finish is desired on flatwork, a trowel can be used. A *trowel*, **Figure 14-17**, is a hand tool used to work with concrete or mortar for masonry construction. A *power trowel*, a piece of motorized equipment used to apply a smooth finish to a concrete surface, can also be used, **Figure 14-18**. Using a trowel creates a nice finish for a basement or other structure

Figure 14-13. The bleed water on top of this concrete pad must be allowed to evaporate before the concrete can be finished.

where the floor may be painted or stained as the final finish. Because they will be slippery when wet, very smooth finishes are used only in areas where the floor will not be wet.

If the flatwork is part of a driveway, livestock lot, or other area where water would make it too slippery with a slick troweled finish, a broom finish can be created. A *broom finish* is a texture created with a stiff bristle broom that is pushed across the slab, leaving behind tiny grooves and rough areas that provide traction. A stamped finish, which is more decorative, is another finishing option. It is created with a stamping tool, **Figure 14-19**, which marks a pattern into the fresh concrete. Stamped finishes are often used with concrete that is colored and makes a pattern that looks like rocks, pavers, or stones. A more specialized tool for stamping a pattern into concrete is a *safety roller*. A safety roller puts a pattern into concrete for ramps, livestock lots, and other places where smooth finishes pose a safety hazard.

When finishing concrete in large slabs where the center cannot be reached from either side, concrete finishers often use knee boards (also called kneeling boards). *Knee boards* are rectangular or round platforms on which the user kneels to distribute body

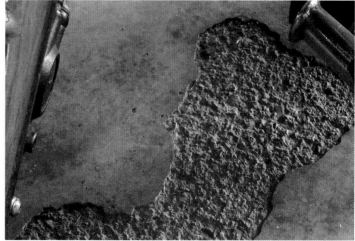

Figure 14-14. Cracks and flaking surfaces are results of cement being mixed with excess water.

362 Agricultural Mechanics and Technology Systems

TFoxFoto/Shutterstock.com

Figure 14-15. These exposed edges are being smoothed by hand with an edger.

weight while moving over a concrete surface during the finishing process. Knee boards allow finishers to spread their body weight over the fresh concrete as they use a float or other finishing tools to ensure no depressions are left in the concrete. *Floor scrapers* are tools used to knock down bumps and lumps of excess concrete or other imperfections left on the top of concrete, **Figure 14-20**. The scraper is also useful in cleaning concrete forms of leftover material. A *floor squeegee*, **Figure 14-21**, is used for removing excess water from a concrete surface and for spreading sealant onto a concrete slab.

Foundation Systems

A foundation is the lowest load-bearing part of a building and is often below ground level. The foundation allows the weight of the building to be spread over a large area of soil, giving it more stability. Foundations also anchor the building to the soil. There are several systems for building foundations. Each has its benefits and specific uses.

T-Foundations

T-foundations are used in areas where the soil may freeze, **Figure 14-22**. The foundation footing is always placed below the average frost line for the

Mark Ross/Shutterstock.com

Figure 14-16. Be sure to form a straight line when cutting expansion joints into fresh concrete.

Copyright Goodheart-Willcox Co., Inc.

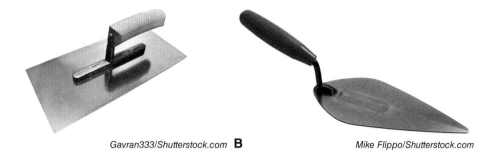

A *Gavran333/Shutterstock.com* **B** *Mike Flippo/Shutterstock.com*

Figure 14-17. Hand trowels come in a variety of shapes and sizes. A—This trowel has a rectangular shape with a wood handle. B—This trowel is narrow with a plastic handle.

soil. The *T* in this foundation is upside down, with the footing forming the cross member on the bottom and the foundation wall going from the bottom to the top. The foundation footing is wider than the foundation wall to add additional support by spreading the weight over a larger area of soil. Generally, this foundation is reinforced using rebar. T-foundations are known as *spread footings* because they spread the weight of the structure over the soil or bedrock below.

Slab-on-Grade Foundations

A *slab-on-grade foundation* is a layer of concrete poured onto a prepared bed that rests directly on the soil. Generally, a slab-on-grade foundation has an edge about twice the thickness of the slab. This edge forms the footing of the foundation, **Figure 14-23**. Slab-on-grade foundations use reinforcing rods in the thickened areas. A slab-on-grade foundation is often used in areas where there is little possibility that the soil will freeze. This foundation is also useful in sandy soils. This system can be used in a heated building in an area where some ground freezing is possible if the building weight is light and the concrete is always inside a heated structure. Follow local codes regarding when this system may be used.

Battered Foundations

In areas where the soil is weak, where earthquakes are likely to occur, or where the structure is so heavy as to overwhelm the soil below, a pile foundation is used. This is often a battered foundation. A *battered foundation* is a base for building that has a flexible pole-like structure (piles) driven or drilled into underlying bedrock or deep into the soil. These piles are formed at an angle from the bottom. The bottom of

Ari N/Shutterstock.com

Figure 14-18. Power trowels are used for finishing large areas of concrete.

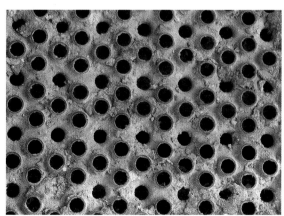

Marek P/Shutterstock.com

Figure 14-19. This perforated plate is used as a stamping tool.

a battered pile is drilled and poured at an angle. This increases the strength of the pile.

Pier Footings

A *pier footing* is a foundation component made of concrete poured in a box shape that is thick enough to spread the weight from a post or structural member into the surrounding soil. Pier footings are generally used in areas where a very heavy building load, such as a chimney, is in one specific location. They are also used where the structure, such as a deck, sits on individual posts or piers. This construction system can be used in areas with no frost or very shallow frost lines. Piers are also useful in situations where the slope of a continuous foundation would be difficult to maintain, such as on a hillside.

Concrete Forms

In order to pour concrete, forms must be prepared. Good forms must contain the heavy, fresh concrete until it is set and can hold a shape. Good concrete forms are:

- Generally made of metal or plywood.
- Smooth or finished because concrete takes the shape of the form.
- Strong enough, with bracing, to hold the heavy, fresh concrete.

Even though for foundations the soil is excavated to receive the concrete, forms often need to be constructed to contain aboveground portions of footers or flatwork. The forms should be dampened before they are filled with concrete, and they should be thoroughly scraped and cleaned after they are removed from around the set concrete. Forms can be reused if cleaned well after each use. When building, all forms should be built with nails or screws coming from the outside into the wood or metal. This is because the concrete will form around the depression left by a screw or nail head and mar the surface of the final product.

Once forms are built, they should be staked on the outside to increase the stability of the forms holding the heavy flowing concrete. A heavy sledgehammer is valuable for driving stakes into the soil to hold the forms in place.

JDPR/Shutterstock.com

Figure 14-20. Floor scrapers help put the finishing touch on concrete.

Winai Tepsuttinun/Shutterstock.com

Figure 14-21. A floor squeegee is used to remove any excess water from the floor before applying sealant.

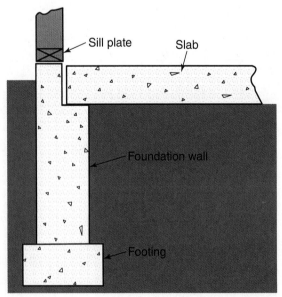

Goodheart-Willcox Publisher

Figure 14-22. Analyzing the local soil will determine how deep T-foundations will be placed.

Insulated Concrete Forms

In some cases, concrete walls need additional insulating properties, especially when used in extreme climates. *Insulated concrete forms (ICFs)* are frames, made with insulating materials on each side, into which concrete is poured. Generally, rebar or other reinforcement materials are part of the system. The forms remain in place as part of the wall to help reduce heat transfer. They can be finished with standard materials. Backfill can be placed against the walls if they are below ground level. An ICF system can be used to create concrete panels that are poured away from the building site and used in modular manufactured buildings.

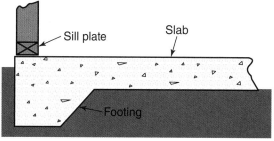

Goodheart-Willcox Publisher

Figure 14-23. Slab-on-grade foundations are generally used in warmer climates.

Flatwork Forms

After completing the foundation system, forms for the concrete flatwork can be built or placed. Flatwork might be a building slab floor, walkway, or driveway. Once the footer is poured and allowed to cure according to local codes (usually 5 to 10 days), the flatwork forms are put in place. **Figure 14-24** shows the overall form as concrete is being poured. To begin the process, excavate the base soil if needed and tamp it to compact the area. Lay out the corner of the slab and prepare a piece of 2″ lumber the length of one edge. If the board is not long enough, splice it with a cleat. This 2″ lumber should be slightly wider than the slab's thickness, allowing it to be slightly below ground level. (A 6″ slab would need a board that is 8″ wide.)

If the slab is thicker than available lumber, plywood is used to build the forms. The plywood is staked in the same way as boards. Putting the edge of the form below ground level allows the concrete to be held cleanly inside the forms. The top edge of the form should be at the same height as the desired height of the concrete slab. A nylon line or string placed at the correct height is helpful in this process. The form lumber should be staked in place every 2′ to hold the excess weight of the fresh concrete. The last stake should be attached to the form lumber and then driven into the soil. This allows the final height to be adjusted to the string with taps of a mallet or sledgehammer.

If the slab is to have square corners, place the first board of the form. Then use the 3-4-5 method (from the Pythagorean theorem) to square the second edge to the first. Set the remaining form boards in a like manner. Continue with the 3-4-5 method in each of the other corners, and follow

Safety Note

Because of the weight of both fresh concrete and cured concrete, placement of concrete often involves heavy lifting. All lifting should be done through the legs and not with the back.

Dmitry Kalinovsky/Shutterstock.com

Figure 14-24. With the form completed, these workers are pouring the flatwork.

the string to keep all boards level to create the form for the slab. Remember to brace all forms at 2' intervals by driving stakes into the soil to hold the form. If the soil has been excavated or is not strong enough to hold the forms, additional bracing is needed. For example, stones may be placed around the boards to support the form.

If the slab is to have edges that are curved lines, the forms must be made of materials that will hold the desired curved shapes. Follow the procedures described above to make sure that the forms are at the proper height to create a level slab.

If reinforcing rods will be used, they are fitted into place after the forms have been secured. Wire mesh, which can be cut to fit the outline of the form, can also be used to reinforce the concrete. For increased strength, a grid of overlapping rebar pieces can be used. The pieces of rebar for the first layer are cut to length and placed in the form. These pieces are overlaid with more rebar pieces going perpendicular to the first layer. The pieces may be tied together with wire. Care should be taken to place the rebar so that it will stay inside the slab area and not be exposed to the surface of the slab.

Flatwork Placement and Curing

Once the concrete forms and reinforcement (if needed) are in place, fresh concrete can be placed (poured) in the forms. Begin by placing several shovelfuls of concrete at intervals under the rebar pieces in the form. This concrete should lift the rebar pieces up 2" to 4" so that it will be held in place in the slab. Remember that all rebar pieces must be entirely encased in the concrete or they will rust and destroy the concrete. Once the rebar has been elevated, start placing concrete mixture into the farthest corner of the form. Shovel the concrete into the corners and other small places before filling the main areas of the form. Make sure the total quantity of the fresh concrete is placed into the form until it is slightly above the rim.

When the truckload of concrete or a large enough area has been poured, use the screed board in a sawing motion across the top of the form. See **Figure 14-25**. Work toward the edge of the fresh concrete to settle the concrete and create a level surface. Continue adding concrete and working in this manner until the entire slab has been poured. If the day is especially warm, the poured concrete may need to be worked with the float soon after it is poured. If the slab is large, the earliest poured concrete may need to be floated while the remaining concrete is being poured. This may require several workers. Some workers will move concrete that is being poured and others will begin the finishing process on the concrete that was poured earlier.

Excess Concrete

A job may call for less than an entire truckload of concrete or slightly less than a full cubic yard. In such a case, other uses may be planned for the concrete that is not needed for the main job. The crew may build pad forms, curb forms, or other forms so the excess concrete can be used. These forms can be filled after the concrete is poured for the main job.

When the concrete truck is empty, move the truck to a location where the remaining concrete can be cleaned from the machine. The location should be somewhat out of the way, where watered-down concrete and aggregate washed from the truck will not cause a problem. Adding lots of water to the concrete allows it to form very weak bonds. This makes the concrete washed from the truck harden into a mass that is weak. It can be broken up easily and removed from the site.

Curing Methods

Concrete curing can go on for as long as there is non-hydrated Portland cement and water available in the mixture. This means that concrete may continue to strengthen for months or even years after placement. However, the initial curing of the concrete can be somewhat controlled by the addition of curing compounds or the use of coverings to hold water in the mixture. A chemical sealant can be applied to concrete immediately after final finish. This sealant keeps available water from evaporating and continues the hydration reaction, giving improved strength after about seven days.

Another curing method involves placing wet burlap or plastic sheeting over the slab and not allowing it to dry so that the hydration process will continue. Using wet burlap requires additional work each day to keep the covering wet. Time and effort is also required to remove the wet burlap or plastic covering material later.

All methods that enhance curing improve early strength of concrete and can improve long-term strength of the slab. One very important curing criterion is that newly placed concrete should not be allowed to freeze for at least one week. Concrete that has been exposed to freezing temperatures in the first 24 hours after placement has about 50 percent of the strength of similarly placed concrete that was not exposed to freezing temperatures.

serato/Shutterstock.com

Figure 14-25. Screed boards are used to help settle concrete right after it has been poured.

CHAPTER 14 Review and Assessment

Summary

- Concrete is often a good choice for building agricultural structures because it is strong, durable, and relatively inexpensive.
- Concrete is made in a step-by-step process that causes a chemical reaction (called hydration) between Portland cement and water. Aggregates, such as gravel or sand, are also used in concrete.
- A slump test may be performed to check the consistency of fresh concrete. Admixtures, such as coloring agents or retarders, may be added to concrete as it is mixed.
- The soil structure will have a great effect on the stability of the concrete. Groundwork may need to be done to prepare the area for a foundation or flatwork.
- When estimating the amount of concrete that will be needed for a project, it is best to measure the volume of the concrete forms. Once a total volume needed is known, concrete can be ordered from a ready-mix company or mixed at the job site.
- A vapor barrier, such as thin plastic sheeting, can be placed under a concrete slab floor to retard the movement of moisture from the underlying soil to the inside of the building.
- Spreaders, rakes, tampers, straightedges, screeds, and floats are some of the tools used to work fresh concrete. A bolt cutter or metal chop saw may be used to cut steel reinforcing rods.
- When concrete is ready for finishing, the edges may be smoothed and expansion joints may be cut into the concrete. A floated finish, slick finish, broom finish, or stamped finish may be applied to the concrete.
- Foundation systems that use concrete include T-foundations, slab-on-grade foundations, and battered foundations. Pier footings may be used to spread the weight from a post or structural member into the surrounding soil.
- Concrete forms are used to contain the heavy, fresh concrete until it sets and can hold a shape. Forms are generally made of metal or plywood.
- Insulated concrete forms (ICFs) are frames made with insulating materials on each side into which concrete is poured. The forms remain in place as part of the wall to help reduce heat transfer.
- Once the footer for a structure is poured and allowed to cure, the flatwork forms are put in place.
- Concrete forms should be set at the proper height to create a level slab and adjusted so that the corners are square. If reinforcing rods will be used, they are fitted into place after the forms have been staked.
- As concrete is being poured, a screed is used to work toward the edge of the fresh concrete to settle the concrete and create a level surface.
- The curing time for fresh concrete may be extended by using a chemical sealant to keep water from evaporating quickly. Wet burlap or plastic sheeting can be placed over the slab, preventing it from drying so that the hydration process will continue.

Words to Know

Match the key terms from the chapter to the correct definition.

A. admixture
B. aggregate
C. battered foundation
D. bleed water
E. broom finish
F. bull float
G. edger
H. exothermic
I. expansion joint
J. float
K. form
L. hydration
M. insulated concrete form (ICF)
N. knee board
O. pier footing
P. Portland cement
Q. power trowel
R. rebar
S. safety roller
T. screed
U. slab-on-grade foundation
V. slump test
W. spread footing
X. straightedge
Y. tamper
Z. vapor barrier

1. A texture created with a stiff bristle broom that is pushed across the floated concrete slab, leaving behind tiny grooves and rough areas that provide traction.
2. A groove cut into the slab that allows the concrete to crack at this point so that it does not crack in a more random, unsightly pattern.
3. An exothermic chemical reaction in which a substance combines with water.
4. A substance that is made from limestone and clay and that hardens as it cures, even under water.
5. A single layer of concrete flatwork with thick edges that serve as footers and that typically rests on a bed of gravel.
6. A type of foundation that uses an inverted *T* shape to spread the weight of the structure over the soil or bedrock below.
7. A tool or material used to press down and settle substances, such as soil or concrete.
8. Something added to concrete (coloring agents, retarders, or accelerators) during the mixing or pouring phase to change the characteristics of the mixture.
9. Water that rises to the surface of concrete during the curing process soon after placement and floating are complete.
10. A tool used to settle aggregate, provide a seal in the concrete, and give a smooth finish to the edges of a concrete slab.
11. A mold or frame (typically made of metal or wood) used to hold fresh concrete until it sets enough to hold the desired shape.
12. A frame made with insulating materials on each side into which concrete is poured and which remains in place as part of the wall.
13. A foundation component made of concrete poured in a box shape that is thick enough to spread the weight from a post or structural member into the surrounding soil.
14. A tool used for stamping a pattern into concrete to create a textured finish.
15. An even piece of lumber (without curves or bends) or an aluminum or magnesium tool similar in shape to a 2″ × 4″ piece of lumber used to strike off concrete level at the top of a form.

16. A material, such as thin plastic sheeting, used to retard the movement of moisture from one area to another.
17. Sand, gravel, or other similar substances added to the mixture of Portland cement to create concrete or masonry units.
18. A tool with a long handle attached to a large flat surface that is moved over the top of fresh concrete to completely settle aggregate, level the surface, and smooth high and low spots.
19. A base for a building that has a flexible pole-like structure (piles) driven or drilled into underlying bedrock or deep into the soil.
20. A reaction or process that releases heat.
21. A tool used to smooth the surface of concrete as part of the finishing process.
22. A rectangular or round platform on which the user kneels to distribute body weight while moving over a concrete surface during the finishing process.
23. A straight piece of lumber used to level fresh concrete and settle the aggregate into the paste.
24. A piece of motorized equipment used to apply a smooth finish to a concrete surface.
25. A method of checking the consistency of fresh concrete by seeing how much a sample of the concrete subsides when the surrounding container is removed.
26. A steel rod used for reinforcement in concrete.

Know and Understand

Answer the following questions using the information provided in this chapter.

1. Which of the following is not a characteristic of concrete?
   A. Relatively inexpensive.
   B. Fire resistant.
   C. Low compressive strength.
   D. Good tensile strength (with reinforcement materials).
2. What are the three distinct phases of concrete hydration?
3. What are some examples of course aggregates that can be used in making concrete?
4. What is the effect of adding an accelerating admixture to concrete?
5. What is the bearing capability of soil and why is this measure important to a builder?
6. What is the best way to estimate the amount of concrete needed for a project?
7. What is the common recipe or mix for creating concrete?
8. Why is it important to know the amount of water in sand that will be used to make concrete?
9. What are traditional and current recommendations for vapor barriers used with a concrete slab?
10. Tensile strength of concrete can be improved by placing wire mesh or _____ in the forms before the concrete is poured.
11. Why should bleed water be allowed to evaporate before final finishing of concrete takes place?
12. What are some examples of flatwork applications where the concrete may have a broom finish?

13. In what areas or situations is a slab-on-grade foundation typically used?
14. Where are pier footings for a foundation generally used?
15. What are some characteristics of good concrete forms?
16. Describe how an overlapping grid of rebar is placed to strengthen concrete.
17. Explain how to begin placing concrete in a form when rebar is used.
18. What are two methods that may be used to control the initial curing of concrete?

STEM Connections and Academic Activities

1. **Science.** Explain the hydration reaction in inorganic compounds using chemical symbols. Why is hydration important for concrete construction?
2. **Technology.** Have you ever heard of see-through concrete? How about concrete roof tile that can eat smog? Admixtures can fundamentally change concrete. Research the latest admixtures and the properties they can give concrete. Are these just interesting ideas or are they being used in the industry? What other properties would improve this age-old building material? Think of three new properties that might take off in the future. Write a short report for your new concrete material and its use.
3. **Engineering.** Use packing peanuts, peat moss, or other recycled products in place of course aggregate in a standard concrete mix and then weigh the resulting product. Next, mix a batch of standard concrete using stone as the course aggregate and weigh the product. How much does each weigh per square inch? Explain what the benefits of lighter concrete would be. Are there any problems with this lighter product as a part of concrete?
4. **Math.** Concrete is bought and used by the cubic yard. A cubic yard is 1 yard high, 1 yard wide, and 1 yard long. How many cubic yards of concrete would it take to fill your classroom, school cafeteria, and gymnasium? Compare the volume of concrete needed to fill these rooms with the volume of concrete it took to build the Hoover Dam.
5. **Social Science.** Concrete is a substance widely used today, but it was also produced thousands of years ago. What is the main difference between today's modern Portland cement–based concrete and the cement used by the ancient Romans? Why has the ancient concrete lasted so long?
6. **Language Arts.** Many ancient structures that were built using early forms of concrete have stood the test of time, including the Pantheon. Write a short story that details the reaction a person who worked on these projects would have to finding their construction still in use in the 21st Century.

Thinking Critically

1. Adding reinforcing rods or admixtures that act as reinforcement increase concrete's tensile strength. Give three examples of situations where concrete might need increased tensile strength.
2. When ordering ready-mix concrete, why is it important to tell the company the exact dimensions of the formed volume, not just how much concrete you want to order?
3. The boss says, "clean tools make more money." Why are clean tools so important for a concrete construction worker?

CHAPTER 15 Masonry

Chapter Outcomes

After studying this chapter, you will be able to:
- Identify masonry tools and equipment used for building projects.
- Explain the importance of following safety procedures in masonry work.
- Identify clay products and concrete products used in masonry building projects.
- Describe types of pattern bonds, mortar joints, bricks, and concrete blocks.
- Discuss the functions of mortar and types of mortar used for masonry projects.
- Discuss activities involved in planning a masonry building project.
- Discuss procedures for building corner leads and walls for masonry projects.

Words to Know

American Society for Testing and Materials (ASTM)	(CMU) corner lead	level	raked joint
	corner pole	line holder	rowlock brick
		margin trowel	rowlock stretcher brick
anchor bolt	course	mason's line	sailor brick
bed of mortar	ears	mason's rule	shiner brick
bond	firing	mortar board	soldier brick
brick	folding rule	mortar box	story pole
brick chisel	green brick	mortar stand	stretcher block
brick hammer	grout	pattern bond	stretcher brick
brick set	header brick	personal protective equipment (PPE)	tile
chisel	hydraulic cement		trowel
closure block	jointer	racking	tuck pointer
concave joint	kiln	racking back the lead	wythe
concrete masonry unit			

Before You Read

Before reading the chapter, review the Words to Know within the chapter. Determine the meaning of each term.

While studying this chapter, look for the activity icon to:

- **Practice** vocabulary terms with e-flash cards and matching activities.
- **Expand** learning with interactive activities.
- **Reinforce** what you learn by completing the end-of-chapter questions.

www.g-wlearning.com/agriculture

Masonry and its associated products are very important to the agriculture industry. The most commonly used building products for agricultural applications are concrete blocks and bricks. A ***brick*** is a molded rectangular block of clay that has been baked in an oven to increase the strength and durability of the material. Concrete blocks can be used for structural purposes, such as foundations, and exterior and interior walls of agriculture buildings. Bricks are most commonly used as a facing material to enhance the exterior or interior appearances of a structure. In the dairy industry, glazed bricks and blocks are often used because they are easy to clean and disinfect. Because of the diversity of masonry units available, contractors or architects can select from an assortment of units when designing agricultural structures.

Masonry construction is a valuable skill that can be used throughout a person's life. Developing knowledge and skills related to masonry takes practice and patience. In the beginning, workers do not need to invest a large amount of money to purchase tools or equipment. Hand tools can be purchased and, if treated correctly, will last a lifetime. Larger tools and equipment can be rented from companies that provide this service.

Masonry Tools and Equipment

Tools and equipment are very important for any building project. Proper application and care of the tools are also important. The worker's safety and the safety of others can depend on selecting the correct tool and using it properly to perform the job. Tools and equipment can last many seasons if they are kept clean and maintained properly. Injuries can occur when a tool or a piece of equipment is not used properly or if the worker is not wearing appropriate protective equipment.

Marshalltown Company

Figure 15-1. Make sure you feel comfortable using the trowel and that it is the proper size for the project.

Trowels

A ***trowel*** is the most commonly used tool to perform masonry work. A trowel is a tool used to apply mortar (a bonding agent) to masonry units, such as bricks and concrete blocks. See **Figure 15-1**. Trowels vary in size and shape, and selection will depend on the activity that is planned. It is important that the worker feel comfortable using the selected trowel. A trowel with a large blade that holds a lot of mortar can be difficult to balance for inexperienced workers. Trowel handles are made in an assortment of materials, such as wood, plastic, or other materials. See **Figure 15-2**.

Marshalltown Company

Figure 15-2. Trowels can be purchased with several different handle constructions.

Two trowels commonly used to perform masonry work are the margin trowel and tuck pointer. The *margin trowel* has a rectangular front edge and a V-shaped heel. See **Figure 15-3**. This trowel is used to hold mortar or fill small areas. It can also be used in conjunction with the tuck pointer trowel to fill in voids between joints in bricks or blocks. The *tuck pointer* is a trowel that has a narrow blade and is used to fit into tight spaces, such as between bricks or blocks. See **Figure 15-4**.

A trowel must be clean to perform properly. After working with any trowel, mortar should be removed so it does not dry or begin the process of rusting the tool. Develop a habit of cleaning your tools at the end of the day. Dry and store tools properly to ensure they are ready for the next use. Depending on the tool, the manufacturer may recommend a light coating of oil on the steel to inhibit rusting or for wood to keep it from drying and cracking.

Marshalltown Company

Figure 15-3. This margin trowel has a wood handle.

Jointers

A *jointer* is a tool used to finish the surface of a newly laid mortar joint between bricks or concrete blocks. Jointers are also called joint tools or finishing tools. A *course* is a complete or continuous row of bricks or blocks laid in a wall, structure, or project. Joints between courses of bricks or blocks may vary in thickness from 5/16″ to 3/4″. To accommodate these variations, jointers are made in different sizes. Jointers are available in different shapes to create rounded or V-shaped joints. They are also made in short and long styles to finish vertical joints or horizontal joints, **Figure 15-5**.

Marshalltown Company

Figure 15-4. A tuck pointer's narrow blade allows you access to small areas unreachable with margin trowels.

Hammers

A *brick hammer* is a tool used to drive nails, chip stone, or break bricks or masonry units. See **Figure 15-6**. The flat head of the hammer is used to drive nails. The other end of the head is drawn out to form a tool for scribing

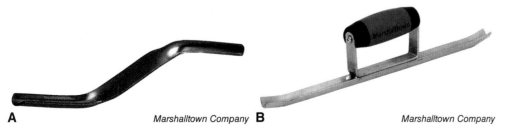

A *Marshalltown Company* B *Marshalltown Company*

Figure 15-5. Choosing the proper jointer depends on the size and thickness of the joints. A—1/2″ and 5/8″ short jointer. B—1/2″ × 15″ half-round sled runner.

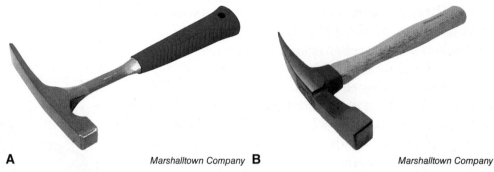

Figure 15-6. These brick hammers have different handles, but perform the same function. A—A mason's brick hammer with a soft grip handle. B—A mason's brick hammer with a wooden handle.

Safety Note

Brick hammers are not intended to be used to strike hardened metal, such as steel stakes. Use an appropriate hammer intended for striking such metal. It is always very important to wear safety glasses when using hammers.

(marking) and cutting bricks, blocks, or stones. It is also used for making grooves in bricks or blocks to create a weak point to aid in breaking or fracturing the unit. The size of a brick hammer and most other hammers is based on the weight of the head of the hammer. Most brick hammers weigh from 12 to 24 ounces.

Measuring Tools

When laying bricks or concrete blocks, there are two common measuring devices that can be used: a mason's rule, **Figure 15-7**, and a retractable steel tape, **Figure 15-8**. A *mason's rule*, sometimes called a *folding rule*, is a measuring device that is 6′ in length and folds for easy storage. It is the standard tool used when working with bricks or blocks. On the front side of the rule is the standard US customary measurement. On the reverse side is a scale for laying bricks or blocks. The back side has special markings that correspond to course heights for various unit sizes and joint thicknesses, depending on the specific ruler selected.

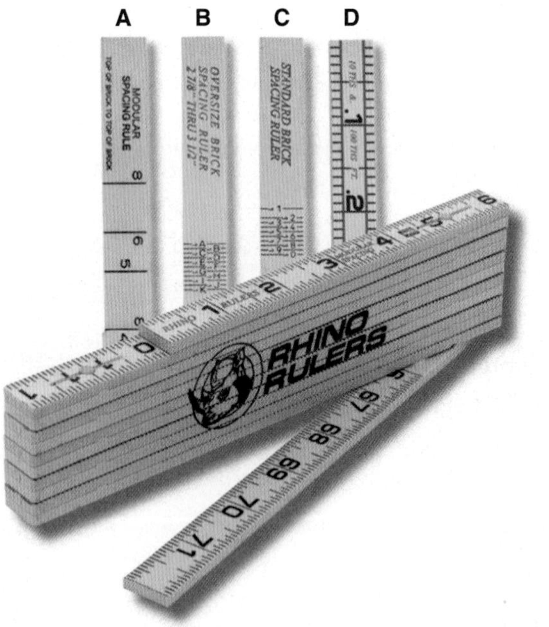

Mortar Boards and Stands

When preparing to lay bricks or concrete blocks, the worker needs to position a stand and mortar board close to the location where the units are to be laid. A *mortar board* is a device, made of plywood or polymer, used to hold mortar when laying bricks or concrete blocks. Plywood has been a common product used to hold mortar for many years. Polymer is a new material being used to manufacture mortar boards. This material is heat resistant and nearly indestructible. An advantage of using a polymer mortar board is that it will not absorb moisture as a wood mortar board can.

A *mortar stand* is a device used to hold and position a mortar board at the correct height when

Figure 15-7. Rulers are made with different scales, depending on the bricks or units being used. A—Modular spacing ruler. B—Oversize brick spacing ruler. C—Standard brick spacing ruler. D—Standard US customary ruler.

laying bricks or blocks. Mortar stands can be erected from concrete blocks set on edge or on end. Metal stands are another option for holding the mortar board at the appropriate height. See **Figure 15-9**. Metal stands are easy to move and can be folded up when the project is done. Most stands have placement holders on the end of the pipes to keep the mortar board from sliding off the stand. Some stands can be turned on end to increase the height of the stand.

Steel Squares

A steel square is a measuring and squaring tool in the shape of a right triangle (90° angle), **Figure 15-10**. The square has two arms. One arm is called the tongue, and the other is called the body. The tongue is 1 1/2″ wide and 16″ long. The body is 2″ wide and 24″ long. The inner and outer edges of the square have been divided into inches and fractions of an inch for measuring. The steel square has many uses, such as drawing straight lines or laying out a 90° angle to create the outline of a brick or block wall.

Levels

A *level* is a tool used to determine whether a surface is exactly horizontal or vertical. When laying bricks or blocks, the level is placed on the top or end of the unit to ensure that the top surface is level (exactly horizontal) and the side or end of the unit is plumb (exactly vertical). It is one of the most delicate tools used in construction. A bubble level (also called a spirit level) has one or more glass vials filled with liquid and an air bubble. The placement of the air bubble indicates whether a surface is level or plumb. Levels can be made of wood, **Figure 15-11**, metal, or a combination of materials. Levels are available in a variety of sizes, from 6′ to less than 1′. The most common levels for laying bricks and blocks are 2′ and 4′ sizes.

Chisels

There are numerous types of chisels used in the masonry field. *Chisels* are hand tools used to scribe (mark) and cut bricks, blocks, and stones. They are made of hardened steel that is able to keep a sharp edge when cutting bricks and blocks. They are also used to cut and shape stone. One of the most common chisels used in the masonry field is the brick set,

Marshalltown Company

Figure 15-8. Retractable steel tapes can be purchased in an assortment of blade lengths and widths. This tape has US customary measurements on one edge and brick spacing on the other.

Goodheart-Willcox Publisher

Figure 15-9. This mortar board and metal stand are used to hold mortar. This stand can be set on end to increase the height.

Johnson Level and Tool Manufacturing Company, Inc.

Figure 15-10. This type of steel square is used in laying out structures and measuring.

Marshalltown Company

Figure 15-11. This 4′ wood level is used in masonry work.

Marshalltown Company

Figure 15-12. A brick set is a wide chisel used to cut and trim bricks.

Marshalltown Company

Figure 15-13. A mason's line is used as a guide to lay blocks between corners.

Figure 15-12. A *brick set*, sometimes called a *brick chisel*, is a type of chisel used to scribe bricks and then cut them to the appropriate length.

Mason's Lines

A *mason's line* is a strong nylon cord used to show the layout of a building or to keep a course (row) of bricks or concrete blocks in a straight line. See **Figure 15-13**. A *line holder* is a device used to keep the mason's line in place when positioned on the bricks or blocks, **Figure 15-14**. One line holder is positioned at each corner. The mason's line is connected to each holder, and then strung under tension to ensure that the line does not sag over its length. The mason's line and line holders work together to help the worker keep the course being laid straight and at the correct height.

Mortar Boxes

Mortar is the material used to *bond* (join together securely) bricks or blocks. A *mortar box* is a container used to hold the ingredients that are mixed to create mortar. Mortar boxes of various sizes are used, depending on the job and number of workers laying blocks or bricks. Mortar boxes are made of metal or polyethylene (poly), **Figure 15-15**. The advantage of using a mortar box made of poly is that mortar does not adhere well to its slippery surfaces. This makes cleaning the box easier.

Masonry Safety

Construction can be a dangerous activity. If workers are not careful and observant to changing conditions or situations, accidents can happen very quickly. Most accidents are caused by unsafe conditions or unsafe actions. Workers who are not concerned about safety can injure themselves or others because of their carelessness. It is important to be trained in the operation of tools and equipment prior to using them, **Figure 15-16**. Through training and the proper use of safety equipment, such as safety glasses, hearing protection, foot protection, and other safety devices, a large number of accidents can be avoided.

Chemicals and hazardous compounds present in masonry products require the use of various types of personal protective equipment. *Personal protective equipment (PPE)* consists of safety devices or clothing used or worn to protect an individual against the health or safety hazards that may be present in a work environment. Portland cement is a prime example of a masonry product that presents potential hazards to an unprotected handler. Certain compounds in Portland cement can cause reactions with skin and eyes. When mixing mortar, you should wear safety glasses to protect your eyes, breathing protection to avoid inhaling the dust, and gloves to protect your skin.

Muriatic acid, used to clean brick masonry, is another product that is hazardous for workers. Muriatic acid can cause serious burns to the skin or eyes and requires the use of safety glasses and appropriate gloves. It is also important to wear a face shield when mixing the solution to prevent the product from splashing on your face or in your eyes, **Figure 15-17**.

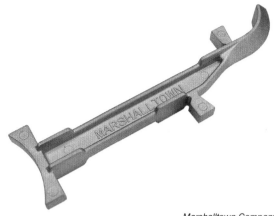

Marshalltown Company

Figure 15-14. Line holders are used to support the mason's line between corners.

Masonry Building Products

Masonry units are divided into three primary categories:
- Clay masonry products (bricks and related products).
- Concrete masonry units (concrete blocks and related products).
- Stone (field stones and cut stones).

In this chapter, clay products and concrete products are discussed.

Marshalltown Company

Figure 15-15. This mortar box has a capacity of 10 cubic feet.

Air Images/Shutterstock.com

Figure 15-16. Students are learning to lay bricks and blocks. Notice that the instructor and students are wearing safety glasses and protective gloves.

Safety Note

Safety glasses, protective gloves, and a face shield should always be worn when mixing or working with cleaning solutions and similar hazardous chemicals. Safety clothing, such as a protective white suit, should be worn to shield clothing and skin from the overspray.

SAE Connection

Masonry and Concrete Work

Block, brick, stone, and rock are very common in construction. Due to their beauty, strength, and long life, masonry products are often the choice for agricultural, residential, and commercial applications. A good mason has a solid career with plenty of work available. The details in this chapter, along with the experience of working in a masonry job, provide a great opportunity for a supervised agricultural experience (SAE) project. An agricultural design and fabrication proficiency or an agricultural services proficiency would also be a good fit for an SAE in masonry or concrete work. The variety of materials and tools available provides countless opportunities for the display of creative talents in this area. Make sure you take plenty of pictures!

goodluz/Shutterstock.com

Structural Clay Products

There are three classifications of structural clay products:
- Solid clay masonry units (bricks).
- Hollow clay masonry units (*tiles*).
- Architectural terra cotta (products used for decoration in construction).

Structural clay products are made from clay or shale. Of the three types, brick will be the one covered in this chapter. Bricks are molded rectangular units of clay that have been baked in an oven to increase the strength and durability of the material. Clay is mined throughout the United States to supply the demand for a multitude of brick styles and color shades.

Manufacturing Brick

The manufacturing process of brick is shown in **Figure 15-18**. The process includes the following steps:
- Mining the raw materials.
- Crushing and storing the mined raw materials.
- Preparing and grinding the raw materials.
- Forming units into bricks or other clay products.

mikeledray/Shutterstock.com

Figure 15-17. A worker is rinsing a brick wall with clean water after scrubbing the surface with a muriatic acid solution to remove excess mortar.

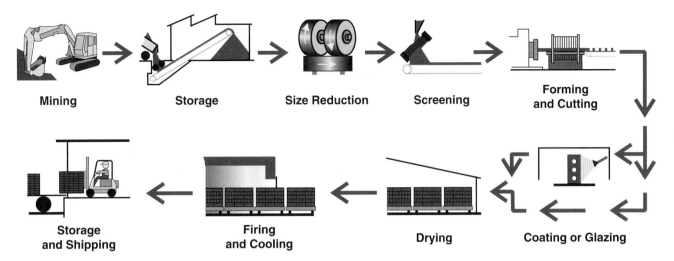

Diagrammatic Representation of Manufacturing Process

Brick Industry Association

Figure 15-18. These are the steps in manufacturing clay bricks.

- Drying units to remove excess moisture.
- *Firing* (heating in an oven) the units and cooling them.
- Storing the products and shipping them to customers.

Making bricks begins with creating a mud-type mixture wet enough to stick together. Then the wet clay mixture is forced through dyes to form a ribbon of clay larger in thickness and width than the final size of the brick. After the ribbon of clay has been formed, it is cut into individual units using wire cutters to create green bricks. A *green brick* is a freshly molded block of clay that has not been heat treated. The bricks are then stacked in cubes to start the drying process. The wet bricks need to dry to a very low moisture content to ensure that they do not deform or crack. After the bricks reach the appropriate moisture content, they are then sent to a kiln. A *kiln* is an oven or furnace where green bricks or other products are heated to a very high temperature. The process is called firing and brings the temperature of the bricks to more than 2,000°F (1,100°C). Firing turns the soft dry clay into units that will withstand extreme pressures and endure severe weather conditions when used as an exterior wall finish.

After the bricks are removed from the kiln, they are stacked in pallets. The bricks are stored until they are purchased. Then they are sent to a construction site to be used as building materials, **Figure 15-19**.

Bonds and Patterns

Mortar is used to bond individual masonry units together to form a masonry wall. A *pattern bond* is the design or arrangement formed by masonry units and the mortar joints on the face of a masonry wall. The

Stephen Finn/Shutterstock.com

Figure 15-19. These bricks stacked on pallets have been delivered to a construction site.

382 Agricultural Mechanics and Technology Systems

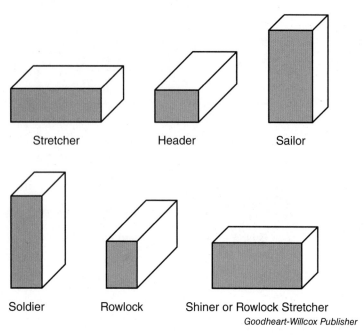

Figure 15-20. Bricks get their names based on how they are laid in a specific position.

overall bond strength of the wall can vary, depending on the pattern bond selected.

Bricks are given names based on how they are positioned in the wall, **Figure 15-20** and **Figure 15-21**. The most commonly laid brick, the *stretcher brick*, is laid flat with the long side exposed in the wall face. The *header brick* is one that is laid flat with the short end of the brick exposed. A *sailor brick* is laid on end with the largest surface of the brick being exposed in the wall face. A *soldier brick* is laid on end with the narrow side of the brick being exposed in the wall face. A *rowlock brick* is laid on the face edge with the end of the brick visible in the wall face. This unit can be used to connect two sections of brick together. A *shiner brick* or *rowlock stretcher brick* is a brick that is laid on edge with the largest surface of the brick exposed in the wall face.

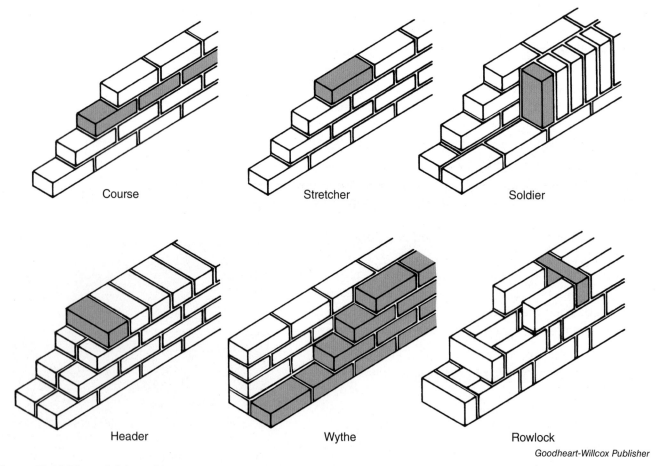

Figure 15-21. These brick positions are used in wall construction.

When masonry units are laid as components in a wall or structure, they are referred to in specific terms. For example, when bricks or blocks are laid in a complete or continuous row, this is called a course. A *wythe* of bricks or blocks is a continuous vertical section of masonry that is one unit in thickness and constructed of a single line of masonry units. The wythe of bricks or blocks may be independent or interlocked with other adjoining wythes of masonry. When different wythes are adjoining, there may be the need to fill the void between them with grout. *Grout* is a fluid cement mixture used to fill gaps or voids between wythes or other areas in masonry construction. Grout bonds the sections together so that they act as one unit. The bonding of wythes depends on the wall type and structural requirements.

Some pattern bonds that create different exterior appearances of masonry walls include running bond, common bond, Flemish bond, English bond, and stack bond. These patterns are shown in **Figure 15-22**.

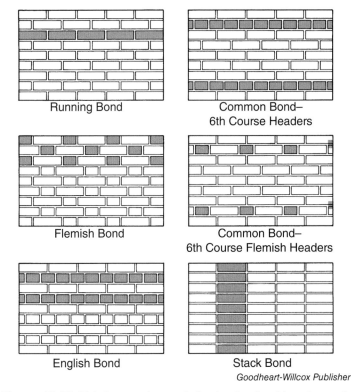

Figure 15-22. This image shows six basic structural bonds used in masonry work.

Mortar Joints

Mortar is the material that bonds the masonry units together as one mass. Mortar joints finish the surface of mortar that is exposed between bricks or concrete block units.

There are many different joint selections available when determining the final finishing of the mortar joint. Finishing or tooling the joint seals it to aid in keeping moisture from entering the mortar. Finishing also compacts the mortar to aid in creating a hard surface.

Two common mortar joints include the concave joint and raked mortar joint. The *concave joint* is a type of mortar joint created when a finishing tool is used to strike the mortar surface and leave an indentation that curves inward in the mortar. This is the most popular type of mortar joint used to finish bricks or concrete blocks. This joint is created by striking off the excess mortar from the joint area with a trowel and then returning the excess mortar to the board. Once the mortar begins to harden, the joint is struck using a concave steel jointing tool. Finishing the joint compacts the mortar and creates a smooth, hard surface finish, **Figure 15-23**.

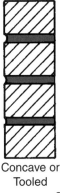

Figure 15-23. This brick joint has a concave indentation in the mortar.

A *raked joint* is a type of mortar joint in which a portion of the mortar is removed to create a joint recessed below the surface of the bricks. The mortar is removed once it begins to set. After removing the mortar, it is important to finish the recessed surface to seal the joint. Finishing the joint will aid in sealing the raked joint from moisture and protect the mortar between the units. **Figure 15-24** shows a completed raked joint after it has been finished using a tuck-pointing tool.

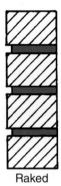

Goodheart-Willcox Publisher

Figure 15-24. This raked brick joint was struck with a tuck-pointing tool.

Concrete Masonry Units

Concrete masonry units (CMUs) are hollow or solid blocks made from Portland cement, aggregates, admixtures, and water. The mixture varies depending on the type of concrete unit being made. Aggregate refers to sand or gravel added to the mixture to create the units. Varying the type of aggregates can change the classification of the CMU, such as to lightweight blocks from standard blocks. Admixtures are substances added to the concrete mixture to change its characteristics. Coloring agents and retarders to slow down the curing time are examples of admixtures that can be used in creating concrete masonry units.

Types of Concrete Blocks

Concrete blocks are usually manufactured as one of four types:
- Solid load-bearing blocks.
- Hollow load-bearing blocks.
- Solid blocks that are not load bearing.
- Hollow blocks that are not load bearing.

The concrete block to select for a specific job will depend on the type of structure being built, its use, and the required strength. For example, a two-story agricultural complex is to be built using concrete block units. Exterior walls will need to support an extreme amount of weight. There will be a number of interior concrete walls that are not load bearing to divide the interior space. In this example, a minimum of two types of blocks would need to be selected.

Exterior walls of the structure should be made of load-bearing concrete blocks because of the strength required. The contractor or architect will need to determine which load-bearing blocks to use, depending on the estimated load that will be applied to the block walls. Either solid load-bearing blocks or hollow load-bearing blocks should be used to build the exterior wall structure. See **Figure 15-25**. Interior walls could be built using concrete units that are not load bearing since the interior walls are not load bearing. In this project, solid blocks that are not load bearing or hollow load-bearing concrete blocks could be used for interior walls.

Sizes and Shapes of Concrete Blocks

Concrete blocks are made in many different sizes and shapes, depending on their intended use, **Figure 15-26**. The selection of the specific unit depends

Thomas Barrat/Shutterstock.com

Figure 15-25. An agricultural building is under construction using concrete masonry units. The exterior walls use load-bearing blocks. The interior walls will use nonload-bearing blocks.

on the strength needed and where the concrete block will be used. Sizes are usually given in nominal dimensions. For example, a standard concrete block is given the nominal dimensions of 8″ × 8″ × 16″. The actual unit measures 7 5/8″ wide, 7 5/8″ high, and 15 5/8″ in length. When a mortar joint of 3/8″ is used with the unit, the combination of the block and the joint will measure 8″ high × 16″ in length.

Blocks are designed to be in modular (standardized) unit lengths. The unit length is divisible by 4″ or modules of 4″. This will aid the mason or contractor when laying out the structure if the overall dimension of the building is also divisible by 4″ or modules of 4″. For example, for a building sized 16′ 8″ × 24′ 0″, both dimensions are divisible evenly by 4″.

Pattern Bonds

The block pattern used in a wall or structure will vary, depending on the appearance desired or the strength created by the pattern. The two patterns commonly used in block laying are the running bond and the stacked bond. See **Figure 15-27**.

Mortar

Mortar is the bonding agent that ties masonry units together to create a strong and watertight structure that will last for decades or longer. The mortar must be strong and durable to keep the structure straight and plumb during extreme weather conditions, such as wind and temperature changes.

386 Agricultural Mechanics and Technology Systems

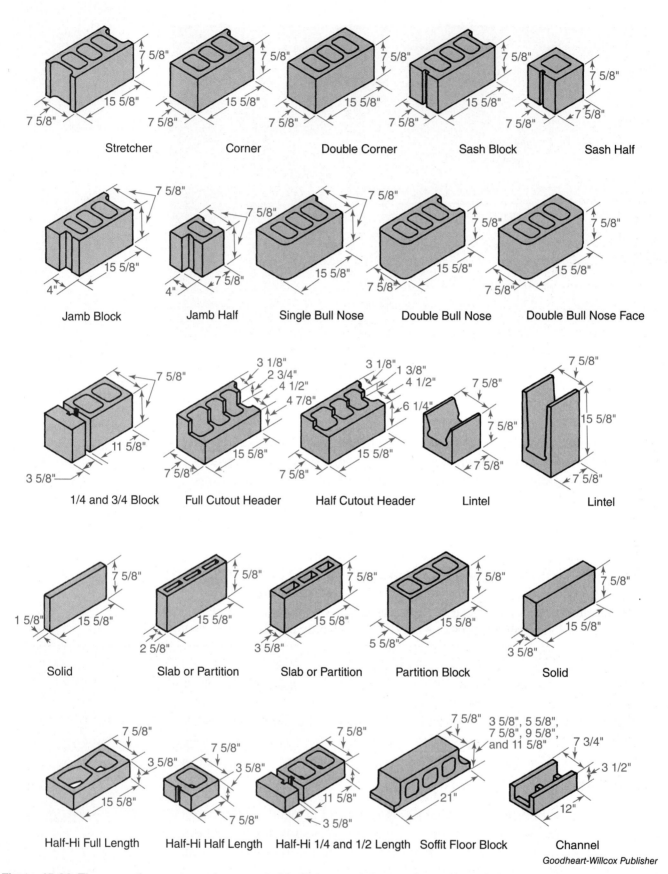

Figure 15-26. These are the most popular concrete block sizes and shapes. Concrete blocks are also available in modular widths of 12″.

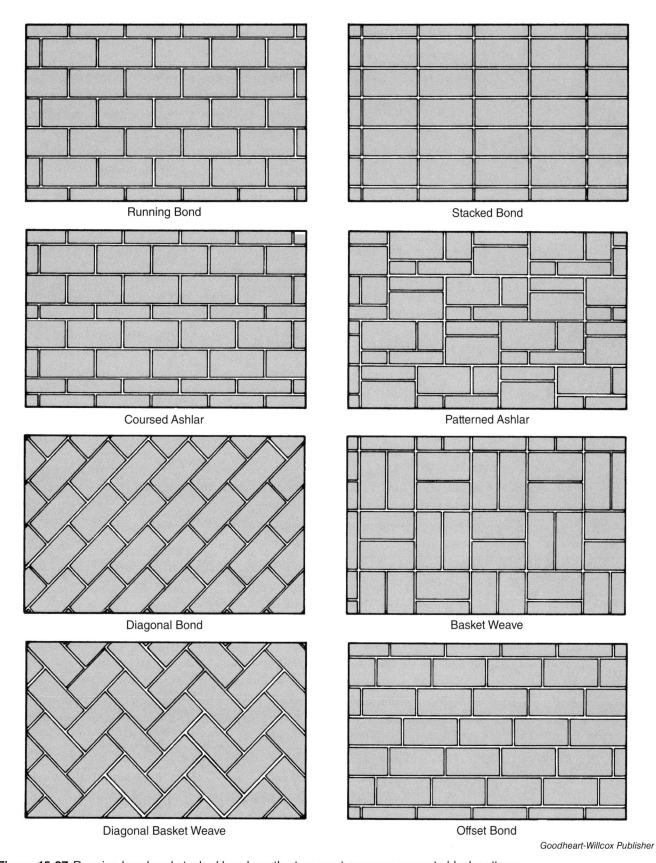

Figure 15-27. Running bond and stacked bond are the two most common concrete block patterns.

Mortar Materials

Mortar is typically made from Portland cement, hydrated lime, and sand. In some mortar mixtures, masonry cement is used instead of Portland cement. Masonry cements usually contain a portion of Portland cement, ground limestone, and other additives. When water is added to the mixture, it turns the dry ingredients into a pliable and workable mixture. The water used in the mixture should be clean and free of visible particles. If the water is drinkable, it is usually acceptable as a component in mortar.

Mortar serves four functions:
- It bonds the units together and seals the area between units.
- It compensates for the variations that may occur among blocks.
- It aids in bonding metal ties or other reinforcement to units to create a structure that acts as one mass.
- It adds character to the wall through the use of different joint finishes and colors.

Types and Properties

The component in mortar that bonds units together is primarily Portland cement. Portland cement is a *hydraulic cement* (a type of cement that will set and harden underwater). The *American Society for Testing and Materials (ASTM)* is an international society that tests materials and develops standards for their use. ASTM has created approximately 12,000 standards; one section pertains to masonry and masonry products. ASTM C150—Standard Specification for Portland Cement discusses eight types of cement. Of those eight, only three are recommended for mortar:
- Type I—cement for general use.
- Type II—cement for moderate sulfate resistance.
- Type III—cement for use when high strength is needed.

Subsets of the three types are type IA, IIA, and IIIA cements. The *A* stands for air entrainment. Air entrainment is the creation of very small air bubbles in concrete. Cements that contain air entrainment additives are not recommended for mortar because the additive reduces the bonds created among the mortar, reinforcement materials, and masonry units.

When selecting lime to add to a mixture, hydrated lime is recommended for mortar. There are two types of hydrated lime: type N and type S. Type S lime is recommended for masonry work because it has characteristics that make it bond well to masonry units.

Selecting Mortar Type

Mortar is made using two common methods. In one method, the contractor selects dry materials that are mixed together to develop a specific mortar based on the strength and properties required for the project. This is performed by selecting the proper Portland cement or masonry cement, sand, lime, additives, and water to develop a mixture that has the needed properties. It is important when developing a mixture that the same products are used for all batches. If the products are changed, this can affect the strength and color from one batch to another.

In another method, the contractor selects a premixed masonry mortar that has the appropriate properties and strengths to meet the needs of the project. Using premixed mortar reduces the time needed to create the mortar. Sand and water are usually the only two ingredients added to the mixture to create this type of mortar. When using premixed mortars, it is important to be consistent and not change brands during a project. This helps to ensure uniformity in color and strength in the batches used to construct the project.

Mixing Mortar

The ingredients in mortar must be combined thoroughly to obtain a uniform and consistent mixture. Mortar can be mixed by hand or by using a motorized mixing machine.

Mixing in a Mortar Box

To mix mortar by hand, measure the correct amount of each of the dry ingredients and place the ingredients in the mortar box. Blend the dry ingredients in the box until the materials are a uniform consistency. Then add water and continue to mix until the mixture reaches the proper consistency. It is important to be precise when mixing to ensure that the mortar is consistent from one batch to the next. Refer to **Figure 15-28**.

Machine Mixing

When mixing mortar by machine, it is important to read and follow the instructions in the manufacturer's manual. Be sure to follow all safety procedures. Start the mixer as directed in the instructions. After the blades are rotating at the proper speed, pour the required amount of water into the drum. Add the cement, sand, and lime or mortar mix to the water being rotated in the mixing drum. Continue to mix until the dry ingredients are completely incorporated into the water to create a uniform mortar mixture. Each type of mortar mixer is somewhat different from others. Because of this, it is important to read the manufacturer's manual even if you have operated a different mortar mixer. Refer to **Figure 15-29**.

> **Safety Note**
> When mixing mortar, it is important to wear safety glasses to protect your eyes, and gloves to protect your hands from chemicals in the mixture.

atyphoto/Shutterstock.com
Figure 15-28. This worker is using a mason hoe to mix ingredients to prepare a batch of mortar.

Planning the Project

It is important to plan before beginning a masonry project. Planning entails activities such as selecting materials, consulting building codes,

> **Safety Note**
> When using a mortar mixer, it is very important to close the safety grill when mixing. The grill protects the operator from the turning paddles in the mixing drum.

Figure 15-29. This mortar mixer is used for large masonry projects.

Marshalltown Company

scheduling deliveries, and identifying storage locations for materials. The type, size, quantity, and color of masonry units to be used for the project must be determined before materials can be purchased. Refer back to **Figure 15-26**, which shows a number of common concrete masonry units that are available for constructing a building. A similar number of options exist for brick units. Bricks will vary in color, texture, size, and shape. It is important to select masonry units that will enhance the project and create a structure that will last for decades.

Building Codes

Building codes are local (or regional) laws that govern how structures should be built or what types of materials should be used. It is important in the planning stages of the project to check with local and state building officials to identify codes that are relevant to the project being considered. One such requirement may be the use of reinforcement materials in footings, walls, or other parts of the structure. Develop a working relationship with the building officials to ensure that code requirements become an integral part of the planning stages of any structure being considered.

Scheduling Deliveries

Planning includes scheduling the delivery of masonry units and mortar materials to the site. It is also important to determine locations where items will be stored until they are needed by the masonry workers. Once the materials are ordered and delivered to the construction site, it is important to protect them from the weather. Rain, for example, may have a detrimental effect on the masonry units and mortar materials. Masonry units and mortar materials should be stacked on pallets and covered with plastic sheets or tarps to keep the products dry. When masonry units are ready to be laid, distribute an appropriate number of bricks or blocks close to the corner to be built. This will aid in reducing the time the worker needs to lay the units.

Fundamentals of Laying Masonry Units

Laying masonry products can be a rewarding experience because workers can look back on their accomplishment with pride and satisfaction. When laying masonry products, it is very important to perform the work with care and a concern for quality because the finished project can last a lifetime.

Basic Operations for Laying Masonry Units

Laying bricks and concrete blocks involves preparation, knowledge of tools and equipment, and the ability to use the tools properly. Laying masonry units can be considered an art because of the skill and finesse required to lay and finish the units properly. It is important to follow standard practices and procedures when laying and finishing masonry units. The methods used affect the structure's longevity, and the finished work will be in view for all to see.

Prior to laying masonry units, a worker needs to develop a number of skills. Basic skills required include loading the trowel, spreading mortar, holding bricks and blocks, adjusting units to be level and plumb, using a mason's line, and finishing joints. Becoming competent at these skills requires practice.

Loading the Trowel and Spreading Mortar

Selecting the correct trowel is an important step for successfully performing masonry work. The worker will need to find a trowel that fits the hand and stays balanced when loading the trowel and spreading mortar. When beginning, select a trowel that is not too long or wide. A standard mason's trowel is approximately 10″. Trowels vary in width, depending on the type selected. Holding the trowel correctly is very important for spreading the mortar properly. To hold the trowel, grasp it in your dominate hand, then wrap your fingers under the handle with your thumb on the top of the ferrule (ring or cap on the end of the handle nearest the blade). See **Figure 15-30**.

Goodheart-Willcox Publisher

Figure 15-30. The correct way to hold a trowel is with the fingers under the handle and the thumb on top of the ferrule.

Loading the trowel can be performed in a number of ways. **Figure 15-31** shows one method of loading or placing mortar on the trowel. Load the trowel with small amounts of mortar at first until you become accustomed to filling the trowel with an appropriate amount of mortar. Continue to practice the steps of loading the trowel until the movements begin to feel natural. The next skill to develop is unloading the trowel in a manner that places mortar uniformly across a surface similar to the size of a brick or concrete block.

Workers also need to be competent at spreading mortar. With practice, you will be able to spread mortar from a trowel uniformly across a surface. Spread mortar with a quick turn of the wrist toward the body, moving the arm away from the point of placement. As mortar is removed from the trowel, tip the trowel to aid in removing the remaining material. Trowel skills and mortar placement techniques will improve as you continue to practice loading and unloading mortar.

Course Height of Masonry Units

There are a number of ways to determine the height of a course when laying masonry units. One method is to use a mason's rule (or ruler) that is

Christina Richards/Shutterstock.com

Figure 15-31. Before loading the trowel, use it to work the mortar into a pile in the center of the mortar board, smooth an area in the mortar, and separate the desired portion of mortar from the smoothed area. To load the trowel, quickly scoop up the mortar and then snap your wrist to set the mortar to the trowel.

Hands-On Agriculture

Practicing Mortar Placement Techniques

On a flat surface, set two concrete blocks on end approximately 3′ apart. Then place a 2″ × 4″ on top of the blocks to create a surface about 4″ wide and 4′ long where mortar can be placed or spread. **Figure A** shows two concrete blocks and a plank lying flat on the recesses of the two blocks. Once the mortar is applied to the surface of the plank, use the tip of the trowel with a back-and-forth motion to create a furrow in the mortar. Once the furrow is completed, remove the mortar from the plank and return it to the mortar board. Continue to practice this procedure to develop the needed skills for applying mortar to create a bed joint or to masonry units to prepare for the next course to be laid.

Goodheart-Willcox Publisher

Figure A. Use a practice station to develop mortar skills by applying mortar to and removing it from a 2″ × 4″.

---

selected for the specific type of bricks or concrete blocks, **Figure 15-7**. As explained earlier, a mason's rule is the standard measuring tool used when laying bricks or blocks. It is usually 6′ in length and folds for easy storage.

Mason's rulers are manufactured according to specific types or brick sizes. The worker will select an appropriate mason's ruler to match the bricks or blocks being laid. The mortar joint will also vary, depending on the layout spacing of the unit and joint selection. A corner pole is another device used to identify course height. The *corner pole* (also called a *story pole*) is a masonry tool with markings used to indicate the top of each course of bricks or blocks. The corner pole is usually made of wood in a size of a 1″ × 2″. The length can be from 4′ to 8′.

Figure 15-32 displays a corner pole being used to determine the height of the first unit being laid in each of the new courses. The second line represents the top of the mortar joint.

STEM Connection

Making a Corner Pole

To make a corner pole, begin by measuring the size of bricks or blocks to be used and determining the mortar joint thickness. For example, if the mortar joint is 3/8″ and the brick is 2 1/4″ in thickness, the total would be 2 5/8″. Measure up from the end of the corner pole until the tape reads 2 5/8″. Mark this location; it will be the top of the first course. Then measure up another 2 5/8″ to mark the top of the second course. To be more accurate, use math to add multiples of 2 5/8″. Continue to draw lines on the corner pole until all courses have been marked. Each line will represent the top of the specific course.

If desired, measure up from each of the course lines 3/8″ to mark a second line. These lines will represent the top of the mortar joints or the bottom of the masonry units (both being the same location).

Layout of Masonry Units on the Footing or Slab

A *corner lead* is a group of bricks or concrete blocks laid several units long and rows high to establish an outside corner of a structure. When preparing for the first course for bricks or concrete blocks, a complete bed of mortar should be placed under the masonry units. A *bed of mortar* is the bonding agent (mortar) placed under masonry units to bond them to the supporting surface. The bed should be as wide as the masonry unit that will be positioned into the mortar.

Position a number of masonry units close to where the corner leads are to be built. The first unit will form the outside corner of the building. Masonry units are laid from the outside corner in both directions, using four to five units. After the first course in the corner lead is completed, a second is started. Again, the first unit is started at the outside corner. The first masonry unit in the second course is laid opposite the first course. This positioning will create an overlapping effect from one course to the next. The last unit in the second course will end prior to the first course, creating a step effect. *Racking* is this process of ending a brick or block course prior to the previous course. This is also called *racking back the lead*.

Corner leads must be constructed carefully because the corners are used to construct the remaining wall between the corners. The process of laying the corners is called laying the leads. A mason's line is stretched between the two corners and used as a guide to lay the bricks or blocks. This will be discussed later in the chapter.

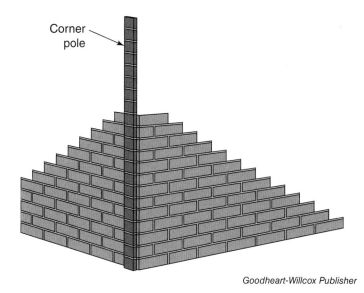

Goodheart-Willcox Publisher

Figure 15-32. This corner pole is in position to mark courses.

Building a Brick Corner Lead

The series of images in **Figure 15-33** displays the process used to build a corner lead using bricks. Building a corner lead using concrete blocks has similar basic steps. Follow these steps to build a brick corner lead:

1. Use a chalk line to mark the outside edge of the bricks or concrete blocks on the footing or concrete slab. Then mark on the concrete where the joint spacing will be located between the masonry units.
2. Lay a mortar bed as wide as the base of the masonry unit. Position the first unit above the intersection of the two chalk lines on the concrete footing that forms an outside corner. When positioning the corner unit, try to keep it level during placement into the mortar bed. The unit should be directly above the lines on the footing.
3. Check the height of the unit using the corner pole or mason's ruler. The height of the unit should be above the first course line to allow for leveling the unit. Place a level across the masonry unit to check that the

Safety Note

When working with mortar, it is important to clean hands and other exposed skin areas that come in contact with mortar. Mortar has chemicals that can burn skin if exposed to it for an extended period of time. Wear long sleeves, long pants, and safety glasses to protect the skin and eyes. If mortar enters the eyes, flush the eyes immediately. Seek medical attention if a reaction occurs with these chemicals.

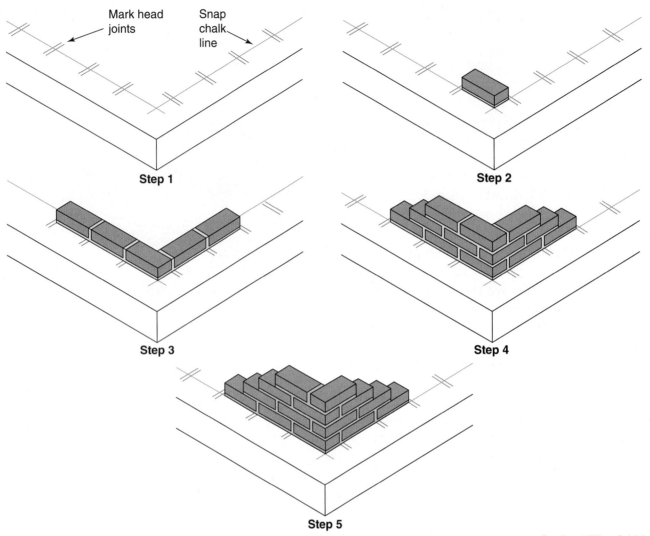

Figure 15-33. This image shows the sequence of laying a masonry corner.

length is level. Then rotate the level 90° to check that the width of the unit is level. Check that the unit is plumb. Recheck the height. Continue to make adjustments, tapping on the unit until it is level, plumb, above the lines on the footing, and at the correct height compared to the corner pole.

4. Position the second unit of each leg into mortar so that it is directly above the marked lines and joint lines. Make sure the second units are level and plumb. Each one should be flush with the top of the corner unit surface and also flush with the exterior edge of the corner unit. Follow the same sequence to lay the third unit in each of the legs.

5. Place the mortar bed for the second course. Position the corner unit for the second course. The corner unit should lie opposite the first course. Check the unit course height with a mason's rule or corner pole. The unit should be flush with the lower course on the outside edge and end. Make adjustments so the unit is level and plumb as was done in the previous course. Place the next units in the course as was done in the previous course. Install the third course following procedures performed in the second course.

6. Place the mortar bed for the fourth course. Set the corner unit and check for height. Ensure the unit is flush with the preceding courses on the outside edge and end. Lay the next units in both legs to complete the corner lead.

For information on completing the brick wall between the corner leads, review procedures found in the *Laying a Block Corner Lead and Wall* section below. The procedures are almost identical.

Laying a Block Corner Lead and Wall

Concrete blocks come in several shapes. Two of the most common shapes are the corner block and the stretcher block. See **Figure 15-34**. A *stretcher block* is the type of masonry unit that is laid between the corner blocks. In preparation for laying, stretcher blocks are positioned on end with the ears pointing up, **Figure 15-35**. The *ears* are the two flanges (ridges) on each end of a stretcher block where the mortar is placed.

Concrete blocks are large units that usually require using both hands for placement. For example, an 8″ × 8″ × 16″ concrete block will weigh 40 pounds or more. Lightweight blocks can weigh 22 to 28 pounds.

Laying the First Course of the Lead

The first course is started by laying the corner block at the intersection of the two chalk lines. The block is checked to see if it is close to the correct height by reading the mason's rule or corner pole. The unit should be slightly above the final height to allow for adjustments so it will be level and plumb.

A stretcher block is laid at the end of the corner block. Just prior to positioning the stretcher block in the wall, mortar is placed on the ears with a trowel. See **Figure 15-36**. Two or three stretcher blocks are placed in each leg of the first course.

Follow these steps for laying blocks for the first course of a corner lead:
1. Determine where the structure or wall is to be laid on the footing or concrete slab. Mark the exterior corner locations on the concrete footing to position the corners of the building or structure.
2. Snap chalk lines between the corner marks to indicate the exterior face of the masonry walls.

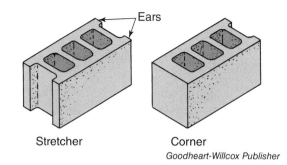

Goodheart-Willcox Publisher

Figure 15-34. A stretcher block and corner block are shown here.

Goodheart-Willcox Publisher

Figure 15-35. Concrete stretcher blocks are positioned on end to make it easy for the worker to access them prior to laying.

Goodheart-Willcox Publisher

Figure 15-36. One of the concrete blocks has received mortar to the ears and is ready to be laid.

3. Mark the position where each of the blocks will be located. Refer back to **Figure 15-33**. The common length of a block is 15 5/8", and a standard joint size can measure 3/8". In this instance, one block and one joint added together will measure 16". This measurement will be marked repeatedly on the concrete to locate block and joint positions. The length of a block plus one joint may change from one project to another, depending on the unit size and the mortar joint selected. When marking the block locations and mortar joints, extend the position lines past the block to ensure they are visible after the blocks are laid. The lines allow the worker to make minor adjustments in the block's position by lightly tapping the block to move it to the correct position.
4. Place an ample amount of mortar to create the bed on the concrete for two to three block lengths. As mortar loses moisture, it becomes more difficult to work with. Therefore, it is important not to spread too much mortar in the beginning. After the mortar bed is laid, create a furrow using the tip of the trowel.
5. Position a number of corner blocks and stretcher blocks close to the location where the corner leads will be built.
6. Select a corner block and lift it with both hands in preparation for it to be laid in the fresh mortar bed. When handling the unit, try to keep the block level. Then position the block directly above the intersection of the two snapped lines on the concrete footing that forms the outside edge of the corner lead.
7. Check to see if the corner block is level and plumb. Place the level over the length of the corner block to begin adjusting the top surface until it is level. Turn the level 90° so it is across the width and perform the same operation. Then use the mason's ruler or corner pole to check the height of the block. Continue to make adjustments to the block until the unit is level and plumb. The side and end of the block should be directly above the chalk lines on the concrete. If mortar restricts the view of the chalk lines, remove the excess mortar on the outside of the unit and return it to the mortar board.
8. Place mortar on the ears of the stretcher block and check to ensure that adequate mortar is in place where the unit will be positioned. Pick up the unit and place it at the end of the corner block. Press the block slightly toward the corner block to form a good joint between the block being laid and the corner block. Check to see if the unit is level and positioned directly above the chalk line. Continue to make minor adjustments until the unit is flush on the top and side with the corner block and the mortar joint is the correct spacing. This is done by looking at the joint marks drawn on the concrete footing. Finally, check again to see if the units are level. Then remove the level from the top and turn it on edge. Place the level against the sides of the blocks to see if they are flush with each other.
9. Check and recheck the corner block to ensure that it is plumb and was not moved when surrounding units were placed against it to form the corner. All other blocks in the lead are positioned based on the corner block, so it is very important the first block is plumb as well as level.

10. Complete the first course of the corner lead by laying the second leg of the corner, **Figure 15-37**.

As blocks or bricks are laid, it is important to cut excess mortar away from the joints. Keep the trowel at an angle to cut the mortar away from the block or brick. This reduces chances of smearing. The mortar that is deposited on the trowel can be returned to the mortar board. Once mortar is smeared into the unit, it becomes very difficult to remove. This affects the final appearance of the project.

Also, use care not to move the blocks or bricks after they have begun to bond to the mortar. Moving the blocks or bricks at this point will break the bond to the mortar.

Christina Richards/Shutterstock.com

Figure 15-37. Once one leg of the corner lead is level and plumb, the worker is setting the first block to begin the second leg of the corner.

Laying the Following Courses of the Lead

A normal corner lead will be four or five courses high. Each leg of the lead should be approximately the same length. For practice, shorter leads can be used to develop skills in building leads. The second course of the corner lead is laid by placing mortar on the corner block leg to be laid first. The corner block in the second course will be positioned in the opposite direction of the first course corner block. Use the same procedures discussed earlier to adjust the corner block so that it is level and plumb. Continue to adjust the corner block until it is at the correct height, level, plumb, and flush vertically on the end and side with the first course.

Follow these steps to build the second and additional courses of a corner lead:

1. Start the second course by placing mortar on the top of the preceding course. The first unit, the corner block, will be laid in an alternating pattern from the first course. Make sure the corner block is as level as possible when positioning it in the mortar. Check the block height against the story pole. Make adjustments as needed until the corner block is flush with the end and side of the previous course, plumb, level, and at the correct height.

2. Lay the stretcher blocks as has been done before, ending the last unit in the course back from the previous course in a stair-step effect. Ensure that the units are level with the corner block and level across the width. Then check to see if the stretcher block is flush on the outside edge with the previous course. Continue this process with the remaining stretcher blocks in both legs.

3. Begin laying the third course by placing the corner block in an alternating pattern to the second course, **Figure 15-38**. Make adjustments to the corner block until it is at the correct height, level, plumb, and flush with the previous courses. Lay stretcher blocks in the third course following procedures used previously.

Christina Richards/Shutterstock.com

Figure 15-38. The worker is positioning the corner block in the next course in the corner lead.

Christina Richards/Shutterstock.com

Figure 15-39. The worker is laying stretcher blocks between the two corner leads. He is using a mason's line as a guide for height and straightness.

4. Begin the fourth course by laying the corner block in an alternating pattern to the third course. Check the corner block for height; lightly tap the unit to make the corner level and plumb. Check to ensure the corner block is flush on the end and side with the previous courses. Check the edge and end to ensure it is straight and plumb. Continue to adjust and recheck until it is at the correct height, level, plumb, and flush. Lay the stretcher blocks in both legs as was done in the previous courses.

Laying the Stretcher Blocks

After the corner leads have been constructed, a mason's line will be strung between the two leads. The mason's line will be used as a guide for height and straightness when laying stretcher blocks. The line should be very tight without noticeable sag. The tension should be uniform each time the line is reset for the next course of stretchers. When laying a stretcher block next to the mason's line, the worker should ensure that it does not touch the line. The block should be one line width away from the mason's line. If blocks are allowed to touch the mason's line, they will cause the line to start to bow out and will affect the straightness of the course.

Follow these steps to lay stretcher blocks between corner leads to build a wall:

1. Place a mortar bed on the concrete footing next to one corner lead. Position the first stretcher block at the end of the corner lead. Check to ensure that the block ends at the mark drawn on the concrete footing. This will ensure the joints between the units are the correct size. Use the mason's line as a guide for height and to keep the block straight between the two corner leads. The first stretcher block should be flush with the outside edge of the corner lead and directly above the snapped line on the concrete footing. The top surface of the newly laid block should be flush with the top of the corner block.

2. Continue to lay stretcher blocks. Work from both corners toward the center of the wall. Check to ensure laid blocks are above the snapped line on the concrete footing and one line away from the mason's line. Check to see that the joints are correctly spaced and that the top of the newly laid block is flush with the previously laid unit.

3. The last block placed in the course is called a *closure block*. When laying this block, place mortar on the ears of both blocks that have been laid. Then slide the closure block between the previously laid blocks to finish the course.

4. Lay stretcher blocks in the following courses using the same procedures as those used in the first course, **Figure 15-39**.

Finishing Joints

Joints are finished when the mortar is thumbprint hard. This means that the mortar has hardened to a point that a thumbprint is left when a thumb is pressed into the mortar. At this point, the joints can be finished to seal against moisture. Select a jointer tool for the type of joint profile desired, such as a concave joint. The jointer tool should be slightly larger than the joint being finished so the tool can make contact along the edges of the block. For example, a joint spacing of 3/8″ would require a jointing tool of 5/8″. When finishing horizontal joints, it would be best to use a sled runner jointer that is at least 22″ in length. Finish the horizontal joints first, then the vertical joints. This procedure will aid in allowing moisture to run off the joints rather than getting caught on the lip of the horizontal joint. After the vertical joints have been finished, use a trowel to remove any burrs that were created when joints were finished, **Figure 15-40**. When the mortar is sufficiently dry, use a stiff masonry brush or rub the wall with a burlap bag to remove dry mortar particles, **Figure 15-41**.

Installing Anchor Bolts

Anchor bolts are specially designed fasteners used to hold wood framing to concrete masonry units, **Figure 15-42**. Anchor bolts are installed in the cores (hollow areas) of the top course. Spacing between anchor bolts can vary, depending on requirements in local or state building codes. Anchor bolts are commonly spaced approximately 4′ apart. The bolts are usually 1/2″ in diameter and 18″ in length so they can extend down through the top two courses of block. When a block wall is laid, a metal lath is often placed under the second-to-last course. Concrete or mortar is then added to fill the top two courses of the cores that contain the anchor bolts. In some cases, a lath is not used and the cores are completely filled with mortar or concrete from the footing to the top of the block wall. This technique will strengthen the wall by tying the footing, concrete blocks, and anchor bolts together to act as one assembly. Check with local and state building codes to determine specific requirements.

Goodheart-Willcox Publisher

Figure 15-40. The worker is using a trowel to remove burrs created during the process of finishing mortar joints. His trowel is at an angle to cut the burrs off the wall.

Goodheart-Willcox Publisher

Figure 15-41. The worker is using a stiff masonry brush to clean the block wall of dried mortar particles.

Goodheart-Willcox Publisher

Figure 15-42. An anchor bolt has been installed in the top course of the cores of the concrete block wall.

CHAPTER 15 Review and Assessment

Summary

- Masonry products are commonly used in agricultural applications. Concrete blocks are used for structural purposes, such as foundations and walls. Bricks are most commonly used as a facing material.
- Tools used for masonry building include trowels, jointers, hammers, measuring devices, mortar boards and stands, steel squares, levels, and chisels. Mason's lines and mortar boxes are also used in masonry construction.
- Many accidents related to masonry work involve tools and hazardous substances. Following safety guidelines and using personal protective equipment can help workers avoid injuries.
- Masonry units include clay products, such as bricks, concrete units, and stones. Bricks are made by firing molded blocks of clay in a kiln.
- Mortar is used to bond individual masonry units together to form a wall. The strength of the wall can vary, depending on the pattern bond selected. Different types of joints are used to finish the surface of the exposed mortar.
- Concrete masonry units (CMUs) are hollow or solid blocks made from Portland cement, aggregates, admixtures, and water. Some blocks are designed to be load bearing in a structure.
- Mortar is typically made from Portland cement, hydrated lime, and sand. Contractors may also use premixed masonry mortar. Mortar can be mixed by hand or by using a machine.
- Planning for a masonry project involves determining the type, size, quantity, and color of masonry units to be used. Building codes for the area should be reviewed during planning.
- Basic skills required for masonry work include loading the trowel, spreading mortar, holding bricks and blocks, adjusting units to be level and plumb, using a mason's line, and finishing joints.
- A trowel is used to apply mortar to masonry units. To determine the height of a course when laying masonry units, a worker may use a mason's rule or a corner pole.
- Masonry units are laid in courses from the outside corner in both directions. The first masonry unit in the second course is laid opposite the first course to create an overlapping effect.
- When building corner leads or walls, adjustments must be made to keep each unit level, plumb, flush with the previous course, and at the correct height.
- Mortar joints in a wall can be finished to seal the joint against moisture. Anchor bolts can be used to hold wood framing to concrete masonry units.

Words to Know

Match the key terms from the chapter to the correct definition.

A. American Society for Testing and Materials
B. anchor bolt
C. bed of mortar
D. bond
E. closure block
F. corner lead
G. corner pole
H. course
I. grout
J. header brick
K. hydraulic cement
L. jointer
M. kiln
N. level
O. line holder
P. margin trowel
Q. mason's line
R. mortar board
S. pattern bond
T. rowlock stretcher brick
U. soldier brick
V. stretcher block
W. tile
X. tuck pointer
Y. wythe

1. The last block or brick laid in the course.
2. A type of hollow clay masonry unit that is fired to harden it.
3. A group of bricks or concrete blocks laid several units long and rows high to establish an outside corner of a structure.
4. An international organization that tests materials and develops standards for their use.
5. A fluid cement mixture used to fill gaps or voids between wythes or other areas in masonry construction.
6. A masonry tool with markings used to indicate the top of each course of bricks or blocks.
7. A term used to describe a surface that is exactly horizontal.
8. A strong nylon cord used to show the layout of a building or to keep a course (row) of bricks or concrete blocks in a straight line.
9. A design or arrangement formed by the masonry units and their mortar joints on the face of a masonry wall.
10. An oven or furnace that is used to heat green bricks to a very high temperature.
11. A complete or continuous row of bricks or blocks laid in a wall, structure, or project.
12. The bonding agent that is placed under the first course of masonry units to support and bond the units to the supporting surface.
13. A brick that is laid flat with the short end exposed.
14. A trowel with a rectangular front edge and a V-shaped heel that is used to hold mortar or fill small areas.
15. A brick that is laid on end with the narrow side exposed in the wall face.
16. A type of trowel with a narrow blade used to fit into tight spaces, such as between bricks or blocks.
17. A continuous vertical section of masonry that is one unit in thickness and constructed of a single line of masonry units.
18. A type of concrete block that is laid between two corner blocks and usually has ears on both ends of the unit.

19. A brick that is laid on edge with the largest surface of the brick exposed in the wall face.
20. A device made of plywood or polymer that is used to hold mortar when laying bricks or concrete blocks.
21. A device used to hold a mason's line in place when positioned on the bricks or blocks.
22. A type of cement that will set and harden underwater, such as Portland cement.
23. To join items or substances together securely by means of an adhesive, heat, or pressure.
24. A specially designed fastener used to hold wood framing to concrete masonry units.
25. A tool used to finish the surface of a mortar joint between bricks or concrete blocks.

Know and Understand

Answer the following questions using the information provided in this chapter.

1. What are two purposes for which concrete blocks are used in agricultural structures?
2. Describe the two types of trowels commonly used to perform masonry work.
3. How is a mason's rule used in laying masonry units?
4. What is the purpose of a brick set?
5. What safety precautions should you take when mixing mortar with Portland cement?
6. What are three classifications of structural clay masonry products?
7. What steps are included in the brick manufacturing process?
8. What are some pattern bonds that create different exterior appearances of masonry walls?
9. How does a concave mortar joint differ from a raked mortar joint?
10. What are some admixtures that may be used when creating concrete masonry units?
11. What are the nominal dimensions and the actual dimensions of a standard concrete block?
12. What are the main ingredients used to make mortar?
13. What steps are involved in mixing mortar by hand?
14. What are some activities that should be done in the planning stage for a masonry project?
15. What are some basic skills that a masonry worker needs?
16. What are two devices that can be used to determine the height of a course when laying masonry units?
17. Explain the process of racking when creating a corner lead.
18. When laying a brick corner lead, how should the first unit placed in the mortar bed be positioned?
19. How is a mason's line used when constructing a block wall?
20. Should vertical or horizontal joints in a wall be finished first? Explain why.

STEM Connections and Academic Activities

1. **Science.** Research the chemical process that causes concrete to set or become hard underwater.
2. **Technology.** Research a new technology that is being introduced to the concrete masonry field or the brick industry. Write a report pertaining to your discoveries. Include the pros and cons of the technology as you see them.
3. **Engineering.** Through research, determine the type of glazed block that would be best used in a dairy or cheese facility. Make note of special tips, such as the fact that light-colored blocks are easier to clean and show fewer water spots after drying. Make selections as listed below and prepare to give reasons for each of your selections.
   - Select a color.
   - Select a width.
   - Select a length.
   - Select a height.
   - Select a manufacturer.
4. **Math.** Determine the dimensions for laying out a three-course corner pole for a concrete block retaining wall. What would be the dimensions for the lines to be drawn on the pole for the top of each joint and top of each course? The joint size is 3/8" and the block is 7 5/8". *Hint: do not forget the top of the first joint!*

Thinking Critically

1. Research the procedures used to lay out the four corners of a building (measuring 10′ × 20′) to ensure that they are square and at 90° angles to each other. Research the procedures to measure diagonally across the corners to determine if the layout is square. Diagonal measurements should be equal.
2. After performing your research for question 1, work in a group with two to four students. Use the data gathered in your research to lay out the building. Keep these points in mind:
   - Stakes should be in each corner.
   - A mason's line should stretch around the perimeter.
   - The end and side measurements should be 10′ and 20′.
   - The diagonal measurements should be the same.
   - Items needed include: 30′ tape measure, four stakes, a mason's line, and a hammer.

CHAPTER 16
Framing Structures

Chapter Outcomes
After studying this chapter, you will be able to:
- Discuss how loads, computer aided design, and building materials affect building design.
- Describe four types of structures used in agriculture.
- Describe five types of framing used for structures.
- List advantages of using trusses in roof construction.
- Describe the main components used in floor construction.
- Describe the main components used in wall construction.
- Identify roof styles used for structures.

Words to Know

arc construction	framing	pole frame construction	square
balloon framing	live load	post	stud
blueprints	load	post frame and truss framing	stud wall construction
bridging	load-bearing member		subfloor
CAD software	modern braced framing	rigid construction	truss
ceiling joist	opening	roof	wall
dead load	plank and beam framing	sheathing	western framing
environmental load		sill	
finishing	platform framing	sill plate	
floor	plumb	span	

Before You Read
Before you begin reading this chapter, consider how the author developed and presented information. How does the information provide the foundation for the next chapter?

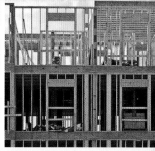

While studying this chapter, look for the activity icon to:

- **Practice** vocabulary terms with e-flash cards and matching activities.
- **Expand** learning with interactive activities.
- **Reinforce** what you learn by completing the end-of-chapter questions.

www.g-wlearning.com/agriculture

The term *framing* is used to describe the basic structural members, or components, of any type of structure or the process of assembling these components. Framing provides the basic skeleton of a building and is usually one of the first steps in construction. Structural members can be constructed from a variety of materials, but most commonly these materials are wood, steel, and concrete. These members support the weight of the building materials and people, animals, and objects that may be in or on the building. Structural members also serve as attachment points for coverings, such as sheetrock or siding. Because of this, framing components are typically not seen. Covering the framing structure is called *finishing*. Some of the most common types of exterior finishing materials include metal, wood, and brick, **Figure 16-1**. However, not all buildings are completely finished. For example, many agricultural structures, such as storage barns, are left unfinished, **Figure 16-2**.

Framing structures and finishing options are selected to fit the purpose a structure will serve. Framing designs should be selected as part of a planning phase prior to construction.

Building Design

Building design is the overall arrangement, function, purpose, use, or appearance of a building. Building design is a career pathway that demands a dedication to mathematics and science. Specialized education and training prepare students for a variety of jobs related to building design. Proven engineering principles must be the basis for any building's design, and these principles are discussed in this section.

AgEd Connection — Home and Community Development

Food, clothing, and shelter have long been considered basic needs of man. All three of these needs are supplied by agricultural products. This chapter covers one aspect of the need for shelter. After studying the material in this chapter, you will have a solid understanding of a process that applies to numerous supervised agricultural experiences (SAE), career development events, and future job opportunities.

The last line of the FFA motto is "Living to Serve," and many service opportunities arise from having a knowledge of planning, designing, drawing, and building or improving shelter. You can take the skills from this chapter and apply them to a home and community development SAE that will not only give a little back to your community but will also provide you with invaluable experience. Discuss with your teacher today the opportunities to "give back" to your community.

©iStock.com/kadmy

Building Loads

A *load* is a stress or weight that is supported by a structure or member. Load factors must be considered during the building design and engineering phase because they affect the strength of a building. They have a direct correlation to how much weight a building can support and the forces it can resist. A *load-bearing member* is a component of a framing system that must support an amount of weight that is important for the strength of a building. Making sure the load-bearing members in a framing structure are correct is an important job for engineers and project managers. Loads are important to consider in design and construction because they will determine the type of structural support needed. A career in engineering is one of many rewarding options in the building and construction industry.

When designing buildings, three basic types of loads must be considered:
- Dead loads.
- Environmental loads.
- Live loads.

Dead Loads

Dead loads are the weights of any materials or components used to build a structure. Items such as boards, shingles, siding, and any other materials used

A

B

C
Goodheart-Willcox Publisher

Figure 16-1. Covering the framing structure is called finishing. A—Framing structures are typically covered during the finishing stage of the construction process with materials such as metal, wood, or brick. B—Wood is a common type of siding for many agricultural barns. C—Brick provides a long service life and is commonly used for siding on houses.

Rob Byron/Shutterstock.com

Figure 16-2. Storage areas in barns are often left unfinished on the inside, which leaves the framing exposed.

Safety Note

Environmental events cannot be predicted reliably. Destruction from a tornado can be disastrous. Seeking shelter in the lowest and most central location of a structure places you in the strongest areas, which are most able to resist a tornado's powerful wind forces.

to construct a building are included as dead loads during building design. When constructing a building, the load-bearing members should support the building itself and any other loads the building needs to withstand.

Environmental Loads

An *environmental load* is a weight or stress caused by a meteorological event (wind, rain, snow, or earthquake) or condition of the surrounding area. Structures should be able to withstand wind storms, heavy rains, the pressure of fill (dirt/gravel) applied around a basement, and many other factors.

Live Loads

A *live load* is the stress caused by or the weight of everything placed on, in, or temporarily attached to a structure, such as machinery, people, furniture, and animals. Live loads are variable stresses or weights that may change over time. A person walking through a room, feed stored in a loft, and a piano placed in a family room are all examples of live loads that need structural support. Live loads are unique because they can change: a person can leave the room, feed can be taken from a storage area, and the piano can be moved to another location.

Computer Aided Design

Computers are commonly used for designing structures. Many software packages are available for computer aided design (CAD). *CAD software* is a program that allows the user to design architectural, engineering, and construction projects on a computer. The user can design a building and can include varied information, such as load calculations, visual images of loads, lists of materials required, and 3-D tours of completed designs. Using CAD software has distinct advantages. Designers can save time, money, and materials; quickly determine material needs; and generate completed designs called blueprints. *Blueprints* are two-dimensional plans that provide all dimensions, measurements, and construction details for a building. Construction companies refer to blueprints for design requirements when completing building projects, **Figure 16-3**.

Building Materials

A large variety of building materials are available for construction projects. As discussed in Chapter 9, *Materials, Fasteners, and Hardware*, the selection of materials depends on several factors, such as local availability, cost, function, strength, and intended life of the structure. Building materials can generally be placed into one of three categories: wood, steel, or concrete.

FranckBoston/Shutterstock.com

Figure 16-3. Computer programs can be used to produce both two-dimensional and three-dimensional drawings. Blueprints are generated from these dimensional renderings.

Structure Types

Four main structure types are used in building design. They are based on the different kinds of building components used in the construction and sometimes on the function of the building. Many types of structures used in agricultural and nonagricultural applications can be included in these four broad categories. Each of these categories includes many different designs or framing types, which are discussed in the next section of this chapter. The four main structure types are:

- Pole frame construction.
- Rigid frame construction.
- Arc construction.
- Stud wall construction.

Pole Frame Construction

Pole frame construction describes buildings that have posts or poles as the main structural component for their walls, **Figure 16-4A**. A *wall* is a vertical component of a building that is used to separate inside and outside environments, divide spaces into smaller rooms, and support the weight of other building components. A *post* is a long, sturdy piece of wood or metal that is set upright in the ground or a structure. These structural posts are also how the building is anchored to the ground. This is either accomplished by inserting the posts into the ground or installing them as part of the concrete foundation. Pole frame construction is very common for agricultural structures. Currently, the most common type of structures for machinery storage, machine shop, workshop, and livestock is pole frame construction. Some residential structures also use pole frame construction. A *truss* is a frame made of beams, bars, or rods and used in roof construction. A *roof* is the external covering of the top of a building that is exposed to the environment. When posts and trusses are used, a large open area inside the structure is the result. This open area is very desirable in agricultural applications. Since the combination of the posts and trusses is self-supporting for all loads, there are many opportunities for using this open space efficiently. Rafters may also be used with this type of construction instead of trusses. Construction is essentially the same, but additional support is required when rafters are used, **Figure 16-4B**.

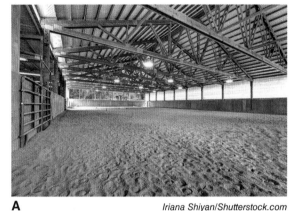
A *Iriana Shiyan/Shutterstock.com*

B *Rick's Photography/Shutterstock.com*

Figure 16-4. A—Pole frame construction takes advantage of the large open space provided by the use of roof trusses. B—Rafters and trusses serve the same function. This picture shows rafters. Roofs made of rafters are built onsite.

Rigid Construction

Rigid construction is a type of structure in which posts and rafters are assembled as a unit and are open the entire span of a structure. *Span* is the full extent of something from end to end. It is similar to pole frame construction. This type of building is usually constructed with a gable roof. Two main differences are the components and the assembly process. A *stud* is a vertical board that is the main structural component of a wall. Two studs or posts are combined with two rafters to form one structural unit stretching from one side of the building to the other. Gussets (braces) are attached with nails, glue, or bolts at the ridge or peak of the roof and at the eave connections. Besides bracing, these are the only framing components, making rigid construction fast and easy to erect. Rigid structures are built using either steel or wood components, resulting in a robust frame. When using steel posts, rafter units can be I-beams, tubing, or steel trusses, **Figure 16-5**. Sizing of wood components is dependent on the loads and spacing of the individual members. An advantage of this type of construction is an interior area free of obstructions very similar to the interior of a structure with pole frame construction using trusses or rafters.

©iStock.com/RachelKathrynGiles

Figure 16-5. Rigid construction commonly uses steel I-beams for posts and rafters.

Arc Construction

Arc construction is a type of structure that uses continuous rafters that are curved from one side of the building to the other. A common type of arc construction using wood or steel construction is the Quonset hut, **Figure 16-6**. When finished, the continuous rafter combines all the functions of walls and rafters that are used in the construction of rigid or pole frame structures. These structures are curved from the point of attachment to the foundation (or a short wall) to the peak. Wooden arc rafters can be laminated into the curved shape. They can also be sawed and assembled to the correct shape on the job site.

Scott Proko/Shutterstock.com

Figure 16-6. A common type of arc construction using wood or steel construction is the Quonset hut. It features a completely curved wall and roof system.

Stud Wall Construction

Stud wall construction is a type of structure in which vertical boards or pieces of metal (studs) are used as the main structural component in walls. This type includes structures that are mainly, but not limited to, residential structures. Stud walls are constructed of boards that are usually 2″ × 4″ or 2″ × 6″. These studs are installed at a recommended spacing that depends on the loads they are designed to support. While wood is the material of choice for studs, steel studs have become common in stud wall construction. See **Figure 16-7A** for an example of steel stud construction. In most applications, studs are spaced every 16″ on center. Stud spacing can be increased if the structure will not be required to support a large amount of

Figure 16-7. Stud wall construction using steel studs, figure A, is compared to figure B showing a stud wall construction using wood studs.

weight for storage or environmental loads. For example, storage sheds for home use many times are constructed with the studs spaced at 24″ on center intervals.

A stud wall contains several important structural components. Not only does a wall need to include studs for supporting a variety of loads, it should always fulfill the functional needs of the structure as well. Spacing, header construction, and placement for windows and doors should all be considered during wall construction. Refer to **Figure 16-7B**.

Framing Types

Understanding the different framing types is important when planning and designing a building for a specific purpose. Having knowledge about a structure's framing is also important when communicating with suppliers, contractors, and owners. A structure is considered framed when all the structural members have been installed. **Figure 16-8** shows a building that has been completely framed. Structural members include floor components (posts, stringers, and joists), walls (studs, top and bottom plates, and headers), trusses or rafters, and any necessary bracing.

While there are numerous building designs, there are five types of framing that are most commonly used in construction. These framing types are:
- Post frame and truss framing.
- Modern braced framing.
- Western or platform framing.
- Balloon framing.
- Plank and beam framing.

Figure 16-8. This structure has been completely framed and awaits sheathing and insulation. Note the pressure treated wood bottom plate that rests on the concrete foundation.

Figure 16-9. Post frame and truss buildings are simple to construct and can be very useful for storage buildings or for shed structures. You can clearly see the posts holding up the roof trusses.

Post Frame and Truss Framing

Post frame and truss framing is a building method that results in large open areas inside a building, **Figure 16-9**. It is the most commonly used type of framing. Many machinery storage, livestock, workshop, and other structures are built using this framing type. The invention of trusses greatly changed the design, usefulness, and open floor space of agricultural buildings. Prior to the mid-1900s, the primary construction type was mainly post and rafter. This type of structure worked well (and still does). However, it limits the usefulness of buildings because rafters require support throughout their span. At a minimum, two support posts are used for the outside walls. Some buildings require more posts, in many cases two more, if additional support is needed. In agricultural equipment applications, this can be a limiting factor because equipment sizes have grown so large during the past few decades. However, these structures are simple to construct and can be very useful as storage buildings or for shed structures. The major advantage of using a truss is the large amount of clear span (open area with no obstructions) in the building, **Figure 16-10**. Trusses require anchor points on each end (the outside walls) but no support in between because the support structure is incorporated into the bracing that is part of the truss. Another advantage is that trusses provide easy installation of ceiling framing and coverings. This is because the bottom of a truss is used as an attachment point for the ceiling in this type of framing.

Figure 16-10. A large clear span is the main advantage of using trusses. Bracing in the roof structure instead of in the floor space below frees up valuable space. In this image, a poultry barn design provides 55′ of open space from one side of the structure to the other.

Modern Braced Framing

Modern braced framing is a building method in which all main structural components are assembled before subflooring is installed. A *subfloor* is a layer of material, such as plywood, that is placed over the floor joists, **Figure 16-11**. These structures use stud walls as their outside wall framing. Each level of the structure is constructed before any subfloor materials are installed. As construction progresses, the floor joists, boards, and any other floor components are installed. When these components are finished, the walls are installed directly to the floor joists. If the blueprints call for additional levels or stories, the construction process is repeated (with floor joists, walls, and then more floor joists on top of the walls). Subflooring is installed later in the construction process after the walls are completed.

Western Framing or Platform Framing

Western framing or *platform framing* is a building method in which each level or story of a building is built as a separate unit. This method is very similar to the modern braced framing method. The main difference is that each level or story of a structure is completed before continuing with additional levels. In this type of framing, joists are installed, subflooring is installed, and then walls are installed. The main advantage of western or platform framing is safety. Carpenters have a solid floor to use as a work and assembly area during construction. See **Figure 16-12**.

Balloon Framing

In *balloon framing*, long studs that extend through two stories of a structure are used, **Figure 16-13**. Wall construction is very similar to that used in modern braced framing and western framing. Since the studs are so long, bracing must be used during construction to make sure the structure does not fall or shift. Floor

Christina Richards/Shutterstock.com

Figure 16-11. Modern braced framing structures are built by installing all framing before any subflooring. Notice the exposed floor joists in the background.

Safety Note

During construction of any building, safety should be considered at all times. Construction tasks include using power tools, lifting heavy objects, and working above ground level. Appropriate safety training and a safety-first mindset are important.

Melinda Fawver/Shutterstock.com

Figure 16-12. Western or platform framing provides a solid floor while other components of the structure are being built. Here, stud walls are constructed with the floor area being used as a workspace.

Figure 16-13. Balloon framing uses long studs that extend through two stories of a structure.

joists are installed by using a ledger board. A ledger board is attached to the inside of the studs at the appropriate height. The joists are then placed on the ledger board and also attached to the sides of each stud. This type of framing allows for quick construction. However, this type of building is not well braced until sheathing is installed. *Sheathing* is material (wood or fibers) attached to framing to strengthen the structure and provide an attachment point for coverings, such as roofing or siding. Until then, the structure is unstable.

Plank and Beam Framing

Plank and beam framing is a building method used to create large open areas inside a structure and to show off exposed wood components, such as support beams, for their visual appeal, **Figure 16-14**. These structures can use studs or posts for wall construction. Architects love the large open areas plank and beam framing provides. Often they will specify architectural details, such as large windows, large doors, or vaulted ceilings in their plans. If this is the case, some of the structural wall members must be removed. If structural components are removed, the strength has to be increased elsewhere to accommodate building loads. With this type of framing, studs will be larger than normal. For example, 2″ × 6″ or larger studs will be used instead of 2″ × 4″ studs. Support beams will be used instead of trusses. Tongue and groove planking may be used instead of plywood sheathing if it will be left exposed. Since this type of framing provides large open areas, many creative design options can be considered.

Figure 16-14. Plank and beam framing can be used to open up an area to natural lighting and highlight natural colors of wood. This is a common type of framing for sunrooms.

Floor Framing

Flooring serves several important functions in a structure. Some of the most important functions include providing an area for storage; adding strength and rigidity to floor framing and the overall structure; and providing an area for the installation of floor coverings, such as carpet, tile, or wood. A *floor* is the lower, supporting surface of a room or building. Floors have structural members that allow them to support structural loads. Structural members can be made from wood, steel, or concrete. Using these materials in

Figure 16-15. A—This huge machinery storage structure has a rigid steel frame. B—This grain storage facility is constructed with concrete.

the floor framing strengthens and reinforces the structure. While all of these building materials have their advantages, wood is typically selected for use in agricultural and residential structures due to lower costs. Wood is also often used for remodeling projects or for additions to an existing structure. In some specialized structures, one material may be used exclusively. Examples include a machinery storage barn with a rigid steel frame, **Figure 16-15A**, and a grain storage facility constructed from concrete, **Figure 16-15B**.

Sills

A floor frame must have an initial mounting location. A *sill*, also called a *sill plate*, is a horizontal board placed on the foundation to which vertical members, such as wall studs, are attached. Many types and styles of foundations are used in modern construction techniques. Foundations are typically solid concrete at the base with concrete blocks or a solid concrete wall added to provide height between the floor framing and the ground. Concrete blocks or solid concrete can be used for one-story structures (only a few blocks high) or for taller structures with a basement. In either structure, a treated wood board will be installed around the perimeter of the building, **Figure 16-16**. Because of its close proximity to the ground, the wood must be treated to prevent rot. All other flooring and the remainder of the structure will rest on the sill.

Figure 16-16. The sill plate is the bottom plate of a wall. The sill plate is used to attach a wall to a proper location inside a structure.

Beams, Girders, and Posts

Floor designs are determined by the physical area of a floor and how much weight or load it is designed to support. Any area of flooring is limited by how much weight the structural members can hold without support. **Figure 16-17**

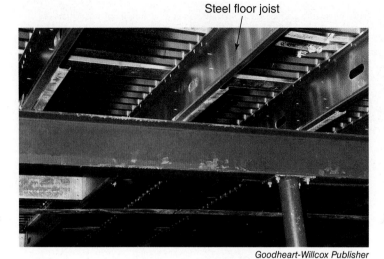

Figure 16-17. These steel floor joists need support to strengthen the floor.

Figure 16-18. One support beam through the middle of the span is used for an open space in this structure.

Figure 16-19. Floor joists are being installed in this house.

shows steel floor joists with a span that exceeds the joist's ability to support the floor without additional support. At some point along a floor joist, the joist will need support to prevent the floor from sagging or bowing. Girders are boards installed to provide an attachment point for other materials or to provide support for the structure. At the point of support, beams or girders are used to strengthen the area and provide support for the overall floor structure. Beams are typically laminated boards, such those made of three $2'' \times 10''$ boards, or a steel I-beam. Either of these components will be supported by a post. As a complete system, a floor would have support on the sill on both sides of the structure and at least one area of support in the middle. In a house of average size with a basement or large open room, there will be one support beam through the middle of the span. See **Figure 16-18**.

Floor Joists

After installation of the sill and any support beams, floor joists are the next components in construction. Joists are the major framing component for floors. They give the floor rigidity, the strength to support weight, and a mounting location for subflooring and flooring. **Figure 16-19** shows how floor joists are installed. The size of a joist is determined by the load it will carry and the distance it will extend. **Figure 16-20** provides guidelines for selecting and sizing joists.

The traditional joist is a wood board, such as a $2'' \times 8''$. It is stood up on its narrow edge and then positioned on the sill. On the outside perimeter, a cover board is used to create a solid area around the floor framing. The cover board also becomes a mounting surface for any sheathing or siding installed later in the construction process.

An alternative to using floor joists is using floor trusses, **Figure 16-21A**, or manufactured I-beams, **Figure 16-21B**. Floor trusses have gained popularity in recent years. Floor trusses offer distinct advantages

Floor Joist Span Data

30 psf live load, 10 psf dead load, deflection <360

Species or Group	Grade	2" × 8"			2" × 10"			2" × 12"		
		12" O.C.	16" O.C.	24" O.C.	12" O.C.	16" O.C.	24" O.C.	12" O.C.	16" O.C.	24" O.C.
Douglas Fir and Larch	Sel. Struc.	16'-6"	15'-0"	13'-1"	21'-0"	19'-1"	16'-8"	25'-7"	23'-3"	20'-3"
	No. 1 & Btr.	16'-2"	14'-8"	12'-10"	20'-8"	18'-9"	16'-1"	25'-1"	22'-10"	18'-8"
	No. 1	15'-10"	14'-5"	12'-4"	20'-3"	18'-5"	15'-0"	24'-8"	21'-4"	17'-5"
	No. 2	15'-7"	14'-1"	11'-6"	19'-10"	17'-2"	14'-1"	23'-0"	19'-11"	16'-3"
	No. 3	12'-4"	10'-8"	8'-8"	15'-0"	13'-0"	10'-7"	17'-5"	15'-1"	12'-4"

40 psf live load, 10 psf dead load, deflection <360

Species or Group	Grade	2" × 8"			2" × 10"			2" × 12"		
		12" O.C.	16" O.C.	24" O.C.	12" O.C.	16" O.C.	24" O.C.	12" O.C.	16" O.C.	24" O.C.
Douglas Fir and Larch	Sel. Struc.	15'-0"	13'-7"	11'-11"	19'-1"	17'-4"	15'-2"	23'-3"	21'-1"	18'-5"
	No. 1 & Btr.	14'-8"	13'-4"	11'-8"	18'-9"	17'-0"	14'-5"	22'-10"	20'-5"	16'-8"
	No. 1	14'-5"	13'-1"	11'-0"	18'-5"	16'-5"	13'-5"	22'-0"	19'-1"	15'-7"
	No. 2	14'-2"	12'-7"	10'-3"	17'-9"	15'-5"	12'-7"	20'-7"	17'-10"	14'-7"
	No. 3	11'-0"	9'-6"	7'-9"	13'-5"	11'-8"	9'-6"	15'-7"	13'-6"	11'-0"

30 psf live load, 10 psf dead load, deflection <360

Species or Group	Grade	2" × 8"			2" × 10"			2" × 12"		
		12" O.C.	16" O.C.	24" O.C.	12" O.C.	16" O.C.	24" O.C.	12" O.C.	16" O.C.	24" O.C.
Southern Pine	Sel. Struc.	16'-2"	14'-8"	12'-10"	20'-8"	18'-9"	16'-5"	25'-1"	22'-10"	19'-11"
	No. 1	15'-10"	14'-5"	12'-7"	20'-3"	18'-5"	16'-1"	24'-8"	22'-5"	19'-6"
	No. 2	15'-7"	14'-2"	12'-4"	19'-10"	18'-0"	14'-8"	24'-2"	21'-1"	17'-2"
	No. 3	13'-3"	11'-6"	9'-5"	15'-8"	13'-7"	11'-1"	18'-8"	16'-2"	13'-2"

40 psf live load, 10 psf dead load, deflection <360

Species or Group	Grade	2" × 8"			2" × 10"			2" × 12"		
		12" O.C.	16" O.C.	24" O.C.	12" O.C.	16" O.C.	24" O.C.	12" O.C.	16" O.C.	24" O.C.
Southern Pine	Sel. Struc.	14'-8"	13'-4"	11'-8"	18'-9"	17'-0"	14'-11"	22'-10"	20'-9"	18'-1"
	No. 1	14'-5"	13'-1"	11'-5"	18'-5"	16'-9"	14'-7"	22'-5"	20'-4"	17'-5"
	No. 2	14'-2"	12'-10"	11'-0"	18'-0"	16'-1"	13'-2"	21'-9"	18'-10"	15'-4"
	No. 3	11'-11"	10'-3"	8'-5"	14'-0"	12'-2"	9'-11"	16'-8"	14'-5"	11'-10"

40 psf live load, 10 psf dead load, deflection <360

Species or Group	Grade	2" × 8"			2" × 10"			2" × 12"		
		12" O.C.	16" O.C.	24" O.C.	12" O.C.	16" O.C.	24" O.C.	12" O.C.	16" O.C.	24" O.C.
Redwood	Cl. All Heart		7'-3"	6'-0"		10'-9"	8'-9"		13'-6"	11'-0"
	Const. Heart		7'-3"	6'-0"		10'-9"	8'-9"		13'-6"	11'-0"
	Const. Common		7'-3"	6'-0"		10'-9"	8'-9"		13'-6"	11'-0"

Goodheart-Willcox Publisher

Figure 16-20. The size of a joist is determined by the load it will carry and the distance it will span.

when compared to boards or I-beams. Floor trusses are manufactured with boards in a design that results in large open areas that can be used to install ductwork for heating and cooling equipment, electrical wiring, plumbing, or any other equipment requirements of the building design. Both the traditional solid board and manufactured I-beam floor framing require drilling holes for wiring or plumbing installation and a separate area or design for ductwork. Costs vary for these options, and a comparison should be part of the design process.

As a general rule, spacing of joists will be the same spacing as the studs in the wall. In most applications, this will be 16″ on center. Spacing can be increased if the floor area will be used for light storage. Heavy storage may require spacing joists closer than 16″ to support the extra weight. Additional columns or posts may also be required for heavy storage.

Stringers

Stringers are sloped boards that provide support and structure for stairs. A minimum of two stringers are needed for any set of stairs. Often, three stringers are used. In applications where greater strength is needed, multiple stringers made of wood or steel are used. In any application, a stringer will be secured on its upper end to a landing. The lower end will be secured to the lower landing area or floor. Support posts are added between these two connection points as needed. Stringers should provide a step or tread that is deep enough horizontally to provide a comfortable step where most of a person's foot will fit. The depth of the tread is called the run. The vertical distance of the stair step, called the rise, should also be comfortable enough so that a person does not have to take too large or too small a step. The rise and run should be designed to provide the distance for the average person's step. **Figure 16-22** illustrates a typical stringer used in a house.

A ©iStock.com/Banks Photos

B TFoxFoto/Shutterstock.com

Figure 16-21. A—Floor trusses (look up at the ceiling) can be used for floor framing instead of boards. This basement floor has a radiant heating system installed. B—Wood I-beams can be used for floor framing instead of boards.

Bridging

Bridging is a brace of wood or steel installed in floor framing to provide extra support and to distribute loads more evenly. While floors can be constructed without bridging, many local codes require them. The use of bridging has become routine for many construction companies. A bridging component is installed every 8′ of length along a floor joist. This spacing provides for rigidity of the floor. It also acts as a mounting location for sheets of subflooring material when using solid bridging. These components are installed perpendicular to floor joists to increase strength. Three types of bridging are commonly used in floor construction: solid bridging, cross bridging, and ledger boards. **Figure 16-23** shows steel bridging.

Subfloors

When all necessary framing is complete in a floor system, subflooring is installed. Two main purposes of subflooring are to add strength and to provide a solid attachment surface for flooring, such as carpet or tile. Subflooring may be installed prior to wall installation, depending on the framing type being used. Subflooring is commonly made from plywood, but it can also be wood boards or concrete, depending on the design, **Figure 16-24**. The choice of subfloor materials can vary and may depend on what is locally available. Whatever subfloor material is selected, it is most likely available in the standard 4′ × 8′ dimension. Some products are manufactured with a tongue and groove system along the 8′ side. The tongue and groove construction adds rigidity and strength to the subfloor.

Wall Framing

After floor framing is completed, walls will be assembled and attached to the structure. The basic components of a wall are shown in **Figure 16-25**. Walls will be constructed with similar components regardless of the wall type or what function they will serve. In stud wall construction, for example, consistent spacing is important for attaching sheathing or hanging sheetrock. In residential applications, the spacing will be 16″ on center. In this spacing, each 4′ × 8′ sheet of sheathing or sheetrock will attach to more than one stud. This attachment further braces the structure and helps keep the walls in the proper shape and alignment.

©iStock.com/Stan Rohrer

Figure 16-22. The stringer provides support for stairs. This figure shows three stringers used for this set of stairs.

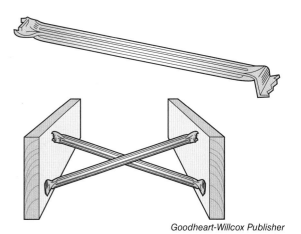

Goodheart-Willcox Publisher

Figure 16-23. Bridging is installed in floor framing to provide extra support and to distribute loads more evenly. Bridging materials can be either wood or steel. This illustration depicts steel bridging.

420   Agricultural Mechanics and Technology Systems

TFoxFoto/Shutterstock.com

Figure 16-24. Plywood subflooring is being installed to the floor framing in this structure.

When any type of wall is installed in a structure, attachment and bracing are very important. Attachment to a floor can be directly to a joist and subfloor. Wall attachment at other points must be performed correctly and securely.

Walls must be square and plumb. *Square* is a term used to describe an element, such as a wall, that is straight horizontally and has sides perpendicular (at a 90° angle) to each other. *Plumb* means that an element, such as a wall, is straight vertically and exactly perpendicular to another component, such as the floor. Consistent use of square subcomponents (foundation, flooring, and walls) makes installation of any additional framing or other components much easier because the materials are typically square when manufactured. During construction, walls will sometimes not be square or plumb and will need adjustments. The following methods can be used to determine whether a wall is square or plumb.

In the first method, a measuring tape is used to measure the diagonal distance of opposite corners. These two measurements should be equal if the wall is square. If the measurements are different, the wall is not square and adjustments should be made. When the wall is square, it must have a temporary brace installed to prevent unintended movement of the components.

The second method is to plumb the wall after the bottom plate is secured to the subfloor. Plumb should be checked to make sure walls are completely perpendicular to the structure on which they

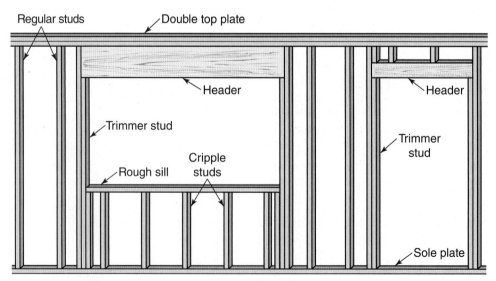

Goodheart-Willcox Publisher

Figure 16-25. Basic wall components include studs, trimmer studs, cripple studs, sill plate, sole plate, header, and top plate.

are installed. To check whether a wall is plumb, a level is typically used against one or more of the studs. The sight is used to make adjustments to the position of the wall. When a wall is determined to be plumb, temporary bracing is installed to hold the wall in place.

Temporary bracing holds framing components in place until more permanent bracing or components can be installed. Sometimes, construction schedules, bad weather, or other unforeseen obstacles delay the project. In such cases, temporary bracing is used for longer periods of time than initially intended. Therefore, temporary braces should be strong and secure enough to hold a building in place until work can be completed. The most secure bracing will extend from the top plate of a wall to the bottom plate of a wall. Some buildings include corner braces to keep structures square and plumb. These braces may be temporary braces that are removed when sheathing is applied. They may also be a permanent part of the structure, **Figure 16-26**.

Christian Delbert/Shutterstock.com

Figure 16-26. Bracing holds the structure level, plumb, and square while the concrete dries. Look closely and you will see that there are many temporary braces located throughout the structure.

Load-Bearing and Interior Walls

Walls will either be load bearing, not load bearing, or have a specific practical function. Interior walls, for example, serve as partitions or dividers between rooms. While these walls extend from the ceiling to the floor just like the exterior walls, their main function is to divide open areas into rooms. Some load-bearing ability is added since the walls have the same attachment points (floor to ceiling) as load-bearing walls, but partitioning is their main function.

A load-bearing wall is a wall that supports a major portion of the building's weight. Load-bearing walls are typically those that are on the outside of a building. If the building is constructed using trusses, there may be only two load-bearing walls because trusses only need support on each end. However, internal walls could also be load-bearing walls, especially if rafters are used for roofing. **Figure 16-27** shows the outside load-bearing walls of a typical house.

Structural members of a load-bearing wall are typically studs, posts, or concrete (blocks or solid concrete). These materials are used to support dead loads,

Christian Delbert/Shutterstock.com

Figure 16-27. The exterior and interior walls of this building are load bearing. Note the use of wood and metal bracing.

environmental loads, and live loads. An inadequate load-bearing member can create structural problems. These problems could range from mild (doors that will not open and shut properly) to severe (roof collapse). Proper planning and design must be performed to eliminate these problems. Construction companies should have engineers on staff who ensure that the materials selected are adequate to support the loads.

Openings

An *opening* is a free space in a wall or other building component that allows passage of people, animals, or objects or where another component, such as a door or window, can be installed. In many structures, openings provide a location to install walk-in doors, windows, garage doors, stairs, and many other practical features, **Figure 16-28**. In many locations, building codes dictate size requirements for openings. For example, the American Disability Association (ADA) recommends that all doors be a minimum of 36″ wide to allow adequate access for wheelchairs or assistive devices for physically impaired individuals. Requirements and codes such as these force builders to design and construct framing structures accordingly.

Jandrie Lombard/Shutterstock.com

Figure 16-28. Openings add to the function of a structure. Framing for openings allows installation of components, such as doors, windows, and ventilation systems. Note the huge headers over these very large openings.

Headers

When an opening is included in a building, the area must be properly braced and structurally supported. Typically, construction for openings requires structural members to be removed. For example, in a machinery storage barn, post spacing is usually several feet apart. In many Morton Buildings™, the posts are set at a 7 1/2′ spacing. If a door that is 20′ wide is needed, two posts need to be removed. This would greatly weaken the structure surrounding the door. To prevent a structural failure, you must install extra bracing and framing systems to reinforce the open area. For a 20′ opening, a header strong enough to replace the strength of the two removed posts would be required.

The most common way to replace strength in a wall is to install a header. Headers are installed to make sure weight held by load-bearing walls is properly supported. The most common type of header for wood construction is made from 2″ × 4″ or 2″ × 6″ boards where the boards are cut to fit the width of the rough opening. A typical header includes two boards that are laminated to a piece of plywood placed between them and an additional board attached to the underside along the longest edge. See **Figure 16-29** for components of a header and **Figure 16-30** for an example of a header supporting a doorway and window openings.

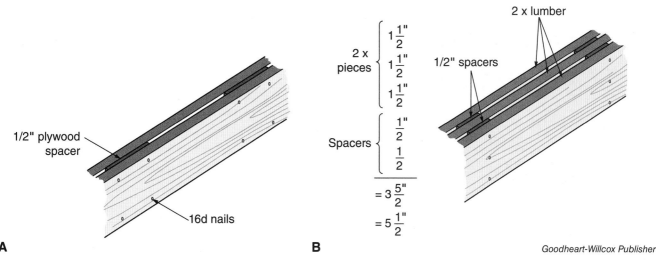

Figure 16-29. A—A header for a 2″ × 4″ stud wall requires two members with ½″ spacers to equal the width of the studs. B—A 2″ × 6″ stud wall requires three members with ½″ spacers between each member.

Headers are available in many sizes. Selection depends on the load the header will support and how wide the opening is. As openings and loads get larger, headers also get larger and may require construction from wider boards and more than two laminated boards. See **Figure 16-31** for an example of an extra-wide door used in a machinery storage barn. Refer to **Figure 16-32** for a chart that highlights the relationship between header spans and the loads they are expected to support.

ChristianDelbert/Shutterstock.com

Figure 16-30. This photo shows multiple examples of basic headers for door and window openings. Note the huge beam on the left that will open up the interior space in this open floorplan.

Marilyn Barbone/Shutterstock.com

Figure 16-31. Farm machinery can be huge and requires large doorways for access. Wide door openings require an extra long header to support the weight of the door as well as wall and roof loads.

While headers are only needed for structural support in side walls, most openings for doors and windows in residential applications are constructed using headers. This construction provides consistent and solid mounting locations for windows and doors. In some specific applications, a header can provide added strength for live loads. An example would be a set of sliding doors installed in the end walls of a machinery workshop or livestock building. In the closed position, the oversize doors need overhead support to stay in proper position. A header must provide the rigid support necessary for proper operation of the doors. See **Figure 16-33**.

Framing Walls

When constructed correctly, wall connections increase the rigidity of a building and provide bracing. Two types of connections are crucial to framing walls: corners and partitions. These connections provide a way for adequately securing two walls together. They also provide a location for attaching finish materials, such as sheathing or sheetrock. To construct corners or partitions, extra studs are needed. These components are

Allowable Header Spans

Header Supporting	Size	Span with 30 lb per sq. ft. snow load		Span with 50 lb per sq. ft. snow load		Span with 70 lb per sq. ft. snow load	
		Building width		Building width		Building width	
		20'	28'	20'	28'	20'	28'
Roof and ceiling	2-2'×4"	3'-6"	3'-2"	3'-2"	2'-9"	2'-10"	2'-6"
	2-2'×10"	8'-5"	7'-3"	7'-3"	6'-3"	6'-6"	5'-7"
	3-2'×8"	8'-4"	7'-5"	7'-5"	6'-5"	6'-8"	5'-9"
Roof, ceiling, and one clear-span floor	2-2'×4"	3'-1"	2'-9"	2'-9"	2'-5"	2'-7"	2'-3"
	2-2'×10"	7'-0"	6'-2"	6'-4"	5'-6"	5'-9"	5'-1"
	3-2'×8"	7'-2"	6'-3"	6'-5"	5'-8"	5'-11"	5'-2"
Roof, ceiling, and two clear-span floors	2-2'×4"	2'-1"	1'-8"	2'-0"	1'-8"	2'-0"	1'-8"
	2-2'×10"	4'-9"	4'-1"	4'-8"	4'-0"	4'-7"	4'-0"
	3-2'×8"	4'-10"	4'-2"	4'-9"	4'-1"	5'-5"	4'-8"

Goodheart-Willcox Publisher

Figure 16-32. Proper sizing of header components will provide adequate support of structural loads. This figure includes the requirements for common headers.

constructed while taking into account the 16″ on center spacing of studs. See **Figure 16-34** for examples of corners and partitions. In addition to the strength added to the wall unit by the connection itself, walls are further secured to each other and strengthened by extending the double top plate to overlap the two connecting walls.

Sheathing

A building is not considered fully braced until sheathing is applied. Sheathing is material used to cover the framing on a building's structure, and it is available in a variety of thicknesses and types. A common type of sheathing is 1/2″ medium density overlay (MDO) panel, **Figure 16-35**. MDO panels are very similar to plywood in dimension and in ability to cover an area. When framing is completed, sheathing is attached directly to the outside of stud walls or to the outside of trusses or rafters. Sheathing serves two main purposes. The first purpose is to cover the framing of a structure and provide the first step of separating the outside environment from the inside environment. The second purpose is to provide an attachment point for any other materials used to insulate or finish a structure. For example, the seams between the sheets of sheathing are sealed to separate the two environments. Then, a synthetic house wrap could be installed before siding is attached. Each of these attaches directly to the sheathing. Roofing materials are installed in a similar method to roof sheathing.

Fabric

In some construction settings, a synthetic fabric wrap can be used to seal seams and increase the effectiveness of insulation. Water vapor can pass through this house wrap material, but water cannot pass through. Air infiltration is also restricted. The ability to dissipate moisture to the outside away from the sheathing improves energy efficiency and control of the internal environment. See **Figure 16-36** for an example of this material on a structure. In some applications, seams between sheathing can be sealed with an air and water barrier material.

Modfos/Shutterstock.com

Figure 16-33. A header must provide the overhead, rigid support necessary for proper function of the sliding doors. The header that supports the large sliding metal door in this photo must be very strong.

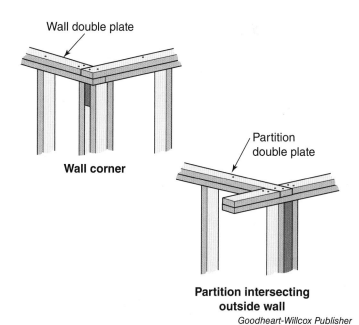

Goodheart-Willcox Publisher

Figure 16-34. Proper connections at corner walls and intersecting walls will ensure a strong overall structure. Double top plates are a key strength feature.

Thinking Green

Insulation

Insulation is one of the most important considerations when designing a structure. Heating and cooling account for 50 percent of energy consumption. Synthetic house wraps dramatically reduce the infiltration of outside air and reduce heating and cooling demands.

Roof and Ceiling Framing

The main function of a roof is to protect the inside of a building from the outside environment. A roof should prevent rain, snow, hail, sleet, and wind events from directly affecting the inside environment of a building. Common roofing materials for residential and agricultural structures include asphalt shingles and steel (metal). However, some owners prefer the visual appeal of wood shingles. A common use of wood shingles is for a house with a log cabin theme or design. Roofing materials have a range of useful life before replacement is needed. In most cases, the life span of a roof ranges from 10 years up to 40 years or more.

A ceiling is the upper interior surface of a room or building. Ceilings can enhance climate and environmental control. Ceilings can be made from a variety of materials, depending on purpose, function, or visual preference of the owner. Ceilings are typically finished during the construction process and are ready to be painted. However, ceilings in some structures, such as basements, farm shops, machinery storage buildings, and livestock barns, may not be installed or finished until a later date.

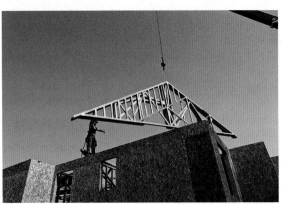

Figure 16-35. Wall sheathing has been installed on this structure, and a roof truss is being lifted into place.

Types of Roofs

Roof types vary widely and depend on several factors. Choosing a roof style may be based on an architectural preference or a purely functional purpose. Refer to **Figure 16-37** for examples of the most common types of roofs. The simplest and most common roof style for modern agricultural buildings is a gable roof.

Rafters and Trusses

Rafters or trusses are framing components that are part of any roof structure. Depending on

Figure 16-36. This house has been wrapped with a synthetic fabric that acts as a barrier to air and water.

the construction style and techniques used, either rafters or trusses will be selected by the architect. When all trusses or rafters are installed, the complete roof structure and associated bracing will be connected and ready for any further roofing material installation, such as asphalt shingles (exterior) or sheetrock (interior).

Rafters are boards that provide the basic framework and slope of a roof. *Ceiling joists* are framing components attached to the bottom of rafters at the point where rafters attach to the outside walls. They are constructed of boards that are typically the same dimensions as rafters. Rafters and joists are typically cut to the proper dimension and length and are installed at the job site.

Trusses are roof structures made of beams, bars, or rods that can be used in place of rafters. They are delivered by truck to a job site where a crane is used to lift them into place. Using trusses speeds the roof construction process because roof trusses are self-contained structural units. Everything needed is integrated into the truss except for any bridging or bracing. The rapid construction possible with trusses limits the time a structure is exposed to the outside environment as compared to building a roof on site with rafters. There are a variety of different types of trusses for different building styles. Refer to **Figure 16-38**. Trusses can be used for a functional purpose (attic room framing) or visual purpose (vaulted or cathedral style ceilings).

Bridging is typically used to fill in spaces between bottom chords. It is also used as a point of attachment for any finishing materials, such as plywood or sheetrock. After bridging is installed, the structure is completely tied together into a strong, integrated assembly.

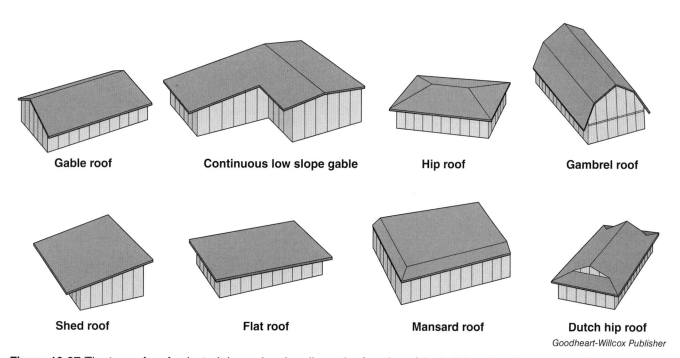

Figure 16-37. The type of roof selected depends primarily on the function of the building. This figure shows the most common roof types used in agricultural and residential applications.

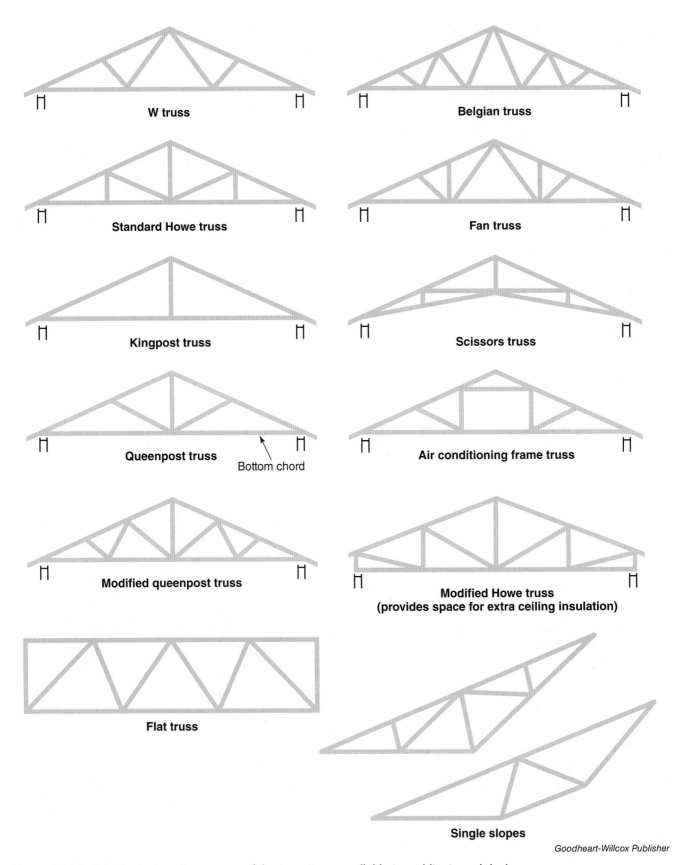

Figure 16-38. This illustration shows some of the truss types available to architects and designers.

CHAPTER 16 Review and Assessment

Summary

- Framing provides the basic skeleton of a building and is usually one of the first steps in construction. Covering the framing structure is called finishing and is commonly performed on residential structures and some agricultural buildings.
- Three types of loads that should be considered during building design include dead loads, environmental loads, and live loads.
- CAD programs can be used to design a building and can include varied information, such as load calculations, lists of materials required, and 3-D tours of completed designs.
- The four types of building structures are pole frame, rigid frame, arc, and stud wall construction.
- The five most common types of framing are post frame and truss, modern braced, western or platform, balloon, and plank and beam framing.
- Flooring adds strength and rigidity to the structure and provides an area for the installation of floor coverings, such as carpet, tile, or wood.
- A floor frame must have an initial mounting location. A sill is a horizontal board placed on the foundation to which vertical members, such as walls, are attached.
- Beams, girders, posts, floor joists, and bridging are used in constructing floors. They provide the strength and rigidity needed to support required loads.
- After floor framing is completed, walls will be assembled. When any type of wall is installed, attachment and bracing are very important. Walls should be square and plumb.
- Wall framing components provide strength to support the weight of building materials and any other stresses the building may encounter. Structural members of a load-bearing wall are typically either studs, posts, or concrete.
- When an opening is included in a building, the area must be properly braced and structurally supported. If posts or studs are removed, a header must be installed to maintain the strength of the wall.
- A variety of roof styles can be used, depending on the owner's preference or the function of the building.
- Rafters or trusses are used to provide roof support and a location for installing roofing materials. Using trusses speeds the roof construction process.

Words to Know

Match the key terms from the chapter to the correct definition.

A. balloon framing
B. blueprints
C. CAD software
D. ceiling joist
E. dead load
F. environmental load
G. finishing
H. framing
I. live load
J. load-bearing member
K. modern braced framing
L. plank and beam framing
M. platform framing
N. plumb
O. pole frame construction
P. post
Q. post frame and truss framing
R. rigid construction
S. sheathing
T. sill
U. span
V. square
W. stud
X. stud wall construction
Y. subfloor
Z. truss

1. The weight of materials used to build a structure.
2. A building method that uses studs that are long enough to extend through two stories.
3. A building method in which all main structural components are assembled before subflooring is installed.
4. A type of construction in which poles or posts are used as the main structural components for walls.
5. A building method in which each level or story of a building is built as a separate unit.
6. The stress caused by or the weight of everything placed on, in, or temporarily attached to a structure.
7. A horizontal board placed on the foundation to which vertical members, such as wall studs, are attached.
8. A vertical board or piece of metal that is the main structural component of a wall in stud wall construction.
9. Material (wood or fibers) attached to framing to strengthen the structure and provide an attachment point for coverings, such as roofing or siding.
10. A frame made of beams, bars, or rods and used in roof or floor construction.
11. A term used to describe an element, such as a wall, that is straight horizontally and has exactly perpendicular sides.
12. A term used to describe an element, such as a wall, that is straight vertically and exactly perpendicular to another component, such as the floor.
13. The basic structural components of a building or the process of assembling these components.
14. A program that allows the user to design architectural, engineering, and construction projects on a computer.
15. A weight or stress caused by a meteorological event (wind, rain, snow, or earthquake) or a condition of the surrounding area, such as dirt pressing against a foundation.
16. A building method used to create large open areas and to show off exposed wood components, such as support beams, for their visual appeal.

17. A building method that results in large open areas inside a building and is the most common type of framing.
18. A type of structure in which vertical boards or pieces of metal (studs) are used as the main structural component in walls.
19. A layer of material, such as plywood, placed over the floor joists.
20. A type of structure in which posts and rafters are assembled as a unit and are open the entire span of a building.
21. Two-dimensional plans that include all dimensions, measurements, and construction details for a building.
22. A framing component made of boards that connects each side of a building where a rafter attaches to a wall.
23. The process of covering the framing structure with a variety of materials, such as metal, wood, or brick.
24. A component of a framing system that must support an amount of weight that is important for the strength of a building.
25. A long, sturdy piece of wood or metal that is set upright in the ground or a structure and often used in agricultural buildings as an attachment point for other structural components.
26. The full extent of something from end to end, such as the complete distance across a building.

Know and Understand

Answer the following questions using the information provided in this chapter.

1. What are three types of loads that must be considered in building design?
2. What are some advantages of using CAD software?
3. What are four main structure types used in building design?
4. What are some examples of buildings made with pole frame construction?
5. What are some advantages of using trusses in roof construction?
6. What is the main advantage of using western or platform framing?
7. What can be done to increase strength and accommodate building loads when using plank and beam framing?
8. What are some materials used for structural members in a floor? What material is typically selected for floors in agricultural structures?
9. How are beams, girders, and posts used for support in constructing a floor?
10. What is the function of stringers in stair construction?
11. What is the purpose of a header in wall framing and how is a header typically constructed?
12. What are the two main purposes of sheathing?
13. What is the most common roof style used for modern agricultural buildings?
14. What are rafters and how are they used with ceiling joists in roof construction?
15. Describe how using trusses speeds the roof construction process and explain how this can be helpful.

STEM Connections and Academic Activities

Example: You are building a house with an attached garage. There will be a storage room in the garage. Since the room will be in the corner, the outside walls will serve as two of the four walls for the room. Therefore, only two walls will be needed in addition to the garage structure. They will be 6' long and 10' long. Stud spacing will be 16" on center and there will be a 3' wide door installed in the 10' long wall.

1. **Science.** Research how much vertical weight can be supported by a common stud with dimensions of $2'' \times 4''$.
2. **Technology.** Draw a set of blueprints for this project. You should concentrate on only an "overhead" view, which will be two-dimensional. You may draw these plans by hand or by using a CAD program. If drawing by hand, use graph paper and draw the plans to scale.
3. **Engineering.** Assemble a wall.
   1. Lay the top and bottom plates on edge and place them close together.
   2. Using a measuring tape, make a mark every 16". This mark will be the center of each stud. Do not forget to mark each end since there will be a stud on the ends of the wall. Since the plates are touching each other, the mark for both the top and bottom plates can be made at the same time.
   3. Place the plates far enough apart to place a stud between them.
   4. Align a stud with each mark so that the center of each stud is aligned with a mark.
   5. Use two to three nails to nail through the plate into each stud.
   6. Attach each stud and the wall is ready to be installed.
4. **Math.** How many $2'' \times 4''$ studs will be needed? Calculate materials for a wall.
   1. Measure the overall length of the wall or wall perimeter. If you measured in feet, you will need to convert the measurement to inches.
   2. Divide the length by the spacing of the studs.
   3. Add one stud to your total number.
   4. Add one stud for each door and one stud for each corner.
   5. The final answer will be the number of studs. If the answer is not a whole number, round up to the nearest whole number.
5. **Social Science.** For the Engineering activity, divide into groups of two, three, or four students. Work together to make appropriate measurements, cuts, and attachments. Make sure to double check each other's work and measurements before making any cuts to boards.
6. **Language Arts.** Write a report describing the procedures you followed, beginning with developing the blueprints. The report should include any procedure you used to complete this task as well as a summary of what you learned from the activity.

Thinking Critically

1. What would be the advantages and, if any, the disadvantages of building with 2″ × 6″ studs as compared with building with the much more common 2″ × 4″ studs. Consider exterior as well as interior walls.
2. What are some of the considerations an architect would need to address when designing an agricultural structure that would house large farming equipment?
3. The gable roof style is the most common roof used in agricultural structures. Why?

CHAPTER 17: Finishing Structures

Chapter Outcomes

After studying this chapter, you will be able to:
- Identify components commonly used in covering and enclosing buildings.
- Describe slopes, underlayment, and construction materials for roofs.
- Describe materials used for covering interior and exterior walls.
- Identify components of a typical window and types of windows.
- List components of doors.
- Explain the purpose of trim for finishing buildings.
- Describe how environmental control and building construction methods affect energy efficiency in a building.

Words to Know

air quality	environmental control	purlin	temperature
build-up roofing	fascia	radiation	thermal pane window
conduction	girder	relative humidity	trim
convection	heat transfer	sash	underlayment
crawlspace	insulation	SEER rating	ventilation
doorjamb	overall rise	shingle	window
door sill	overhang	siding	window frame
energy efficient	prehung door	soffit	wrap

Before You Read

Write all of the chapter terms on a sheet of paper. Highlight the words that you don't know. Before you begin reading, look up the highlighted words in the glossary and write the definitions.

While studying this chapter, look for the activity icon to:

- **Practice** vocabulary terms with e-flash cards and matching activities.
- **Expand** learning with interactive activities.
- **Reinforce** what you learn by completing the end-of-chapter questions.

www.g-wlearning.com/agriculture

Finishing is the final stage of a construction project. Finishing includes roofing, insulating, installing wall coverings, trimming, and painting. Finishing is an important part of the construction process for several reasons. One of the main reasons is that the finishing process separates the outside environment from the inside environment. Any structure will have as its main goal the control of the internal environment, which begins by separating the outside environment from the inside environment. *Environmental control* is managing conditions, such as heat, light, and air moisture, inside a structure or area. A residential or workshop area that is well insulated may have a high degree of control. Simple storage structures with only a roof to prevent rain or wind from damaging agricultural equipment or commodities, **Figure 17-1**, have limited environmental control. The level of environmental control varies depending on the purpose and function of the building.

Another reason for finishing structures is the look, or the visual appeal, desired by the owner. Personal, company, or organizational preferences vary widely as do finish options. Personal preferences, local building codes, building purpose or function, and many other reasons can influence the design and final appearance of a structure. Considerable expense and time may be required during the design and construction process to create a structure the owner will be satisfied with. While any part of a structure can be modified after completion, there will be additional time and cost involved. Changes are more easily performed during framing than in the finishing stages of construction. In many instances, construction of agricultural buildings will be coordinated according to the style, color, and spacing of other buildings on the property. While farmsteads and agricultural businesses are not typically sold on a regular basis, resale value is increased when structures are visually appealing.

©iStock/StraightShooter

Figure 17-1. Some agricultural barns provide adequate protection by having a roof and one or more walls left open to the outside environment.

Covering and Enclosing Structures

To properly enclose a structure and increase its useful life, a covering system must be installed. Several different types of covering products are available. These products increase the rigidity of the framing system and provide a starting point for installing insulation and other materials. *Insulation* is material that slows the movement of heat from warm areas to cool areas.

Sheathing

Sheathing is material that is attached to and covers the main framing structures of a building. In most cases, sheathing is some form of wood product, such as plywood or oriented strand board (OSB), **Figure 17-2**. Most of these products are applied as 4′ × 8′ sheets after framing is completed. Sheathing can also include fiberboard or planking, although these materials are not as commonly used. In post frame and truss buildings, the most common type of agricultural buildings, the sheathing consists of girders (for walls) and purlins (for roofs). Girders and purlins are used to connect posts or trusses together in the same way 4′ × 8′ sheets connect and reinforce studs in stud wall construction. *Girders* attach to each post horizontally and provide attachment points for the outside wall covering, which is typically metal or wood. See **Figure 17-3**. *Purlins* are boards used to connect trusses or rafters, and provide an attachment point for roofing materials. See **Figure 17-4**. Some structure designs include metal strapping that acts as bracing. The strapping is installed in corners or in all outside walls before the girders are installed. Girders and purlins are covered in detail later in this chapter.

Wrap

Wrap is a synthetic material installed over sheathing to provide an additional barrier to separate inside and outside environments. This material is usually very thin and resembles bubble wrap or fabric, **Figure 17-5**. Wrap material provides a small level of insulation, but its primary function is to help prevent air movement out of or into the building. When the wrap has been installed, the structure is ready for the outer covering.

TFoxFoto/Shutterstock.com

Figure 17-2. Sheathing provides an initial covering for a building and also provides a location to attach siding, wrap, or other construction materials. OSB is being installed on the building in this image.

Siding

Siding is material, such as brick, vinyl, or wood, that is placed on the outside walls of a building. Siding serves two main purposes: visual appeal and environmental control. Siding is a great way to add color, style, and an architectural flavor to any building. There are many siding options available to create a desired effect. Owners or builders should choose carefully during the planning and design phase of the project. If there are any local building codes, they should be considered. There are various colors and textures available for any type of siding option. As with framing materials, options may depend on local

©iStock/Ca2hill

Figure 17-3. Girders are a form of sheathing used in many agricultural buildings. Girders are the boards installed horizontally.

Figure 17-4. Purlins are installed to connect trusses and provide a location to install roof materials.

Figure 17-5. Wrap materials are used to increase the effectiveness of insulation.

Figure 17-6. Brick is a common material used for siding on houses. However, it is not a common choice for agricultural buildings.

availability. Local suppliers should be consulted prior to construction.

Another purpose of siding is to provide a barrier between the inside and outside environments. Preventing rain, snow, sleet, hail, and any other weather events from entering into a structure is very important. There are four main types of siding:

- Brick siding.
- Metal siding.
- Vinyl siding.
- Wood siding.

Brick Siding

Brick is a material that is similar in strength to concrete and is available in a wide choice of textures and colors. It is also available in several sizes, but the 3″ × 6″ brick is the most common. Each brick is placed individually by a professional bricklayer in successive courses on the side of a building to create a solid covering. Brick is a common material for siding. Typically, it is not used for agricultural structures due to cost of materials and labor as well as its limited repair options. Primarily, brick is used in residential applications where appearance is an important consideration. See **Figure 17-6**.

Metal Siding

Metal siding is a very common material for finishing agricultural structures due to its long service life, durability, and ease of repair. Many colors are available, and metal siding can be used for both roofing and siding. See **Figure 17-7**.

Vinyl Siding

Vinyl siding is the least durable of siding choices, but it is also the lowest in price. While vinyl will last a long time, extreme or even hostile environments found in some agricultural applications make vinyl siding more appropriate for residential use. See **Figure 17-8**.

Wood Siding

Wood siding is a common choice for many agricultural buildings due to its high durability, ease of repair, and long life. Wood accepts virtually any kind of finish. It may be painted or left untreated, allowing for a natural, weathered look. See **Figure 17-9**.

Roofs

Roofs provide protection for the interior of a structure from the outside environment. They also provide a variety of finishing, material, and appearance options as well. A roof is a part of a building that is not typically replaced very often. When installed properly with the right choice of materials, a roof should last at least 20 years before replacement is needed. Using high quality roofing materials can provide a lower overall cost for a roof in the long run. Weather damage to a roof is one of the major causes of deterioration. Inspections should be performed routinely to check for weather damage or any other deterioration problems that prevent a roof from performing adequately.

Roof Slopes

Roofs must have an appropriate grade or slope for adequate drainage to occur, **Figure 17-10**. Even roofs that are considered flat will have a slight amount of slope (approximately 1" per foot of length) to allow for drainage. In some designs, roof slope may have an impact on appearance or building use, such as the amount of usable space in an attic. However, the main function of a roof's slope is for proper drainage.

Slope is expressed as rise (increase in vertical distance) per foot of run (horizontal distance covered). As noted in blueprints and building plans, these two values are always expressed in inches, **Figure 17-11**. This basic definition of slope can also be used to describe the slope of floors, roads, drainage ditches, waterways, or other structures that must have a slope for proper function or drainage. To further describe slope, *overall rise* is the measurement of how many inches or feet of rise are included in a roof from eave to peak. In other applications, such as building roadways, installing concrete floors, or laying drainage tile, this same measurement may be called *drop* or *fall* when used to describe the slope of the area. In any application where slope is an important consideration, the main goal always is efficient drainage.

Underlayment

Underlayment is a material, such as tar paper, that is installed before final roofing materials to help prevent water from leaking into the building. Tar paper, the most common underlayment material,

Goodheart-Willcox Publisher

Figure 17-7. Metal siding is commonly used for agricultural buildings.

Bonita R. Cheshier/Shutterstock.com

Figure 17-8. Vinyl siding is a common siding option for residential applications.

Goodheart-Willcox Publisher

Figure 17-9. Wood is a common siding material for many agricultural buildings.

Figure 17-10. A roof's slope helps provide proper drainage.

comes in rolls. Roofers unroll the tar paper, lay it down, and staple it to the roof sheathing when OSB or similar material is used, **Figure 17-12**. Installation should begin at the eave or lower area of a roof. Each subsequent strip of underlayment will be higher up on the roof than the one before. This technique ensures that the layers of tar paper will overlap and prevent water from contacting the sheathing.

Roof Materials

Roof materials must provide durability and adequate service life. They must resist severe conditions, such as heat in summer sun and accumulated snow in winter. Roofing is directly affected by the service environment, which may include rain, snow, sunshine, wind, hail, and other environmental factors. *Temperature* is the degree or amount of heat in something, such as the air, an object, or a substance. The temperature difference between weather seasons, or even temperature fluctuations during a single day, causes roofing materials to expand and contract. Any movement due to expansion or contraction can affect roofing material fasteners. The effect on fasteners can be observed in roofs that are several years old. For example, screws or nails will occasionally become loose or come out of a metal roof. These screws or nails must be replaced or reattached.

A roof should provide protection from the outside environment by having an adequate water seal. For example, during a rain storm, water collects on a roof and then runs off. While the water is on the roof, the roofing material must prevent the water from leaking into the building. Leaking water can cause numerous problems within a structure. If the leaky area is not repaired, it can lead to major problems and structural issues. There are different methods for creating and maintaining a good seal, depending on the type of material used for the roof. Common seal methods are discussed in each of the three roof material sections that follow. The three main types of roofing material are:

- Metal roofing.
- Asphalt roofing.
- Wood roofing.

Metal Roofing

Metal roofs have gained popularity in recent years for use with agricultural and other types of buildings. Service life is comparable with other roofing materials. Differences include color availability, appearance,

Inches of Rise	Slope
3	3:12
4	4:12
6	6:12
8	8:12
12	12:12
18	18:12
24	24:12

Goodheart-Willcox Publisher

Figure 17-11. This chart includes the rise in inches per 1 foot (12 inches) of run for roof construction. As rise increases, the slope also increases. As inches of rise decrease, roof slope and drainage also decrease. A roof with inadequate slope will not drain properly.

SAE Connection

SAEs and CDEs—The Finishing Touch

A roof without shingles or a wall without siding would be like a car without tires. All components must be in place before a product can be considered finished. All of the finish components of a building, such as insulation, roofing, siding, and other materials, come together to bring completion to the construction project. The finishing process can look slow, especially compared to the speed in which wall and roof framing is completed. Nevertheless, framed roofs and walls are not complete, and it is very important to include those finishing elements that provide protection for the structure and give it a finished look.

The same principle applies to your classroom experience in agricultural education. Supervised agricultural experience (SAE) projects and career development events (CDEs) are vital components of agricultural education, and along with classroom/ laboratory activities, they provide the finishing touch for your education. Take the time today to research opportunities for SAEs and CDEs. Also, don't forget to fill out the proficiency application that best fits your SAE program.

Brandt Bolding/Shutterstock.com

and sound transmission. See **Figure 17-13** for common colors. Metal is currently the material most commonly used with agricultural buildings. Metal roofing panels come in stock thicknesses of 18, 20, 22, and 24 gauge with widths of 27″ through 46″. One supplier offers a panel that is 42″ wide and available in 8′, 10′, and 12′ lengths. The same supplier offers a panel 27″ wide in those same lengths. However, the panel can be up to 20′ long.

Metal roofing panels are secured with either nails or screws, depending on the manufacturer or the building's design. Since fasteners punch a hole in the roofing when installed, a rubber washer is used to provide a tight seal around the fastener (nail or screw), **Figure 17-14**. Care must be taken during installation to ensure that fasteners are properly installed and are not too loose or too tight. If the rubber washer is deformed during installation, there is a greater chance for a leak to develop.

TFoxFoto/Shutterstock.com

Figure 17-12. Underlayment helps to protect a roof structure. Note how one layer overlaps the one that came before.

-Taurus-/Shutterstock.com

Figure 17-13. A wide variety of colors and style options are available for metal roofing and siding materials.

-Taurus-/Shutterstock.com

Figure 17-14. Proper installation of fasteners helps to ensure a leak does not occur at the attachment point.

Asphalt Roofing

Asphalt roofing is made from the same kinds of asphalt materials that are used as pavement for roads. These materials are soft (pliable) when warm and brittle at cold temperatures. ***Build-up roofing*** consists of alternating layers of rolled asphalt and roofing tar. This material is used primarily for roofs that are flat (those with a slope of 1/4″ per foot to 1/8″ per foot). The lifespan of the roof depends on the number of built-up roofing layers. To ensure a proper bond, heat is applied to the roof. Heat causes the asphalt and tar to meld together and form a waterproof bond. See **Figure 17-15**.

A *shingle* is a roof covering made of asphalt or wood and installed in overlapping layers. Asphalt shingles are similar to rolled asphalt roofing; however, they are manufactured and packaged in 1′ × 3′ strips. Some shingles have tabs, which means that the shingle is spilt into sections on one side. If the shingle has tabs, each tab will be 12″. These are called 3-tab shingles. Shingles made without tabs are solid across the exposed edge. A hole is punched in a shingle during installation in the same way as with metal roofing material. Nails are most commonly used to attach shingles to a roof. Since a hole in the material is created, leaks are possible. However, the attachment point is covered by the next row of shingles. Exposure of a typical 12″ tall shingle is 5″. The configuration offers adequate protection from weather for a roof with an average slope. See **Figure 17-16**. A second measure of protection when installing shingles is a tar strip on the back side of each exposed edge. This tar strip helps to seal the shingles to the previous layer or row. Shingle installation should only be done in weather warm enough for the tar strip to become sticky and seal to the previous row of shingles.

Thinking Green

Roof Shingles

Through advances in manufacturing technology, asphalt shingle products are now more environmentally friendly than ever. Computerized equipment has improved roofing production efficiencies, resulting in less material waste. Shingles removed from roofs can be recycled and used to pave roads across America. Solar reflective shingles are now available and can help reduce energy use.

Dimitri Kalinovsky/Shutterstock.com

Figure 17-15. Build-up asphalt roofing consists of alternating layers of rolled asphalt roofing and liquid tar. Heat is used to seal each layer properly.

Wood Roofing

Wood shingles are a popular roofing option and provide a very distinct appearance, **Figure 17-17**. Wood shingles have a long service life and can last up to 50 years if installed properly and used in the right environment. The main durability issue for the life of a wood shingle roof is water damage. The shingles should be kept as dry as possible. A key requirement for a long service life is that the slope of the roof is steep enough to provide adequate drainage. This is obviously a challenge in wet weather. However, a roof with a steep slope will drain water quickly. Wood shingles do not have a sealing option as do metal or asphalt materials. Roofs using wood shingles rely on adequate underlayment, steep slopes for quick drainage, and a greater shingle overlap to prevent water from getting into attachment points.

Overhang

Overhang is the part of a roof that extends beyond the edge of the outside wall along the roof line. Overhang provides some protection to the structure at the point where the roof members attach to the walls. This area also contains trim (discussed later in this chapter) and often has fascia and soffit. *Trim* is a nonstructural building component installed in a structure to

Christina Richards/Shutterstock.com

Figure 17-16. To prevent leaks in an asphalt shingle roof, nails should be placed where they will not be exposed and above the tar strip. Note how the shingles overlap.

Figure 17-17. Wood shingles are another option for residential roofing material.

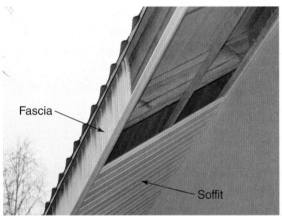

Figure 17-18. This image shows an unfinished roof overhang including the fascia board and the area for soffit.

Figure 17-19. Vented soffit allows air to flow through and up into the attic area.

enhance appearance (such as a baseboard or crown moulding) or to help seal areas where water or air can penetrate a building. *Fascia* is trim made from vinyl, metal, or wood that is installed on the vertical edge of the overhang on a building. *Soffit* is trim placed underneath the roof structure in the space between the fascia and the edge of the building. See **Figure 17-18**. Some of the soffit is usually vented, **Figure 17-19**, which helps increase airflow in attic areas.

Girders and Purlins

Most modern farm and agricultural buildings are constructed using similar techniques. The most common type of design for agricultural buildings is the post frame and truss. This type of building uses posts in combination with trusses or rafters to provide the main framing of a structure. Often, the spacing of these structural components does not match the width of the materials applied for either the roof or siding. As a result, a solid surface necessary for the installation of siding or roofing is lacking for a few feet. To provide a consistent location for installation and provide additional bracing, girders and purlins are used.

Girders and purlins essentially accomplish the same functions—to add strength to the structure and provide a place to attach other materials, such as shingles. Their location, spacing, and orientation varies depending on where they are needed and building requirements or codes. Girders are used on walls and attach to each post in the post frame and truss structure. They are installed horizontally and at a vertical spacing of approximately 24″ on center, although manufacturers' design requirements can differ, **Figure 17-20**. Girders can also be used to support building components, such as a loft, floor, or roof. See **Figure 17-21**. Upon the installation of girders, a building will be braced. Corner braces are usually added also. However, when sheets of metal siding are installed, a greater amount of bracing, support, and rigidity is achieved. These three things are important for any building to support itself and any loads that may be encountered. Girders used for siding are generally 2″ × 4″ boards, which are attached to posts. These boards allow the siding to be attached using either screws or nails. Since sheets of siding are on the outside of the building, they are exposed to any weather events and should function as a barrier to the outside environment. Because each screw or nail used to attach the siding makes a hole in the metal, a proper seal in these areas is crucial. A leak can ruin a building in a short time.

The best location for a fastener is in the middle of any girder. During the installation process, a chalk line is very useful for marking the location for a fastener. When chalk is applied, the installation team has a clear mark to follow when fastening siding to the girders, **Figure 17-22**.

Purlins provide the same basic function as girders. They are installed across trusses or rafters and provide an attachment location for metal roofing. Purlins can be attached using the 3 1/2″ edge or, for added strength, the 1 1/2″ edge of a board. In northern climates, the 1 1/2″ edge will be used due to greater snow loads. Purlins are 2″ × 4″ or larger, depending on strength and load requirements. See **Figure 17-23**.

Walls

Walls provide several functions for a structure. As discussed in Chapter 16, *Framing Structures,* one of the main functions of a wall is to provide strength in a variety of load-bearing situations. Another important purpose of wall is to separate the inside environment from the outside environment and to help limit heat transfer. *Heat transfer* is the movement of warm air into an area with cool air. To help maintain an

©iStock.com/zhengzaishuru

Figure 17-20. Girders are horizontal boards that function as sheathing for walls of agricultural buildings.

Goodheart-Willcox Publisher

Figure 17-21. Girders are used to support several different types of areas in a building. This image shows how a laminated girder is used to support a floor for a loft in a farm shop. Girders for roof support are at the top of the image.

David Desoer/Shutterstock.com

Figure 17-22. A chalk line, such as the one shown here, can be used to mark the middle of a girder before installing siding.

Figure 17-23. Purlins connect trusses and are installed on their 1½" edge in this building.

environment inside a structure, insulation is installed when required. Heat transfer and insulation are discussed in greater detail later in this chapter. Wall construction components allow insulation to be easily installed. They also provide an attachment surface for the easy installation of finishing materials, such as metal, plywood, or sheetrock. Depending on the desired look, finish materials can be painted as a last step to complete a project. Walls are classified as either exterior or interior.

Exterior Walls

Exterior walls are those that are located along the perimeter of a building. The majority of building loads are supported by exterior walls. These walls also directly contact the outside environment. Proper finish procedures, such as wrapping, covering the outside surface with sheathing, and installing siding or brick are extremely important. See **Figure 17-24**.

Interior Walls

Interior walls are those that are located in the interior of a structure. They may also be called partition walls since they divide a structure into smaller spaces. In some applications, such as when roof trusses are used, interior walls are not load bearing. In cases such as these, their function is primarily to divide the inside space into rooms, **Figure 17-25**. Insulation is not usually required in interior walls to control the environment. However, insulation may be installed to reduce sound transfer between adjoining rooms, which is a secondary function of insulation.

Materials for Walls

Material selection for finishing interior wall surfaces depends on the function and the desired look. Materials for finishing the interior of a building include many of the same types of materials used to finish the exterior of a building. The most common materials for finishing the interior walls of a building are wood, metal, or sheetrock. Since the interior of a building is either lived in or used more extensively, the materials should be selected specifically based on their intended use. When finishing with any material, keep in mind that framing

Figure 17-24. Proper finishing procedures, such as installing brick, are important for exterior walls.

Chapter 17 Finishing Structures 447

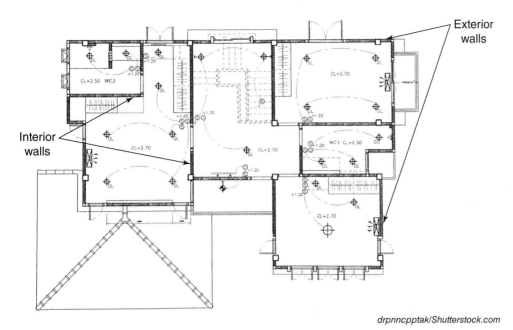

drpnncpptak/Shutterstock.com

Figure 17-25. Interior walls have a main function of dividing a building into smaller rooms.

components will be covered and can no longer be seen or accessed unless special access is installed.

The interior of confined feeding facilities, such as ones used on chicken farms, are typically lined with a wood material, such as oriented strand board (OSB). OSB is a material that is manufactured in sheets that are typically 4′ × 8′ and similar to plywood. OSB is constructed of wood strands that are glued and pressed into various thicknesses. Plywood is also constructed of wood, but it is made by gluing together plys or layers of wood instead of wood particles, **Figure 17-26**. OSB and plywood materials provide durability in heavy service or aggressive use areas and at a more affordable cost than some other materials. While these materials can be painted, there is no specific advantage to painting other than visual appearance. Installation is simplified since there are no other steps in the finishing procedure. In other applications, such as in a repair shop, a painted wood surface may be desirable.

An interior lined with corrugated metal may be desirable in applications where a more detailed, enhanced, or specific appearance is needed or wanted. See **Figure 17-27**. Metal that is used for covering the outside of a structure may also be used for interior spaces. In some structures, such as a post frame and truss building, girders are needed on the interior of the walls to provide an area for installation of the wall covering.

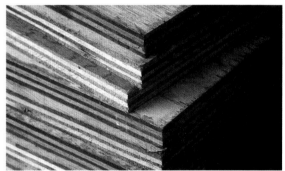

Noppharat46/Shutterstock.com

Figure 17-26. Plywood is a material that is used for a variety of finishing options. The outside layer can be manufactured with a variety of qualities or appearances.

©iStock.com/amandine45

Figure 17-27. Building interiors finished with metal have a unique look.

448 Agricultural Mechanics and Technology Systems

Ernest R. Prim/Shutterstock.com

Figure 17-28. Sheetrock provides an affordable finishing option in residential and office areas.

Sheetrock provides a very affordable option for finishing the interior of a building. However, the drawbacks of sheetrock include a lower durability and less resistance to moisture compared to steel and wood. For this reason, sheetrock is primarily used in residential or office applications, **Figure 17-28**.

Windows

A *window* is an opening in a wall or roof that may be fitted with glass or another covering, allows light and air to pass through, and provides a view of the outside. Windows are used in buildings for several reasons. Windows let natural light into a building, allow airflow when open, and enhance the look or architectural style of a building. Many options are possible when considering window installation. Size, design, style, and the number of windows should be considered during the planning stages of a construction project.

Window Components

Windows are usually shipped from the manufacturer fully assembled and ready for installation. Although the assembly stage had already been completed when the window arrives, knowing the parts of a window is important.

Window Sash

A window *sash* is the panel or bars in a window that hold the glass in place. The type or name of a window is determined by the sash and the way the sash functions. Sashes may be installed to stay in a fixed position, slide, or swing on hinges.

Window Frame

A *window frame* is the part of a window that provides the structure for the sash to work properly and allows the window to be installed into a building. See **Figure 17-29** for a diagram of window components. A window frame consists of a sill and jambs. A window sill is a horizontal component that is located on the bottom of a window. A sill is installed or shaped so that water will shed properly to prevent

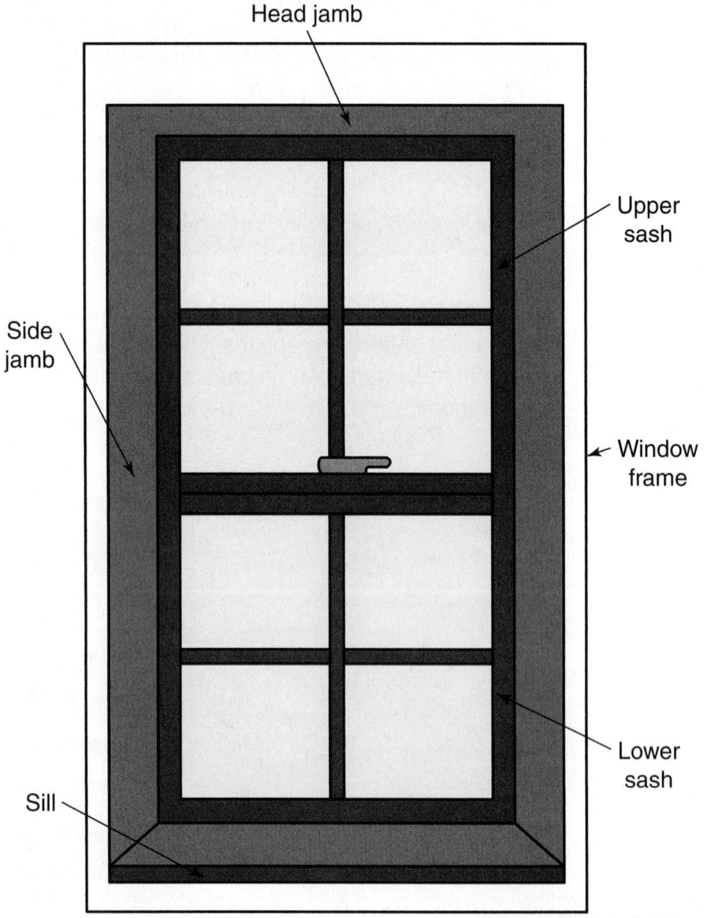

Goodheart-Willcox Publisher

Figure 17-29. This figure shows components of a typical window.

Copyright Goodheart-Willcox Co., Inc.

leaks inside the building. A head jamb is the top of a window frame. Side jambs are the vertical frame components. Depending on the type of window, a groove or track may also be installed on the inside of one of the frame components to slide or open by a hinge.

Window Types

A variety of window styles, sizes, and designs are available for architects and designers to choose from. Many windows have optional design features, such as decorative grids, folding parts for cleaning, or screens to keep out insects when windows are open. Selection of windows should be made during the design stage of a project. The wall framing process is affected by the framing needed for windows. For this reason, decisions regarding windows should be made early in the planning process.

Fixed Windows

Fixed windows are those that do not open or close. When secured, these windows include large panes of glass and are installed as one unit. See **Figure 17-30**.

Single and Double Hung Windows

Single and double hung windows are a popular choice. They originated in England and have been in use in North America since the days of the first colonial settlers. These windows have at least one sash that slides vertically to open and close. A single hung window is a window that has the top sash fixed and the bottom sash slides. A double hung window is made so that both the top and bottom sash can slide. In newer windows, both sashes will also release and rotate to the inside of the structure, which helps make cleaning easier. See **Figure 17-31**.

Casement Windows

A casement window is a window unit that is hinged vertically and typically swings outward. These windows are usually operated by a hand crank or lever. See **Figure 17-32**.

Sliding Windows

Sliding windows are those that can be moved horizontally. Essentially, these windows operate the same as single and double hung windows except they slide horizontally instead of vertically. See **Figure 17-33**.

Maria Dryfhout/Shutterstock.com

Figure 17-30. Fixed windows, such as those in the front of this barn, do not open or close. In structures such as this one, large areas of glass can be unobstructed for viewing the outdoors or letting light into a room.

©iStock.com/beright

Figure 17-31. This double hung window allows both the top and bottom sash to open. These windows are also equipped with sashes that fold inward for easier cleaning.

Figure 17-32. Casement windows swing on hinges vertically.

Figure 17-33. A sliding window opens by one or both sashes sliding horizontally.

Figure 17-34. Awning or hopper windows operate on a hinge at the top or bottom of the window.

Awning and Hopper Windows

Awning windows and hopper windows are designed to fold on a hinge as they are opened. The hinge is attached at the top or bottom of the window and allows the window to fold to open and close. See **Figure 17-34**.

Skylight and Roof Windows

Skylights and roof windows are installed in the roof of a building. See **Figure 17-35**. These windows allow extra light into a room and allow heat to be vented to the outside. These windows can also allow warmth from the sun to enter the building. A disadvantage of any type of window installed in a roof is the potential for leaks and subsequent water damage to a structure as the roofing material deteriorates with age.

Glass in Windows

Windows use glass as a primary component. A piece of glass in a window that a person can look through is sometimes called a window light or pane. Depending on the design of the window, there will be one or more lights. The thicknesses of glass in windows varies, depending on how large the windows are and the style or function of the window.

Safety glass is used when required by building codes or when added safety is a concern. Most windows for residential use do not contain safety glass and present a safety concern when broken. Ordinary window glass remains in large and sharp pieces when broken. Safety glass shatters into countless small pieces when broken and makes these windows safer to clean or work with.

Thermal pane windows are windows that have two or more glass panes separated by a space filled with a vacuum, air, or gas. R-value (the measure of resistance for heat transfer) of windows is important because windows are the primary way heat is lost from the inside of a building. Single pane glass windows are constructed by using only one light of glass. By trapping air or gas between two panes of glass, heat transfer from one side of the window to the other is reduced. Some thermal pane windows have argon gas in the space between panes. Argon gas is used because it transfers heat at a slower rate than air. Windows may also have materials that reflect heat to keep heat inside a building in winter and repel heat from the sun in the summer. See **Figure 17-36**.

Thinking Green

Thermal Pane Windows

Thermal pane windows have two or more glass panes separated by a vacuum or a gas. Since it is difficult to obtain a vacuum between glass panes, the space between is filled often with insulating gases. This reduces the heat transmission. During summer, the windows block heat entering into the house. During winter, heat from inside is prevented from going outside. This helps save energy used to cool and heat the house.

Installing Windows

Proper installation of windows allows them to function as designed and prevents air or moisture from entering the structure around the windows. Most windows are installed after sheathing and wrap materials have been attached to the building. Windows are installed from the outside of the structure. After initial attachments, other items, such as flashing, trim, and caulk, are used to help seal the attachment areas and installation points to prevent wind and water from penetrating past the window unit and getting into the wall or inside the structure. See **Figure 17-37**.

ariadna de raadt/Shutterstock.com

Figure 17-35. Skylights or roof windows are located in a roof to allow natural light into a building.

Doors

Doors allow passage from the outside and through the interior rooms of a structure. Door installation is in many ways similar to window installation. Proper fit during the installation allows doors to operate properly and have a long service life. Door components consist of the door and the door frame. The frame contains a *door sill*, which is found at the bottom of the unit. The sill is the part that a person steps on or steps over when walking through a door. A *doorjamb* is the door frame component that the perimeter of a door rests against when closed. There is a jamb across the header area and vertically on each side of the door. The sill or lower jamb may be adjustable to make the best fit after the door unit is installed. Proper fitting and adjustment of a door is crucial to ensure that sticking, hanging, or drafts are not an issue. See **Figure 17-38**. Most doors used in agricultural or residential applications are *prehung doors*. Prehung doors are purchased as a complete unit with a door and frame with hinges already installed. Installing the unit consists of placing

Coprid/Shutterstock.com

Figure 17-36. Thermal pane windows contain air space and components to reduce heat transfer. The space between the window panes can be filled with a gas that slows heat loss in winter and heat gain in summer.

TFoxFoto/Shutterstock.com

Figure 17-37. Many windows are installed from the outside of a structure and require attachment areas to be properly sealed. Note how the house wrap material was installed before the window.

Dimitri Kalinovsky/Shutterstock.com

Figure 17-38. Doors must be fitted and adjusted properly when installed.

the unit in the rough opening of a structure from the outside. The exterior casing will then rest against the sheathing and final attachment and adjustments can begin.

Trim

Typically, trim is the last component attached when finishing a building. Functionally, trim is one more way to prevent water from getting into the framing of a structure. Trim will seal seams between components, such as the seam between a roof and walls. Another important purpose of trim is to enhance the appearance of a room or the outside of a building. Trim is available in a variety of widths and finishes (painted or stained). Trim can be selected according to the owner's preference or for the desired architectural look. See **Figure 17-39** for examples of various types of trim.

Energy Efficiency

As energy costs continue to rise, energy efficiency has become increasingly important. Conserving energy is important in homes, offices, farm shops, livestock buildings, and many other types of buildings. *Energy efficient* is a term used to describe a low use of natural resources to heat, cool, use appliances, or operate equipment inside a building. A building that is energy efficient will result in lower operating costs and help conserve resources. One aspect of conserving energy is efficient use of electricity. Many appliances have a rating for energy efficiency. This rating is printed on information tags attached to new appliances. The **SEER rating** (seasonal energy efficiency ratio) is a measure that describes the energy efficiency of electrical equipment or appliances. It can be used to estimate the monthly or annual cost of operating electrical appliances or equipment. The higher the SEER rating, the more energy efficient the device.

Environmental Control

Energy efficiency can be achieved by using energy efficient appliances or equipment and through the use of insulation in buildings. Environmental control is the driving force behind structure designs and insulation plans. The main purpose of environmental control is to

Figure 17-39. There are many types of trim available. A—Composite trim is being installed around a window before siding is installed. B—Metal trim is available in colors that complement the siding.

manage elements, such as temperature, *relative humidity* (the amount of water vapor in the air), or air quality. *Air quality* is a measure of the degree to which air is free of pollution or other harmful substances. Total control is not possible due to required air exchanges or heat losses. However, control can be improved by using insulation effectively. Control of environmental factors inside a building can be done either for comfort, such as in a home or office, or for a specific function. In agricultural buildings, for example, humidity must be monitored in confined feeding facilities. Humidity in the air combined with natural building moisture, respiration and body heat produced by animals, and moisture produced by cooling or watering systems increase the moisture levels and make breathing difficult for animals. If not corrected, animals may not be able to breathe and, in extreme or uncontrolled conditions, will die. For these reasons, confined feeding facilities have exhaust fans to control humidity. These fans can also help manage other factors for the living conditions of animals, such as air temperature and air quality. Environmental control varies depending on the use and function of the building. Four factors affect environmental control:

- Building construction.
- Heat transfer.
- Insulation.
- Ventilation.

Building Construction

During the construction stages of a building, care must be taken to properly seal all areas of the structure from the outside environment. In agricultural applications, this includes sealing the area to the point where mice and other vermin cannot easily enter. Properly wrapping or sealing the structure is also important. This sealing process is performed after any sheathing is installed. The sealing material functions as an initial barrier to the outside environment. Wrapping or sealing may function as the sole barrier if insulation is not used inside the structure. The main purpose of properly sealing a structure is to enhance the ability of the structure to maintain the desired internal environment. Any unnecessary airflow or temperature migration will affect the overall environment and make controlling it more difficult. Methods and materials used in construction, such as sealing and wrapping a structure, can influence whether the building will be airtight (able to keep air from passing through). An airtight building can have more efficient environmental controls than buildings that are not airtight. Air spaces, cracks, or areas where building components are not attached well can all become problem areas. These areas should be addressed by using glue, caulk, or tight fitting components if sealing and wrapping do not seal these gaps properly.

Heat Transfer

Typically, when framing is complete or nearly complete, insulation is installed. Insulation will be covered with sheathing, sheetrock, or other coverings to finish the structure. Insulation slows down or interrupts undesirable heat transfer—the movement of warm air into an area with cooler air. Proper management of heat transfer is critical to maintaining environmental control inside a building. There are three types of heat transfer that can affect environments: conduction, convection, and radiation. See **Figure 17-40**.

Conduction is a type of heat transfer that occurs through contact. As two parts, components, or materials touch each other, the warmer component transfers heat to the cooler one. An example of conduction would be a hot burner on a stove transferring heat to a skillet placed on the burner.

Convection occurs when heat is transferred from one area to another through a gas or liquid. When discussing the control of inside environments, convection through a gas (air) is of greatest concern.

Managing air temperature is a major part of controlling the comfort level of living spaces in residential settings. When adjusting a thermostat for optimum comfort, temperature is the condition that is affected.

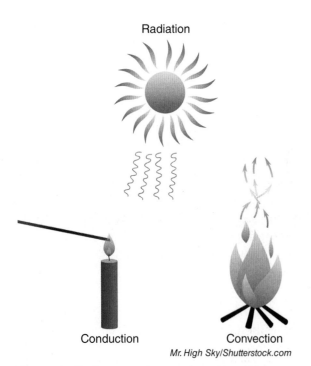

Mr. High Sky/Shutterstock.com

Figure 17-40. Heat can be transferred by radiation, conduction, or convection.

Special characteristics of air are considered when designing and building. Today, the standard inside ceiling height for many residential structures is 8′. However, many older homes and public buildings were originally constructed with ceilings taller than 8′. Before air conditioning systems came into common use, many buildings used 10′ or 12′ ceilings to help keep indoor temperatures cooler in warm seasons. Since warm air rises, room temperature in these buildings was cooler in the living or work areas during warmer parts of the year. Many renovations in older buildings include adding false ceilings to lower the ceiling height. A typical false ceiling includes a support system installed to suspend ceiling materials, such as lightweight panels, and lower the height of a ceiling, **Figure 17-41**. Having this lower ceiling height means that the room has a smaller amount of air to heat or cool. Another option is to use ceiling fans to circulate air. This option often is used in modern construction when a vaulted or cathedral ceiling is used, **Figure 17-42**.

stockofnepal/Shutterstock.com

Figure 17-41. A false ceiling can provide a way to economically cover a ceiling and to lower a ceiling if needed.

In agricultural use, a tall ceiling is sometimes used to help control the living environment for animals. Some structures, such as equine stall barns or poultry barns, commonly have taller ceiling heights in the living areas to move warm air away from animals. By allowing warm air to move up and away from the animals, greater control of the living environment can be achieved. This is especially important in confined feeding operations. Difficult breathing conditions for 30,000 chickens, for example, can be a very big problem for a person managing a flock, **Figure 17-43**.

Radiation is a transfer of heat energy through heat waves traveling from an emitting heat source to a person or object. A camp fire, heat lamp, and the sun's rays are examples of heating by radiation.

Insulation

Insulation plays an important part in environmental control inside a building. Insulation is used to fill spaces or voids in areas of a building, such as a wall or ceiling, that separate the inside environment from the outside environment. Examples of voids are the spaces between studs, posts, rafters, or trusses, **Figure 17-44**. Insulation resists heat transfer by preventing air movement. When the insulation material traps air inside wall spaces or voids, heat transfer is greatly reduced. A building that is well insulated will have less change in the inside temperature when

Phong Krit/Shutterstock.com

Figure 17-42. Ceiling fans are used to help circulate air, especially when ceilings are open, vaulted, or cathedral style.

Figure 17-43. Tall ceiling heights are useful in confined feeding operations to help move heat up and away from animals in their living areas.

the outside environment changes. The trapped air maintains a low amount of heat transfer provided there are no air leaks, drafts, or breezes to move the air. Proper construction of a building will keep air infiltration problems at a minimum.

Insulation Requirements

Some materials are better insulators than others. Each type of insulating material has a unique resistance to heat transfer. Materials are rated by their resistance values for heat transfer (R-value). These R-values are important to consider when designing a structure. The greater the R-value, the better the material will insulate the structure. R-value can vary by the material's thickness (value per inch) or by the material as manufactured (the inherent insulating qualities of the material itself). For materials that have a value per inch, there is an advantage to filling as much space with insulation as possible. For example, if stud walls are 3 1/2″ thick (the actual thickness of a standard 2″ × 4″ stud), the full 3 1/2″ of space should be filled with insulation to provide maximum insulating values. As energy costs continue to rise, builders may choose to use more insulation. Constructing stud walls using 2″ × 6″ boards allows for 2″ of additional insulation. While this number may seem small, the cost savings in heating and cooling quickly pays for the extra construction costs. Many agricultural buildings in the post frame and truss design are built using 2″ × 6″ posts, which provides approximately 6″ inches of insulation area in the walls.

Figure 17-44. A—Spaces or voids inside walls or ceilings provide a place for insulation. B—Adding insulation will make a structure more energy efficient and keep heat transfer to a minimum.

Ceilings and windows are typically the areas of greatest heat loss in a structure, **Figure 17-45**. Because heat rises, ceilings require more insulation than other parts of a structure to keep warm air from escaping. As structures age, they may need more insulation installed in the ceiling space. This addition may be needed due to a small amount of insulation used during construction (especially with older structures) or due to insulation settling over time and causing a reduced area for trapping air space. R-value requirements for ceilings should be approximately double that of walls to provide the most efficient management of heating and cooling resources.

©iStock.com/gabort71

Figure 17-45. The windows in this infrared image show heat loss.

Types of Insulation

Insulation is manufactured in a variety of forms. The choice of a particular type is dependent on a variety of factors, including location used, function, and ease of installation. There are five main types of insulation:

- Loose fill insulation.
- Batt or blanket insulation.
- Rigid insulation.
- Reflective insulation.
- Foam-in-place insulation.

Safety Note

Always wear a respirator and enough clothing to cover exposed skin when working with insulation.

Loose fill insulation is usually sold in bulk (large quantities) in a bale or a bag. The actual insulation material is not formed or attached to anything as with other types of insulation. In the bag or bale, there are only particles of the insulation material. See **Figure 17-46**. This material is made of rock wool, fiberglass, or cellulose. Installation includes pouring, placing by hand, or using a machine to blow the material in place. This type of material and installation works well for insulating ceiling or attic areas since the material can easily fill any void area.

Insulation that is manufactured as batt insulation or blanket insulation is usually sold in rolls. This type of insulation is usually 16″ or 24″ wide, although wider rolls are available. Batt or blanket insulation is available either with or without a facing. The facing acts as a vapor barrier and also makes it easier to attach the insulation on the inside edge of wall studs. See **Figure 17-47**. Where a vapor barrier is not desirable, the unfaced insulation should be used. This type of

Ozgur Boskum/Shutterstock.com

Figure 17-46. Loose fill insulation can be easily placed in areas with voids, such as an attic.

Christina Richards/Shutterstock.com

Figure 17-47. Batt or blanket insulation is usually manufactured in rolls and can be easily cut to length and installed.

insulation is more solid than loose fill. Each roll is one solid piece of material. This quality helps builders to place or install the insulation in certain applications. When different lengths are required, the rolls can be cut to the appropriate length prior to installation.

Rigid insulation is used to add an extra layer of insulation in locations such as on concrete foundations, under concrete flooring, or under roof decking or siding. Rigid insulation is solid and more durable than other types of insulation. It may also be called insulation board. This type of insulation is manufactured in sheets and can be cut easily for installation. See **Figure 17-48**.

Reflective insulation is made with a reflective foil on the outside surface, **Figure 17-49**. The function of this foil is to reflect heat and prevent unwanted heat transfer into a building. This insulation works well in summer months to reduce the amount of radiant heat transferred into a building by the sun. However, this type of insulation is not as useful during winter months in buildings located in cold climates because warmth from the sun is desired. When covered with dust or corrosion, reflective insulation loses its insulating value.

Foam-in-place insulation is applied as a liquid foam and then cures or dries into a solid. Some amount of expansion occurs with any foam-in-place insulation.

Sima/Shutterstock.com

Figure 17-48. Rigid panels provide an extra layer of insulation where flexible insulation is not suitable.

Installation includes simple methods, such as spraying from aerosol cans that can be bought from many home improvement stores. The foam is simply applied to an area in which insulation is needed, such as cracks around windows. In more complex application methods, equipment is used to mix two or more chemicals to make the final foam product just before application. Necessary equipment includes hoses, pumps, and application guns. See **Figure 17-50**.

Ventilation

Ventilation is the movement of fresh air into a building or controlled environment. Proper ventilation in a building is important for controlling air quality and environmental conditions, such as humidity. Ventilation is needed in three main areas of a structure: the crawlspace; the main area used for living, working, or storage; and the attic. The *crawlspace* is an area below a building that allows access to the underside of a structure. Agricultural buildings typically do not have a crawlspace, but residential houses sometimes do. These areas may contain ductwork, electrical wiring, plumbing pipes, or fixtures for other utilities. Crawlspaces can be used for installation, service, or replacement of any of these systems or equipment. Airflow is managed by using vents around the perimeter of the crawlspace. See **Figure 17-51**. These

Osgur Boskun/Shutterstock.com

Figure 17-49. Some insulation materials are made with a layer of a reflective material.

vents are necessary to assist in keeping the area under a house dry. Vents are either manually operated or equipped with an automatic system. This system will open the vents when temperatures rise and close them when temperatures fall. Because a crawlspace is not heated, open vents in cold weather will cause higher heating costs. Vents should only be opened in warm weather when drying conditions are favorable.

Another area requiring airflow is the area inside a structure. Whether the use is residential or agricultural, proper airflow is important for air quality and moisture management. A thermostat inside a house will measure the temperature of the air surrounding the thermostat. When a heating or air conditioning unit is activated by the thermostat, air is cycled through the ductwork from the unit and then back through the air return ductwork. As temperature is managed, moisture and other air quality factors are improved as air is exchanged. In agricultural applications, such as confined feeding structures, air is exchanged at regular intervals. In poultry houses, for example, exhaust fans are turned on approximately every 30 seconds. These fans help remove ammonia, moisture, and other harmful elements, **Figure 17-52**.

Ventilation is also important is in attic or ceiling areas. Since heat rises, an attic area can collect a large amount of heat during warm months. In addition, a rise in temperature results from the sun shining on a roof. This area needs to be managed adequately to prevent heat stresses to the building. Simple airflow systems contain vents to allow air flow through the attic or ceiling space. For buildings with a gable-style roof, the vents would be located in the walls on each end of the gable, **Figure 17-53**. Airflow can be enhanced by using soffit vents, turbine vents (commonly called whirlybirds), cupolas, and ridge vents, **Figure 17-54**. If greater control of airflow is needed, powered fans can also be installed. A common use of a powered fan is in residential applications where a fan is installed in a cupola. In agricultural settings, a combination of vent systems is used unless greater control of the air movement is needed.

In any type of building, the surroundings greatly affect the needed airflow management and also heating and cooling costs. In the warm seasons, roofs that are at least partially shaded by trees or other buildings will not need as much airflow management as those exposed to full sun.

©iStock.com/joebelanger

Figure 17-50. Foam-in-place insulation is being sprayed between the studs of this wall. After application, the foam expands and cures.

Goodheart-Willcox Publisher

Figure 17-51. Vents around the perimeter of a crawlspace allow air to flow underneath a structure.

Christopher Meder/Shutterstock.com

Figure 17-52. Ventilation is achieved in a confined feeding building by using powerful exhaust fans.

Goodheart-Willcox Publisher

Figure 17-53. Gable vents allow air to flow freely in attic areas.

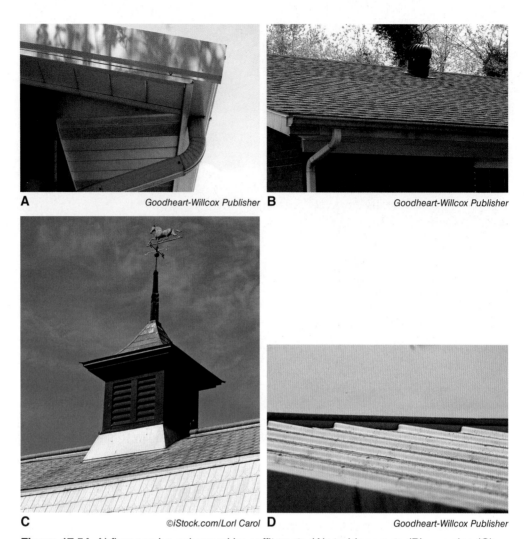

Figure 17-54. Airflow can be enhanced by soffit vents (A), turbine vents (B), cupolas (C), and ridge vents (D).

CHAPTER 17 Review and Assessment

Summary

- Finishing is the final stage of building a structure. In the finishing process, structures are covered or enclosed to increase the usefulness of the building.
- Finishing the exterior walls of a building includes installing sheathing, building wrap material, and siding. Siding may be made of brick, metal, vinyl, or wood.
- A roof provides protection for the interior of a structure from the outside environment. Roofs must have a grade or slope for adequate drainage to occur.
- An underlayment is used under roofing materials made of wood, asphalt, or metal. An overhang provides some protection to the structure at the point where the roof attaches to the walls.
- Purlins and girders act as sheathing to which metal roofing or siding can be attached.
- Walls provide strength in a structure, block the outside environment, and divide an area into rooms. Interior walls are most commonly finished with wood, metal, or sheetrock.
- Windows let natural light into a building, allow ventilation, and enhance the look of a building. A variety of window types are available.
- Thermal pane windows have two or more glass panes separated by a space filled with a vacuum, air, or gas. They have a high insulating value compared to single pane windows.
- Proper fit during the installation allows doors to operate properly and have a long service life. Door components consist of the door itself and the door frame.
- Trim is a nonstructural building component installed to enhance appearance or to help seal areas where water or air can penetrate the building.
- A building that is energy efficient will result in lower operating costs and help conserve resources.
- The purpose of environmental control is to manage elements such as temperature, relative humidity, and air quality. Four factors that affect environmental control include building construction, heat transfer management, insulation, and ventilation.
- Insulation resists heat transfer by preventing air movement. Types of insulation include loose fill, batt or blanket, rigid, reflective, and foam-in-place insulation.
- Ventilation systems are important for a building to have adequate air exchange underneath structures in crawlspaces or basements, in living or usable areas, and in attic or ceiling spaces.

Words to Know

Match the key terms from the chapter to the correct definition.

A. build-up roofing
B. conduction
C. convection
D. crawlspace
E. door sill
F. energy efficient
G. environmental control
H. fascia
I. insulation
J. overall rise
K. overhang
L. purlin
M. radiation
N. relative humidity
O. sash
P. SEER rating
Q. shingle
R. siding
S. soffit
T. thermal pane window
U. trim
V. underlayment
W. ventilation
X. window frame
Y. window
Z. wrap

1. Managing conditions, such as temperature, humidity, and light, inside a structure or area.
2. A type of asphalt roofing that is made from alternating layers of asphalt and roofing tar.
3. A board that is part of roof framing and is used to connect trusses or rafters and function as sheathing.
4. The amount of water vapor present in the air.
5. Material, such as brick, vinyl, or wood, that is placed on the outside walls of a building.
6. The part of a window (sill and jams) that provides the structure for the sash to work properly and allows the window to be installed into a building.
7. The transfer of heat by contact between a warm object and a cool object.
8. The transfer of heat through heat waves traveling from an emitting heat source to a person or object.
9. The transfer of heat through a gas or liquid.
10. A type of roof covering made from asphalt or wood that is installed in overlapping layers.
11. Nonstructural building components installed in a structure to enhance appearance or help seal areas where water or air can penetrate a building.
12. A window that contains two or more panes of glass to reduce heat transfer.
13. An opening in a wall or roof that may be fitted with glass or another covering, allows light and air to pass through, and provides a view of the outside.
14. Material, such as tar paper, that is installed before final roofing materials to help prevent water from leaking into the building.
15. A synthetic material installed over sheathing to help prevent air movement out of or into the building.
16. The movement of fresh air into a building or controlled environment.
17. A descriptive term for something that has low use of natural resources or electricity to heat, cool, run appliances, or operate equipment.
18. The part of the roof that extends beyond the edge of the outside wall and along the roof line.

19. Building trim made from vinyl, metal, or wood that is installed on the vertical edge of the overhang on a building.
20. A measure that describes the energy efficiency of electrical equipment or appliances.
21. A material that slows the movement of heat from warm areas to cool areas.
22. Building trim placed underneath the roof structure in the space between the fascia board and the edge of the building.
23. The door frame component located on the very bottom of a door frame.
24. The measurement of how many inches or feet of rise are included in a roof from eave to peak.
25. The area below a building that allows access underneath a structure.
26. Panels or bars in a window that hold the glass in the window.

Know and Understand

Answer the following questions using the information provided in this chapter.

1. List examples of a finishing process in a building.
2. What is sheathing and when is it applied to a structure?
3. What are four types of siding used in building construction?
4. What is a major function of a roof and how long should a properly installed roof last before needing replacement?
5. How is roof slope expressed, and why is having an appropriate roof slope important?
6. What is the most common underlayment material used for roofs and how is it installed?
7. What are the three main types of roofing materials used today?
8. How are metal roofing panels secured to the roof, and why is having a tight seal important?
9. What are two types of asphalt roofing, and how do they differ?
10. What is a key requirement for a long service life for wood shingles?
11. What purpose do girders and purlins serve in a framing structure?
12. What are two important functions of walls in a structure?
13. What type of wall supports the majority of building loads?
14. What are the most common materials for finishing the interior walls of a building?
15. Describe the purpose and parts of a window frame.
16. How does a single hung window differ from a double hung window?
17. Why is the R-value of a window important?
18. Why is proper fitting and adjustment important when installing doors?
19. What are five types of insulation used in buildings?
20. Ventilation is needed in what three main areas of a structure?

STEM Connections and Academic Activities

1. **Science.** Roofing tar is essential to help properly seal a roof. Research the appropriate temperature for roofing tar to seal shingles on a roof. Explore the reason tar becomes sticky at higher temperatures.
2. **Technology.** Use a smartphone and scan the QR™ code on any piece of lumber or plywood. Examine the technical information and specifications and determine if there are any acceptable or restricted uses.
3. **Engineering.** Plywood is a material that is commonly used for sheathing and subflooring. In order to perform under certain loads (such as the weight of building materials, wind, or snow) plywood must have the ability to hold the weight. The panel rating provides information on appropriate use of plywood for both roof sheathing and subflooring. Note the span rating (identification index) in the following image.

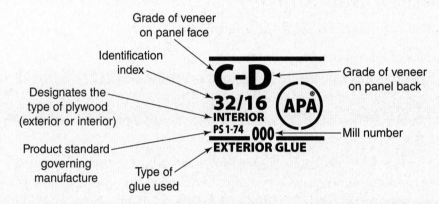

A rating of 32/16 means that when used as roof sheathing, the maximum distance between supports (rafters or trusses) is 32″. When used as subflooring, the maximum distance between floor joists is 16″.

Example: You are building a machinery storage building and the trusses are spaced at 24″ on center. Would the above plywood with the span rating of 32/16 be usable for this building?

4. **Math.** Calculate the amount of metal needed to install a roof for a machinery storage shed. The building is 40′ wide × 60′ long. The distance from the peak of the roof to the eave is 22′. Each piece of metal is approximately 3′ in width and can be cut to exactly 22′. How many pieces will be needed for the whole roof?
5. **Social Science.** Divide into groups of four students. Subdivide your group into pairs and research a type of roofing material for the machinery building above. One group should research metal and one group should research asphalt shingles. Find out specifics for this type of roofing material, such as cost, warranty, service life, and color options. Compare and contrast each pair's findings.
6. **Language Arts.** Research the types of plywood available to consumers. Write a report that discusses the differences in the various types. Give an example use for each one.

Thinking Critically

1. Using heat loss and heat gain as your only considerations, what are the advantages and disadvantages of a south-facing window?
2. Consider the movement of air through a wall. Describe how the air would move. What effect does moisture and temperature have on air movement? What would excessive moisture do if it were trapped in a wall? How does a wall vent excess moisture?
3. What are some of the technologies architects use to make a roof waterproof?

CHAPTER 18
Specialized Agricultural Structures

Chapter Outcomes
After studying this chapter, you will be able to:
- Identify the different types of greenhouse structures and roof styles.
- Discuss basic plant health needs in greenhouses.
- Discuss basic plant health needs in hydroponic systems.
- Identify the different types of aquaculture systems.
- Explain the differences among hydroponics, aquaculture, and aquaponics.
- List the advantages of prefabricated structures.
- Define and give examples of earthen structures.

Words to Know 📱

aquaculture	filtration	hydroponics	specialized agricultural structure
aquaponics	flow-through aquaculture system	pest	sprinkler system
aquatic plant	framing	plant disease	transplanting
automatic watering system	greenhouse	photosynthesis	ventilation
controlled environment agriculture (CEA)	growing environment	polyethylene	water bed
cooling system	growing medium	pond system	water quality
covering	growth cycle	prefabricated agricultural structure	water system
drip watering system	gutter-connected greenhouse	propagation	
earthen structure	headhouse	recirculating system	
		rigid panels	

Before You Read
Review the chapter headings and use them to create an outline for taking notes during reading and class discussion. Under each heading, list any Words to Know within that section. Write two questions that you expect the chapter to answer.

While studying this chapter, look for the activity icon to:
- **Practice** vocabulary terms with e-flash cards and matching activities.
- **Expand** learning with interactive activities.
- **Reinforce** what you learn by completing the end-of-chapter questions.

www.g-wlearning.com/agriculture

©iStock.com/jjdc23

There are many types of structures in agriculture. The most common types have stud or post frame and truss construction. These structures are discussed in Chapters 16 and 17. Other structures that are designed for a specific use or function may have different material requirements and characteristics. These less common structures are described in this chapter.

Specialized Agricultural Structures

Specialized agricultural structures are those that are not common to every farm or agriculture business. However, they have a very specific function in a sector of the agriculture industry. Greenhouses, for example, are not used in every farmstead, but their use in plant production makes them very valuable for some agriculture businesses. The total number of greenhouse, nursery, and floriculture farms has grown to more than 50,000. The market value of the products from these farms has grown to $14 billion annually. This value is approximately 12% of the market value of all crops produced in the United States. While the market value has grown since 1997, there has been a 22% decrease in the number of farms involved in this area. In floriculture, the number of farms declined, but the square footage of greenhouses and other structures increased to over 870 million square feet in 2012.

In addition to greenhouses, several other types of specialized structures are used in agriculture. The most common are:

- Hydroponic structures
- Aquaculture structures
- Aquaponic structures
- Prefabricated structures
- Earthen structures

Figure 18-1. Greenhouses are often equipped to control and enhance the growing environment. A—Commercial greenhouses often use specialized irrigation equipment. B—Fans are used for air circulation as well as temperature control.

Greenhouse Structures

A *greenhouse* is used to start plants from seed and provide a consistent environment for plants to grow. Greenhouses can be basic structures with no equipment, or they can include many types of equipment to control or enhance the *growing environment* (temperature, humidity, moisture, etc.). See **Figure 18-1**. Many greenhouses use *controlled environment agriculture (CEA)*, a system in which growing environments are enhanced through controlled conditions. One major advantage of a controlled environment is that optimum growing conditions can be provided, which removes the variability of the outdoor environment. To increase the effectiveness and production of a greenhouse, some factors—such as location, water quality, type of structure, and equipment—should be considered.

470 Agricultural Mechanics and Technology Systems

Location and Orientation

Because a greenhouse utilizes the sun as a heat source, its efficiency depends partly on its location and orientation. To be efficient, it must be designed and constructed in the proper position to take advantage of the sun's rays. However, while the sun can usually provide a heat source during daylight hours, supplemental heat systems are often needed to maintain temperatures at night and on cloudy days. See **Figure 18-2**. While a heater can be installed to help warm or maintain air temperature, excessive use of heaters can increase operating costs.

Effective placement of a greenhouse can reduce the amount of supplemental heat needed. The best location for a greenhouse is generally where the sun can be best utilized. Ideally, greenhouses should be constructed on a site that has little or no obstruction by trees or other buildings. See **Figure 18-3**.

jan kranendonk/Shutterstock.com

Figure 18-2. Heaters are used to help maintain temperatures inside greenhouses. These heaters help keep the temperature inside the greenhouse stable when sunlight is not available or sufficient to maintain the required temperature.

Orientation is also an important part of placing a greenhouse. A greenhouse's orientation is its position related to the sun's movement through the sky. Plants need sunlight to perform photosynthesis. Proper orientation of the greenhouse allows them to use maximum sunlight.

The proper orientation varies depending on the part of the United States in which a greenhouse will be built. The dividing line for orientation is around 40° latitude. In locations above 40°, a greenhouse should be positioned with the ridge of the roof running east and west. In winter months especially, plants generally get sunlight throughout the day as the sun travels through the sky from east to west. Below 40°, greenhouses are better positioned with the ridge of the roof running north and south. In these areas, where sunlight is more intense, plants grow better when sunlight is less intense during part of the day. This can be accomplished by orienting the buildings from north to south or by planning the location so that trees or buildings shade the greenhouse at intervals.

Water Quality and Availability

Water is a crucial natural resource for a greenhouse system because plants must have adequate moisture. Since greenhouses remain warm, usually much warmer than outside temperatures, adequate water for plants to grow should be closely monitored. There are several methods for watering plants in a greenhouse. These methods

PavelSvoboda

Figure 18-3. Greenhouses should be placed in an area free of trees and other structures that might obstruct the sun.

472 Agricultural Mechanics and Technology Systems

A

gtflour/Shutterstock.com

B

Filip Ristevski/Shutterstock.com

C

muph/Shutterstock.com

Figure 18-4. A—Sprinkler watering systems provide water from above the growing area. B—Drip watering systems provide water directly to the soil. C—Automatic watering systems can make watering plants more efficient.

range from individualized watering to automatic systems.

The type of *water system* (method and equipment involved delivering water to plants) used in a greenhouse depends on the budget, the size of the operation, and the specific water needs of the plants being grown. The simplest, and least efficient, type of water system involves using a watering hose. Plants in this system may be in various types of pots or trays, depending on the type and size of the plants.

Sprinkler systems are watering systems in which water is applied from above the plants. See **Figure 18-4A**. Another type of system is the *drip watering system*. These systems have water lines that apply small amounts of water at the soil level instead of over the whole plant or growing area. See **Figure 18-4B**. *Automatic watering systems* can also be used. Automatic systems apply water according to moisture needs or by timers. Automated systems can result in a considerable savings on labor to perform watering tasks. See **Figure 18-4C**.

In some agricultural applications, *water beds* are constructed to hold a few inches of water. Styrofoam trays containing seedlings float on top of the water. Holes in the bottom of the trays allow water to moisten the soil as needed. The water provides moisture as needed for plants but can also contain nutrients. Large numbers of plants can be managed using this method. For example, the tobacco industry often uses this method.

Water should be installed in a greenhouse even for minimal use. *Water quality* and availability are important aspects of managing a greenhouse. The water supply should be free of any contaminants or particulates that would negatively affect plant health. Enough water should be available to meet the needs of the plants being grown.

The location of a water source should be considered when planning a greenhouse.

Copyright Goodheart-Willcox Co., Inc.

A well can provide water directly to a greenhouse in certain areas of the United States. Water can also be installed from public systems. Since a consistent and reliable water source is needed, rainwater collection systems are typically not used. However, they are a possibility in some applications and in some areas of the country.

Both the additives in public water systems and naturally occurring particulates can negatively affect plant health in some plants. In more precise water quality needs, close monitoring and water *filtration* (removal of anything that would negatively affect plant health) should be used.

Greenhouse Styles

A variety of greenhouse styles is available for many different applications in agriculture. The most basic greenhouses are simple in construction and do not contain any technology for nutrient application or air exchange. More complex greenhouses include automated systems for monitoring and adjusting the growing environment.

Technology Selection

The use and productivity of a greenhouse should be considered when planning construction or expansion. Technology can make greenhouse management more efficient and effective, but it is expensive to install. Careful consideration should be given to what level of technology is appropriate for a specific greenhouse application. Technology includes lighting, exhaust fans, ventilation, cooling systems, water systems, and automation. As a complete system, these components enhance not only the growing area, but also management practices and procedures.

Artificial lights provide an adequate amount of lighting at night and during the day when natural sunlight is not available. A variety of lighting options and controls allow light conditions to be customized to meet the needs of individual greenhouse locations and functions.

Ventilation is the process of exchanging air. Air enters the growing area through vents and may be filtered. Exhaust fans pull air from the growing area and exhaust it to the outside air. *Cooling systems* are included in greenhouses to keep the growing area from getting too hot. The air conditioning systems used in residential housing are typically not practical for use in greenhouses due to the heat load in the growing area. Since most plants require warmer temperatures for proper growth, a high level of cooling is not required. However, heat management is important to maintain an adequate growing environment.

A common system used to cool greenhouses relies on evaporative cooling and consists of an exhaust fan and a cooling pad. The exhaust fan draws air into the greenhouse and through the water-soaked cooling pad. The air traveling through the pad is cooled as water evaporates from the pad. The cooled air then flows through the greenhouse. See **Figure 18-5**. This type of cooling system is not suitable in high-humidity environments, since it introduces water into the growing environment. Other systems used to cool greenhouses include misting, shading, and increasing ventilation.

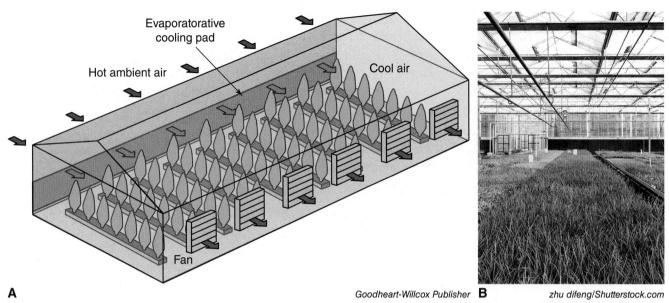

Figure 18-5. Cooling pads are a common way to help cool the growing environment in a greenhouse. A—Ambient air is cooled as it travels through the water-soaked cooling pads. B—The brown wall at the far end of the greenhouse is the evaporative cooling system.

Water systems provide needed moisture to plants as they grow. These systems increase the efficiency of watering chores. Water systems range from simple to very complex. In more complex systems, water is routed through water meters and branch circuits to supply variable amounts of water for plants requiring different amounts of moisture. See **Figure 18-6**.

Any of these processes can be automated to help manage and control growing conditions. If the main purpose of a greenhouse is to start plants from seed and grow them to an appropriate transplant size on a small scale, then a lower level of technology may be needed, depending on plant variety. In warmer climates, less technology may be possible. However, if a greenhouse will be used in a full-time nursery, and especially if plans include off-season production, more technology is required.

Roof Structure

A variety of roof styles is available for greenhouse construction. Each type of roof structure has advantages for specific greenhouse applications. Specific use, environmental control, initial cost, and expansion options are a few of the factors that impact the style used and narrow the choices during the planning stages. **Figure 18-7** illustrates some of the major types of roof structures for greenhouses.

Single vs. Multiple Structures

One of the main decisions affecting which roof structure will be used is whether that structure will be

Figure 18-6. A water control system in a large greenhouse can provide more precise control of moisture requirements for a variety of plants.

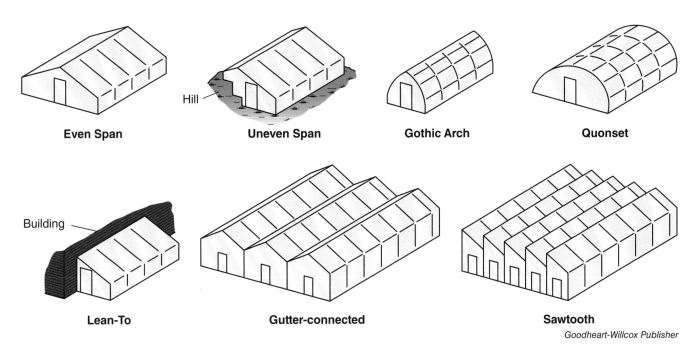

Figure 18-7. Greenhouse roof styles.

a stand-alone unit or if there will be multiple structures. If multiple units will be added, then a gutter-connected style may be a good option.

A single greenhouse provides a growing environment under one roof structure that can be small or large in square footage. A *gutter-connected greenhouse* consists of two or more single greenhouses joined together to increase the square footage of the growing area. Gutter-connected greenhouses share a roof because all the structures are attached. Although a single greenhouse can be very functional, a gutter-connected structure has some distinct advantages. See **Figure 18-8**. One of the biggest structural advantages is the option for future expansion. Because outside walls are shared between the individual structures, an additional greenhouse area can be easily added. The next structure can be constructed by connecting it to an existing structure after site preparation has been completed.

Another advantage of connected multiple structures is that shared outside walls increase energy efficiency. Outside walls in greenhouses are not insulated, as they are in other agricultural structures, and they tend to lose energy easily. As an outside wall is eliminated, since the area is shared between two structures, one complete wall between two greenhouses is no longer exposed to the outside environment. This benefit provides greater control of the inside environment.

Figure 18-8. Gutter-connected greenhouse design allows for easier expansion. This provides growers more options for increasing the square footage of their growing area.

Also, if beneficial for the plants being grown or the organization of the greenhouse, the shared wall can be removed in the future to have a more open area. From a functional perspective, shared areas inside a gutter-connected greenhouse provide greater environmental control and better use of equipment. One control system and reduced equipment needs greatly simplify these types of greenhouse structures. See **Figure 18-9**.

Materials

Greenhouse structures are simple in their design and construction compared to some other agricultural structures. Construction materials for basic greenhouse frames include wood and steel. Wood is the least expensive option and can provide easy construction. However, due to water use for plants and moisture in the controlled environment, wood does not last very long. Although treated lumber can help extend the life of lumber, decay and rot due to the environment will cause a shorter service life.

Photodiem/Shutterstock.com

Figure 18-9. Control systems provide easier management of larger greenhouse irrigation, ventilation, and heating and cooling systems.

In situations in which lower cost is a requirement, or when replacement or growth is planned, using wood framing may be the best option. However, when permanent structures are needed or when long-term use is planned, steel framing is a better choice. Steel structures for greenhouses are typically galvanized so that rust is not a problem. These structures allow more permanent planning and use. Although there are several optional components and subsystems, materials used to build the main structure fall into two main categories: the frame and the covering.

The Greenhouse Frame

Framing for a greenhouse is the main support for the structure. The frame must support the weight or forces of the construction materials, as well as any forces caused by weather, such as snow or wind. The simplest greenhouses have a frame that consists of arched materials to which a covering, such as plastic, is attached. See **Figure 18-10**. Since one of the main functions of a greenhouse is to allow light to enter the growing area, framing should obstruct as little sunlight as possible. In most greenhouses, framing components are galvanized steel and are as small as specifications allow. They are designed to provide adequate strength without obstructing the sunlight. In larger, more complex greenhouse systems, framing may include steel I-beams, hinges

lumokajlinioj/Shutterstock.com

Figure 18-10. Some of the simplest greenhouses include a frame and a covering.

for opening panels, and other functional components. See **Figure 18-11**.

The Greenhouse Covering

When framing for a greenhouse structure is complete, a *covering* is installed. Coverings form a barrier between the inside and outside environments. They do not insulate well because as much sunlight as possible must be allowed to enter the growing area. Because *photosynthesis* (the process by which plants create food from carbon dioxide and water) is crucial for maximum plant growth, in a simple greenhouse, the non-insulating covering extends over both the roof and wall components.

In some applications, however, a headhouse is needed. A *headhouse* is an area of a greenhouse that is used for sales, instruction, storage, potting, or other tasks. These areas are typically not used for growing plants and do not have the same sunlight requirements as the growing environments. These are best placed on the side of a greenhouse that gets the least amount of sunlight. Alternatively, they can be separated from the greenhouse by a breezeway or hallway so that sunlight can still enter most of the wall where the connection is located.

Covering materials are generally determined by use and cost. Plastic provides the lowest cost of any material, allows sufficient sunlight into a growing area, and provides environment barriers. *Polyethylene* is one of the most common plastic materials used for greenhouse covering. See **Figure 18-12**. However, periodic replacement is needed due to damage, age, or weather. If replacement is needed during a production season, valuable time and energy can be lost.

Glass is another material that can be used as a greenhouse covering. Glass is an excellent material to allow sunlight and heat to pass into a growing area. Although replacement costs are extremely high, it provides a long service life. See **Figure 18-13**.

Rigid panels are another choice for covering a greenhouse. Rigid panels are made of materials such

Zick Svift/Shutterstock.com

Figure 18-11. Framing for larger greenhouses can include steel trusses, I-beams, and other hardware for installed equipment.

Bahadir Yeniceri/Shutterstock.com

Figure 18-12. Polyethylene is one of the most common greenhouse coverings due to its low cost and suitable properties.

val lawless/Shutterstock.com

Figure 18-13. Glass is a material option for long-term greenhouse coverings.

AgEd Connection: Greenhouse Maintenance and Repair

Greenhouses are used in many areas of agriculture. We are all familiar with flowers grown in greenhouses, but did you know that a variety of agricultural crops are grown in greenhouses too?

Greenhouse designs range from simple to complex. Some greenhouses have plastic coverings, while others are made from glass. Many have plumbing systems that supply water for irrigation and electrical systems that provide power for lighting, irrigation control, and heating.

Students with supervised agricultural experience (SAE) programs who work with flowers, nursery plants, or agricultural crops grown for either all or part of their life in greenhouses will benefit from agricultural mechanics skills if they are required to maintain or repair these specialized structures. These skills will also help enhance proficiency applications in floriculture or nursery landscape, and can be useful in preparing for career development events (CDEs) in these areas.

Solis Images/Shutterstock.com

Marina Lohrbach/Shutterstock.com

Figure 18-14. Rigid panels made of polycarbonate or acrylic are often used for long-term coverings in greenhouses due to their strength and light weight.

as polycarbonate or acrylic. These materials have some of the same benefits as the other materials, but they also have some distinct advantages. They are stronger than glass, and also lighter. Many greenhouses that are used long-term in a business use rigid panels. See **Figure 18-14**.

Plant Health

The main purpose of a greenhouse is to provide an optimum growing environment for plant growth or production. While a proper growing environment is crucial for this, plant health is equally important. Disease or pest issues can greatly affect plant health and, in worst cases, can kill a complete crop or business. Plants should be inspected regularly for signs of problems, and preventive or corrective measures should be taken as needed. The most important areas affecting plant health include:

- Air quality
- Moisture
- Nutrients
- Pest and disease control

Air Quality and Temperature

The quality of the air in the growing area is important to plant health. Air that is relatively free from contaminants, such as dust and other hazards, provides a clean environment for plants to carry out the process of photosynthesis and grow or produce well. Air quality is therefore one of the most important factors to help plants grow at their greatest potential.

Consistent and proper temperature is also important for plants to grow. One of the main benefits of environmental controls in a greenhouse is the ability to provide more optimum and constant temperatures compared to outdoor environments. Heat systems, cooling systems, and circulation fans help to keep the growing environment optimal.

Moisture

Plants must have moisture (water) to be healthy and productive. The amount of moisture needed varies by plant type, so each species should be appropriately researched and provided with the correct amount. Too much or too little water can harm the plants. Some plants may actually die if they receive an incorrect amount of water, so proper moisture control is very important.

There are several ways to deliver water to plants inside a greenhouse. One of the simplest is to use a water hose with a nozzle attachment. See **Figure 18-15**. This method provides an opportunity to inspect the individual plants as water is applied, allowing greenhouse workers to identify any plant health issues. Other types of watering systems are more automated and can provide faster completion of watering chores. Sprinkler or drip watering systems provide moisture to many plants at the same time. Care must be taken that water is placed appropriately so that every plant receives water. Automatic controls can be added to sprinklers or drip systems to increase the efficient use of time for managers or workers. See **Figure 18-16**.

Tyler Olson/Shutterstock.com

Figure 18-15. Manual watering systems in a greenhouse are the simplest and least expensive to install. They also allow workers to inspect plant health.

Veychko/Shutterstock.com

Figure 18-16. Automatic watering systems make watering tasks more efficient.

Nutrients

Proper amounts and types of nutrients are another important factor for plant health. Nutrients can either be applied through the soil, where they are absorbed

Figure 18-17. Nutrients can be added to automatic watering systems to efficiently supply both moisture and fertilizer needs.

Figure 18-18. A growing medium for seedlings should provide adequate nutrients for root growth.

Figure 18-19. Transplanting involves starting plants in one growing area and later transferring them to another. This picture shows tomato plants that were started as seedlings being transplanted to larger pots for additional growth and later sales.

by root systems, or on top of the plants, where they are absorbed through the leaves or other foliage. Nutrients can be applied using systems similar to those used to supply moisture.

Broadcasting fertilizer (by mixing and using a watering can, for example) is effective and may work well with some greenhouses and for some fertilizer needs. In other cases, nutrients may be added to the watering system. This provides greater efficiency, especially when the system is an automated or inline system. See **Figure 18-17**. This type of application is used primarily for supplemental nutrients because the most common way for plants to absorb nutrients is through the root system.

An important part of plant *propagation*, or plant cultivation and growth, is using the correct growing medium. *Growing medium* is a substance that is used to start and grow plants and should include proper nutrients for the growth cycle of plants. Contents of a growing medium may be different at various stages of plant growth. See **Figure 18-18**. A *growth cycle* describes a beginning point of plant propagation through an end point. In some cases, a growth cycle may start with seed germination and continue through adult stages. However, it can also be a shorter amount of time. For example, the growth cycle might start with seed germination and end when the plants are ready for transplanting. In this case, plants may not be fully grown when the growth cycle ends. A new growth cycle begins when plants are transplanted.

Transplanting is a process of removing a plant from a growing area, pot, or other container and re-potting or placing that plant into a bigger pot or final growing location. See **Figure 18-19**. A common example of transplanting is found in the landscaping industry. Ornamental flowers, trees, or other plants are grown in pots large enough to allow plants to grow to a marketable size. After the plants reach this size, consumers or landscaping companies buy the plants and transplant them into decorative beds. Having the proper nutrients in each location (greenhouse, pots, landscaped area) should be considered because the availability of these nutrients is important for plant health.

Pest and Disease Control

Two of the biggest problems with plant health that must be monitored and corrected quickly are pests and diseases. An infestation of *pests* (insects or animals) or *plant diseases* (infections caused by harmful organisms) can quickly cause plant health to decline to a point of limited or no return to healthy status. If a complete greenhouse or plant variety is affected, a business could lose a large profit. For a farm business, an entire crop may be lost.

Cleanliness is one of the best ways to help control some basic health issues. Clean work or growing areas, sterilized growing containers, and controlling traffic into a growing area are all important items to consider to limit the spread and control any issues. See **Figure 18-20**. Organic or chemical applications should be used appropriately to control pest or disease infestations.

Vlad Teodor/Shutterstock.com

Figure 18-20. Plant health is monitored throughout the growth cycle. Even seedlings should be monitored for disease or pests.

Hydroponic Systems

Hydroponics is a system for growing plants in a water solution. See **Figure 18-21**. Water quality is essential in hydroponics because the plants depend entirely on the water for both nutrients and moisture. In fact, one of the distinct advantages of hydroponics is everything the plants need to grow and be healthy is included in the water solution. Everything needed is calculated, measured, and given to the plants through the water. With traditional soil systems, adjustments must be made to nutrients based on what is already contained in the soil or potting mixtures.

Wasu Watcharadachaphong/Shutterstock.com

Figure 18-21. Hydroponics is a growing system in which a water solution, rather than soil, is used to supply nutrients.

Hands-On Agriculture

Building a Terrarium

A terrarium is basically a small greenhouse that provides a growing environment for plants. The construction techniques for a basic terrarium are outlined below. Keeping a terrarium in plain view will allow you to see plant growth and respiration during the growing process.

1. Assemble the following supplies: a glass jar (one commonly used in canning is sufficient), plastic wrap, rubber band, soil or other growing medium, plants appropriate for the container's size.
2. Place soil or growing medium in the glass jar and add the appropriate amount of moisture.
3. Place the plants in the growing medium.
4. Cover the opening of the jar with plastic wrap and secure it with a rubber band.
5. After a few hours you will notice moisture begin to collect on the inside of the growing environment. The moisture will stay inside the container and provide adequate moisture for the plants.
6. Place the container in a sunny location to watch your plants grow. Check the plants during the day to make sure they do not get too hot. If they show signs of wilting or stress, move them out of direct sun.

Many of the advantages of growing plants in a greenhouse are also advantages of hydroponics. For example, a significant advantage of using hydroponic systems is that crops can be grown in off seasons. In climates where growing conditions outdoors are not favorable in cooler or winter months, a hydroponic system can extend a growing season and make plant or vegetable production more profitable. See **Figure 18-22**. One drawback to hydroponic systems is that they are more expensive than traditional soil systems due to specialized equipment, techniques, and plant nutrients required.

Hydroponic structures vary widely in size and complexity. Structures designed for plants that remain in the system until they are harvested are generally more complex than those designed for plants that will eventually be transplanted to a different growing system. This is because structures used for seasonal growth and harvest must provide greater control of the growing environment.

The main component of any hydroponic system is the container for the water solution. Containers are available in a variety of configurations, including large tanks, beds, plastic barrels, reservoirs, and grow trays. Depending on the system application, other components may include a filtration system, a circulating system, a nutrient pump, an air pump, growing medium, and pots or trays.

Aquaculture Systems

Aquaculture is a system for growing and harvesting animals and plants that normally grow in water. Aquaculture systems provide a concentrated water-based environment to allow them to be grown more efficiently and in more manageable settings than in their natural habitats. Some of these systems are small enough that they can be placed indoors in structures similar to greenhouses. Others are outdoor systems in specially designed ponds or other earthen structures.

ssguy/Shutterstock.com

Figure 18-22. Hydroponics can extend growing seasons even in cold climates.

Aquatic animals commonly raised in aquaculture systems are shrimp, catfish, and tilapia. See **Figure 18-23**. Examples of plant aquaculture include watercress and kelp. The difference between plant aquaculture and hydroponics is that aquaculture is a growing system for *aquatic plants*—plants that can only grow in water—whereas hydroponic systems are designed to grow plants that grow naturally in soil.

Types of Aquaculture Systems

The design of an aquaculture system depends on cost, space, production of the system, and the requirements for species being grown. The systems can be installed in either indoor or outdoor settings. The three types of systems are:

- Flow-through systems
- Pond systems
- Recirculating systems

Pan Xunbin/Shutterstock.com

Figure 18-23. Aquaculture systems allow plants or animals, such as fish or shrimp, to be grown in a controlled water environment.

Flow-Through Systems

Flow-through aquaculture systems provide a constant flow of fresh water through the growing area. Since a constant flow is required, a consistent and reliable water source must be used. If water quality is acceptable, streams, wells, or natural springs can be used. If not, other, more reliable sources are used. For example, the flow-through system can be designed to work with an existing local municipal water supply. Waste collection areas are placed at strategic areas within the system so that as the water flows, waste can be collected easily. See **Figure 18-24** for an example of a flow-through system.

Vlada/Zhikhareva/Shutterstock.com

Figure 18-24. Flow-through aquaculture systems have a continual flow of fresh water through the system.

Pond Systems

A *pond system* is similar to many ponds or lakes found in agricultural settings today in which the water supply is contained in one location. See **Figure 18-25**. Basic construction includes digging and shaping the soil so that there is a depression in the ground and filling it with water. When this type of system is used, at some point, waste that falls to the bottom is cleaned. The cleaning interval depends on factors such as the frequency of use and depth of the pond.

tangencial/Shutterstock.com

Figure 18-25. A pond system is an earthen structure, such as a natural or constructed lake or pond. Fish farms typically have multiple ponds separated by earthen banks.

Recirculating Systems

Recirculating systems are designed to reuse the water in an aquaculture system. To increase water quality in the growing environment, a filtration system is added to the system. This requires more complex plumbing and equipment than other types of systems, but the growing environment is consistently cleaner and can be more easily regulated.

Pan Xunbin/Shutterstock.com

Figure 18-26. Recirculating aquaculture systems filter water for reuse. These systems allow a greater control of the growing environment.

Because greater regulation of water quality is possible, animal population can be maximized. Increasing the population allows a greater cost return because more of the growing area can be utilized. Recirculating systems are a good alternative if a good water supply is not available or if discharge areas are not feasible, as required by other systems. See **Figure 18-26** for an example of a recirculating aquaculture system.

Location

Locations for aquaculture systems should be thoroughly evaluated prior to deciding on a location or breaking ground for a facility. One of the biggest determining factors for installing an aquaculture system is a good-quality water supply (water supply that is free of contaminants that harm growth). Many surface water supplies are contaminated with chemicals, parasites, or other harmful substances. A reliable, clean supply of water needs to be available because water is the animals' living environment. Especially in systems that flow or recirculate, water must be available as needed for the health and success of the system. Wells or natural springs may provide quality water and should be considered. However, any water source identified for aquaculture should be tested to make sure it can provide a water source of good enough quality for the system.

The physical location of an aquaculture facility should also be considered. Transportation of animals can be difficult and can affect their health and well-being. Shorter transport times for processing help to ensure a fresh harvest and product for the intended market.

Aquaponics

Aquaponics is a combination of aquaculture and hydroponics in which both plants and aquatic animals are grown in the same system. See **Figure 18-27**. As an example, both plants and fish may be grown together in the same environment or enclosure.

There is a mutual benefit for the plants and fish in this type of growing environment. One of the issues with aquaculture systems is that the waste produced by the aquatic animals must be cleaned from the growing area. The solids must be removed using a filtration system or other means in order to maintain a healthy growing environment. The waste can then be used for fertilizer, composting, or other uses. While the waste has a value in aquaculture systems, the cost of waste removal can be eliminated with an aquaponics system. Fish waste provides fertilizer for the plants, and in return, the plants clean the water for the fish. Aquaponics allows two commodities (plants and fish) to be harvested from one system.

Joseph Sohm/Shutterstock.com
Figure 18-27. Aquaponic systems provide a growing environment for plants and aquatic animals in the same system.

Prefabricated Structures

Prefabricated agricultural structures are structures built at one location and then transported to another location for installation and use. Portions of buildings that are transported separately and then assembled into a complete structure on site are also considered prefabricated structures. A common example of a prefabricated structure is an outdoor storage building. See **Figure 18-28**. An example of a structure that has been built in pieces and then assembled on site is a pre-manufactured home.

James R. Martin/Shutterstock.com
Figure 18-28. Outdoor storage buildings are prefabricated structures that are ready to use upon delivery and placement on a property.

Warren Price Photography/Shutterstock.com

Figure 18-29. Larger prefabricated structures are built and delivered in smaller pieces and then assembled on site before they can be used.

Many pre-manufactured homes, especially those over a certain width, are built in two pieces. Each half is transported separately. The halves are then fitted and attached together to make a complete structure on site. See **Figure 18-29**. Other examples of prefabricated structures include grain storage bins, concrete panels, Quonset buildings, water storage tanks, irrigation systems, and many others.

Transportation of prefabricated structures includes proper planning of routes, obtaining permits, and adequate equipment for delivery. One of the biggest factors limiting the transportation of a structure is its height. Power lines, stoplights, bridges, and other overhead structures are built to an economical and effective height. Any prefabricated structure must be planned and built within local and state requirements to prevent damage to these structures during transport.

Another limitation is weight. Roads are able to support only a certain amount of weight before they start to break down. Wheels and tires used during transport can damage road surfaces if more weight is applied than the road surface is rated to support.

Earthen Structures

Earthen structures are those that are made from natural components, such as soil, sod, and topography. See **Figure 18-30**. Common earthen structures include holding ponds, lakes, canals, terraces, waterways, diversion ditches, and spillways. Each type of earthen structure is designed and constructed for a specific purpose. Some of the common purposes include water flow control, manure holding areas, rainwater collection, and erosion control. Information and design for many of these structures can be obtained through local university extension offices or from the Natural Resources Conservation Service.

Stu49/Shutterstock.com

Figure 18-30. Earthen structures are made from the earth and are used to control water in some way. In this picture, a drainage ditch provides a location for water to be drained and removed from a field to prevent flooding.

CHAPTER 18 Review and Assessment

Summary

- Greenhouses can provide a controlled environment to propagate plants.
- Greenhouses can have many different roof styles, including even span, uneven span, gothic arc, Quonset, lean-to, gutter-connected, and sawtooth.
- Greenhouses can be constructed as single or multiple structures.
- The two basic parts of a greenhouse are the framing and the covering.
- Basic plant health in a greenhouse requires proper moisture, nutrients, disease and pest control, air quality, and temperature control.
- Hydroponic systems allow plants that are traditionally grown in soil to grow in a water solution that contains all of the nutrients they require.
- Aquaculture systems allow the growth of some animals, such as shellfish or catfish, in a controlled environment.
- The three main types of aquaculture systems are flow-through, pond, and recirculating systems.
- An aquaponic system is a combination system that involves growing plants and animals in the same space.
- Prefabricated agricultural structures are those that are constructed off-site and then moved to the site where they will be used.
- Earthen structures are constructed as part of a landscape and may include natural features such as soil and topology.

Words to Know

Match the key terms from the chapter to the correct definition.

A. aquaculture
B. controlled environment agriculture (CEA)
C. covering
D. drip watering system
E. earthen structure
F. flow-through aquaculture system
G. greenhouse
H. growing environment
I. growing medium
J. gutter-connected greenhouse
K. hydroponics
L. headhouse
M. pest
N. plant disease
O. polyethylene
P. pond system
Q. prefabricated agriculture structure
R. propagation
S. recirculating system
T. rigid panels
U. specialized agricultural structure
V. sprinkler system
W. transplanting
X. ventilation
Y. water quality
Z. water system

1. A growing area for plants or animals in which the growing environment is controlled to enhance growth or production.
2. The characteristics of an area where plants are grown, including factors affecting plant health, such as temperature, humidity, and moisture.
3. An aquaculture system in which water is filtered and recirculated through the system.
4. A system for growing and harvesting animals that normally grow in water.
5. A structure that is made from natural components such as soil, sod, and topography.
6. Manufactured panels that are stronger, lighter, and longer lasting than other greenhouse covering materials.
7. A material that separates the inside environment of a greenhouse from the outside environment.
8. The characteristics of the available water supply.
9. An area of a greenhouse that is used for instruction, storage, or other function, rather than for growing plants.
10. A type of watering system that applies water through water lines at the soil level.
11. Structures used for specific agricultural applications that are less common than other areas of the agricultural industry.
12. An agricultural structure used to start and maintain plants in specialized growing environments.
13. An aquaculture system that has a constant flow of fresh water through the system.
14. Multiple greenhouses joined together in one structure, sharing common walls.
15. The method and equipment used to deliver water to plants.
16. Unwanted insects or animals in a growing environment.
17. A material commonly used to cover a greenhouse to divide the growing area from the outside environment.
18. The process of exchanging air in a growing environment using vents and exhaust fans.
19. An aquaculture system in which the water supply is contained in one location.
20. A system for growing plants that would traditionally be grown in soil in a water solution.
21. Plant cultivation and growth.
22. The process of moving a plant from one growing area to another.
23. Agricultural structures that are built at one location and transported to another location for installation and use.
24. An infection caused by an organism that can negatively affect plant health and can cause plant death.
25. A substance or material used to start and grow plants.
26. A type of watering system that applies water above plants and resembles rainfall.

Know and Understand

Answer the following questions using the information provided in this chapter.

1. What name is given to structures in the agricultural industry that are less common in design or function?
2. A(n) _____ is an agricultural structure that is used to provide a consistent environment for plants to grow.
3. What is the name used to describe growing environments that are enhanced for optimum growth?
4. Why should the orientation of a greenhouse structure be considered during planning and construction?
5. Discuss the importance of water availability and water quality in a greenhouse.
6. What are the advantages of installing a gutter-connected greenhouse?
7. The term _____ is used to describe growing plants and animals together in the same system.
    A. aquaponics
    B. aquaculture
    C. hydroponics
    D. hydroculture
8. Name some advantages of incorporating technology in greenhouses.
9. What characteristics should framing and covering materials have?
10. *True or False?* The biggest factor in determining the location for an aquaculture facility is a good-quality water supply.

STEM and Academic Activities

1. **Science.** Heat transfer properties are basically defined as the amount of heat transferred from one object to another. In greenhouses, heat transferred through a covering is important because that heat is used to maintain temperatures in a growing environment. The heat transfer value of various covering materials is called the *heat transfer coefficient*. Knowing the heat transfer coefficient of various materials allows an evaluation of which materials have the best heat transfer ability. Collect the following supplies: a digital temperature probe and three drinking glasses made of different materials (glass, plastic, and Styrofoam, for example). On a sunny day, go outside and turn one of the glasses upside down on a level surface with the temperature probe underneath. Allow the sun to shine on the glass for 1 or 2 minutes and record the temperature reading. Repeat this procedure for each glass. Make sure each glass is exposed to the sun for the same amount of time when taking the temperature reading. Which material allows the greatest amount of heat transfer?
2. **Technology.** Explore your local home improvement store. You may want to travel there in person, or you could also search online if the company has a website. Search for heaters and examine their specifications. Record the BTU value of at least three heaters and also the square footage the heaters are rated to heat. BTU is an abbreviation for "British Thermal Unit." This value is used to express the energy required to raise one pound of water one degree Fahrenheit.

3. **Engineering.** Using the information you found for the heaters above, determine if each heater is adequate for the space in the following example. You decide to build a specialized agricultural structure with the dimensions 28′ × 40′. Compare the square footage ratings of the heaters you researched in Activity #2 with the square footage of the structure. Are any of these heaters adequate? If not, how many of each type would be needed to adequately heat the structure?

4. **Math.** Calculate the exposed surface area of a greenhouse. The dimensions of the greenhouse are shown in the following diagram.

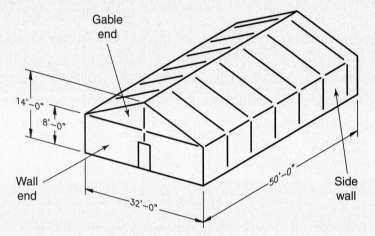

Calculate the surface area of the following areas:
a. Outside walls each end
b. Outside walls each side
c. Each gable end
d. Both sides of roof
e. Total area

5. **Social Science.** What temperature makes you comfortable? Ask or determine the thermostat setting in your home. The next day in class, form groups of three students. In these groups, compare the temperatures from your homes and discuss whether you feel too warm or too cool at each temperature, and why.

6. **Language Arts.** Research the different types of fuels used to heat structures. These structures could be homes, workshops, factories, or specialized agricultural structures. Prepare to discuss with your classmates the following information:
a. Name of the fuel
b. BTU value
c. Availability (can you buy the fuel in any area or is the supply limited?)
d. Cost
e. Is the fuel practical for consumers or businesses to use?

Thinking Critically

1. You are working as a greenhouse employee. While applying fertilizer as part of your normal duties, you notice that plant health is not very good in one area. What would you do?
2. While working the weekend shift at a local nursery, you notice that some plants look stressed from lack of water. How would you report this issue to the manager?
3. A local elementary school has contacted you to bring a science class to your aquaculture facility for a field trip. In your business, you have outdoor catfish ponds and hydroponically grown vegetables. Create an outline of what you would include in a presentation to the students. What activities would you do with these students to enhance their learning experience?

CHAPTER 19
Fencing and Livestock Structures

Chapter Outcomes

After studying this chapter, you will be able to:
- Discuss the differences among the types of fences.
- Describe the steps in planning for construction of a fence.
- Estimate materials for a fencing project.
- List the materials used to construct a fence.
- Identify the various types of wire and explain their uses.
- Explain how a fence charger should be connected to a fence.
- Discuss the function of gates and gaps in fences.
- Discuss the significance of corner posts.
- Explain how to set posts.
- Explain how to install fencing materials.
- Identify tools commonly used to install fences.
- Describe the proper maintenance of various types of fences.

Words to Know

barbed wire	gap	post	t-post
braided wire	gate post	post-and-rail fencing	welded wire fence
cattle guard	hardware cloth	post driver	wire fabric
chain-link fence	high-tensile wire	poultry wire	wire net
corner post	insulator	powered auger	woven wire
electric fence	livestock	ratchet tensioner	X-post fencing
electric fence charger	material list	smooth wire	
fencing material	paddock	splicing	
footing	perimeter fence	survey	

Before You Read

Before reading the chapter, skim the photos and their captions. As you read, determine how these concepts contribute to the ideas presented in the text.

While studying this chapter, look for the activity icon to:

- **Practice** vocabulary terms with e-flash cards and matching activities.
- **Expand** learning with interactive activities.
- **Reinforce** what you learn by completing the end-of-chapter questions.

www.g-wlearning.com/agriculture

Fences have been used for a very long time in agriculture. Going back centuries, fences have been essential for families that used animals or animal products, such as eggs or milk, for food. Fences contained the animals and kept out predators. This chapter will discuss the different types of fences used in the agriculture industry. It will also detail the installation, inspection, and maintenance of these fences.

Fencing Applications

Modern fencing is used for a variety of reasons. Some fences are used primarily to visually enhance an area, while others have a specific function. Decorative, ornate, and painted or vinyl fences definitely increase visual appeal of a property, but they may also have a specific function. See **Figure 19-1**. Functional fences are used either to keep animals in a confined area, such as cattle in a grazing pasture, or to keep animals out of an area such as a garden, to prevent plant or produce destruction.

One of the most common agricultural uses of fencing is for *livestock* (animal) farms. The average livestock farm has approximately 420 acres. As an industry, livestock farms use a total area of around 900 million acres each year. In 2012, product value from farms involved in animal production in the United States was over 185 billion dollars. Farms specifically devoted to cattle ranching and cattle farming totaled 112 billion dollars. With the value and investment in this specific industry, proper protection and containment of livestock is essential.

Miao Liao/Shutterstock.com

Figure 19-1. Many fences can visually enhance property while also being functional.

Fences in livestock applications provide important safeguards for farmers and ranchers. A *perimeter fence* (a fence that entirely encloses an area for growing livestock) contains the livestock and limits the animals' living areas to the land owned by the producers. See **Figure 19-2**. Most states have laws that require livestock owners to contain their animals. However, a few states in the Midwest area of the United States still permit open-range grazing. Before building a livestock operation, it is important to understand the applicable state and local laws concerning fencing.

Containment protects both the producers and the animals in their care. If allowed to wander beyond their intended areas, livestock can cause damage to other crops in the area, damage neighbors' crops, or cause a driving hazard for passing motorists. Fencing also protects animal health by keeping the animals away from areas too close to traffic, other animals that may be dangerous, and other hazardous areas.

MBoe/Shutterstock.com

Figure 19-2. Perimeter fences contain animals in specific areas on farms.

Another reason for using fencing in agriculture is to keep animals out of certain areas. In a garden or a produce business, for example, animals can quickly destroy a crop or even a whole year's harvest. In these areas, a fence is installed to protect the plants to keep animals out of the area. The animals are considered a pest or nuisance in this situation.

Types of Fencing

The type of fence used depends on a variety of factors, including the types of animals involved, the topography of the land, soil condition, and strength needed. The intended use of the fence determines the materials needed and their durability. Fences can primarily be placed in one of three categories:

- Agricultural fences
- Privacy and security fences
- Aesthetics fencing

Agricultural Fences

The primary purpose of an agricultural fence is to create a boundary, or a border, to control access to certain areas. See **Figure 19-3**. One of the main functions of an agricultural fence is to control animal access either by containing a group of animals or by preventing their access to an area. Proper fencing can keep animals safe, including pets and farm or ranch animals. The type of fencing used is determined by the type of animal, material availability, material cost, and other factors.

Joe Belanger/Shutterstock.com

Figure 19-3. Boundary fences provide a simple boundary between farm areas or property lines.

Post-and-Rail Fences

Post-and-rail fencing is often used for visual appeal but also does a good job of controlling animal access. See **Figure 19-4**. These fences are simple in design and are made from either wood or vinyl. They consist primarily of posts to support the fence and rails, which are installed horizontally between the posts. When assembled properly, a post-and-rail fence can enhance property borders or fence lines while controlling livestock access. If animals do not respect the boundary set by this type of fence, an electric wire can be added to better contain them.

Cynthia Farmer/Shutterstock.com

Figure 19-4. Post-and-rail fencing consists of vertical and horizontal elements. These components can be attached as shown in this image or by screws, nails, or other hardware.

X-Post Fences

X-post fencing is typically used only when soil is too hard for other types of fencing to be installed. All other types of fencing utilize installing a post in the ground and the fence material is then attached to the post. An X post is constructed in an X shape and sits on top of the ground. See **Figure 19-5**. Rails are then attached to the posts. Unless specifically required by soil conditions, this type of fence is not used in modern fencing practices due to the amount of material required and lower strength when finished.

Jen duMoulin/Shutterstock.com

Figure 19-5. X-post fencing is installed by making an X to form a post on top of the ground. No digging is necessary.

Electric Fences

Electric fences are metal fences that use a small amount of electricity to shock and, as a result, discourage animals from trying to penetrate the fence. In agricultural applications, they have the same functions as any other fence: either to contain animals or to keep them out. Technically, any type of fence—agricultural, privacy/security, or aesthetic—can be electrified. However, electric fences are most often used for agricultural and security purposes.

Privacy and Security Fences

Privacy fences restrict viewing into or out of the fenced area. They are commonly used in neighborhoods where houses are close together. See **Figure 19-6**. In addition to providing privacy, they can help keep children and pets safe by containing them within a specific area. Many privacy fences are aesthetically pleasing as well.

romakoma/Shutterstock.com

Figure 19-6. Privacy fences block an area from view.

Privacy fencing typically is not used in agriculture. Farm buildings, tree lines, or windbreaks are many times used for privacy in agricultural settings when more than just enclosing a backyard or social area is desired.

The purpose of a security fence is to restrict access into or out of the fenced area. Businesses often use a fence to contain products intended for sale. For example, seed and fertilizer businesses, home improvement stores, and machinery equipment manufacturers use security fencing to keep unauthorized people out of the area when the business is closed. See **Figure 19-7**. In more sensitive situations, security cameras or security guards are used in addition to fencing to help ensure the security of an area.

Aesthetic Fences

Fences that are installed purely for visual appeal are aesthetic fences. While this type of fence can be used for multiple purposes, a primary focus is the visual enhancement provided. See **Figure 19-8**. Aesthetic fences are made from a variety of materials, including wood, vinyl, steel, brick, and rock. Final designs are based solely on the preferences of the property owners.

Planning Fence Construction

An important part of any construction project is planning. In many projects, more time is spent planning a project than in actual construction. Planning includes creating a visual plan for design and placement of the fence, comparing fencing materials, and considering the costs of various options.

Obtaining a Survey

A *survey* is a procedure to establish exact points, locations, boundaries, and property lines (legal boundaries between two properties). A survey is sometimes required by local ordinances before a fence can be constructed. Even when not required by law, a survey often provides a great starting point

Johnny Habell/Shutterstock.com

Figure 19-7. Security fences are used in a variety of settings to control or limit access to certain areas.

Jorge Salcedo/Shutterstock.com

Figure 19-8. Aesthetic fences provide visual appeal.

Figure 19-9. A survey is useful for establishing fences, especially if they will be used as a property line.

for a fencing project. See **Figure 19-9**. Although land or property owners know the approximate property or boundary lines, an official survey provides a legal description of their location. If the owner does not obtain a survey and places a fence outside a property line, the resulting issues may complicate relationships with neighbors and delay fence construction.

Surveys involve determining exact locations on the earth's surface. In addition to determining property lines, surveys also provide exact coordinates to locate buildings and roads. In some cases, they are used to guide site preparation and elevation of a build site. Survey points are marked with permanent rods or other markers so they can be found whenever necessary. When property is sold or a construction project is planned, for example, accurate point location not only makes the survey accurate, but helps to ensure that property owners retain legal ownership of a specific area.

Determining Fence Lines

A fence line is used to establish a boundary between neighboring properties or between separate areas within a property. Fence lines can follow a surveyed line, the contour of the land, or a location for a specific function.

Surveyed fence lines are often used to identify the location of property boundaries. These fences may remain intact for many decades whether or not they contain livestock. Other boundary markers, such as an oak tree, a concrete pillar, or a creek, may be used as additional markers, because they typically will not move or be removed over a period of many years. When a legal property line or boundary marker is established, it is sometimes used for generations. In areas where natural markers are scarce, surveyed fence lines become more important.

Career Connection

Soil Conservation Technician

Job description: Soil conservation technicians work closely with soil conservationists and natural resource managers to development resource management and conservation plans. These plans help prevent soil erosion and protect our natural resources.

Education required: Bachelor's Degree in Soil Science or Natural Resources.

Job fit. This job might be for you if you love protecting our natural resources. You should also enjoy working outdoors.

Another use of surveyed fence lines is to accurately divide land into smaller sections called ***paddocks***. In livestock operations, paddocks play a key role in rotational grazing practices to ensure the health and productivity of animals and also forages (natural grasses and plants eaten by livestock). Dividing an area of land into separate paddocks allows grazing operations to manage grazing areas more efficiently. These grazing areas are then rotated for use every two or three days.

A temporary fence is often used to divide land into paddocks or grazing areas. Temporary fencing has fence posts and wire, for example, that can easily be moved to another location. See **Figure 19-10**. Temporary fencing may be left in one location for several years or even decades, but it can be moved at any time to redesign pastures.

Goodheart-Willcox Publisher

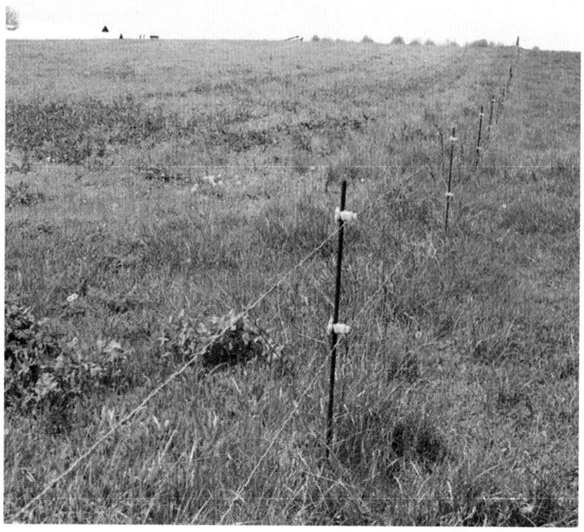

Figure 19-10. Temporary fences may be left in one location for several years, but they also can be easily moved to another location.

Fences can also be installed according to the contour of the land. See **Figure 19-11**. In locations where there are hills, valleys, or other natural topography that would make surveys impractical, fences may follow the border of a creek, ridge, embankment, yard, or other features. In these applications, fence lines follow the contour of those features instead of a prescribed path. In many of these cases, surveys are not performed.

Material Estimation

Determining the materials needed for a fence is an important step in the planning process. Having the proper materials on hand is important both for budgeting and for making efficient use of time and resources. One of the first

Brad Whitsitt/Shutterstock.com

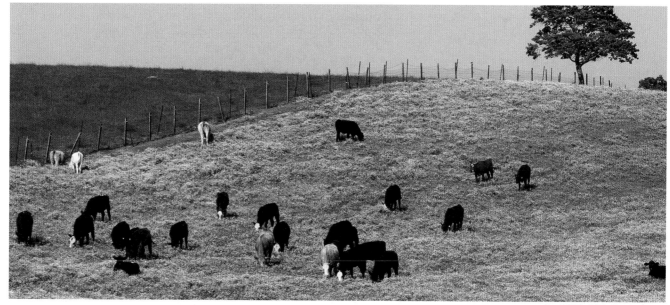

Figure 19-11. Fences are sometimes installed to follow land features.

decisions to make in estimating fence materials is to determine the function of the fence, which allows an appropriate type of fence to be selected. After the type of fence is decided, materials can be determined.

Because more than one type or style of fencing may meet the property owner's needs, a cost comparison may be beneficial. Cost comparisons allow a direct comparison of material quality and cost of various fencing options for planning purposes. Any project should include a cost comparison to help plan the budget.

When the style and type of fencing are finalized, a *material list* can be developed. A material list is a list of all the materials needed to complete a project. Before the material list can be completed, the proposed enclosure must be measured. This is because the number of posts needed is based on post spacing around the perimeter, and wire is bought in rolls that contain a specific length of fencing material. Most fencing estimates are based on linear (straight) measurements because most fences are installed in straight lines, but even if a curved fence is planned, a linear measurement is still needed. In this case, a measuring wheel is a useful tool. This tool includes any changes in topography or curves as measuring is conducted.

After the material list is developed, the installation is planned. Having an adequate installation plan allows the fencing project to proceed smoothly and efficiently. The plan should include not only a schedule for purchasing materials, but also a list of tools needed and an estimate of the labor required to complete the job. Sources of tools and labor are also included in the plan.

Materials

Materials for fencing are chosen based on their intended use. If a fence fails, the landowner may experience profit or harvest losses, loss of privacy, or significant costs for replacement. The proper materials are therefore important to allow the fence to function as needed. Materials for most types of fencing include posts and fencing materials. Additional items, such as hardware and electric chargers, are needed for some fence types.

Goodheart-Willcox Publisher

Figure 19-12. Wood fence posts have been used to support fences for many generations.

Types of Posts

Posts are basically the framework and support for a fence. They are commonly made of wood, steel, or vinyl. They can be either temporary or permanent.

Wood Posts

Wood posts have been used for many generations in farming applications. Wood has always been readily available in most areas. Many wood posts today are manufactured to be perfectly round or square. However, in the past, and even today in some applications, wood posts have sometimes been made from trees or tree limbs, as shown in **Figure 19-12**. Wood posts usually last approximately the same amount of time as fencing materials, so when a fence needs updating, the posts are updated as well. If a harder wood is used, then posts may last longer.

Steel Posts

Steel has become a popular material for fence posts due to long service life and an ability to be reused. Some of the most common types of steel fence posts include steel pipe, t-posts, and temporary posts. Steel pipe requires more installation time and effort, so it is sometimes a second choice as a material unless cost or availability makes steel an advantage. See **Figure 19-13**.

Steel *t-posts* are commonly used today. See **Figure 19-14**. When driven in the ground, t-posts are very sturdy and have excellent service life. A post driver is required to install t-posts, however.

Vinyl Posts

Vinyl posts are primarily used in applications where aesthetics are important. Vinyl fencing material is used with these vinyl posts to provide a consistent appearance. See **Figure 19-15**. Since the strength of vinyl is lower than that of wood or steel, vinyl posts are typically not used where high durability is needed unless other fencing options are also included. For example, vinyl fencing is not strong enough to withstand use in horse pastures. In order to achieve

Hank Shiffman/Shutterstock.com

Figure 19-13. Steel pipe is an effective material for fencing.

Goodheart-Willcox Publisher

Figure 19-14. T-posts are commonly used to support agricultural fences.

Shane Wilson Link/Shutterstock.com

Figure 19-15. Vinyl posts and fence materials provide a distinctive appearance while also providing a boundary.

502 Agricultural Mechanics and Technology Systems

Goodheart-Willcox Publisher

Figure 19-16. Temporary fence posts can be installed easily when constructing or moving a fence.

THPStock/Shutterstock.com

Figure 19-17. Corner posts provide extra strength for supporting a fence. They are used at any beginning, end, or angled attachment in a fence line.

Simon Laprida/Shutterstock.com

Figure 19-18. Smooth wire is made from mild steel and provides a barrier. It is also used in some cases for other functions, such as bracing.

the look of vinyl, an electric fence may be used with the vinyl fence to keep horses away from the fence and prevent any damage or escape.

Temporary Posts

Temporary fence posts are manufactured from round steel, fiberglass, or plastic. See **Figure 19-16**. A big advantage of temporary posts is their easy installation. Unless the soil is hard or dry, these temporary posts can be pressed into the desired location by hand. Some of these posts include a small piece of metal that a person's foot can press on to help install the post.

Corner Posts

Corner posts are posts that provide a starting point for a fence. They add extra strength when the fencing material is stretched into place. Fencing extends from the corner posts at a 90° angle unless a boundary requires a different position or angle. See **Figure 19-17**. Corner posts serve two purposes. The first is to have a post in the corner of a fence when the angle changes. Fence posts used between corners (in a straight line, for example) are mainly used to hold the fence off the ground in the proper position. When a section of fence wire is stretched, there is a large amount of pressure on corner posts due to the tension required on the wire.

Wire Fencing Materials

Fencing material is the part of a fence that provides a barrier or decorative design in a completed fence. Material options for fencing and related fasteners vary widely. Examples include wire, wire fabric, rails, and solid panels. Certain options are more appropriate for use with certain types of animals.

Wire

Smooth wire is made from mild steel. See **Figure 19-18**. In most cases, smooth wire is used only as a barrier in locations where larger openings between the wires are not a concern. Smooth wire can also be used as bracing for corner posts, gates, or gaps. It can be electrified if needed.

Several different types of stranded wire are available to meet various functions and needs. The most commonly used type of stranded wire in agriculture is *barbed wire*. See **Figure 19-19**. Barbed wire is constructed with two or three strands of smooth

Copyright Goodheart-Willcox Co., Inc.

wire with metal barbs arranged at regular intervals. This type of wire is typically used with larger livestock, such as cows, due to the strength and durability of the wire. Service life is also an advantage.

Barbed wire is an example of *high-tensile wire.* High-tensile wire has a high tensile strength, which means that it is stronger and has better resistance to being broken. Barbed wire forms a strong barrier. Like smooth wire, it can be electrified.

Wire Fabric

Wire fabric is a material made by twisting, welding, or linking wires together to form a fence material. Wire fabric can be made of a variety of wire sizes. Several types of wire fabric are used in agriculture. Each one provides a specific advantage for containing animals such as cattle, goats, pigs, or sheep, as well as smaller animals. Wire fabric is installed vertically and is attached to posts.

Hardware cloth is a type of welded wire made with very small, square openings. See **Figure 19-20**. This type of wire can be used as a fence material, but it is also sturdy enough to build cages for small animals such as rabbits or birds.

Poultry wire, also called *chicken wire,* is a small-gauge wire twisted to form a repeated hexagon shape. See **Figure 19-21**. When installed, this material provides a barrier that will keep small birds, such as chickens, contained in a pen. It is also used to keep pests such as rabbits out of garden areas.

Some wire fabric has larger rectangular or square openings. *Welded wire fence* and no-climb wire fence are two examples of this type of fence. They are made from heavier gauge wire and can be used where smaller openings are not a necessity, such as for medium-sized animals such as hogs. See **Figure 19-22**.

Chain-link fence is commonly used as a border around yards, kennels, parks, ball fields, and many other areas. Larger gauge wire is wound and formed together in a chainlike pattern. Although not common, chain-link fence can be used with some livestock. Chain-link fencing is made from

Eddie J. Rodriquez/Shutterstock.com

Figure 19-19. Barbed wire is the most common type of wire in agricultural applications.

frantab/Shutterstock.com

Figure 19-20. Hardware cloth has small openings and is used to form barriers or cages for small animals.

jurgenfr/Shutterstock.com

Figure 19-21. Poultry wire can be used to provide a boundary for several types of animals, including chickens.

Federico Rostagno/Shutterstock.com

Figure 19-22. Welded wire fence is used in many applications including containing medium-sized animals such as hogs.

Laurie Dugdale/Shutterstock.com

Figure 19-23. Woven wire fencing material is used in combination with barbed wire on this farm.

a larger gauge wire and can withstand livestock pressure due to strength and the ability to stretch.

Woven wire has the largest openings of any of the wire fabric options. This type of wire is most commonly used with larger livestock such as cattle and horses. See **Figure 19-23**. As with other types of wire, the openings provide an adequate barrier to keep animals contained inside the boundary. Barbed wire or an electrified single strand of smooth wire is sometimes used along with woven wire to better contain livestock.

Electric Fencing Materials

Two of the main differences between electric fencing and other types of fencing materials are the types of wire used and the added equipment of a fence charger.

Wire for Electric Fences

Theoretically, any type of metal fence can conduct electricity. However, not every material can do it as efficiently or as effectively as wire. Using wire allows the shock to be concentrated. The type of wire most commonly used with electric fences is barbed wire. *Braided wire* and *wire net* are made of a combination of polyester fibers with small-gauge smooth wire interwoven. This type of wire is very portable and provides greater flexibility and visibility than barbed wire. Braided wire is commonly used in horse pastures. See **Figure 19-24**.

A *Goodheart-Willcox Publisher* B *Ekaterina Kondratova/Shutterstock.com*

Figure 19-24. A—Electric fences provide both a barrier and discouragement to animals that try to get through a fence. Once an animal receives the low-voltage shock, it avoids the fence in order to avoid being shocked again. B—Wire braid is often used with horses to increase visibility for them.

Chargers

Electric fence chargers, or fence boxes, supply the voltage necessary to make electric fences work. See **Figure 19-25**. Chargers generally operate on common power supplies of 120 volts. During operation, electricity flows through wires continually in a circuit. See Chapters 20, 21, and 22 for more information about electric circuits. The pathway for electrical flow in a fence is through the fencing wire. If anything, including animals, humans, plants, or anything else, comes in contact with the fencing wire, electricity flows through that object instead of the wire. Livestock that comes in contact with the wire receives an electric shock. The shock experienced is enough to deter the animals from going through the fence, but is not enough to injure them.

Fence chargers are rated by the miles of fence they can electrify. For a charger to work properly and perform as rated, the fence must be clean and contain no unnecessary grounding. A short circuit in a wire, grass, trees, weeds, and a wire touching a metal post due to a broken insulator are all examples of undesirable grounding. These situations prevent the electricity from flowing and being effective further down the fence.

Goodheart-Willcox Publisher

Figure 19-25. Electric fence chargers use a regular 120-volt power supply to electrify fences.

Power sources for fence chargers are usually electric outlets in a barn or service area. However, in recent years, solar-powered fence chargers have become popular. See **Figure 19-26**. A small solar panel is used instead of an electrical outlet to power the charger. This option allows greater flexibility for charger location and use. Chargers can be installed very close to a fence, although most require protection from weather events and still need to be installed in a protected area.

Gates and Gaps

Gates are necessary in fences to provide access to enclosed areas. At some point in a length of fence, an opening is necessary for access to a pasture, field, paddock, or other area of land. Gates provide a strong structure in a fence opening that can be closed or opened to allow access. Access may be needed for a number of reasons. One example is adding or removing livestock from a field. See **Figure 19-27**.

©iStock.com/Peter Burnett

Figure 19-26. Solar-powered fence chargers have gained popularity because they can be installed directly in a fence.

David Hughes/Shutterstock.com

Figure 19-27. Gates provide an access point through a fence while also providing a barrier to prevent animals from going through the opening.

Gates can also be smaller to allow access for people in a livestock facility, or they can be larger to allow equipment to pass through an area.

Two types of gates are used with livestock: wire gates and manufactured gates. Wire gates are called *gaps*. Depending on the need for control, gaps are typically constructed using one or two strands of wire. They may or may not be electrified. See **Figure 19-28**.

Goodheart-Willcox Publisher

Figure 19-28. Gaps perform the same function as gates and are usually made of wire.

Manufactured gates are made from steel, aluminum, or wood. If the materials are available, gates can be manufactured on a farm. Manufactured gates are typically hinged for easy use, but they can also be installed in a fixed position. Many cattle working facilities use hinged gates to help control the path of animals when they are being medicated or when their health is evaluated.

Installation

Although there are many methods of installing a fence, some common goals and procedures apply to any installation. At a minimum, a fence should be stable; it should not fall down or come apart after it is installed. It should also function as intended to limit or control access to the areas it surrounds.

Footings

Footings provide underground support for fence posts to keep them stationary. They are needed for posts that will support major sections of a fence. Common places for footings include the fence's starting and ending points, fence corners, and places where the topography changes. Footings consist of holes dug into the ground into which the post will be inserted. After the post is inserted in the hole, fill dirt or concrete is used to fill the space. See **Figure 19-29**. If fill dirt is used, it should be packed or tamped (compacted) tightly around the post to hold it securely in place. Proper compaction around a post provides the strength necessary to support the weight and tightness of a fence and also keeps the post in the proper position throughout its service life. Posts that require a footing are typically those that have a significant footprint. Many wooden posts or round steel posts require footings.

Setting Posts

Setting posts is a time-consuming part of fencing installation, so an efficient method for this part of the process is an advantage. Some fences for larger livestock operations may be many miles in length, so an easy method of setting posts is important to complete a fencing job quickly.

Dmitry Trubitsyn/Shutterstock.com

Figure 19-29. Footings are important to provide the strength for a post to support a fence. This image shows a concrete footing being placed to secure a post.

Goodheart-Willcox Publisher

Figure 19-30. A manual post driver is used to install steel fence posts.

The method used to set a post depends on the type and function of the post. Many types of temporary posts and steel t-posts are driven into the ground with a manual or powered post driver. See **Figure 19-30**. A tractor-mounted post driver may be used if conditions and post material (such as wood) permit. Driving a post is more difficult when soil conditions are extremely dry or hard. In these conditions, an X post may be the best option. If more strength is required or if a consistent appearance is needed, each post will need a footing and must be checked for plumb.

Corner posts require more time for installation because they support the tension of a fence. They often require a footing because they must be well anchored so they are not pulled out of alignment while supporting and maintaining tension. Corners are best constructed using multiple posts that are well braced.

Installing Fence Materials

After the posts are set, the fencing material is installed. If a wire fence is being used, the wire will be stretched so that it does not sag or droop between fence posts. When wire is stretched, proper attachment is important to make sure the fence will remain tight when finished.

Installing Wire Fences

For a barbed wire fence, installation begins by attaching an *insulator* to a corner post. The insulator has two functions: it prevents electrical flow and provides a connection for the fencing material. See **Figure 19-31A**. After the insulator is attached to the post, the end of a wire is attached to the insulator. Enough wire should extend past the insulator so that the wire can be bent back and wrapped around itself to be properly secured. See **Figure 19-31B** for an illustration of this procedure.

Figure 19-31. A—A beginning or ending attachment of a fence includes an insulator at a corner post. This image shows a single-strand barbed wire fence and a two-strand gap. B—Wrapping a wire around itself provides a secure connection when installing a fence. Three attachments were made for this fence: two for the two-strand barbed wire fence and one for the single strand for the gap.

After the initial attachment is secured, the wire is unrolled to the length of a fence section to the next connection point. When the roll or wire supply has been completely used, the fence can be continued by splicing. *Splicing* is a method of overlapping wire from the old and new roll by approximately 8 to 12 inches and wrapping the two wires together around themselves in a method similar to starting a new connection to a corner post. Splicing is used both to continue a fence during initial installation and to repair broken sections of fencing.

After a wire is stretched to the next connection point, the second connection is made. This ending connection is the same as the beginning connection to a post. However, when the second connection is made, tension is applied to the wire. More tension is needed over longer distances, but in general, the tension should be high enough that the fence does not sag or droop between posts. The remaining step in the procedure is to attach the wire to each of the posts in the fence line, between the two connection points, using an insulator. A wire fence is complete at this point unless an electric fence is being installed. For an electric fence, the charger needs to be installed, connected, and tested. See **Figure 19-32**.

Goodheart-Willcox Publisher

Figure 19-32. Insulators secure fencing material and also prevent electrical flow into a post when used with electric fences. In this image, the top and bottom wires are not electrified. The middle wire is electrified and is secured with an appropriate insulator.

Installing Rail Fences

If a rail fence is being installed, the goal is basically the same as for a wire fence. The posts are set first, and after the posts are set, the bulk of the work has been accomplished. Installing rails involves attaching the rails at each post. Hardware such as screws, nails, or specialty holders may be required for this procedure. For steel pipe, a welder may be required. After all the rails have been attached, the fence is complete. Some rail fences also use electric wire to provide greater visibility or to discourage livestock from crossing the boundary. If an electric wire is added, then the procedure discussed earlier in this section should be used.

Gates and Access

Gates that are properly hung will swing freely, open and close with ease, and not droop or drag on the ground as they are used. As with wire fence, proper installation of posts is crucial for gates to function properly. *Gate posts* are posts that support the full weight and function of a gate. They must be properly installed and braced so that the gates function well. Many of the same requirements and procedures for setting corner posts in a fence are applied to gate posts. Gates are typically secured in the closed position by a chain or latch. See **Figure 19-33**.

Figure 19-33. Swinging gates provide easy access through a fence or livestock working facility.

A *cattle guard* is a permanent structure that is installed in the ground. It allows a vehicle to drive over it or a person to walk across it, but serves as a barrier to livestock. See **Figure 19-34**. Cattle guards are sometimes used in place of gates to provide an access point into a field. When a cattle guard is used, no gate or gap is used; there is a permanent opening in the fence. The crossing is made from round steel pipe, which is very difficult for cattle to walk on. Due to this difficulty, cattle cannot go through the opening. Human access is greatly improved because no gate or gap needs to be opened in order to check on livestock or fence conditions, mow pastures, or perform other chores. Vehicles can be driven through these openings without stopping.

Figure 19-34. A cattle guard provides an opening in a fence for equipment to access a field while still providing a barrier to contain animals.

SAE Connection — Entrepreneurship

Few people in the workforce today are self-employed. One reason for this is that entrepreneurs assume a certain amount of risk that many people are unwilling to take. An example of this risk is the investment of personal finances in the startup of a business. If the business is not successful, the entrepreneur loses money.

Building fences for other people is an entrepreneurship supervised agricultural experience (SAE) project that has less risk than some others. Startup costs are low because there are relatively few tools required. You may already have experience in putting up fence that you could put to use for others and generate some income at the same time. The layout skills taught in agricultural mechanics, along with basic construction skills, will play a useful part in a fence-building business. From farm fencing to a backyard privacy fence, technical knowledge and an eye for detail will be a good start to a potential career. Just look around at the fences in your neighborhood. Fences are common but there is still plenty of room for more. Ask your teacher today about SAE opportunities in this area.

Tools and Equipment

Fence installation and repair require relatively few tools compared to some other areas of agriculture. Several tools are available, however, to make the procedures more efficient. Below are some of the most important tools for fencing.

Post Hole Diggers

There are two types of post hole diggers. A manual post hole digger is a tool that consists of two shovels connected at a pivot point. It is operated by the physical strength of a person. See **Figure 19-35**. Manual post hole diggers have been used for a number of years to prepare a footing for a post. Since there is a pivot above the shovels, dirt can be held and lifted out of a hole after being cut or dislodged from the sides of a hole. While this tool works very well, it is best used when the number of holes to be dug is limited.

The second type of post hole digger is a *powered auger*. Powered augers are used to drill a hole quickly. They can be mounted on tractors or skid-steer loaders, or they can be held manually between two people. See **Figure 19-36**. When a large number of holes are needed for a fence project, powered augers are more efficient than manual post hole diggers because they can dig the required holes more easily and quickly.

digitalreflections/Shutterstock.com

Figure 19-35. A manual post hole digger provides an efficient way to dig post holes.

vallefrias/Shutterstock.com

Figure 19-36. Powered augers are useful when digging a large number of post holes.

Goodheart-Willcox Publisher

Figure 19-37. Post drivers are used to drive fence posts into place when a footing is not needed.

Goodheart-Willcox Publisher

Figure 19-38. A variety of tools are useful for installing fences.

Post Driver

Post drivers are used to install posts without using a footing. A post driver is commonly used with steel t-posts and with some wooden posts. They are used to simply drive the post into the ground. See **Figure 19-37**. A manually-operated driver is lifted over a post and slammed against the top of the post in a downward motion. Hydraulic drivers are typically tractor-mounted and operate on the same principle. With mechanical and hydraulic forces, the tractor-mounted driver provides more force to drive a post into place. Tractor-mounted drivers are typically used only with wooden posts.

Miscellaneous Tools

Fencing tools also include general tools such as hammers, wire cutters, and pliers. Fencing pliers have specific uses in fencing installation, such as driving nails or staples and holding or pulling wire. Fence stretchers, gloves, and a fencing wagon are also helpful tools for fencing projects. See **Figure 19-38**.

Inspection and Maintenance

Proper maintenance and service are important so that livestock is properly contained in an enclosed area. Unless excessive storm damage, animal activity, deterioration due to age, or other major factors occur, a fence will function well and have a long service life. However, an effective farm or ranch manager inspects fences periodically. Especially when other animals, such as deer, are active, an inspection is necessary to make sure fences remain secure and that they are not broken due to age or from livestock or other animals pressing against them.

The most effective inspection is performed by walking along the fence and visually examining connections, insulators, fence tension, and any other areas or factors affecting a fence. See **Figure 19-39**. If long areas of fence are used, an ATV, utility vehicle, or other farm vehicle may be used.

Sometimes fences lose their tension over time due to stretching. Maintenance for a stretched fence includes retightening wires by pulling and rewrapping the connection at a starting or ending insulator. A *ratchet tensioner* can be used to retighten or increase the tension easily. A ratchet tensioner is a device installed in a wire fence that is used to increase fence tension using a wrench or ratchet. See **Figure 19-40**.

For electric fences, an important part of fence maintenance is to keep any plants from growing through the fence. Plants can provide a pathway to ground, shorting out the electric current. Visual inspections are also needed to find and repair any broken wire, insulators, or posts.

Post-and-rail fences must also be checked periodically. Maintenance is needed when a rail becomes detached due to age, rotting, or livestock or machinery damage.

Depending on climate, a wire fence may last a few decades. Replacement is needed when the steel wire rusts and deteriorates to a point that the wire breaks and splicing any breaks no longer is possible.

Goodheart-Willcox Publisher

Figure 19-39. Periodic inspection of fences helps ensure that fences remain secured and identifies needed repairs. Maintaining appropriate fence tension provides an effective barrier.

Goodheart-Willcox Publisher

Figure 19-40. The ratchet tensioner in the lower-right corner of this image is used to install and adjust fence tension.

CHAPTER 19
Review and Assessment

Summary

- Fences are essential in agriculture to contain animals and to keep out predators.
- In agriculture, fences are used to mark property lines, establish a perimeter, and divide areas into paddocks.
- Fences can also be used in residential and commercial applications to protect children, supply privacy, provide security, or visually improve property.
- Electric fences carry a small electric charge to discourage animals from trying to penetrate the fence.
- An installation plan for building a fence provides an opportunity for budgeting, comparing materials, and lining up resources such as tools and labor.
- Materials used in fencing vary widely, but wood, steel, and vinyl are most common.
- A temporary fence can easily divide an area into smaller grazing areas called *paddocks*.
- Posts support fence materials and provide an anchoring point for fence materials that need tension when installed.
- Corner posts are used at the start or end of a fenced area and provide more strength for supporting fence tension.
- The most commonly used type of wire in agricultural applications is barbed wire.
- Gates and gaps provide access for equipment, animals, and people needing to enter or exit a fenced area.
- Footings are used with posts that need to be strongly anchored.
- Insulators are used to attach electric fence materials to fence posts.
- Specialized tools and equipment are available for installing and repairing fences.
- Periodic inspections of fences are necessary to repair any damaged or weak areas and keep them functioning properly.

Words to Know

Match the key terms from the chapter to the correct definition.

A. barbed wire
B. braided wire
C. cattle guard
D. corner post
E. electric fence
F. electric fence charger
G. fencing material
H. footing
I. gap
J. gate
K. gate post
L. high-tensile wire
M. insulator
N. paddock
O. perimeter fence
P. post
Q. post-and-rail fence
R. post driver
S. poultry wire
T. powered auger
U. ratchet tensioner
V. smooth wire
W. splicing
X. survey
Y. welded wire fence
Z. x-post fencing

1. A type of wire commonly used with poultry and specifically chickens.
2. Type of wire made of a combination of polyester fibers and small-gauge smooth wire that is commonly used in electric fences.
3. A type of fence material that is made from mild steel and has no barbs, twists, or woven patterns.
4. A fence that uses a small electric shock to discourage animals from breaking through the fence.
5. The part of a fence that provides a barrier between posts.
6. A device that is installed at ground level and functions as a gate because it is difficult for cattle to cross, yet it is easy for vehicles or tractors to cross.
7. An opening in a fence spanned by one or two strands of wire attached to a post on both sides of the opening.
8. A device installed in a wire fence that is used to increase the fence tension.
9. The underground support for a post that is used to hold the post firmly in place.
10. A type of wire fence made from heavier gauge wire for use with larger animals such as hogs.
11. The most commonly used type of wire in agriculture.
12. The component of an electric fence that supplies the necessary voltage.
13. A fence component that spans the opening in a fence to provide a point of entry; usually hung on hinges.
14. A fence component that prevents electrical flow and attaches a fence to a fence post.
15. A procedure used to establish exact point, locations, boundaries, and property lines.
16. A fence that is installed all the way around an area where livestock are grown.
17. A post that supports the full weight and function of a gate.
18. A type of fence that has posts for vertical components and boards, or rails, for horizontal components.
19. A type of fencing that is installed entirely aboveground in areas where the soil is very hard.
20. A tool used to mechanically drive posts into the ground; can be either manually operated or tractor mounted.

21. The component that provides the basic framework for a fence.
22. A tool used to drill a hole in the ground quickly using a mechanical power source such as an engine, tractor, or skid-steer loader.
23. A smaller portion of land that is divided from a larger property and used as living areas for livestock.
24. A fencing procedure in which the ends of two wires are overlapped approximately 8-12 inches and wrapped together; used for repairing broken fences or installing new fence.
25. Wire that is made to have high strength and resist being broken.
26. A post that starts or ends a fence and provides extra strength when fencing material is installed.

Know and Understand

Answer the following questions using the information provided in this chapter.

1. Name two specific uses for a fence.
2. What is the primary use for a fence in agricultural applications?
3. Which of the following does not allow easy access through a fence?
   A. Gate.
   B. Gap.
   C. Cattle guard.
   D. Corner post.
4. What are the most common fence post materials?
5. A(n) _____ is an animal grazing area that is a portion of a larger area with boundaries established by a temporary fence.
6. Describe the advantages of creating a fencing installation plan.
7. What is a primary advantage of using a temporary fence?
8. *True or False?* Agricultural fences made of barbed wire cannot be electrified.
9. What is the primary reason for using an electrically charged fence?
10. What is the name of a post used for supporting the tension of a fence?

STEM and Academic Activities

1. **Science.** Research how solar panels produce electricity. Solar panels have become popular in recent years to power electrical equipment including fence chargers. Obtain a solar panel used for powering a fence charger and set up the equipment on a sunny day. Using a voltmeter or multimeter, measure how many volts the panel is producing.
2. **Technology.** Go to www.arcgis.com and retrieve an aerial image of a farm. If you haven't used this website before, you will need to create an account. When you log in, click on *map/create a map* and then find a farm (your home farm or one close to where you live). Click on *basemap* and choose *imagery*.
   a. Use the *measure* tool to determine how many acres are open and available for grazing on the property you found.

b. Use the *distance* tool to determine the length of a fence that will be installed around the perimeter of the property. Make sure your fence location does not include forests, bodies of water, roads, or any other areas not desirable for cattle to live. Your final answer should be in miles.

3. **Engineering.** Designing and organizing a fence is important for livestock farms to best utilize forages and to provide adequate nutrients and daily weight gains. Design the field you identified from the computer exercise in Activity #2 for a cattle herd that contains 50 head of cattle. According to the Natural Resources Conservation Service, if the forages are average in production and growth, approximately 1.5-2 acres are needed for each animal. Is your property capable of supporting all 50 head? If not, how many can it support?

4. **Math.** Calculate materials needed to build the fence you identified and planned from the above questions. For the perimeter fence, you will install steel t-posts at 12-foot spacing. The fence material will be three-strand barbed wire. Standard rolls of barbed wire contain approximately 1,320 feet of wire.

Example answer:

Posts:

Total perimeter length determined from the online exercise is 4,515′. Dividing by 12′ post spacing = 377 posts.

Wire:

4,515′ × 3 strands = 13,545′

13,545′/1,320′ per roll = 10.26 or 11 rolls minimum would be needed

5. **Social Science.** Are there any neighbors to the property you identified in Activity #2? What would be the impact on neighbors and on the farm if the field were changed to livestock production? Gather into groups of four and discuss the problems, challenges, and solutions for this situation.

6. **Language Arts.** In groups of three, research any zones or restrictions associated with the properties each group member has chosen. Summarize the findings and make an oral report in front of the class on what was found and any challenges discovered.

Thinking Critically

1. Research the difference in cost between meats that are mass produced on farms and meats that are produced locally or for personal use. You may either make visits to grocery stores and meat markets or find appropriate online markets. What difference did you find and to what do you attribute the difference?

2. Would you prefer to grow your own food (either meat or plants) or to buy your food? What challenges besides fencing and protecting the products would you have if you decided to grow your own food?

3. There are urban agriculture systems and techniques available that can be used in small spaces. Research these systems and techniques to determine if they will work at your home.

CHAPTER 20
Electric Theory

Chapter Outcomes
After studying this chapter, you will be able to:
- Identify principles of electric wiring and wiring terminology.
- Define the terms and principles of electricity.
- Estimate electric needs and loads.
- Discuss the principles of alternating current and direct current.
- Identify sources of electric power.
- Demonstrate the use of various meters.
- Explore alternative power systems, including solar and wind.
- Demonstrate the use of various meters designed to help people work safely around electricity.

Words to Know

alternating current (AC)	ground conductor	Ohm's law	step-down transformer
ampere (amp)	ground fault circuit interrupter (GFCI)	photovoltaic cell	step-up transformer
battery		polarity	switch
circuit	halogen	receptacle tester	three-phase power
circuit breaker	hertz (Hz)	resistance	voltage
conductor	incandescent	single-phase power	voltage detector
continuity	insulator	single-pole, double-throw (SPDT) switch	voltage drop
direct current (DC)	load		watt
electric current	multimeter	single-pole, single-throw (SPST) switch	
electricity	neutral conductors		
fuse	ohm	split-phase power	

Before You Read
As you read the chapter, record any questions that come to mind. Indicate where the answer to each question may be found, whether it is in the text, from your teacher, in another book, on the Internet, or by reflecting on your own knowledge and experiences. Pursue the answers to your questions.

While studying this chapter, look for the activity icon to:

- **Practice** vocabulary terms with e-flash cards and matching activities.
- **Expand** learning with interactive activities.
- **Reinforce** what you learn by completing the end-of-chapter questions.

www.g-wlearning.com/agriculture

This chapter will introduce you to the basics of electricity. *Electricity* is a type of energy that consists of charged particles, such as electrons. When a charge builds up in one place, it is known as *static electricity*. When it flows along a path, it is called *electric current*. In other words, electric current is the flow of electrons.

How do we use electricity? In agriculture, tools and equipment convert the power of electricity into light, heat, sound, and mechanical motion.

Conductors and Insulators

Conductors are materials that readily support electron flow. In a conductor, electrons move from one atom to the next at approximately the speed of light (186,000 miles/second). As an electron from the first atom moves to the second, another electron is passed to the third atom, which in turn forces the transfer of yet another electron to the fourth, and so on until an electron leaves the conductor at the other end. Imagine a tube filled with marbles. The tube represents the conductor and each marble represents an electron along a conductor. As a marble is pushed into the left end of the tube, all of the other marbles (electrons) move one place to the right, and the one on the right end is expelled from the conductor, **Figure 20-1**.

Types of conductors include gold, silver, and copper. Gold is the best conductor, while copper is the most common. Aluminum is also commonly used as a conductor.

Compare the gold and copper atoms shown in **Figure 20-2**. What are the similarities? Each has a single valence electron. A *valence electron* is loosely held in the outermost ring of the atom. Single valence electrons of metal atoms have relatively weak bonds and can readily move from one atom to another.

Electrons carry a negative charge (–). When one electron moves to the outer ring of a neighboring atom, its negative charge repels other negatively charged valence electrons, causing an electron from that atom to jump to the next atom, and so on. The movement of these "free" electrons constitutes electric current. The amount of current (in amps) represents the amount of electricity flowing across a conductor at a given point.

Insulators are materials that resist the free movement of electrons from one atom to another. The valence electrons of insulators are held much more tightly to the nucleus of the atoms than those of conductors. While insulators are not useful for transferring current, they are critical in the management

Goodheart-Willcox Publisher

Figure 20-1. Just like marbles in a tube, an electron pushed into the end of the conductor causes each successive free electron to be moved to the right, with the electron on the opposite end being expelled from the conductor.

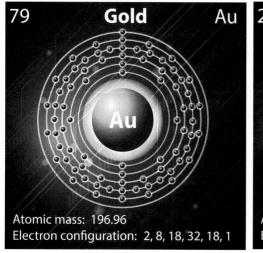
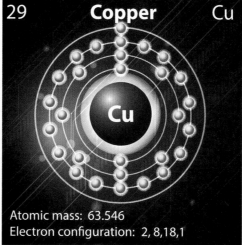

Figure 20-2. Gold and copper atoms have a single electron loosely held in their outermost ring.

of electric power. We use insulators to cover conductors to restrict the flow of current to intended circuits. Insulated coverings prevent unintentional human contact with energized conductors, which could result in an electric shock.

Common insulating materials include glass, porcelain, rubber, and many plastics, **Figure 20-3**. Often, hand tools designed for working around electricity are built with extra insulation in the handles to prevent shock if inadvertent contact is made with a live conductor. Electric power tools have insulation around the conductors and are often constructed with insulated plastic handles and housings to prevent human contact with electric current.

Electric Circuits

The power of electricity can seem mysterious, but it is not magic—it is mechanical. Electricity is predictable and follows a logical path called a *circuit*. A complete circuit flows from the power supply through conductors and a load and back to the power supply. For our purposes, a *load* is that part of the circuit that does the work. Power tools, lights, and electrical equipment are all examples of loads. Loads convert electric energy into other usable forms of energy. We can use a switch to control when electric power is delivered to a load.

Figure 20-3. Electric insulators are often color-coded to aid in routing power through complex systems.

AgEd Connection: Electrical Systems

A basic understanding of electricity and electrical components will prove useful in a variety of situations. FFA members participating in an agricultural technology and mechanical systems career development event (CDE) will have an opportunity to demonstrate their understanding of electricity by performing various problem-solving and skill activities. They may be required to wire a circuit, calculate amperage for an electric motor, assemble electrical components, or perform numerous other activities associated with electric theory and electrical materials.

Supervised agricultural experience (SAE) programs in the area of agricultural mechanics repair and maintenance may involve working in a job where you must troubleshoot electrical components and determine the best method of repair. Designing and installing the electrical system for a livestock facility would be a great match for an agricultural mechanics design and fabrication SAE. If electricity is an interest of yours, ask your agriculture teacher today how you can gain electrical experience for your SAE program.

itman__47/Shutterstock.com

Amps, Watts, Volts, and Ohms

To understand electric circuits, you need to understand the forces involved and the units of measurement used to describe them. Electric current is measured in *amperes (amps)*. One ampere is approximately 6,240,000,000,000,000,000 electrons passing through a given point on a conductor per second.

Voltage is the potential energy between two points along an electrical conductor. If electricity flowing through a conductor is compared to water flowing through a hose, the amount of water would represent the current, or amps, and the pressure of the water would be analogous to the voltage.

Now imagine putting a nozzle on the hose to restrict and concentrate the force of the water. In electrical terms, the restriction to flow is known as *resistance* and is measured in *ohms*. All loads have resistance. Ohms are represented by the omega symbol (Ω). *Ohm's law* defines the relationship between voltage, amperage, and resistance with the mathematical formulas:

$$\text{Voltage (V)} = \text{Current (I)} \times \text{Resistance (R)} \quad (V = I \times R)$$
$$\text{Current (I)} = \text{Voltage (V)} \div \text{Resistance (R)} \quad (I = V \div R)$$
$$\text{Resistance (R)} = \text{Voltage (V)} \div \text{Current (I)} \quad (R = V \div I)$$

The *watt* is the standard unit of electric power; it is the rate at which work is done. The relationship between power, current, and voltage can be expressed mathematically by the following formula: Power = Voltage × Current ($P = V \times I$). This formula, which is known as Watt's law, can be rearranged mathematically to allow you to find any of the three components. Additional math allows you to calculate resistance as well using a derivative of this formula. See **Figure 20-4**.

Loads

In agriculture, electrical loads convert the power of moving electrons into light, heat, sound, and mechanical motion. Lighting is commonly incandescent, fluorescent, or more recently, light-emitting diode (LED). Electric heaters may be used for warming products, livestock, people, or plants. Communication devices and alarms convert electricity into sound and sometimes light as well. Electric motors convert electric energy to rotary motion that may be converted to linear motion through different mechanical methods to power machinery and tools.

Lighting

Traditional electric lighting is based on the use of *incandescent* bulbs. Incandescent lighting uses a tungsten resistor coil supported in a vacuum tube. An incandescent lightbulb has a thin, often frosted or colored glass envelope. Inside the glass is a nonreactive gas such as argon or nitrogen, which helps preserve the tungsten filament in the center. Electric current passing through the filament heats it to about 4500°F (2482°C). The tungsten gets white-hot at that heat level and emits a great deal of visible light. Incandescent lighting is still the most common lighting for DC applications on agricultural vehicles. See **Figure 20-5**.

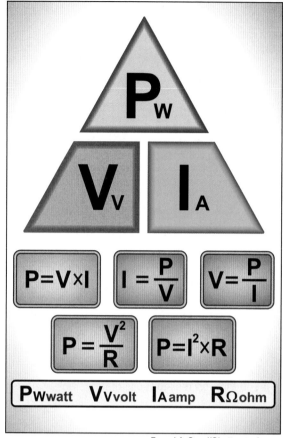

Fouad A. Saad/Shutterstock.com

Figure 20-4. Mathematic relationship between power (in watts), voltage (in volts), current (in amps), and resistance (in ohms).

STEM Connection

Using Watt's Law

Suppose a blow dryer is needed in a barn to groom livestock. The dryer is rated at 1680 watts when powered by 120 V of AC electricity. What amperage is required to run the blower? We can plug the values we have into the Watt's law formula to get the answer. Referring again to **Figure 20-4**, use the version of the formula that solves for current: I = P/V. 1680 watts ÷ 120 V = 14 amps. With this information, we can determine which circuit in the barn can handle the load of the dryer.

Goodheart-Willcox Publisher

In another example, a poultry facility has a main light circuit with 18 light receptacles. The current 60 watt bulbs do not provide enough light. What wattage bulbs could we use as replacements if the circuit is rated to operate at 15 A and 120 V? Plug the values we have into the formula, P = V × I; 120 V × 15 A = 1800 watts. Divided equally among the 18 receptacles, we can safely use up to 100 W bulbs to increase the level of lighting in the poultry facility.

524 Agricultural Mechanics and Technology Systems

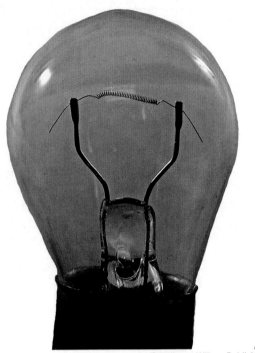

Goodheart-Willcox Publisher

Figure 20-5. In an incandescent bulb, current is passed through resistor coils of tungsten supported in a glass tube.

A *halogen* lamp has a tungsten filament encased inside a bulb that is much smaller than an incandescent bulb. The bulb is made of quartz, because it is so close to the filament that it would melt if it were made of glass. The bulb contains a gas from the halogen group. This gas combines with tungsten vapor. If the temperature is high enough, the halogen gas combines with tungsten atoms as they evaporate and redeposits them on the filament. This redeposition process extends the life of the filament. It is possible to run the filament hotter and obtain more light per unit of energy. See **Figure 20-6**. Because the quartz envelope is so close to the filament, it is extremely hot compared to an incandescent bulb.

Introduced in the 1930s, fluorescent bulbs function by ionizing mercury vapor in a glass tube. The ionized gas emits particles at ultraviolet (UV) frequencies. The UV light causes a phosphor coating on the inside of the tube to glow, releasing standard visible light. Fluorescent lights are energy-efficient for lighting indoor spaces, **Figure 20-7**. The small amount of mercury in each bulb makes responsible disposal of worn-out bulbs an important environmental issue.

LED lamps are composed of single or multiple light-emitting diodes. The larger electrode inside the LED bulb is the cathode (–). The cathode acts as a heat sink and support for a tiny LED semiconductor crystal. A *heat sink* is a device that absorbs heat, often protecting more delicate surrounding parts. The smaller electrode is the anode (+), which is connected by a fine wire (not a filament) to one side of the LED crystal. See **Figure 20-8**. LEDs are diodes, which allow electricity to flow in only one direction. Therefore, they must be installed with the correct polarity. The cathode must be connected to the negative side of the circuit (ground) for the light to illuminate.

A *NorGal/Shutterstock.com*

B *NorGal/Shutterstock.com*

Figure 20-6. Halogen lighting is used in applications where a bright focused light is desired. A—Football stadium lighting. B—Halogen headlamp.

Goodheart-Willcox Publisher

Figure 20-7. The fluorescent lighting in this shop is much more efficient than attempting to illuminate the same space with incandescent lamps.

The filaments of traditional incandescent bulbs stretch, weaken, and eventually fail. LED lamps avoid this weakness by not using filaments. LED lights are initially more expensive to install than traditional incandescent lights, but they last longer because they stand up to vibration more effectively than incandescent bulbs. The life expectancy of an LED is about six times that of an average incandescent bulb. LED lighting also consumes significantly less power than incandescent lighting, drawing as little as one-tenth the power of comparable incandescent bulbs.

Heat Applications

The agricultural uses of electrically generated heat are usually limited to fairly small, concentrated applications. Electricity is generally not an economical source of heat for heating large, open spaces. Water heaters and heat lamps for young animals are examples of common electric heating loads used in agriculture. Basically, an electric heating element is a type of conductor that converts electric energy to heat. The element itself heats up as electric current passes through, **Figure 20-9**.

Motion Applications

Electric motors are used to convert electric energy to mechanical motion useful for numerous tasks in

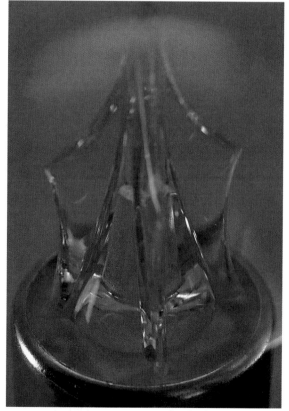

Goodheart-Willcox Publisher

Figure 20-8. The yellow-orange part of this LED assembly is the diode. The crystalline structure on top is designed for maximum diffusion of the generated light.

Goodheart-Willcox Publisher

Figure 20-9. Heat generated by the movement of electrons flowing across the coiled conductors of this dryer heating element is transferred to a stream of air driven by a blower motor.

agriculture. Much the same way that generators use motion and magnetic fields to produce electricity, electric motors use electricity to create magnetic fields that produce motion, **Figure 20-10**. Electric motors may be small enough to drive the hard drive of a computer or large enough to exert tremendous force.

Switches

A *switch* is any device that is used to interrupt the flow of electrons in a circuit, stopping or redirecting the flow of electricity, **Figure 20-11**. Switches are used to stop and start the supply of electric power to a load. They are either completely on (closed) or completely off (open). The simplest type of switch physically connects electrical conductors together and pulls them apart at a contact point. Electricity tends to arc, or spark, when the contacts are moved. In this way, switches may experience wear, eventually leading to a need for replacement.

Switches are always installed in the hot side of a circuit to minimize the length of an energized conductor when the switch is open. This helps to

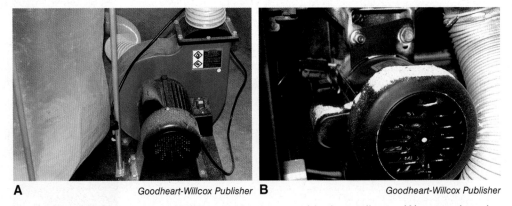

A *Goodheart-Willcox Publisher* B *Goodheart-Willcox Publisher*

Figure 20-10. Electric motors may be exposed, as on this dust collector (A), or enclosed in a housing or cabinet (B), as is this table saw motor.

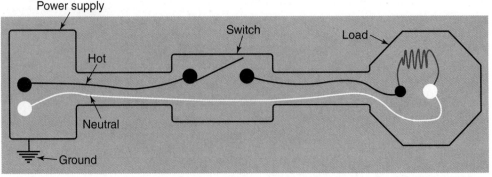

Goodheart-Willcox Publisher

Figure 20-11. A complete circuit includes a power supply, conductors, a load, and a switch.

prevent electrocutions and shorts to ground that could harm or kill. If a switch were located on the neutral side of a circuit, a load could be live but not operating. If someone were to touch the load or any live part of it while being in contact with ground, electric current would flow through the person. This is why a switch must be on hot conductors, not neutral conductors.

Common switches found in household circuits for operating lights, fans, and other loads come to rest in either the *off* or *on* position. This means that when the switch is off, it stays off. When it is on, it remains on. However, some machines and equipment operate only when a switch is engaged and automatically shut off when pressure is released on the switch, **Figure 20-12**. This type of switch is known as a *momentary switch*.

A typical light switch has two positions and one set of contacts, known as a *single-pole, single-throw (SPST) switch*. Another common type of switch is the *single-pole, double-throw (SPDT) switch*, **Figure 20-13**. SPDT switches have one common terminal and two opposing terminals that can connect to the common. SPDTs may be used in pairs to operate a load, such as lighting, from two separate locations. In this application, each SPDT operates as a three-way switch, **Figure 20-14**.

Protecting the Circuit

Electric circuits can be damaged in two basic ways: physical forces from the outside can damage insulators and conductors, and surges of excess current

Goodheart-Willcox Publisher

Figure 20-12. The trigger switch on this power drill is an example of a momentary switch. When the trigger is engaged, the drill is on, and when the trigger is released, it is off.

Goodheart-Willcox Publisher

Figure 20-13. The SPST switch on the left is marked *On* and *Off*. The SPDT switch on the right has no markings because contacts inside the switch are connected to conductors in both switch positions.

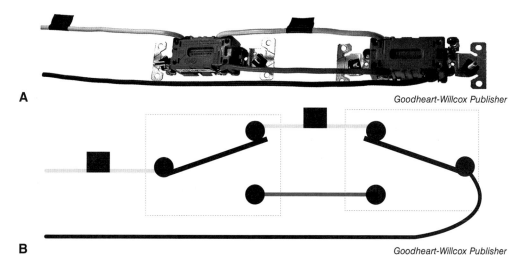
Goodheart-Willcox Publisher

Figure 20-14. A—Two SPDT switches are used together to connect a three-way circuit. B—Diagram of a three-way circuit.

Figure 20-15. The effectiveness of the insulation on this conductor has been compromised by rodent activity.

can overheat electrical system components. Examples of external physical force might be gnawing rodents, invasions of insects, flooding, or damage by people making modifications to a building, **Figure 20-15**. To protect electrical systems from physical damage, conductors are enclosed as much as possible in protective covers that reduce chances of damage to the insulation and the conductors inside.

Fuses

Electrical surges can be caused by lightning, faulty equipment, and physical damage to electrical systems. To protect from surge, the most basic circuit protection device is the *fuse*. A fuse is simply a very thin wire enclosed in an insulating casing that is connected like a switch into the supply side of a circuit, **Figure 20-16**. When a circuit is in use, all current flows through the fuse. As a result, the fuse is exposed to exactly the same current as every other point along the circuit. The fuse is engineered to disintegrate if the current climbs too high; the resulting heat burns up the wire, "blowing" the fuse. Sacrificing the fuse breaks the circuit before the excess current can damage more expensive components within the circuit. The drawback of fuses is that they work only once. Every time a fuse functions to protect a circuit, it has to be replaced with a new fuse.

Circuit Breakers

A *circuit breaker* serves the same purpose as a fuse by breaking a circuit if the current rises to an unsafe level. Their advantage is that circuit breakers can be used over and over again. The common circuit breaker consists of a simple switch governed by an electromagnet. When current rises too high, the electromagnet actuates to open the switch. This shuts off the electricity.

Figure 20-16. Several different types of fuses are shown here. Common to all fuses is an element that is designed to fail safely if there is a current overload. Fuses are a weak link wired in series in an electric circuit.

In another type of circuit breaker, a bimetallic strip is constructed of two different metals that expand at different rates. When the strip is heated, the heat causes the strip to bend by a predictable amount. In a circuit breaker, the hot wire is connected to the two ends of the switch. When the switch is in the *on* position, electricity flows from the input terminal through the bimetallic strip, up to the moving contact, across to the stationary contact, and to the output terminal. A surge that increases the current heats the bimetallic strip. When the current reaches an unsafe level, the bending strip pulls down a metal lever connected to the switch linkage. The entire linkage shifts, tilting the moving contact away from the stationary contact to break the circuit. See **Figure 20-17**.

Goodheart-Willcox Publisher

Figure 20-17. Heat buildup from excessive current causes the bimetallic strip to bend just enough to trip the breaker, releasing the contact points and cutting power to the circuit.

Typically, fuses or circuit breakers are located at the service entrance or at equipment disconnects. These locations afford the most protection to larger sections of the electrical system and the building structure.

Ground Fault Circuit Interrupter

Circuit breakers are designed to protect equipment and buildings. *Ground fault circuit interrupters (GFCIs)* are engineered to protect humans from electric shock. See **Figure 20-18**. According to the National Electrical Code, a *ground fault* is a conducting connection, intentional or accidental, between any electrical conductor and any conducting material that is grounded or that may become grounded. Electricity will always seek a path of least resistance to the earth or ground. In a ground fault, electricity has found a path to ground, but typically it is a path the electricity was never intended to follow.

A GFCI constantly monitors the current in both the neutral wire and the hot wire in a circuit. Under normal operating conditions, the current in both wires should be exactly the same. If somebody accidentally contacts the hot wire or a ground fault occurs, the current level surges in the side of the circuit making the best contact to ground. The GFCI reacts to differences in the amount of current as small as 4 to 5 milliamps on either side of the circuit by breaking the circuit immediately, preventing electrocution. Unlike a circuit breaker, a GFCI does not have to wait for current to build up heat. It reacts much more quickly.

GFCIs should be installed in every circuit where appliances and power tools are used in close proximity to water. GFCI devices are typically included in convenience outlets, but may also be incorporated into circuit breakers and portable plug-in equipment,

StevenRussellSmithPhotos/Shutterstock.com

Figure 20-18. Ground fault circuit interrupters, or GFCIs, are designed to reduce the chances of serious injury from electric shock.

Goodheart-Willcox Publisher

Figure 20-19. This circuit breaker is intended to control a circuit in an environment that is subject to frequent high levels of moisture.

Figure 20-19. Locating a GFCI closer to the locations of predictable contact with electricity and moisture provides a quicker reaction time to protect people from electric shock.

Generally, it is a better idea to protect a freezer or refrigerator located in a garage or shop by connecting the frame directly to a grounded conductor rather than using a GFCI because a surge of current could trip a GFCI without warning. The loss of power could cause the loss of the contents. Note that the NEC contains a list of locations and circuits that should be protected by a GFCI.

Alternating Current

Electricity supplied by power companies is *alternating current (AC)*. The power source is constantly alternating, or switching, *polarity*. Negative voltage goes to positive and positive voltage switches to negative. Changing polarity changes the direction of electron flow. In an AC circuit, the electrons regularly switch their direction of flow in measurable cycles. During the first half of a cycle, electrons flow in one direction, and in the second half, they flow in the opposite direction. The number of cycles per second is an AC signal's frequency, which is measured in units called *hertz (Hz)*. See **Figure 20-20**.

In the United States, electricity is uniformly supplied at 60 Hz, or 60 cycles per second. At 60 Hz, the power to lights and other loads is at a level of zero 120 times each second. This happens very quickly, and our eyes become adjusted to this rate of switching current, so we do not notice the effect on the lights. Travelers from the United States visiting Europe may be aware that the lights flicker slightly. European power companies regularly supply power at 50 Hz. The difference is recognizable to someone conditioned to 60 Hz power delivery.

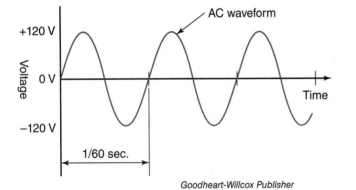

Goodheart-Willcox Publisher

Figure 20-20. This AC waveform is a graphical representation of an AC signal over time. Note that the signal alternates polarity during each cycle and that there are 60 cycles per second.

Hot, Neutral, and Ground

Within an AC circuit, conductors are considered to be either hot, neutral, or ground. The hot conductor is the wiring that is energized (has voltage). Sometimes hot conductors are referred to as being *live*, **Figure 20-21**. Hot wires are the portion of the path or circuit that delivers power from the source to the load. Hot conductors are typically color-coded black or red, but they may be other colors in some applications.

Neutral conductors have voltage and carry a current only when the circuit is complete. Neutral conductors in properly functioning circuits have no charge when the load is not operating. The standard color code for a neutral conductor is white.

A *ground conductor* is often included in electric cable as a bare, noninsulated conductor, **Figure 20-22**. If it is insulated, green is the standard color used to indicate a ground wire. Ground conductors are not intended to be part of an AC circuit. Instead, they are a failsafe. If the circuit has a problem or fails, the ground conductor allows the system to fail safely, carrying the current to the earth where it is safely dissipated.

Goodheart-Willcox Publisher

Figure 20-21. Using a voltage indicator to check for a live conductor. The red indicator light signifies a hot circuit.

Goodheart-Willcox Publisher

Figure 20-22. In this common conductor, the hot wire is black, the neutral wire is white, and the ground conductor is bare.

Phase

We have said that AC current changes polarity 60 times a second (60 Hz), and that a typical power supply provides 120/240 volts at the service entrance of a building. To complete the circuit to a common power tool such as a drill, a single hot wire is required to supply 120 V of current, and a neutral wire completes the circuit back to the power source. This is an example of *single-phase power*, one hot and one neutral conductor. Only one AC signal travels on these conductors.

Some equipment requires more power than can be efficiently delivered through a typical single hot wire. Many of these larger pieces of equipment or appliances are set up to operate on 240 V of power. To supply 240 V of AC power, two hot conductors are used without a neutral conductor, **Figure 20-23**. The power is directed from hot to hot. This works because the phases of the two hot connectors are offset by 180°. Imagine that when the voltage on line A is positive, the voltage on line B is negative, and 60 times a second they switch. See **Figure 20-24**. Equipment powered in this manner is typically referred to as *split-phase power* and rated as 240 V AC single-phase equipment.

A different setup is often used in industry, schools, and other places where a large amount of power is consumed.

Goodheart-Willcox Publisher

Figure 20-23. This plug connects a welding machine to the AC power source. What other equipment or appliances might use this type of plug?

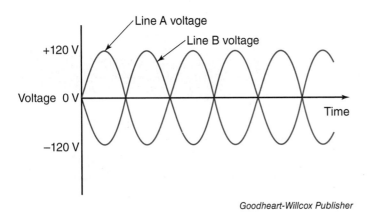

Figure 20-24. In split-phase power, there are two AC signals. When one signal is positive, the other is negative.

Three-phase power uses three hot conductors with three separate AC signals. The phases are offset by 120°. See **Figure 20-25**.

Three-phase power is more efficient than single-phase, but the wiring is more complex and can be configured several different ways. The services of a competent professional electrician should be used when installing or performing repairs on three-phase circuitry.

Direct Current

Electricity that flows in only one direction is called *direct current (DC)*. Direct current is used in motor vehicles, battery-operated equipment, and welding. In older motor vehicles, generators produced DC electricity. Today, AC-producing alternators, in conjunction with other components, convert the AC to DC.

A *battery* is a storage device for an electric charge. In a complete DC circuit, electricity flows from a negatively-charged terminal, through conductors, through a load, and through more conductors to a positively-charged terminal. Batteries use chemical reactions to charge and discharge. See **Figure 20-26**. Batteries are recharged by pushing an electric current through them backward. This reverses the chemical reactions to replenish chemical fuel. Over time, by-products from these reactions slow the reaction process, reducing the efficiency of the chemical reactions in storing and releasing energy. Worn-out batteries should be disposed of responsibly because batteries contain heavy metals, such as lithium, lead, cadmium, and nickel, which can be harmful to the environment. State and federal laws govern the recycling and disposal of batteries.

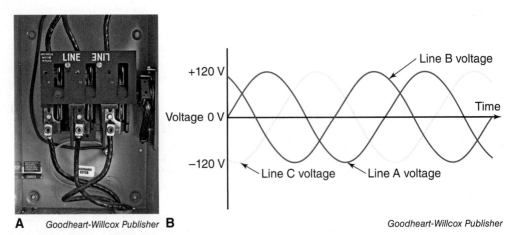

Figure 20-25. A—The three side-by-side black conductors in this service disconnect box indicate the power supply is three-phase. Only qualified individuals with the proper training should be allowed to repair or modify this circuit. B—In a three-phase circuit, the AC signals are phased at 120°.

Sources of Electric Power

In agriculture, there are several ways to acquire useful electric power. Most electric power is converted from mechanical energy using machines, known as *alternators* and *generators*, that pass coiled conductors through a magnetic field in order to induce the flow of electrons.

Alternators

In most cases, power-generating alternators are driven by some type of fuel-consuming apparatus, **Figure 20-27**. Most large power plants are fueled by coal, natural gas, or nuclear power. Some power plants use the energy of moving water to drive their alternators. These sources depend on geographic locations with constant moving water, so their numbers are limited.

Generators

Generators convert mechanical energy to AC electricity by moving a conductor through a magnetic field. See **Figure 20-28**. The armature of a generator consists of conductors (bundled wires) that are spun between two or more energized electromagnets (field coils) attached to the generator case. The electricity moves through the armature and flows through carbon brushes to the commutator. The commutator distributes the charge to the loads (parts of the electrical circuit that resist the flow of current) and reverses the charge into the field coils. This reversal strengthens the magnetic field, which in turn makes the armature generate even more electricity.

The field coils require a ground to generate a magnetic field. Improving the connection to ground increases the current in the field coils, strengthening the magnetic field. Regulation of the generator output is achieved by controlling the direct connection of the field coils to ground or by grounding through a resistor that limits the current's connection to ground.

Wind Energy

The movement of wind may be used to drive windmills that convert motion to electric energy. The numbers of these facilities is also limited by geography. To be effective, they can be used only where a constant wind can be predicted. It takes a large amount of

S1001/Shutterstock.com

Figure 20-26. Because lithium is a highly reactive element, a great deal of energy can be stored in its atomic bonds when it is used to make lithium-ion batteries. This lightweight battery is used to power an electric fence energizer.

James Marvin Phelps/Shutterstock.com

Figure 20-27. This power plant converts coal into electric energy.

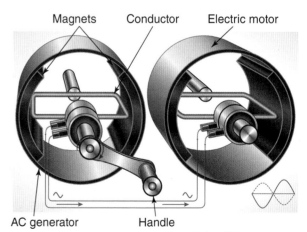

Fouad A. Saad/Shutterstock.com

Figure 20-28. Turning the handle on this generator moves the conductor through the magnetic field created by the magnets. This produces electricity to run the electric motor.

Sarah Fields Photography/Shutterstock.com

Figure 20-29. Wind farms in west Texas and other states use huge turbines to generate clean, renewable energy.

wind to make useful amounts of electricity. Electricity-generating windmills are typically located in wind farms made up of many windmills working together to produce commercially useful electric current, **Figure 20-29**.

Solar Energy

The energy of the sun may also be used in a couple of different ways to generate electricity. Large solar-collecting facilities built in desert areas combine thousands of mirrors or heat-collecting troughs to concentrate the sun's energy. The resulting heat is used to generate steam that turns turbine-driven alternators. Solar energy can also be harnessed by *photovoltaic cells* that absorb photons of light and release electrons that can be used as electricity, **Figure 20-30**. For more information about how photovoltaic cells work, refer to Chapter 22, *Wiring DC Circuits*.

Like wind energy, solar power requires a great number of energy conversion devices in order to be useful on a large scale. Solar power plants require large areas of dedicated acreage in locations where plentiful sunlight is predictable.

Fabio Alcini/Shutterstock.com

Figure 20-30. The solar panels on the roofs of the buildings on this hog farm supplement electric energy purchased from the local electric cooperative.

Distribution of Electric Power

Electricity in the United States is sold by the kilowatt-hour (kW-h), a unit of energy equal to a constant delivery of 1000 watts (W) over the period of 1 hour. A power tool rated at 1000 W operating for 1 hour uses 1 kilowatt-hour of energy. A 100 W lightbulb operating for 10 hours also uses 1 kilowatt-hour.

Alternators in power plants typically generate 15,000 volts (V) of electric power. This is passed through a *step-up transformer* that increases the voltage of the electricity leaving the power plants to as much as 750,000 V. This high voltage is necessary for moving large quantities of power over long distances. Electricity is sent from the power station to substations that use *step-down transformers* to reduce voltage to levels of 25,000 or fewer volts. This is the electricity in the typical residential power line that is routed to homes and businesses. The final step in delivery is sending the electricity through another step-down transformer that reduces the voltage to common usable levels of 120 V and 240 V, **Figure 20-31**. The power company's system is connected to the building at the service entrance. A meter at the service entrance records the kilowatt-hour usage, **Figure 20-32**.

Voltage Drop

All conductors and loads have resistance. Loads are intended to have resistance. Conductors are not intended to exert resistance on a circuit. Any amount of resistance will decrease the potential energy, or voltage, of a current as it flows through the resistance. This decrease results in a loss of energy called *voltage drop*. An electric load is an intended voltage drop. A poorly sized conductor or a poorly made electrical connection is an unintended voltage drop.

Voltage drop becomes a significant issue when electricity has to be transmitted over long distances. Electric companies typically provide 120/240 VAC at 60 Hz from the transformer. If the service entrance is several hundred feet or farther from the transformer, a voltage drop must be considered when planning electrical systems. Using long extension cords or locating equipment long distances from the service entrance can have the same effect. For example, an exhaust fan motor rated to operate at 12.5 amps and 115 volts will burn up and be ruined after a few months of running at a lower voltage of 105 volts created by a voltage drop problem. Voltage drops should be kept to less than 5% of rated voltages; 2% to 3% is even better.

How do we deal with voltage drop? The power company is not going to supply current at a higher voltage to compensate. Instead, it is necessary to choose a conductor with lower resistance. By increasing the diameter of the conductor, you can reduce the resistance over distance. The result is a lower voltage drop. Tables of recommended conductor sizes for required amperages and lengths of run are available, **Figure 20-33**.

Goodheart-Willcox Publisher

Figure 20-31. A step-down transformer is necessary to reduce the voltage supplied to homes and businesses to 120/240 VAC.

Safety around Electricity

Electricity can be dangerous, even fatal. Only qualified persons should work on or repair electrical systems and connections. The Occupational Safety and Health Administration (OSHA) defines a qualified person as one who has received adequate training. Nothing is a substitute for understanding the nature of electricity when faced with a need to service electrical systems.

A common misconception is that a shock of 15,000 volts is more likely to kill than a shock of 50 volts. The true measure of shock's intensity is the amount of current (amperes) forced through the body, not

Goodheart-Willcox Publisher

Figure 20-32. This meter at the service entrance of a building records the kilowatt-hour usage.

536 Agricultural Mechanics and Technology Systems

Goodheart-Willcox Publisher

Figure 20-33. Several voltage drop calculators are available online. Smartphone apps are available as well.

Goodheart-Willcox Publisher

Figure 20-34. The indicator light on this voltage detector indicates the presence of electric voltage when the probes are connected between a conductor and a confirmed path to ground.

the voltage. Any electrical device connected to a typical residential circuit can, under certain conditions, transmit a fatal current. Typically, electric currents between 0.1 and 0.2 amperes are lethal. Any current over 10 milliamps (0.01 amp) can possibly produce a painful to severe shock. The actual resistance of the body varies, depending on the points of contact and the skin condition. Dry skin can have 500 times the resistance of wet skin. The best way to avoid shock and electrocution is to keep electrical current confined to predictable paths and to not touch energized conductors. It is a good idea to assume that any conductor associated with an electric circuit is hot unless testing proves that it is not.

Tools for Electrical Safety

There are a number of tools and devices designed to keep people safe when working around electricity. However, even the very best tools and testing equipment cannot replace training, education, and good sense. To safely harness the power of moving electrons, take the time to really understand how electricity works.

A *voltage detector* is a device that determines if voltage is present in a circuit, **Figure 20-34**. Older two-lead styles must actually make physical contact with a conductor and a known ground. A small lamp glows to indicate a hot conductor. Used with a known good hot conductor, this two-lead voltage tester can also check the function of questionable ground conductors.

Newer electronic models of voltage detectors light up and emit an audible beep when placed in proximity of a hot conductor. See **Figure 20-35**. Electronic voltage detectors offer the advantage of detecting voltage through normal insulation materials without the need to expose a hot conductor.

A *receptacle tester* is a device that plugs into a receptacle or outlet to verify that circuits are wired properly. Simply plug the device into a convenience outlet and read the combination of indicator lights according to a chart attached to the tester, **Figure 20-36**.

Most *multimeters* can measure voltage, current, and resistance. One of their many functions is determining whether voltage supplies are correct. Checking loads for proper resistance according to the manufacturer's specifications can often identify faulty equipment. Using the multimeter to check for *continuity* can save much time in diagnosing problems

Goodheart-Willcox Publisher

Figure 20-35. A no-contact voltage detector allows you to check a circuit or conductor to be sure it is free from voltage before exposing a bare conductor.

in electric circuits. A circuit that has continuity is complete (not broken) and allows the flow of electricity. Some multimeters have a visual as well as an audible indication of continuity, **Figure 20-37**.

Safe Practices

Using the proper tools and personal protective equipment (PPE) is also critical to maintaining your own safety. When working around electricity, use only insulated hand tools that are designed to be used near live or possibly live conductors. Insulated tools are built with extra insulation to protect against accidental contact. These tools should be marked with the voltage rating of the protection provided. Of course, the best plan is to be

> **Safety Note**
> An electronic tester is powered by small batteries. Check the batteries to make sure the tester is in good working order before relying on its readings.

Goodheart-Willcox Publisher

Figure 20-36. Simply plugging this receptacle tester into a convenience outlet and observing the combination of lighted indicator lamps allows you to determine the functional condition of an electrical circuit.

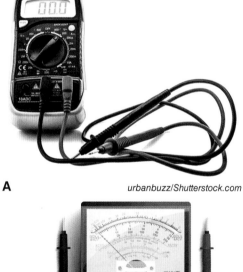

A

urbanbuzz/Shutterstock.com

B

Sirapob/Shutterstock.com

Figure 20-37. A—A digital multimeter. B—An analog multimeter.

Career Connection

Electrician

bbernard/Shutterstock.com

Job description: Electricians work on a variety of activities, including installation and maintenance of electric power, lighting, and control and communication systems in homes, businesses, and industry. In agriculture, this includes working on animal and plant environments as well as processing facilities for food and other agricultural products. Being an electrician requires skills in science, math, and engineering. People in this career work closely with people in many other careers in agriculture and related industries.

Education required: Most electricians have graduated from high school or attained a General Educational Development (GED) certification. They enter the field through an apprenticeship that typically lasts two to four years. Training in technical schools is often included as part of the process. The final step in the apprenticeship is to be recognized as a journeyman electrician. Upon completion of apprenticeship training, individuals may be licensed by their state as electricians.

Career success: This job may be a fit for you if you enjoy working with technical information, have the ability to pay close attention to detail, and like working with tools.

sure the power is shut off before working. Extra insulation on tools should be considered a failsafe measure, **Figure 20-38**.

Gloves and other personal protective equipment are available for work around electric systems, **Figure 20-39**. Like insulated hand tools, they are rated as to the level of protection. PPEs used for electrical work should be carefully inspected before each use and considered a last line of defense against electrical hazards.

maxim ibragimov/Shutterstock.com

Figure 20-38. Some needle-nose pliers are insulated as a failsafe measure for electrical work.

dcwcreations/Shutterstock.com

Figure 20-39. Even though all electrical work should be performed only after shutting off the power, gloves with special insulation rated for electrical work, used in combination with insulated tools such as this screwdriver, help to protect workers from accidental contact with electrically charged conductors.

CHAPTER 20 Review and Assessment

Summary

- Electricity is a type of energy that consists of charged particles, such as electrons. Electric current is the flow of electrons along a path.
- Electric current represents the amount of electricity flowing across a conductor.
- Insulators impede the movement of electrons from one atom to another.
- A complete circuit flows from the power supply through conductors and a load and back to the power supply.
- Ohm's law can be used to find the amount of current, voltage, or resistance, as long as two of the three values are known. Power can also be calculated using Ohm's law: $P = V \times I$.
- Lighting is commonly provided by incandescent, fluorescent, or light-emitting diode (LED) sources.
- Types of switches include momentary; single-pole, single-throw; and single-pole, double-throw. A pair of SPDT switches can be arranged as two three-way switches.
- Fuses protect a circuit only once; circuit breakers can be used multiple times.
- Ground fault circuit interrupters (GFCI) protect humans from electric shock.
- The three types of power used in alternating current are single-phase, split-phase, and three-phase.
- In alternating current, the electrons regularly switch direction of flow in measurable cycles.
- Direct current is electricity that flows in only one direction.
- Generators and alternators are used to produce electric power.
- Solar energy and wind energy are alternative sources for electric power.
- The kilowatt-hour is the unit of measure for electricity sold in America. One thousand watts operating for one hour equals one kilo-watt hour. This same value (1 kW-h) can be calculated from other wattage values used over different lengths of time.
- Voltage drop becomes an issue when electricity is transmitted over long distances.
- Tools for working safely with electricity include voltage directors, receptacle testers, and multimeters.

Words to Know

Match the key terms from the chapter to the correct definition.

A. alternating current
B. ampere (amp)
C. circuit
D. conductor
E. direct current (DC)
F. fuse
G. ground fault circuit interrupter (GFCI)
H. multimeter
I. ohm
J. receptacle tester

1. A device that can measure voltage, current, or resistance in a circuit.
2. Electricity that flows in only one direction.
3. The unit of measure for electrical resistance.
4. Device that plugs into a convenience outlet to verify that circuits are wired properly.
5. Power source that constantly switches polarity back and forth between negative and positive.
6. The closed, logical path electricity follows.
7. Any material that readily supports electron flow.
8. Device engineered to protect humans from electric shock.
9. A very thin wire enclosed in an insulating casing that is connected like a switch into the supply side of a circuit.
10. Unit of measure for electric current.

Know and Understand

Answer the following questions using the information provided in this chapter.

1. Gold and copper are excellent conductors of electricity because atomically they each have a single _____ electron that is easily transferred across atoms.
   A. charged
   B. covalent
   C. neutral
   D. valence
2. Insulators are not useful for carrying current because they _____ the flow of electrons; instead, insulators are useful for controlling electricity.
   A. increase
   B. restrict
   C. redirect
   D. supplement
3. Electric current in which the power source constantly switches from negative to positive and back again is _____ current.
4. Which two alternative electrical power production methods require large areas of dedicated acreage to harness significant amounts of power?
   A. Solar and wind.
   B. Fuel cell and natural gas.
   C. Water wheels and tide turbines.
   D. Geothermal and coal.
5. A 100-watt lightbulb operating for 20 hours uses _____ kilowatt-hours of electricity.
6. A(n) _____ is the part of the circuit that does the work. Examples include power tools, lights, and electrical equipment.

7. _____ conductors become live when the circuit is complete.
8. The frequency of an AC signal is measured in _____, or cycles per second.
9. As a failsafe measure, switches are always installed on the _____ side of an electric circuit.
10. All loads have _____ measured in _____, which are represented by the symbol (Ω).
11. To compensate for _____ over long transmission distances, we should choose larger conductors with a lower resistance.
12. A(n) _____ is a weak link engineered into a circuit to serve as a failsafe in case of a current overload.
13. _____ are designed to protect buildings and equipment, while _____ are engineered to protect people from electric shock.
14. The true destructive measure of an electric shock's intensity is the amount of _____ forced through the body.
15. Use a multimeter to see if a circuit has _____, indicating that it is not broken and will allow the flow of electricity.

STEM and Academic Activities

1. **Science.** Investigate the periodic table and compare Bohr models of different elements. Prepare a short presentation explaining which elements make the best conductors and why.
2. **Technology.** Compare the costs of installing and using LED lighting to that of incandescent lighting in a simple structure such as a livestock barn. How long would it take to make LEDs more economical?
3. **Engineering.** Study the way a circuit breaker operates. Draw diagrams depicting other ways to incorporate this technology into useful items.
4. **Math.** Use a multimeter to measure the voltage in different areas of your school. Collect and record the data. Organize your data into graph form. Can you detect areas affected by voltage drop?
5. **Social Science.** Imagine that the price of copper rose to equal that of gold. What would happen to the use of electricity as a power source?
6. **Language Arts.** Look at the formulas for Ohm's law and for calculating electric power. Conduct research to determine where the symbols and words used in the formulas originated.

Thinking Critically

1. Lightning results from a buildup of static electricity. Why, then, has it been impossible thus far to harness lightning as a source of electric power?
2. Many farmers and ranchers use electric fencing to discourage farm animals from leaving their assigned fields or pastures. Why does the electric fence deliver a shock to animals that come in contact with it?

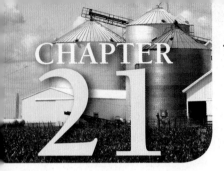

CHAPTER 21 Wiring AC Circuits

Chapter Outcomes

After studying this chapter, you will be able to:
- Plan installations using local codes and National Electric Code guidelines.
- Understand the construction and purpose of service entrance components.
- Perform service on a breaker box.
- Differentiate between series circuits and parallel circuits.
- Select appropriate circuit wiring materials and supplies.
- Differentiate among various types of enclosures and connectors.
- Install electrical devices such as receptacles and switches.
- Connect solid single conductor wires to devices such as switches and receptacles using screw terminals.

Words to Know

American Wire Gauge (AWG)	hardwired	parallel circuit	single-strand insulated conductor
armored cable	hot bars	pigtail	splice
cable ripper	junction box	raceway	strip gauge
conduit	multiple-strand insulated conductor	receptacle	tap
dielectric gel	National Electric Code® (NEC)	rigid conduit	terminal connection
distribution panel		Romex®	underground feeder (UF)
EMT conduit	neutral bar	screw terminal	wire nut
grounding rod	nonmetallic sheathed cable (NMC)	series circuit	wire stripper
handy box		service disconnect	
		service meter	

Before You Read

As you read the chapter, record any questions that come to mind. Indicate where the answer to each question may be found—in the text, from your teacher, in another book, on the Internet, or by reflecting on your own knowledge and experiences. Pursue the answers to your questions and share the answers with your classmates.

While studying this chapter, look for the activity icon to:

- **Practice** vocabulary terms with e-flash cards and matching activities.
- **Expand** learning with interactive activities.
- **Reinforce** what you learn by completing the end-of-chapter questions.

www.g-wlearning.com/agriculture

Alternating Current (AC) circuits are used to power homes, schools, businesses, and many other facilities that use electrical power. AC circuits run the lights and other electrical components that are commonly used in these areas. Electricians design and install the wiring in these facilities according to the requirements of the National Electrical Code (NEC) and are required to have training and certification in order to perform electrical work.

National Electric Code and Other Regulations

The National Fire Protection Association (NFPA) is the recognized authority on electrical safety. The NFPA publishes NFPA 70, or the *National Electric Code®*, which is commonly referred to simply as the *NEC*. See **Figure 21-1**. The NEC is adopted in all 50 states.

The NEC serves as the definitive guide for proper electrical installation and inspections to prevent people and physical property from damage by electrical hazards. All aspects of design and installation of electrical systems are covered by this resource. Local building codes enacted and enforced by cities and other government entities are either written according to or reference complete adoption of the NEC. Electricians must study, understand, and be able to pass testing over the NEC to become licensed. The NEC may be purchased in book form or accessed for free in a read-only format through the NFPA website. The NEC is updated every three years.

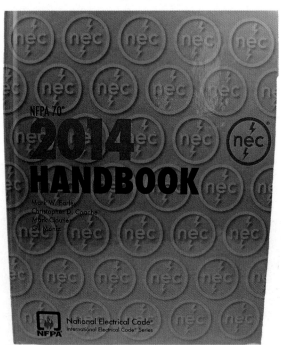

Goodheart-Willcox Publisher

Figure 21-1. The NEC is published and regularly revised by the NFPA. It is the authority for everything electric.

Service Entrance

The service entrance is the point in the electrical system where the ownership and responsibility for maintenance of components changes from the local power supply company to the consumer. The power company typically is responsible for everything on the supply side of the service entrance. The consumer must maintain everything past the connection point. In most cases, the power company will not connect to a service entrance until the service entrance has been checked and certified safe by a licensed electrician. All service entrances are not the same, but common components include a meter, a circuit breaker box, a main switch, a direct connection to ground, and rigid tubing, including a weatherproof entrance for conductors, **Figure 21-2**.

Service Entrance Components

The *service meter* measures the amount of electric current for which the power company charges the customer, **Figure 21-3**. Service meters may be analog or digital and measure billable electricity in kilowatt-hours. Older service meters required monthly reading by a power company technician. Newer meters are equipped to send a wireless signal directly to the power company. Meters should be serviced only by authorized persons.

The electric supply cables pass through rigid tubing known as *conduit* to protect them from damage prior to entering the main service panel. Conduit is galvanized metal or plastic tubing that is designed to provide a *raceway*, or protected path, for electrical conductors.

If the supply lines are overhead, the conduit entering the service panel or meter enclosure is extended into a mast or riser that keeps power company supply lines safely overhead. At the top of the mast is a component called a service head (also called a weatherhead). See **Figure 21-4**. The service head is angled and sealed to prevent moisture from entering the electrical system. A thick gasket seals around the

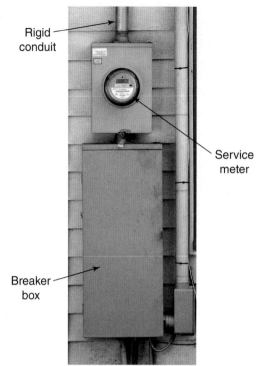

Goodheart-Willcox Publisher

Figure 21-2. Typically, each service entrance includes a service meter, a breaker box, a service disconnect, a direct ground connection, and rigid conduit topped with a service head.

A — *Goodheart-Willcox Publisher*   B — *Goodheart-Willcox Publisher*

Figure 21-3. In many jurisdictions, unauthorized access to meter connections is illegal. A—This meter is locked to prevent unauthorized access. B—The tag on this meter is designed to discourage tampering.

Figure 21-4. The service head entrance angle positions power cables so that rainwater drips off the cable insulation instead of entering the electrical system.

Figure 21-5. The main circuit breaker may also serve as the service disconnect for an electrical service entrance.

conductors to keep out moisture and reduce damage to the insulating covering on the wires. Conductors are positioned with large loops to direct drips away from the service head and to reduce stress from the weight and movement of conductors in the wind.

The *service disconnect* is a manually operated switch or other means of disconnecting the electric power to the entire system. The service disconnect may be located in the system before or after the service meter connection, depending on the local requirements. Several different styles of service disconnects are common. The service disconnect is often in the form of a main breaker at the top of a breaker panel, **Figure 21-5**. Another common type of disconnect, often used with air conditioning units, has a removable bridge that connects across blade-type connectors, **Figure 21-6**. The service disconnect may also be in a separate enclosure with an external lever operating a set of blade-type switches, one for each hot conductor. See **Figure 21-7**.

The *distribution panel* is the enclosure that protects the circuit breakers and main electrical connections. All branch circuits are connected at this point. The distribution panel is also known as a breaker box.

Each service entrance must be connected, or bonded, directly to ground. Usually this requires connecting a heavy copper conductor from the service distribution panel to a copper-clad steel *grounding rod* driven deep into stable earth. The clamp used to connect the ground conductor to the ground rod is engineered for this purpose. Common grounding rods are 8′ to 10′ long and 5/8 in diameter. See **Figure 21-8**. In some instances, local codes allow grounding connections to underground water piping.

The Breaker Box

Inside a typical breaker box, the two incoming hot conductors (the 120 V supply lines from the power company) are connected to screw terminals called lugs. A *screw terminal* is a connector that uses screw pressure to secure a wire or cable to another electrical device. These terminals connect either to a main service disconnect or directly to a pair of parallel strips of metal called *hot bars*. One terminal of each 120 V breaker connects to one of the hot bars; 240 V breakers connect to both hot bars. Hot conductors connect to the output terminals of the circuit breakers and distribute power to the individual circuits, one breaker per circuit.

The neutral conductor from the power supplier is connected by a screw terminal to a metal bar containing a row of smaller screw terminals. This *neutral bar* is connected to every neutral from all of the branch circuits, making the neutral continuous and common to all circuits. In the United States, the neutral bar is required to be connected to the ground connection and bonded to the metal enclosure of the distribution panel. To be bonded means that the electrical component is securely connected to a grounded conductor. Some breaker boxes have a separate ground bar that in US applications should be electrically bonded to the neutral bar. See **Figure 21-9**.

Safety Note

AC devices will operate in the absence of a neutral conductor if the device is connected to a hot wire and a ground wire. This is an unsafe situation. The ground wire should be energized only when there is an electrical system failure. The purpose of the ground conductor is to direct electric energy that has strayed from an intended circuit safely to the ground, away from people and livestock.

Distribution panels designed for outside installation have two covers. One is an inside cover that exposes only the switch levers of the breakers. The other is an outside cover or door that closes to protect the entire panel. Panels designed for indoor installation have only the inner panel that covers the electrical connections. The circuit breakers are attached to the box by a hooking mechanism. The mechanism attaches to two parallel steel bars, located one on each side of the rear wall of the box, and to one of the hot bars in the center of the box.

To attach a breaker, first connect the hot wire(s) leading to the circuit to the screw terminal on the breaker. Next, insert the hook on the rear of the breaker under the stationary bar and rock the breaker inward to push it into secure contact with the hot bar(s). Breakers that control a 120 V circuit contact only one of the hot bars, while 240 V breakers look like two 120 V breakers joined together and connect to both hot bars at the same time.

Goodheart-Willcox Publisher

Figure 21-6. Pulling the handle of this service disconnect removes a bridge across the two hot conductors, instantly cutting the power to the circuit.

Goodheart-Willcox Publisher

Figure 21-7. The blade-type contacts make secure connections in this service disconnect that also incorporates fuses to protect the circuit from current overload. The three hot conductors indicate that this is a three-phase circuit.

Goodheart-Willcox Publisher

Figure 21-8. Like an iceberg, the bulk of this 5/8" × 8' copper-clad grounding rod is hidden from view.

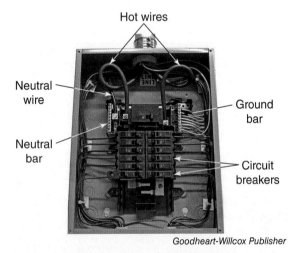

Goodheart-Willcox Publisher

Figure 21-9. It is important to learn the correct names of the components inside an electrical distribution panel.

Safety Note

For reference and for safety during maintenance, each breaker in a distribution panel should be labeled to identify where the circuit is located. Always shut off the main panel before removing the inside cover of a distribution panel. Never perform service or maintenance on an energized circuit. Switch the circuit breaker to *Off* to de-energize a circuit for service; never rely on simply turning off a switch. Even after switching a breaker to *Off*, use a voltage detector or a multimeter to ensure that the power is disconnected. See **Figure 21-10**.

Types of Circuits

A properly connected complete circuit flows from the power supply through conductors and a load or multiple loads and through other conductors back to the power supply. In a *series circuit*, multiple loads are arranged so each load is a conductor necessary for the function of the entire circuit. See **Figure 21-11**. If a load in a series circuit fails, such as a light burning out, the circuit is broken and no other loads will function. Wiring devices in series is usually limited to strings of cheaply constructed holiday lights or to circuits in which the failure of one device is intended to shut down the circuit as a safety measure. Wiring switches, fuses, and circuit breakers in series with the rest of the circuits they control or protect is an intentional use of the nature of series circuits.

In a *parallel circuit*, each load is wired separately so that the failure of one load does not have an adverse effect on the other loads. See **Figure 21-12**. In most household and commercial circuits, multiple loads are wired in parallel. These typical circuits are also protected by circuit breakers wired in series.

SAE Connection Electricians

Most electricians are busy with lots of work and earn good pay. If electrical circuits interest you, why not explore the potential job opportunities for electrical work in your area? Ask your agriculture teacher to recommend local electricians that you could job shadow or work for to gain valuable experience to use in your supervised agricultural experience (SAE) program. The experience that you gain could be a jump start for a great career.

Planning AC Circuits

An AC circuit begins at the service panel, passes through a circuit breaker, extends out over a hot conductor to a load (or loads), and returns via the common neutral wiring to the service panel. Selecting the correct wire is one of the first steps in planning an AC circuit. Most residential and commercial applications use copper wiring. Aluminum was once popular for household wiring, but it is no longer installed as residential wiring due to an excessive number of problems creating fire hazards. Copper wire conductors differ in three major ways: the size of the conductor, the structure or arrangement of the conducting material, and the type of insulating or protective covering.

Goodheart-Willcox Publisher

Figure 21-10. Always shut off the main panel before removing the cover. Never service or perform maintenance on an energized circuit.

Wire Size

Wire is sized according to the *American Wire Gauge (AWG)* system. Increasing the amperage a circuit is required to handle requires the use of a larger diameter wire. As wire diameter increases, the numeric designation of the wire gauge decreases. See **Figure 21-13**. For example, a 6 AWG conductor is much larger and carries more amperage than a 12 AWG conductor. **Figure 21-14** shows some common amperage/gauge requirements.

Residential wiring is made up of *single-strand insulated conductors* or *multiple-strand insulated conductors*. Single-strand wiring consists of a single, solid conductor. Multiple-strand (multi-strand) wiring includes several strands of wire twisted together into a single conductor. Single-strand conductors are the most common in residential wiring systems because they are less expensive to manufacture and easier to connect to electrical devices. Multiple-strand wires are suited for applications where the conductors may experience flexing or vibration. For applications such as welding cable, where extreme flexibility is required, a large number of very small wires are bundled together to increase flexibility, **Figure 21-15**.

ulegundo/Shutterstock.com

Figure 21-11. In a series circuit like this one, if one lightbulb (load) burns out, the circuit is no longer complete, so all of the lights go out.

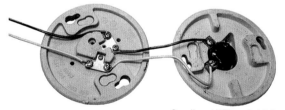

Goodheart-Willcox Publisher

Figure 21-12. When one light in this parallel circuit burns out, the remaining light continues to function.

Wire Colors

Common conductors are insulated with thermoplastic high-heat nylon (THHN) or thermoplastic high-heat water nylon (THWN) coverings. THHN wire is available in several colors that are used to identify their purpose in a circuit:

- Black wires—These wires should always be used for hot wires. They may connect to a switch, outlet, or light. Black wires should never be used to connect a neutral or ground.

AWG/Diameter in Inches	Maximum Amperage (single conductor)	Maximum Amperage (7 to 24 conductors)
20 / .032	6.0	3.5
18 / .040	9.5	4.9
16 / .051	20	7.0
14 / .064	24	10.5
12 / .081	20	12.0
10 / .100	30	18
6 / .170	55	38
4 / .200	70	49

Goodheart-Willcox Publisher

Figure 21-13. AWG–Amperage Rating Chart.

Amperage	Gauge Requirement
200 A service entrance	00 AWG
150 A service entrance	0 AWG
50 A arc welder	6 AWG
30 A air conditioning unit	10 AWG
20 A kitchen appliances	12 AWG
15 A lighting circuit	14 AWG
(Some local codes require 12 AWG minimum for all circuits.)	

Goodheart-Willcox Publisher

Figure 21-14. Gauge requirements for common amperage levels.

Safety Note

An exception to the color standards listed is the use of a white wire as a second hot wire. In this case, the white wire should be marked with a flag of black or red electrical tape to indicate that it is not neutral.

- Red wires—These wires are also used as hot conductors, especially when two hot wires are used in the same circuit.
- White wires—Neutral wires are always white.
- Yellow or blue wires—These wires may be used as hot wires on motor controls or as links between switches or from switches to loads.
- Green and bare wires—Green and bare wires should only be used as ground connections. Under normal conditions, ground conductors do not carry current.

According to NEC regulations, whenever a flagged white conductor is used to connect a switch, the conductor should bring power to the switch, not away from the switch. See **Figure 21-16**.

Romex® Conductors

Conductors in residential wiring systems may be single insulated conductors or, more commonly, multiple insulated conductors bundled together in a common protective sheath. *Romex*® is a type of wiring commonly used in residential applications. Romex® conductors can be used in damp environments, such as barns and other animal facilities. This type of wiring is categorized by the NEC as *nonmetallic sheathed cable (NMC)* or *underground feeder (UF)*.

NMC conductors consist of two or more insulated conductors bundled in a nonmetallic sheath, **Figure 21-17**. The outer protective sheathing on NMC cable is nonconducting and flame- and moisture-resistant. NMC cable is easily cut with lineman's pliers or diagonal cutting pliers. Cable cutters may be used to cut heavier gauge conductors. During installation, NMC must be securely fastened to the structure with supports spaced no farther than 4′ apart. When using staples to fasten the cable, take care not to puncture, cut, or crush the protective insulation.

Nonmetallic cables are identified according to the number and size of individual conductors inside the outer sheathing. See **Figure 21-18**. For example, a 12-2 NMC contains three 12 AWG wires, including two insulated conductors, one insulated hot wire, and one insulated neutral wire; the third conductor is a bare ground wire. A 10-3 NMC includes two hot wires and one neutral insulated 10 AWG wire, along with a bare ground. What would be inside a Romex® cable with the markings 14-2 NMC?

Underground feeder (UF) cables are similar to NMC cables except that UF cables have a solid plastic core. UF conductors are used for buried cable and similar applications, such as power to submersible water pumps.

Armored Cable

NMC conductors are not permitted by the NEC in commercial construction—a more secure form of protection is required. *Armored cable*, which has a flexible metallic sheathing that allows for extra protection, can be used instead. Several types of armored cable are available, **Figure 21-19**. In *bonded armored cables*, a bare ground wire remains in constant contact with the metal armor. The ground wire and armor together function in emergency grounding. Armored cable, like NMC, should be installed with supports at a maximum of 4′ apart.

Conduit

For many applications, NMC or armored cable does not offer adequate protection for electric conductors. In these cases, a conduit is required to enclose

Goodheart-Willcox Publisher

Figure 21-15. The diameter of multiple-strand copper wire must be greater than that of solid wire to carry the same amperage. This is due to the small gaps between the individual strands of multiple-strand wire, which allow heat to be retained more readily.

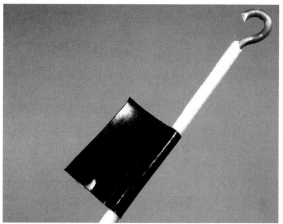

Goodheart-Willcox Publisher

Figure 21-16. A white wire that is used as a hot wire is marked with a flag of black or red electrical tape to alert others that it is not a neutral carrier.

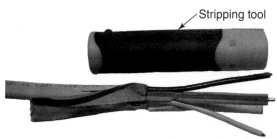

Goodheart-Willcox Publisher

Figure 21-17. Romex® NMC cable contains up to four insulated conductors and a bare wire for grounding. The stripping tool is a plastic tube with a small angled blade protruding to the inside.

552 Agricultural Mechanics and Technology Systems

Goodheart-Willcox Publisher

Figure 21-18. These markings indicate that this Romex® cable contains two 14-gauge insulated conductors and a bare ground conductor. NM-B refers to nonmetallic cable with a "B" heat rating of 194°F (90°C). This cable is rated to safely carry a maximum peak voltage of 600 V.

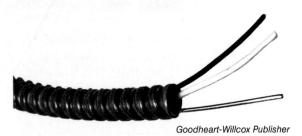

Goodheart-Willcox Publisher

Figure 21-19. Type AC armored cable contains up to four separately insulated conductors as well as a bare wire. The bare wire, in conjunction with the armor shell, serves as a safety ground conductor.

Safety Note
Wear approved eye protection and leather gloves when cutting and bending conduit to reduce chances of injury.

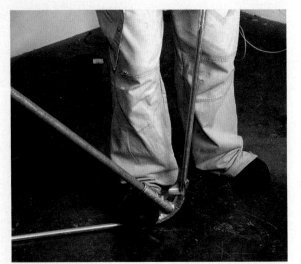

Lisa F. Young/Shutterstock.com

Figure 21-20. A tool used for bending EMT conduit.

conductors within the structure. Conduit is often thin-walled steel tubing that is used to protect electric circuits, but it may be constructed from rigid or flexible plastic tubing as well.

Two types of galvanized steel conduit are common. *Rigid conduit* is similar to water pipe and is thick enough to be threaded to accept threaded fittings. *EMT conduit*, made of electrical metallic tubing, is thin-walled steel tubing that can be bent to follow the structure of buildings, **Figure 21-20**.

Special tools and additional skills are required for the installation of metal conduit. EMT conduit comes in 10′ lengths and is easily cut with a 32-tooth hacksaw. Using a tubing bender made for EMT allows bending at convenient radii without decreasing the internal diameter of the conduit. When cutting metal conduit, be sure to cut pieces cleanly and leave no sharp edges on the ends of conduit that may damage the insulation covering the conductors. Bends in conduit should be smooth enough so that pulling wires is not difficult.

PVC (polyvinyl chloride) tubing can also be used for conduit. PVC tubing is typically gray to distinguish electrical tubing from water pipes. Another common conduit is electrical nonmetallic tubing, which is thin-walled corrugated tubing that is moisture- and flame-retardant. These flexible blue tubes are often used in new residential and commercial construction because the conduit is easy to work with simple tools, **Figure 21-21**. Buildings with more than three stories are required to have the electrical system enclosed in conduit rather than exposed flexible electric cable.

Machinery and equipment permanently installed in a building often are connected using a liquid-tight flexible metal conduit (LFMC), which resembles an armored cable covered with a plastic waterproof coating. See **Figure 21-22**. LFMC conduit is not only durable and watertight, but it also protects the enclosed wire from damage due to vibration, occasional shifting, and other physical hazards.

Enclosures and Connectors

Specific types of electrical boxes are designed to enclose wire connections for particular applications. The boxes provide convenient access to electrical components. The types of electrical boxes that can be used in specific applications are regulated by the NEC and local building codes.

Junction Boxes

According to the NEC, all electrical connections in an AC circuit must be made within an approved enclosure. A *junction box* is an enclosure designed to protect connections between electrical wires. Junction boxes are available in many sizes and shapes. They may be constructed of steel or plastic material, **Figure 21-23**.

Some junction boxes are surface-mounted on the building structure, leaving cables or conduit attached to the boxes somewhat exposed. Other junction boxes are flush-mounted, with only an inspection cover exposed. The rest of the enclosure is concealed, along with the wiring within the wall structure. It is usually best to nail flush-mounted boxes to the building frame and run wiring before installing wall coverings such as paneling or sheetrock.

All junction boxes should be closed with the proper cover plates—not using a cover plate defeats the purpose of protecting connections within an enclosure. The NEC limits the number of conductors permitted in each enclosure based on the size of the enclosure, gauge of the wires, and other items in the box, such as cable clamps and devices.

Electricians and plumbers often work in close proximity on job sites because they are installing building components in similar places inside of wall frames. Cooperation is necessary to ensure that the installation of electrical systems and plumbing components do not interfere with one another.

Boxes That Enclose Electrical Devices

Some junction boxes are designed not only to contain wire connections but also to provide a mounting place for electrical devices, such as switches and convenience outlets. There are many types of these switch enclosures, or boxes. They may be made of galvanized steel, aluminum, or various plastic materials.

A *handy box* is an enclosure that is press-formed from steel, giving it rounded, smooth corners. Handy boxes can be flush- or surface-mounted and are the correct size for most 120 V switches and outlets. The side walls of a handy box are covered with knockouts (prepunched circles) that must have the centers removed for connecting cables or conduit to the enclosure. To open knockouts, remove the center pieces by prying with a pair of lineman's pliers, **Figure 21-24**.

Flegere/Shutterstock.com

Figure 21-21. Flexible electrical nonmetallic tubing is easily cut and put into place.

Goodheart-Willcox Publisher

Figure 21-22. The liquid-tight flexible metal conduit (LFMC) used to protect wiring to this compressor resists damage from vibration and occasional flexing.

Goodheart-Willcox Publisher

Figure 21-23. Junction boxes are available in many shapes, sizes, and materials.

Figure 21-24. Lineman's pliers are used to remove the slugs from junction box knockouts.

Thinking Green

The center pieces removed from the knockouts are called slugs. These slugs should be recycled along with other metal scraps. A surprisingly large number of slugs can be left over from a single wiring job.

Junction boxes and device enclosures designed for mounting outdoors or in damp environments have features to prevent damage from moisture. A gasket that seals the cover plate to the box helps prevent moisture from compromising electrical connections. Two basic types of hinged secondary covers protect convenience outlets from moisture. The simplest type closes access to the outlet when it is not in use. The other type consists of a bubble-like door that covers the device but still allows it to be used when covered. This second type may also be used to cover outdoor switches. See **Figure 21-25**.

Safety Note

Do not open more knockouts than will be used—open holes leave connections exposed. In addition, use caution to avoid being cut by sharp edges when working with metal junction boxes.

Connecting Enclosures

When cables or sections of conduit join to junction boxes, they must be securely fastened. Some junction boxes are formed with built-in clamping mechanisms; others require additional parts known as cable connectors and conduit connectors. See **Figure 21-26**. Cable connectors clamp onto the outer sheathing of cables and extend into knockout holes with threaded nipples that are locked into place by tightening thin locknuts. The NEC allows only the insulated conductors and the bare ground wire to be inside the enclosure. The outer sheathing of the cable must be removed and only a maximum of 1/2″ is allowed in the enclosure. Conductors should be trimmed to a length of 6″ to 8″ inside the enclosure. See **Figure 21-27**. Conduit connectors work similarly to cable connectors, clamping securely to the outside of conduit and locking in place with a threaded fastener.

It is a good idea to secure cable or conduit connectors to the junction box before installing the box to the building structure because of limited space after installation. When positioning a cable connector, keep in mind that in its locked position, the screws that tighten to the cable sheath need to be accessible.

Hardwiring

When a load is directly connected to the building circuit without an easy means of disconnecting it from the circuit, it is considered to be *hardwired*. Large machinery that is too big to be easily moved or must remain in a specific location for safety reasons is often hardwired to the building's electrical system. See **Figure 21-28**.

Figure 21-25. This outdoor-rated electrical box protects both the connections of the conductors and the electrical device.

Safety Note
When hardwiring equipment, it is essential to connect, or bond, the metal frame of the machine directly to a grounded conductor.

It is very common to place hardwired equipment on a dedicated circuit, or one that carries no other loads. Smoke detectors and fire alarms are typically required by code to be hardwired and installed on dedicated circuits to isolate them from other electrical equipment that may experience problems and to reduce tampering. Can you think of any other common appliances that should be connected to a dedicated circuit?

Electrical Devices

Electrical devices used in common circuits can be classified as loads, switches, and receptacles. As previously discussed, loads convert energy to do work, and switches provide a way to break the current flow, thus controlling the electric power delivered to a load. *Receptacles* provide a means of conveniently connecting to or disconnecting a load from a circuit. Two common examples of receptacles are light sockets and convenience outlets, **Figure 21-29**.

Receptacles

Receptacles vary in structure and configuration based on the amperage they are designed to deliver. Light socket receptacles, like those found in a closet or in a garage, are porcelain or plastic structures into which a lightbulb is screwed into place. The most common receptacle is the standard duty 15 A, 120 V duplex receptacle into which the plug from a power cord is inserted, **Figure 21-30**.

Duplex Receptacles

Closely examine a duplex receptacle, in which two outlets are combined into one receptacle. There are four screw terminals for connecting conductor wires and a fifth green terminal that is reserved for the ground wire. Notice that the terminals connected to the side of the device with longer slots are silver in color, while the terminals connected to the shorter slots are bronze. The hot (black wire) conductor should always be connected to the bronze screws, and the neutral (white wire) conductor should always be connected to the silver screws. The two outlets are configured to be connected together, but they may be separated

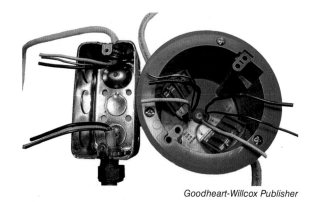

Goodheart-Willcox Publisher

Figure 21-26. Compare the handy box with the cable connector and rigid conduit connector installed to the plastic box with integral clamping mechanisms. Which would be quickest to install? Which would be easiest to service?

Goodheart-Willcox Publisher

Figure 21-27. Conductors should be trimmed to a length of 6″ extending beyond the face of the enclosure. Gripping the conductors in one hand while leaving a little extra is a good estimate.

by removing the two tiny brass tabs that connect the terminals on each side, **Figure 21-31**. This makes it possible to wire one outlet hot all of the time and have the other controlled by a switch. The grounding terminal of the receptacle is connected directly to the metal frame of the device. The large D-shaped holes that accept the grounding pin of the plug are also connected to the frame.

GFCI Receptacles

Code requires that circuits that are near to, or may be exposed to, moisture be protected through a GFCI. A GFCI receptacle can be used for this purpose. GFCI receptacles must be wired according to specific guidelines in order to function correctly.

Examine the back side of a GFCI receptacle—five screw terminals should be visible. The ground terminal should be connected to a bare or green wire. A pair of terminals—one brass and one silver—are marked *line*. See **Figure 21-32**. The line terminals connect to power from the service panel. If there is only one cable in the box, use the line terminals. Only the GFCI receptacle is protected against ground fault when connected in this manner. To protect the entire circuit with a single GFCI, connect outgoing power leading to the remainder of the circuit to the load terminals. In this configuration, the GFCI monitors the balance of current flow and protects the entire circuit.

Heavy-Duty Receptacles

Equipment and appliances that operate at increased amperage requirements must be plugged into circuits equipped with receptacles capable of delivering higher amperages. These devices are referred to as heavy-duty receptacles. Higher-amperage receptacles have heavier-duty internal parts and different blade configurations to help workers distinguish their capacity at a glance. See **Figure 21-33**. A plug designed to carry 120 V electricity cannot be inserted into receptacles designed for 240 V current. Likewise, a 30 A 240 V connection is very different from a 50 A 240 V connection. The different types of plugs and receptacles engineered for different amperage outputs reduce the chances of overloading standard circuits with heavy-duty loads.

Goodheart-Willcox Publisher

Figure 21-28. Moving this air compressor between jobs would be impractical, and it must be vented to the outdoors. It is therefore hardwired and fixed in a permanent location.

Goodheart-Willcox Publisher

Figure 21-29. The convenience outlet and light socket are examples of electrical receptacles. Each provides a mechanism for connecting a load to a circuit.

Goodheart-Willcox Publisher

Figure 21-30. Standard 15 A duplex receptacle.

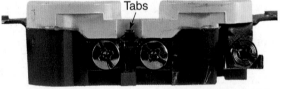

Goodheart-Willcox Publisher

Figure 21-31. The tabs between the two hot terminals and the two neutral terminals on this duplex receptacle may be removed with pliers to operate each outlet independently.

Switches

When you think of a switch, it is easiest to visualize a single-pole single-throw (SPST) arrangement in which the switch is either *Off* or *On*. The job of the switch is to break the circuit before electricity reaches the load, **Figure 21-34**. The switch is located on the hot side as a fail-safe measure, reducing the number of energized conductors when the switch is in the *Off* position. The conductor connected from the power source to the switch is always hot. The conductor leading from the switch to the load is called a switch leg, and it is hot only when the switch is in the *On* position.

Examine an SPST switch. The two screw terminals are bronze, signifying that they should be connected to a hot (black) conductor. The switch lever is marked *On* and *Off*, indicating that power passes through the switch in only one position. The grounding terminal is directly attached to the metal frame of the device.

In 240 V circuits, there are two hot wires, requiring a different type of switch. Both conductors

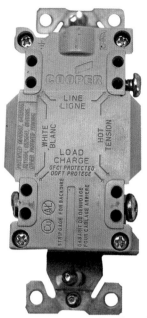

Goodheart-Willcox Publisher

Figure 21-32. A GFCI receptacle can provide protection in one location or to an entire circuit. The line terminals are for power in, and the load terminals are for power out.

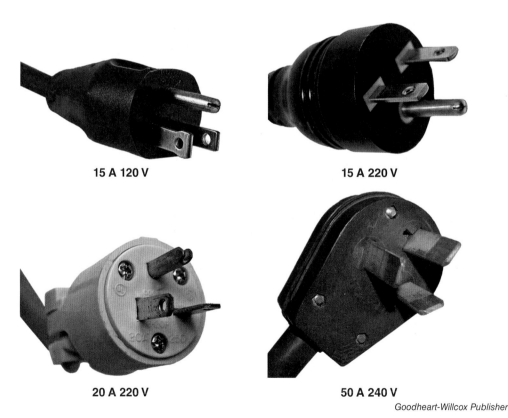

Goodheart-Willcox Publisher

Figure 21-33. Having different plug/receptacle configurations for different amperages helps prevent accidental connection of equipment to circuits not designed to accommodate heavier loads.

Goodheart-Willcox Publisher

Figure 21-34. Switches are always located on the hot side of a circuit, where they break, or control, the flow of electricity to the load.

are simultaneously interrupted using a double-pole single-throw (DPST) switch. In this type of switch, a single movement of the switch lever operates two side-by-side switches within the same device. Most 240 V equipment is engineered with the switch as part of the machine, eliminating the need to mount big DPST switches in building circuits.

It is often advantageous to be able to control an electric load from more than one location. For example, in a large room with doors at both ends, one should be able to switch the lights off and on from either door. A pair of three-way switches is used for this purpose. Three-way switches have three screw terminals. A single screw terminal on one side is the common terminal (single-pole), and a pair on the opposite side are the traveler terminals (double-throw). See **Figure 21-35**.

When a pair of three-way switches is installed, the common terminal on the first switch is connected to the hot (black) conductor. The common terminal is connected to a conductor (switch leg) leading to the load. The two switches are joined together with two conductors (travelers) connected between the traveler terminals on each switch. See **Figure 21-36**. No matter what position the two switches are in, only one of the travelers is energized at any point in time. When the two switches are synchronized to the same traveler, the current is delivered to the load.

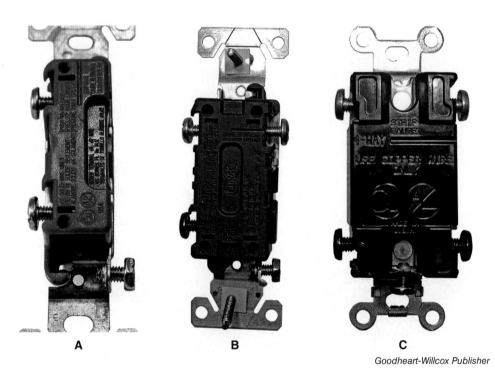

Goodheart-Willcox Publisher

Figure 21-35. Comparison of switches. A—The SPST switch has two hot terminals. B—The three-way switch (SPDT) has one common terminal and two traveler terminals. C—The four-way switch has four traveler terminals and is always used between two three-way or four-way switches.

A four-way switch has no common terminals, only four travelers. These double-pole double-throw (DPDT) switches are designed to be connected to the travelers between a pair of three-way switches, **Figure 21-37**. Theoretically, any number of four-way switches could be installed between a pair of three-way switches to create a circuit with many different control points.

Figure 21-36. Three-way switches. Which one of the travelers is energized? Are the switches in position to energize the load?

Connecting Wires

Electrical connections must always be tight and secure. The effectiveness of the connection depends on having adequate surface-to-surface contact between conductors so that moving electrons can make the transition without arcing (the current jumping a gap in the circuit). Connections can be classified as splices, taps, or terminal connections.

Wire-to-Wire Connections

Splices are connections from wire to wire when one section is not long enough. It is best to avoid having to make splices by planning ahead.

A *tap* is another type of wire-to-wire connection. One wire is connected along the length of another in order to create a branch in another direction. Taps are very common in AC circuits. A short piece of wire tapped into another and leading to a device in the same enclosure is known as a *pigtail*, **Figure 21-38**.

The most common method of connecting wires in AC circuits is to use solderless twist-on connectors called wire nuts. A *wire nut* is a plastic or ceramic cone with a tapered, spiral metal insert that grips the ends of wires to make a connection. When a wire nut is twisted onto the stripped and twisted-together ends of wires, the conductors are drawn into the metal spiral and

Safety Note
According to the NEC, all connections within an AC circuit must be made inside protective enclosures, such as junction boxes or switch boxes.

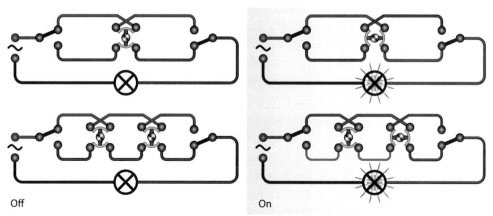

Figure 21-37. Two of the conductors connected to a properly installed four-way switch are always energized. Any number of four-way switches may be installed between two three-way switches.

Figure 21-38. The short piece of wire is a pigtail that is to be connected to the terminal of a fixture.

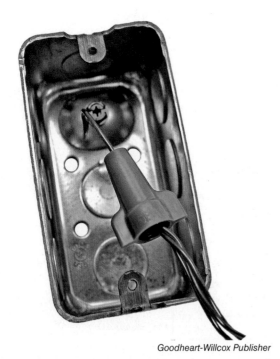

Figure 21-39. The pigtail formed by passing a short piece of wire through the hole in this grounding wire nut is bonded to the rear of the box. Note the green grounding screw in the prethreaded hole in the rear of the box.

compressed together inside. Wire nuts have external grooves or wing-shaped extensions molded into their exterior plastic shells to make them easier to twist into place. The direct wire-to-wire contact and wire contact with the metal insert work together to maintain electrical continuity at the connection.

Several types of wire nuts are available. One type intended for use in wet environments is filled with a dielectric gel and has an attached skirt-like gasket that seals around the connected wires. A *dielectric gel* or grease is a nonconductive, silicone-based substance designed to seal out moisture and prevent corrosion on electrical connectors. Another common type of wire nut has a hole for one wire to protrude through the end of the connector, creating a pigtail from the connection. This type of pass-through wire nut is green and is used for grounding devices or enclosures, **Figure 21-39**. On wire gauges thicker than 10 AWG, crimp-on connectors or connectors using setscrews or clamps are used in place of wire nuts because large-diameter wires are too stiff to be reliably connected with this method.

Terminal Connections

Terminal connections are points at which wire conductors terminate and are connected to electrical devices. Every wire in a service panel is attached at a terminal connection. Switches, receptacles, and other devices are connected to circuits using terminal connections. Solid, single-wire conductors are easily bent into a hook shape and secured under the heads of common screw terminals.

Terminals such as those found on the neutral bar in the service panel directly compress the conductor under the tip of the set screw. Others use small clamping devices to grip the wire. Many switches and receptacles have terminals in the back of the device where the bare end of a connector may be pushed into a spring-loaded clamp. Electric motors and equipment are often equipped with blade-type connectors that require wires to be terminated with crimp-on or solder-on attachments to fit snugly to the blade terminals. Similar crimp-on attachments are used when working with multiple-strand conductors and screw terminals. These attachments have rings, hooks, or forks designed to be tightened into place under screw terminals.

Stripping Insulation

Before electrical connections can be made, the insulation must be removed from the ends of the conductor. In the case of nonmetallic sheathed Romex® cable, first the outer sheath must be removed, and then each individual wire must be stripped. Remember that the NEC states that no outer sheath material is permitted inside enclosures and that 6″ of free conductor should extend beyond the face of the junction box before connections are made. If the box is 2″ deep, 2″ plus 6″ means that the removal of 8″ of outer sheath is required.

The best and safest way to remove outer sheathing is with a tool called a cable ripper. A *cable ripper* cuts the outside sheath for removal while minimizing damage to the insulation on the conductors inside. Several varieties of cable rippers are available, **Figure 21-40**. The most basic type has a tiny, sharp, knife-like blade that is guided through the sheathing.

The next step is to remove a little bit of the insulation from the ends of the conductor wires. The goal is to strip just enough to make a good connection and not leave any bare conductor exposed past the connection.

Quality switches or standard duplex receptacles have a *strip gauge*, which is two parallel lines with a space between them molded into the plastic housing. See **Figure 21-41**. The space is about the same width as the diameter of the conductor wire. The strip gauge may or may not be marked. It is usually about 9/16″ long and is the length of wire that should be stripped to fit correctly around the screw terminal when bent properly. These strip gauges also work for stripping wires that are to be connected using wire nuts. Devices with stab-in terminals usually have another strip gauge on the back that is a bit shorter for wires that are to be pushed into the back terminals. See **Figure 21-42**.

A *wire stripper* is a tool that resembles a pair of pliers with a series of openings fitted to different gauges of wire. Most stripping tools include a cutter, and sections of the jaws may also be configured for crimping attachments, **Figure 21-43**. Use a wire stripper to cut the insulation at the proper length indicated by the strip gauge, and use the jaws of the stripper to pull the waste piece of insulation straight off the end of the wire. Be sure to use the correct place in the jaws to match the gauge of the wire. Close the jaws into place and then rotate the stripping tool a couple of degrees around the wire to completely cut the insulation without

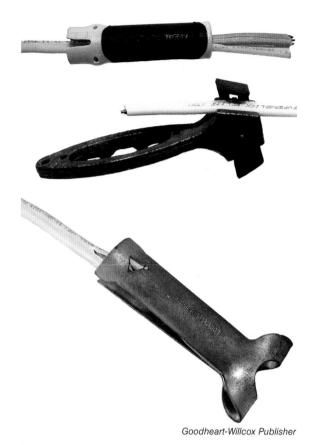

Goodheart-Willcox Publisher

Figure 21-40. Examples of cable rippers.

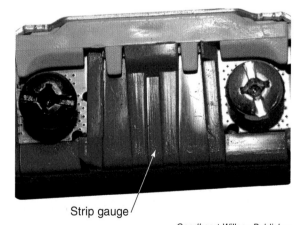

Goodheart-Willcox Publisher

Figure 21-41. Stripping conductors according to this strip gauge exposes just enough wire to make a secure connection without leaving excess of bare metal.

Figure 21-42. The second strip gauge on the back of this device indicates the correct length of bare conductor needed to fit the push-in terminals.

nicking the wire. See **Figure 21-44**. To prevent a reduction in the amperage-carrying capacity of the wire, do not allow the jaws to nick or cut into the conductor. Properly dispose of waste pieces of sheathing and insulation.

Notches formed in the corners of many electrical devices are designed to be used as wire strippers. Examine the two notches in **Figure 21-45**. Notice that they are different sizes to accommodate two different gauges of wire.

Safety Note

Remember to disconnect the power to a circuit at the service panel before beginning installation. Use a voltage tester or multimeter to double-check that the power is off before performing any electrical work.

Installing a Device

When connecting solid single conductor wires to devices such as switches and receptacles, it is preferable to use the screw terminals. When properly connected, screw terminals offer a more secure connection with a greater amount of surface area in contact at the connection point than stab-in terminals with spring-loaded locking mechanisms. Use the following procedure:

1. Strip the conductor wires back to the length indicated by the strip gauge on the side of the device.
2. Grip the end of one of the conductor wires with needle-nose pliers and roll the pliers in your hand, forming a tight loop. Move the tip of the pliers to grip the wire directly opposite of the end of the newly formed loop and bend the wire back sharply approximately 10° to form a shape similar to a shepherd's hook, **Figure 21-46**.
3. Repeat Steps 1 and 2 to form hooks in the other conductor and the ground wire.
4. One wire at a time, place the hooked end clockwise around the correct screw terminal. Use a screwdriver to tighten the screw snugly against the loop—this causes the hook to close securely around the screw. Placing the hook around the screw counterclockwise forces the hook to open when the screw is tightened.

Figure 21-43. Several options are available for stripping insulation from conductors. Each of these wire strippers has a graduated series of sharp openings in the jaws for different gauges of conductors. The wire stripper on the right also has notches for crimping various connecting attachments to the ends of conductors.

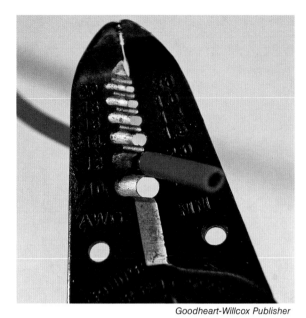

Goodheart-Willcox Publisher

Figure 21-44. Selecting the correct size opening in the wire stripper to match the gauge of the conductor neatly removes the insulation without damaging the conductor.

Goodheart-Willcox Publisher

Figure 21-45. The sharp-edged slots on the tab of this switch can be used to strip the insulation from the conductor. The slot on the left is intended for 12-gauge conductors. The narrower slot on the right is configured for stripping 14-gauge wires.

Goodheart-Willcox Publisher

Figure 21-46. Placing the hooked end of the wire clockwise around the screw terminal allows screw pressure to tighten the loop around the shaft of the screw.

Goodheart-Willcox Publisher

Figure 21-47. A connected receptacle ready to be positioned into the box.

5. Repeat the process for the other two wires. When properly connected, only bare conductor wire should be engaged by the screw terminal, yet the bare portion of the conductor should not extend beyond the edge of the terminal area of the device.
6. Tighten any unused screw terminals to reduce chances of unintentional contact when the wires and device are fitted into the box.
7. Neatly fold the wires into the rear of the box as the device is positioned to be secured with screws to the frame of the enclosure. Be sure that the bare ground wire is folded into the rear of the box first and that any unused screw terminals are tightened in to reduce chances of accidental grounding. See **Figure 21-47**.
8. Check the function of the device. If good, install the appropriate cover plate.

Hands-On Agriculture

Removing Cable Sheathing

1. Slip the stripping tool over the cable to approximately 6″ to 8″ from the end (**Figure A**).

A

2. Press the stripping tool downward to engage the knife tip into the outer sheathing (**Figure B**).

B

3. Slide the stripping tool along the cable to the end, splitting the outer insulation.

4. Peel back the sheathing to expose the individual conductors (**Figure C**).

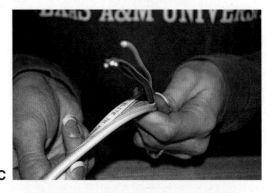
C

5. Use cutting pliers to trim away the waste sheathing (**Figure D**).

D

6. Cut away any paper or string material that may also be inside the cable (**Figure E**).

E

Goodheart-Willcox Publisher

CHAPTER 21 Review and Assessment

Summary

- The National Electric Code® (NEC) is the definitive guideline for proper electrical installation and inspections.
- Typical components of a service entrance are a meter, a circuit breaker box, a main switch, a direct connection to ground, and rigid tubing, including a weatherproof entrance for conductors.
- In a series circuit, multiple loads are arranged so each load is a conductor necessary for the function of the entire circuit. In a parallel circuit, each load is wired separately so the failure of one does not have an adverse effect on the other loads.
- Copper wire conductors differ in three major ways: the size of the conductor, the structure or arrangement of the conducting material, and the type of insulating or protective covering.
- Romex® nonmetallic sheathed cable (NMC) is commonly used in residential applications. NMC conductors are not permitted by the NEC in commercial construction; armored cable may be used instead.
- Receptacles vary in structure and configuration based on the amperage they are designed to deliver. The most common receptacle is the standard duty 15 A 120 V duplex receptacle into which the plug from a power cord is inserted.
- Switches are always installed on the hot side of a 120 V circuit. The job of the switch is to break the circuit before electricity is connected to the load.
- Electrical connections can be classified as splices, taps, and terminal connections.
- Before electrical connections can be made, the insulation must be removed from the ends of the conductor. Cable rippers and wire strippers are tools used in this process.

Words to Know

Match the key terms from the chapter to the correct definition.

A. armored cable	F. handy box	K. screw terminal
B. cable ripper	G. hardwired	L. series circuit
C. conduit	H. junction box	M. service meter
D. dielectric gel	I. parallel circuit	N. splice
E. distribution panel	J. pigtail	O. wire nut

1. Device used to measure electric power consumption.
2. Rigid tubing used to protect electric cables.
3. Connection from wire to wire when one section is not long enough.

4. Commonly called a breaker box; all branch circuits are connected here.
5. Electrical connector in which a wire or cable is connected to a device by screw pressure.
6. If one load fails, every other load on this path will lose power.
7. Each load on this path is connected independently of all other loads on the same path.
8. Enclosure to protect electric connections that must be made outside of the distribution panel.
9. Cable that has a flexible metallic sheathing that allows for extra protection; often used in commercial construction.
10. Method of connecting large equipment directly to an electric circuit without an easy means of disconnecting it from the circuit.
11. Nonconductive, silicone-based substance designed to seal out moisture.
12. Plastic or ceramic cone that grips the ends of wires to make a connection.
13. Short piece of wire connected to another conductor used to connect a device in the same enclosure.
14. Tool used to split outer insulation layers for removal.
15. Enclosure that has rounded, smooth corners and is used to enclose wire connections and electrical devices.

Know and Understand

Answer the following questions using the information provided in this chapter.

1. A service _____ typically includes a meter, a circuit breaker box, a main switch, a direct connection to ground, and rigid tubing, including a weatherproof entrance for conductors.
2. The enclosure that protects the circuit breakers and main electrical connections is known as the _____.
3. All of the neutral conductors in a breaker box are connected securely to the _____, which is also required to be bonded to a ground connection.
4. What are the three major ways in which copper wire conductors differ?
5. A flag of black electrical tape on a white conductor in a circuit indicates that the wire is being used as a(n) _____ conductor.
6. What are the center pieces removed from knockouts called?
7. Hardwired equipment is often placed on a(n) _____ circuit.
8. In a duplex receptacle, the hot conductor should always be connected to the _____-colored screws.
9. The conductor leading from a switch to the load is called a(n) _____, and it is hot only when the switch is in the *On* position.
10. Installing _____ switches between two three-way switches allows the location of an unlimited number of controls on a single circuit.
11. A wire nut that is filled with a dielectric gel and has an attached skirt-like gasket that seals around the connected wires is intended for use in _____ environments.
12. When forming a shepherd's hook on the end of a conductor, in which direction should you wrap it around the screw terminal before tightening the screw?

STEM and Academic Activities

1. **Science.** Carefully disassemble a common duplex receptacle. Formulate a hypothesis concerning the relative value of using push-in type connections versus screw terminals to connect to the conductor wires. Present your hypothesis to the class.
2. **Technology.** Use a multimeter to trace the open and closed paths of electricity through wired models of three-way and four-way switch assemblies. Demonstrate how an infinite number of four-way switches can function between two three-way switches.
3. **Engineering.** Draw diagrams depicting different arrangements possible in electric circuits. Build compact models of these circuits using wiring boards or wall structures.
4. **Math.** Use the NFPA website to look up the rule for the number of conductors in a junction box. Compare two or more example junction boxes by calculating the volume of each to determine the allowed maximum number of connectors. Explain why this limitation exists.
5. **Language Arts.** Why is the nomenclature of electrical components an important part of the NEC? What would happen if, like many tools, electrical components had different names in different parts of the country?

Thinking Critically

1. Compare the following two types of boxes:
   A. A handy box with a cable connector and a rigid conduit connector installed.
   B. A plastic box with integral clamping mechanisms.
   Which would be quicker to install? Which would be easier to service? State the reasons for your answers.
2. In what situations should you use a flagged white conductor to connect a switch rather than using black or red for hot?

CHAPTER 22
Wiring DC Circuits

Chapter Outcomes

After studying this chapter, you will be able to:
- Identify applications for DC circuits.
- Identify sources of DC power.
- Explain the different types of electrical loads found in agriculture.
- Identify types of switches.
- Wire and repair electrical connections.
- Wire an agricultural trailer.
- Troubleshoot DC circuits.

Words to Know

array	limit switch	photovoltaic module	shorted
blade connector	lineman splice	positive ground system	shrink tube
bullet connector	loom	rectifier	solenoid
diode	normally closed (N/C) switch	relay	SPDT (single-pole double-throw) relay
duty cycle		rheostat	
electrical tape	normally open (N/O) switch	Scotchlok® connector	tap splice
end splice		screw terminal	toggle switch
in-line fuse	photovoltaic cell	sensor	wiring harness
in-line splice			

Before You Read

Before reading this chapter, review the objectives. Based on this information, write down two or three items that you think are important to note while you are reading.

568   Copyright Goodheart-Willcox Co., Inc.

While studying this chapter, look for the activity icon to:

- **Practice** vocabulary terms with e-flash cards and matching activities.
- **Expand** learning with interactive activities.
- **Reinforce** what you learn by completing the end-of-chapter questions.

www.g-wlearning.com/agriculture

In 1882, Thomas Edison's company started the world's first electric power distribution system. The company provided 110 volts of direct current (DC) to customers in New York. Due to significant advantages of AC over DC in power distribution, nearly all commercially distributed electric power is now AC. However, DC electricity is still an important power source for many applications, including agriculture. DC electric energy is commonly produced by generators that convert mechanical power to electric power and by solar cells that convert light energy directly into electricity. See **Figure 22-1**. Alternating current can be converted to direct current through several types of devices.

Laura Stone/Shutterstock.com

Figure 22-1. The generators used on older tractors produce DC electricity by converting mechanical energy to electric energy.

DC Circuits

Direct current flows in a constant direction, from a negative charge to a positive charge. This is different from alternating current, which changes polarity within each cycle. Direct current is used to charge batteries, which in turn are the power supplies for many devices and purposes. The positive side of a DC electric circuit is connected to the positive terminal of the battery and is considered to be hot. The negative side is the ground side.

On most mobile vehicles and equipment, the ground side of the circuit is connected to the metal frame and the negative terminal of the battery. See **Figure 22-2**. Some older tractors and cars were factory-wired in the opposite manner and had a *positive ground system*. This system worked because DC electricity can operate most loads in either polarity; however, generators and charging systems for positive ground systems must also be configured in reverse.

Most automotive electrical systems are based on a 12 V DC electric current. Tractors and other agricultural machines use DC circuits for starting, lighting, and monitoring systems. Nearly all electronic systems use DC current. Many portable power tools and appliances are operated on power supplied by DC batteries. See **Figure 22-3**.

Relatively low-voltage DC systems can provide safe power for outdoor lighting and communication systems. Monitoring systems in greenhouses often rely on low-voltage DC power.

Goodheart-Willcox Publisher

Figure 22-2. The negative, or ground, terminal of the battery is connected directly to the frame of this utility vehicle.

Sources of DC Power

Sources of DC electricity include alternators, generators, and solar power. Generators convert mechanical energy to AC electricity by moving a conductor through a magnetic field. For DC applications, the AC is then rectified, or converted to DC electricity. In addition, the AC current produced by an alternator can be converted to DC electricity.

Converting AC Current to DC Current

Modern tractors and automotive applications use alternators or generators to generate electric power to charge the batteries of DC electric systems. See **Figure 22-4**. An alternator or generator produces AC power that must be converted to DC electricity. Converting AC to DC is accomplished by an electronic device called a *rectifier*.

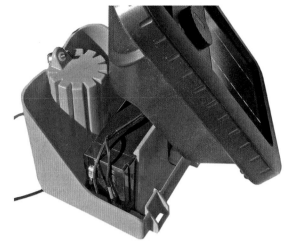

Goodheart-Willcox Publisher

Figure 22-3. The battery inside this solar fence energizer stores power collected from the sun to send pulses of DC current through a containment fence.

Rectifiers use diodes to convert alternating current into direct current. *Diodes* are electric gates that allow current to pass in one direction but not the other. What happens to the other portion of the AC energy if only half is allowed to pass through the diodes? Rectifiers are engineered to dissipate excess energy in the form of heat and are often equipped with conductive fins that radiate excess heat energy into the air. Rectifiers may also incorporate a transformer that steps the voltage either up or down. See **Figure 22-5**.

Most electronic devices are powered by DC current. Batteries of all types provide smooth DC power. Electronic machines, such as personal computers, that plug in to common 110 V AC outlets usually incorporate a power supply that is an AC-to-DC converter. This converter is basically a rectifier/transformer component.

A *Goodheart-Willcox Publisher*

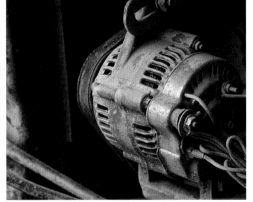
B *RMIKKA/Shutterstock.com*

Figure 22-4. Alternators provide AC power that is passed through a rectifier to convert the energy to DC electricity used by the vehicle. A—Alternator on a utility vehicle. B—Alternator on a tractor.

AgEd Connection: DC Circuits

Direct current (DC) circuits are used to power most electronic devices like computers, phones, and the electrical components of automobiles. Maybe the inside of a computer is of interest to you or the fact that a computer can be used to control the components of a large facility like a cotton gin or even a food processing facility. With such variety comes a huge job market. What potential is available locally for you to gain experience with DC circuits or components? Talk with your agriculture teacher and find out today.

michaeljung/Shutterstock.com

Goodheart-Willcox Publisher

Figure 22-5. This cell phone charger incorporates a transformer to step the voltage down and a rectifier to convert AC to DC electricity.

Goodheart-Willcox Publisher

Figure 22-6. A photovoltaic module is a group of solar cells mounted in a frame and electrically connected to each other. How many individual cells are built into this module?

Solar Energy

A *photovoltaic cell* converts light directly into DC electricity at the atomic level. Certain materials exhibit a property known as the *photoelectric effect*, which causes them to absorb photons of light and release negatively charged electrons. A basic photovoltaic cell, also called a *solar cell*, captures these free electrons, resulting in a useful electric current. Solar cells are made of semiconductor materials, such as silicon. When light energy hits the solar cell, electrons are freed from the atoms in the semiconductor material. The free electrons are captured in the form of DC electricity. This electricity can then be used to charge a battery or to power a load.

A *photovoltaic module* is a group of solar cells mounted in a frame and electrically connected to each other. See **Figure 22-6.** Each module supplies electricity at a certain voltage, typically 12 V. The amount of current produced depends directly on the quantity of light reaching the module. Multiple modules are wired together to form an *array*. See **Figure 22-7.** The larger the surface area of an array, the more light it can absorb, and the more electricity it produces. Photovoltaic modules and arrays can be connected in either series or parallel circuits to produce a variety of voltage and current combinations.

Solar power is used in agriculture to charge batteries, provide light, and power essential equipment. Photovoltaic arrays are often more economical than

STEM Connection

Batteries

A battery is an electrochemical device that stores and produces electricity. It contains two plates: one positive and the other negative. When it is fully charged, there is a significant difference in the voltage potential between the positive and negative plates. When a load is placed across the positive and negative terminals of the battery, electrons (current) flow from the negative terminal to the positive terminal in an attempt to equalize the potential difference. As the current flows, the potential difference between the terminals decreases and the battery begins to run down. Charging the battery restores the difference in voltage potential between the two plates.

Battery design and construction is often the limiting factor in advancing the design and engineering of tools and equipment. When initially introduced, cordless power tools typically lacked the power and endurance required for professional use. As a result of the development of battery technology, particularly economical lithium ion batteries, cordless power tools are now the norm on jobsites.

For many years, lead-acid batteries have been the standard for storing and providing power to vehicle electric systems. However, lead-acid batteries are too bulky to be effective for efficient electric-powered vehicles. The US Department of Energy is encouraging the development of more economical, compact, and efficient battery technologies for electric vehicles. How do you believe the development of more efficient batteries will affect agricultural equipment?

---

installing new electric lines to supply electricity for electric fences, lights, and water pumps in remote locations. Solar systems require no fuel and have no moving parts, so they are more convenient than fuel-powered generators. They are also economical for remote livestock water supply, pond aeration, and small irrigation systems.

Electrical Loads

A load is a device that uses electrical power and creates resistance. Examples of loads are a vehicle's ignition system and lights. In agriculture, DC electricity is used to power a variety of different loads, including lighting, electric motors, and other devices.

Lighting

Typically all of the lighting on agricultural tractors, trucks, and other vehicles is powered by 12 V battery systems. Handheld battery-powered portable lights and flashlights are also important tools for people working in agriculture. Outdoor lighting systems that are located close to the ground, moisture, or

Iakov Filimonov/Shutterstock.com

Figure 22-7. Multiple modules are wired together to form an array.

Figure 22-8. This self-contained landscaping light uses solar energy to recharge a DC battery that powers the LED lamp.

Figure 22-9. This landscape lighting system uses a rectifier and step-down transformer to supply safe electrical power to a series of outdoor lamps.

Figure 22-10. The starter on this old Ford tractor is an example of a DC motor in practical use.

people are frequently operated using a low-voltage DC power supply. See **Figure 22-8** and **Figure 22-9**. Lights used in DC applications include incandescent bulbs, halogen lamps, and light-emitting diodes.

Safety Note

Use of LEDs for vehicle lighting is increasing due to greater potential for safety. Incandescent lights have to heat up, taking about a quarter of a second to attain 90% full brightness. LEDs respond almost instantaneously when a vehicle's brake lights are applied to provide a brighter, more focused illumination.

Halogen lighting is used in applications where a bright, focused light is desired. In DC applications, halogen lamps are used in vehicle headlights.

Electric Motors

DC power can be used to drive electric motors for many different operations. See **Figure 22-10**. Common applications include starters for large, fuel-driven engines; pumps for fuel, livestock water, or cooling; and fans. DC motors are often used in heavy-duty industrial applications because they can maintain a constant RPM at varying loads.

The construction of electric motors closely resembles that of generators. See **Figure 22-11**. Similar components work together to do the opposite job of a generator. An electric motor uses magnetic fields to convert electrical energy to rotating mechanical energy.

Small DC-powered electric motors are used to drive cordless power tools. These motors deliver power efficiently, but must be protected from overheating. The *duty cycle* of an electric motor is the amount of time it can safely operate in relation to the time it needs to rest or cool down. Cordless tools often have an overload protection switch that opens when the tool becomes too warm. The switch shuts down the power to the motor, preventing damage from overheating. Tools with a low duty cycle cut out more quickly. Other DC electric motors, such as pump or fan motors, also commonly have overload protection.

Starter motors have heavy-duty windings to provide the capacity for large current loads. Starters typically operate on short bursts of 12 V DC current at amperages of over 250 amps. These short bursts use the stored energy of a battery rather rapidly, so starters need to be efficient. Switching mechanisms known as electromagnetic *solenoids* deliver power to starters and

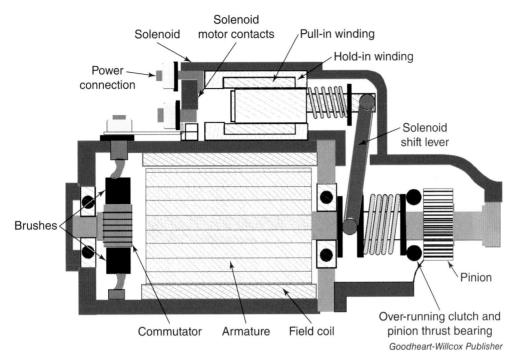

Figure 22-11. The construction of electric motors closely resembles the way generators are assembled.

shift the starter gear mechanism to engage the flywheel of the larger engine. See **Figure 22-12**.

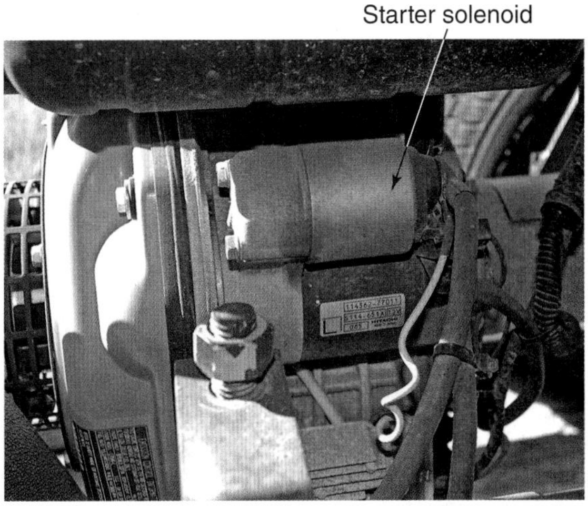

Electromagnets

DC electromagnets are important not only for the function of generators and electric motors; they are also used for trailer braking systems and switching solenoids. Brakes and solenoids, like motors, convert electric energy to motion or mechanical energy.

Electric Brakes

Small- to medium-size agricultural trailers are often equipped with drum brakes actuated by electromagnets. Applying pressure to the brake pedal of the pulling vehicle sends DC to the electromagnets within each brake drum. Each magnet is attracted to the inside face of the rotating drum. As the drum turns, it pulls the magnet, activating a lever arm that spreads the brake shoes outward into the drum. Friction between the brake shoes and the drum slows the vehicle. When the electric power is stopped, the magnets release and springs return the brake shoes to their normal positions. See **Figure 22-13**.

Figure 22-12. The starter solenoid on this wood chipper allows cranking energy to be supplied without passing the full current through the ignition switch.

Electromagnetic Solenoids

Electromagnetic solenoids use a controlled electromagnetic field created by passing current through a coiled wire to convert electricity to linear motion. Current passing through the coil causes a steel or iron armature

Electromagnet
Goodheart-Willcox Publisher

Figure 22-13. Applying pressure to the brake pedal sends current to the electromagnet within the brake drum.

to become an electromagnet. The resulting magnetic force is used to move the armature a short linear distance. This short movement may be harnessed to open or close pneumatic or hydraulic valves, operate mechanisms such as door locks, or control the position of engine components. Solenoids also allow remote operation of large electric switching mechanisms on starter motors and winch controllers. See **Figure 22-14**.

Fence Controllers

Electric fence controllers use step-up transformers to increase the voltage and drop the amperage. See **Figure 22-15**. Current (amperage) is the part of an electrical charge that can cause muscle contractions and even cause death. The controller, also known as an *energizer* or *charger*, sends energized pulses through a fence wire that must be completely insulated from the supporting fence posts. High enough voltage provides an intense shock, which encourages animals to stay away from the fence, without resulting in injury or damage.

Both DC and solar-powered fence chargers use battery power. These fence chargers are used for fencing remote locations. Solar-powered fence chargers offer the benefits of automatically recharging the battery using power from the sun. Battery life depends on how often the fence is grounded. In other words, a charged fence carries a potential for shock but uses minimal energy until the fence wire is grounded by animals or weeds making contact. An animal receives a stimulating high-voltage pulse of short duration on contact with the fence. With contact, the battery drains faster than if the wire remains untouched. Animals learning to avoid a fence touch it more often at the beginning of training and less often as time goes by. See **Figure 22-16**.

Goodheart-Willcox Publisher

Figure 22-14. This electrically operated solenoid used to remotely shut down the fuel flow on a diesel engine. When the electromagnet is energized, the steel shaft is drawn into the cylinder, which in turn pulls a valve lever attached to the end of the linkage.

Switches

Switches are used to break, or disrupt, a DC circuit. Some switches are on/off switches; others are multipositional. Some include a resistor to limit current. Circuit switches are typically located on the hot (positive) side of the circuit. Switching mechanisms may be operated manually, triggered by the changing position of machine parts, or triggered by rising or falling levels of consumables or products in storage receptacles, such as fuel in a tank or grain in a storage bin.

On/Off Switches

Simple switches have two positions: *On* and *Off*. The circuit is either complete or broken. **Toggle switches** work like a typical light switch. When placed in the *On* position, toggle switches remain in that position until manually moved to the *Off* position. See **Figure 22-17**.

As explained in Chapter 20, *Electric Theory*, a momentary switch remains in operating position only while the switch is physically triggered. Momentary switches are either **normally open (N/O) switches** or **normally closed (N/C) switches**, and some switches are wired to operate either way. An example of an N/O momentary switch is the horn button on a vehicle. When the button is pushed, the switch closes the circuit and the horn sounds. The brake lights on vehicles are also controlled by N/O momentary switches.

A *limit switch* on a grain bin is an example of an N/C momentary switch. When the bin is full, the switch is triggered, interrupting the control circuit that operates the motor driving the auger that fills the bin. In a greenhouse, there might be an N/C switch connected to a float in the cooling system reservoir. If the water level drops too low, the switch connecting the pump control circuit disconnects power to prevent the pump from overheating. See **Figure 22-18**.

Goodheart-Willcox Publisher

Figure 22-15. A solar-powered fence controller uses a photovoltaic module to convert light energy to DC electricity, a battery to store the charge, and a step-up transformer to provide a safe but intense shock.

marikond/Shutterstock.com

Figure 22-16. Livestock touch an electric fence more at the beginning of training and avoid it as time passes.

Figure 22-17. This simple toggle switch has two positions—*On* and *Off*. It is used to interrupt the circuit and is placed in the hot side of the circuit under most circumstances.

BACHTUB DMITRII/Shutterstock.com

Figure 22-18. This float-operated sensor switch from inside a diesel fuel tank sends a signal to the fuel gauge to indicate the fuel level.

Goodheart-Willcox Publisher

Figure 22-19. Most ignition switches have three or four positions. Typically, ignition switches incorporate a momentary switch with one or two on/off switches. Ignition switches may or may not incorporate a key for security.

Multipositional Switches

Not all switches are confined to *Off* and *On* positions controlling a single component. Multiposition switches have multiple terminals that can be used to operate several different components. They may also be connected to a single component that is designed to operate at different levels. A tractor ignition switch is an example of a multiposition switch that controls multiple components. See **Figure 22-19**. In the first position, only the lights and radio communication will operate. In the second position, the engine and essential systems will run. The third position functions as a momentary switch and controls the starter motor.

Some multiposition switches use a *rheostat*, or adjustable resistor, to vary the resistance in an electric circuit. See **Figure 22-20**. A rheostat performs tasks such as dimming lights and controlling the speed of motors. The resistance used in these applications is typically a result of passing the current through different sections of coiled metal wire.

A *sensor* is a switch that is triggered by a means other than an operator's intentional manipulation of a device such as a handle or knob. Sensors may be affected by the position of a piece of equipment, the presence of movement, or the location of materials. The float in a tractor's diesel fuel tank is attached to a multiposition switch using a rheostat. As the float moves, the rheostat limits the amount of current passing through the sensor. The changing current level is interpreted by the fuel gauge as the current fuel level.

Relays

A *relay* is an electromagnetic switch that is connected to two circuits—a primary controlling circuit and a secondary controlled circuit. Using a relay allows the secondary switch, which draws a large amount of current, to be controlled by a very small switch, wiring, and a solenoid in a primary circuit that only draws minimal current.

A relay commonly used in DC circuits on vehicles is the *SPDT (single-pole, double-throw) relay*. This relay consists of a coil and terminals (primary circuit), one common terminal, one N/C terminal, and one N/O terminal (secondary circuit). See **Figure 22-21**. Power in the primary circuit operates the solenoid, which in turn mechanically moves a switch in the secondary circuit to control power to a load. When the coil of the relay is not energized, the common terminal and the N/C terminal have continuity. When the coil is energized, the common terminal and the N/O terminal have continuity.

The coil of a common SPDT relay draws very little current (less than 200 milliamps). The amount of current that passes through a relay's common, N/O, and N/C contacts is as high as 30 A or 40 A.

Diodes are sometimes used across the coil of a relay to provide a path for current when the current path to the relay is switched off. This allows the coil field to collapse without generating a voltage spike that could damage switch or relay contacts.

Fuses

Many DC electric circuits used in agriculture have fuses that protect the circuit from overloading. As described in earlier chapters, fuses are failsafe mechanisms, meaning that a fuse allows the system to fail safely without starting a fire or damaging expensive components.

Different types of fuses are available. Buss fuses contain elements that can be viewed through glass. The thin strip of metal in the center of the fuse is the weak link within a circuit, and melts if exposed to excessive current. See **Figure 22-22**. Blade fuses are less expensive to manufacture than buss fuses. Their design allows for a compact arrangement in central fuse blocks in order to locate most of the fuses used on a piece of equipment in an easily accessible location. See **Figure 22-23**.

The size of a fuse depends on the size of wire and the capacity of other components in the circuit. The fuse should be designed to fail at a current level smaller than maximum current limit of the next weakest component. The same fuse should not be replaced repeatedly without determining the cause of the overload, nor should a burned fuse be replaced with a fuse with a higher amperage rating. Remember that fuses are intended to be a failsafe—if they are not the weak point in the circuit, they have no function.

Figure 22-20. A coiled resistor inside this rheostat switch is moved in and out of the circuit by turning the control knob. Multiple contact points along the coil provide different levels of resistance.

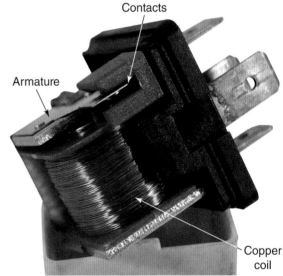

Figure 22-21. A primary circuit flowing through the copper coil operates the electromagnet in this relay. The electromagnet moves the contacts to control electricity flowing through a secondary circuit.

Figure 22-22. Buss fuses.

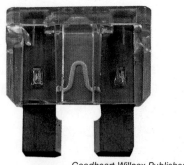

Figure 22-23. Blade fuses.

All fuses may be conveniently mounted in a common fuse box that serves as the distribution point for multiple circuits in a piece of equipment. Most large equipment manufacturers leave extra slots in the central fuse block for additional equipment. However, it is occasionally necessary to use an in-line fuse holder. See **Figure 22-24**. A fuse may be placed individually as an *in-line fuse* to protect a single circuit, such as the starting circuit on a tractor.

Some applications using DC motors employ circuit breakers as overload protection for the motor. Circuit breakers can be reset and reused. Chapter 20, *Electric Theory*, describes circuit breakers in more detail.

Making Wire Connections

Insulated wiring is used as a conductor to move electrical power throughout a DC system. The AWG size of the electrical wiring must be heavy enough to handle the intended load without overheating. Single-strand wire is useful for wiring stationary installations and can be easily shaped and tucked out of the way around machinery. Bundled wire (multiple-strand wire) has a slightly larger overall diameter than a solid wire of the same AWG due to the small gaps between the strands, **Figure 22-25**.

Bundled wire is preferred to single-strand wire because current travels on the outer surface of the wire, so a bundle of smaller wires can carry more current than a solid wire of the same size. Bundled wire can withstand vibrations and flexing better than single-strand conductors. If vibration or other movement of the equipment will cause flexing of the conductor, a multiple-strand conductor is a superior choice.

Poor connections create excessive resistance and are the most common type of electrical problem. Electrical connections and connectors must be designed to maintain good contact when exposed to the conditions at the sites of their use.

Types of Wire Connections

Many types and styles of connectors are available for making connections in DC electrical circuits. The connectors may be permanent or designed to be conveniently disconnected. Connections in stationary applications, such as greenhouses or livestock facilities, may be made with wire nuts similar to those used in AC circuits. Mobile applications generally experience higher levels of vibration and other movement, requiring more secure methods of connecting conductors. Crimp-on connectors offer relatively

Figure 22-24. Examples of in-line fuse holders.

permanent connections by being mechanically compressed to wire conductors. Common connection types are as follows:

- *Bullet connectors* are round connectors used to connect wires to wires. They come in male/female pairs and are usually crimp-on. These connectors provide a somewhat temporary connection, allowing for simple repairs. See **Figure 22-26**.
- *Blade connectors* are also male/female pairs used to connect wires to wires. Blade connectors are flat and often crimp-on. Many electrical components are set up with male blade connectors and are attached to wires terminating in a female blade connector. See **Figure 22-27**.
- *Screw terminals* of several types are commonly used to make electrical connections in DC circuits. The most basic screw terminal connection requires that the wire conductor be wrapped clockwise around the screw. As the screw is tightened, the conductor is compressed between the screw head and the terminal. Alternatively, devices such as ring- or fork-type crimp-on connectors may be used to complete the junction. See **Figure 22-28**. Another type uses a screw-driven compression device to grip the connector and press it securely to the terminal. See **Figure 22-29**.

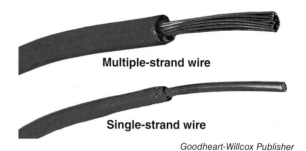

Figure 22-25. Multiple-strand and single-strand conductors.

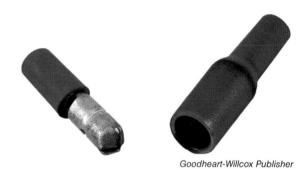

Figure 22-26. Crimp-on bullet connectors are made in male/female pairs.

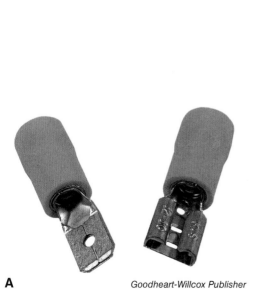

Figure 22-27. A—Blade connectors come in male/female pairs. B—A female blade connector used separately as a connection to a component designed with a built-in male blade connector.

Figure 22-28. Screw terminals. A—A conductor may be wrapped around and tightened into place on a screw terminal. B—Ring-type crimp-on connector. C—Fork-type crimp-on connector.

Figure 22-29. A lug-type grounding connector uses a screw to exert pressure on a connection block. The conductor wire is gripped with great force to ensure a positive connection.

Figure 22-30. Scotchlok® connectors are placed directly over insulated conductors. Pressing the slotted blade into place cuts the insulation and completes the electrical connection between two conductors.

Figure 22-31. An end splice.

- *Scotchlok® connectors* are made for quick repairs without the need for specialized electrical tools. They consist of a plastic frame and a slotted metal blade that is pressed into the conductors with a regular pair of pliers. See **Figure 22-30**.
- Splicing involves stripping the insulation from the ends of the wire and twisting the wires in a prescribed manner to achieve a secure electrical connection.
  - In an *end splice*, the wires that are pointing the same direction are securely twisted together, **Figure 22-31**. The splice is completed by insulating the bare conductor and securing the wires to prevent loosening the junction through exposure to movement. An end splice is suitable for stationary locations where plenty of room is available.
  - An *in-line splice* connects wires that point toward each other. The splice should be only marginally thicker than the unspliced conductor. Properly executed splices retain most of the strength of the original. The splice must be completed with an insulating cover and may be secured with solder or a crimping device. There are several versions of in-line splices. See **Figure 22-32**.
  - In a *tap splice*, a conductor is joined to another conductor by connection to a wire in a circuit without breaking or cutting the original wire to make the connection.

Soldering

Soldering may be preferred as a more permanent solution for applications that are exposed to vibration or bad weather. Solder can also enhance screw terminal connections. Multiple-strand wire can be soldered together at the tip prior to attachment to a screw connector. Bonding the strands allows more torque

Figure 22-32. Two versions of an in-line splice.

from the connector and a more secure connection. See **Figure 22-33**.

Solder may be used to attach connectors similar to crimp-on connectors to a conductor. Soldering is less likely to damage a conductor and offers less resistance than crimped connections. See **Figure 22-34**.

Wire splices exposed to considerable vibration may be soldered to ensure a secure connection. When soldering wires, use caution not to damage insulation. Apply just enough solder to thoroughly bond the connection. Soldering connections may slow the wiring process, but for some applications the extra security is well worth the effort. See **Figure 22-35**.

Protecting Wire Connections

Electrical connections need to be insulated and protected after assembly. Electrical connections that are exposed to the outdoors also must be watertight. Many connectors incorporate an integral insulator—if the connector is correctly installed, the insulation is in place.

Splices and noninsulated connectors can be insulated in several ways. Heat-sensitive *shrink tube* is a plastic sleeve that must be slipped onto a wire before a connection is made. After the connection is secure, the shrink tube is slid into place over the joint and shrunk by applying heat from a heat gun or by careful exposure to a small open flame. See **Figure 22-36**.

Electrical tape has been used for many years as a standard insulating material for electrical connections in DC applications. Not all electrical tape is the same—different types, assorted colors, and varying qualities are available. Common electrical tape is black, is made from vinyl, and has a shiny finish on one side and an adhesive on the other.

Electrical tape is pressure-sensitive. It bonds better to itself if stretched a little as it is applied. Tearing from the roll instead of cutting with scissors stretches the end of the tape, causing it to adhere better than a cut end. In most applications, several layers of tape are necessary to provide adequate electrical insulation.

Figure 22-33. Applying a small amount of solder to bond the individual wires of a multiple-strand conductor stiffens it sufficiently to enable a sound connection under a screw terminal.

Figure 22-34. Application of rosin core solder to attach a ring terminal connector. Heat applied to the barrel of the connector melts the solder, and capillary action draws it into the joint.

584 Agricultural Mechanics and Technology Systems

Goodheart-Willcox Publisher

Figure 22-35. Soldering enhances the security of an in-line tap splice. The blue wire is being tapped into the black conductor without breaking the main conductor.

Goodheart-Willcox Publisher

Figure 22-36. Applying shrink tube insulation with a heat gun.

Apply the layers in one continuous wrap without allowing the tape to bunch up. See **Figure 22-37**.

For wiring applications such as agricultural trailers and other mobile equipment, tape alone is often not enough protection for exposed wiring. Plastic *loom* is used as an added layer of protection for wires and bundles of wire known as *wiring harnesses*. Loom is available in several forms, **Figure 22-38**. A convoluted or corrugated (ridged or grooved) black plastic type with a single split down one side is offered in an assortment of diameters. The corrugation provides a measure of strength and has an insulating effect, protecting enclosed wires from external heat sources. Another type is a thick plastic preformed loom in a long spiral tube. This type of loom can be wrapped around a bundle of wire while allowing multiple connections along the length of the bundle. Loom is often secured using electrical tape and cable ties.

Lineman Splice

The *lineman splice*, or Western Union splice, was developed during the introduction of the telegraph to mechanically and electrically connect wires that were subject to considerable stress. The wires are twisted together in a way that causes the electrical connection to tighten as the conductors pull against each other. See **Figure 22-39**.

This type of splice was originally designed for solid rather than stranded conductors, but it works for both types. The lineman splice is a general-purpose splice suitable for joining wires of similar diameter (a maximum difference of 4 AWG). This type of splice has excellent mechanical strength, small size, regular (in-line) shape, and very low resistance. Short and long versions can be executed—use a longer splice to accommodate the flow of solder.

To make a lineman splice, use your fingers to twist the ends of two wires together three-quarters of a turn each counterclockwise. Next, tightly twist the ends three to five complete turns using needle-nose pliers. Push the cut-off ends close to the center wire to make a smooth junction.

Soldering a Wire Splice

The following is a procedure for wire-to-wire soldering, such as repairing a wiring harness:

1. Before splicing, cut and strip the wires. If they are stranded, twist them snugly so there are no frayed bits in the way.

2. Put a length of heat-shrink tubing about 2″ long over one of the wires. Move it out of the way so you can use it to seal the joint when you are finished. See **Figure 22-40**.
3. Make a tight lineman splice.
4. Heat the bottom of the splice with the soldering iron and apply solder at the top. Allow a thin layer of solder to coat each wire surface; this is *tinning*. As the solder melts, it will flow around the wires, resulting in an excellent mechanical and electrical connection. See **Figure 22-41**. Use as little solder as needed to bond the connection. Do not make a big, drippy ball. See **Figure 22-42**.

Safety Note

Take appropriate safety precautions when soldering. Always wear approved eye protection—liquid solder and fluxes can pop out of the work zone. A soldering iron remains hot for some time, so set it down in a safe area so it does not damage a surface or burn someone. Work in a well-ventilated area—metal fumes and vapors from acidic fluxes should not be inhaled. After soldering is completed, wash your hands and clean working surfaces to remove residues.

5. To limit melting of insulation, clip it at each end with alligator clips to act as a heat sink.
6. When the connection has cooled, slide the heat shrink tubing over the connection. Use a butane lighter or heat gun to carefully apply just enough heat to shrink the tubing.

Goodheart-Willcox Publisher

Figure 22-37. Apply layers of electrical tape in one continuous motion to bind multiple conductors into a wiring harness. The resulting bundle is stronger than the individual wires.

Goodheart-Willcox Publisher

Figure 22-38. Convoluted loom is easily slipped into place over a bundle of individual conductors. The loom protects the wiring harness from damaging external forces.

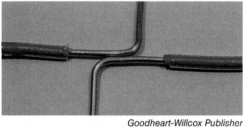
Goodheart-Willcox Publisher

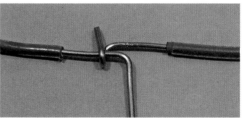
Goodheart-Willcox Publisher

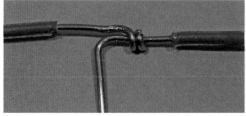
Goodheart-Willcox Publisher

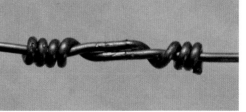
Goodheart-Willcox Publisher

Figure 22-39. Steps to complete a lineman splice.

Figure 22-40. Heat-shrink tubing must be slipped over a conductor prior to making a splice. It should be set back far enough to stay away from the heat of soldering.

7. Multiple splices in the same wiring harness should be offset in order to limit the diameter of the wiring harness. Do not put them all in the same place.

Wiring an Agricultural Trailer

Before wiring a trailer, first determine what components on the trailer need wiring. How many and what type of lights are required? Is the trailer heavy enough to need trailer brakes? Are other electric systems needed? Other considerations include the intended use of the trailer, whether the trailer will be exposed to a harsh environment, the budget available for components, and the time available for wiring work. Consider these variables before selecting and installing the electrical system of a trailer.

Figure 22-41. Heat rises and allows the liquid solder to flow between adjoining surfaces of the connection.

Career Connection

Agricultural Machinery Engineer

Job description: Agricultural machinery engineers work on a variety of activities, including designing equipment, developing biofuels, improving conservation, planning animal environments, and developing better ways to process food. Being an agricultural machinery engineer requires skills in science, math, and engineering. People in this career work closely with people in many other careers in the agricultural industry.

Education required: Educational requirements vary, but most agricultural engineers enter the field with a four-year bachelor's degree related to agricultural equipment or other area of agricultural engineering.

Job fit: This job may be a fit for you if you enjoy working with technical information, have the ability to visualize your ideas, and like working with tools.

When planning the wiring system for a new trailer, it is a good idea to start by drawing a wiring diagram or map of the different conductors needed to operate the trailer's electrical system. Drawing the diagram helps you understand how the DC current will flow. Use colored pencils, markers, or a CAD program to develop a usable diagram. A good wiring diagram simplifies troubleshooting when there is a problem. See **Figure 22-43**.

Figure 22-42. Use only enough solder to provide for a strong smooth bond. More is not always better.

Components

Every trailer wiring system must include a minimum of three circuits, all connected to a common ground. One circuit controls the running lights that identify the trailer in low light conditions. Another circuit connects the left turn and stop signals. The third circuit connects the right turn and stop signals. Other circuits may be included for brake controls, auxiliary lighting, or extra equipment.

Required Lighting

Federal and state regulations govern lighting requirements for agricultural trailers. Be sure to check the legal lighting requirements in your location. Refer to the National Highway Transportation Safety Administration (NHTSA) website for the most current federal specifications.

According to the NHTSA, trailers of all sizes must have one red stop lamp and one red or amber turn signal lamp mounted on each side of the rear of the trailer, as far apart as is practicable. A white license plate light must be mounted above or on the side of the license plate. Trailers 30′ and longer must also have amber intermediate side marker clearance lamps on each side of the trailer.

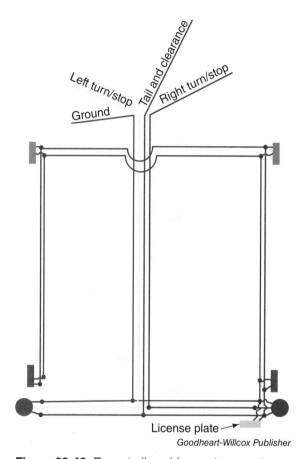

Figure 22-43. Every trailer wiring system must include a minimum of three circuits, all connected to a common ground. Can you identify the three critical circuits in this wiring diagram for a small utility trailer?

Trailers 80″ and wider have additional front and rear lighting requirements. A minimum of two red rear clearance lamps must be mounted at the widest point on or near the rear, facing rearward, and as high as practicable. A group of three red rear identification lamps must be centered on the rear of the trailer, horizontally spaced 6″ to 12″ apart, facing rearward, and placed as high as practicable. Two yellow front clearance lamps must be mounted symmetrically at the widest point on the front or near the front, facing forward, and as high as practicable. See **Figure 22-44**.

Trailers may also have additional lighting, such as working lights or backup lights. Typical trailer lighting systems are sold with incandescent lamps. Selecting LED lamps instead will drastically reduce the amperage load

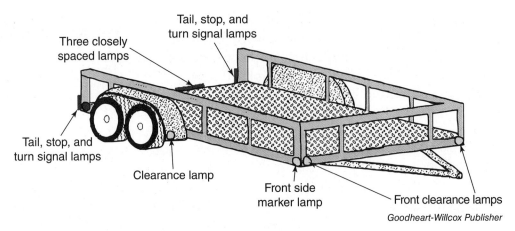

Figure 22-44. A 16′ trailer should have two red tail, stop, and turn signal lamps on the rear; one white license plate lamp; two red side-facing clearance lamps; and yellow front clearance lamps. All trailers 80″ or wider need three closely spaced red lamps in the center of the rear. Clearance marker lamps are needed to identify the widest point of the trailer.

on the towing vehicle's electrical system. When replacing incandescent turn signal lamps with LEDs, it is often necessary to install additional resistors in the circuit so flashing or blinking mechanisms will function properly.

Trailer Brakes

For trailers with a gross vehicle weight (GVW) of more than 1,500 pounds, many states require a dedicated braking system and a breakaway mechanism. This mechanism, located on the tongue of the trailer, activates the trailer brakes in the event the trailer separates from the towing vehicle. Some states require that brakes be installed on each axle; others do not.

Electric brakes usually provide drivers with automatic and manual control for trailer brakes. Electric brakes require the tow vehicle to be equipped with a controlling device synchronized with the towing vehicle's braking system and wiring for electrical power. See **Figure 22-45**. The control box should be installed in the towing vehicle within reach of the driver so it can be manually or automatically applied.

Any trailer equipped with brakes should also have a breakaway system that will stop and hold a runaway trailer for a minimum of 15 minutes. Electric brakes require an auxiliary power supply (battery). Emergency trailer braking systems include a battery for auxiliary power and a breakaway switch that separates into two parts if the trailer accidentally becomes unhitched during transport. The switch, mounted on the trailer frame, is connected to the towing vehicle with a short length of wire rope or cable. The battery can be recharged with a solar recharger or using energy from the towing vehicle's electrical system. See **Figure 22-46**.

Figure 22-45. A trailer brake control unit mounted under the dashboard. This unit energizes the trailer brakes in response to changes in motion of the towing vehicle or by manually manipulating the lever on the front of the unit.

Safety Note
Batteries in breakaway systems should be checked on a regular schedule to ensure that the system will function in an emergency.

Figure 22-46. Emergency trailer braking system components. A—Breakaway switch on trailer tongue. B—Breakaway switch on trailer.

Conductors

Most manufacturers adhere to SAE J560, a standard for wiring colors set by the Society of Automotive Engineers. However, some are a little different. It is always best to follow the standard wiring code so that anyone accessing the wiring system will know which color wire operates each component. See **Figure 22-47**.

Wiring must be of adequate size for the circuit. AWG 14 or AWG 16 is adequate for all but the longest circuits. Remember that adding length and loads (lamps) to a circuit increases the resistance and the required amperage. For wiring brake circuits and complex tail lamp circuits, heavier AWG 10 or AWG 12 is a safer choice to handle the higher amperage loads. Trailer wiring is subject to road vibrations and other factors that mechanically stress conductors. Using heavier conductors allows for some damage from this stress, such as a few breaking strands at stress points. All wiring for agricultural trailers should be multiple-strand. Wiring harnesses can consist of single wires bundled together or prebundled multiconductor cables. See **Figure 22-48**. Multiconductor cables are usually easier to route than bundles of single connectors.

Grounding

The white conductor in a trailer wiring harness is the ground connector. It is common to find older

Standard Trailer Wiring Color Code

Color	Component
Green	Right turn, stop, and hazard lamps
Yellow	Left turn, stop, and hazard lamps
Brown	Tail, clearance, and identification lamps
Black	Clearance, side marker, and license lamps
Blue	Electric brakes
Red	Battery (always hot)
Orange	Backup/auxiliary power
White	Ground, connected to towing vehicle

Figure 22-47. Standard trailer wiring color code.

Figure 22-48. This flat, 4-wire multiconductor cable for wiring trailers includes a white ground conductor, brown stop and tail conductor, yellow left turn conductor, and green right turn conductor. How is this color coding system different from the system for AC wiring?

590 Agricultural Mechanics and Technology Systems

Figure 22-49. The steel frame of this trailer serves as the grounding conductor for the lighting. This grounding method works but is unreliable and not recommended.

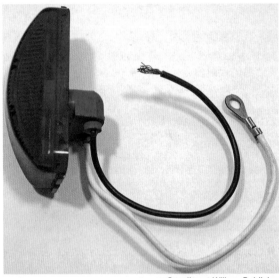

Figure 22-50. A self-contained replacement lamp. The black conductor on the pigtail is connected to the hot. The white conductor is connected to a ground conductor or the metal frame of the equipment.

Figure 22-51. Diagonal pliers are used to cut the end of a cable tie.

trailers that use the steel frame of the trailer as the grounding conductor. This method of grounding works, but it is unreliable. See **Figure 22-49**. An insufficient ground connection is a likely explanation for a trailer being pulled down a road with its lights blinking for no apparent reason. The lights blink because the ground current is routed through the trailer hitch, which by design must be loose for turning. To prevent this blinking, connect all electrical loads directly to a dedicated, white conductor that connects to the towing vehicle's ground conductor.

Making the Connections

Wiring connections on a trailer can be crimp-on, screw terminal, spliced, or soldered. Most trailers are wired using a combination of methods. Since agricultural trailers are commonly subjected to harsh conditions, connections need to be secure, waterproof, and durable.

Trailer lamps typically have screw-on stud terminals, or pigtails, with insulated plug connectors on one end and short (6″ to 10″) wires that must be spliced into the wiring harness on the other end. See **Figure 22-50**. Crimp-on ring or hook terminals work well with screw connections. Pigtails can be spliced in with crimp-on connectors, by wire splicing, or by soldering. Using heat shrink insulation over pigtail connections aids in securing the connections and provides water resistance.

Securing the pigtail loop to the trailer frame with a nylon cable tie reduces the effects of vibration and shaking. Cable ties are also used to secure bundled sections of the wiring harness. Clipping the loose end of cable ties with a pair of diagonal wire-cutting pliers after installation results in a more professional appearance. See **Figure 22-51**.

Trailer Wiring Connectors

Examine the commonly used 4-wire connector, **Figure 22-52**. Notice that the female connector is attached to the pulling vehicle. The male connector is for the trailer. The different male/female pole on each connector is the ground. Whenever the ignition is on, the vehicle's electrical circuit is energized. Protecting the connectors in this manner prevents exposing live circuitry that could contact metal objects, causing a short circuit when a trailer is not connected. This

Trailer Connection
Goodheart-Willcox Publisher

Pulling Vehicle Connection
Goodheart-Willcox Publisher

Figure 22-52. Flat, 4-wire connections are common on smaller trailers that are not heavy enough to require brakes. This connection provides the minimum number of circuits required to operate trailer lighting.

principle is true for all types of trailer connectors—the connector on the pulling vehicle should always be the female connector and the trailer should have the male connector.

The basic 4-pole wiring system is for running lights and signal only. This system should be used only on very lightweight utility trailers. A 6-wire connector has two additional terminals. One is for operating electric brakes. The other is an always-hot, direct 12 V source that is usually intended for working lights, but may be used for anything else requiring electric power. See **Figure 22-53**. A 7-terminal connector offers an additional backup circuit. The auxiliary (backup) circuit provides an extra 12 V source that is typically powered by dedicated batteries and fused separately. See **Figure 22-54**. Using this circuit for lights and small loads does not affect the starting or operating circuits of the pulling vehicle.

Trailer Connection
Goodheart-Willcox Publisher

Pulling Vehicle Connection
Goodheart-Willcox Publisher

Figure 22-53. A 6-terminal trailer connection system includes the four standard conductors, an extra line to control the brakes, and an accessory control circuit that is usually used for additional lighting.

Trailer Connection
Goodheart-Willcox Publisher

Pulling Vehicle Connection
Goodheart-Willcox Publisher

Figure 22-54. A 7-terminal trailer connection system includes the conductors found in a 6-conductor system and an extra always-hot conductor for accessories used while the trailer is parked.

Routing and Connecting Wires

When wiring a trailer, tuck all wires in and away from anything that could rub or catch on them, such as other parts of the trailer, potential trailer loads, roadside brush, and inquisitive calves if you are parked in a cow pasture. Keep trailer wiring neat and protected. Sagging wires can catch on brush and other things, causing extensive damage. See **Figure 22-55**.

Routing wires through conduit or inside hollow frame members results in a clean appearance and helps to protect the wiring. Use grommets or extra insulation to protect wires where they pass through holes in metal or across rubbing points. Grommets, formed from rubber or plastic materials, are sized to fit snugly into round holes in metal. The grommets completely cover the sharp edges of the holes. There should be no potential pinch points—on top of axles, springs, or other moving parts that could damage wires. Holes should be drilled, not burned with a torch, through angle iron and other

Goodheart-Willcox Publisher   *Goodheart-Willcox Publisher*

Figure 22-55. Wires hanging below the frames of utility trailers represent accidents waiting to happen. Unprotected wiring can easily snag on road hazards or vegetation when the trailer is used off paved roadways.

frame members where wire will pass through. See **Figure 22-56**. Remove all cutting and drilling burrs from the edges of the metal where wires will be routed.

All wiring connections should be sealed and watertight. Holes on the bottom of frame members should be left open for drainage. Other holes can be sealed to prevent water from filling the tubular frame. Cable ties or clips that attach to the frame can be used with trailers with channel or angle iron construction to keep wiring tight inside the frame and out of the way. An even better solution is to use small tubing as conduit to completely enclose as much of the wiring as possible. See **Figure 22-57**.

Troubleshooting DC Circuits

Electricity is logical—it flows on a prescribed path, or circuit. When something is wrong with an electric circuit, the path of the electrons has been disrupted or redirected. Fixing the problem is a matter of finding the disruption. If a load such as a trailer lamp does not work, the circuit is either broken or *shorted* (prematurely grounded).

A broken circuit no longer has a complete path from the positive terminal of the battery through a load or loads and back to the negative terminal of the battery. Common causes of a broken DC circuit on agricultural equipment are loose connections, burned-out bulbs, and burned fuses. Another common cause is exposed conductors that have been damaged.

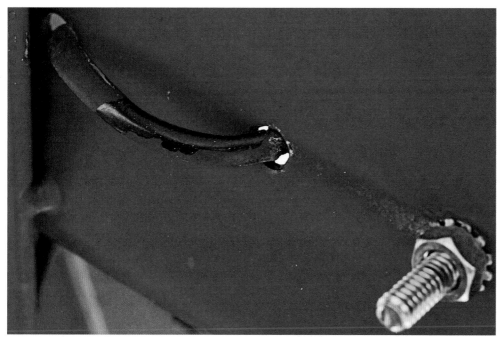

Goodheart-Willcox Publisher

Figure 22-56. Sharp burrs around the hole used for wiring are eroding the insulation from the conductor. If left unchecked, what will be the result of this erosion?

Figure 22-57. Lengths of tubing are added to trailer frames to provide secure conduits that protect the wiring from damage.

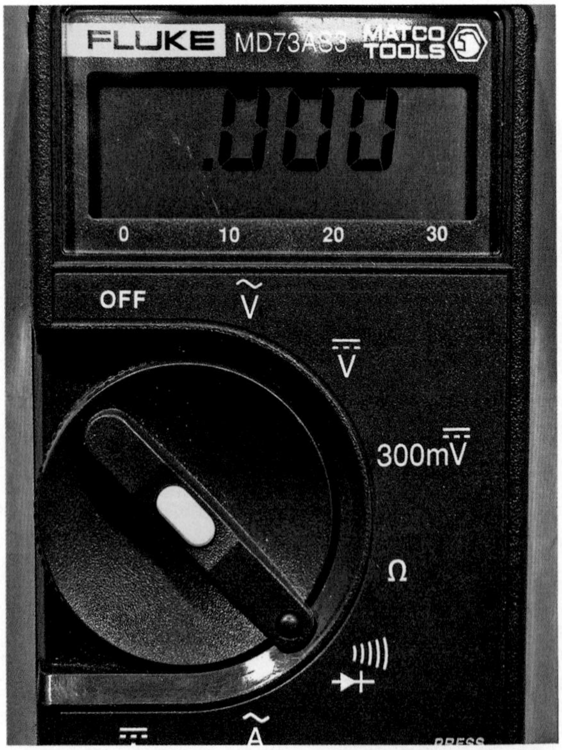

Broken circuits are easily identified with a multimeter set to detect continuity. See **Figure 22-58**. Check for continuity by placing one probe of the multimeter in the circuit's terminal at the trailer-to-vehicle connector and the other on the ground terminal or the trailer frame. If no continuity is detected, check the fuses; bulbs; splices; and places where conductors are not protected by loom, conduit, or the frame of the equipment.

Burned fuses often indicate that a short circuit caused the fuse to overload from excessive amperage. If the multimeter indicates continuity on a nonworking circuit, a short circuit is likely. Check for water-filled light housings and inspect conductors where they enter conduits or frame members and may have damaged insulation. Failure of some components may result in a short circuit; however, most failed components result in a broken circuit.

Figure 22-58. The *.000* reading on this multimeter indicates that the tested circuit has complete continuity with minimal impedance.

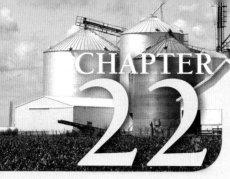

CHAPTER 22 Review and Assessment

Summary

- DC electric energy is commonly produced by alternators and generators that convert mechanical power to electric power and by solar cells that convert light energy directly into electricity. Electricity produced by alternators and generators must be converted from AC to DC.
- In agriculture, DC electricity powers a variety of loads, including lighting, electric motors, and other devices.
- Some switches are confined to *Off* and *On* positions that control a single component. Others are multipositional switches, which may have multiple terminals operating several different components.
- Many DC electric circuits used in agriculture have fuses that protect the circuit from overloading.
- The insulated wiring that serves as a conductor to move electrical power throughout a DC system must have an AWG size heavy enough to handle the intended load without overheating.
- Methods of making connections in DC electrical circuits include bullet connectors, blade connectors, screw terminals, Scotchlok® connectors, splicing, and soldering.
- In a lineman splice, the wires are twisted together in a way that causes the electrical connection to tighten as the conductors pull against each other. This splice is suitable for joining wires of similar diameter.
- Federal and state regulations govern lighting requirements for agricultural trailers.
- Any trailer equipped with brakes should also have a breakaway system that can stop and hold a runaway trailer for a minimum of 15 minutes.
- The trailer connector on the pulling vehicle should always be the female connector and the trailer should have the male connector.
- All wiring connections should be sealed and watertight.
- Broken circuits are easily identified with a multimeter set to detect continuity.

Words to Know

Match the key terms from the chapter to the correct definition.

A. array
B. diode
C. electrical tape
D. end splice
E. limit switch
F. lineman splice
G. loom
H. photovoltaic cell
I. photovoltaic module
J. rectifier
K. relay
L. rheostat
M. shorted
N. shrink tube
O. tap splice

1. A group of interconnected solar cells grouped together in a single frame.

2. A collection of modules grouped together for the collection of solar energy.
3. Electronic device used to convert AC to DC.
4. Gateway that allows electric current to pass through in one direction only.
5. Switch operated by the position of equipment or materials rather than manually.
6. Adjustable resistor that may be used to control light intensity.
7. Electromagnetic switch that allows control of a high-amperage circuit through a low-amperage switching mechanism.
8. Wire-to-wire connection in which exposed wires are pointing the same direction before being twisted together.
9. Insulating protector that requires heat to protect bare wires and connections.
10. Insulating protector that can be used to cover bare wires and connections without the addition of heat.
11. Used as an added layer of protection to cover wires and wire harnesses.
12. Wire-to-wire connection in which parallel lines are pointing in opposite directions before being twisted together.
13. Condition in which a conductor makes contact to ground prematurely or in an unintended location.
14. Wire-to-wire connection in which exposed wires are perpendicular to each other before twisting together.
15. Device that converts light directly to DC electricity at the atomic level.

Know and Understand

Answer the following questions using the information provided in this chapter.

1. DC electrical circuits are used in agricultural machines for starting, _____, and monitoring systems.
2. To convert mechanical energy to DC electricity, generators move a conductor through a(n) _____ field.
3. Rectifiers use _____ to convert AC to DC electricity.
4. The electrons freed from the _____ material in a solar cell are captured in the form of DC electricity.
5. In most cases, the lighting on agricultural vehicles is powered by 12 V _____ systems.
6. In applications where a bright, focused light is required, _____ lighting is used.
7. DC electric motors are commonly used in heavy-duty applications in which a constant _____ must be maintained under varying loads.
8. On trailers equipped with electric brakes, an electromagnet initiates a motion causing friction between the brake _____ and the brake drum.
9. DC and solar-powered fence chargers are powered by a(n) _____.
10. The multipositional switch that operates a tractor starter motor only when the key is physically held in position is an example of a(n) _____ switch.

11. The _____ circuit of a relay typically draws much less current than the secondary circuit.
12. The size of a fuse for a given circuit depends on the capacity of the other components in the circuit and on the size of the _____ used in the circuit.
13. Bullet and blade wire connectors come in _____ pairs that are designed for ease in making connections.
14. What is the purpose of using plastic loom on parts of a wiring harness?
15. According to NHTSA regulations, all trailers must have one red stop lamp and one red or amber _____ lamp mounted on each side of the rear of the trailer, as far apart as is practicable.
16. A burned fuse is often an indication of a _____ circuit in a lighting circuit on an agricultural trailer.

STEM and Academic Activities

1. **Science.** Reflex reflectors are often incorporated into the design of equipment lighting. Research the characteristics of these reflectors and write a brief report.
2. **Technology.** Compare several different methods of joining electric wiring for DC applications. Construct a table illustrating the benefits and drawbacks of each method.
3. **Engineering.** Study the function of a trailer brake controller. Draw and present to the class a wiring diagram that illustrates how energy is applied to the trailer brakes.
4. **Math.** Examine a typical 16′ to 20′ utility trailer. Determine the amperage required to illuminate all of the tail and clearance marker lights if the trailer is equipped with standard incandescent lighting. What would the amperage need to be for the same system equipped with LED lighting?
5. **Social Science.** Investigate the historical development of tractor electrical systems. When did the conversion from using generators to using alternators occur? Why did this happen?
6. **Language Arts.** With a partner, review the section headings in this chapter and choose a topic. Prepare a presentation in which you and your partner explain this topic to the class and answer questions from your classmates. Include visual aids in your presentation.

Thinking Critically

1. Research the changes in battery technology. How have these changes impacted battery life, the materials used to manufacture batteries, and how power is delivered from the battery to a device? What improvements to current battery technology do you think would be beneficial for agricultural applications?
2. What methods can you think of to remember the color coding system for flat, 4-wire multiconductor cable for wiring trailers? How is this color coding system different from the color coding for AC wiring?

CHAPTER 23: Electric Motors and Controls

Chapter Outcomes

After studying this chapter, you will be able to:
- Explain the advantages and disadvantages of electric motors as a power source.
- Explain how electric motors operate.
- Identify common components of an electric motor.
- Identify common types of electric motors.
- Interpret a motor nameplate.
- Select an electric motor for a given agricultural application.
- Install an electric motor.
- Identify and install basic electric motor controls.
- Demonstrate proper electric motor maintenance.
- Describe how to work safely using electric motors.

Words to Know

bearings	electromagnet	slip
brushes	encoder	split-phase motor
bushings	induction	stator
capacitor-start motor	insulation classes	stepper motor
commutator	inverse-time breaker	three-phase motor
coupling	motor housing	universal motor
drive ratio	repulsion-start motor	variable frequency
drive shaft	rotor	drive
electrolytic capacitor	servo motor	

Before You Read

Arrange a study session to read the chapter aloud with a classmate. At the end of each session, discuss any words you do not know. Take notes of words you would like to discuss in class.

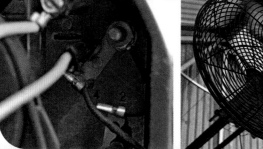

While studying this chapter, look for the activity icon to:

- **Practice** vocabulary terms with e-flash cards and matching activities.
- **Expand** learning with interactive activities.
- **Reinforce** what you learn by completing the end-of-chapter questions.

www.g-wlearning.com/agriculture

Electric motors are widely used in agriculture, providing power for things such as fans, pumps, mixers, power tools, and other equipment with moving parts. See **Figure 23-1**. Often, when pneumatic or hydraulic power is used to operate agricultural equipment, an electric motor is first used to drive a pump.

Electric motors are typically limited to use in operations where electrical power is constantly and reliably available. Usually, this means AC (alternating current) power from the power grid or a generator or DC (direct current) power supplied by a generator or a battery source. AC motors can be constructed to operate on single-phase or three-phase power. Electric motors can be constructed to operate at many different voltages. The introduction of lithium-based battery packs has elevated the use of electric motors to operate certain cordless landscaping equipment, such as mowers, blowers, and chain saws. See **Figure 23-2**. These tools lack the power of comparable tools powered by internal combustion engines. Battery-powered DC electric motors are beginning to see regular use for small vehicles that operate within a limited range between charging stations. Economical batteries of sufficient capacity to operate large motors for a useful length of time have not yet been developed. They remain too expensive and too heavy to be justified in all but a few vehicular applications. See **Figure 23-3**.

Advantages of Electric Motors

Electric motors have quite a few advantages over other sources of mechanical power:

- Electric motors are efficient and economical. Common electric motors operate at an efficiency rating between 75% and 95%. Very little energy is wasted to heat and friction. Even a small 1/2 hp (horsepower) electric motor can do the work of five people. Operating continuously, an electric motor requires no breaks.

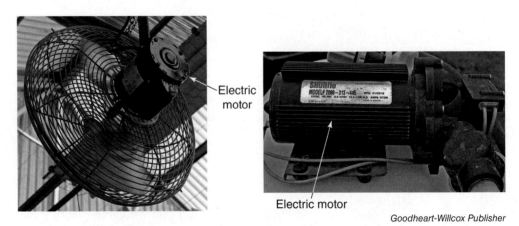

Goodheart-Willcox Publisher

Figure 23-1. Electric motors are used for many purposes in agriculture. High-volume fans are used to circulate air in greenhouses and to cool stalled livestock. Electric pumps are used to move everything from water to molasses.

- Electric motors require only a reliable source of electricity to run. They do not depend on fossil fuels or a clean air supply. Properly configured electric motors may be operated when completely submerged in liquid.
- Compared to internal combustion engines, electric motors have few moving parts. This reduces wear points and maintenance needs.
- Quiet operation is a huge advantage when working with livestock and when communication between workers is important.
- Electric motors are simple to operate and dependable. Little special training is needed to switch on a motor. Motors operate in a wide range of climatic conditions. Changes in outside temperature have little effect on starting or running efficiency.
- When properly installed and periodically serviced, an electric motor can provide many years of reliable operation. It is not uncommon for today's agricultural mechanic to encounter working electric motors that have been in service for 40 years or more.
- A variety of available switches and controls makes the automation of electric motors very common. In many applications, such as water pumps and greenhouse equipment, electric motors are rarely, if ever, operated manually. Timers and sensors control these operations.
- Electric motors make working conditions less hazardous than other power sources. They have fewer moving parts, no hazardous exhaust fumes, and no hot exhaust ports. In addition, electric motors operate relatively quietly, reducing the risk of noise hazard.

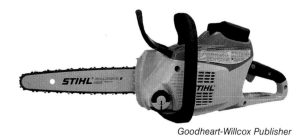

Goodheart-Willcox Publisher

Figure 23-2. Improvements in the construction of lithium-ion batteries have made tools such as this battery-operated chain saw possible.

Goodheart-Willcox Publisher

Figure 23-3. Battery-operated carts are used in agriculture to transport people, equipment, and supplies. This cart is used to provide tours at an agricultural facility. Livestock at the facility are not disturbed by its quiet operation.

Electric Motors and Magnetism

Electric motors change electrical energy into mechanical energy. Motors produce a rotational force that may be measured in speed (rotations per minute, or RPM) and in intensity (torque). Most motors use an electric current passing through coils to produce a set of electromagnets with corresponding magnetic fields that interact with each other, resulting in torque on a rotating part.

Remember that each end of a magnet is called a *pole*. These magnetic poles are identified as north (N) and south (S). A fundamental rule of magnetism is that like poles repel each other and unlike poles attract. To get an electric motor to work, engineers take advantage of this rule.

Figure 23-4 illustrates the basic principle of motor operation. The moving bar magnet in the center rotates because its N pole is attracted to the S pole of the stationary magnet and its S pole is attracted to the N pole of the stationary magnet. Just as the N pole of the moving magnet reaches the S pole of the stationary magnet, the polarity is switched by changing the direction of current flow through the coils. Reversing the polarity changes the direction of the magnetic forces, causing the center magnet to rotate for another half turn.

Some motors use permanent magnets, but most motors use electromagnets. An *electromagnet* is a magnet in which the magnetic field is produced by an electric current. Whenever electric current passes through a conductor, the current creates a magnetic field. When a conductor is coiled around an iron core, a strong magnetic field is produced. Most electromagnetics used in motors consist of coils of copper conductors (called *windings*) wrapped around iron cores.

Some windings are energized by direct connection to an electrical power source. Other windings are energized by moving in a magnetic field. When a conductor moves through a magnetic field, current is generated in the conductor. This process is called *induction*. The induced current then creates a magnetic field. Thus, an energized winding produces a magnetic field. This magnetic field can induce a magnetic field in another winding, and the interaction of these two magnetic fields can result in motor rotation. The force of a motor's rotation is influenced by the amperage of the input current and the strength of the magnetic fields.

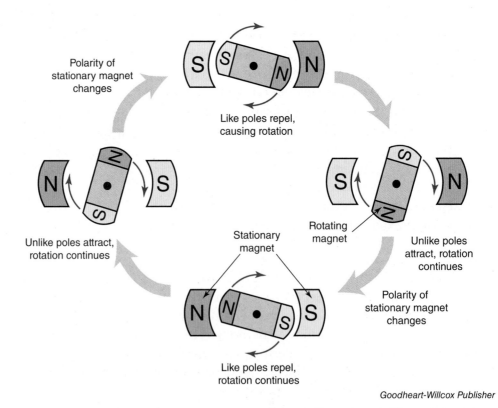

Goodheart-Willcox Publisher

Figure 23-4. The basic principle of electric motors is that like poles of different magnets repel one another and opposite poles attract. The magnets may be permanent magnets or electromagnets.

Electric Motor Components

Most electric motors have some common components. See **Figure 23-5**. The stationary portion of the motor mechanism is called the *stator*. The rotating portion of the motor is called the *rotor*. Both the stator and rotor include magnets or electromagnetics in their construction.

In some motor types, the windings of the stator are connected to a *commutator*, which is composed of thin strips of conductors that are aligned side by side around the diameter of the drive shaft. Insulating material such as mica is used to isolate these strips from each other and the drive shaft. In some motors, a set of carbon *brushes* rubs across the commutator to transfer electrical energy to the rotor winding.

All electric motors have a frame or *motor housing* that encloses and supports the other components of the motor. A solid *drive shaft* rotates in the center of each motor and is the part that transfers mechanical power from the motor to other equipment. Drive shafts must have some method of attachment to other equipment and must be supported in the frame. A pair of *bearings* or *bushings* provides this support.

Bushings are single-piece sleeves constructed of a material with a lower coefficient of friction than the drive shaft or housing. Often, they are formed from a soft alloy, such as bronze. See **Figure 23-6**. Bushings are found in light-duty motors that are not expected to experience heavy loads. Electric motors equipped with bushings may require periodic lubrication to minimize wear and extend the life of the bushings. Bushings typically attach to the shaft or the support and are engineered to wear before the metal parts they protect.

The drive shafts of heavy-duty motors and those designed for harsh conditions are usually supported by bearings consisting of multiple parts. Common bearings include an inner race and an outer race that serve as tracks for a set of rollers. Bearings are made from very hard steel alloys to resist wear. Rollers may be balls, cylinders, or very fine cylinders known as needles. See **Figure 23-7**.

Some older motors are engineered with lubrication points connected to passages that lead to the internal bearings. Larger equipment may use automatic oilers with oil reservoirs that require periodic filling. Most modern bearings are internally lubricated for the life of the bearing and sealed to contain the lubricant and keep out contaminants.

Disassembled Motor

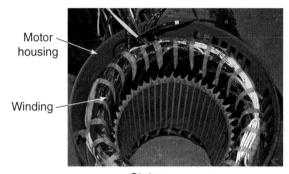
Stator

Rotor

Goodheart-Willcox Publisher

Figure 23-5. All electric motors have certain parts in common, including a housing, a drive shaft, bearings or bushings, windings, and a fan or other method of dissipating excess heat.

Figure 23-6. A bronze bushing is often used to reduce friction and control wear between a motor drive shaft and support. When installed, the bushing fits tightly to either the support or the drive shaft, limiting the friction to the outside or inside surface of the bushing.

Figure 23-7. This open roller bearing has inner and outer races surrounding a set of roller balls separated by a cage. A bearing fits tightly to the shaft and its support in order to contain the friction and wear surfaces within the internal structure of the bearing.

Types of Electric Motors

Several different types of electric motors are used in agriculture. They may be classified based on the input current and voltage as well as the use of AC or DC power. The method of starting is often the most important factor used to differentiate electric motors. Physical size and output horsepower may also be used as a method to group motors. The construction design of the motor housing, mounting method, and driveshaft configuration can also be used to classify electric motors. Whatever the classification method, electric motors should be specifically suited to the intended load and the unique environmental conditions presented by agricultural applications.

DC Motors

Small DC motors often use permanent magnets to create a stationary field. Two other styles of DC motors are series-wound and shunt-wound. Each uses an armature with brushes and a commutator. However, the stator uses windings instead of permanent magnets to create the field. See **Figure 23-8**. A series-wound motor has the stator windings connected in series with the armature, and a shunt-wound motor has the stator windings connected in parallel with the armature winding. Series-wound motors are more common. DC motors have a tendency to run at dangerously high speeds under no-load conditions. Such speeds could mechanically damage the motor. Series-wound motors should not be installed using drive belts for this reason. If the belt breaks, slips, or becomes dislodged from the pulley, the motor could overspeed and be severely damaged.

Two main types of DC motors are found in agriculture. These are brushed and brushless motors. Electric motors containing brushes are commonly used as starting motors for internal combustion engines, to power electric winches, and to power cordless portable power tools. Within brushed motors, the electric current is conducted along a solid path. The electrical energy is transferred from the stator to the rotor using a commutator that rubs against stationary brushes. The operating speed of DC motors may be controlled by using an electrical resistance in the circuit supplying power to the field or the armature. In some designs, varying the operating speed has little effect on the output torque.

SAE Connection

Electric Motors

Electric motors and controls are found everywhere in agriculture. From pumping water to driving augers, electric motors perform various jobs safely and efficiently. The fact that moving electrons have the capability to "make wheels turn" may be fascinating to you. If designing and building electric motor components or repairing these same components stirs your thoughts, then take the opportunity to research potential job experience in this area.

Electric motors provide a cleaner alternative to powering agriculture. What current agricultural processes use electric motors and controls? Which agricultural activities that currently use other power sources could be switched to electrical power using an electric motor? How much energy is saved by using electric motors and controls? What environmental impact does the use of electric motors have? Each of these questions would be a great foundation for a research SAE. Ask your agriculture teacher how you can get started today.

Franco Nadalin/Shutterstock.com

Brushless DC motors use an electronic controller to perform the function of the commutator. Brushless DC motors are typically used to operate cooling fans in computers and other electronic controllers, some small battery-operated power tools, and remote control applications. Brushless motors offer the advantages of noise reduction, improved efficiency, and reduced maintenance. Brushless motors cannot be used in dirty environments or service where excessive shock may affect the sensitive controller circuitry, which limits their application in agricultural mechanics.

Fouad A. Saad/Shutterstock.com

Figure 23-8. The stator in a wound DC motor uses electromagnets instead of permanent magnets to produce a magnetic field.

Split-Phase Motors

Split-phase motors are used for easy-to-start loads. See **Figure 23-9**. They are commonly used to power fans, centrifugal pumps, and in applications where the amount of load increases as the speed increases. Split-phase motors are powered by single-phase AC power. The input current is split between two sets of windings in the stator: a main set for running the motor and a second set for starting. The starting winding is mounted at 90° to the running winding. The magnetic field generated by the running winding does not rotate. Instead, it pulsates with maximum strength as the motor turns at 0° and 180°. When the two sets are compared, the running winding has a lower resistance than the starting winding. The magnetic field in the starting winding develops first, with the

Goodheart-Willcox Publisher

Figure 23-9. One of the most commonly used electric motors is the split-phase induction motor. They are economical and are used to drive easy-to-start loads.

Goodheart-Willcox Publisher

Figure 23-10. Alternate starting winding and running winding in the stator generate a rotating magnetic field that begins the rotation in a split-phase motor.

field in the running winding following slightly behind. This results in the rotating magnetic field needed for movement. See **Figure 23-10**. After the motor rotation reaches about 75% of its full speed, a centrifugal switch inside the motor opens, cutting power to the starting winding and diverting the full current flow to the running winding. See **Figure 23-11**.

Split-phase motors have the advantages of being relatively inexpensive and of being able to reach full speed fairly quickly. Operation is at an approximately constant speed. A four-pole motor powered by 60 Hz AC power turns at about 1725 RPM under full load. Split-phase motors can be easily reversed. Loads that require the motor to overcome significant inertia or conditions that require repeated starting and stopping are not suited for split-phase motors. Because of the higher resistance in the starting windings, spilt-phase motor circuits must be wired with larger conductors than necessary for running the motor.

Capacitor-Start Motors

Similar to split-phase motors, capacitor-start motors are engineered with both running and starting windings. *Capacitor-start motors* can easily be identified by the electrolytic capacitor mounted on the outside of the motor housing. See **Figure 23-12**. An *electrolytic capacitor* causes the voltage in a circuit to lag behind the current by a predetermined degree of the phase. In practice, the capacitor is wired in series with the starting winding to create a stronger rotating magnetic field than is possible in a split-phase motor. The more powerful magnetic field at startup gives the capacitor-start motor the ability to overcome the initial inertia of heavier loads such as agricultural pumps, compressors, and other processing equipment. As with split-phase motors, when the motor approaches operating speed, a centrifugal switch cuts power to the starting winding.

Repulsion-Start Motors

In *repulsion-start motors*, the stator winding is connected directly to an AC power source. The rotor is connected to a commutator and brush assembly, similar to the construction of a DC motor. The commutator and brush assembly is connected to create a complete circuit using only some of the rotor winding. See **Figure 23-13**.

In repulsion-start motors, power to the rotor is supplied by induction from the stator windings. When alternating current is supplied to the stator winding, it induces a strong electromagnetic force (emf) in the rotor. The starting force is powerful enough to start heavy loads such as those found on grain elevators, conveyors, and augers. As the rotor turns, the portion of the rotor winding with induced current changes.

Similar to other induction motors, once a repulsion-start motor reaches about 75% of its full speed, a centrifugal switch short circuits the segments of the commutator. The motor begins running as an induction motor. In some models, the centrifugal switch mechanism also lifts the carbon brushes out of contact with the commutator to reduce wear. Typically, repulsion-start motors used in agriculture are bulky and constructed using a heavy-duty frame and mounting system. Repulsion-start motors are seldom used today on new equipment, but their solid construction has kept a considerable number in service for many years.

Three-Phase Motors

The stator of the *three-phase motor* has a solid, laminated-steel magnetic core with slots in the inner surface. The phase windings are placed in the slots of the magnetic core and are separated by insulation. Because the waves of three-phase power are synchronized so that each phase peaks in power every 60°, the magnetic field produced is rotating and no separate starting windings are needed. Larger three-phase induction motors do use a starting circuit that switches the configuration of the three-phase power input. The switch may manually shift from start to run or work automatically on a timer.

Three-phase motors can be constructed lighter than single-phase motors of similar output, and they operate more efficiently than single-phase motors. Motors connected to three-phase power have an even energy draw on the electrical system, resulting in little to no light flickering when the motor is started. Three-phase motors typically cost less than single-phase motors, but the initial installation costs of three-phase power supplies may override any savings. The actual use of three-phase motors is often limited to industrial or other applications where a high power requirement is likely. In agriculture, three-phase equipment is most likely to be stationary and used for powering compressors, high-capacity pumps, large blower motors, and other production and processing equipment. See **Figure 23-14**.

Goodheart-Willcox Publisher

Figure 23-11. As a split-phase or capacitor-start motor reaches about 75% of its operating speed, centrifugal force on a slip disc located on the motor drive shaft causes a switch to open, disconnecting power to the starting winding.

Goodheart-Willcox Publisher

Figure 23-12. A capacitor-start motor is easily identified by the hump-shaped shell containing the capacitor mounted on the exterior of the motor housing.

Figure 23-13. In a repulsion-start motor, the brush assembly aligns the carbon brushes to make contact with the alternately wound coils on the rotor. Shifting the position of this bracket a few degrees clockwise or counterclockwise reverses the direction of rotation.

Figure 23-14. This 100 hp three-phase electric motor is mounted in the pit below a feed mill where it drives the feed-grinding hammer mill. A tightly sealed construction protects it from damage in the extremely dusty environment.

Some motors are wired to be able to operate on three-phase or single-phase power. Basically, in single-phase mode these motors do not use one of the sets of three-phase windings. The result is a warmer operating temperature and a reduction in power.

Universal Motors

Universal motors are often used for corded portable power tools such as electric drills, saws, and sanders. These motors can operate on single-phase AC or DC power. The universal motor is basically a series-wound DC motor that is designed to operate on AC power as well as on DC power. The design characteristics of a universal motor determine whether it turns faster using AC or DC power—very few operate at exactly the same speed on both. The starting torque is always higher when the motor is started using DC power. Just like DC motors, a universal motor has a tendency to run at a dangerously high speed under no-load conditions and should not be used in belt-driven systems or otherwise allowed to overspeed.

Stepper and Servo Motors

Stepper and servo motors are used to actuate and control linear motion. *Stepper motors* typically are brushless. They use 50 to 100 magnetic poles, with polarity generated either by coils or by permanent magnets. Stepper motors move incrementally using electrical pulses released by an electronic controller.

Servo motors have only four to eight poles that must be controlled by an *encoder*, which is a more complex controller than that used for stepper motors. The ability of stepper and servo motors to rotate incrementally for controlled distances makes them ideal for applications in which the position of machinery or parts of machines is controlled by a computer-driven application. Examples include automatic egg turners, greenhouse controls, robotics, CNC machinery, and many machines used in the processing of agricultural products. See **Figure 23-15**. Both stepper and servo motors are available in models that either operate on AC or DC power. DC-operated positioners tend to be less powerful and are limited to smaller applications. Electronic controllers used with servo motors also incorporate sensors that provide feedback to the controller indicating the exact motor position.

Motor Nameplates

Important information needed to select a motor for a given application can be found on the motor nameplate. See **Figure 23-16**. The type of data provided by the motor nameplate is standardized by the National Electrical Manufacturers Association (NEMA), an organization that develops and promotes electrical manufacturing standards. The National Electrical Code provides detailed information concerning the required markings on motor nameplates. Different motors have different nameplates, and each is matched to the important characteristics of the specific motor. Nameplates are located prominently on the outside of the motor housing. When possible, motors should be mounted to equipment so that the motor nameplate is accessible with a minimum of disassembly required to obtain service information. Motor nameplates contain the following information:

- Manufacturer. This information provides the company's name and address to help locate engineering specifications on the motor.
- Horsepower. This provides the full-load horsepower rating for the motor.
- Specification or serial number. This number helps identify parts and warranty information regarding the motor.
- Frame. This number reflects the mounting bolt hole dimensions and locations as well as the drive shaft configuration. Standard frame numbers have been adopted by NEMA to be consistent among different manufacturers. This information is necessary to determine if motors can be interchanged.
- RPM. The number of revolutions per minute is measured when a motor is turning at full load under the rated voltage and frequency. The full-load speed of induction motors is somewhere between 96% and 99% of the no-load speed. The difference is known as the motor's *slip*.
- Phase. This tells if the motor is designed to operate on a single-phase or a three-phase electric power supply.
- Insulation class. *Insulation classes* are designated in order of their thermal capacities by A, B, F, and H. The higher the code letter, the greater the heat capacity.

Goodheart-Willcox Publisher

Figure 23-15. The precision of a servo motor makes it useful on mechanisms designed for positioning machinery and materials, such as this CNC drive mechanism.

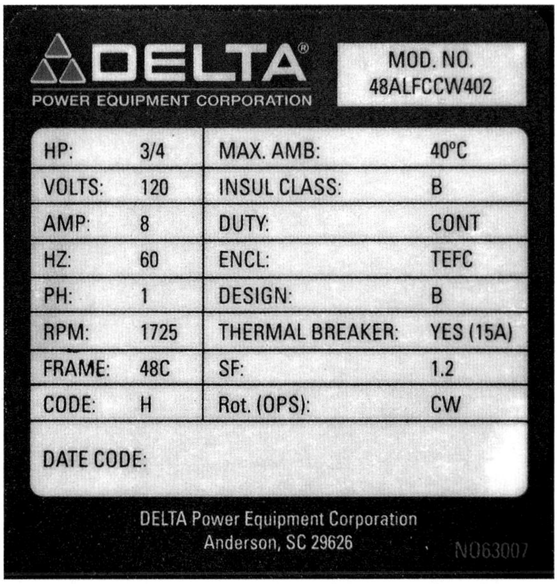

Goodheart-Willcox Publisher

Figure 23-16. Motor nameplates provide information needed for electric motor installation, service, and replacement.

- Service factor. The service factor indicates the percentage of overload the motor will handle. A motor with a 1.0 service factor will not safely operate above its rated horsepower. A motor with a 1.10 service factor will operate with infrequent loads at 10% above its rated horsepower. Operating above rated horsepower results in excess heat buildup that will shorten the motor's expected life.
- Hertz. In the United States, this is 60 cycles per second and indicates the operation frequency of the electrical power supply for which the motor is designed.
- Voltage. This indicates the voltage rating at which the motor is designed to operate most efficiently. Electric motors are engineered to operate within a plus or minus 10% tolerance of this value. Many motors can be wired to accommodate multiple voltage ranges, such as 110/220/440 V. This is achieved by attaching the input power to different terminals connected at selected points along the motor windings.
- Amperage. The number provides the amount of current that the motor is expected to draw under full load. Motors that will operate in multiple voltage ranges have different amperage draws in each configuration.
- Duty rating. Sometimes called *duty cycle*, this rating indicates whether the motor can be used continuously or only intermittently. Intermittent duty has a maximum amount of time that the motor may be operated under load, as opposed to being at rest.

Selecting Electric Motors

When choosing an electric motor for a particular application, several factors must be taken into consideration. First, a motor should have the ability to drive its load without being overloaded. A motor that is larger than needed will be inefficient; optimum efficiency is achieved when an electric motor operates between 75% and 100% of its rated load. Using a belt-and-pulley or chain-and-sprocket drive system to connect a motor to the load will cause a power loss of approximately 7%. This must be taken into account to select the motor. Select a motor horsepower that ensures that, once the load is started, the motor will operate near or somewhat below the rated horsepower on the nameplate. An electric motor to be used in place of an existing small internal combustion engine should supply 2/3 to 3/4 as much power as the rating of the small engine. When replacing another electric motor, refer to the motor nameplates to ensure a good match.

An electric motor must be able to reliably start the load. Some loads require a higher starting torque in order to begin turning them from a resting position and accelerate them to the appropriate operating speed. Select the type of motor whose horsepower and starting torque are most appropriate. The more starting power a motor has, the more it is likely to cost.

Always consider the location and environment in which an electric motor will be working. A motor mounted inside a covered machine housing on a machine permanently located in a climate-controlled building (such as a table saw) requires much less protection than a motor mounted to a portable unit expected to be exposed to wet outdoor conditions (such as a steam cleaner). Indoor-rated motors are often constructed with an open, skeleton-like frame that allows air circulation for cooling. Other motors are enclosed in protective housings with internal fans that force cooling air through passages in the motor housing. Motors designed for service in conditions complicated by severe dust or submersion in liquid are completely sealed to keep out the elements. See **Figure 23-17**. Submersible water pumps use sealed motors that are cooled by contact with the water as it passes over the surface of the motor housing.

Goodheart-Willcox Publisher

Figure 23-17. This electric motor used to drive a submersible pump in a greenhouse cooling system is completely sealed to protect it from the moisture.

Installing Electric Motors

An electric motor performs work and serves as the load in the circuit. In addition to the motor, a complete motor circuit includes a circuit protection device, a means of disconnecting the motor, a controller, and possibly an overload protection device. Circuit protection is commonly found in the service entrance in the form of fuses or circuit breakers. Protection from overheating due to short circuit is provided to the circuit components. Controllers may be simply a switch or more complex devices that control factors such as the speed of the motor or the time for which motor is allowed to operate. Electric circuits including motors that operate equipment near humans or animals in damp environments may also incorporate a GFCI device for protection against electric shock. Many electric motors are constructed with internal mechanisms to prevent overheating.

Circuit Protection

Like all AC circuits, motor circuits are required to include overload protection to prevent excess amperage from building up damaging heat in the circuit components. For single-phase and three-phase motors, *inverse-time breakers* or time-delay fuses are recommended to provide this protection. Unlike standard circuit breakers and fuses, these devices will handle a brief overload for a predetermined time (enough for motor starting) before breaking the circuit. AC motors typically have a higher starting current than is required to keep the motor running. The amperage rating for the circuit must be large enough to carry the starting current of the motor. To accommodate starting current demands, the maximum size for a time-delay fuse is 1.75 times the full-load motor current. The maximum size for an inverse-time breaker is 2.50 times the full-load motor current. Always use a device with the smallest rating that permits the motor to start and operate properly. When using thermally protected electric motors, the circuit breaker or fuse may have a higher amperage rating than the conductor ampacity because the motor overload device protects the conductors from overheating.

Figure 23-18. Timers save energy and wear on equipment. This manually set timer allows an electric motor to operate for a maximum of one hour before the timer must be reset.

Often, electric motor circuits include a timing device that limits the amount of time the motor operates. Limiting operating time saves energy and wear on the equipment. Ventilation systems, pumps, and agitation devices are commonly controlled with a timer. Timers may need to be manually reset, **Figure 23-18**, or they may be fully automatic and programmable.

Overload Protection

During starting and overloads, the motor draws more current than when delivering its rated horsepower. Motor overloading can be caused by several conditions, including exposure to excessive torque, low input voltage, and loss of or imbalance of input currents. Overloading will damage the motor if sustained over a long time. A three-phase motor may continue to run on two phases, but damaging heat will rapidly build up inside the motor windings. Similarly, a motor wired for 240 V may continue to turn for a time, although slower and hotter, if supplied by a single 120 V hot conductor. Overload protection devices are commonly built in as an integral part of an electric motor. Thermal sensors inside a motor can detect heat from overload conditions and automatically open the circuit to switch off the power supply to the windings to prevent the heat from damaging the motor. Motors that have this capability are marked "thermally protected" on the motor nameplate. Some will reset themselves when operating temperatures return to acceptable levels. Others must be manually reset. See **Figure 23-19**.

Motor Controls

The current required to start many electric motors demands that switching devices be more heavily constructed than typical switches used for lighting. Conductors and other circuit components are sized to limit voltage drop, which can cause motors to overheat or perform poorly. The NEC prescribes the appropriate diameter of conductor needed to limit the voltage drop based on required current, voltage, and type and length of conductors.

Electric motor circuits should have an individual means of disconnecting the load and controller from all ungrounded (hot) wires in order to properly shut down the circuit to service the motor. The means of disconnect must be within sight and within 50′ of the controller and motor. In many situations, a standard receptacle and plug combination serves as the service disconnect. You simply unplug the motor for service.

Figure 23-19. An internal sensor and switch on this thermally protected motor cut the power when excessive heat is built up. After a cool-down period, the switch may be reset by pressing the red button.

Stationary equipment requires a separate disconnect device. Disconnect devices completely cut all power to the motor until they are manually reconnected or reset. The prominently located emergency stop button may be part of a service disconnect if activating it triggers a mechanical disconnection of the power supply from the motor. Many disconnects are designed to be locked with a lockout/tagout system to prevent someone from accidentally energizing a piece of equipment that is being repaired or waiting on service. See **Figure 23-20**.

Switches can be used in motor circuits but must be rated greater than or equal to the motor horsepower rating. The motor circuit switch must also have a current rating greater than or equal to 1.15 times the motor full-load current rating. Switches for electric motors must be designed to carry significant amperage loads. They commonly use an electromagnetic relay to energize the motor circuit.

In addition to simple switches for starting and stopping the motor, motor controllers may include thermostats, variable-speed controllers, or timers. The NEC requires that controllers be located within sight and within 50′ of the motor.

The most common output speed of AC electric motors is 1725 to 1750 RPM. Slower or faster fixed speeds can be achieved through different motor designs. Variable motor speeds are commonly developed by means of an electronic controller. One type of motor controller is the *variable frequency drive*, or VFD, which electronically manipulates the AC frequency. The change in frequency results in a change in the speed of rotation in three-phase motors. See **Figure 23-21**. A rheostat may be also used for controlling motor speed. Whenever a motor controller is used to determine motor speed, the motor must be specifically designed for this use.

Mounting

There are many different ways to mount electric motors, including direct drive and the use of belts, chains, or gears. Electric motors may be clamped in place using a form-fitting cradle as part of the framework of the equipment or bolted securely to a flat spot or platform designed for this purpose. Integral mounting of the electric motor in the main frame or body of a machine is used in most handheld power tools and other equipment such as simple ventilation

Holes align in off position for lockout/tagout device

Goodheart-Willcox Publisher

Figure 23-20. Conveniently mounted on the base of an electric metalworking machine, this switch is used to disconnect the power supply before the machine is serviced. The holes around the perimeter of the switch line up when in the off position in order to accommodate a lockout/tagout device.

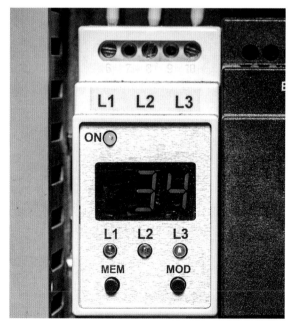

Kalabi Yau/Shutterstock.com

Figure 23-21. A digital display on a variable frequency drive shows the exact frequency of the input current.

fans. Mounting systems may be designed to hold a motor tightly in position with absolutely no movement, or they may include components that allow for slight vibration or twisting due to varying torque loads. Often, the position of the motor is adjustable to establish correct tension on a drive mechanism such as a belt or chain. Adjustment mechanisms may be set to allow the motor to slide or pivot in order to apply tension to a drive belt. Some systems incorporate an adjusting bolt, others use spring tension, and some rely on the mass of the motor and gravity to maintain a constant tension on drive components.

Direct-Drive Mounting Systems

A positive, no slip, connection of the motor to the equipment is known as *direct drive*. In a direct-drive mounting system, the motor-driven parts rotate at the speed of the motor. Some direct-drive systems connect the drive shaft of the electric motor to the equipment through a ***coupling***. The coupling absorbs shock from starting and stopping and vibration from slight misalignment between the motor and equipment. See **Figure 23-22**. Many pumps are powered using direct-drive mounted motors. Servo motors used to position equipment are usually mounted using direct drive.

Gear-Box Systems

In situations where a direct-drive mounting is the best choice, but equipment needs to be operated at considerably less than motor speed, a mounting system through a gear box may be used. A set of two or more matched gears mounted in a solid framework can effectively alter the output speed of rotation and change the orientation of the rotation. Reducing the speed has the added effect of increasing the torque. Gear boxes can be equipped with a clutch that may be used to disengage power from the equipment or simply to absorb shock within the system.

Bean Coupling   **Jaw Coupling**

Goodheart-Willcox Publisher

Figure 23-22. Direct-drive couplings are often designed to absorb vibration and to allow for miniscule differences in motor-to-machine alignment.

Chain-Drive Systems

A system of chain and sprockets similar to those used on a bicycle may be used to connect an electric motor to equipment. A chain-and-sprocket system creates a connection almost as positive as a direct drive. See **Figure 23-23**. Chain drives are simple, but often produce more noise and require more maintenance than other drive systems. Using different sprockets of difference sizes allows the machine to be operated at slower or faster speeds than standard motor output. The sprocket mounted to the motor drive shaft is called the *drive sprocket* and the sprocket mounted to the equipment is the *driven sprocket*. The size of the drive sprocket compared to the driven sprocket is in the same ratio as the speed of the motor compared to the speed of the driven machine. If a drive sprocket turning at 1725 RPM has 10 teeth and a driven sprocket has 15 teeth, the speed of the driven sprocket will be two-thirds (10/15) that of the motor, or 1150 RPM. Another way to look at the relationship is that every time the drive sprocket makes a full rotation, the driven sprocket makes 10/15, or two-thirds, of a rotation.

Goodheart-Willcox Publisher

Figure 23-23. A chain-and-sprocket system provides a positive connection from the motor to driven components. An idler sprocket is used to adjust and maintain tension in the chain.

Belt-Drive Systems

Typical belt-drive systems use V-shaped or flat belts and pulleys to transfer motion from an electric motor to other machine components. The drive pulley must be securely attached to the motor drive shaft in a manner that prevents rotation on the shaft. Drive shafts are engineered with flat sides or square notches made to accept square keys that lock into similar notches cut into the hole of the pulley. Setscrews apply tension to assist in keeping pulleys in position on drive shafts. Similar to sprockets, the size relationship between pulleys can be used to determine the speed of equipment. The ratio is based on the diameters of pulleys rather than the number of teeth. For example, a 2″ drive pulley will turn half as fast as a 4″ driven pulley.

A system using a belt and pulleys typically has a predictable degree of slippage. A small amount of slippage may be desirable to protect machine components from the shock of variable loads. When a belt slips too much, the machine becomes inefficient and excessive wear is likely. Slippage is reduced by maintaining the correct belt tension by adjusting the motor position and possibly by providing extra tension using an idler pulley. Idler pulleys do not turn equipment. Their only purposes are to maintain the belt position and tension.

Belt tension is measured and checked by deflecting the belt from its normal path. See **Figure 23-24**. The

Goodheart-Willcox Publisher

Figure 23-24. Machine manufacturers provide specifications for checking belt tension. When the correct pressure is applied, a correctly tensioned belt should deflect a specified distance.

Goodheart-Willcox Publisher

Figure 23-25. This flat belt from a timing mechanism is equipped with teeth that fit into the notches machined on the pulleys to provide a positive drive that is resistant to slippage and loss of timing.

manufacturer of the equipment provides maintenance specifications in the owner's manual that indicate where to check belt tension and how much deflection should be present when a predetermined force is applied.

Some belt-and-pulley systems use flat belts with notched teeth formed on the contact side of the belt and corresponding notches machined into the pulleys. See **Figure 23-25.** These systems function with the positive connection of a chain drive combined with the quieter operation of a belt.

STEM Connection

Belts and Pulleys

Pulleys and belts have two uses: to transfer power from one shaft to another and to increase or reduce speed or torque. If the transfer of power is all that is needed, then two pulleys should be of the same diameter. Most of the time, the engineer also wants to take the opportunity to trade speed for torque, or vice versa. If a small pulley is connected to a large pulley and the small pulley is the driver, the large pulley will turn more slowly but with increased torque or rotational force. However, if the large pulley is the driver, then the smaller pulley will rotate faster than the larger driving pulley but with less torque.

The pitch diameter of a pulley is not the exact outside or inside diameter. In fact, the pitch diameter can be difficult to measure directly. If you cut a V-belt and look at the end, you'll see a row of fibers near the outside surface. This is the tension-carrying part of the belt; the rest of the belt exists only to carry the forces from the pulley to and from these fibers. The pitch diameter of any pulley is measured at these fibers. If you think about this for a moment, you will see that the pitch diameter of a pulley depends not just on the pulley itself, but on the width of the belt. If you put a B-series belt on an A-series pulley, it will ride higher than usual, increasing the effective pitch diameter. The pitch diameter equal to the pulley diameter only when the belt rides with the fiber core even with the outside diameter of the pulley.

The ratio of the pitch diameters of two paired pulleys is called the **drive ratio**, the ratio by which torque is increased and speed is decreased, or vice versa. Power is the product of speed and force, or in the case of things that spin, speed and torque. Pulleys do not affect power. When they increase torque, it is at the expense of speed, and vice versa.

To calculate pulley speed, divide the diameter of the motor driver pulley by the diameter of the driven pulley. The result is the drive ratio in revolutions per minute, (RPMs). For example: a 6″ motor driver pulley with a 2″ driven pulley results in a 1:3 drive ratio. In other words, the driven pulley will make three complete revolutions for every 1 revolution of the motor driver pulley. If the motor turns at 1000 RPM, the driven pulley will turn at 3000 RPM.

An easy way to measure the circumference of a belt is to roll it along a flat surface, then measure the distance you have traveled when you return to the same point on the belt. On a machine where the belt is missing, you can run a string around the pulleys and measure that for an approximate length. Mathematically, belt length is calculated by adding two times the distance between the pulley centers to 1/2 of the circumference of each pulley measured at the pitch diameter.

When designing belt drives, remember that belts come in specific standard lengths, and pulleys come in specific standard pitch diameters. You cannot just arbitrarily select dimensions and hope to find such components.

Changing the Direction of Rotation

When installing an electric motor, sometimes the motor's rotation will be in the wrong direction for the application. In this situation, it is rarely feasible to simply mount the motor pointed in the opposite direction. The next logical solution might be to use a different motor that turns the other way, but this can be an expensive fix and is likely not necessary. Many motors can easily be modified to switch the direction of rotation.

In split-phase and capacitor-start motors, the starting winding and the running winding are connected in parallel. When the connection on the running winding is reversed, the rotation of the motor is also reversed. See **Figure 23-26**. Many motors are engineered to simplify this operation. First, remove the inspection plate usually located on one end of the motor. A wiring diagram is often included indicating the positions of the blade-type connections for clockwise or counterclockwise rotation. See **Figure 23-27**. Adjustments for operating the motor on different voltages are also made by following the wiring diagram. Three-phase motors may also be reversed by switching connection terminals on the windings. Always follow the manufacturer's wiring diagram when making these adjustments.

In repulsion-start motors, the direction of induced current in the armature conductors depends on the position of the brushes. Changing the direction of motor rotation is as simple as loosening a setscrew and rotating the ring or frame on which the brushes are mounted a few degrees one way or the other and tightening the setscrew.

> **Safety Note**
> Always remember to first disconnect the power before making any changes to an electric motor.

Electric Motor Maintenance

Most electric motors have very basic maintenance needs. Maintenance includes controlling exposure to dust, moisture, friction, and uneven loading. Cleanliness is the key to proper maintenance of electric motors. Motors that are not designed to withstand exposure to moisture must be mounted, used, and stored where exposure to the elements is minimized. Complete

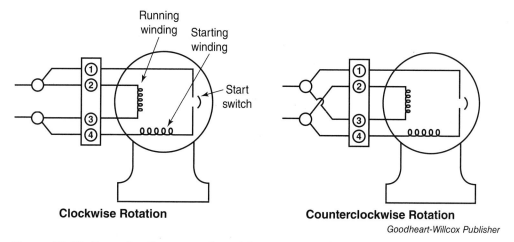

Goodheart-Willcox Publisher

Figure 23-26. Reversing the connection of the power supply to the running winding changes the direction of rotation in split-phase and capacitor-start motors.

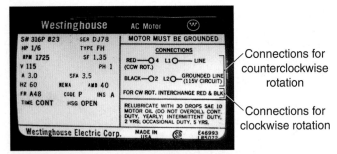

Wiring Diagram on Motor Nameplate

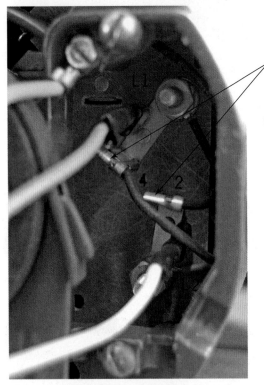

Motor Wiring Terminals

Goodheart-Willcox Publisher

Figure 23-27. The directions for reversing this split-phase motor are found on the motor nameplate. To change direction, switch the red and black wires by moving the blade connectors on terminals #2 and #4.

Safety Note
After inspection, always return guards to their proper position.

maintenance involves cleaning, lubrication, and attention to motor mounting systems.

Dust and debris may need to be removed to reduce the likelihood of the motor overheating or experiencing premature wear. Dust should be removed from both the outside and inside of motors. The conditions in which a motor is operated determine the frequency of service. See **Figure 23-28**. A layer of dust on the outside of an electric motor acts as an insulator, reducing the motor's ability to dissipate heat. Contaminants collecting inside a motor contribute to the degradation of electrical components and increase friction and wear on moving parts. Some types of dust can also become a fire hazard. Motors may be cleaned by wiping with a rag and a noncombustible solvent and by using compressed air at or below 40 psi. Using higher air pressures can damage internal insulation.

Some electric motors have bearings or bushings that require lubrication on a regular schedule. Others are designed with sealed bearings that need no lubrication after leaving the factory. Always refer to the operator's manual for the electric motor to determine the service needs and the proper service intervals. Equipment operated by an electric motor must also be lubricated regularly to keep excess friction from creating loads that exceed the motor's design. Overlubrication can be as destructive to an electric motor as underlubrication. Stray amounts of oil used for lubrication can severely damage the internal insulation separating motor windings. Excess lubrication also attracts dust and other debris, holding it in lumps that can interfere with motor function and airflow needed to cool the motor. Always apply lubricants according to the manufacturer's recommendations and wipe away any spills.

When servicing electric motors, inspect the condition of mounting and power transfer components. A loose mounting can cause vibration, which can lead to damage or rapid wear of the motor or other system components. Belts and pulleys or chains and sprockets should be properly aligned and adjusted to manufacturer's specifications. Misalignment or improper tension puts undue strain on motor bearings or bushings.

Figure 23-28. This three-phase motor drives the main feed mixes in a mill. Due to very dusty conditions, the motor requires frequent service to keep the collected dust layer to a minimum.

Often, problems with mounting or motor alignment are accompanied by unusual noises coming from operating machinery. For example, a worn bearing may make a high-pitched grinding, squalling howl. A loose belt may develop an audible thumping sound. Knowing what proper operation

Career Connection

Industrial Machinery Mechanic

Job description: In the field of agriculture, industrial machinery mechanics maintain and repair stationary and mobile equipment and machinery used in the processing of agricultural products. This includes equipment such as conveying systems, production machinery, and packaging equipment. Industrial machinery mechanics must be familiar with many different types of machines. Knowledge about electric wiring, electric motors, hydraulics, pneumatics, and mechanical mechanisms is essential. In this occupation, workers must follow safety precautions and use protective equipment. Employment is usually full-time with opportunities for overtime. Often, night or weekend shifts are part of the job.

Education required: Most industrial machinery mechanics have graduated from high school or attained a GED. They may attend training programs in a technical school for a year or more. Most also receive on-the-job training specific to the company. Regular updates to training are needed as equipment is updated in a facility.

This job may be a fit for you if: You are interested in how things work, you have the ability to pay close attention to detail, and you like working with tools.

sounds like and paying close attention while the machinery is running can help to catch many problems while repair is still possible.

Safely Using Electric Motors

Handle all electrical components associated with electric motors using respect and caution. Electrocution is one of the leading causes of occupational fatalities in the United States. Understanding electricity is essential to safely harnessing its power. Motors present a distinct set of hazards which can be managed by observing the following basic safety rules:

- Wear adequate eye protection (OSHA z87.1) whenever working around moving machinery.
- Always refer to manufacturer's specifications for the application, installation, adjustment, and service of electric motors.
- Always disconnect an electric motor from the power source before making adjustments, cleaning, or repairing. In many applications, this may include following lockout/tagout procedures.
- Protect any moving parts from accidental contact by people, animals, or materials using guards or shields. Guards and shields should always be replaced when removed for service. See **Figure 23-29**.
- Be sure that wiring always includes a ground connection to the equipment. Replace damaged wiring or cords.

Goodheart-Willcox Publisher

Figure 23-29. This grain auger drive unit has shielding installed around belt drive components to protect people and animals from contacting the moving parts.

CHAPTER 23 Review and Assessment

Summary

- Electric motors are used in agriculture because of their efficiency, convenience, safety and quiet operation.
- Electric motors are driven by forces of magnetism through the process of induction.
- The construction of various types of electric motors results in many choices regarding the capabilities and operating characteristics of the motors.
- Electric motors are required to have nameplates that identify the motor's operating characteristics. Always refer to the nameplate and operator's manual for the proper motor selection.
- Proper installation of electric motors requires knowledge of electric circuits, mounting, and attachment of drive components.
- Maintenance of electric motors involves cleanliness, lubrication, and attention to details concerning the attachment to other machinery.
- Safety when working with or around electric motors includes dealing with electricity and moving parts.

Words to Know

Match the key terms from the chapter to the correct definition.

A. bearings
B. brushes
C. bushings
D. capacitor-start motor
E. coupling
F. drive shaft
G. drive ratio
H. electromagnet
I. induction
J. repulsion-start motor
K. servo motor
L. split-phase motor
M. stepper motor
N. three-phase motor
O. universal motor

1. Uses a strong electromagnetic force to start and rotate heavy loads.
2. Device used in direct-drive mounting systems to connect the motor to a load.
3. Divides incoming AC power between a starting and a running circuit.
4. The process of moving a conductor through a magnetic field to generate current in the conductor.
5. Uses a device wired in series with the starting circuit to increase its electromagnetic force.
6. A magnetic field produced by an electric current.
7. Mostly used to operate stationary equipment due to the required input current.
8. Used to reduce friction; include races and rollers.
9. Used to reduce friction; single-piece construction.
10. Constructed using 50 to 100 magnetic poles.

11. The amount by which torque is increased and speed is decreased, or vice versa, between two connected pulleys.
12. Point of attachment for sprockets or pulleys.
13. Series-wound motor designed to operate using AC or DC power.
14. Transfer energy through the rotating commutator.
15. Must be controlled by the use of a controller known as an *encoder*.

Know and Understand

Answer the following questions using the information provided in this chapter.

1. The introduction of _____-based battery packs has elevated the use of electric motors to operate certain cordless landscaping tools.
2. Electric motors change electrical energy into _____ energy.
3. The force of a motor's rotation is influenced by the _____ of the input electric current and the strength of the magnetic fields.
4. _____ are used to support the rotating shaft of a heavy-duty electric motor.
5. Physical size and output _____ may be used as a method to group motors.
6. In a series-wound motor, the stator windings are connected in series with the armature, and in a(n) _____-wound motor, the stator windings are in parallel with the armature winding.
7. A four-pole split-phase motor powered by 60Hz AC current will turn at about _____ RPM under full load.
8. Once single-phase induction motors reach a designated speed, a(n) _____ switch disengages the starting circuit to allow the motor to operate on induction alone.
9. _____, which list critical motor data, are located on the outside of the motor housing where they are easily visible.
10. A belt-and-pulley system consumes about _____ of the power produced by an electric motor.
11. Circuit protection devices for motor circuits incorporate time-delay components that allow brief periods of overload needed for motor _____.
12. _____ protected electric motors have overload protection devices built into the construction of the motor.
13. Switching devices for electric motors often use an electromagnetic _____ to energize the motor circuit in order to handle significant current loads encountered during starting.
14. A variable frequency drive can be used to monitor and control the _____ of a three-phase electric motor.
15. Increasing the size of the driven pulley while using the same drive pulley will _____ the RPM of the equipment.
16. Belt _____ is measured and checked by deflecting the belt from its normal path at a given point using a specified amount of force.
17. Reversing the connections to the running windings will change the direction of rotation in split-phase and _____-start electric motors.
18. An excessive layer of dust on the outside of an electric motor is like a blanket that reduces the motor's ability to dissipate _____.

19. _____ can lead to buildup of lumps of debris that interfere with proper operation of an electric motor.
20. Equipment using a belt and pulleys should be equipped with _____ to prevent accidental contact with the moving parts.

STEM and Academic Activities

1. **Science.** Research Faraday's experiments in 1831 involving magnetism and induction. Prepare a presentation describing the original experiment for the class.
2. **Technology.** Remove the inspection plate from a split-phase or a capacitor-start motor, and use a multimeter to find the terminals corresponding to the two sets of windings. With the dial set to continuity, touch different pairs of terminals with the probes until two continuous windings are located. Use the ohms setting to determine which set of terminals connects to the starting winding and which set connects to the running winding. The start winding will exhibit a higher resistance.
3. **Engineering.** Locate two different machines that use an electric motor connected with a belt-and-pulley drive system and the owner's manuals for each. It may be necessary to locate the owner's manuals online. Using the specifications provided, check the tension on the drive belts. If the tensions are not correct, get your instructor's permission to make the needed adjustments.
4. **Math.** A 2″ driver pulley is connected to an electric motor that turns at 1750 RPM. This drives an 8″ pulley connected to a fan that is 36″ in diameter. The distance between the centers of the two pulleys is 24″.

    Sketch the belt and pulley setup.

    A. What is the pulley drive ratio?
    B. What is the operating speed of the fan?
    C. What is the RPM of the outer perimeter of the fan?
    D. What is the length of the drive belt?
5. **Language Arts.** For mechanics, it is important to communicate important information efficiently in ways that are standardized and easily shared. Motor nameplates are an example of this principle. Make a poster that can be displayed in class explaining all of the information found on a motor nameplate.
6. **Social Science.** Write an essay that explains how the use of electric motors for pumping water has changed the lives of humans. What about other common applications? What uses for electric motors do you see in the future?

Thinking Critically

1. Imagine that you are designing a wood chipper that will be powered by an electric motor. How much horsepower do you need and what type of electric motor will you select? Explain your answer.
2. Perform a survey of the electric motors used in your home. How many can you identify? How would life change if suddenly you had no access to electric power?

CHAPTER 24
Plumbing Design and Installation

Chapter Outcomes
After studying this chapter, you will be able to:
- Understand the sources of plumbing codes adopted by local governments.
- Identify the components of a water supply system.
- Describe the effects of gravity and water pressure in a plumbing system.
- Explain the functions of a drain-waste-vent system.
- Explain the operation of a septic system.
- List tools commonly used by plumbers.
- Describe the three basic plumbing pipe materials and their variations.
- Identify different types of pipe fittings and explain their purpose.

Words to Know ⤤

acrylonitrile butadiene styrene (ABS)	cross-linked polyethylene (PEX)	liquid layer	soldering
annealing	drain-waste-vent (DWV) system	primer	solvent cement
black pipe	fitting	polyvinyl chloride (PVC)	trap
bladder	fixture	scum	valve
burr	flux paste	septic system	water supply system
chlorinated polyvinyl chloride (CPVC)	galvanized pipe	sewage	well
compression ring	leach field	sewer gas	
		sludge	

Before You Read
After reading each section (separated by main headings), stop and write a three- to four-sentence summary of what you just read. Be sure to paraphrase and use your own words.

While studying this chapter, look for the activity icon to:

- **Practice** vocabulary terms with e-flash cards and matching activities.
- **Expand** learning with interactive activities.
- **Reinforce** what you learn by completing the end-of-chapter questions.

www.g-wlearning.com/agriculture

Elena Elisseeva/Shutterstock.com

Plumbing installations are an important part of modern agricultural systems, **Figure 24-1**. Plumbing systems perform the following functions:
- Supply fresh water.
- Remove wastewater and other waste elements.
- Bring gas or water into a heating system.
- Transfer milk from a milking station to the processing area.

Knowing how to choose the best materials, fixtures, and location for plumbing systems is an important skill. Although some plumbing work can be done by novices, most jobs should be performed under the guidance of a licensed plumber in order to comply with local zoning and health department regulations.

A
kezza/Shutterstock.com

B
GETIENjones/Shutterstock.com

Figure 24-1. Agricultural plumbing. A—Water storage. B—Irrigation pipes.

Plumbing Codes

Plumbing codes are laws adopted by local governments that detail how plumbing installations should be completed. These codes are based on three sources:
- International Plumbing Code (IPC). This set of rules and procedures, developed by the International Code Council (ICC), sets minimum regulations for plumbing systems and components. These codes are the basis for laws on plumbing installation in 35 states.
- Uniform Plumbing Code (UPC). This set of rules was developed by the International Plumbing and Mechanical Officials and is the basis for the laws in most states not using the IPC.
- Local or state level government agencies, in conjunction with local plumbers.

The various code systems are similar in requirements, but all local codes must be consulted prior to beginning plumbing projects to ensure that the correct procedures will be followed. To determine what codes your town or state uses, you can visit your town or state government's official website. Local rules can also be found by calling the local health department, building code and inspection service, local zoning officials, or local government officials.

Plumbing Systems

While most plumbing involves moving potable water in and wastewater out, plumbing is also found in other agriculture systems, including greenhouses, irrigation, gas piping, milk handling, and manure handling. See **Figure 24-2**. These systems all have the same basic piping. They work

Figure 24-2. Examples of water systems in agriculture. A—Irrigation system. B—Milking system.

on the principle that liquids and gases move under pressure or by gravity toward the lowest point in the system. Plumbing systems consist of two basic systems: the water supply system and the drain-waste-vent (DWV) system.

Water Supply System

The *water supply system* includes the water source and all of the pipes and fixtures that bring the water from the source to the final point of use. *Fixtures* are permanent parts of a plumbing system that use water or drain the system. Examples of fixtures are showerheads, toilets, faucets at sinks, and hose bibs (outside faucets to which hoses are attached).

Water moves easily toward gravity, and it can be made to move uphill under pressure or by creating a low pressure area above it. Many rural water systems consist of water pumped from a private *well* (a deep, manmade hole in the ground from which water can be pumped) or storage containers and a gravity-fed water removal system. See **Figure 24-3**.

When water is pumped up into a storage tank or water tower, **Figure 24-4**, it has

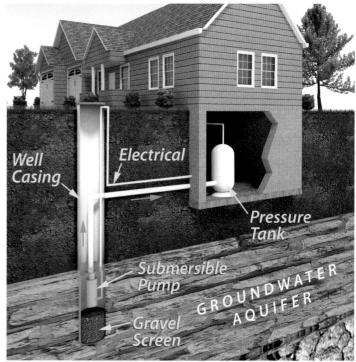

Figure 24-3. Water to this structure is supplied by a private well.

Figure 24-4. Water towers provide water in many communities.

potential energy. When it flows down, water has about 43 psi of pressure for each 100′ above ground level. Most plumbing systems require between 30 psi and 50 psi of pressure to meet the needs of water fixtures.

Pressure for a water supply system can also be produced using a pump and pressure tank. The pump brings water up from the groundwater, and the pressure tank holds the water. A pressure tank has a water-holding *bladder*, a soft bag that fills with water or air and expands. The bladder is inside the sealed tank, which also contains air. When the tank bladder fills with water pumped from the well, it squeezes the air in the surrounding tank and applies this pressure to the water. The water is then held until needed by a fixture in the system. When the valve is opened, the high pressure in the tank pushes water out of the bladder, supplying the pressurized water needed by the supply system.

DWV System

The *drain-waste-vent (DWV) system* consists of the pipes and fixtures that remove water, waste, and gases from the plumbing system and maintain atmospheric air pressure within the system. The DWV system often handles toxic or unsanitary waste products. In addition to piping and fixtures, this system includes traps, vents, and containment areas.

The waste removal portion of the system is often a gravity system that allows wastewater and the solids carried in wastewater to move to the lowest point in the system. The venting portion of the DWV system has a dual function—to remove sewer gas from the system in a safe manner and to maintain air pressure in the system. *Sewer gas*, a toxic, noxious gas, results from the decomposition of sewage. *Sewage* is waste material, especially urine and feces.

From each fixture, a vent stack or a pipe leads to outside the structure. This vent allows air pressure to remain relatively constant in the system. If a DWV system did not maintain air pressure, a vacuum would form behind the water each time a fixture drains water. The vacuum would cause water (and waste) to be pulled back into the fixture. The vent also allows sewer gas to escape the structure. See **Figure 24-5**.

In a gravity-fed wastewater system, all of the outlet lines run down from the fixtures. The angle is carefully calculated (about 1/4″ per foot of horizontal pipe) for waste pipe. In a gravity-dependent waste system, an angle that is too shallow causes water and solids to move slowly, creating clogs. If the angle is too steep, the water runs quickly away from the solids,

Figure 24-5. Vent pipes help prevent water (and waste) from being drawn back into a fixture and allow sewer gas to escape.

leaving a clog. In a good plumbing system, all the water and the waste solids are carried out of the system together. The septic system (a system for cleaning wastewater) or municipal sewage system must be the lowest point in the drain system so gravity can work.

Plumbing Traps

A plumbing *trap* seals water into a vented pipe to stop sewer gas from entering a plumbing fixture. Since sewer gas is foul-smelling and may even be toxic, traps are important parts of plumbing systems. Without a trap, a drain pipe could not hold water, and an open pipe would lead directly to the rotting waste.

A trap works by a combination of air pressure and gravity. See **Figure 24-6**. With a trap in place, the water flows down the pipe until the weight of the water above is equal to the weight of the air pressure on the other end of the trap, holding some water in the trap. This water forms a water seal to keep out odors.

SPb photo maker/Shutterstock.com

Figure 24-6. A plumbing trap prevents sewer gas from flowing back into a fixture.

Adding a small amount of additional water at a slow speed allows the water to stay in the bowl. Adding a large amount of water quickly—as in flushing or pouring a bucket full of water quickly into the fixture—allows the bowl to empty quickly, removing waste along with the water.

Septic System

A *septic system* is designed to clean the wastewater from individual residences. The system consists of a septic tank, a leach field, and a series of connecting pipes. See **Figure 24-7**. Wastewater goes into the septic tank, where it forms these three layers:

- *Sludge*. A mixture of solids and liquids. This is the layer that is broken down by microorganisms. Most organic solids and some inorganic and/or heavy material make up this layer.

A *trufero/Shutterstock.com* B *Christian Delbert/Shutterstock.com*

Figure 24-7. Typical septic system components are shown here. A—Septic tank. B—New leach field being installed.

- *Liquid layer.* The water, urine, and liquid waste products that are above the sludge layer and quickly leave the septic tank.
- *Scum.* The top layer, which includes very lightweight solids, fats, and other waste materials.

In the septic tank, the sludge layer is constantly broken down by bacteria that enter the system from the digestive tracts of the system's users. As the sludge is broken down into ever finer particles, the sludge layer slowly leaves the system as wastewater. The liquid layer leaves the system when the liquid is above the outlet pipe for the tank. It then moves into the leach field.

The *leach field* is a series of perforated pipes placed in covered, gravel-lined trenches downstream from the septic tank. Water drains into the trench and continues to filter into the local groundwater. The leach field must be below the frost line in areas where temperatures are low enough to freeze the soil. The soil must have enough water-percolating ability to remove all the water from the system in a timely manner. Water standing at the surface is a sign of a malfunctioning system, and the problem must be fixed immediately.

The scum layer often causes problems in the septic system and should be kept to a minimum. To minimize the scum layer, food waste, toxic chemical waste (including cleaning agents and drain cleaners), and other nondegradable items should not be put into the system. Examples of items that should not be put into the system are paper (other than designated toilet paper), tissues, facial or baby wipes, personal hygiene products, cloth, petroleum-based products, paint, stains, grease and oils, coffee grounds, and household waste.

Septic tanks should be inspected routinely about once a year. A professional carefully vents the accumulated sewer gas before looking into the system to determine the depth of the sludge and scum layers. If the sludge layer is more than one-third the depth of the septic tank or blocks the outlet pipe, the tank should be pumped to remove accumulated solids that cannot be broken down by microorganisms.

Pumping is done by a professional because toxic sewer gas is a health hazard and must be disposed of in an approved manner. The frequency with which a system needs to be pumped depends on the number of people using the system, how much waste is removed, how much water is used, and other factors. Pumping should be done before a problem is apparent (soggy area in the lawn, backed-up plumbing fixtures, or fixtures through which water does not flow readily). Failing to fix the system in a timely manner may result in the need for costly repairs.

Plumbing Tools

The tools commonly found in a plumber's toolbox include those used by most agricultural mechanics, such as an adjustable wrench, tongue-and-groove pliers, and various screwdrivers and pliers. Plumbers also use several specialty tools. These tools include:

- A pipe cutter, **Figure 24-8**, is placed around the pipe at the location to be cut. The cutting wheel is tightened until it just touches the pipe. The

pramual jermjun/Shutterstock.com

Figure 24-8. Pipe cutters work on most pipe materials.

cutter is then rotated around the pipe once, then tightened and rotated again, continuing until the pipe has been cleanly cut. Some ratcheting pipe cutters are designed for use on plastic pipe as well as metal pipe.

- A plunger, **Figure 24-9**, is used to pull out or break up clogs. The plunger is fitted over the opening of the pipe to be cleared and gently depressed until a seal forms around the pipe. The plunger is then pulled up quickly to form a vacuum in the pipe, dislodging the clog or breaking it up and allowing it to flow out of the pipe.
- A pipe wrench, **Figure 24-10**, is generally used to hold pipes or turn threaded pipes. Pipe wrenches are often used two at a time, one to hold the stationary pipe and one to rotate the other pipe.
- A hand auger, **Figure 24-11A**, is also known as a *plumber's snake*. The tool is cranked into a clogged drain and through the clog, breaking it up and clearing the drain.
- A closet auger, **Figure 24-11B**, is a special-use hand auger. It has a covering to protect the porcelain fixture when the tool is used on a toilet bowl.

Andrey_Popov/Shutterstock.com

Figure 24-9. This plumber is using a plunger to unclog a sink. A plunger works best when the drain contains some water.

modustollens/Shutterstock.com

Figure 24-10. The jaws of a pipe wrench are deeply grooved and movable, allowing for a strong grip on various pipe sizes and material.

SAE Connection

Plumbing

We all need water. The use of water in homes, businesses, and agricultural applications requires plumbing supply and waste systems that bring in fresh water and effectively dispose of the resulting wastewater. The design, installation, maintenance, and repair of these systems requires a specific knowledge and skill set that comes from training and experience. If a career in plumbing is appealing to you, research the training requirements and discuss with your agriculture teacher how you can gain valuable experience working with a local plumber. Remember to record your experience, hours, and pay so that you can complete the

Monkey Business Images/Shutterstock.com

application for an Agricultural Mechanics Design and Fabrication proficiency. A plumbing SAE can lead to a profitable career and give you valuable experience to get you started.

- A basin wrench, **Figure 24-12**, is used to reach the nuts on a faucet on the underside of a sink basin. It has a long, thin shaft to reach the work area and a firm gripping jaw.
- A tubing cutter, **Figure 24-13**, is used to cut soft, flexible copper tubing and other easily cut pipe, such as rigid copper.
- A flaring yoke can be used to put a flared end on soft copper tubing. A flaring tool is used with a flaring yoke to press into the copper tubing and create a flared or bulged end on the pipe.
- A crimping tool can be used to crimp a *fitting* (a small part used to join, turn, or otherwise change the system) on a hot water supply line.
- A propane torch, **Figure 24-14**, is a canister of propane with a torch fitting that regulates the size and quality of a flame. The flame is used to add heat to a copper pipe when soldering a fixture.

A

photosync/Shutterstock.com

Aleksei Golovanov/Shutterstock.com

Figure 24-13. This tubing cutter is designed to be used in tight spaces.

B

David Spates/Shutterstock.com

Figure 24-11. Augers are available in a variety of configurations. A—Hand auger (plumber's snake). B—Closet auger.

digitalreflections/Shutterstock.com

Figure 24-12. Basin wrench are designed to access nuts under a sink basin without removing the sink.

Dan Kosmayer/Shutterstock.com

Figure 24-14. A propane torch is commonly used to melt solder when joining copper pipe and fittings.

Thinking Green

Water Conservation

Many people in the United States face water restrictions due to short- or long-term shortages, but everyone has a stake in conserving fresh water. As fresh water becomes scarcer, the cost of using it increases. Water conservation involves the following:
- Using less water to accomplish the same activity, such as installing low-flow toilets.
- Reusing gray water (household wastewater, except that from toilets) for other tasks, such as watering household plants.
- Using something instead of water, such as cleaning a driveway with a broom instead of a hose.

All fixtures (toilets, faucets, showerheads, and others) are labeled to allow consumers to make informed choices based on their flow rate. Flow rate is stated in gallons per minute (gpm) at a peak pressure measured in pounds per square inch (psi) or, in the case of toilets, bidets, and urinals, gallons per flush (gpf).

A plumber can determine the pressure of a plumbing system and adjust it with a pressure-reducing valve (PRV), if necessary. Higher water pressure means more water is pushed through the system faster, so reducing the pressure saves water. In general, pressures over 60 psi are harmful to plumbing systems and should be lowered. All fixtures in a system should have the same or very similar pressure ratings. This allows the system to function at an adequate pressure.

Water conservation also involves being aware of where water is used or misused and taking steps to correct these issues. The following are some steps that can be taken to conserve water:
- Install and properly maintain the most efficient fixtures and appliances available.
- Stop running faucets while lathering hands, shaving, and brushing teeth.
- Turn off faucets completely when not in use.
- Stop all leaks in plumbing systems. Examples are toilets that run longer than necessary to refill or faucets that drip (1 drip per second equals about 8 gallons of wasted water per day).
- Install and maintain landscaping that requires less water.
- Install and use a rainwater collection system for nonpotable water needs.
- Install and use a water recycling system.

Plumbing Pipe

Three basic types of plumbing pipe are commonly used by agriculture mechanics: plastic, copper, and steel. Currently, the most common and cost-effective pipe is made of plastic. This pipe is used for both supply and DWV pipe. Copper piping was commonly installed in the last 50 years and thus is found in many existing systems. Copper is preferred over steel pipe, but because of its cost and the need for skill in assembling the system, it is quickly being replaced by plastic in most modern systems. Many agricultural locations still have plumbing features that rely on steel pipe for either supply or DWV pipes. Each pipe material has benefits and drawbacks.

Plastic

Plastic is more cost-effective than metal piping. It is generally more resistant to corrosion and takes less skill to install. The four basic types of plastic pipe material are PEX, PVC, CPVC, and ABS.

Cross-Linked Polyethylene (PEX)

Cross-linked polyethylene (PEX) is a flexible, hard, thermoset plastic that can be used to supply both hot and cold water in a plumbing system, **Figure 24-15**. The following are benefits of PEX:

- Easy to install. Even a novice can use push-on fittings.
- Reliable—rarely leaks.
- No soldering or hazardous pipe cement is needed for installation.
- Easy to merge with existing copper when expanding or repairing systems.
- Long lasting.
- Sustains less damage from freezing. PEX may withstand up to six or seven freezing cycles without breaking.

Drawbacks of PEX tubing as water supply piping include:

- Degrades in sunlight and develops pinholes. All installations must be shielded from sunlight.
- Can be penetrated by insects. All components should be protected from insect damage.
- Fittings (generally brass) are expensive. However, fewer total fittings are required, because PEX is flexible and can be bent for many directional changes that would require a fitting in a rigid pipe system.

The following is a procedure for crimping or cinching PEX tubing to a system. Obtain a PEX fitting (brass or stainless steel), a PEX crimp ring, a crimping tool, and a length of PEX tubing from your instructor. Perform the following steps:

1. Place the ring over the end of the PEX tubing and push the PEX tubing onto the fitting until it is seated in the shoulder of the fitting, past the grooves.
2. Draw the crimping ring over the tubing onto the fitting so it is 1/8″ to 1/4″ from the end of the tubing (in the fitting's grooves) and place the crimping tool over the ring.
3. Squeeze the tool firmly and hold until the ring crimps tightly onto the fitting, making a watertight seal. If you are using a ratcheting crimping tool, ratchet until it is tight. See **Figure 24-16**.

PEX tubing is flexible and, like many plastics, expands and contracts as the temperature changes. Therefore, the line must be plumbed loosely enough in the system

Goodheart-Willcox Publisher

Figure 24-15. PEX water supply lines. Note that the PEX is color-coded. Red is for hot water supply and blue is for cold water applications.

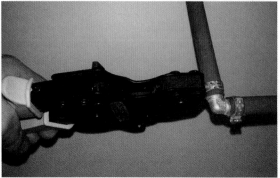

Doug Lautenschlager

Figure 24-16. This fitting is being installed with a ratcheting PEX crimping tool. A ratcheting tool makes the job easier, especially in tight spaces.

to allow for this contraction. A loop in long runs (a 4″ loop for 1/2″ tubing) yields enough to contain the distortion in the plastic. Long vertical runs, as well as long horizontal runs, need to be supported with hangers. Hangers are spaced between 32″ and 6′. It is recommended that horizontal PEX runs be supported every 32″. Since PEX tubing expands and contracts with changes in temperature, hangers must not be too tight. The tubing should be able to move easily in its support.

Polyvinyl Chloride (PVC)

Polyvinyl chloride (PVC) is generally used for DWV pipes, **Figure 24-17**. PVC is rigid and not heat-tolerant enough to be used in water supply pipes. Benefits of using PVC include:
- Light weight.
- Low cost.
- Low maintenance.
- Generally inert to chemical and biological agents.
- Easily joined together with solvent cement or a fusion system.
- Generally fire retardant (unless exposed to direct flames).

Drawbacks of PVC use include:
- Good ventilation is required when cement is used to join piping.
- Cannot be used for hot water supply applications.
- Will burn if exposed to open flames.

Chlorinated Polyvinyl Chloride (CPVC)

When chlorine is added to polyvinyl chloride, the result is *chlorinated polyvinyl chloride (CPVC)*. See **Figure 24-18**. CPVC offers the same benefits as PVC but is much more flexible. In addition, CPVC is heat-tolerant up to 200°F (93°C), allowing it to be used for hot water supply lines.

Drawbacks of CPVC include:
- Emits toxic chlorine gas when exposed to flames.
- Good ventilation is required when cement is used to join piping.

The following is a procedure for making PVC or CPVC connections. Obtain a length of PVC pipe, a joint fixture, and the correct *primer* (cleaner and surface preparation liquid) and solvent cement from your instructor. *Solvent cement* is the adhesive that bonds the plastic parts together. **Perform the following steps:**

1. Gently dry-assemble the fixture onto the pipe.
2. Mark the correct location of the fixture (oriented in the direction the pipe needs to turn). This is best accomplished with a small level and a permanent marker.

Goodheart-Willcox Publisher

Figure 24-17. PVC pipe is commonly used for drain and vent lines.

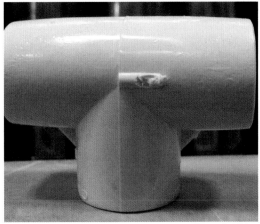

Goodheart-Willcox Publisher

Figure 24-18. CPVC tee fitting. CPVC pipe and fittings can be used for hot water supply lines.

Note: If during dry assembly the pipe is pushed too far into the fixture and is difficult to remove, use a hammer to gently tap a block of wood placed on the lip of the fixture to dislodge the pipe.

3. Cut the pipe to length with a fine-toothed saw, hacksaw, or a PVC pipe saw.

Note: Measure carefully before cutting. Pipe will not fit completely into the fitting until solvent cement is applied, so take this distance into account when you are measuring for correct length. On a 1-1/2" pipe, the distance might be an additional 3/8" at each joint.

4. Make sure the cut is square, and then clean any *burrs* (small bits of metal stuck to the end of the pipe) from the inside or outside of the cut.
5. Apply purple primer to both the inside of the fixture and the outside of the pipe.
6. Apply solvent cement to both the inside of the fixture and the outside of the pipe.

Note: Do not over-apply solvent cement. Overapplication can cause protrusions on the interior joint that impede water flow.

7. Quickly press the pipe into the fixture and turn the pipe a one-quarter turn to line up the marks made earlier. The pipe must be seated and lined up within approximately 10–15 seconds. After 15 seconds, solvent cement sets and will no longer move. Hold the pipe 10–15 seconds longer to allow the solvent to set completely to ensure watertightness.
8. Wipe any excess solvent off the outside with a cloth. Check for watertightness.
9. Return tools, materials, and equipment to their proper places.

Safety Note
When connecting pipe, always wear eye protection. Work in a well-ventilated area or wear an approved organic vapor respirator when working with solvent cement.

Acrylonitrile Butadiene Styrene (ABS)

In plumbing applications, *acrylonitrile butadiene styrene (ABS)* is used only for DWV pipes. Because of its low resistance to heat, it cannot be used in hot water applications. This plastic is a cost-effective thermoset plastic which is often used in commercial plumbing applications. See **Figure 24-19**. Compared to PVC, ABS offers the following advantages:

- Lower cost.
- Easy to install—no primer needed.
- More impact resistant.
- More cold tolerant.

Drawbacks of ABS include:

- Degrades with exposure to UV light.
- Deforms with heat.

Assembling ABS plastic pipe involves a procedure similar to that for PVC or CPVC. The only difference is that primer is not required before the solvent cement is applied.

Goodheart-Willcox Publisher
Figure 24-19. ABS plastic pipe in underground use.

Copper

Copper pipe, generally called *tubing*, is available in rigid and soft (flexible) styles, **Figure 24-20**. As copper is pulled into a long tube shape, it becomes hard. The copper must be annealed to make it soft again, adding to the cost of flexible copper tubing. *Annealing* is the process of heating a metal and letting it cool slowly, making the material tougher and more flexible.

The benefits of rigid copper pipe include the following:
- Long lasting.
- Noncorrosive (is not affected by chemical reactions that weaken the metal).
- Relatively easy to join together.
- Works with other piping systems.
- Requires little or no maintenance.

Drawbacks to rigid copper include the following:
- Relatively expensive.
- Pitting (pinholes) under certain circumstances, such as with acidic water or improperly grounded electrical wiring.

Aksenenko Olga/Shutterstock.com

Figure 24-20. Rigid copper pipe was used for plumbing applications during much of the latter half of the twentieth century.

Soft copper tubing is used for water supply lines. Because it is flexible, the tubing can be bent around obstacles, requiring fewer joints than rigid copper pipe. Soft copper tubing has the same drawbacks as rigid tubing.

Copper tubing is joined using three basic methods:
- Soldering.
- Compression fittings.
- Flare fittings.

Occasionally, a crimped fitting is used in commercial applications. However, this is rare in residential and farmstead applications due to the cost and time involved in the process.

Soldering

Soldering is a process that joins two metal pieces by heating and flowing a different metal around the joint. Soldering pipes is often referred to by plumbers as "sweating" the pipes. Soldering is a necessary skill for working with copper piping, **Figure 24-21**. The technique varies depending on whether the plumbing installation is new or existing. With new installations, there is no water to exclude from the joint. In an existing system repair, excluding water is important so the fixture heats enough to melt the solder.

borzywoj/Shutterstock.com

Figure 24-21. Soldering is the most common method of joining copper pipe and fittings.

The following is a procedure for soldering a new joint. To perform this activity, obtain a length of pipe from your instructor and one 90° elbow fitting.

Perform the following steps:
1. Cut the pipe into two lengths with a tubing cutter.
2. Remove any burrs on the inside or outside of the pipe with an emery cloth or fine sandpaper. Careful cleaning of the fitting and the pipe with the emery cloth is critical to ensure good adhesion of the solder to the copper.
3. Finish cleaning the fitting and pipe with *flux paste*, a slightly abrasive cleaner, to remove any remaining oxidation.

Safety Note

Always wear standard safety eye protection when working with metal and propane torches. Follow all fire safety procedures and make sure a working fire extinguisher is present.

638 Agricultural Mechanics and Technology Systems

> **Safety Note**
> Wear fireproof clothing and heat-protective gear when working with open flames. Always turn off a propane torch before setting it on a work surface. Torches can easily tip.

4. Score the pieces slightly with fine sandpaper or emery cloth to prepare the surface for soldering.
5. Put the pipe into the fixture. Hold the solder on the far side of the fixture at the joining point of the pipe and the fixture.
6. Heat the fixture on the near side of the joint with the cone of the propane torch flame (the tip of the cone is the hottest point of the flame) until the melting point of the solder is reached. See **Figure 24-22.** Move the flame gently around the joint without touching the solder to evenly heat the joint.
7. When the melting point of the solder is reached, the solder quickly melts and disappears into the joint. Make sure that all parts of the joint get solder, then quickly remove the heat source and the solder wire. Solder should be pulled into the joint, not pushed into the joint.
8. Wipe any excess solder from the joint with a soft cloth. Check that the entire joint is filled with solder and is leakproof.

> **Safety Note**
> Protect flammable materials that will be subject to the torch flame with a heat shield or fire protection blanket. When soldering copper pipe, have a fire extinguisher handy. A bucket of ice water is also helpful. Be mindful of hot solder drips when doing overhead repairs.

To solder a joint in an existing plumbing run, water must be excluded and the surrounding material must be protected from open flame. First, drain the system completely and dry the joint. A propane torch can be used to drive water out of the pipe if the proper precautions are taken.

As in the procedure for soldering a new joint, remove any burrs, clean the joint and the pipe, and apply flux to both pieces before beginning to heat the fixture. When soldering is complete, visually inspect the soldered joint before adding water back to the system. If the joint is not sealed completely, the entire process must be repeated. Removing water before each attempt is a time-consuming process, and leaks can damage surrounding material.

Compression Fittings

When a soldered joint is not possible, a compression joint can be used, **Figure 24-23**. A compression joint is less sturdy than a soldered joint and may not remain leak-free as long. However, it has the advantage of being able to join two different pipe materials, such as copper and PEX.

A compression fitting consists of a compression nut that fits over one pipe end and a *compression ring* that fits onto the second pipe. When the ring

Vladimir Breytberg/Shutterstock.com

Figure 24-22. When soldering a copper joint, heat the fitting until it is hot enough to melt the solder.

Kuchina/Shutterstock.com

Figure 24-23. This plumber is connecting water pipes using compression fittings.

Copyright Goodheart-Willcox Co., Inc.

is fitted into the nut and tightened, it forms a seal by squeezing the ring into the compression space. This type of joint does not require thread seal tape (plumber's tape) or other sealants. Avoid overtightening the nut, which can deform the seal and cause leakage. A compression ring (ferrule) may need to be replaced if it has been deformed by overtightening.

Flared Fittings

If it is not possible or desirable to solder a joint in copper tubing, a flared joint can be made. A flared joint is more labor-intensive than a soldered joint, but it generally lasts as long. The general procedure for making a flared joint is as follows:

1. Select the proper flare fitting, correctly sized for the pipe and the fixture.
2. Place the flare nut over the end of the pipe to be flared. Check to be sure it is facing the correct direction, with the threads toward the end of the pipe.
3. Place the pipe, which has been deburred and cleaned with an abrasive cloth, into the flaring yoke in the correct size hole at the proper depth. For most flare yokes, this means flush with the surface of the yoke vise, but refer to the manufacturer's directions.
4. Tighten the yoke to hold the pipe securely.
5. Add the flaring tool to the yoke. Position the flare over the end of the pipe in the clamp. Begin turning the handle of the flaring tool until it is snugly pushed into the pipe.
6. Release the flaring tool and yoke vise. Check that the flare is round and without burrs.
7. Pull up the flare nut and insert the fixture piece into the flare nut. Tighten completely, forming the fitting in a leakproof joint.
8. Check the joint for leaks and continue assembling the system's fittings and fixtures.

Safety Note:
Follow all local codes regarding flare fittings in gas pipe systems. Local codes may require a professional plumber to monitor or complete the work. Always make sure gas is off and evacuated from the system before attempting to repair. Always wear eye protection when working with metal.

Steel Pipe

Steel pipe comes in two basic forms—galvanized pipe and black pipe. Both types have fallen out of favor as pipes for water supply plumbing, but they may remain in service in some old farmsteads.

Galvanized Pipe

Galvanized pipe is zinc-coated steel pipe, **Figure 24-24**. This pipe was used extensively for both supply and DWV piping until the mid-1960s because it was long-lasting, relatively inexpensive, and easy to install with threaded fittings. Galvanized pipe is not installed today, however, due to several serious drawbacks. Once the zinc

designbydx/Shutterstock.com

Figure 24-24. Galvanized pipe can be found in older plumbing installations.

coating is compromised, the pipe quickly corrodes and fails. The zinc coating, when heated, is toxic to breathe. Buildup inside the pipes can leach lead or other hazards into the water supply. In addition, galvanized pipe is heavy to install.

Black Pipe

Black pipe is black because of the carbon formed when the iron is manufactured into pipe, **Figure 24-25**. This type of pipe is no longer used for water supply except in irrigation applications. However, it is often used for pressurized gas supply lines, such as those that deliver natural gas and propane.

Black pipe is durable and performs well under pressure. It is easy to assemble and less costly than aluminum or copper. However, it is heavy and corrodes easily, ruling it out as a general water supply pipe.

withGod/Shutterstock.com

Figure 24-25. Black pipe nipples. A pipe nipple is a prethreaded section of galvanized pipe.

Pipe Fittings and Valves

Pipe fittings are junctions that allow a pipe to change direction, size, or material. Pipe fittings are generally made of the same material as the connecting pipe, except in the case of connecting one pipe material to another. Fitting configurations are either male (they fit inside the pipe) or female (pipe fits into them). Fittings are also used to join pipe that has been cut for repair and to join pipe from different directions. Various types of fittings are shown and described in **Figure 24-26**.

Pipe Fittings		
Shape	**Material**	**Function**
Tee	Rigid or soft copper, brass (in PEX systems), PVC, CPVC, ABS, steel	Allows water to enter the system and branch off in two directions from the same source.
Elbow (ell) including 90°, 60°, and 45°	Rigid copper, brass (in PEX systems), PVC, CPVC, ABS, steel	To make a 90°, 60°, or 45° turn in a rigid pipe.

(Continued)
Goodheart-Willcox Publisher

Figure 24-26. Various fittings and their functions.

Pipe Fittings *(Continued)*

Shape	Material	Function
Wye (Y) *Anita Potter/Shutterstock.com*	Rigid copper, brass (in PEX systems), PVC, CPVC, ABS, steel	Most often used in DWV systems as a point to clean out a system while allowing water to continue to flow.
Cross *Gumpanat/Shutterstock.com*	Rigid copper, brass (in PEX systems), PVC, CPVC, ABS, steel	Used when supply comes into a system and is diverted in three directions, such as in an irrigation system. Occasionally used in waste systems when three waste streams meet and leave by one outlet pipe.
Coupler, reducer *everst/Shutterstock.com*	Rigid copper, PVC, CPVC, ABS	Used to change from one size pipe to another. Restricts water flow near an outlet in a supply system to increase pressure at a fixture. Couplers can be used in pipes of the same size when a line has been cut and needs a fixture to repair the gap.
Union *Goodheart-Willcox Publisher*	Rigid copper, PVC, CPVC, ABS, steel	Used like a coupler to repair pipe, change from one pipe material to another, or change sizes. This three-part fitting can be easily disassembled to leave two ends free for other fixtures or additional repairs.
Nipple *Flegere/Shutterstock.com*	Rigid copper, PVC, CPVC, ABS, steel	A short pipe, often threaded on one or both ends, that is used to repair a cut pipe or replace a fixture that is being bypassed. Generally has male configurations on both ends.
Cap *niwat chaiyawoot/Shutterstock.com*	Rigid copper, brass (in PEX systems), PVC, CPVC, ABS, steel	Seals off the end of a pipe for either DWV or supply lines. Sometimes used as a location to clean out a system or add on in the future.
Plug *Sukpaiboonwat/Shutterstock.com*	Rigid copper, brass (in PEX systems), PVC, CPVC, ABS, steel	Seals the end of a pipe with a male configuration.

Some fittings are also fixtures. *Valves* allow water to leave the system, restrict the flow of fluids in the system, or divert water to another part of the system. Valves come in a variety of styles to accomplish different purposes. Valves commonly used by the agriculture industry are shown in **Figure 24-27**.

Valves		
Valve	**Description**	**Function**
Gate valve 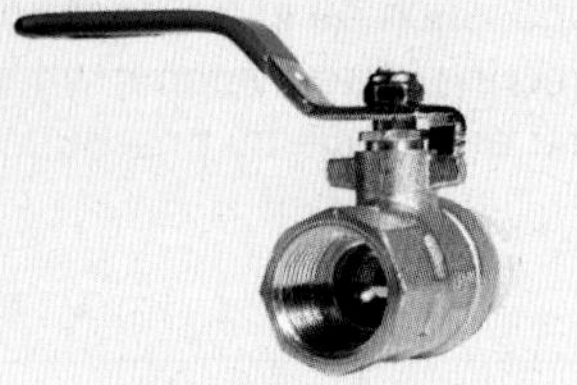 *yevgeniy11/Shutterstock.com*	A gate, or restrictive plate, is turned or pushed from one side of the valve to the other.	A slow close/open valve that is either entirely open or closed and is not used to restrict flow.
Globe valve  *cameracantabile/Shutterstock.com*	Recognized by its rounded "globe" area that contains a restricting mechanism.	A slow open/close valve used on hose bibs and other locations that require no leakage under a variety of pressures.
Ball valve 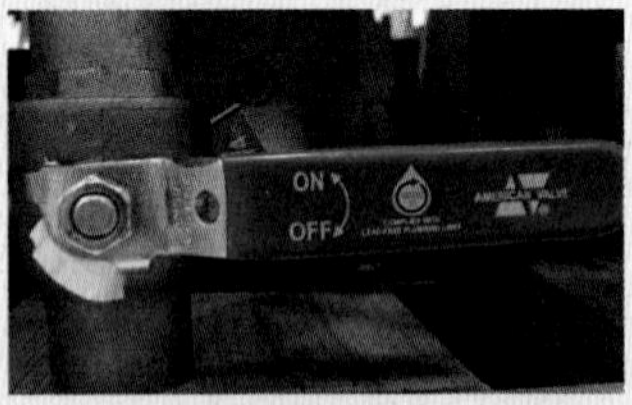 *Goodheart-Willcox Publisher*	Has a slightly rounded middle where a ball mechanism stops the flow of water when it is turned 90°.	A fast close/open valve that slightly restricts the water going through the valve but does not restrict the flow of fluid in a partially open or closed position. This valve is used where immediate control is needed, such as in food processing and chemical management systems.
Butterfly valve  *Nadezda Murmakova/Shutterstock.com*	Has a thin plate that can turn from the narrow side toward the direction of flow to the plate, completely stopping the flow when turned 90°.	Used in air, fluid, or steam systems because it can be opened or closed quickly.
Check valve 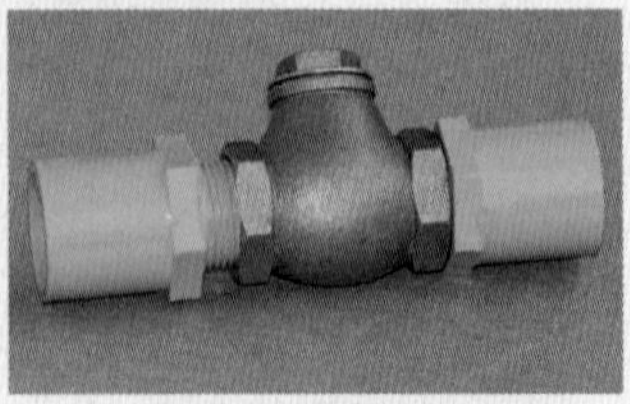 *bonga1965/Shutterstock.com*	A one-way, two-part system that contains the flow until a certain threshold is met, and then allows fluid to flow in one direction only. Fluid is met with a closed valve when trying to move in the opposite direction.	Used in systems such as waste exclusion, chemical handling, and food processing.

Goodheart-Willcox Publisher

Figure 24-27. Various plumbing valves and their functions.

CHAPTER 24 Review and Assessment

Summary

- Most local plumbing codes in the United States are based on the International Plumbing Code or the Uniform Plumbing Code.
- Most plumbing systems involve moving potable water in and wastewater out.
- The two basic parts of a plumbing system are the water supply system and the drain-waste-vent (DWV) system.
- Plumbing traps prevent sewer gas from coming back up through a plumbing fixture and into a residence.
- A septic system cleans the wastewater from individual residences by means of a septic tank, a leach field, and a series of connecting pipes.
- Plumbers use tools used by most agricultural mechanics, as well as specialty tools designed specifically for working with plumbing systems.
- Three basic types of plumbing pipe are commonly used by agriculture mechanics: plastic, copper, and steel.
- The four basic types of plastic pipe material are PEX, PVC, CPVC, and ABS.
- Pipe fittings are junctions that allow a pipe to change direction, size, or material.
- Valves control the flow of fluid from and within pipes.

Words to Know 📖

Match the key terms from the chapter to the correct definition.

A. annealing	D. fitting	H. septic system
B. cross-linked polyethylene (PEX)	E. fixture	I. sludge
	F. polyvinyl chloride (PVC)	J. soldering
C. drain-waste-vent (DWV) system	G. scum	K. water supply system

1. A system designed to clean the wastewater from individual residences.
2. The process of heating a metal and letting it cool slowly, making the material tougher and more flexible.
3. A type of plastic pipe that can be used to supply both hot and cold water in a plumbing system.
4. The part of a plumbing system that includes the water source and all of the pipes and fixtures that bring the water from the source to the final point of use.
5. A permanent part of a plumbing system that uses water or drains the system, such as hose bibs and showerheads.

6. The part of a plumbing system that removes water, waste, and gases from the system and maintains atmospheric pressure within the system.
7. The top layer in a septic tank, which consists of very lightweight solids, fats, and other waste materials.
8. A mixture of solids and liquids that forms the bottom layer in a septic tank.
9. A small part used to join, turn, or otherwise change a plumbing system.
10. A process that joins two metal pieces by heating and flowing a different metal around the joint.
11. A type of rigid plastic pipe that is often used in DWV systems.

Know and Understand

Answer the following questions using the information provided in this chapter.

1. In agricultural plumbing systems, liquids and gases move under _____ or by _____ toward the lowest point in the system.
2. In a DWV system, what two functions does the venting system serve?
3. What would happen when a fixture drains water if air pressure were not maintained in a DWV system?
4. The _____ field of a septic system must be below the frost line in areas where temperatures are low enough to freeze the soil.
5. To minimize the scum layer, what items should not be put into a septic system?
6. Which of the following is *not* a characteristic of PEX pipe?
   A. It is easy to install.
   B. It must be shielded from sunlight.
   C. It is reliable and long lasting.
   D. Pipe cement is required for installation.
7. CPVC emits toxic _____ gas when exposed to flames.
8. ABS pipe is used only for _____ pipes in plumbing applications.
9. Which type of copper tubing—rigid or soft—is used for water supply lines?
10. When soldering copper pipes in an existing system, you must exclude _____ so the fixture heats enough to melt the solder.
11. What are the two basic forms of steel pipe?
12. What is the function of valves in a plumbing system?

STEM and Academic Activities

1. **Science.** Given that the atmosphere on the moon is much thinner than that on Earth and gravity is much lower, would it be possible to build a plumbing system on the moon that would flush using current Earthbound technology? Explain your answer.
2. **Technology.** Research recent technological advances in plumbing that save water, improve sanitation, or contribute to the environmental protection in other ways. Choose a technology and give a report to the class using presentation software.
3. **Engineering.** Design a rainwater collection system that would contain both potable and nonpotable water. How would you solve the problem of the first water from the rain washing off debris from the surfaces? How could this water be incorporated into an existing plumbing system?
4. **Math.** In a DWV system, the standard slope in the drain system is generally 1/4" per foot as a minimum and 3" per foot as the maximum slope to allow both water and waste solids to leave the drain pipe. If the drain leaves your bathroom 27' from the exit end of the house, what would be the minimum and maximum fall for this bathroom drain if it starts 8' above the floor and moves to the outside wall at the correct slope?
5. **Social Science.** In what ways is the wide adoption of the International Plumbing Code beneficial? Is the IPC the basis for plumbing laws in your state?
6. **Language Arts.** Prepare flash cards for the different types of plumbing tools described in this chapter. On the front, draw a color image of the plumbing tool. On the back, write the name of the tool, along with a description and a list of uses. With a partner, use the cards to quiz each other on the plumbing tools.

Thinking Critically

1. If you were building three new buildings—a house, a livestock barn, and a greenhouse—what pipe material would you choose for each building, and why?
2. What personal behaviors can you change in order to reduce your consumption of water?

CHAPTER 25
Farm and Landscape Irrigation

Chapter Outcomes
After studying this chapter, you will be able to:
- Briefly describe the history of irrigation in the United States.
- Identify sources of irrigation water.
- Explain the need for water conservation in agricultural irrigation systems.
- Identify irrigation equipment components.
- Describe irrigation systems used for landscape and crops.
- Understand the process of irrigation system design.
- Explain how to manage and repair an existing irrigation system.
- Identify potential irrigation problems and solutions.

Words to Know 📄

aquifer	evapotranspiration	polyethylene pipe	sprinkler head
backflow preventer	furrow irrigation	reclaimed water	sump pump
basin irrigation	grading	reservoir	surface irrigation
border irrigation	hose traveler system	retention pond	swale
center-pivot sprinkler system	irrigation	runoff	thread tape
	irrigation controller	slope	turf rotor
detention pond	irrigation zone	soil structure	water cycle
drip irrigation	lateral line	soil texture	water pressure
erosion	main line	spray pattern	xeriscaping

Before You Read
Find the list of terms at the beginning of the chapter. Write what you think each term means. Then look up the term in the glossary and write the textbook definition.

While studying this chapter, look for the activity icon to:

- **Practice** vocabulary terms with e-flash cards and matching activities.
- **Expand** learning with interactive activities.
- **Reinforce** what you learn by completing the end-of-chapter questions.

www.g-wlearning.com/agriculture

Elena Elisseeva/Shutterstock.com

United States agriculture plays an instrumental role in the world food supply. With a world population of over 7 billion, farmers worldwide are producing more than ever before. Irrigation plays a vital role in the ability of farmers to produce the resources that we consume on a daily basis. *Irrigation* is the application of water to crops to aid in plant growth. Irrigation also plays a critical role in producing high-quality turf for residential lawns, athletic fields, and commercial buildings, **Figure 25-1**.

connel/Shutterstock.com

Figure 25-1. Irrigation provides water to crops and the landscape when rainfall is not sufficient.

Irrigation History

Egyptians were the first to use irrigation in approximately 6000 BCE. By diverting the water from the flooding Nile and Euphrates rivers, ancient Egyptians could provide irrigation water for their crops for over two months. Today, over 56 million acres of farmland in the United States are irrigated. The majority of this irrigated land is dedicated to grain and oilseed crops, vegetables, greenhouse and nursery crops, and hay crops. United States farmers apply an average of 128,000 million gallons of water per day to their crops. The majority of this water is applied in areas that receive too little rainfall to promote proper plant growth. For example, low rainfall totals in the western United States require more irrigation to maintain crop production. See **Figure 25-2**.

Dust Bowl

The Dust Bowl of the 1930s had a devastating impact on agriculture in the United States. A combination of drought and poor agricultural techniques turned once fertile prairie soil into loose sand and dust, **Figure 25-3**. As a result of the Dust Bowl, farmers realized that new strategies and tactics were needed to produce enough food to feed a growing country. Universities and government agencies began to conduct research to improve agricultural practices. New conservation practices, such as erosion control, irrigation, and crop rotation, were developed as a result.

Colorado River Storage Project

In 1956, the Colorado River Storage Project provided water to once-barren land in the western United States. Through a collection of water storage dams, states such as Colorado, New Mexico, Utah, Wyoming, Arizona, California, and Nevada were provided a consistent water source to apply to fertile soil. The combination of reliable water supply and fertile soil quickly led to successful crop production and an increased food supply.

Chapter 25 Farm and Landscape Irrigation 649

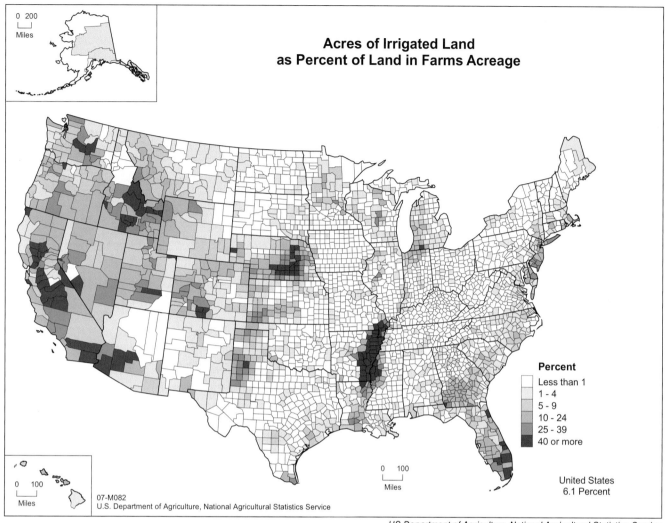

Figure 25-2. The percentage of irrigated cropland in the United States is shown here. Note the higher concentration of irrigated land in the western states.

Figure 25-3. Farm equipment buried in sand and dust in rural South Dakota during the Dust Bowl.

Green Industry

The growth of the Green Industry has greatly increased the demand for irrigation in the United States. The Green Industry includes businesses such as landscapers, florists, garden centers, golf courses, sod producers, and Christmas tree growers. Unlike traditional agriculture, the Green Industry produces a crop or service solely for ornamental purposes. Clients of these businesses demand a high-quality product. As a result, the Green Industry relies heavily on irrigation to provide the required crop quality. See **Figure 25-4**.

Water Sources

Water from a variety of sources provides drinking water for humans and irrigation water for crops. Water sources include rain, city reservoirs, backyard streams and rivers, and underground water. The available water source determines the method used to irrigate crops and landscape.

The Water Cycle

The *water cycle* is the continuous circulation of water on, above, and below the earth through evaporation, condensation, precipitation, and transpiration.

Water ultimately drains and is stored in large bodies of water. The water is returned to the earth's atmosphere through the process of evapotranspiration. *Evapotranspiration* is a combination of water loss from large bodies of water (*evaporation*) and water lost through plant growth processes (*transpiration*). Evapotranspiration returns water to the atmosphere. The moisture accumulates through the process of condensation. The moisture eventually returns to earth in the form of precipitation. **Figure 25-5** illustrates the water cycle.

Reservoirs

A reservoir, **Figure 25-6**, is a common source of water for residents of cities. Landscaped areas, such as athletic fields, business properties, home lawns, and golf courses, are often located in heavily populated areas and rely on a reservoir for irrigation water. A *reservoir* is a manmade or natural lake that provides a water supply. Water from the reservoir is used for human consumption as well as for irrigation.

©iStock/pedrosala

Figure 25-4. Irrigation is needed to keep a golf course in pristine condition.

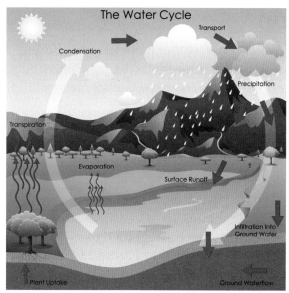
stockshoppe/Shutterstock.com
Figure 25-5. The water cycle.

Dchauy/Shutterstock.com
Figure 25-6. Reservoirs supply water for drinking, crop irrigation, and recreation.

Groundwater

Farmers of row crops are often located too far from densely populated areas to access reservoir water for irrigation Another water source is a well. A well is a manmade hole dug into the earth to reach an aquifer. An *aquifer* is an underground layer of rock or sand that is saturated with water. Wells vary greatly in their depth and water supply. Finding underground water (groundwater) can be challenging and expensive.

When wells cannot keep up with the demand for irrigation water, well water can be pumped into a pond. Ponds provide additional water to supply irrigation requirements. Water from the well is pumped to refill the pond. Excess water from irrigation drains and ultimately returns to the pond.

Thinking Green

Reclaimed Water

In order to conserve drinking water, reclaimed water (recycled water) is being increasingly used for irrigation, as well as for industrial processes. **Reclaimed water** is wastewater that is captured and treated. The treatment removes solid debris, breaks down organic materials, removes microorganisms, and filters and disinfects the water. Reclaimed water is a sustainable and cost-effective water supply. For irrigation purposes, the salts and nutrients contained in reclaimed water should be considered. For example, the need for additional fertilizers could be reduced.

CTR Photos/Shutterstock.com

One example of a reclaimed water system is in the town of Cary, North Carolina. The reclaimed water is used to irrigate lawns and planting beds, saving up to 20 million gallons of drinking water a day during the summer months.

Water Conservation

Earth's water supply is finite. Water conservation methods and efficient irrigation techniques must continue to be developed in order to balance the water demands of crops and humans. Researchers and scientists are working to develop ways to reduce water consumption.

Farmers and landscapers pay close attention to water consumption. A thorough understanding of a crop's cultivation requirements, soil conditions, and basic irrigation techniques helps to ensure that irrigation water is applied efficiently.

Irrigation systems are designed so that water can be reused and recycled as much as possible. In severe drought conditions, local and state officials can limit water consumption. Irrigation systems may be permitted to run only at certain times of the day on certain days of the week. See **Figure 25-7**.

Runoff

No irrigation system allows the crop to use 100% of all the water applied. *Runoff* is the draining away of water that plants are unable to absorb. See **Figure 25-8**. When too much water is applied to crops, the water carries away topsoil as well as any fertilizer and nutrients that are on the ground surface. This can be an issue especially if the field is located close to a river or stream. Agricultural runoff often contains excess fertilizers and pesticides that can be detrimental to surrounding streams and bodies of water and the plants and animals that live there.

Slope determines the direction in which runoff will flow. *Slope* is the slope or angle of the land from horizontal, called the *lay* of the land. A minimum slope of 1% is required to drain water. If the lay of the land does not efficiently drain water, *grading* (changing the slope of the land) may be required.

Mykola Mazuryk/Shutterstock.com

Figure 25-7. The effects of drought. What industries besides agriculture are affected by drought?

Steve Collender/Shutterstock.com

Figure 25-8. Runoff from a cornfield.

Erosion

Erosion is similar to runoff but causes more visible damage. *Erosion* is the washing away of soil caused by excess water or wind. The three main types of erosion are sheet, rill, and gully erosion. *Sheet erosion* is the removal of soil in thin layers by raindrops. Sheet erosion commonly goes unnoticed because it occurs slowly over long periods of time.

Have you ever noticed a large amount of water trying to drain after a thunderstorm? The channeling of water during the draining process is called *rill erosion*. Rill erosion is easily identified by the channels created by the draining water. *Gully erosion* is caused by the removal of soil along drainage lines.

Soil texture plays a vital role in the soil's susceptibility to erosion. *Soil texture* is the size of the individual soil particles. In soils with extremely small soil particles, such as clay, water takes longer to infiltrate through the ground. While the water is waiting to enter the ground, soil is often washed away. In soils with larger particles, such as sand, the water is easily able to pass through the soil. However, the particles are so light that they can easily be carried away by wind.

Soil structure is the way that individual particles of soil stick together. The tighter the soil sticks together, the more water the soil can hold, and the less often irrigation is needed. Soil that sticks together loosely holds less water and needs to be irrigated more frequently.

Retention and Detention Ponds

In agricultural and landscape settings, runoff drains into a series of runoff management basins. Retention ponds and detention ponds capture and filter excess irrigation water and excess rainwater. A *retention pond* holds a permanent pool of water. This water ultimately drains into a larger body of water and is then returned to the water supply after it is filtered. See **Figure 25-9**. A *detention pond*, also called a *dry pond*, is a flood control mechanism. It serves as a spot for water to drain and later evaporate. It contains water only for brief periods after rainfall or application of irrigation water.

Farmers manage excess water to control runoff in crop production with a series of swales. A *swale* is a type of ditch that drains excess water (runoff) away from crops, **Figure 25-10**.

Jwwaddles/Shutterstock.com

Figure 25-9. Retention ponds contain a permanent supply of water. They are useful for filtering water as it runs off fields, lawns, and parking lots

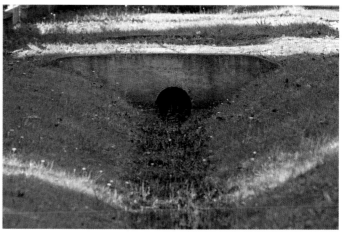

Nneirda/Shutterstock.com

Figure 25-10. Swales provide a source of drainage in agronomic settings. Swales allow water to properly drain and be recycled.

Rainwater Harvesting

In rainwater harvesting, rainwater is collected and stored for future use rather than allowed to run off. Rainwater harvesting can provide irrigation water to farmers and landscapers who manage properties with a large square footage of roofs or paved surfaces. For example, Metrolina Greenhouses in Huntersville, North Carolina, harvests rainwater to irrigate over 100 acres of greenhouses. Just one inch of rainfall provides Metrolina with over 3 million gallons of irrigation water. The rainwater is collected from the roofs and then diverted to a storage pond that becomes a water source for plants. The water is filtered through a series of screens and filters before it reaches the storage pond. Vegetation around the storage pond filters the solid material from the water. The water is filtered again before returning to the greenhouse.

Rainwater cisterns are also used to harvest rainwater on a commercial and residential scale. Water stored in these cisterns can be pumped to irrigate plants as needed, **Figure 25-11**.

Steven Frame/Shutterstock.com

Figure 25-11. Rainwater cisterns are used by homeowners to collect rainwater for use in their landscaping. Water is then pumped to where it is needed.

Xeriscaping

Proper plant selection and planting time can be more important than the type of irrigation system. *Xeriscaping* (pronounced ZEER-escaping) is a landscaping method that uses water-efficient plants. Plants are grouped in zones according to their water usage. Plants that require larger amounts of water are used sparingly and are grouped together. In xeriscaping, crops are irrigated early in the morning to allow the plants to use the water efficiently in the photosynthesis process. Plants are watered thoroughly and deeply to ensure the development of a deep root system. The actual amount of water needed varies depending on the climate and type of crop.

Irrigation Equipment

An irrigation system consists of several components that move water from its source to the location needed by the crop. A pump mechanically moves water from the well to the location needed by the irrigation system, **Figure 25-12**.

A *backflow preventer* prevents water pumped by the irrigation system from returning to the

Deyan Georgiev/Shutterstock.com

Figure 25-12. Irrigation pumps vary in type and size depending on the application.

groundwater in the well or to water in the main irrigation line. The water can travel through the backflow preventer in one direction, but it is prevented from going backward in order to protect the water source from contamination.

Irrigation Lines

Pumping equipment delivers the water supply from the source to the piping equipment. A *main line*, **Figure 25-13**, is a large-diameter pipe that is used to carry water to lateral lines. Water is under constant pressure in main lines. A *lateral line* is a secondary irrigation line that carries water from the main line to the sprinkler head or emitter. Lateral lines are under pressure only when the irrigation system is in use. Water pipes in agricultural and landscape settings are commonly made of PVC, galvanized steel, PEX, or polyethylene. Each piping material has its own advantages and disadvantages.

FooTToo/Shutterstock.com

Figure 25-13. Main irrigation lines are larger in diameter than lateral lines.

PVC Pipe

PVC (polyvinyl chloride) pipe is the most commonly used pipe in irrigation systems. PVC pipe is cost-effective, lightweight, and corrosion-resistant. See **Figure 25-14**. It is available in a large variety of diameters and lengths but can be easily cut to the desired length. PVC pipe is often buried underground but can be used aboveground as well. Equipment and machinery can easily damage PVC pipe. PVC can be primed and glued together. However, the glue that is used to bond PVC pipe often does not last as long as the pipe itself, creating leaks. See Chapter 24, *Plumbing Design and Installation*, for more information about joining PVC and other types of pipe.

©iStock/koosen

Figure 25-14. PVC pipe is easy to connect and can be cut to specific lengths.

Galvanized Pipe

Galvanized (zinc-coated steel) pipe is also used in irrigation systems. Properly installed and maintained, galvanized pipe has an average lifespan of 25 to 50 years. Over time, the zinc can wear off the pipe and cause the pipe to rust. Galvanized pipe is more rigid than PVC pipe and is less likely to break when in contact with machines and equipment. However, the additional strength of galvanized pipe makes it more difficult to work with.

656 Agricultural Mechanics and Technology Systems

Praiwun Thungsam/Shutterstock.com

Figure 25-15. Galvanized pipe, common in older irrigation systems, rusts over time as the zinc coating wears off.

Pawel G/Shutterstock.com

Figure 25-16. PEX pipe is installed with a special crimping tool and crimping rings.

Le Do/Shutterstock.com

Figure 25-17. Polyethylene pipe is soft and flexible. It is commonly used in low-pressure irrigation systems.

Galvanized pipe is screwed together to create strong connections with other pipe. See **Figure 25-15**.

PEX Pipe

Many homes and irrigation systems are now being plumbed using PEX (cross-linked polyethylene) pipe. PEX pipe is flexible and is usually white, but can also be red, blue, or orange. It must be buried or installed under a structure because the sun's ultraviolet rays can degrade the pipe. PEX pipe is commonly sold in smaller diameters and is ideal for water-saving irrigation systems. The pipes are clamped together with a series of crimping rings and a crimping tool. See **Figure 25-16**. If not installed properly, leaks can soon arise at the rings.

Polyethylene Pipe

Polyethylene pipe is soft and flexible like PEX pipe; however, it can be used aboveground. Polyethylene pipe is usually black in color, **Figure 25-17**. The pipe is often joined together with barbed couplings (connecting devices) and hose clamps. Polyethylene pipe is common in low-pressure irrigation systems such as drip and trickle-tube irrigation.

Pipe Fittings

Irrigation pipe is joined together with pipe fittings. Pipe fittings control the layout of the pipe as well as water pressure and flow rate. See **Figure 25-18**.

Irrigation Systems

Sprinkler systems vary greatly, depending on the types of plants or crops and the size of the area to be irrigated. For example, a residential landscaping irrigation system is typically set up much differently from a crop irrigation system on a large farm.

Copyright Goodheart-Willcox Co., Inc.

Pipe Fittings Used in Irrigation

Image	Pipe Fitting	Usage
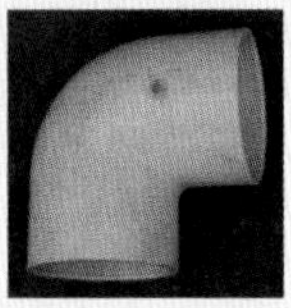 *Goodheart-Willcox Publisher*	90° elbow	Making a 90° turn with pipe. A 90° turn can make irrigation systems tight and compact, but water pressure is greatly increased when such sharp turns are made.
 *Goodheart-Willcox Publisher*	45° elbow	Making a 45° turn with pipe. A 45° turn is more gradual and does not have as much pressure build up as a 90° turn
 *Goodheart-Willcox Publisher*	Pipe coupling	Joining two pieces of pipe without changing the direction of flow.
 *Goodheart-Willcox Publisher*	Pipe bushing	Reducing pipe size. A pipe bushing is screwed or glued into one end of a pipe. A smaller pipe is then glued into the bushing.
 *Goodheart-Willcox Publisher*	Pipe reducer	Reducing pipe size. A pipe reducer is attached to the outside of the pipe.
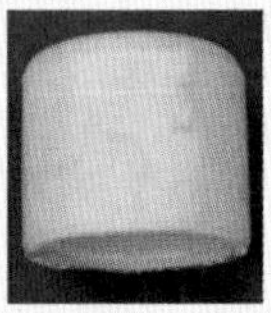 *Goodheart-Willcox Publisher*	Pipe cap	Closing the end of a pipe. A pipe cap is not easily removed.
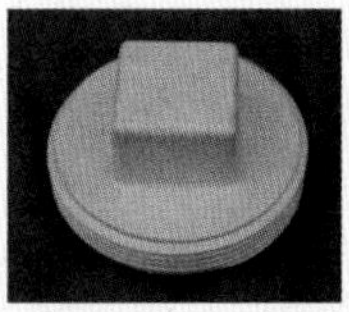 *Goodheart-Willcox Publisher*	Pipe plug	Closing one end of a pipe. Easily screwed on and off with a pipe wrench.

(Continued)
Goodheart-Willcox Publisher

Figure 25-18. Types of pipe fittings commonly used in farm and landscape irrigation.

| Pipe Fittings Used in Irrigation *(Continued)* |||
Image	Pipe Fitting	Usage
Goodheart-Willcox Publisher	Ball valve	Cutting off the water supply to a lateral or main line. A simple way to prevent water from flowing through certain pipes.
Goodheart-Willcox Publisher	Pipe union	Joining two pipes without changing the direction of flow. Can be easily connected and disconnected without damaging the pipes.

topseller/Shutterstock.com

Figure 25-19. Pop-up sprinkler heads have very few moving parts.

nvelichko/Shutterstock.com

Figure 25-20. Gear-driven rotary heads are larger and have more moving parts than pop-up sprinkler heads.

Sprinkler Irrigation

As the name implies, sprinkler irrigation involves the use of a sprinkler head to provide water to a crop or landscape. A *sprinkler head* is a device attached to an irrigation line that dispenses the water. The four main categories of sprinkler heads are pop-up, gear-driven rotary head, impact rotary head, and large turf rotor.

Pop-Up Sprinkler Heads

Pop-up sprinkler heads are commonly used in small landscape irrigation settings. An advantage of pop-up sprinkler heads is their ability to be concealed. Water pressure causes the sprinkler head to pop up and dispense water. Water is sprayed in a certain direction depending on the nozzle type. Nozzle types can be easily adjusted and interchanged to change the spray direction. Depending on water pressure, pop-up sprinkler heads can typically dispense up to 2.5″ of water per hour. See **Figure 25-19**.

Gear-Driven Rotary Heads

Gear-driven rotary heads are commonly used in commercial landscape settings. These sprinkler heads, like the pop-up type, are concealed when not in use. Gear-driven rotary heads are driven by a series of gears inside of the sprinkler head. The gears allow for a quick adjustment of spray pattern and arc. See **Figure 25-20**. A *spray pattern* is the area that one sprinkler head can cover. Gear-driven rotary heads provide more adjustment from a single nozzle and are better suited for large landscape areas. A disadvantage of both gear-driven rotary heads and pop-up sprinkler heads is that they are easily broken when run over by a lawnmower or car.

Impact Rotary Heads

Impact rotary heads are most commonly used in agronomic settings. They cover large areas (a radius of up to 150′) with a continuous stream of water. Impact rotary heads are often installed above the ground and can apply up to 2″ of water per hour. Impact rotary heads are noisy and are not common in landscape settings. Nurseries commonly irrigate using impact rotary heads, **Figure 25-21**.

Turf Rotors

Turf rotors are commonly used on large, grassy areas, such as athletic fields and golf courses. Turf rotors allow landscapers to disperse a large amount of water in a short time. Turf rotors are also easy to install. Their convenience allows landscapers to provide irrigation water to areas that may be inaccessible by other irrigation systems.

Large turf rotors can cover a radius of up to 200″ and dispense up to 80 gallons of water per minute. Large turf rotors attach directly to a lateral line. They also can be easily removed.

Agronomic Irrigation

Farmers irrigate hundreds or even thousands of acres. In order to produce high-quality crops, the amount of water applied to the crop must be controlled. Farmers who cannot justify the expense of costly irrigation controllers irrigate their crops with simpler, yet effective, methods. Types of irrigation commonly used in agronomic settings are hose traveler systems, center-pivot irrigation, drip (or trickle) irrigation, and surface irrigation.

Goodheart-Willcox Publisher

Figure 25-21. Impact rotary heads are noisy. They are commonly installed above the plants being irrigated.

Figure 25-22. A hose traveler system can shoot water up to 1/2 mile throughout the crop.

Figure 25-23. A—Center-pivot irrigation is an efficient way to irrigate large acreage. B—Center-pivot irrigation provides water in a circular pattern, as seen from this aerial view.

Hose Traveler Systems

In a *hose traveler system*, a cart equipped with a high-pressure irrigation nozzle is pulled by a tractor through the field to irrigate the crops, **Figure 25-22**. Hose travel systems are also referred to as *water cannons*, *hose reel systems*, or *traveling gun systems*.

Hose traveler systems are capable of dispensing water up to 1/2 mile and work best on rectangular fields. Disadvantages of this system are that installation is labor-intensive and the grower may need to move the cart to provide water to the entire field.

Center-pivot sprinkler systems pivot and move in a circular pattern across the field, **Figure 25-23**. In this system, the sprinkler heads are attached to several large trusses on wheels. The *truss* is a series of metal segments that are used to provide stability to the boom of the sprinkler system. When the sprinkler first begins to move, it is on the perimeter of the field. With each pass, the sprinkler moves closer to the center of the field. Center-pivot irrigation systems are commonly 1/4 mile in length. The system is commonly powered by an electric motor or water-driven gears.

Drip Irrigation

Drip irrigation, also referred to as *trickle irrigation*, is a popular irrigation technique in the fruit and vegetable industry as well as among ornamental plant producers. See **Figure 25-24**. Drip irrigation uses water efficiently. The water drips out of an emitter at a slow pace, ensuring that the plants are able to take up most or all of the water supplied. Runoff is greatly reduced. A disadvantage of this system is that the emitters must be located in close proximity to the plants. This greatly reduces their application in an agronomic system.

A micro-sprinkler system is a form of drip irrigation system. The micro-sprinkler system actually sprays water on the plant; however, water is still used efficiently and conserved.

Surface Irrigation

Surface irrigation is the process of flooding entire rows or fields of crops, **Figure 25-25**. Gravity and the slope of the field cause the water to spread across the field. Surface irrigation is a simple but effective method of watering large areas. The disadvantage of this method is that the grower cannot target the water needs of a specific group of plants within the field.

Virtually every agronomic crop can be watered by some type of surface irrigation system. Types of surface irrigation include basin, border, and furrow.

The most common and efficient type of surface irrigation is basin irrigation. *Basin irrigation* involves pumping water into the field to flood the crop for a brief period of time. Gravity and the slope of the land distribute the water to the entire field. Water is typically pumped from an irrigation pond. Small berms along the perimeter of the field surround the water and keep it in the desired field.

Border irrigation requires an evenly sloping field. Water is pumped along the border of the field. Gravity and the slope of the land carry the water throughout the field.

Furrow irrigation is similar to border irrigation, except that the water is pumped into furrows between crop rows. The crops themselves are not flooded, only the furrows. Another difference is that the amount of water supplied to each furrow can be varied, allowing greater control over irrigation to different parts of the field.

Tortoon Thodsapol/Shutterstock.com

Figure 25-24. Drip irrigation is an efficient and sustainable method of watering crops and landscape plants.

Irrigation System Design

Before the first piece of pipe is purchased and the first trench is dug, the irrigation system is designed by an experienced landscape contractor or plumber. Proper irrigation design ensures that the irrigation system has enough water supply and water pressure to work properly and cover the intended area. Irrigation system prints use symbols to indicate the components of the system and their locations. Common irrigation symbols are shown in **Figure 25-26.**

noramin.s/Shutterstock.com

Figure 25-25. Crops flooded by surface irrigation must be able to withstand wet soil for long periods.

Irrigation Symbols	
Sprinkler	•—•—•
Main line	—·—·—·—
Lateral line	————————
Control box	⊗ ⊠
Water meter (water source)	[M]
Backflow preventer	-N-
Throw area of sprinkler	(fan symbol)

Goodheart-Willcox Publisher

Figure 25-26. Irrigation symbols.

TUM2282/Shutterstock.com

Figure 25-27. Pressure dials indicate the amount of water pressure in irrigation lines.

Designing an efficient and effective irrigation system requires knowledge of basic concepts of water and water pressure. Water flow is the passage of water through the irrigation system. Twists and turns of irrigation pipe have an impact on water flow. *Water pressure* is the force that water applies to its surroundings. Water pressure is measured in pounds of pressure per square inch (psi). Water pressure is determined by a number of factors, including the slope and elevation of the land, water pipe diameter, and water flow rate. In an effective irrigation system, a uniform water pressure is maintained. Pressure gauges measure the water pressure in the system. Water pressure that is too high for the designed system can cause the irrigation system to break, especially at pipe fittings. See **Figure 25-27**.

Water pressure decreases as water flows farther away from the source. Water friction caused by the motion of the water against the irrigation pipe, also causes a reduction in water pressure. In order to maintain pressure, pipe size is often reduced using either a pipe reducer or pipe bushing.

Water Calculation

The first step in irrigation system design is to calculate the amount of water needed by the crop. Much research and testing has been conducted to determine accurate moisture requirements for virtually all agronomic and landscape crops. For example, in North Carolina, fescue requires approximately 1.5″ of water per week.

Precipitation rate is the rate, or speed, at which water is applied over a specific area. Precipitation rate is measured in inches of water per hour. The precipitation rate is used to calculate the specific amount of time the irrigation system needs to run to provide sufficient moisture to the crop. Water that moves through the irrigation system is measured in gallons per minute (gpm) or gallons per hour (gph).

Not all crops require the same amount of water. The same crops grown in different regions also require varying amounts of water. The moisture requirements of the crop, along with the geographic location, play a crucial role in determining moisture content of the harvested crop. Watermelon has higher moisture content than field corn. More water is required to produce a watermelon than a single ear of corn. As another example, it takes approximately one gallon of water to produce a single almond, but approximately 25 gallons of water to produce one ear of corn. See **Figure 25-28**. Sufficient moisture levels must be maintained in order to produce optimal crop levels.

Irrigation Controller

The system design also includes the specification of the control equipment used to manage the amount of water applied and the irrigation schedule. *Irrigation controllers* allow farmers and landscapers to control the duration of irrigation, as well as the time of day the irrigation water is applied to the landscape or crop. See **Figure 25-29**. An irrigation controller operates *irrigation zones*, a group of sprinklers that is turned on to water a particular area.

Christopher Boswell/Shutterstock.com; Polarpx/Shutterstock.com

Figure 25-28. Irrigation allows farmers to meet the water requirements of a variety of crops in different climate conditions.

Goodheart-Willcox Publisher

Figure 25-29. Irrigation controllers, also called *time clocks*, allow users to easily program the duration of irrigation and the area to which water is to be applied.

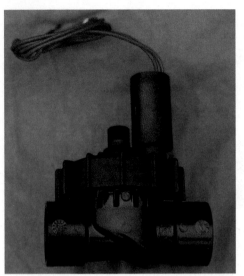

Goodheart-Willcox Publisher

Figure 25-30. Solenoid valves connect the irrigation controller to the zone. Each zone has one solenoid.

Irrigation controllers are attached to solenoid valves. A solenoid valve electrically controls a water valve, causing it to open or close. The irrigation controller sends an electrical charge to a solenoid. The charge opens or closes the solenoid, which turns the water on or off. See **Figure 25-30**.

Irrigation systems supplement rainfall, so their use may not be necessary during rainy periods. Many irrigation controllers are installed with a rain sensor. A rain sensor measures the rainfall and indicates the amount to the irrigation controller. If the rainfall is over a certain amount, the irrigation system automatically turns off for that day. When managed properly, an irrigation controller and rain sensor increase the efficiency of any irrigation system.

All irrigation controllers must be connected by electrical wire to the zones controlled by the timer. The irrigation wire is connected to the solenoid. Irrigation controllers may control up to 16 zones. Once installed, the irrigation controller timer must be programmed correctly by following the manufacturer's instructions. The newly programmed irrigation controller should be closely monitored to ensure that it is operating properly.

Finalizing the Design

Once the moisture requirement of the crop has been analyzed and a consistent source of water identified, the irrigation system is designed. Irrigation systems should always be drawn to scale. The design should indicate the location of the main lines, the location of lateral lines, and the throw radius of sprinkler heads.

Each sprinkler distributes the full amount of water to 2/3 of its diameter. In order to ensure that all areas receive equal amounts of water, this overlap should be included in the design. Sprinkler heads should be spaced to allow for a 2/3 overlap of the throw radius.

The water source, irrigation pipe size, power of the pump, and distance from the source of water are all factors that determine the amount of water pressure needed for the irrigation system to function. It is recommended that an experienced plumber or irrigation contractor determine the requirements that work best for the particular situation.

Installation Safety

When installing irrigation systems in a large trench, make sure the trench is secure and stable before entering. Remove excess water from a trench before installing or repairing pipe.

Because irrigation systems involve water under pressure, moving parts, and electricity, safety is a primary concern. Before working with any electrical system, cut off the electricity at the breaker box. Before repairing irrigation lines, turn off the water to make the repair safer and cleaner.

Always wear personal protective equipment (PPE) when working on an irrigation system. PPE should include closed-toe shoes, safety glasses, gloves, a hardhat, and a safety vest. Avoid wearing loose clothing that could become entangled in moving parts. Employers may also recommend additional PPE.

Irrigation System Management

Once the irrigation system is installed, it must be properly managed and maintained for optimal operation. Sprinklers must continuously be adjusted to ensure proper coverage areas. Nozzles must be cleaned on a regular basis to remove debris. Sprinkler heads and nozzles can be easily adjusted using an irrigation key or a small screwdriver, depending on the make and model.

Output of irrigation systems can be measured by placing rain gauges in select areas of the landscape. Soil moisture sensors can be used to measure the specific amount of moisture in the soil. Soil moisture sensors should be placed at representative intervals throughout the field or landscape. A soil moisture sensor meter can be used to read moisture levels at all of the sensors.

Winterizing Irrigation Lines

Freezing temperatures can cause water in irrigation lines to freeze and expand, resulting in breakage of irrigation lines and fittings. See **Figure 25-31**. Therefore, irrigation lines must be properly winterized before the system is exposed to freezing temperatures. Winterizing is especially important on PVC pipe, as it can easily break as a result of freezing water.

All water must be drained from the lines by turning off the water supply to the irrigation system and draining the remaining water from the irrigation lines. The process of draining irrigation lines depends on the layout of the irrigation system. Irrigation lines are commonly drained by cutting off water to the lateral line supplying water to the irrigation zone. The release valve connecting the irrigation zone to the lateral line should be opened to drain the irrigation water. If the irrigation system is needed during the winter, the valves must be closed and the water supply turned on again. After use, all lines must be redrained before the pipes are exposed to freezing temperatures.

Goodheart-Willcox Publisher
Figure 25-31. This broken pipe must be repaired before the nursery plants can receive any water.

Repairing Irrigation Pipe

Irrigation pipes are connected through a series of couplings and fittings. Methods of attaching fittings to pipe vary. Regardless of the type of pipe, the pipe must be cut at a straight angle and installed properly to the fitting.

> **Safety Note**
> When cutting pipe, wear safety glasses and gloves.

PVC Pipe

PVC pipes are glued together. When the pipe is cut, a clean, straight cut must be produced. Small burrs of plastic formed during cutting must be removed before the pipe is glued. The burrs of plastic can easily be removed by wiping the lip of the freshly cut pipe with a clean rag.

Before the pipe is glued together, the pipe and the fitting must be primed and cleaned. The purpose of the primer is to remove debris from the pipe and create a porous surface to which the glue can bind. See **Figure 25-32**. The inside of the fitting and the outside of the pipe should be covered evenly with the primer. Prime pipe only in the area the fitting will cover. Once the pipe and fitting are cleaned and primed, apply the glue in the same manner in which the primer was applied. Spread the glue evenly across the entire perimeter of the pipe and fitting. Once the pipe and fitting are covered in glue, fit the two pieces together. See **Figure 25-33**.

PVC fittings are sold with a raised lip inside the fitting. All pipe material should be pushed as far as the pipe will go into the fitting to ensure a secure fit. Once the pipe is inside the fitting, rotate the pipe 90° to ensure a proper fit. Then allow the pipe to dry for 15 minutes.

PVC pipe that is 1″ or more in diameter is not flexible enough for sections to be cut out and replaced. The pipe will not bend enough to correctly install the pipe fittings. Broken sections of PVC pipe can be repaired with a slip coupling. A slip coupling is composed of two sections of PVC pipe and rubber gaskets. It expands to fit between the pieces of broken pipe and is glued to the ends of the broken pipe. The slip coupling serves as a bridge to repair the broken section. See **Figure 25-34**. The larger end of PVC pipe is glued to one side of pipe. The smaller end of PVC pipe is then slid far enough to glue to the other end of the pipe.

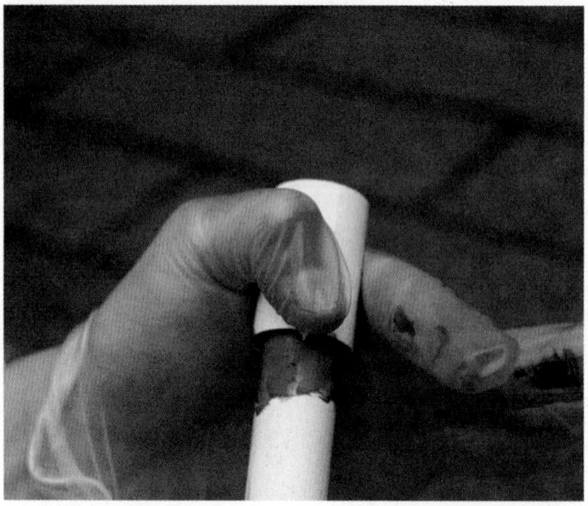

Goodheart-Willcox Publisher

Figure 25-32. Primer should be applied on both the pipe and the fitting that is being attached.

Figure 25-33. The fitting should easily slide over the pipe.

PEX Pipe

PEX pipe is quickly and easily attached together. It is crimped to a coupling or a fitting

Goodheart-Willcox Publisher

Figure 25-34. A quick slip repair coupling.

using a PEX crimping tool. To prevent a leak at the fitting, it is recommended that two crimp rings be installed on each end of the fitting. If the irrigation system continues to leak at the fittings, the crimping tool may not be attaching the rings tightly enough.

Galvanized Pipe

Galvanized pipe is screwed together with threads and thread tape. *Thread tape* is a nonsticky, stretchy white tape that provides a tight fit when joining galvanized pipe. The thread tape should be wrapped in a clockwise direction around the male end of the pipe. If the tape is not wrapped in a clockwise direction, it will begin to unravel when the new fitting is installed.

Galvanized pipe repairs can also be made with a pipe nipple. A pipe nipple is a prethreaded section of galvanized pipe. Pipe nipples are sold in a variety of lengths and diameters. If the repair cannot be made using a pipe nipple, a piece of pipe must be cut and threaded to the exact length required. Many hardware stores offer free cutting and threading of galvanized pipe.

Polyethylene Pipe

Polyethylene pipe is joined together using heat and barbed couplings and fittings. The pipe must be cut straight and be free of burrs before it is installed to a coupling or fitting. Polyethylene should be heated with a torch to make the pipe soft enough for correct installation of the fittings, **Figure 25-35**. Most couplings or fittings have a mark or notch that indicates how deep the coupling should be inserted. See **Figure 25-36**.

For added security, hose clamps are used to secure the pipe to the fitting. The hose clamps should be installed tightly on each side of the fitting, **Figure 25-37**.

Goodheart-Willcox Publisher

Figure 25-35. Polyethylene pipe can be heated with a propane torch.

Goodheart-Willcox Publisher

Figure 25-36. The pipe should not be pushed too far on the coupling or fitting. A raised lip on the fitting indicates the correct depth.

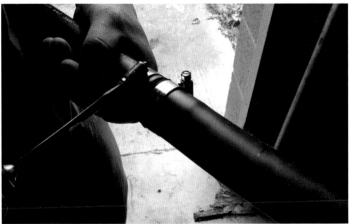

Goodheart-Willcox Publisher

Figure 25-37. Hose clamps ensure that the fitting is secured tightly to the irrigation pipe.

Career Connection

Irrigation Repair Technician

Job description: Irrigation repair technicians are involved in design and repair of irrigation systems for commercial landscape companies. Work includes repairing broken lines, calibrating sprinkler heads, designing irrigation systems, and installing new irrigation systems.

Education required: Associate's or bachelor's degree in Horticulture Science.

Job fit: This job may be a fit for you if you enjoy working outside and getting your hands dirty. You must also enjoy working with people and have good problem-solving skills.

sima/Shutterstock.com

Irrigation System Troubleshooting

Regardless of the type or design of the irrigation system, problems occur sooner or later. Troubleshooting skills save time and frustration. Quickly solving the problem also saves water and prevents damage to the farm or the landscape. Locating problems can be difficult when pipes and fittings are buried underground, **Figure 25-38**. The following are a few examples of irrigation problems and solutions.

Problem: The irrigation controller will not turn off the irrigation zone.

Solution #1: The solenoid for the zone is not receiving the signal to close the valve. First, check to see if the solenoid is correctly wired to the irrigation

Goodheart-Willcox Publisher

Figure 25-38. This underground pipe fitting has broken, causing a leak in the greenhouse.

wire that connects to the irrigation controller. The problem could also be due to a split in the wire that runs from the solenoid to the irrigation controller. A wire and valve locator is a handy tool that can be used to help locate the break in the irrigation wire. The wire and valve locator emits a tone when held close to the buried irrigation wire. It stops emitting the tone when held over the broken wire.

Solution #2: The solenoid valve is stuck open. The solenoid valve works through a series of rubber diaphragms and seals. Debris can easily enter the solenoid and cause the solenoid to malfunction. Before replacing the solenoid, take it apart and clean the rubber diaphragm. The solenoid is held together by a series of screws located around its perimeter. After cleaning, if the solenoid still cannot be controlled by the irrigation controller, it is likely that the solenoid will need to be replaced.

Problem: Sprinkler heads are not covering the intended irrigation area.

Solution #1: The sprinkler heads and nozzles need adjustment. Adjust sprinkler heads and nozzles on a routine basis to ensure that they are in optimal working condition. Properly calibrated sprinkler heads help to ensure that the irrigation system is working efficiently. Commercial and residential landscapes in urban settings are subject to scrutiny regarding the use of water. Municipalities can penalize or fine property owners for improperly tuned irrigation systems.

The best time to adjust sprinkler heads is when the irrigation system is running. Adjust sprinkler heads by zone. Each sprinkler head is tuned differently, depending on the brand and model of sprinkler head. Most sprinkler heads require a dedicated tool to adjust the sprinkler head. Other irrigation heads can be adjusted with a small flat-head screwdriver. Properly calibrated sprinkler heads are shown in **Figure 25-39**.

It can be difficult to adjust the sprinkler heads by walking or driving back and forth from the irrigation controller to the irrigation zone. An

Rashid Valitov/Shutterstock.com

Figure 25-39. Sprinkler heads must be fined-tuned to work efficiently.

irrigation remote controller makes irrigation system maintenance much easier. It plugs into the irrigation controller and allows the user to remotely turn on irrigation zones. The remote control can operate the control box from up to two miles away. Operators can simply turn on a specific irrigation zone and then proceed to adjust each sprinkler head as needed.

Solution #2: Water pressure is reduced due to a broken pipe or leaky fitting. Water pressure over time will cause pipe fittings to degrade. PVC pipe can be damaged by vehicles and other construction equipment. A buried PVC irrigation line rarely survives the weight of a large piece of equipment driving over it. Sprinkler heads do not fare well when they come in contact with a vehicle tire.

Replace broken irrigation pipes or fittings as soon as possible. A trench shovel is useful to dig down and expose broken irrigation pipe. See **Figure 25-40**. After digging, pump excess water out of holes with a *sump pump* (a manual or electric portable pump used to remove water).

Figure 25-40. A trench shovel is narrower than a common shovel, allowing the user to easily dig around irrigation lines.

Solution #3: Water pressure is reduced due to clogged or dirty irrigation lines. Over time, irrigation lines can become clogged with dirt, rocks, trash, algae, and other debris that can easily enter the exposed end of the pipe. The debris clogs nozzles and increases the amount of friction in the pipe, reducing water pressure. Because drip irrigation systems are in direct contact with the ground, soil can easily clog the openings, **Figure 25-41**. Flush irrigation lines on a routine basis to ensure that they are clean. Irrigation lines are commonly flushed at the beginning and end of each growing season. Also flush the lines after repairs are made.

Problem: Ground within an irrigation zone is wet and soggy.

Solution #1: A repair of the lateral line is needed. Irrigation pipe commonly leaks around couplings and other pipe fittings. Water pressure, combined with turns in the irrigation pipe, can cause couplings to burst or simply wear out.

Vadym Zaitsev/Shutterstock.com

Figure 25-41. Drip irrigation systems can easily become clogged because the drip holes are so close to the soil.

To repair a broken lateral line, first cut off the water supply and then dig down to expose the broken pipe. Be careful not to damage the pipe with your shovel. Turn on the water supply. If there is a leak, you should be able to quickly identify it. Once you have identified the location of the leak, repair the pipe.

Solution #2: The slope or drainage is improper. Excess water drains to the lowest point in the field. In a newly designed irrigation system or one where recent grading has occurred, the slope could be the source of the problem. If no grading has occurred and the spot has previously not caused concern, the slope and grade are probably fine.

Solution #3: The sprinkler heads are not calibrated correctly. Over time, sprinkler heads require adjustment. Irrigation systems are designed to overlap to ensure complete coverage. If the sprinkler heads are not calibrated correctly, excess overlap can occur.

CHAPTER 25
Review and Assessment

Summary

- In the last 80 years, universities and government agencies have worked to improve architectural irrigation practices.
- Water from a variety of sources, including reservoirs, ponds, rivers, groundwater, and reclaimed water, is used to irrigate crops.
- Irrigation systems are designed so water can be reused and recycled as much as possible.
- When too much water is applied to crops, the water carries away topsoil as well as any fertilizer and nutrients that are on the ground surface.
- An agricultural irrigation system consists of many components, including a pump, backflow preventer, irrigation lines, and control equipment.
- Water pipes in agricultural and landscape settings are commonly made of PVC, galvanized steel, PEX, or polyethylene.
- Four main categories of sprinkler heads are pop-up, impact rotary head, gear-driven rotary head, and large turf rotor. The different types are suited to different applications.
- Irrigation methods commonly used in agronomic settings are hose traveler systems, center-pivot irrigation, trickle or drip irrigation, and surface irrigation.
- Water pressure in an irrigation system is determined by a number of factors, including slope and elevation of the land, pipe diameter, and water flow rate.
- In an effective irrigation system, a uniform water pressure is maintained.
- Irrigation system design takes into consideration the amount of water needed by the crop as well as the physical components and how they will be controlled or timed.
- Irrigation lines must be properly winterized before the system is exposed to freezing temperatures.
- When repairing irrigation pipes, a series of couplings and fittings may need to be adjusted or installed. The pipe must be cut at a straight angle and installed properly to the fitting.
- Always wear personal protective equipment (PPE) when working on an irrigation system.

Words to Know

Match the key terms from the chapter to the correct definition.

A. backflow preventer
B. basin irrigation
C. center-pivot irrigation
D. detention pond
E. drip irrigation
F. hose traveler system
G. irrigation zone
H. lateral line
I. main line
J. polyethylene pipe
K. reservoir
L. retention pond
M. spray pattern
N. surface irrigation
O. swale
P. turf rotor

1. A common form of surface irrigation in which the crop is flooded for a brief period of time. A pump, gravity, and the slope of the land are used to distribute the water from a single location.
2. A pond used to capture and filter excess irrigation and rainwater. It serves as a flood control mechanism and may dry out completely.
3. The area that one sprinkler head can cover.
4. A type of irrigation system that uses a cart equipped with a high-pressure irrigation nozzle.
5. An irrigation system device used to ensure that water pumped by the irrigation system is not returned to the ground water in the well, or water in the main line.
6. A large-diameter pipe that carries water to the lateral lines of an irrigation system.
7. A type of ditch used in production agriculture that drains excess water away from crops.
8. A group of sprinklers that is turned on to water a specific area.
9. An irrigation system in which water is emitted at a slow pace that ensures the plants area able to take up the water.
10. Agricultural irrigation system in which the sprinkler system moves in a circular pattern across the field.
11. A soft, flexible irrigation system pipe that can be used on top of the ground; usually black.
12. A manmade or natural lake that is used for a water supply.
13. A removable sprinkler head used on large grassy areas such as athletic fields and golf courses.
14. A pond used to capture and filter excess irrigation and rainwater; holds a permanent pool of water.
15. The process of flooding entire rows or fields of crops. The process uses gravity and the slope of the field to disperse the water.
16. A secondary irrigation line that carries water from the main line to the sprinkler head or emitter.

Know and Understand

Answer the following questions using the information provided in this chapter.

1. What is irrigation?
2. What event caused universities and government agencies to conduct research to better understand crop techniques and principles?
3. Briefly describe the water cycle.
4. When a well cannot keep up with the demand for irrigation water, the well water can be pumped into a small, open body of water called a(n) _____.
5. Treated wastewater that is usable for watering landscape and planting beds is called _____ water.
6. What is the difference between a retention pond and a detention pond?
7. The most common type of pipe used in irrigation systems is _____ pipe.
   A. galvanized
   B. PEX
   C. PVC
   D. aluminum
8. What is the purpose of a rain sensor in an irrigation system?
9. Which type of sprinkler head is commonly used for irrigation by nurseries?
10. List the four main categories of sprinkler heads.
11. List three types of irrigation systems common in agronomic settings.
12. A(n) _____ is a tool that can be used to help locate breaks in irrigation wire.
13. Water pressure is measured in _____.
14. How is precipitation rate used when an irrigation controller is being programmed?
15. In order to join two pieces of PVC, the pipe should be cleaned with _____ and bonded with _____.
16. Why must thread tape be wrapped in a clockwise direction around the male end of galvanized pipe?
17. What PPE should be worn by someone who is repairing irrigation lines?

STEM and Academic Activities

1. **Science.** Investigate the physical characteristics of the soil in a local grassy area. Write a report describing the characteristics of the soil and how this would affect water requirements of a crop.
2. **Technology.** Conduct research to determine how technology is helping farmers and landscapers to use water more efficiently. Write a brief report on your findings.
3. **Engineering.** Select a crop and design an irrigation system suitable for that crop. Draw the irrigation system to scale. Include the location of the couplings, irrigation heads, and nozzles.
4. **Math.** Calculate how much it would cost to install an irrigation system that is 100' long if pipe costs $1.19 per foot. Include a sprinkler head every 20'. Each sprinkler head costs $10. Add $15 to account for primer, glue, and couplings. Research current labor costs and include the cost of labor in your total.
5. **Language Arts**. Research *dowsing*, a technique that has been used for hundreds of years to locate groundwater sources. Participate in a class discussion regarding the history and scientific validity of dowsing.
6. **Social Science.** Select a civilization from the past and research the irrigation practices of the time period. Compare and contrast the ancient methods to the common irrigation practices of today.

Thinking Critically

1. You are the foreman of a landscape crew and have been called to the jobsite to solve a problem. The irrigation controller will not turn off Zone 5 of the irrigation system. All other zones are working properly. Zone 5 is now flooded and is wasting water. Describe possible solutions for this issue.
2. You have been asked by a local farmer to recommend an irrigation system for his tomato crop. The farmer is environmentally conscious and wants to install an irrigation system that is as sustainable as possible. Which type of irrigation system would you recommend and why?
3. While installing an overhead irrigation system in a field of vegetables, you notice that the sprinkler head at the very end of the irrigation line is not operating properly due to lack of water pressure. As the installer, what could you do to increase the water pressure for the end sprinkler head?

CHAPTER 26
Environmental Controls

Chapter Outcomes

After studying this chapter, you will be able to:
- Explain the need for environmental controls in agricultural structures.
- Discuss heat transfer, psychrometrics, and control instruments as they relate to managing building environments.
- Discuss environmental controls related to housing livestock.
- Describe environmental controls used for grain storage.
- Describe environmental controls used for perishables and other crop storage.
- Explain how refrigeration is used to maintain storage temperatures for perishable crops.
- Describe how evaporative cooling is used for environmental control in animal housing or greenhouses.
- Describe ventilation and exhaust systems used in agricultural structures or work areas.

Words to Know

dew point	humidity	psychrometric chart	thermodynamics
dry-bulb temperature	humidistat	psychrometrics	thermometer
enthalpy	hyperthermy	sensible cooling	thermostat
environment	hypothermy	sensible heat	volume
evaporative cooling	latent heat	sensible heating	wet-bulb temperature
heat	manometer	specific humidity	
homeostasis	mechanical	thermal equilibrium	
homeothermic	refrigeration	thermal radiation	

Before You Read

As you read the chapter, record any questions that come to mind. Indicate where the answer to each question can be found: within the text, by asking your teacher, in another book, on the Internet, or by reflecting on your own knowledge and experiences. Pursue the answers to your questions.

While studying this chapter, look for the activity icon to:

- **Practice** vocabulary terms with e-flash cards and matching activities.
- **Expand** learning with interactive activities.
- **Reinforce** what you learn by completing the end-of-chapter questions.

www.g-wlearning.com/agriculture

Elena Elisseeva/Shutterstock.com

All structures are designed and built with the end use in mind. A home is designed to provide comfort and protection from the elements. A bank is designed for security and customer service, and a school is designed to create a comfortable learning environment. An *environment* consists of the surroundings or conditions that affect animals, plants, or objects. Each one of these structures has specific systems and design elements to create the desired environment. The same principles apply to agricultural structures. Each structure must be designed for its intended use. As with any business, structures that are properly built and maintained are vital to an agricultural business. Agricultural structures are used for purposes such as:

- Housing animals in healthy environments.
- Storing animal feed and bedding materials.
- Storing crops and maintaining their quality.
- Safely containing or processing waste materials.
- Storing agricultural equipment.
- Providing safe work areas.
- Providing healthy growing space for plants.
- Securely storing chemicals.
- Providing clean processing and holding areas.

To fulfill these purposes, the environment within the structures must be properly controlled and maintained, **Figure 26-1**. The elements or conditions that require control include temperature, moisture, light, dust, odors, noise,

Aumsama/Shutterstock.com

Figure 26-1. Using a heat lamp to warm the area for these piglets is an example of controlling the environment.

airflow, and air quality. This chapter covers the design factors and equipment used to create and maintain environments in agricultural structures as well as those scientific theories relevant to indoor environmental control.

Thermodynamics

Before being able to control an environment, you must understand how the nature of heat and energy affects an environment. *Heat* is a form of energy that makes atoms move. *Thermodynamics* is the branch of science dealing with the relationships between heat and other forms of energy. Simply put, a thermodynamic system is a system that interacts and exchanges energy with the area around it. Thermodynamics has several basic principles and concepts.

- *Thermal equilibrium* is a balance of temperature between two bodies. For example, if two bodies at different temperatures (one hot and one cold) are in physical contact, the temperature of both bodies will change. The hot body will tend to become cooler, and the cold body will tend to become warmer. Eventually, both bodies will reach the same temperature. They will be in a state of thermal equilibrium.
- The law of conservation of energy states that energy can neither be created nor destroyed, but it can be changed from one form to another. For example, a toaster changes electrical energy into heat energy.
- Natural heat flow means that heat tends to move from warm areas to cool areas. For example, if one end of a metal rod is heated, heat will flow from the hot end to the cool end. See **Figure 26-2**.

Heat Transfer

All matter contains some thermal energy (heat energy) that causes atoms to move. More heat causes more movement of atoms. Less heat causes less movement of atoms. Heat transfer is the movement of this thermal energy from one object to another object with a different temperature. The objects may be two solids, a solid and a liquid or gas, or two liquids or gases. The objects may also be living organisms or their surrounding environments.

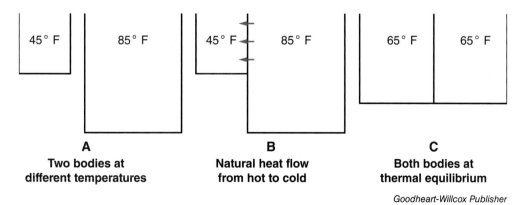

Goodheart-Willcox Publisher

Figure 26-2. This figure shows natural heat flow. A—Two bodies at different temperatures. B—Natural heat flow from hot to cold. C—Both bodies at thermal equilibrium.

Heat energy will move between objects through conduction, convection, or thermal radiation:

- Conduction—the flow of heat through a substance by the molecular vibrations in that substance, **Figure 26-3A**.
- Convection—the movement of heat from one place to another by way of a liquid or gas, **Figure 26-3B**.
- *Thermal radiation*—the transfer of heat by electromagnetic waves, **Figure 26-3C**.

It is important to understand heat transfer when designing, installing, and maintaining environmental control systems. The many factors that affect heat transfer will help determine the best types of construction materials, mechanical systems, and controls to be used. For example, construction materials can be thermal insulators or thermal conductors. Building materials can absorb, reflect, or transmit thermal energy. For instance, dull black paint will absorb radiant heat energy from the sun. However, a shiny aluminum surface will reflect a high percentage of the radiation. Greenhouse glass or polyethylene surfaces allow some heat waves to pass through them, transmitting the heat to the inside of the building, **Figure 26-4**.

Psychrometrics

Understanding psychrometrics is required to manage building environments successfully. *Psychrometrics* is the study of science involving the thermodynamic properties of air and water vapor. Air is a complex mixture of gases, water vapor, and heat. The amount of water vapor in the atmosphere is called *humidity*. Management of air and water vapor can be accomplished through heating, cooling, and ventilation systems. The seven physical properties of air used in psychrometrics are:

- *Specific humidity*—the moisture content of air as a measure of the actual amount of water held in the air in the form of vapor.
- Relative humidity—the ratio of the maximum amount of water the air could hold if it were fully saturated to the amount of water that is in the air. It is expressed as a percentage.
- *Dry-bulb temperature*—the temperature of air measured with a standard *thermometer* an instrument used to measure temperature. Dry-bulb temperature does not take moisture content into account. It measures

A *Andor Bujdoso/Shutterstock.com*

B *Natan86/Shutterstock.com*

C *Aphelleon/Shutterstock.com*

Figure 26-3. A—An example of conduction is when a metal rod heated on one end transfers heat to the other end through molecular vibration. B—With this blow dryer, heat is transferred through the movement of heated air, which is called convection. C—Heat rays travel across space to warm the surface of the Earth through radiation.

only sensible heat. *Sensible heat* is energy that changes the temperature of a substance with no change in the physical state of the substance.

- *Wet-bulb temperature*—the temperature of air measured with a standard thermometer that takes into account the moisture content of the air. Wet-bulb temperature takes into account both sensible heat and latent heat. *Latent heat* is energy that changes a substance's physical phase from one state to another (liquid, gas, or solid). The higher the moisture content, the more heat energy in the air. The drier the air is, the greater the evaporation of moisture, leading to a greater cooling effect.
- *Dew point*—the temperature at which the moisture in the air begins to condense or form droplets that are too large to remain suspended in the air. The sensible temperature at which air is fully saturated is 100 percent relative humidity. Note that since cold air has less heat energy, it cannot hold as much water vapor as warm air can hold.
- *Enthalpy*—the total heat energy in the air expressed in Btu per pound of air. This includes the sensible heat of dry-bulb temperature and the latent heat of wet-bulb temperature.
- *Volume*—the amount of space occupied by a substance or object. In this case, volume refers to a pound of air at standard atmospheric pressure. It is expressed in cubic feet per pound of air. The volume may decrease if atmospheric pressure increases. Volume may increase with an increase in temperature and humidity.

Psychrometric charts are used to understand air and its manipulation. The *psychrometric chart* serves as a graphical representation of the seven physical properties of air, **Figure 26-5**. A psychrometric chart is designed to be used with any two known properties of the air. The remaining five unknown properties can be determined using a stated point to reference them on the chart. It is common to find high-temperature, normal-temperature, and low-temperature charts. Heating and ventilation rates may also be estimated using the chart.

A — *Foto by M/Shutterstock.com* B — *Foto by M/Shutterstock.com*

Figure 26-4. A—Greenhouse glass or polyethylene surfaces allow some radiation wavelengths to pass through them, transmitting the heat to the inside of the building. B—The amount of radiation allowed to enter the greenhouse may be limited through the use of coverings, such as this green netting.

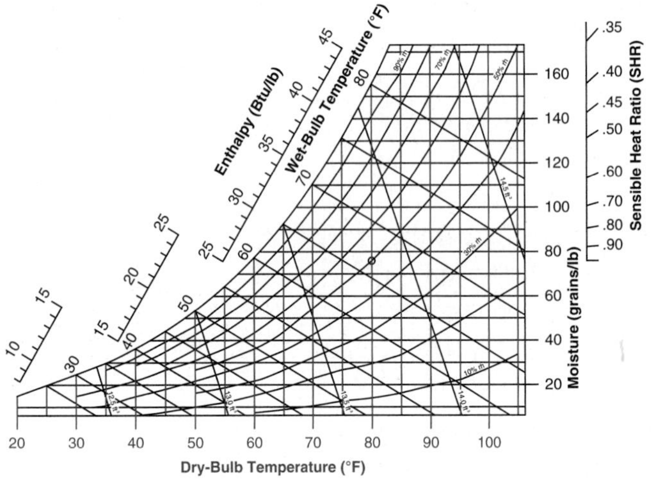

A

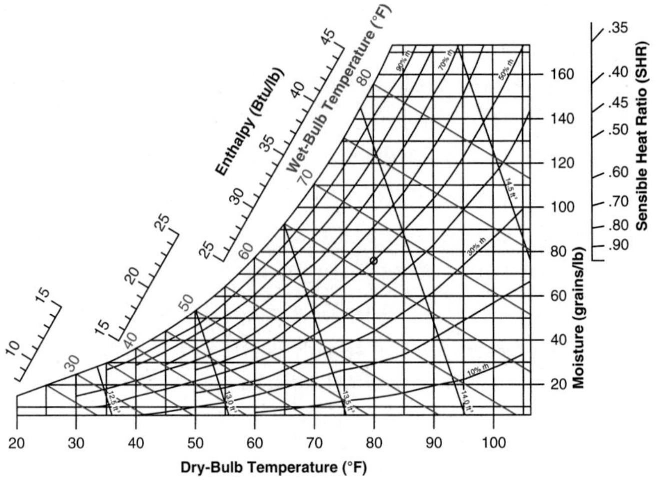

B

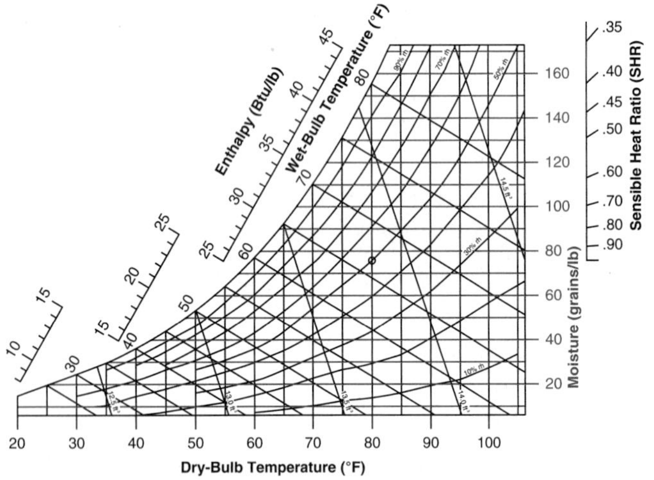

C

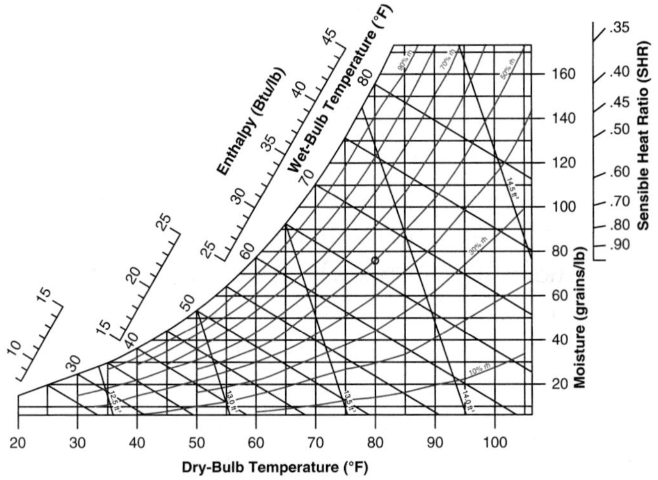

D

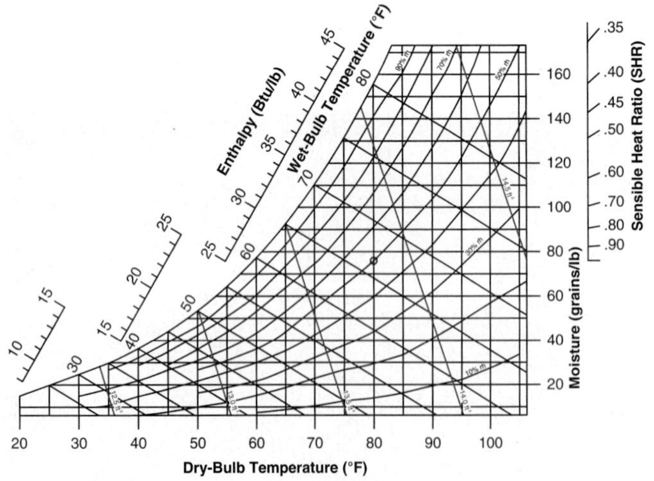

E

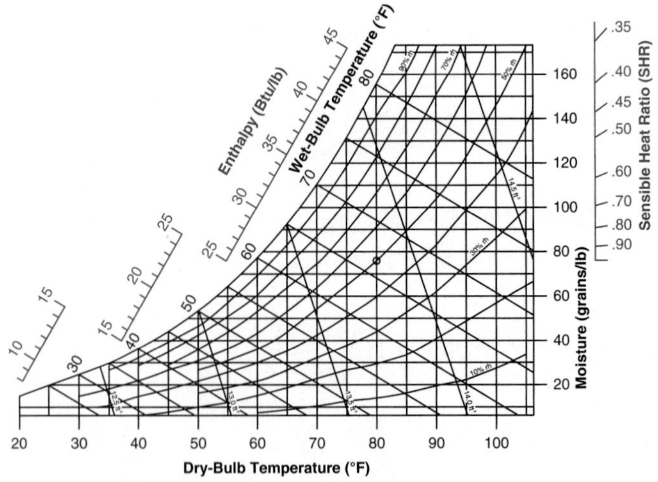

F

Goodheart-Willcox Publisher

Figure 26-5. A—The lines on this psychrometric chart represent constant dry-bulb temperatures. B—These lines represent constant wet-bulb temperatures. C—These lines represent constant specific humidity levels. D—These lines represent constant relative humidity levels. E—These lines represent constant specific volumes. F—These scales are used to determine the total heat energy in the air sample and the ratio of sensible heat to latent heat.

AgEd Connection: Environmental and Natural Resource CDE

As our world population increases and our land mass remains unchanged, the age-old challenge of producing enough food keeps getting more difficult. Food storage, animal housing, waste processing, and efficient plant growth are concerns that are met, at least in part, by environmental control. If research into the ways temperature, moisture, light, and other factors influence food production is interesting to you, take some time to investigate the content of this chapter.

An environmental and natural resources CDE will allow you to gain competitive experience in this area. Ask your agriculture teacher how you can get started with an environmental supervised agricultural experience program.

Steve Boyko/Shutterstock.com

Environmental Controls and Instruments

A thermometer (an instrument used to measure temperature) can be analog or digital and can use a variety of sensing methods. Thermocouple probes are one of the most widely used electronic temperature sensors. Different types of thermocouples can be selected for different applications.

Common temperature scales on thermometers are Celsius, Fahrenheit, and Kelvin. Converting between these scales can be accomplished using formulas. Some thermometers display temperature in Fahrenheit and Celsius or use a button to switch the scales being displayed. A thermometer used for a specific purpose may need to be of a certain build for a given environment, **Figure 26-6**.

In some situations, it is not safe or convenient to measure temperature by being in close contact with an area or an object. This is when an infrared thermometer should be used to measure temperature from a distance, **Figure 26-7**.

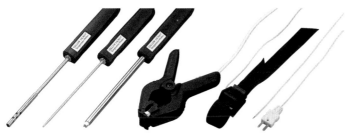
Ritchie Engineering Co., Inc. – YELLOW JACKET Products Division

Figure 26-6. A variety of temperature probes are used with electronic thermometers.

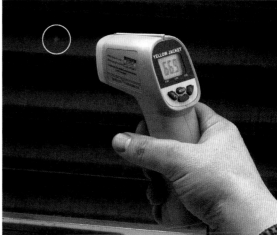
Ritchie Engineering Co., Inc. – YELLOW JACKET Products Division

Figure 26-7. An infrared thermometer allows temperature to be taken from a safe or convenient distance.

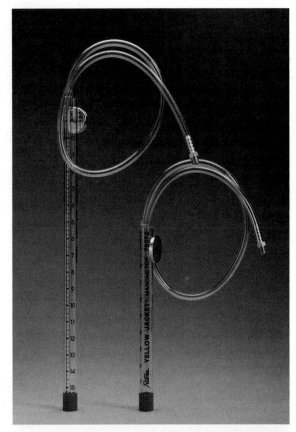

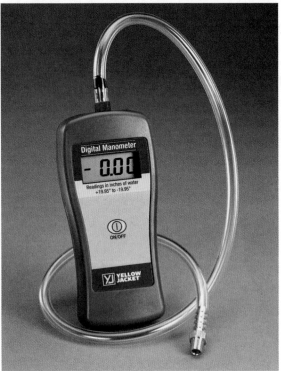

Ritchie Engineering Co., Inc. — YELLOW JACKET Products Division

Figure 26-8. Analog and digital manometers are used in different low-pressure measurement applications.

A *thermostat* is a device used to control a heating or cooling system. A thermostat senses temperature or pressure and switches a heating or cooling system on or off to maintain a desired temperature. Sensing devices vary by system. The thermostat is different from a thermometer in functionality. A thermostat serves as an automatic control switch, while a thermometer serves as a measuring instrument.

A *humidistat* is a control device that senses the relative humidity of the air. These devices are used to control humidifiers and dehumidifiers to maintain different environments, such as a greenhouse or climate-controlled warehouse. These devices rely on hygrometers much like a thermostat relies on a thermometer.

A *manometer* is an instrument used to measure very low pressure in air, gas, or liquid (generally new atmospheric pressure). Some manometers are used to determine the difference between two pressures, such as the pressures before and after an air filter in a ventilation system. Some are filled with fluid and marked with gradients on the fluid housing. Others may have digital readouts. Typical units of measure are inches of mercury or inches of water, **Figure 26-8**.

A variety of alarms and devices monitor different variables in an agricultural structure. They may be programmed to alert personnel in the event of a temperature change or a power outage. Typical alarm systems autodial telephone numbers to notify an individual of the problem. Some devices provide remote access that makes it possible to call the system using a telephone. The caller can check the status of all monitored conditions and make program changes. Technology updates have given increased access and control from remote locations by using the Internet to access the system. The purpose of using these tools is to detect problems before they become disasters.

Housing Livestock

Structures for livestock must be designed for the type of livestock being housed. For example, newborn piglets generate less heat than an adult cow. The piglets will likely require a supplemental heat source for their survival. The adult cow may require a means of keeping its environment cooler.

Body Temperature

Most livestock are *homeothermic*. This means they maintain a nearly constant core body temperature. The maintenance of this internal temperature, despite external changes, is called *homeostasis*. A simple example of homeostasis is your own body's ability to maintain a temperature around 98.6°F (37°C), regardless of the ambient (surrounding) temperature. However, prolonged exposure to less than ideal circumstances (too hot, cold, dry, or wet) will interfere with a homeothermic body's ability to maintain its internal temperature. Livestock are susceptible to hypothermy and hyperthermy. *Hypothermy* is when the body temperature falls below its norm. When the body temperature rises above its norm, the condition is called *hyperthermy*.

The goal for livestock structures is to keep the temperature in a range that minimizes stress on the animals. This can be accomplished by maintaining the appropriate temperature range for the specific livestock being housed and avoiding extreme temperature and humidity changes. If animals are too hot, their production will be reduced. In some cases, such as in swine production, overheating can be fatal. Young and immature animals may also suffer from overheating. For large animals, cool temperatures typically do not affect production as much as hot temperatures. Young or immature animals are affected more by cold temperatures than are mature animals, **Figure 26-9**.

Type of Animal	Animal Temperature	Optimum Temperature Range
Swine		
Lactating sow	60°F (15.6°C)	50°F–70°F (10°C–21°C)
Litter (newborn)	95°F (35°C)	90°F–100°F (32.2°C–37.8°C)
Litter (3 weeks)	75°F (24°C)	70°F–80°F (21°C–26.7°C)
Pre-nursery pigs (12–30 lb)	80°F (26.7°C)	75°F–85°F (24°C–29.4°C)
Nursery pigs (30–75 lb)	75°F (24°C)	70°F–80°F (21°C–26.7°C)
Growing-finishing hogs	60°F (15.6°C)	50°F–70°F (10°C–21°C)
Gestation sow and boar	60°F (15.6°C)	50°F–70°F (10°C–21°C)
Dairy		
Cows	50°F (10°C)	45°F–70°F (7.2°C–21°C)
Calves	70°F (21°C)	45°F–80°F (7.2°C–26.7°C)

Goodheart-Willcox Publisher

Figure 26-9. This chart shows temperature optimums and ranges for confinement for swine and dairy animals.

One of the primary ways livestock can reduce body heat is through their respiratory tract (by panting). Animals that are exposed to drafts may lose body heat at a more rapid rate than those that are not exposed to drafts. Animals can also lose body heat by coming into contact with cool surfaces, such as a concrete floor.

Air Quality and Ventilation

Livestock housing must be ventilated to remove heat and moisture given off by the animals as they breathe. The harmful gases that can develop in the building should also be removed. Excess moisture or too little moisture can create conditions that are favorable for respiratory diseases. High humidity can be damaging to equipment and machinery by creating a corrosive environment. Humidity also has the effect of increasing the enthalpy (total heat content) in a livestock building. Air that is high in humidity will feel hotter than air that is the same dry-bulb temperature but with lower humidity. High air temperatures can lower production. This requires bringing cooler fresh air into the building. Depending on climate and the type of livestock, supplemental heating may also be needed because the animals may lose more heat than they are able to produce. This is called a heat balance, and it is dependent on building features, air conditions, climate, and livestock type. Animals will have different metabolic rates and respiration rates that affect their housing needs. A livestock building ventilation system provides for moisture and odor control in winter and for heat control the rest of the year. See **Figure 26-10**.

Type of Animal	Surrounding Temperature	Heat Produced (Btus/hr)
Swine		
Sow and litter	60°F (15.6°C)	1200
Sow and litter	70°F (21°C)	1100
Pre-nursery pig (12–30 lb)	80°F (26.7°C)	75
Nursery pig (30–75 lb)	75°F (24°C)	125
Growing pig (75–150 lb)	60°F (15.6°C)	250
Finishing hog (150–220 lb)	60°F (15.6°C)	350
Gestating sow (325 lb)	60°F (15.6°C)	550
Boar (400 lb)	60°F (15.6°C)	700
Dairy		
Cow	50°F (10°C)	3220
Calf	70°F (21°C)	600

Goodheart-Willcox Publisher

Figure 26-10. This chart shows body heat produced by swine and dairy animals under certain surrounding temperatures.

Mechanical Ventilation for Animal Housing

Mechanical ventilation is created using fans, thermostats, and building openings for air exchange. Axial flow fans or centrifugal fans are commonly used to exchange air in a building with fresh air from outside, **Figure 26-11**. Fans can be used in a positive pressure system where air is blown into a building. In a negative pressure system, fans blow air out of the building, and fresh air is drawn in through inlets on the opposite side. Negative pressure is typically more common. Young animals, such as dairy calves, may benefit from a positive pressure system if they need low ventilation capacity. Ceiling fans and circulation fans mounted inside the building are used to improve uniformity of air conditions. Mechanical ventilation provides higher control over temperature and airflow within the facility than natural ventilation.

Installation expense, operating cost, and the amount of heat loss during winter should be considered when selecting a mechanical ventilation system. A supplemental heating system could be needed to keep the animals sufficiently warm when ventilating to control for humidity during cold temperatures.

Fans that cycle on and off should be equipped with louvers. Louvers are commonly used where control is needed to maintain air temperature and humidity levels in housing that is designed for animals that are sensitive to cold and drafty conditions. Examples include swine farrowing and nursery buildings, lambing sheds, warm dairy cattle barns, calf housing, and milking parlors. Louvers prevent the entry of cold air and other objects during periods when the fan is not operating. They should also have protective guards or screening to prevent injuries to animals or to people who work around them.

Budimir Jevtic/Shutterstock.com

Figure 26-11. An axial fan is often used in livestock buildings.

Fan Designs and Operation

A ventilation fan causes air to move through the building. Their designs can vary by blade pitch, number of blades per fan, operating speeds, and required motor horsepower. When a fan motor turns the blades, the blades move through the air at an incline. The inclined surface passing through the air pushes the air forward. The air is scooped up by the leading edge of the fan blade and set in motion as it passes over the surface. Most fan blades are mounted on an inclined angle of less than 90°. The motor constantly turns the blades in a single direction to direct air out the front of the fan fixture.

A basic axial fan is a propeller/impeller that is mounted on a shaft within a housing. The housing reduces the drag effect on the tips of the blades. The shaft can be directly mounted on the motor or rotated with a pulley system using a belt. The housing provides direction for the airflow. An axial fan gets its name by the way it blows air across its axis in a linear direction. The blades of these fans force air to move parallel to the shaft, **Figure 26-12**.

Centrifugal fans work by throwing air away from its blades. Centrifugal fan blades can be straight, backward, or forward curved. The blade wheel is mounted within a housing, **Figure 26-13**. A centrifugal fan provides high-velocity airflow but with lower volume than an axial fan. An axial fan provides lower velocity but higher volume of airflow than a centrifugal fan.

Supplemental Heating for Animal Housing

The heat produced by animals can help to warm a building. However, young animals may not produce enough heat when buildings are ventilated to remove excess moisture. Additionally, high heat loss through poor insulation and drafts may require supplemental heat. Four basic types of heating systems used for livestock housing are overhead radiant heating, floor heating, unit or space heating, and make-up air heating.

Some producers think that a combination of systems is a good option in certain housing types. A thermostat controls temperature settings based on management decisions to improve animal health and productivity. Alarms and monitors can be used to alert the manager if heating units have failed or temperature settings have fallen below acceptable limits.

Radiant heaters can be used in feeding areas for young animals or in sleeping areas for large animals instead of bedding material. Electric resistance cable or boiler pipes can be buried in concrete to provide floor heating, **Figure 26-14**. Temporary heating pads can be placed on top of

worradirek/Shutterstock.com

Figure 26-12. Axial fans can be used in a large greenhouse to provide ventilation.

ra3rn/Shutterstock.com

Figure 26-13. As the motor turns the blade wheel in this centrifugal fan, air that is drawn in from the side is thrown outward. Its flow is directed by the housing.

existing flooring. Both overhead and floor heating can be used in areas for young animals. One may be removed as the animals become acclimated a few days after birth. Animals may move closer or farther away from overhead heaters, depending on their temperature needs.

Wall- or ceiling-mounted unit heaters (either vented or unvented) can be used to heat and circulate building air. These heating units typically create heat by electric resistance or fuel-fired furnaces. A large disadvantage is high maintenance need for these units. An accepted practice is to place the heating units so that they blow along the coldest wall in order to intercept cold drafts and reduce radiant heat losses to the wall. When more than one heater is used, they are often arranged to create a circular air motion within the room.

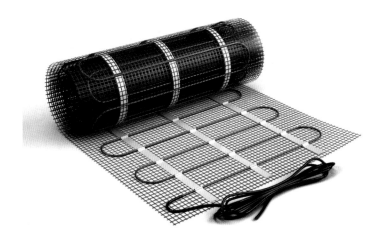

Sashkin/Shutterstock.com

Figure 26-14. To provide floor heat, electric heating cables can be installed under flooring or within poured concrete.

Make-up air heaters use direct gas-fired burners to heat incoming fresh air brought in during ventilation of the building. They are typically located outside the livestock building. Adjusting the burn rate of fuel in the burner maintains a constant temperature as the air leaves the blower. These heaters are sized carefully to prevent overrunning the cold weather exhaust fans.

Efficient heaters that are unvented provide a uniform building temperature, but they may add more water to the air through the combustion process. The extra moisture must be considered in determining the appropriate cold weather ventilation rate.

Some rooms of a building designed for people, such as offices, mud rooms, or laundry areas, may use forced-air furnaces in combination with central air conditioning to heat and cool air that is recirculated through the area.

Thermostats controlling heaters and ventilation fans should be located away from ventilation air inlets and direct sunlight. They should also be protected from animal contact. A recommended practice is to place furnace and fan controls in the center of the room approximately 4′ off the floor. Use an accurate thermometer for adjusting the heater controls, and read from this same thermometer to check the accuracy of fan thermostats.

Building Design Considerations

Systems with electronic controls are not always needed to maintain conditions in a building. Energy savings and reduction of building costs can be accomplished by using natural ventilation and managing sun exposure. Buildings may be constructed to use prevailing winds and sun location to help heat and cool the structure. These types of building may serve to provide shelter for animals or equipment during bad weather for short periods of time. Sheds with open fronts facing away from prevailing winter winds may benefit from the cooling effects of summer air currents. A general rule is to locate the long dimension of a structure east to west to allow maximum sun exposure on the south. To minimize late afternoon summer sun exposure, end walls should face toward the west. A roof overhang can keep out summer sun that comes from a high angle.

Grain Storage

Stored grain requires careful management of environmental conditions to inhibit germination, mold growth, and insect activity. Moisture, temperature, and maturity of the grain must be monitored to maintain the best environment for long- and short-term storage, **Figure 26-15**. Harvesting grain that has matured and is at the appropriate moisture content can reduce the need for drying.

Aeration is used to maintain cool temperatures throughout the grain. The low temperatures and uniform airflow limit hot spots in the grain where mold and insects may create grain quality problems. See **Figure 26-16**.

Corn Temperature	**Corn Moisture Content**					
	13%	**14%**	**15%**	**16%**	**17%**	**18%**
40°F (4.4°C)	150.0	61.0	29.0	15.0	9.4	6.1
50°F (10°C)	84.0	34.0	16.0	8.9	5.3	3.4
60°F (15.6°C)	47.0	19.0	9.2	5.0	3.0	1.9
70°F (21°C)	26.0	11.0	5.2	2.8	1.7	1.1
80°F (26.7°C)	15.0	6.0	2.9	1.6	0.9	0.9

Goodheart-Willcox Publisher

Figure 26-15. This chart shows maximum storage time in months for shelled corn based on 0.5 percent maximum dry matter loss (calculated on the basis of USDA research at Iowa State University).

Bukhanovskyy/Shutterstock.com

Figure 26-16. Aeration is used to maintain cool temperatures in grain storage buildings.

Grain Dryers

Moisture content recommendations for storage can vary between 13 percent and 16 percent, depending on the crop. Grain dryers may be used to remove excess moisture from grain after it is harvested. Drying grain can be accomplished through high-temperature or low-temperature natural-air drying or layer drying. Natural-air drying is limited by the outside air temperature and moisture content. The equipment needed for natural-air drying includes a bin with side walls less than 18′ high. The bin should have fully perforated flooring and a fan with a capacity of 0.5 to 1.0 cfm/bu (cubic feet of air per minute per bushel). There should be exhaust vents on top of the bin that will provide at least 1 square foot of vent area per 1,000 cfm of airflow. Layer drying is similar to natural-air drying, but wet grain is added to the bin in layers as it dries, **Figure 26-17**.

For grain that is harvested at slightly higher moisture content, high-temperature dryers can be used to dry the grain quickly. The common dryer types include batch-in-bin, recirculating bin dryer, continuous-flow bin dryer, and column dryers, **Figure 26-18**. Grain stirring devices can be used to uniformly dry grain.

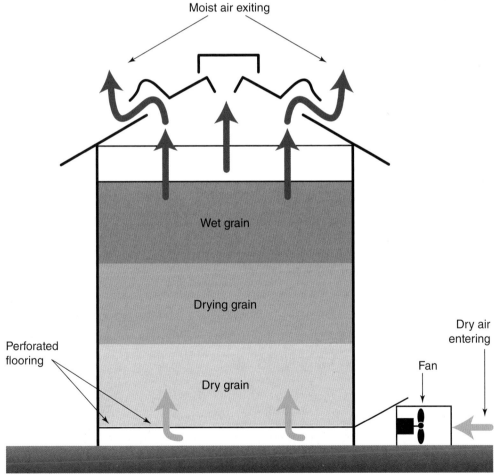

Goodheart-Willcox Publisher

Figure 26-17. This diagram shows the layer drying process. As grain dries, more wet grain is added to the top of the stack.

Figure 26-18. This continuous-flow column grain dryer is used for wheat, barley, maize, and similar grains and seeds.

High-temperature drying is not limited by the outside air temperature and moisture content. However, high-temperature drying requires more energy, which increases costs associated with production and limits the profit return when the grain is sold. Grain that is dried using high temperatures must be cooled to outdoor temperatures soon after drying. However, rapid cooling causes stress cracks in kernels that lead to breakage in handling. These conditions may lower grain quality premiums when sold. Grain breakages increase fines (broken pieces), which creates conditions for insects and mold to become active.

Using a combination of drying methods may reduce energy costs and lower the likelihood of breakage. Dryeration is a process where hot grain is removed from the dryer with a moisture content slightly higher than desired for storage. It is then transferred to a different bin to temper (rest) without airflow. This allows the moisture content to equalize. After being tempered, a cooling fan is turned on. This cools the grain and drys it to the final moisture content before it is moved to final storage, **Figure 26-19**.

Sensor Cables

Temperature and humidity sensor cables may be installed in the bin to monitor grain. This helps managers detect conditions that may lead to spoilage or shrinkage of grain before either happens. These cables can be connected to an alarm system or controller that will activate fans or heaters, depending on the situation. Portable handheld probes can also be used to monitor grain in storage.

Figure 26-19. A combination grain drying installation may be used to reduce energy costs and lower the likelihood of breakage.

Hazards Related to Grain Storage

Spoiled grain can form clumps in the bin, making handling difficult. Workers may need to remove or break up the clumps. Toxic gases can be given off by the spoiled grain, posing a breathing hazard for workers. A grain bridge is an area of grain in a bin that appears solid. However, the grain underneath has settled or fallen away. This can create a dangerous situation for workers who might stand on the grain bridge, fall through, and be entrapped in the grain.

Grain dust is highly flammable. Dust exposed to sparks or excessive heat can explode or catch fire. Equipment used near grain should be monitored to avoid excessive heat buildup or sparks. Electrical equipment designed to prevent explosions may be needed when working with grain dust in confined areas. Grain cleaners can be used to reduce the amount of dust and fines in stored grain.

> **Safety Note**
>
> Grain dusts and manure gases can present explosion hazards. Follow OSHA grain dust safety recommendations found in various OSHA factsheets and OSHA standard 29 CFR 1910.272. Also refer to the information and recommendations found in DHHS (NIOSH) Publication Number 90-103, *Preventing Deaths of Farm Workers in Manure Pits*.

Perishables and other Crop Storage

Agricultural products, such as fresh fruits and vegetables, are living and biologically active. Storage design considerations must account for how to maintain product qualities, such as appearance, texture, moisture level, nutritional content, and flavor. Fruits and vegetables will continue to respire long after harvest. This results in the conversion of starches into sugars and then sugar into oxygen, water, and heat. This leads to degradation of product quality and resistance to microbial activity.

Temperature and relative humidity should be considered when designing storage atmosphere conditions for fresh fruits and vegetables. Relative humidity should be monitored and controlled in storage. A hygrometer or a sling psychrometer should be used to monitor humidity. A variety of methods can be used to regulate conditions of storage air for fruits and vegetables. These methods include:

- Operating a humidifier in the storage area.
- Regulating air movement and ventilation in relation to storage room load.
- Maintaining refrigeration coil temperature within 2°F of the storage room air temperature.
- Using moisture barriers in the insulation of the storage room or transport vehicle and in the lining of the packing containers.
- Wetting the storage room floor.
- Using crushed ice to pack produce for shipment.
- Sprinkling leafy vegetables, cool-season root vegetables, and immature fruits and vegetables with water.

Storage temperature recommendations for cool-season vegetables range from 32°F to 35°F (0°C to 2°C.) Warm-season crops storage temperatures range from 45°F to 55°F (7°C to 13°C). A recording thermometer can be helpful in determining whether storage facilities are maintaining ideal conditions. The thermometer used for monitoring should not be the thermostat controlling the refrigeration equipment.

Ethylene gas is generated from the ripening process of fruits and vegetables. Some crops are sensitive to this gas and will ripen quickly in its presence. Ethylene gas will build up in the storage area and must be controlled or removed to keep crops from overripening. Media containing potassium permanganate can be placed in the storage building or a scrubber can be installed to reduce or eliminate ethylene gas. This typically requires using a controlled atmosphere storage facility. These facilities are gas tight in order to maintain a 3 percent to 5 percent oxygen level and a 0 percent to 5 percent carbon dioxide level. Workers should not enter these facilities without the proper training and equipment due to oxygen deficiency. Carbon dioxide levels should be monitored and maintained at the recommended levels. Carbon dioxide can be removed with a caustic soda scrubber, a water scrubber, fresh hydrated high calcium spray lime, or a carbon dioxide scrubber. Nitrogen generators are used to reduce the oxygen level quickly.

Refrigeration

Refrigeration is commonly used in agriculture to maintain storage temperatures for perishable crops. *Mechanical refrigeration* is a process that uses the principles of various scientific gas laws to absorb heat into a refrigerant to produce a low temperature. Mechanical refrigeration systems include several key components, including a compressor, evaporator coil, condenser coil, metering device, and refrigerant. Many systems also use blower fans to increase heat transfer.

Refrigerant is a chemical that exists in the forms of vapor and liquid. As it circulates through a refrigeration system, refrigerant will boil into a vapor and condense into a liquid multiple times in different parts of the system. As it boils, it absorbs heat. As it condenses it expels heat. Refrigerant boils or condenses based on pressure applied to it.

A mechanical refrigeration system is divided into two sides: the high side and the low side. When a system is operating, the high side is under high pressure, and the low side is under low pressure. It is the pressure and temperature of a refrigerant that determines whether the refrigerant absorbs heat or expels heat. Refrigerant under low pressure will be at low temperature and can absorb heat. Refrigerant under high pressure will be at high temperature and can expel heat, **Figure 26-20**.

A compressor compresses (squeezes) incoming refrigerant vapor to a high pressure as it enters the high side. This hot refrigerant circulates through the condenser coil and loses its heat to ambient air. Often, condensers are located outside. Outside air blown by a fan over the coil helps to increase heat transfer. As refrigerant in the condenser cools, it will change into a liquid after it has expelled enough heat.

Liquid refrigerant on the high side flows through a metering device into the low side of the system. A metering device does two important tasks. It passes refrigerant into the evaporator, and it reduces refrigerant pressure to

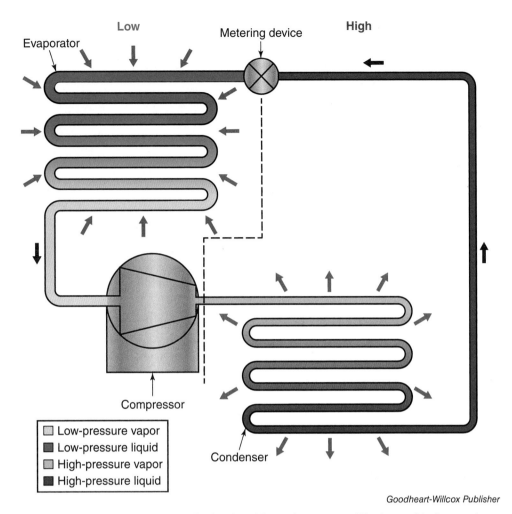

Figure 26-20. This is a diagram of a basic refrigeration system. The legend indicates the pressure and physical state of the refrigerant throughout the system.

low pressure. There are different types of metering devices. All use a small orifice that the liquid under high pressure has to squeeze through.

Between the metering device and the compressor is an area of low pressure. This is because the compressor is creating suction and pulling the refrigerant through the system. As liquid refrigerant circulates through the evaporator, some of it absorbs heat and boils into a vapor. Air moved by a blower fan increases heat transfer with the evaporator. As heat is absorbed into refrigerant in an evaporator, temperature of the area drops. Refrigerant gauges installed on the high and low side of a system can be used to identify problems within the system, **Figure 26-21**.

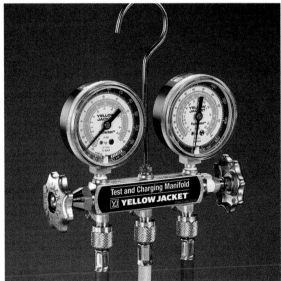

Figure 26-21. A gauge manifold is used to measure pressure in this mechanical refrigeration system.

STEM Connection

Latent and Sensible Heat

Heat is a form of energy. Temperature is the degree or intensity of heat. Latent heat is energy that changes a substance's physical phase from one state to another (liquid, gas, or solid). Sensible heat is energy that changes the temperature of a substance with no change in physical state. Perform the following experiment to better understand the difference between latent heat and sensible heat.

1. Place ice cubes in a beaker.
2. Apply heat with a hot plate or other appropriate heat source.
3. Record the temperature changes of the water at regular intervals as the ice cubes melt.
4. Record the temperature of the water when the ice is completely melted.
5. Record the temperature changes of the water at regular intervals for several minutes after the ice cubes have completely melted.

What was the water temperature while it contained ice cubes? The water temperature should have been 32°F (0°C) until the ice was completely gone. The change of state from ice to water requires heat, yet the water temperature did not change while it contained ice cubes. This is an example of latent heat causing the change of state.

What was the water temperature after the ice was completely melted? The water temperature should have begun rising after the ice cubes were

TomeK K./Shutterstock.com

completely melted. This is an example of sensible heat causing the temperature change.

Sensible heating is adding heat to air without changing its humidity ratio. Examples include heated-air grain drying or heating a room of air during the winter. **Sensible cooling** is removing heat from the air while keeping a constant humidity. An example would be air passing over a cooling coil with a surface temperature above the dew point temperature of the air. **Evaporative cooling** is a process in which water vapor is added to the air and lowers the dry-bulb temperature without altering the amount of heat in the air. This type of cooling is most effective in hot, dry climates. Cooling and dehumidifying lowers both the dry-bulb temperature and humidity during air conditioning.

Safety Note

The electrical supply for some refrigeration equipment can carry enough electrical current to cause serious injury and even death. Therefore, lockout/tagout procedures should be practiced when performing electric motor repairs or replacing electrical equipment components used in refrigeration systems.

Evaporative Cooling

In climates with low relative humidity, evaporative cooling may serve as an economical option for cooling animal housing or greenhouses. When water evaporates, heat energy is absorbed into the water vapor, reducing the dry-bulb temperature but raising humidity. The air gives up heat energy to evaporate water.

The basic components of an evaporative cooling system include a water pump, water tank, cooling pads, transfer lines, and a fan. The fan draws in air from outside through the unit. The water pump moves water from the tank and distributes it over the cooling pads. Air passing over the saturated cooling pads promotes the evaporation of water. Evaporation lowers the temperature of the incoming air.

Other evaporative cooling systems may spray animals with misters and use circulation fans to generate breezes that evaporate the excess water. This evaporation lowers the surface temperature of the animals and cools them much like sweating cools a person.

Ventilation and Exhaust Systems

Fresh air is necessary for a variety of environments in agriculture. Fresh air can be displaced by combustible dust, toxic fumes, or dangerous level of gases that should not be inhaled by humans or animals. Bringing in fresh air and exhausting stale air is necessary in multiple environments found in agriculture.

Manure Storage Ventilation and Exhaust Systems

Buildings used to house dairy cattle and swine may have slotted floors so that manure and urine can be stored in pits built underneath. These manure pits can create hazardous atmospheres. Workers should not enter these areas without proper training and unless safety measures have been taken. Livestock buildings with manure pits can benefit from proper ventilation. Ventilation can be used to remove gases generated from the decomposing manure. This reduces odor levels and permits more uniform air distribution in the building. It also helps warm and dry the floor. Pit ventilation is designed to control for moisture and odor. Fans are used to accomplish this. Recommended pit ventilation rates will vary by species and age of the animal. There are winter rates and summer rates, depending on the comfort needs of the animals.

Safety Note

Confined spaces, such as a manure pits, silos, or grain storage areas, are some of the most dangerous environments found on an agricultural work site. These environments present entanglement and engulfment hazards. Asphyxiation from toxic gas is also a danger. Avoid entering these confined spaces unless doing so is absolutely necessary. OSHA outlines several recommendations for entering confined spaces when necessary.

Fume Exhaust Systems

Adequate ventilation is needed to protect workers when work activities produce hazardous fumes and particulates. Welding, painting, grinding, sanding, and similar work that may produce fumes or particulates should be done in well-ventilated areas. Paint booths and welding stations, **Figure 26-22**, are two areas in which mechanical ventilation is commonly used. Exhaust air leaving the area should be filtered to prevent the spread of pollutants into the environment, and fresh air should be brought in. Manometers can be used to check the static pressure deferential to determine when filters should be changed.

Painting Areas

Painting areas should have smooth, bright interiors that can be easily cleaned. This will help prevent a buildup of paint, which could create a fire hazard. The size of paint booths can vary, depending on the equipment that is commonly painted. Small booths can be used to paint parts.

Spray paint booths may have these components:
- Compressed air filter, pressure-reducing valve, and manifold for sprayer operation.
- Filters to remove particulates.
- Inlet for fresh air.
- Ventilation fans.
- Exhaust ducting.
- Lights.
- Entry door.
- Equipment/overhead door.
- Fire suppression systems.

Baloncici/Shutterstock.com

Figure 26-22. This welding workstation has an overhead exhaust system.

Welding Areas

Welding areas should provide ventilation to remove welding fumes and gases away from welders. This can be accomplished through natural or mechanically forced movement of fresh air. When welding outdoors or in open work spaces, there may be adequate ventilation if drafts are created by natural air currents. Welders should use proper positioning to keep their heads out of the released gases or particulates. Forced-air ventilation systems should be designed to remove fumes and gases from the welder's breathing zone at an adequate rate.

Mechanical Ventilation Systems

Mechanical ventilation systems may be portable or centralized. These systems can be either low vacuum or high vacuum. Some units may have extraction arms or overhead hoods to remove the fumes and gases. Extraction arms on portable or centralized units should be located close to the source of the fumes to remove the maximum amount of fumes and gases away from the welder's breathing zone. Portable or mobile systems can be moved to various locations in the laboratory if jobs are too large to fit underneath a ventilation hood. Downdraft tables offer a unique option to draw the welding fumes down and away from the welder rather than up past the breathing zone (as most ventilation units do).

CHAPTER 26
Review and Assessment

Summary

- Agricultural structures may need to be controlled for temperature, moisture, light, dust, odors, noise, airflow, and air quality.
- Understanding heat transfer, psychrometrics, and environmental instruments and controls is helpful in managing building environments successfully.
- Livestock structures should be managed to keep the temperature and humidity in a range that minimizes stress on the animals. Mechanical and natural ventilation is used to remove heat, moisture, and harmful gases.
- Mechanical ventilation is created using fans, thermostats, and building openings for air exchange. Supplemental heat for livestock housing may be provided using overhead radiant heating, floor heating, unit or space heating, and make-up air heating.
- Stored grain requires careful management to inhibit germination, mold growth, and insect activity. Moisture, temperature, and maturity of the grain must be monitored and controlled.
- Temperature and relative humidity should be considered when designing storage atmosphere conditions for fresh fruits and vegetables.
- Refrigeration is commonly used in agriculture to maintain storage temperatures for perishable crops.
- Evaporative cooling may serve as an economical option for cooling animal housing or greenhouses in some areas.
- Bringing in fresh air and exhausting stale air is necessary in multiple environments found in agriculture.
- Ventilation can be used to remove gases generated from the decomposing manure in livestock buildings.
- Welding, painting, grinding, sanding, and similar work that may produce fumes or particulates should be done in well-ventilated areas. Portable or centralized mechanical systems can be used to provide ventilation.

Words to Know

Match the key terms from the chapter to the correct definition.

A. dew point
B. dry-bulb temperature
C. enthalpy
D. environment
E. evaporative cooling
F. homeostasis
G. homeothermic
H. humidity
I. humidistat
J. hyperthermy
K. hypothermy
L. latent heat
M. manometer
N. mechanical refrigeration
O. psychrometric chart
P. psychrometrics
Q. sensible cooling
R. sensible heating
S. specific humidity
T. thermal equilibrium
U. thermal radiation
V. thermodynamics
W. thermometer
X. thermostat
Y. volume
Z. wet-bulb temperature

1. The surroundings or conditions that affect animals, plants, or objects.
2. Energy that changes a substance's physical phase from one state to another (liquid, gas, or solid).
3. Removing heat from the air while keeping a constant humidity ratio.
4. A balance of temperature between two bodies.
5. Adding heat to air without changing its humidity ratio.
6. The amount of water vapor in the atmosphere.
7. A process in which water vapor is added to the air and lowers the dry-bulb temperature without altering the amount of heat in the air.
8. The temperature at which the moisture in the air begins to condense or form droplets that are too large to remain suspended in the air.
9. The maintenance of a nearly constant internal temperature, despite external changes.
10. A process that uses the principles of various scientific gas laws to absorb heat into a refrigerant to produce a low temperature.
11. The branch of science dealing with the relationships between heat and other forms of energy.
12. The volume amount of space occupied by a substance or object.
13. The moisture content of air as a measure of the actual amount of water held in the air in the form of vapor.
14. An instrument used to measure temperature.
15. A graphical representation of the physical properties of air.
16. A condition is which body temperature is above the norm.
17. The temperature of air measured with a standard thermometer.
18. The total heat energy in the air expressed in Btu per pound of air.
19. A control device that senses the relative humidity of the air.
20. Maintaining a nearly constant core body temperature regardless of the surrounding environment.

21. A condition is which body temperature is below the norm.
22. An instrument used to measure pressure in air, gas, or liquid or to determine the difference between two pressures.
23. The study of science involving thermodynamic properties of air and water vapor.
24. The transfer of heat by electromagnetic waves.
25. The temperature of air measured with a standard thermometer that takes into account the moisture content of the air.
26. A device used to control a heating or cooling system.

Know and Understand

Answer the following questions using the information provided in this chapter.

1. What are some purposes for which agricultural structures are used?
2. For agricultural structures, what environmental elements or conditions may require control?
3. What is natural heat flow, and what is one example of natural heat flow?
4. What are three methods by which heat is transferred?
5. How can air and water vapor be managed in agricultural structures?
6. What are the seven physical properties of air used in psychrometrics?
7. What are three common temperature scales used on thermometers?
8. What is a thermostat, and how does it differ from a thermometer?
9. How can overheating affect livestock?
10. Why is it important to ventilate livestock housing and other agricultural structures?
11. What factors should be considered when selecting a mechanical ventilation system?
12. What basic types of heating systems are used for livestock housing?
13. What are some examples of how building design can be used to provide natural ventilation and manage sun exposure?
14. What factors must be monitored to maintain the best environment for grain for storage?
15. What are some hazards related to grain storage?
16. What methods can be used to regulate conditions of storage air for fruits and vegetables?
17. Explain how refrigerant changes form to absorb and expel heat as it circulates through a refrigeration system.
18. What are the basic components of an evaporative cooling system and how does the system work?
19. How can livestock buildings with manure pits benefit from ventilation?
20. What are some examples of work for which adequate ventilation is needed to protect workers from hazardous fumes and particulates?

STEM Connections and Academic Activities

1. **Science.** Perform an experiment that will test different types of insulation. Develop hypotheses on which is the most effective insulation type. Be sure to have a control sample (no insulation) to compare your results with. Construct a framed wooden stud wall using 16′ of 2″ × 4″ lumber and 1/2″ plywood. You will need tools and fasteners to construct the wall. The dimensions of the wall should be 4′ in length and 1′ in height. You will space the studs 16″ on center. Fasten the boards together to form a rigid rectangular box using screws. Fasten a piece of 1/2″ plywood to the back side of the wall with screws. This will provide backing to keep the insulation and heating coil in the wall. Acquire 60′ of roof deicing cable and place the cable in between the studs. You may drill holes in the studs to pass the cable through each one. In each of the three spaces between the studs, place the insulation on top of the deicing cable. Each space should have a different type of insulation. Place another piece of plywood on the front of the wall and fasten it with screws. Energize the deicing cable to begin heating the inside of the wall. Use an infrared thermometer to measure the temperature of the front of the wall and collect your data. Discuss the results with your teacher.
2. **Technology.** Perform the following test to determine the accuracy of at least three different types of thermometers (digital, type K, and analog).
    - Begin by filling a large container to the top with ice cubes.
    - Add water to just below the top of the ice.
    - Stir the mixture and wait for two minutes.
    - Insert the thermometer's stem or probe 2″ into the center of the ice bath and stir gently, keeping the stem surrounded by the ice cubes and moving constantly.

    Do not let the thermometers rest against the ice or you will get a low reading. Do not let the thermometers rest against the container or you will get a high reading. An accurate thermometer should read 32°F (0°C). Of the thermometers tested, which seemed most accurate?
3. **Engineering.** Design and build a chicken egg incubator that can maintain the following environmental conditions and hatch a dozen eggs.
    - **Temperature:** The eggs need to be kept at 99.5°F (37.5°C) at all times; just one degree higher or lower for a few hours can terminate the embryo.
    - **Humidity:** From 40 to 50 percent humidity must be maintained for the first 18 days; 65 to 75 percent humidity is needed for the final days before hatching.
    - **Ventilation:** Egg shells are porous, allowing oxygen to enter and carbon dioxide to exit. Incubators need to have holes or vents that allow fresh air to circulate so the fetuses can breathe.

    Test your design by monitoring the environmental conditions for the duration of an incubation time frame.

4. **Math.** Practice unit conversion between temperature scales. Use thermometers to measure the temperature of a room in one scale and convert this temperature to other scales commonly used. Common temperature scales include Celsius, Fahrenheit, Rankine, and Kelvin. Formulas for conversion between scales are:
   - Temperature in Rankine = Temperature in Fahrenheit + 460
   - Temperature in Kelvin = Temperature in Celsius + 273
   - Temperature in Celsius = 5/9 (Temperature in Fahrenheit – 32)
   - Temperature in Fahrenheit = 9/5 (Temperature in Celsius + 32)
5. **Social Science.** Complete a research report on the history of refrigeration and its impact on the American agricultural and food industry. Report on the dates and the advancements of various refrigeration technology with resulting impacts on agriculture. To supplement this activity, read Upton Sinclair's novel, *The Jungle*, and discuss the need for sanitary refrigeration methods in the meatpacking industry.
6. **Language Arts.** Create a visual display to communicate the three different heat transfer methods using agricultural structure examples. The display must use written as well as visual communication methods. As a service project, present the display to other students, such as elementary school students or other students not enrolled in agriculture, to help them learn more about the sciences used in agricultural mechanics.

Thinking Critically

1. Investigate three different fans used for ventilation in agricultural buildings (greenhouses, livestock barns, or poultry barns) in your local area. Use the Internet to gather information about the performance of the fans. Compare and contrast the fans based on the following:
   - Quantity of air delivered at different static pressures.
   - Energy efficiency (CFM/Watt).
   - Cost for purchase.
   - Cost of operation.
2. Examine an existing livestock barn or greenhouse for temperature differences in various zones within the structure. Place thermometers around the building room and record temperatures throughout the day. Identity areas that have significantly higher or lower temperatures by comparing the average building temperature with each zone temperature. What methods could be used to better manage temperatures in the structure?

CHAPTER 27
Fundamentals of Metalworking

Chapter Outcomes
After studying this chapter, you will be able to:
- Explain how advances in metalworking have led to new technical developments and processes in agricultural mechanics.
- Identify the ferrous and nonferrous metals most commonly used in agricultural mechanics.
- Create a project plan for a metalworking project.
- Identify and use assorted tools for measuring and marking metal.
- Select an appropriate cutting process for a specific project.
- Describe various methods of shaping metals.
- Explain how threaded holes are created in metal parts.
- Fasten metal parts using a variety of methods.
- Describe the soldering process.
- Explain how various heat treatments affect metals.

Words to Know

abrasive	ductility	jig	solder
annealing	ergonomic	machine bolt	soldering
bill of materials	eutectic	machine screw	springback
blind rivet	flux	malleability	tap
cap screw	forge	mandrel	tempering
cold working	forging	pipe roller	thread locking compound
computer numeric control (CNC)	formed-in-place gasket	pneumatic hammer	thread series
conduit bender	galvanized	power shear	tinning
cut list	heat-treating	punching	toughness
drawing	hydraulic pipe bender	quenching	wing nut
	jam nut	slitting chisel	

Before You Read
Arrange a study session to read the chapter aloud with a classmate. At the end of each section, discuss any words you do not know. Take notes of words you would like to discuss in class.

While studying this chapter, look for the activity icon **to:**

- **Practice** vocabulary terms with e-flash cards and matching activities.
- **Expand** learning with interactive activities.
- **Reinforce** what you learn by completing the end-of-chapter questions.

www.g-wlearning.com/agriculture

Lone Wolf Photography/Shutterstock.com

Throughout history, advances in metalworking have been crucial to the development of more productive and efficient agricultural practices. Ten thousand years ago, as agriculture was first being developed all over the globe, tools such as hoes and sickles were made from naturally occurring materials such as wood, stone, and bone. The development of metalworking skills gradually resulted in the replacement of natural materials with metals, first with copper or perhaps iron from meteors, then bronze, manmade iron, and finally steel. Metal tools made early farmers more efficient and productive. Blacksmiths became important members of every farming community, **Figure 27-1**. They possessed the knowledge and skills to make metal tools and repair them when they dulled or were damaged.

In addition to better materials, better-engineered tools also enabled farmers to be more productive. In 1837, a blacksmith named John Deere living in Grand Detour, Illinois, improved on the design of conventional plows by reusing a broken sawmill blade to make a plow from polished steel instead of cast iron, **Figure 27-2**. Deere's steel plow easily cut through the roots of the prairie grass and the soil slid right off its polished metal moldboards. Because farmers had to stop less to clean their plows, Deere's design was more efficient than his competitor's. In this way, developments in metalworking and agriculture have gone hand-in-hand since antiquity.

Metals Commonly Used in Agriculture

For the agricultural mechanic, metalworking generally consists of cutting, shaping, or fastening metal parts. Metal is crucial to the construction of tools, equipment, and structures. Metals commonly found in agricultural settings include cast iron, steel, aluminum, copper, and copper alloys.

Ferrous Metals

Ferrous metals are metals consisting primarily of iron. Wrought iron is almost pure iron and is not commonly used for structural purposes today. Cast iron is an alloy of iron containing a large percentage (above 2.0%) of carbon. Cast iron is very hard and resistant to wear, but relatively brittle. It has limited applications in heavy equipment bodies, housings, and some components. Agricultural mechanics are more likely to make repairs to cast iron parts than construct new parts using iron. Cast iron is easily machined once the hard surface scale has been removed.

Steel is an alloy of iron with a lower percentage of carbon than cast iron (between 0 and 2.0%). Mild steel has low

McCarthy's PhotoWorks/Shutterstock.com

Figure 27-1. Blacksmiths were important members of the farming community. Only they possessed the knowledge and skill to make and repair metal tools.

carbon content (less than 0.30%). Mild steel is versatile and easy to form, cut, and drill. It is strong enough to withstand considerable wear and stress. Mild steel is routinely used to frame buildings and equipment, **Figure 27-3**. Mild steel is also commonly used in thin sheets as coverings for buildings and cabinets for storing and housing equipment.

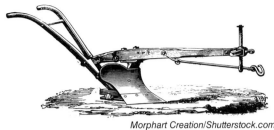

Morphart Creation/Shutterstock.com

Figure 27-2. An illustration of a plow similar to the original John Deere polished steel plow.

Higher-carbon steels (between 0.30% and 2.0% carbon) are harder and less easy to work with than mild steels. Higher-carbon steels can be heat-treated and sharpened to form cutting edges. Stainless steels contain high quantities of chromium and nickel. Stainless steels are more difficult to work than mild steel, but are extremely resistant to corrosion. This makes them ideal for many agricultural applications.

Steel is available in a variety of structural shapes well suited to the needs of agricultural construction. Channel iron, which is used as the main support for many structures, is typically designated by its overall width. The length of a channel's legs is proportional to its width. Structural angle, or angle iron, is L-shaped and specified by the length of each leg, as well as the thickness of the legs. Bar stock is solid material extruded or roll-formed into numerous shapes, including flat, square, and round. Pipe and tubing are hollow and are very common in the construction of agricultural equipment. Tubing is designated by its outside dimensions, and pipe is designated by its inside diameter. See **Figure 27-4.**

Nonferrous Metals

Nonferrous metals include any metal not composed primarily of iron. Agricultural mechanics need to know how to work with aluminum, copper, and copper alloys.

Konzeptm/Shutterstock.com

Figure 27-3. A steel agricultural building.

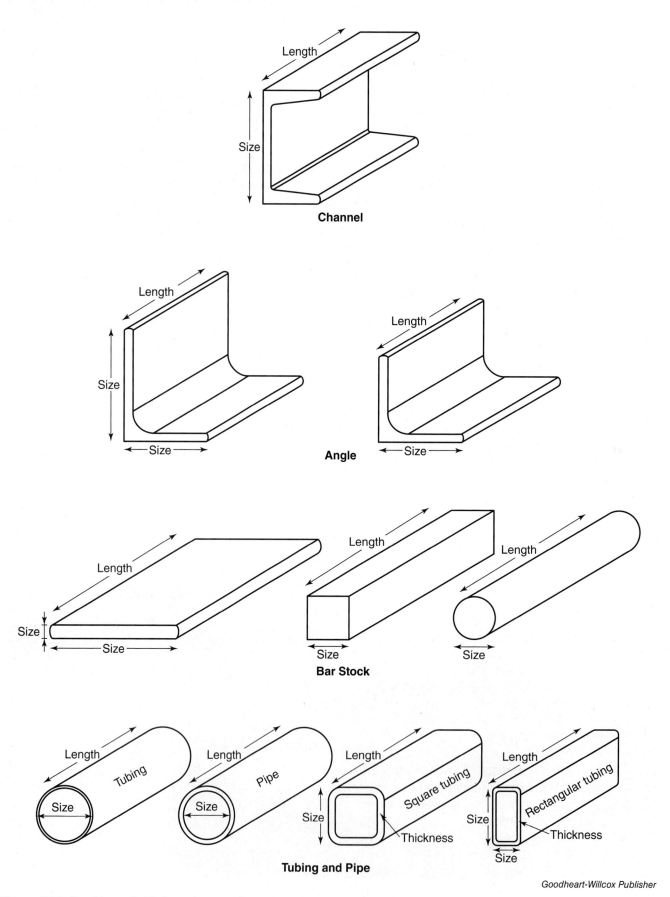

Figure 27-4. Steel is available in various configurations, including channel, angle, bar stock, pipe, and tubing.

Aluminum is strong, lightweight, and resistant to corrosion. It is well suited to constructing aircraft frames, automobile bodies, and buildings. Pure aluminum is relatively soft and weak. Most aluminum has been alloyed to increase its strength and toughness. Aluminum is easily machined, formed, and welded.

Copper is widely used for electrical wiring, plumbing fixtures, and radiators. The softness of copper makes it easy to machine or form into parts. Brass and bronze are alloys of copper. Brass is an alloy of copper and zinc. Bronze is an alloy of copper and tin or copper and aluminum. Both brass and bronze are stronger than copper and are easily machined, but bronze is harder and more expensive than brass. Both are used in agriculture for many types of valves and bodies of control mechanisms. See **Figure 27-5**.

Marcel Derweduwen/Shutterstock.com

Figure 27-5. A variety of plumbing fixtures made from copper and its two major alloys, brass and bronze.

Designing and Planning Metal Projects

Whenever taking on a new project, one should always have a plan. A good plan includes a list of needed materials, tools, and time required. A good plan also allows for enough space to complete the project and an

SAE Connection
Metalworking

Do you have a metals laboratory in your school? Do you have access to metalworking equipment at home? As you study this chapter, research potential projects that can be created from metal. Steel utility trailers, aluminum animal pens, stainless steel fryers, and hundreds of other items can be designed and built from a variety of metals.

Discuss with your agriculture teacher the potential projects that could be created in your school laboratory. Most metalworking projects can be completed in a short amount of time and would make a good supervised agricultural experience (SAE) program for someone interested in metalworking. Also, many local machine shops or metal fabrication shops commonly employ full or part-time student labor which would make a really good placement (SAE) for you to gain valuable experience that just might be the start of a great career

TunedIn by Westend61/Shutterstock.com

estimate of any costs. For repair jobs, the plan may be simple and involve only replacing broken or worn parts. For construction projects, a plan needs to be more complex. Agricultural mechanics should aim to design projects that fit a client's specific needs. The simplest designs are often the best. Complex machines with simple components are easier to make reliable and they are easier to repair.

Sketches and Drawings

Begin construction plans with an image, a sketch or *drawing*. The complexity of the project dictates the detail required in the sketch or drawing. Photos of similar ideas and pencil sketches are used to bring ideas together and contemplate design elements. More detailed plans use scale measurements and dimensions to develop drawings that may be used to guide the construction process. Computer-aided drafting and design (CAD) software automatically includes the scale and dimensioning in drawings. CAD drawings are easily modified, saved, and duplicated. See **Figure 27-6**.

Bill of Materials

Complete drawings should be accompanied by a *bill of materials*—a list of all the materials needed for a project. Using a spreadsheet program helps keep a bill of materials well organized. A complete bill of materials should include quantities, descriptions, and costs of everything needed to complete the project, **Figure 27-7**. Often suppliers are included when resources need to be obtained from multiple sources. Having a list of everything needed at the start of a project reduces the chance for unpleasant surprises in the middle of a project.

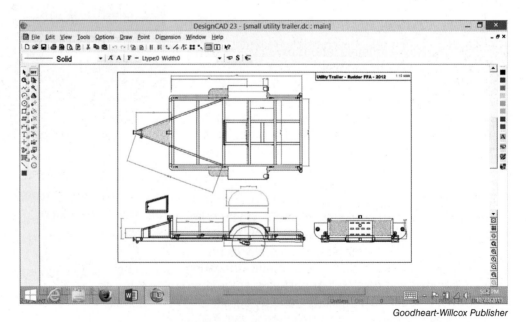

Goodheart-Willcox Publisher

Figure 27-6. Two-dimensional CAD programs that will meet most of the needs of an agricultural mechanic are available at minimal cost.

Ranch Bench

Bill of Materials

Quantity	Size	Material	Location	Cost
1	48″	1″ × 2″ rectangular steel tubing	Horizontal front of seat	
4	46″	1″ × 2″ rectangular steel tubing	Horizontal seat supports and top and bottom of backrest frame	
2	19″	1″ × 2″ rectangular steel tubing	Vertical backrest supports	
2	25″	1″ × 2″ rectangular steel tubing	Sides of seat frame	
2	23″	1″ × 2″ rectangular steel tubing	Sides of backrest frame	
2	22″	1″ × 2″ rectangular steel tubing	Front legs	
2	22¾″	1″ × 2″ rectangular steel tubing	Rear legs	
2	26¼″	1½″ × ¼″ steel strap	Supports for armrests	
2	29½″	2″ × 6″ lumber	Armrests	
1	¼ sheet	24″ × 46″ × ¾″ plywood	Seat bottom	
4	46″	1″ × 6″ pine car siding	Backrest	
	Varies to personal preference		Seat cushion	
	Varies to personal preference		Seat covering	
6	Std	Worn horseshoes	Decorative on sides of seat frame	
8	1¼″ × #10	Screws w/ nuts	Mounting hardware for backrest	
4	¼″	Self-attaching countersunk nuts	Mounting hardware for seat	
4	1½″ × ¼″	Cap screws w/ lock washers	Mounting hardware for seat	
4	2½″ × ¼″	Carriage bolts w/ nuts and lock washers	Mounting hardware for armrests	
	Varies to personal preference		Upholstery hardware for seat	
1	Quart	High gloss urethane	Finish for backrest	
1	Quart	Paint	Finish	
			Total Investment	

Goodheart-Willcox Publisher

Figure 27-7. A bill of materials is a kind of shopping list of all the items needed to complete a project.

Metal stock is generally sold in standard shapes and dimensions. These common sizes are efficient for transport and commerce. Structural steels are usually sold in 20′, 24′, and 40′ lengths. Sheet metal is commonly marketed in 4′ × 8′ or 5′ × 10′ sheets. These common sizes must be considered when developing a bill of materials.

For example, if 12′ of 2″ × 2″ angle iron is needed for a project, it is necessary to acquire a 20′ section. Twenty feet of 2″ × 2″ angle iron would be included in the bill of materials. The remaining 8′ section is technically waste, but it could be reused in a future project. This would allow for the recovery of the excess cost.

Cut List

Many times a bill of materials is accompanied by a detailed list, called a *cut list*, of the different lengths of each material required for the project. A cut list includes the quantity, dimensions, and position of each cut piece. Using a cut list, most or all of the parts can be gathered, cut, and set aside at the beginning of a project. Sometimes development of a cut list reveals the need to purchase additional stock.

As an example, if two 12′ sections and one 16′ section of 2″ × 2″ angle iron are needed, and it is only available in 20′ lengths, then three 20′ lengths need to be purchased. Although the three sections total 40′ of angle iron, if the sections cannot be welded together from shorter sections, the pieces cannot be cut and assembled from two 20′ lengths. Whereas a bill of materials may include waste or extra material, the cut list should include only the exact specifications of materials needed for construction.

Order of Operations

A project plan should also include a list of procedures, along with the needed tools and associated safety precautions. Some procedures must be performed before others can be started. A hole needs to be threaded before a bolt can be tightened into it, and joints must fit properly before being fastened. The key is to think about the steps that must be completed before you begin working. Doing so helps avoid a situation in which the next step would have been much easier if you completed it earlier in the process. To be efficient, have a good plan and follow that plan.

An order of operations can be formal or informal. A formal plan may include every step, tool, and safety measure needed. For simple projects, a few notes in an outline will suffice.

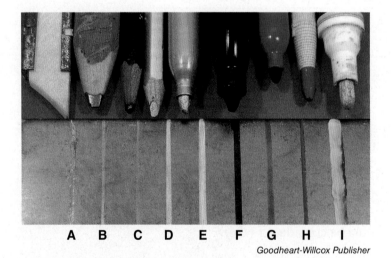

Goodheart-Willcox Publisher

Figure 27-8. Selecting the best marking device can have a pronounced effect on the precision of the work. A—Soapstone. B—Carpenter's graphite pencil. C—Standard graphite pencil. D—Argent pencil. E—Silver permanent marker. F—Black permanent marker. G—Blue permanent marker. H—Blue marking crayon. I—Yellow paint marker.

Marking Metal

Before metal parts can be cut, drilled, shaped, or otherwise modified for use, they need to be measured and marked. The precision required in the finished part and the type of metal dictate the choice of measuring and marking devices. Typical construction using metals involves close tolerances. Even on large projects, small errors in measurement can cause major problems.

When choosing a marking method, carefully consider the width or size of the mark. Will the mark itself be wider than the allowed tolerances? Soapstone, graphite pencils, and marking crayons are common and effective markers, but they can leave marks over 1/8″ wide. See **Figure 27-8.** If tolerances must be held to within thousandths of an inch, as with machinery with moving parts, these markers would be inadequate. When cutting parts for a barn frame or a trailer, however, these relatively wide marks may be just fine.

For temporary marking on mild steel, a sharpened soapstone or argent pencil will leave lines sufficient for sawing and bending. Aluminum and stainless steels cannot be marked easily with soapstone. Multisurface permanent markers with black or blue ink work well for slick, shiny surfaces.

Temporary markings can be accidentally removed by handling the workpiece. A prick punch can be used to make more permanent markings. Prick punches can be used to mark the location of holes or to mark a dotted line. A scratch awl or a scribe can be used on most metals to mark very thin precise lines. Use a guide or a straightedge when scribing marks. This ensures that the mark is correctly placed without creating stray marks.

Cutting Metal

Always be sure that the metal being cut is softer than the cutting tool. Twist drills, saw blades, cold chisels, and the blades of shearing equipment can be easily broken or prematurely worn out in a just a few seconds when used on hardened stock.

Aluminum, copper, and their alloys are typically soft and easily cut. Higher carbon and stainless steels can have sufficient hardness to cause concern. Several testing methods may be used to determine the precise hardness of metals. The agricultural mechanic usually does not need to be specific, and a relative estimate of hardness is sufficient.

One common method of estimating hardness is to use a center punch. Position the center punch on the metal and hit it with a hammer. Then study the depth of penetration. Do this several times. Try to be consistent with the force applied to the hammer. Test a piece of metal that is known to be soft enough to cut and test a piece of leaf spring from a truck or trailer. Leaf springs are too hard for most tools and will only be slightly marked by the punch. Compare the punch marks in the test pieces to the original series of marks in the material that needs to be cut. Very hard steel will damage the point of the punch. See **Figure 27-9**.

A second method of determining the hardness is to use the corner of a file to make several light filing strokes on the metal to be cut. If the file easily cuts into the metal, standard cutting tools should suffice. If instead the file noisily chatters across the surface, this is an indication that the material is too hard to cut with standard tools.

Securing Your Work

Before attempting to cut or drill metal parts, you must secure them to prevent unwanted or unexpected movement. Large pieces may be bulky enough to be

Goodheart-Willcox Publisher

Figure 27-9. Very hard steel can damage the point of the punch. The punch on the right has been deformed by the steel.

714 Agricultural Mechanics and Technology Systems

Goodheart-Willcox Publisher

Figure 27-10. A student standing at a workbench that has a comfortable height. Working at a stable workbench of proper height is more ergonomic than working on the floor.

secure with no additional effort, but even a large trailer can roll slightly when enough force is applied. Before working on large rolling equipment, always block the tires and secure it from moving.

Two things should be kept in mind when securing smaller objects for work. First, they must not move while you are cutting or drilling. Second, they should be in a convenient, *ergonomic* position for working. Ergonomics is the science of work and a human's relationship to that work. If a task is ergonomic, it is comfortable to perform and is unlikely to cause stress or injury. For metalworking, it is usually more ergonomic to work using a stable workbench of proper height than to work on the floor. See **Figure 27-10**.

Using a Vise

Vises and clamps are the primary means of securing small- and moderate-size parts for cutting, drilling, and shaping. A machinist's vise is an excellent tool for gripping smaller metal parts with flat parallel sections. The larger the vise, the more powerful its grip and the more stability it offers.

Most machinist's, mechanic's, and utility vises are equipped with removable jaws that have a textured gripping surface. For parts used in rough work, this results in a better grip. For parts with a more finished surface, these textured jaws can leave undesirable marks. Soft inserts can be used to protect parts from damage, **Figure 27-11**. Thin sheets of brass and scrap pieces of wood both work well.

Use a pipe vise or a vise with V-shaped notches in the jaws to hold round objects. Try to position objects in a vise so that the force applied is in line with the main screw that tightens the vise. This ensures that the majority of the force will be applied perpendicular to the jaws.

Using Clamps

When objects are too large for a vise or no vise is available, clamps can be used to secure metal parts to a workbench or other suitable, stable structure. Choose the type of clamp based on the clamping force required and the shape of the workpiece. C-clamps offer great clamping force, but they are limited by their size and shape. Bar-type clamps work like C-clamps, trading slightly lower grip strength for increased speed of operation. Locking jaw clamps are fast and work well, especially when two or more clamps are used.

Birgit Reitz-Hofmann/Shutterstock.com

Figure 27-11. A toolmaker's vise with soft plastic inserts to protect the work from the hard steel jaws.

Copyright Goodheart-Willcox Co., Inc.

When using clamps to secure metal workpieces, use the following suggestions:

- Be sure that the clamping surface is adequate. Is it stable? Will performing this work risk damage to the structure or related components? Is the working height ergonomic?
- Select a clamp or clamps that will provide security without damaging the part or the clamping surface.
- Clamps should be clean and free from excess lubricants.
- Position clamps where they will provide the least interference with the metalworking operation while still providing adequate support. Using multiple clamps that can be repositioned during the operation is often necessary.
- Clamp the metal part where the cut, hole, or bend will be beyond the edge of the clamping surface to keep metalworking tools from damaging the workbench. Working close to the clamp and the edge of the support will reduce vibration and noise.
- If it is impractical work over the edge, use a sacrificial scrap of wood between the metal part and the workbench.
- Protect delicate surfaces from clamping damage with scrap pieces of wood or metal plate that will distribute the force from the clamp over a larger surface area. See **Figure 27-12**.
- Be sure to clean and return all clamps and other tools to their proper place when the job is completed.

Goodheart-Willcox Publisher

Figure 27-12. The scrap pieces of wood will distribute the force from the C-clamp over a larger surface area and protect the surface of the part from damage.

Sawing Metals

Many different methods are available for cutting metal. The choice of cutting tool depends on the shape and type of metal to be cut, as well as the size of the job.

Hacksaws for Cutting Metal

Small sections of metal tubing or rod can be cut with a hacksaw. Hacksaws are readily available and inexpensive. They are effective if used properly and limited to small jobs. Be sure to match the pitch, or teeth per inch (TPI), to the material. Mild steel can often be cut with an 18 TPI blade. A smoother cut can be obtained on softer metals if a finer pitch is used.

When using a hacksaw, always make sure the metal part is securely held in a vise or clamped to a workbench. The closer the cut is to the stable edge of the vise or table, the less the part will vibrate. Minimizing vibration makes the cutting action more efficient and reduces the energy lost to wasted motion. Thin sections of sheet metal tend to flex very easily. They may be clamped tightly to a backing piece of sacrificial wood that serves to support the metal through the saw strokes.

Procedure for Using a Hacksaw

- Always use a sharp hacksaw blade. When a blade gets dull or damaged, dispose of it and replace it with a new blade.
- Blades should be installed tightly in the frame of the saw with enough tension to prevent the blade from flexing during the cutting process.
- The cutting action of a hacksaw occurs only on the forward stroke. See **Figure 27-13**. Do not apply any downward pressure on the backstroke. When cutting, apply firm pressure on the cutting stroke and run the saw as far as the blade allows, getting the full length of the blade cutting. Avoid pushing so hard that the blade flexes too much. Let the sharpness of the blade do the work.
- Using a small amount of light oil will reduce friction, cool the blade, and help to clear metal particles from the teeth.
- A sawing motion generated by moving the whole body (instead of just the arm) engages more muscle groups. This makes the effort of cutting easier and reduces fatigue.
- Cutting at a moderate pace is less tiring than trying to cut too fast.
- When the saw nears the end of the cut, slow down and apply less pressure.
- Support the workpiece as it is cut free; be careful of sharp burrs created by the cutting action.

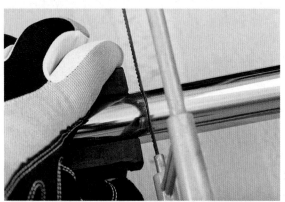

vvoe/Shutterstock.com

Figure 27-13. The cutting action of a hacksaw occurs only on the forward stroke.

Cutting Metal with Power Saws

Several different types of power saws can be configured for cutting metal. As a general rule, stationary machines do a better job than portable saws. When selecting a cutting method, keep in mind the motion of each type of saw and the way it may affect the metal. Circular saw and band saw blades move in one direction. Reciprocating saw blades move back and forth very quickly. Very thin sheet metals may be damaged by the aggressive action of reciprocating saws.

Cold saws are heavy-duty stationary tools designed for cutting metal, **Figure 27-14**. They incorporate a gear reduction mechanism to slow the rotational speed of the blade and a liquid coolant

Goodheart-Willcox Publisher

Figure 27-14. Cold saws work much like miter saws for cutting metal. Note the use of the machine's vise to securely hold the metal work material. A fence or miter gauge may be used as an additional guide.

system to reduce heat damage caused by friction. Cold saws have sturdy vises built into the body of the saw to keep material absolutely still while being cut. They are capable of very precise cuts and, as the name implies, the material does not become hot enough to alter the physical properties of the metal. Blades for cold saws are constructed of complex high-speed steel or tipped with sharpened tungsten carbide inserts brazed to the teeth.

Some metal cutting saws, like cold saws and certain horizontal band saws, have mechanisms to cool metal parts as the cut progresses. See **Figure 27-15**. It is important to keep the reservoirs on these machines filled with the proper cutting fluids. No matter which saw is used, always match the blade type, pitch, and speed of the saw to the metal being cut. This information can often be found in machinist's handbooks.

Worakit Sirijinda/Shutterstock.com

Figure 27-15. This horizontal band saw uses cutting fluid to cool the part and the blade as the cut is made.

Cold Chisels

Prior to the advent of power tools, a great deal of metalworking was completed using cold chisels. A cold chisel is a sharpened wedge that is driven into the metal using a hammer, **Figure 27-16**. The term "cold" means that the metal is cut without being heated. Cold chisels can cut and clean the surface of metal in areas where other tools are inefficient or cannot reach.

When metal is cut with a cold chisel, it is first compressed in the area of the cut. As the chisel continues, the metal shears and separates. Chisel cuts are generally not smooth or precise. Cold chisels are available in a variety of shapes and sizes for making different types of cuts. *Slitting chisels* are used for cutting thin sheet metal, **Figure 27-17**.

Pneumatic Hammer and Chisel

A *pneumatic hammer*, also known as an *air hammer* or *air chisel*, is a power tool that may be used instead of some cold chisels for cutting metal. A pneumatic hammer is similar to a jack hammer. The power of compressed air makes pneumatic hammers much more effective than chiseling by hand. The chisel bit is driven into the material thousands of times per minute. Different tasks can be performed by changing the bit. Chisel bits are used for cutting. Other bits with rounded or flat tips can be used for shaping metals.

Goodheart-Willcox Publisher

Figure 27-16. A cold chisel is a sharpened wedge that is used to work metal at room temperature.

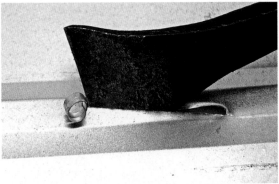

Figure 27-17. Cutting sheet metal with a slitting chisel. Chisel cuts are rarely smooth or precise.

Figure 27-18. Engage the locking mechanism around the pneumatic hammer bit.

Safety Note

Pneumatic hammers are extremely powerful tools. Remember to follow these safety precautions when using them:
- OSHA-approved (Z87.1) impact-resistant eyewear must be worn at all times in conjunction with a full face shield.
- Pneumatic hammers are loud. Ear protection must be worn at all times. Try to set up where others will be least affected by the noise.
- Wear vibration-resistant gloves to help isolate the hands from the extreme motion of the tool.
- Inspect bits before each use. Do not use bits with visible cracks. The extreme forces that pneumatic hammers exert can shatter damaged bits.
- When changing bits, always disconnect the air supply first. Be sure to properly engage the locking mechanism around the new bit. See **Figure 27-18**.
- Always be aware of the position of the tool, workpiece, and other people in the area. The force pneumatic hammers exert can cause the tool or workpiece to jump violently.
- Do not operate the tool unless the bit is in contact with the work. If the retaining mechanism fails unexpectedly, the bit can become a projectile.

Shearing Metals

Sheet metal and small wires or bars can be cut using a shearing action similar to the way scissors cut paper. When metal is sheared, the force of a set of straight blades stresses the metal to a point beyond the shear strength of the material. The clearance between the blades is usually between 5% and 40% of the thickness of the metal being cut. A tighter clearance yields a superior finish on the cut edge, but requires more energy to complete the cut. If there is too much clearance between the blades, the metal will not shear cleanly. Instead it may only deform the metal or result in a rolled edge or an edge with excessive burrs. The clearance on higher-quality shears can be adjusted by tightening the connector at the pivot point of the blades. See **Figure 27-19**.

Punching metal is technically a shearing process. Punching metal, much like a hole punch for paper, uses a punch and die. As the metal is pushed by the punch through the die, a shearing action separates the center piece from the edges of the hole, **Figure 27-20**.

Manual Shearing Tools

Tin snips and aviation snips work well for light cutting of sheet metal. Using them is almost like using scissors to cut paper or cloth. For heavier cutting tasks, snips with a compound lever system in the handles will

Figure 27-19. The clearance on a pair of metal-cutting shears can be adjusted by tightening the connection at the pivot point of the blades.

multiply the squeezing effort. Aviation snips are used to cut sheet metal for covering buildings or roofing. They can be straight or angled to the left or right. The handles are color coded for ease of identification. Yellow-handled snips cut straight, red-handled snips cut to the left, and green-handled snips cut to the right, **Figure 27-21**.

Bolt Cutters

Bolt cutters incorporate tremendous leverage to cut fasteners and mild steel bar stock of similar size. Bolt cutters also make quick work of cutting livestock panels for fencing. The cutting jaws of bolt cutters are two hardened, chisel-shaped blades designed to pinch through metal. Placing an object as deep into the jaws as it will go before pulling the handles maximizes the leverage and force exerted by the cutting jaws. Attempting to use bolt cutters on hardened materials, such as quality padlocks, can quickly damage the blades beyond usefulness. See **Figure 27-22**.

Power Shears

One of the cleanest methods of cutting sheet metal is to use a power shear. *Power shears* function just as smaller handheld tools but with the advantage of electric or hydraulic power. Large stationary units will cut full 4′ × 8′ sheets of metal in perfectly straight lines in a single motion. Smaller handheld units cut using multiple strokes. These lightweight tools can be used with a straightedge to cut long, straight lines or manipulated to cut intricate curves. See **Figure 27-23**. Still other power shears have massively strong jaws designed for cutting sections of steel bar stock.

All power shears are rated based on the thickness and types of metal they will cut. These ratings should not be exceeded. Most power shears are not made to cut plastics or papers. Attempting to cut soft materials will likely foul the cutting edge. Hardened metals can also be damaging to shears. Once a shear blade is deformed or chipped, its cutting efficiency is severely limited.

Nibblers are handheld power tools that cut sheet metal by making a continuous series of punches. They function by pushing little bits of material with a reciprocating punch through a fixed die. Power shears and nibblers may operate on electric power, or they may be pneumatically powered. A portable shear or nibbler is an excellent tool to cut sheet metal for siding buildings or roofing. Nibblers are especially well suited for cutting corrugated materials.

Goodheart-Willcox Publisher

Figure 27-20. Metal may be punched by forcing a punch through the part, as in this hand-operated sheet metal hole punch.

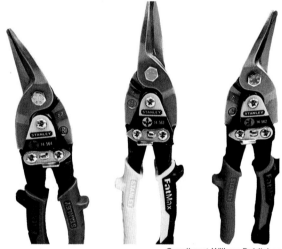

Goodheart-Willcox Publisher

Figure 27-21. The handles of aviation snips are color-coded. Yellow cuts straight, red cuts to the left, and green cuts to the right.

Safety Note
When using a bolt cutter to trim a short piece, wear eye protection and be aware that the smaller piece may be expelled from the cutter with significant force.

Steve Collender/Shutterstock.com

Figure 27-22. Attempting to use bolt cutters on hardened material such as quality padlocks can quickly damage the blades beyond usefulness.

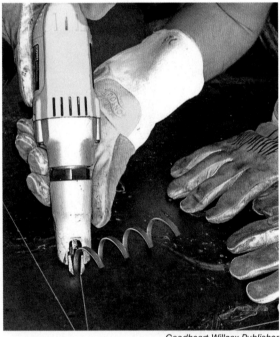

Goodheart-Willcox Publisher

Figure 27-23. Handheld power shears are efficient at cutting thin sheet metal.

Power Shear and Nibbler Rules

Follow all the power tool safety rules and these recommendations for safely using portable power shears or nibblers:

- Power shears are tremendously powerful tools. The blades of these machines will easily pass through flesh and bone. It is extremely important to keep fingers and hands away from the cutting edges.
- The forces exerted by power shears can create pinch points between the stock and the machine parts well away from the cutting action. Be aware of what may move when the machine is engaged.
- Always wear protective leather gloves when handling metal cut with a shear. The edges of the metal may have razor-sharp burrs. These burrs need to be removed before the piece can be handled safely. Sharp burrs can be removed by lightly using a file or a disc sander to smooth the edge.
- Work in a clean area with adequate clearance to safely handle long sections of sheet metal.
- Keep the tool cutting edge sharp, clean, and well lubricated.
- Support the material as close as possible to the cut to reduce vibration. Be aware of what is located just below the cut.
- Use both hands, and do not overreach. Plan your cuts from beginning to end so you are not ending a cut in a hazardous position. It is better to start in a less comfortable position and work toward a more comfortable stopping point than to progressively become more uncomfortable as the cut progresses.
- Do not drag power cords or air supply lines across the sharp edges of cut materials.

Cutting with Abrasives

Metal parts are often cut with *abrasives*. In fact, when unidentified steel parts need to be cut, using an abrasive method may be the best choice. See **Figure 27-24**.

Abrasives work by grinding or wearing away material through friction. The edges of the individual grains of an abrasive wheel serve as microscopic cutting edges. The most common abrasive materials are aluminum oxide, silicon carbide, and zirconia. Aluminum oxide is most often used for mild steel. Zirconia is better for alloyed steels.

Abrasive methods of cutting generally produce quite a bit of heat in the metal. Cutting fluids can be used to control excess heat during cut-off operations. For some hardened metals, using an abrasive cut-off saw or a handheld abrasive cut-off tool may be the only option. Other metal-cutting tools simply may not be hard enough.

Abrasive Cut-Off Saws

Abrasive cut-off saws, also called *chop saws*, look and operate much like miter saws. Abrasive cut-off saws use a thin abrasive blade to cut through metals. See **Figure 27-25**. They are very useful for cutting the structural steel commonly used in the fabrication of agricultural buildings and equipment, but they should never be used to cut lumber. The pivot point on a chop saw is usually stable and moves only in a plane perpendicular to the base. Cutting angles are achieved by changing the position of the vise that holds the metal during a cut.

A great deal of friction is generated by the abrasive wheel or blade of the chop saw, which tends to heat the metal more than other mechanical methods of cutting. This excess heat and the sharp metal slivers on cut ends require the use of leather gloves to handle metal parts after a cut. The excess heat created can also make abrasive cutting unsuitable for some jobs.

fluke samed/Shutterstock.com

Figure 27-24. Using an abrasive cut-off saw may be the best choice for cutting many metals.

apiguide/Shutterstock.com

Figure 27-25. Abrasive cut-off saws are great for cutting steel structural materials on a job site.

Shaping Metal

Metals are ductile, malleable, and tough. A metal with high *ductility* can be drawn or stretched into thin flexible wires. *Malleability* is the ability of metals to be hammered or rolled into flat sheets, **Figure 27-26**.

aaltair/Shutterstock.com

Figure 27-26. The malleability of steel allows it to be flattened into sheets in a rolling mill.

Figure 27-27. Toughness allows metals to be bent and deformed without breaking.

Figure 27-28. A farrier using a hammer and anvil to shape a horseshoe.

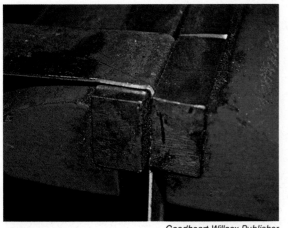

Figure 27-29. Flat bar being bent to a 90° angle in a vise.

High *toughness* means a metal can withstand tearing or breaking when bent or deformed, **Figure 27-27**. Understanding these properties allows agricultural mechanics to shape metals for many important uses.

Forging

Forging is a forming process that involves changing the shape of a metal part through the application of force. Forging is an important process in the manufacturing of metal agricultural goods. Iron, steel, aluminum, and brass are all commonly forged. Forging can be performed cold, but is most often performed hot. Forging alters the microscopic grain structure within the metal, often strengthening the part.

Traditional blacksmith forging involves heating an iron or steel bar in a *forge* (a furnace for heating metals) to an orange or red heat, then pounding the piece with a heavy hammer against an anvil. Metal heated in the forge can also be twisted and bent into detailed forms. Uniform heat is key to successful forging. Traditionally charcoal was used to heat forges, but modern forges are usually heated with propane. Some farriers still use traditional forging methods for shaping and fitting horseshoes, **Figure 27-28**.

Bending Metal

Not all metal parts need to be heated to be bent into useful shapes. Bending sheet metal and bar stock is commonly performed cold (at room temperature). A common method of bending small metal parts is to grip them in a vise and use leverage or a hammer to create a bend just above the jaws of the vise. Leverage can be gained from the part itself if it is long enough. A short section of pipe or a large adjustable wrench may be used for leverage as well. Leverage is often enough to start a bend, but to get a sharp bend, hammering is necessary.

Most machinist's and mechanic's vises incorporate a small anvil as part of the frame. Round, square, and flat steel bars are easily shaped using a vise as a mandrel. See **Figure 27-29**. A *mandrel* is a tool used as a form or guide for bending metal parts. Using a mandrel allows parts to be bent into smooth curves without undesirable creasing or kinking. Bends of other shapes are possible by using the tapered horn and other surfaces of an anvil.

It is often worth the investment in time and materials to construct a jig for bending multiple copies of the same part. A *jig*, much like a template, is a pattern used to accurately guide the formation of a part. Using a jig makes the work repeatable. For instance, if a hay feeder required twenty 2' long steel bars bent in the center at a 35° angle, a jig would ensure that all of the bars were uniform in size and shape. See **Figure 27-30**.

Working with metal while cold can result in a stronger finished product. *Cold working* is the process of forming, bending, or hammering metals well below their melting point to improve their strength and hardness. Cold working is commonly used to harden nonferrous metals such as aluminum and copper.

Goodheart-Willcox Publisher

Figure 27-30. These parts were formed with a jig to ensure that they were all of uniform size and shape.

Bending Pipe and Tubing

Many times it is necessary to bend pipe or tubing in the construction and repair of agricultural equipment or facilities. Several tools are available for bending different types of pipe or tubing without the use of heat. A simple *conduit bender* consists of a semicircular mandrel, a catch that holds the tube in place, and a handle that acts as a lever to force the tube around the mandrel. See **Figure 27-31**.

Conduit benders work well on light-gauge steel tubing and copper tubing, but for heavier-gauge steel pipe and tubing, a *hydraulic pipe bender* will usually yield better results. A hydraulic pipe bender may be stationary or portable and can be powered by a hand-operated, electric, or pneumatic pump. See **Figure 27-32**. Most hydraulic benders have interchangeable mandrels, or shoes, and rollers to accommodate different sizes of pipe.

Hydraulic pipe benders work by using hydraulic pressure behind the mandrel to force the pipe between a pair of rollers. As the bend progresses, the pipe is wrapped around the curve of the mandrel. The size of the mandrel controls the radius of the bend. Some hydraulic pipe benders are equipped with attachments for bending square tubing and angle iron.

Manufacturers of pipe benders provide specific directions for bending prescribed angles in pipe. Following these directions carefully will give the best results. See **Figure 27-33**. Be aware that different batches of pipe can differ enough in their physical

Appleton Electric Co.

Figure 27-31. A manual conduit bender uses the leverage provided by a long handle to bend the tubing.

Dmitry Kalinovsky/Shutterstock.com

Figure 27-32. A hydraulic pipe bender is capable of shaping a range of different sized pipe and tubing to bends with precise radii.

Nominal Pipe Size	Set-Back	Center Line Radius	Minimum Stub Length
½"	1.125	3.250	7.500
¾"	1.500	4.500	8.750
1"	1.875	5.750	10.500
1¼"	2.375	7.250	12.000
1½"	2.750	8.250	13.250
2"	3.250	9.500	15.000

Goodheart-Willcox Publisher

Figure 27-33. Pipe bending chart showing the set-back, center line radius, and minimum stub length for several nominal pipe sizes.

properties to affect the way the pipes bend. Check the measurements frequently when forming multiple copies of the same part.

A *pipe roller* is used to create long curves in metal parts. Hydraulic or screw pressure is applied similarly to the way a pipe bender functions. Additional force is applied to roll the pipe lengthwise during the bending process. This creates bends with a much larger radius than possible with a simple bender. Feeders for round hay bales are often constructed using rings formed with a pipe roller. See **Figure 27-34**. The mandrel of the roller is completely round and is rotated to drive the pipe or tubing through the machine. Pipe rollers can be powered or manually operated. A similar type of roller is used to form steel plates into large curved parts.

When steel is bent cold, it has a tendency to spring back to shape. The type of steel influences the amount of *springback*. Steel needs to be bent a little past the desired shape to account for the springback. When working with an unknown variety of steel, perform this over-bending in small, incremental steps until the desired shape is achieved. If making multiple parts of the same shape, pay close attention to how far each piece must be over-bent to get the correct shape.

Grinding Metal

Grinding is a particularly useful method for shaping the edges of metal parts. Grinding may be part of the preparation for other processes, or it may be used to finish and smooth the effects of other metalworking

Goodheart-Willcox Publisher

Figure 27-34. Pipe roller forming a hay ring.

processes. Grinding is essential to preparing joints for metal-joining processes such as soldering and welding. See **Figure 27-35**. For some hard steels, like tool steel, grinding is the only recommended method of shaping the metal without the use of additional heat.

The heat caused by grinding can be considerable. Use caution when grinding. Metal that is too hot to safely touch looks just like cold metal. When using a bench grinder, cool small parts or tools by frequently quenching them in water. Steel goes through a series of color changes as the temperature increases. At around 400°F, it turns a light straw color. Between 540°F and 575°F, it becomes a dark purplish-blue color. See **Figure 27-36**. Reaching the blue color is an indication that tool steel has been overheated, potentially degrading the effects of any heat treatment.

Goodheart-Willcox Publisher

Figure 27-35. Grinding is often used to grind beveled joints for welding.

Milling Machines

Milling machines, also called *mills*, are stationary power tools that work in the same manner as routers. They use rotating bits or cutters to remove material from metal and other dense materials to manufacture parts. Mills are used for smoothing perfectly flat surfaces, cutting slots in many configurations, and shaping hard materials into intricate shapes. Mills look much like heavy-duty drill presses and may be used for boring holes with extreme depth and positioning precision. See **Figure 27-37A**. They are equipped with sophisticated vises that are designed to precisely feed material into the rotating cutter. The first mills had to be manually fed, but improvements in technology have led to the development of automatic feed systems that control feed rates and depths.

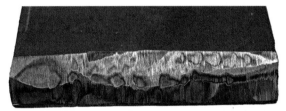

Goodheart-Willcox Publisher

Figure 27-36. Care must be take not to overheat the metal while grinding. The dark blue coloring on this tool steel shows where it was overheated.

Safety Note
Grinding produces sparks and heat. Care should be taken to direct sparks away from people or flammable materials.

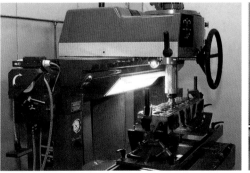
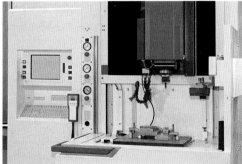

A *Funnycreature/Shutterstock.com* B *Baloncici/Shutterstock.com*

Figure 27-37. Milling machines use cutters and bits similar to those used in a router, but they are engineered to cut away harder materials. A—This mill is being used to resurface the head of an engine. B—Today's CNC milling machines have limitless possibilities. Computers control nearly every aspect of the milling process.

The newest milling machines use *computer numeric control (CNC)* to manage all of the aspects of a machining operation, **Figure 27-37B**. When using CNC machines, the operator positions the part in the vise (often using a jig for repetitive work), selects the file containing the computer codes that drive the mill, and turns on the power. Computer-generated machine codes direct the speed and position of the bit and operate the motion of the vise holding the part to be shaped.

Many mills also incorporate a liquid coolant system to control heat buildup and allow for faster machining rates. Due to their size and cost, milling machines are typically restricted to specialty shops and manufacturers. Many shops offer custom services for occasional milling needs during equipment repairs.

Sharpening and Dressing Metal Tools

When the cutting edge or working end of many tools becomes dull or damaged, they can be resharpened with a grinding wheel. Cold chisels and many knives are sharpened by beveling both sides of the cutting edge. Wood chisels, plane irons, and shovels are sharpened on one side only. Tools such as standard screwdrivers have flat edges with square shoulders.

Before sharpening a tool on a bench grinder, adjust the tool rest to the correct angle. Loosen the nut that holds the tool rest and adjust the angle of the rest so that the desired angle will be ground with the tool lying flat against the tool rest. See **Figure 27-38**. For a cold chisel, the two beveled faces together should form a 60° angle. A wood chisel is ground to a much flatter angle. Check to make sure that the front edge of the tool rest is no more than 1/8″ from the face of the grinding wheel. Tighten the locking nut to hold the tool rest in place.

Next, turn on the grinder and allow it to reach full speed. Place the tool on the tool rest. Bring one of the surfaces of the cutting edge into light contact with the face of the grinding wheel and move the tool slowly back and forth across the entire working surface of the grinding stone. Complete one beveled surface, then the other.

Frequently quench the cutting edge into a container of cool, clean water. See **Figure 27-39**. Watch for the color change. When a straw color appears, it is time to cool. Cooling before the straw color appears is okay. Allowing the blue color to appear indicates that the edge is damaged. Examine the beveled surfaces for

Goodheart-Willcox Publisher
Figure 27-38. Cold chisels can be sharpened on a bench grinder.

Goodheart-Willcox Publisher
Figure 27-39. When grinding a chisel, frequently quench the cutting edge into a container of cool, clean water to keep it from overheating.

a smooth, uniform finish. Smoothing, or dressing, the surface on a finer stone or on a sander is desirable for some edges.

Drilling Metal

Whether using a handheld power drill or a drill press to cut round holes in metal, several factors must be considered, and most of the same steps must be followed.

- Select a bit engineered for drilling the specific type of metal.
- Use the appropriate PPE—eye protection rated Z87.1.
- Do not wear loose clothing, and tie back long hair.
- Mark all locations for holes with a center punch. See **Figure 27-40A**.
- Secure the workpiece; use a vise or clamp for smaller pieces. See **Figure 27-40B**.
- Using a pilot bit to start a larger hole makes the hole easier to place precisely and reduces friction and heat buildup on the larger bits that follow.
- A few drops of light oil reduce friction on the bit and help to move waste chips from the hole. See **Figure 27-40C**.

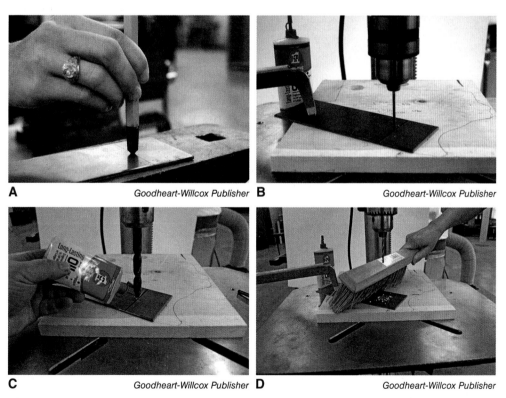

A *Goodheart-Willcox Publisher* B *Goodheart-Willcox Publisher*
C *Goodheart-Willcox Publisher* D *Goodheart-Willcox Publisher*

Figure 27-40. A—Before drilling, mark the hole location with a center punch. The punch mark not only identifies the location of the hole, but it also guides the bit. B—This workpiece is secured with a C-clamp. Note the backing board, which reduces the chance of the bit binding as it passes through the metal piece. C—Lubricating oil reduces friction. D—Remove chips from the work area using a brush. Never use compressed air to clear the chips.

- Apply just enough pressure to the bit that it is continuously removing material whenever the bit is in contact with the work. If using a handheld power drill, be sure to keep the drill perpendicular to the desired hole; do not allow the angle to vary.
- Moving the rotating bit in and out of the completed hole a couple of times will clear most of the burrs from the edges of the hole.
- Use a soft-bristled table brush to remove sharp chips from the work area before handling parts. Be careful: newly drilled holes will likely have sharp edges. See **Figure 27-40D**.

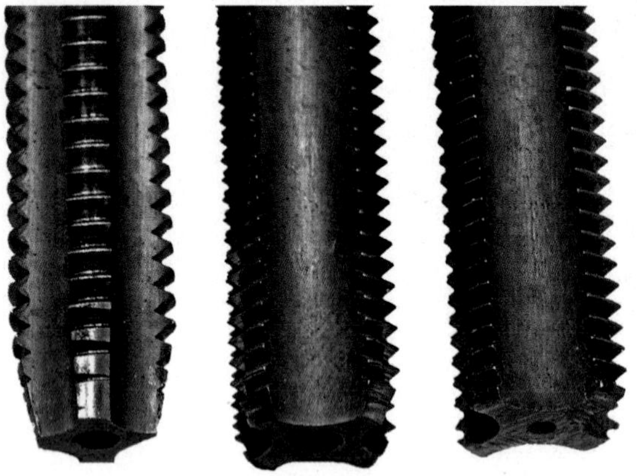

Goodheart-Willcox Publisher

Figure 27-41. Taper, plug, and bottoming taps are used to thread drilled holes.

Threading Metal Parts

When constructing new equipment, holes must be threaded using a ***tap***. If the hole will pass completely through a part, then a taper tap and a tap wrench are all that is required. A taper tap is tapered on about the first ten threads. The tapered section only cuts a partial thread and reduces the force needed to start the tap in the hole. If the hole is blind, or does not pass all of the way through a part, then the tapered tap must be followed by a plug tap and a bottoming tap. See **Figure 27-41**. Plug taps have a shorter beveled section than tapered taps, with only about the first five threads being partial cuts. The bottoming tap is straight and cuts threads all the way down.

Tapping a Hole in Metal

Follow these steps to tap a hole in metal:

1. Drill the proper size hole for tapping. Use a drill bit selection chart to choose a drill bit to match the size of the tap.
2. Insert the square end of a taper tap into a tap wrench and secure it.
3. Secure the part to be tapped in a vise or clamp attached securely to a workbench.
4. Position the end of the tap in the hole while holding the tap wrench directly over the back end of the tap.
5. Turn the wrench clockwise while applying downward pressure to start the tap. Use a machinist's square to check that the tap is entering the hole in a true perpendicular orientation.
6. Continue turning the wrench from the center until the tap starts to bite into the metal.
7. When the tap begins to pull itself into the hole, use both hands to slowly turn the tap wrench by applying even force to each handle. Adding a few drops of cutting oil lubricates the tap to reduce friction and helps to clear chips.

8. Make one to two full turns forward. Then back the tap out (reverse rotation) one half turn to break and clear waste chips. Failure to clear the chips will cause excess friction and damage the threads and possibly the tap. See **Figure 27-42**.
9. Continue alternating between one to two full turns forward and one half turn back until the tap reaches the bottom or passes through the hole and turns freely.
10. Slowly back the tap out of the hole and clean all chips from it before storing it properly. Use caution because chips are sharp.
11. For blind holes, repeat the process with a plug tap followed by a bottoming tap.

Goodheart-Willcox Publisher

Figure 27-42. Backing out the tap clears the chips. Failure to do so can damage the threads and the tap.

Cleaning Existing Threads

Sometimes taps are used during repair work to clean existing threads instead of cutting entirely new threads. Running a tap through a previously threaded hole can clear away corrosion and reshape damaged threads. The same principle is often used with used bolts. Damaged or corroded threads on a bolt or machine screw can be restored by chasing the thread with a die made for cutting threads. See **Figure 27-43**.

Hands-On Agriculture

Drilling Holes

Drilling holes in work material is a basic skill required for many jobs in agriculture. For this exercise, you will need an electric drill, a selection of different size drill bits, scrap wood or metal pieces, clamps or a vise, and measuring and marking tools suitable for the scrap material at hand.

The first step is to understand a few basic rules for drilling holes.

- Holes are located by finding (measuring) and marking centers with a center punch.
- Distance between holes is commonly measured from hole center to hole center.
- Space between holes is measured from hole edge to hole edge. See **Figure A**.
- Holes near the edge of work material should be placed so that the edge of the hole is at least the distance of the diameter of the hole from the edge. For example, the center of a 1″ hole should be placed at least 1 1/2″ from the edge of the material.

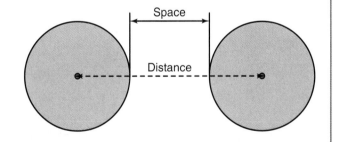

Get your instructor's permission to drill two holes in your scrap material. Wear proper safety apparel and follow the drill safety rules. Mark the location of all holes and drill a 1/2″ hole in your scrap material, leaving exactly 1/2″ between the edge of your hole and the edge of the material. Now drill a second 1/2″ hole exactly 1 1/2″ from the first. Show you work to your instructor for evaluation.

Figure 27-43. A die can be used to chase the threads on a rusty bolt.

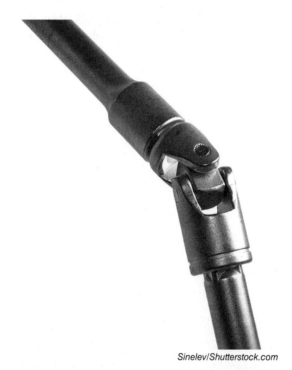

Figure 27-44. The bearings of this U-joint are held by a press fit.

Figure 27-45. Two bolts. The bolt on the top has fine threads and the bolt on the bottom has coarse threads.

Assembling Metal

There are many ways to assemble metals. The methods we choose depend on several factors. For example, what is to be attached: metal to metal, metal to wood, or metal to some type of plastic? What type of stresses will the joint be subjected to? Is the object stationary, or is it a mobile piece of equipment? Will the joint be permanent, or will it need to be disassembled periodically? Can the joint be accessed with a welding process? For which fastening methods are the skills and resources available?

Press Fitting

Steel parts are sometimes fastened simply by a close fit. Mechanical parts, bearings, bushings, collars, and finials may be fitted to round or square shafts using a pressed fit. In a press fit, the parts fit so closely that once in position they do not move. A press fit usually requires mechanical or hydraulic pressure to force the parts together. See **Figure 27-44**. Another method of press fitting involves heating the outer piece of an assembly prior to fitting. Once cooled, it becomes locked in place on the inner part due to natural contraction.

Threaded Fasteners

Threaded fasteners such as screws, nuts, and bolts rely on the wedging action created by a screw thread to clamp parts together. Threaded fasteners are available in both inch-based Unified Thread Standard and metric (ISO Metric) *thread series*. The most common inch-based thread forms are the Unified National Coarse (UNC) and Unified National Fine (UNF) thread series. UNC threads are general-purpose threads. UNF threads have more threads per inch than UNC threads, **Figure 27-45**. UNF threads are used when vibration is expected, such as on power-driven equipment. Fasteners of different thread series are not interchangeable.

Threaded fasteners are available in many different configurations and many different materials, including steel, stainless steel, brass, and even nylon. It should be noted that the definition and differentiation of many threaded fasteners are not always agreed upon.

Machine screws are 1/4″ in diameter or less and are used in agriculture to secure parts in machines, tools, and equipment of all kinds. *Cap screws* are found on assemblies requiring both precision and a more finished appearance. See **Figure 27-46**. Instead of using a nut, cap screws pass through a clearance hole on one part and screw into a threaded hole in another part.

Machine bolts are used to secure parts that do not require close tolerances and are often furnished with a matching nut. They are used to construct equipment and structures that may need to be dismantled occasionally for service or repairs. Carriage bolts can be seen in the assembly of tillage equipment. They are also used for attaching flooring to trailer beds.

Wing nuts are fasteners used on equipment that must be serviced or adjusted frequently. The two prominent "wings" on a wing nut can be gripped with the fingers. See **Figure 27-47**. To secure a nut in place, a *jam nut* can be torqued tightly against the first nut. The friction and tension between the two nuts holds both securely.

Self-tapping sheet metal screws are engineered to cut threads into predrilled holes. Often the tip is tapered and notched to aid in chip removal during thread cutting. Self-drilling sheet metal screws are formed with a short twist drill on the tip. They bore their own hole and tap it with threads.

Goodheart-Willcox Publisher
Figure 27-46. Cap screws are used when precise fits need to be kept during assembly, such as on this engine housing.

Nonthreaded Fasteners

Nails and rivets lack screw threads and are classified as nonthreaded fasteners. Sheet metal can be attached to frameworks using roofing nails. Roofing nails are engineered to punch through sheet metal and penetrate into wood frameworks. Typically, the sheet metal and roofing nails are *galvanized*. Galvanization protects the steel sheet and nails from corrosion with a coating of zinc. Aluminum roofing nails are available for use with aluminum siding.

Phillip Bird LRPS CPAGB/Shutterstock.com
Figure 27-47. Wing nuts can be gripped by hand and are used when an assembly must be serviced or adjusted frequently.

Solid rivets create very reliable permanent assemblies. Steel rivets are used in the construction of some agricultural equipment, such as attaching knife sections to the sickle bars used in combines and hay mowers. See **Figure 27-48**. Setting a solid rivet requires access to both sides of a joint. In agricultural applications, rivets are often set cold with a hammer and anvil. Large-diameter steel rivets may need to be heated.

To set a solid rivet, use a rivet that extends 1/8" to 3/16" beyond the back side of the joint. Then position the head on the face of the anvil. Use the flat side of a ball-peen hammer to upset the tail of the rivet. Round the edges to an oval shape with the round side of the hammer. See **Figure 27-49**.

Figure 27-48. The knife sections have been riveted to this sickle bar.

Figure 27-49. Images of solid rivets before and after being set.

Blind rivets consist of a tubular rivet and a solid pin. Blind rivets are used when the joint is accessible from only one side. The rivet assembly is inserted into a hole drilled through the parts and a rivet gun is used to pull the pin head into the rivet body. The pin causes the tail of the rivet to flare against the reverse side of the joint, **Figure 27-50**. The pin is designed to snap off and can sometimes fall out. The resulting hollow rivet may not be as strong as a solid fastener.

Adhesives

Adhesives used in the assembly of metal components can be classified into two distinct types. Structural adhesives are used to bond metal and other parts, and machinery adhesives are used to improve the seal between parts or increase the holding force of fasteners.

Structural Adhesives

Structural adhesives are intended to be used as the primary means of fastening parts. As the adhesive cures, a permanent bond forms, sealing the joint and protecting it from corrosion. Epoxies designed for repairing cast metal parts have been around for many years. Modern equipment design trends include the use of very thin metals and plastics that are difficult to secure in place with traditional fasteners, but they are well suited to assembly using sophisticated structural adhesives.

Machinery Adhesives

Machinery adhesives are widely used to provide a seal between faces of housings or as formed-in-place gaskets. A sealant is necessary whenever aluminum is attached directly to other metals to prevent galvanic corrosion. *Formed-in-place gaskets* are made by dispensing a bead of adhesive onto the contact surfaces of two parts. When the parts are assembled, the gasket material is form-fitted to the joint, sealing all parts of the connection. As the gasket cures, it adheres to both sides of the joint and forms a flexible sealing layer between the parts.

Thread locking compounds prevent threaded fasteners from loosening and protect against corrosion of the connection. The compound also helps to lubricate the fastener as it is torqued into place. If the fastener must be removed, the thread locking compound releases after the application of a predetermined amount of torque.

Surface Preparation for Adhesives

Most adhesives will not bond through surface contaminants. The bonding surfaces must first be cleaned of all oxidation, moisture, paints, or lubricants.. Many adhesives require cleaning the part with a quick-drying

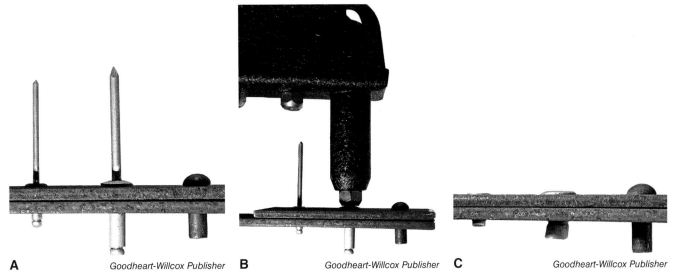

Figure 27-50. Blind rivets are used when only one side of the joint is accessible. A—Two types of blind rivets and a solid rivet for comparison. B—A rivet gun is used to set blind rivets. C—Blind rivets after being set. The pins used to set the rivets have been removed.

solvent prior to application. When adhesives fail, it is usually the result of improper cleaning or using the wrong adhesive for the job.

Soldering

Soldering is a joining process in which metal parts are bonded together by melting a filler metal with a lower melting point than the base metals into the joint. At its most basic, soldering is like gluing with molten metal. Soldering differs from welding in that the base metals do not melt or fuse together. Soldering is an essential skill for a variety of agricultural work, most commonly plumbing and electrical work.

Soldering Alloys

During the soldering process, the filler material, molten *solder*, flows over the surface of the base metal and fills the joint. Ideally, a layer approximately .002″ to .007″ thick of soldering alloy remains between the parts of a completed joint, **Figure 27-51**. Alloys used for soldering typically have a melting range from 190°F to 840°F (88°C to 449°C). Plumbers often use an alloy of 50% tin and 50% lead (50/50 Sn/Pb). Electrical work is often done with a 60/40 ratio of tin to lead. The tin content of a solder is always given first. Commonly available tin-lead solders melt at around 400°F (204°C).

Most alloy mixtures used for soldering are completely liquid at one temperature and completely solid at a different, lower temperature. This creates a range of temperatures at which the solder is in a plastic, or easily moldable, state. A few alloys, such as 63/37 solder, melt and solidify at exactly the

Safety Note
Always read adhesive labels thoroughly before selecting a product.

Figure 27-51. A soldered joint of two copper fittings. Only a thin layer of solder is necessary to form a soldered joint.

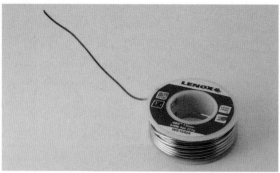

Figure 27-52. Lead-free soldering wire made of a tin/copper (Sn/Cu) alloy.

Figure 27-53. Flux being applied to a copper pipe prior to soldering. The flux helps clean the metal, creating better adhesion and a stronger joint.

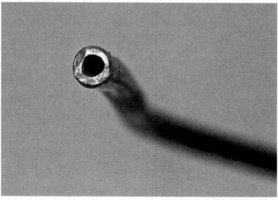

Figure 27-54. The hollow tip of flux-cored soldering wire.

same point. These are known as *eutectic* alloys and are much easier to work with. A range of alloys with different melting points and working properties are available.

Lead is highly poisonous if consumed or inhaled. Lead-free solders are used in most food-grade applications and in industry to reduce hazardous lead waste pollution. Lead-free solders are typically alloys of tin and antimony, silver, zinc, or indium. They may also contain small amounts of copper, bismuth, or aluminum.

Solder is usually found in the form of a rolled wire, **Figure 27-52**. Plumbers sometimes use heavier bars of solder.

Soldering Fluxes

Successful soldering requires the use of a *flux*, a substance that removes metal oxides and protects the metal from oxidizing during the soldering process. This cleaning action by the flux creates better adhesion at the joint and improved conductivity, **Figure 27-53**.

There are two principal types of flux: acid-based and rosin fluxes. Acid-based fluxes are used for metal assembly and plumbing applications. Rosin-based fluxes are used in electric and electronic applications where corrosive acid fluxes and the vapors released from soldering could damage the circuitry. Due to the corrosive nature of fluxes, excess flux should be cleaned from soldered materials on completion of the joining process.

Flux-cored solder is available for many applications, **Figure 27-54**. If the joints are small and not likely to be heavily stressed, this may be adequate. Large joints and critical junctions should be prepared with a separate fluxing step prior to the application of the alloy. Fluxes typically are supplied in a paste or liquid form and are applied with a small bush.

Heat for Soldering

Soldering requires a heat source. Although most solders are workable in the 400°F (204°C) range, the heat source needs to capable of consistently holding a temperature somewhere between 600°F and 700°F (316°C and 371°C). Heat should always be applied to the parts and not directly to the solder. Try to apply heat below the joint and allow it to be conducted through the material to the joint.

Soldering can be accomplished with a soldering iron, soldering gun, or small torch. Soldering irons are composed of heated metal tips attached to insulated handles. See **Figure 27-55**. Soldering irons can be heated with electricity or a small torch.

Soldering guns are pistol-shaped, electric soldering tools. Unlike electric soldering irons, which require time to warm up and then remain heated, soldering guns heat up rapidly when the trigger is pulled, but cool quickly if the trigger is released.

A variety of tip styles are available for both soldering irons and guns. Choose a tip with a chisel shape. Surface contact is key to heat transfer. Tips with flat surfaces make better contact than round or conical tips in most applications.

For some applications, such as joining copper pipe or attaching bolsters to a knife handle, an open flame supplied by a small, air-aspirated torch may be used to provide heat for soldering. See **Figure 27-56**. Always heat the metal, never the soldering alloy.

Preparing a Soldered Joint

Surface preparation at the joint is critical to successful bonding. All possible contaminants, such as dirt, paint, oils, or waxes, must be removed prior to soldering. Mechanical cleaning, brushing, scraping, or sanding should be done prior to using flux for chemical cleaning. Merely stripping the insulation from electrical wiring is often not enough. The oxidation on the wires must also be removed. Once parts are cleaned, do not touch bonding surfaces with bare fingers to avoid contamination.

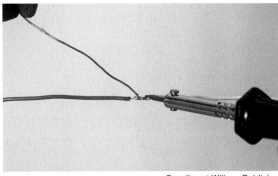

Goodheart-Willcox Publisher

Figure 27-55. An electric soldering iron being used to join two wires.

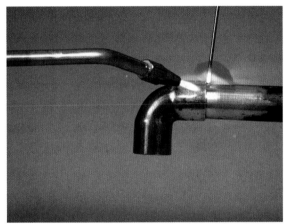

Goodheart-Willcox Publisher

Figure 27-56. A small torch being used to solder a copper fitting to a copper pipe.

Closely fitted joints yield superior results. Liquid solder flows by capillary action between the fitted parts. Soldering alloys are not particularly strong materials. The strength of a soldered joint depends on keeping the amount of bonding material to a minimum. Excess filler will weaken a soldered joint. A splice between wires should be twisted for a strong, close joint prior to the application of solder. Multiple strand wire should be twisted together prior to soldering.

Joints commonly require clamping to secure them in position for soldering. Clamps need to be positioned to hold without interfering with the heat transfer. A machinist's vise is not normally used for holding soldering jobs because the mass of the cast iron or steel vise will act as a heat sink, absorbing the heat needed to properly solder the joint. Instead, clamps should be used judiciously to keep parts in close contact. Special clamping devices are available for soldering small, difficult-to-hold joints.

Tinning

The tip of the soldering iron or gun must be tinned prior to use. *Tinning* is the process of applying a thin coating of soldering alloy to a surface prior to the joining process. Most soldering tips are a copper core surrounded by iron. Liquid solder will stick to the iron. A thin layer of soldering alloy protects the tip from rusting and ensures maximum heat transfer.. See **Figure 27-57**. Wiping the hot, freshly-tinned tip with a damp sponge will remove excess alloy and contaminants.

Tinning is achieved by applying only enough solder to form a thin surface layer. A second pass to add additional solder builds the joint. Smaller joints, such as splicing small-diameter wires, may be tinned while in contact with one another. Heating the parts to temperature, then touching the solder to the joint will cause the alloy to melt and flow over the surfaces, Figure 27-55.

Large soldered joints should be tinned prior to joining. Tinning ensures that the soldering alloy will flow completely throughout the joint.

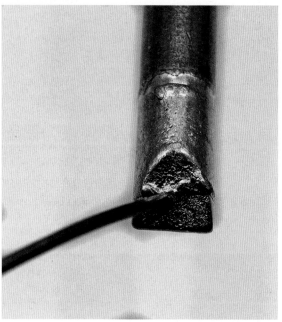

Goodheart-Willcox Publisher

Figure 27-57. Tinning the tip of a soldering iron helps protect it against corrosion.

Completing the Soldered Joint

Once the parts are properly positioned, apply heat to the base metal. If using a soldering iron or gun, position the tip where the flat, tinned face will make as much contact as possible with the base metal. Heat from the underside and allow the heat to rise into the part.

Watch the solder on the tip of the iron. When the proper soldering temperature is reached, the alloy on the tip will become molten and shiny. At this point, touch the end of the soldering alloy wire to the base metal just above the tip of the iron. Allow a small amount of solder to melt and be drawn into the joint. Continue adding solder to the joint until the entire joint is enveloped with the molten alloy.

Be careful not to add too much solder. Excess solder can create lumps or drips on the joint. Remove the heat source before gravity causes the molten solder to pool at the lower part of the joint. Recognizing this usually requires some practice. A rough joint can sometimes be smoothed with the reapplication of heat, but this can also release the bond. Allow cooling time before releasing parts from the clamps, especially when using non-eutectic solders.

At the end of a soldering job, wipe the tip clean with a damp sponge, cover the tip with solder, wipe it again, and then unplug the iron. This will flush and re-tin your tip, protecting it from corrosion while in storage.

Safety Note

When Soldering

Observe the following safety precautions when soldering:
- Always wear the proper personal protective equipment (PPE). OSHA-approved (Z87.1) goggles or glasses are a must.
- Always wear covered shoes. Molten solder can drip easily.
- Under certain conditions, a respirator and heat-resistant gloves may be needed.
- Work in a well-ventilated area. Be aware that too much cross breeze can interfere with uniform heating of parts.
- Handle parts below the heated area—heat rises!
- Use pliers when possible for handling small parts.
- Keep a damp sponge handy for cleaning the tip and wiping excess solder.
- Always place a hot soldering iron where the tip will not come into contact with unintended items such as power cords.
- Do not use an extension cord.
- Regularly check the power cord for burns or cracks. Replace worn cords before using.
- Make sure that the power cord is not positioned where it could be pulled off the workbench.
- Do not drop or handle the iron roughly. Heating elements can crack or break easily, especially when hot.
- Do not allow the iron to rest at operating temperatures for extended periods. This will reduce the working life of the tool.
- Occasionally remove the tip and lightly tap the barrel of wire-wound heater irons to remove loose oxides that form on the coil.
- When using flux cored soldering wire, crimp the end of the wire when finished to prevent the flux from leaking.

Heat-Treating Metal

The heat and repeated application of force metals endure during metalworking operations can change the physical properties of the metal. Cold forming a metal can make a metal harder and stronger, but also more brittle. Often this is a welcome side effect of the primary metalworking process. Sometimes, though, the changes are unwanted. *Heat-treating* is the controlled heating and cooling of metals to obtain certain desirable changes to their physical properties.

Steels with higher carbon contents exhibit the most pronounced effects from heating and cooling. *Quenching*, or rapidly cooling a metal in water or oil, hardens steels enough be sharpened and hold a cutting edge. See **Figure 27-58**. However, this also increases the brittleness and causes internal stresses within the metal.

Tempering is done after quenching to reduce brittleness or hardness. To temper steel, it is heated and held at a relatively high temperature and then

Goodheart-Willcox Publisher

Figure 27-58. Quenching a piece of heated steel in water. Quenching hardens the steel, but also makes it more brittle.

Figure 27-59. Annealing is accomplished by heating a part and allowing it to cool slowly. In this example, burying the hot file in lime insulates the part and allows the metal to cool slowly.

immediately cooled. The exact temperature depends on the particular metal and intended application. This information can be found in metalworking handbooks and steelmaker's catalogs. Tempering releases the internal stresses caused by quenching and reduces brittleness. The steel loses some hardness, but this is offset by the increased toughness and impact resistance.

Annealing reduces the hardness of metals so they can be cut or machined more easily. Annealing steel requires heating the part to a high temperature and allowing it to cool very slowly. For smaller operations, an agricultural mechanic can bury a part in a bucket of agricultural lime, ground limestone, or calcium carbonate. See **Figure 27-59**. Lime is a very good insulator and will hold heat in a part for several hours. Nonferrous metals can also be annealed.

CHAPTER 27 Review and Assessment

Summary

- Metalworking is key to the search for better ways of doing things and looking for ideas to repurpose old tools into new equipment in agriculture today.
- Metal used in agriculture comes in many shapes and forms suited to many different uses.
- A project plan for metalworking should include a drawing, bill of materials, cut list, and a list of procedures.
- Before metal parts can be cut, drilled, shaped, or otherwise modified for use, they need to be measured and marked.
- The specific type of metal and the required precision for the project are the major determining factors when selecting a cutting process.
- Metals may be shaped using a variety of hot and cold methods.
- Taps are used to thread holes in metal parts.
- Metal parts are commonly joined together using fasteners.
- Soldering can be used to bond metal parts securely for certain applications.
- Heat-treating processes such as quenching, tempering, and annealing are used to change the characteristics of metals.

Words to Know

Match the key terms from the chapter to the correct definition.

A. annealing	E. forging	I. mandrel
B. cap screw	F. galvanized	J. tinning
C. ductility	G. jam nut	K. welding
D. flux	H. malleability	L. wing nut

1. Cleaner used to reduce oxides on metal surfaces.
2. Serves as a locking device.
3. Guide for bending metal parts.
4. Designed to fasten materials without tools.
5. Forceful process of compressing and changing the shape of metals.
6. Ability to be stretched or drawn.
7. Coated with zinc.
8. Ability to be hammered or rolled into flat sheets.
9. Applying a thin coat of solder to a metal surface.
10. Fastener that threads directly into a metal part.

11. Softening process for high-carbon steel.
12. Joining metal by melting.

Know and Understand

Answer the following questions using the information provided in this chapter.

1. A complete project plan begins with some type of sketch or drawing and also includes a detailed _____ of materials.
2. *True or False?* Although soapstone is commonly used to mark aluminum, it will not satisfactorily mark mild steel.
3. A(n) _____ can be used to determine the relative hardness of metals prior to cutting or drilling.
4. Cutting or grinding close to the location of clamps or supports will reduce _____ and noise.
5. The individual _____ of an abrasive wheel serve as microscopic cutting edges.
6. _____ hot high-carbon steel in oil or water hardens the steel by locking the grain structure of the metal.
7. An indication that heat from grinding friction has overheated tool steel is a color change in the surface of the metal to a(n) _____ shade.
    A. orange
    B. yellow
    C. blue
    D. green
8. The first step in sharpening tools on the bench grinder is to adjust the _____ to the correct angle.
9. Rivets bind two or more metal parts with a head on one side of a hole and an upset or expanded tail on the other; _____ rivets can be used when only one side of the joint is accessible.
10. *True or False?* Wing nuts are designed to be used on equipment that must be serviced frequently.
11. When aluminum parts are attached to other metals, a(n) _____ is often used to reduce the chance of damage from galvanic corrosion.
12. Heat draws molten solder through _____ action between closely fitted parts.

STEM and Academic Activities

1. **Science.** Gather several different types and grades of metals. Use a grinder to perform spark tests. Prepare a demonstration for the class explaining how a spark test can aid in identifying metals.
2. **Technology.** Compare several methods that might be used to test the effectiveness of different ways of fastening metal parts.
3. **Engineering.** Closely examine a trailer or similar piece of agricultural equipment. Imagine that you were going to build the object from raw materials. What would be your order of operations? What special equipment or tools would be required?
4. **Math.** Closely examine a trailer or similar piece of agricultural equipment. Take measurements and develop a complete bill of materials that would be required to build it as a project. If you purchased standard lengths of materials, how much leftover material would you have after building the project?
5. **Social Science.** For many years, almost every town had a blacksmith. This occupation was essential to the well-being of other businesses in town. Why has this changed? What other occupations can you predict will follow the way of the local blacksmith?
6. **Language Arts.** Design and fabricate a replacement metal battery tray for a garden tractor. The main part of the tray should be rectangular, slightly larger than the footprint of the battery, with a 0.75" lip around the perimeter. Measure the battery, and then layout and cut the form needed from 16 gauge mild steel sheet. File the edges of the sheet metal to remove burrs, and then bend the edges to form the perimeter lip. Drill four holes in the tray for mounting hardware. When your tray is complete, write detailed step-by-step instructions that someone else could use to fabricate an identical tray.

Thinking Critically

1. You are tasked with designing a metal frame for a feeder constructed using a 55-gallon plastic drum split vertically for the trough. What material will you choose to build the frame? Explain the reasons for your choices.
2. Justify the following statement: The strength of a bond between metal parts is based on the surface area engaged in creating the bond.

CHAPTER 28
Oxyfuel Welding, Cutting, and Brazing

Chapter Outcomes

After studying this chapter, you will be able to:
- Explain the processes of oxyfuel welding and oxyfuel cutting.
- Identify the characteristics of gases used with oxyfuel torches.
- Safely handle and store gas cylinders.
- Safely work with oxyfuel equipment.
- Properly light and shut down an oxyfuel torch.
- Differentiate between types of oxyacetylene flames and their applications.
- Use a cutting torch.
- Weld with an oxyacetylene torch.
- Compare and contrast brazing and braze welding.
- Use a multiflame heating tip.

Words to Know

accelerant	cylinder cap	liquefied petroleum gas (LPG)	propane
acetylene	dross	manifold system	reverse flow check valve
braze welding	ethyl mercaptan	mild steel	single-piece torch
brazing	feather	multiflame heating tips	starving
carburizing flame	ferrule	neutral flame	torch body
cutting attachment	flashback	oxidizing flame	torch wrench
cutting lever	flashback arrestor	oxyfuel cutting	working pressure
cutting table	hose barbs	oxyfuel welding	
cutting tip	kindling temperature	pressure regulator	
cylinder			

Before You Read

Write all of the chapter terms on a sheet of paper. Highlight the words that you do not know. Before you begin reading, look up the highlighted words in the glossary and write the definitions.

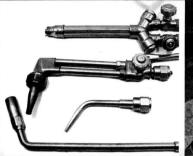

While studying this chapter, look for the activity icon to:

- **Practice** vocabulary terms with e-flash cards and matching activities.
- **Expand** learning with interactive activities.
- **Reinforce** what you learn by completing the end-of-chapter questions.

www.g-wlearning.com/agriculture

Lone Wolf Photography/Shutterstock.com

An oxyfuel torch mixes a compressed fuel gas with oxygen to create a concentrated flame that can be directed and controlled for the purposes of heating, cutting, or bonding metals. As early as 1776, the principle of oxyfuel cutting was understood; however, the fuel gases available did not burn hot enough to be efficient. It was not until 1836 that Edmund Davy discovered *acetylene*, a hydrocarbon gas that burns with a bright flame. The first practical use of acetylene was to generate light, **Figure 28-1**. Around 1900, the first successful acetylene torch was developed. World War I and World War II created a need for better metalworking technologies, particularly in the shipbuilding industry. The use of oxyacetylene developed quickly during this time.

Oxyfuel Welding and Cutting

Oxyfuel welding is a process of joining metals through the introduction of heat generated by burning gases. The fuel gas (typically acetylene) is mixed with the correct proportions of oxygen in the mixing chamber of a torch to produce a flame of just over 5700°F (3149°C). This level of heat makes it possible to bring metals such as cast iron, steel, copper, and aluminum to a molten state. In the molten state, two pieces of like metals can be fused in a welded joint. The combustion of oxygen and acetylene produces CO_2, which shields the molten weld pool from oxidation and other contamination. If filler metal (additional similar metal) is added to the joint, the resulting bonded joint can be stronger than the original base metal. See **Figure 28-2**.

Oxyfuel cutting uses acetylene and oxygen to preheat metal to a bright orange color. This color indicates *kindling temperature*,

Goodheart-Willcox Publisher

Figure 28-1. Coal miners in the past used lamps consisting of an acetylene generator and a small torch. The lamps were handheld or hung from a bracket on the miner's hat.

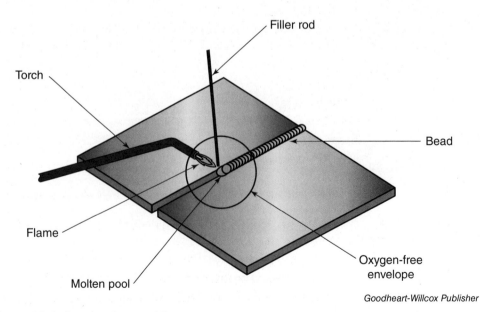

Goodheart-Willcox Publisher

Figure 28-2. Oxyacetylene welding.

approximately 1600°F (871°C), at which the steel readily reacts with oxygen. Then, a blast of pure oxygen is used to burn away the preheated metal. See **Figure 28-3**. The heat for the cutting process is derived from the burning of the base metal. Since oxyfuel cutting is an oxidation process, it is effective only on metals that are easily oxidized at this temperature, such as mild steel and low alloy steels. Oxyfuel cutting can be used to cut metal thicknesses from less than 1/8″ up to 12″.

Oxyfuel can be used for fusion welding, brazing, cutting, and bending. It can also be useful for flame hardening, hardfacing, soldering, and stress relieving, as well as for freeing rusted fasteners. This chapter explains the basic concepts of oxyfuel welding and cutting and describes basic equipment and safety precautions.

Goodheart-Willcox Publisher

Figure 28-3. The oxyfuel cutting process burns a cut into ferrous metals.

Chemistry of Burning Gases

Several fuel gases are commonly used with oxyfuel torches. Acetylene (C_2H_2) is the most popular and most widely used. Other common fuel gases are *propane* (C_3H_8), propylene (C_3H_6), butane (C_4H_{10}), MAPP gas (C_3H_4), and compressed natural gas (CH_4), or CNG. Fuel gases must be mixed with adequate amounts of oxygen for effective burning.

Oxygen

Oxygen (O_2) is naturally found in the atmosphere—the air we breathe consists of approximately 21% oxygen. Oxygen is not flammable; however, the presence of oxygen is required to support any burning process. Oxygen is an *accelerant*. When pure oxygen is mixed with a fuel gas, the temperature of the resulting flame and the combustion rate are increased. About two and one-half parts of oxygen per one part acetylene (2.5:1) is necessary to achieve complete combustion of acetylene. Typically, the torch provides a ratio of oxygen to acetylene of about one to one. The other one and one-half parts of O_2 are pulled from the atmosphere. The chemical reaction, or oxidation, of acetylene is represented by the formula $2C_2H_2 + 5O_2 = 4CO_2 + 2H_2O + heat$. As the fuel gas and oxygen are combined, carbon dioxide and water vapor are produced through the oxidation reaction. Because heat is produced during the oxidation process, it is considered an *exothermic* reaction.

Acetylene

Acetylene is a colorless gas produced by submerging calcium carbide in water. It can be identified by its distinctive, garlic-like odor. Acetylene is preferred as a fuel gas for welding because its molecular formula causes it to burn at higher temperatures than other options. The inner cone of an acetylene flame burns at almost 5600°F (3093°C). Acetylene can also be used for heating, cutting, and brazing.

SAE Connection

Metalworker

The oxyfuel processes allows for the design and fabrication of metal for a variety of purposes, from the construction of certain projects to the repair or maintenance of others. As you work in the agricultural mechanics lab of your school and learn how to safely use oxyfuel equipment, you may develop an interest in this area that can lead to a lucrative career as a metalworker. Metalworkers design, fabricate, and repair a variety of metal products.

Take the time to investigate potential jobs in your community and ask your agriculture teacher how you can get started gaining valuable experience in this area today. Remember to keep up with each new skill that you learn and your hours invested so you can accurately record this information when completing the appropriate proficiency application for your supervised agricultural experience (SAE) program.

gyn9037/Shutterstock.com

Safety Note

Propane and LPG are colorless and odorless gases. As a safety measure, a small amount of **ethyl mercaptan** is added as a scent marker. This compound emits a distinctive, strongly disagreeable odor for the purpose of leak detection.

Propane

Propane produces very high heat, but the temperature of the inner cone of the flame is not as hot as that of acetylene. The flame temperature of propane is about 5200°F (2871°C). Although not suitable for welding, propane and propane mixes are efficient for cutting and other heating needs when the proper torch is used. Propane is often mixed with similar gases, such as butane and/or propylene. These mixtures are sold as *liquefied petroleum gas (LPG)*.

Propane differs from acetylene in that it requires a relatively high oxygen-to-fuel gas ratio (4.3:1) to produce its maximum flame temperature. When propane is used as a fuel gas, oxygen cylinders need to be refilled or exchanged more often. In most locations, the lower cost of propane more than makes up for the difference.

Oxyfuel Equipment

A basic acetylene gas torch setup consists of a pressurized *cylinder* of acetylene, a pressurized cylinder of oxygen, pressure regulators for both tanks, gas hoses, the torch handle where fuel gas and oxygen are mixed, and the torch head attachment. See **Figure 28-4**. The assembly must be fastened securely to a building structure, mounted to a vehicle, or secured to a mobile cart. The valves on top of oxygen cylinders are set up with a male connection using standard clockwise threads. Fuel gas cylinder valves for both acetylene and propane have female connectors with counterclockwise threads. Cylinder valves also include pressure- and heat-sensitive emergency pressure release devices. These devices vent gases in a fire to prevent the cylinder from exploding. When cylinders are not in use, the valves should be protected from accidental damage with metal *cylinder caps* that fit over the soft valves and are screwed to the top of the cylinders.

Oxygen Cylinders

Standard oxygen cylinders are seamless steel tanks with wall thicknesses of at least 1/4″. When full, the most common size cylinder contains approximately 250 cubic feet (ft^3) of oxygen at a pressure of 2200 pounds per square inch (psi). Smaller cylinders have a lower capacity, but are also pressurized at 2200 psi. See **Figure 28-5**.

The orifice, or opening, at the top of the cylinder is about the size of a pencil lead. High-pressure gas accidentally released through this hole can cause an unsecured cylinder to become a violent projectile. A pressure relief device is incorporated into the valve to vent excessive pressure caused by overfilling or exposure to high temperatures, such as in a fire. The valve on an oxygen cylinder is made of brass and is constructed with two seats, or seals. It should always be turned all the way open or all the way closed. See **Figure 28-6**.

Goodheart-Willcox Publisher

Figure 28-4. Parts of an oxyfuel torch assembly.

Goodheart-Willcox Publisher

Figure 28-5. Cylinders for compressed oxygen are available in a variety of sizes. Each is compressed to about 2200 psi. Oxygen is sold based on the number of cubic feet of the uncompressed gas.

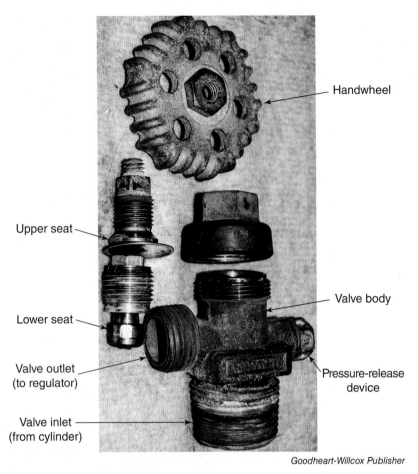

Figure 28-6. A disassembled oxygen cylinder valve. Note the two seats that seal pressure when the cylinder is closed or opened all of the way.

Acetylene Cylinders

Acetylene becomes explosively unstable when pressurized above 15 psi. To counteract this characteristic, acetylene is stored in a cylinder filled with a uniformly porous mass. See **Figure 28-7**. The pores of this mass are saturated with acetone. As acetylene is pumped into the cylinder, it becomes dissolved in the acetone. Because the pores separate the acetylene and acetone mixture into millions of tiny pockets, the acetylene is prevented from forming large pressurized concentrations. In this way, acetylene may be pressurized to around 250 psi in a full cylinder. A full cylinder is not completely filled to the top—a little reserve space is always left to absorb expansion from changing temperatures. Because the acetylene gas has to evaporate, or boil, out of the acetone solution, it may be safely drawn from the cylinder only at a rate not to exceed 1/7th of the cylinder capacity per hour. For example, the withdrawal rate from a 300 ft^3 cylinder should not be greater than 43 cubic feet per hour (cfh).

If the cylinder is placed on its side and then set upright again, it takes some time (as much as two full days) for the acetone to drain out of the dip tube in the valve and be redistributed. If the torch is operated during this period, acetone will be expelled from the cylinder, reducing the cylinder capacity and potentially damaging equipment. Acetylene cylinders should be kept upright at all times.

Acetylene cylinder valves should be opened no more than three quarters of a turn to make sure that an emergency shutdown can be performed in one quick motion. Two common styles of acetylene cylinders are used in the United States. One type has a permanent rounded handle, or handwheel, on the cylinder valve. The other type incorporates a valve that must be operated with a 3/8″ square wrench. See **Figure 28-8**. Some acetylene cylinders have a protective steel ring surrounding the cylinder valve.

Figure 28-7. Acetylene cylinders are packed with a uniformly porous mass to prevent the formation of large, volatile, pressurized concentrations of acetylene gas within the cylinder.

Figure 28-8. Different types of acetylene valves are available for various applications. Some jurisdictions require a specific valve type as a standard to promote safety.

Propane Cylinders

Propane and LPG cylinders are simple tanks of strong, light-gauge steel. Most LPG cylinders include a steel ring around the cylinder valve that protects the valve and acts as a handle for moving the cylinder. Common propane cylinders used with outdoor grills and recreational vehicles, **Figure 28-9**, may be used with a properly equipped oxyfuel torch. The valves on most cylinders of this type have two ports—one that accepts the quick connect female fittings found on propane cookers and one that accepts the left-hand threaded male fittings used on fuel gas pressure regulators.

When propane gas is compressed, it forms a liquid in the cylinder. Like acetylene cylinders, propane cylinders should be kept in the vertical position so only vaporized fuel is drawn from the container. Drawing fuel from an improperly positioned cylinder allows liquid propane to enter the regulator, hoses, and torch. Propane cylinder valves must be opened completely during use to ensure an adequate flow rate.

Propane cylinders should be stored in well ventilated areas. The density of propane causes it to be significantly heavier than air, resulting in a tendency for leaking gas to settle and collect in low areas rather than dissipate into the atmosphere. In fact, this is true for all varieties of LPG.

Figure 28-9. A common propane cylinder valve inlet accepts both the traditional male threaded connections found on many torch regulators and the female threaded connectors common on outdoor appliances, such as propane grills.

Regulators

An adjustable gas *pressure regulator* that controls the pressure of gas flowing toward the torch is essential. Regulators are constructed of soft brass or bronze because these materials have resistance to sparking and minimal susceptibility to corrosion. Typical regulators are equipped with two pressure gauges. The first measures the gas pressure in the cylinder, and the second indicates pressure in the hose and torch, or *working pressure*.

Many regulators function in two stages. See **Figure 28-10**. The first stage provides a higher pressure than is required by the torch. As the cylinder empties, the first-stage pressure drops gradually but remains higher than needed for torch operation. The second stage steps down the pressure and allows adjustment to specified levels required for proper torch operation. As a result, the pressure delivered to the torch remains constant, despite the decreasing cylinder pressure.

Simpler oxyfuel regulators have only a single stage, which results in a gradual decrease in outlet pressure as the cylinder is emptied. However, single-stage regulators are less costly. For many agricultural operations, the trade-off is acceptable. Only periodic readjustment of the working pressure is required. See **Figure 28-11**.

Regulators are engineered for specific gases and should not be interchanged. The fittings supplied with most units generally prevent connection to the wrong cylinders. Adapters should never be used to install the regulators on a cylinder of gas other than those it is designed for. The cylinder valve should be purged, or "cracked" open slightly before a regulator is attached to it. This is necessary to clear any contaminants from the passageway. Crack the valve quickly, releasing only a minimum amount of gas before closing it again. See **Figure 28-12**. In addition, the regulator's pressure adjusting screw should be backed out to the point that no pressure is applied to the internal diaphragm. This reduces the chances of damaging the regulator.

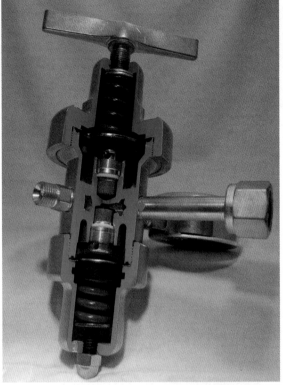

Goodheart-Willcox Publisher

Figure 28-10. A two-stage gas regulator reduces and controls the pressure of torch gases. Dual diaphragms and chambers provide a stable working pressure at the torch.

Goodheart-Willcox Publisher

Figure 28-11. Single-stage regulators use an economical single diaphragm and chamber to regulate gas pressure.

Goodheart-Willcox Publisher

Figure 28-12. Do not stand in front of the gas flow when purging the cylinder valve.

Safety Note

Do not stand in front of the gas flow when purging the cylinder valve. Position your face, hand, arm, and all other body parts away from the front of the pressure regulator while turning the handwheel slowly to release pressure into the regulator.

Regulator fittings are made of soft brass. Engage the threads, being careful not to cross-thread the connection. Lightly jiggle the regulator to allow the connection to be tightened most of the way. Complete the connection by tightening with a *torch wrench*, which is made to fit all of the fittings associated with an oxyfuel torch outfit. See **Figure 28-13**. The short length of this wrench and the sharpness of the grip are designed to purposefully limit the amount of torque that can be applied, preventing overtightening of brass fittings.

Goodheart-Willcox Publisher

Figure 28-13. A torch wrench has jaw openings for all of the common fittings found on a torch assembly.

When opening a cylinder valve, stand to the side of the regulator. Oxyfuel pressure regulators have pressure relief devices to prevent buildup of dangerous pressure. Cylinder pressure gauges are engineered with solid-front, safety-back construction. If exposed to excessively high pressure, the light-metal safety back is designed to pop off to relieve the pressure. The blast of venting gas can be violent. It is possible for old or improperly maintained oxygen regulators to ignite. Even a new oxygen regulator can burn out if the cylinder valve is abruptly turned on. The quick blast of high-pressure oxygen from the cylinder into the regulator chamber recompresses the gas inside the regulator, resulting in rapid exothermic heat buildup.

Safety Note

An oxygen cylinder should always be opened slowly to protect the regulator and the operator.

Hoses

Oxyfuel hoses are designed for use with specific gases and pressures. Hoses are usually color-coded to prevent improper use. Red is used for acetylene and other fuel gases, and green is used for oxygen. Black hoses are usually reserved for inert gases and compressed air.

Torch hoses consist of a tube and protective cover. Oxyfuel hoses can be found in three grades and four types. The three grades are as follows:

- Grade R is for acetylene only. The inside tube and outside cover are not flame-resistant and not oil-resistant.
- Grade RM is also for acetylene only. The inside tube is neither flame-resistant nor oil-resistant. The outside cover is flame-resistant and oil-resistant.
- Grade T can be used for most fuel gases. The inside tube and outside cover are flame-resistant and oil-resistant.

The four types of oxyfuel hoses are as follows:

- Type L is single-line, designed for light-duty service.
- Type S is single-line, used for standard-duty service.
- Type H is single-line, used for heavy-duty service.
- Type VD is bonded double-line (most often used with oxyfuel rigs).

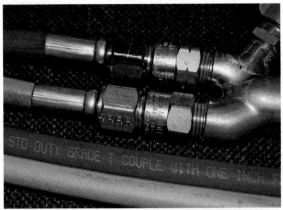

Figure 28-14. Torch hoses are typically colored green for oxygen and red for fuel. A grade T hose setup allows the safe use of any fuel gas.

Grade T hose should be used with any fuel gas other than acetylene, to ensure compatibility and user safety. See **Figure 28-14**. Grade T hose also has a self-extinguishing tube and cover that provides external and internal protection from conditions such as contact with hot slag, flashbacks, or inadvertent exposure to fire.

Grade R offers neither internal nor external flame protection. Grade R hoses are a little less costly than grade T hoses; however, the advantage of using grade T hoses is significant. The Compressed Gas Association (CGA) recommends using grade T hose for all oxygas torch applications.

Commercially manufactured torch hose assemblies include fuel gas line connectors with left-hand threaded brass nuts that are notched to identify them as designed to carry fuel gas. Standard right-hand threads are used for connecting oxygen hoses. Hose connectors have tapered, ringed nipples, known as *hose barbs*, which are inserted into the ends of the hose and secured by crimped-on brass collars, or *ferrules*. Torch hoses should never be used to suspend a torch or be pulled in such a manner that would cause the hose to bend or kink sharply. Damaged oxyfuel hoses can be repaired by cutting away the damaged section and attaching new connectors on the remaining good hose. Repairs must be made with the correct fittings and the proper tools. See **Figure 28-15**.

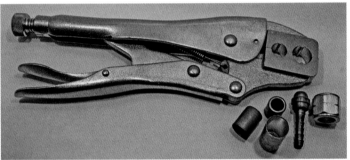

Figure 28-15. Brass connectors on the ends of torch hoses are crimped securely into place. This hose repair tool is used to crimp the sleeve to the outside of a hose that is slipped over the internal barb.

Torches

Oxyfuel torches mix oxygen and the fuel gases inside the torch. The torch operator controls the ratio of the gases by opening and closing the torch valves. The operator manually ignites the oxyfuel mixture at the torch tip, and the flow of gas through the torch sustains the flame. Adjusting the control valves determines the shape and intensity of the flame.

Torch assemblies vary according to the intended use (cutting, welding, brazing, or heating). Most manufacturers offer *light*, *medium*, and *heavy-duty* torches. The term *duty* in this case does not necessarily indicate durability. Torches built by reputable manufacturers are engineered to take decades of hard use—at least one manufacturer offers a lifetime warranty. *Duty* refers to physical size and the ability to produce gas flow rates capable of supporting specific attachments or tip sizes.

Heavy-duty *single-piece torches* are available that are engineered strictly for cutting. Single-piece torches range from compact units less than

12" long, designed for police or military breaching of barricades, to torches over 6' long that are used in steel salvage facilities. The advantage of a single-piece torch is its simplicity. Fewer valves and connections result in a sturdy, trouble-free design built for a single purpose—cutting steel. See **Figure 28-16**.

By far the most common type of oxyfuel torch is the two-piece torch, which consists of a *torch body*, or handle, that can be fitted with a variety of attachments for specific jobs. Attachments for cutting, welding, and heating are often included in basic kits with two-piece torches. See **Figure 28-17**. Attachments can be obtained for various specific heating or welding applications.

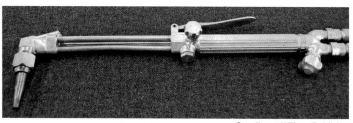

Goodheart-Willcox Publisher

Figure 28-16. A single-piece cutting torch is a sturdy tool dedicated to cutting steel.

Torch Bodies

Torch bodies consist of a handle that is textured to increase grip, a fuel control valve, and an oxygen control valve. The torch body has fittings at one end so oxygen and fuel gas hoses can be attached. Passageways for oxygen and fuel gas run through the handle, from the fittings to the other end of the torch body. The control valves regulate gas flow through these passages. Attachments are threaded onto the other end of the torch body and fitted at this connection into at least two seats. The gas mixing chamber is not in the torch body. Instead, the fuel gas and oxygen are prevented from mixing until they reach the mixing chamber in the attachment.

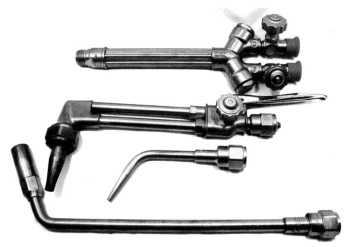

Goodheart-Willcox Publisher

Figure 28-17. Two-piece torches have interchangeable attachments that can be used to perform a variety of cutting, heating, and welding jobs.

Welding and Brazing Tips

Welding and brazing tips are simple angled tubes with a specifically shaped and sized orifice to produce a working flame of the desired size and shape. The mixing chamber is located at the base of the attachment. A mixture of fuel and oxygen flows through the length of the tip toward the opening on the end. On a tip designed for welding or brazing, a single orifice at the end of the tip provides the concentrated flame that is required for control of the heat used in metal joining processes. The flame on a welding tip is entirely controlled by the size of the tip and by the metering valves on the torch body. Different size tips may be selected using charts provided by each torch manufacturer. As the overall size of the tip increases, the size of the orifice also increases.

Welding generally requires more heat for the same thickness of metal than brazing. The thicker the base metal is, the more heat is required.

Safety Note

Never mix torch bodies and attachments from different manufacturers. Even if the threaded connections seem to match the actual seating or sealing, surfaces within the connections are made to different specifications and may be dangerously incompatible.

754 Agricultural Mechanics and Technology Systems

Safety Note

Cutting tips vary by manufacturer. See **Figure 28-18**. It is an unsafe practice to use tips in torches for which they are not designed.

Increasing the heat requires choosing a larger tip. All sizes of tips, when set to the proper flame, burn at approximately the same temperature. However, larger tips supply a bigger flame, resulting in a greater volume of heat.

Cutting Attachments

Cutting attachments allow the torch to burn through many ferrous metals. The tip of a cutting torch has a center cutting orifice, through which pure oxygen is pushed, surrounded by a ring of smaller orifices that supply oxyfuel mix for the preheat flames. *Cutting tips* are replaceable and are attached to the head of the torch with a large retaining nut that keeps the tip seated tightly. Cutting attachments include two additional valves for controlling the oxygen—a rotary valve for setting the preheat oxygen level, and a *cutting lever*–operated valve that opens and closes the oxygen flow through the cutting orifice.

When a cutting attachment is connected to the torch body, the oxygen control valve on the torch body is opened completely to bypass control to the control valve on the cutting attachment. The cutting lever should be treated as an on/off valve. It should either be completely closed or completely open. When operating the cutting lever, depress it fully in one smooth motion. Avoid working it like the throttle of a motor vehicle.

The mixing chamber may be located near the valves or closer to the flame in or near the replaceable tip. Cutting attachments, depending on the location of the mixing chamber, have two or three tubes extending from the valve assembly to the head.

Cutting Tips

Cutting tips designed for acetylene are made from solid pieces of copper alloy. The passageways, or tubes, extend through the length of the tips. The center orifice is the cutting orifice, and the outer orifices are the preheat orifices. There can be as many as eight preheat flames on tips designed for heavy cutting, or as few as one on flattened tips designed for cutting thin sheet steel with minimal distortion from excess heat. In practice, a preheat flame should immediately precede the cutting orifice during the cutting process. Typically, using a tip with larger orifices enables the torch operator to cut thicker sections of steel. The outer ring of orifices flow with mixed fuel and oxygen for the preheat flames, while the center orifice directs the cutting jet of pure oxygen.

Propane and other alternate fuels require the use of a two-piece tip. Two-piece cutting tips are formed from

Goodheart-Willcox Publisher

Goodheart-Willcox Publisher

Goodheart-Willcox Publisher

Figure 28-18. Acetylene cutting tips are designed with different sealing seats and output orifices.

Copyright Goodheart-Willcox Co., Inc.

a solid brass core having a central passageway for the cutting stream of oxygen and parallel slots or grooves surrounding the central tube. The second part of the tip is a smooth, often chrome-plated tube that fits snugly over the center tube, creating a ring of passages from the parallel slots. The density of the alternate fuels is the reason for this more open construction. See **Figure 28-19**.

Replaceable cutting tips control the capacity of a cutting torch. A tip with larger passageways and bigger orifices produces more heat, resulting in the ability to cut through thicker materials. All torch manufacturers supply tip selection charts. These charts are specific to each manufacturer and are not interchangeable. They indicate the thickness of steel that may be cut with each size cutting tip, as well as the recommended oxygen and fuel pressures for each operation. See **Figure 28-20**.

Safety Note

Always follow the manufacturers' cutting tip recommendations. It is unsafe to attempt cutting thicker steel simply by increasing gas pressures.

A large tip with too much capacity for the equipment can starve the flame, causing overheating and a possible flashback. Most cutting tips are suitable for a range of material thicknesses. Choose a tip that will accommodate most of the common sizes you will encounter. Not having to change cutting tips frequently can save time and wear on the seating surfaces of the torch.

A damaged seating surface on either the tip or the head can create a hazardous condition, resulting in a fire or flashback. This can potentially damage the cutting attachment. If the seating surface of a tip becomes damaged, do not use the tip. Some tip styles use replaceable O-rings to provide the seal, while others seal against precision-ground, unrepairable metal seats.

Multiflame Heating Tips

Multiflame heating tips, often referred to as *rosebuds*, are welding attachments with bulbous tips having a ring, or multiple rings, of orifices. Each orifice emits a separate flame that combines with the others, resulting in a high-volume heating tip capable of supplying the large amounts of heat necessary for shaping metal parts. See **Figure 28-21**.

Goodheart-Willcox Publisher

Figure 28-19. Cutting tips designed for propane and other dense LPG fuels are constructed in two pieces. A series of machined flutes directs the preheat gas mixture.

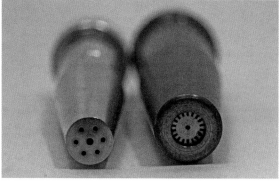

Goodheart-Willcox Publisher

Figure 28-20. Follow the manufacturer's tip selection chart when choosing torch tip sizes and setting regulator working pressures.

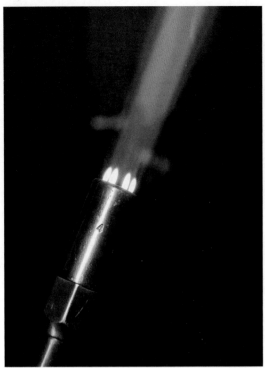

Figure 28-21. A multiflame heating attachment (rosebud torch) can be used to heat metal for bending and other purposes.

Figure 28-22. Some high-capacity heating torches require fuel to be supplied at rates that exceed the safe withdrawal rate of a single cylinder. A manifold system connects several cylinders for a higher capacity.

Rosebuds require large volumes of fuel gas and oxygen. Large multiflame tips may require a higher volume of gas than can be safely drawn from a single acetylene cylinder without exceeding the 1/7th of cylinder capacity limitation. The necessary volume can be provided with an alternative fuel gas for high-volume needs. Propane is effective for large heating jobs. Alternatively, two or more cylinders can be connected together in a *manifold system*. In this system, gas is drawn from several cylinders at once, **Figure 28-22**. Manifold systems can also be used when it is desirable to store fuel and oxygen cylinders outside of a building, with torch setups connected through a network of plumbing.

Safety Devices

A *flashback* is the recession of a flame into the tip or all of the way back into the mixing chamber of the oxyfuel gas torch. A flashback usually produces a distinctive hissing or squealing noise; if this happens, immediately shut off the torch. In severe cases, the flashback can travel back through the hose to the regulator, causing a fire at the fuel cylinder. Because the oxygen working pressure is typically higher than the fuel gas pressure, a flashback is most likely to occur in the fuel line. *Reverse flow check valves* allow gas to flow in one direction only. They can be installed on torch hose connections to reduce the chance of backflow of one gas into the hose of the other in the event of a sudden drop in pressure in one hose. However, reverse flow check valves are not designed to stop flashbacks. *Flashback arrestors* contain a stainless steel mesh filter designed to stop the progress of flashback traveling through gas passages. See **Figure 28-23**. Flashback arrestors may be built into torch or regulator fittings. However, for full protection, flame arrestors should be installed on both ends of torch hoses.

Safety

Oxyfuel torches are reliable and safe when operated in accordance with the manufacturer's instructions. An oxyfuel torch is one of the most versatile pieces of equipment at our disposal; it also is potentially the most dangerous. Oxyfuel is powerful enough to present a hazard to everyone in the immediate vicinity. Safety when using oxyfuel equipment means keeping yourself safe from injury and taking precautions to ensure that your coworkers remain protected.

The following list of precautions may seem long at first, but if you understand the process of using an oxyfuel torch, each precaution is logical and easy to remember.

Clothing and PPE

- Always wear the proper personal protective equipment (PPE). OSHA-approved (Z87.1) goggles or glasses are a must. A face shield may be required. Protective eyewear should have, at minimum, a shade 4 or filter lens.
- Wear gauntlet type, heavy leather welding gloves and covered leather shoes to help prevent weld burns and injury. Wearing a leather apron, welding chaps, or welding jacket further reduces the risk of burns.
- Do not wear clothing made of synthetic, petroleum-based fibers such as rayon or nylon while using a torch. Always avoid rolled or frayed cuffs.
- Do not saturate your clothing with oxygen or fuel gas. Before and while lighting the flame, keep the torch pointed away from your face and body. The presence of pure oxygen dramatically increases fire intensity, causing flammable materials like clothes to burn easily.

Storing Oxyfuel

- Do not allow anyone to smoke near the oxyfuel torch assembly. Smoking near leaking gas could provide ignition and cause a fire or an explosion.
- Store oxygen cylinders away from acetylene cylinders. At a minimum, a noncombustible wall at least 5′ high should separate oxygen and acetylene cylinders in storage.
- Chain cylinders securely in an upright position to a wall or cart.
- If cylinders will not be used for an extended period, or if they are being moved, remove the pressure regulators and keep safety caps in place to prevent damage to cylinder valves.
- Cylinders must be secured in an upright position for transport, prior to use, and during use. Do not operate the torch with the cylinders on their side.
- Close the cylinder valves and release pressure from the hoses before leaving the work area.
- If you are the first to enter a room or building and detect the odor of leaking fuel gas, leave immediately and inform your instructor or supervisor. Do not touch any light switches. A tiny spark could be enough to ignite the leaking fuel.

A *Goodheart-Willcox Publisher*

B *Goodheart-Willcox Publisher*

Figure 28-23. Reverse flow check valves and flashback arrestors are often included in the same device. A—Reverse flow check valves allow gas to flow in one direction only. B—Flashback arrestors have a stainless steel screen that extinguishes a flame traveling into a torch system.

Using Oxyfuel Equipment Safely

- Always report damaged equipment to your instructor or supervisor. Do not use equipment if its safety is questionable.
- Reverse flow check valves and flash arrestors should be installed on oxygen and acetylene lines.
- It is unsafe to mix torch parts from different manufacturers. Parts that seem compatible can be dangerous combinations.
- Never use petroleum-based lubricants on oxyfuel equipment. Oxygen can reduce the kindling temperature of oil and grease to room temperature. Petroleum products can ignite spontaneously when compressed in the presence of concentrated oxygen.
- Check all connections for leaking gases using a commercial testing solution or nonpetroleum-based soapy water.
- Before disconnecting a regulator from a cylinder, always release all pressure from the regulator.
- Regulators are delicate and must be handled carefully. Use the proper wrench to attach them. Never use pliers or a pipe wrench.
- Any hoses with leaks, worn spots, or burns must be replaced or repaired before use.
- Use only fittings that are bronze or brass (less than 65% copper) to repair torch hoses—never use copper. Copper in the presence of acetylene can form copper acetylide, a red, powdery substance that is highly explosive.

Using the Torch

- Always work in a well-ventilated area free from flammable debris or other items susceptible to damage from fire.
- Always have a fully charged fire extinguisher handy.
- Open the valve on an acetylene cylinder no more than three-fourths of a turn so it can be closed quickly in case of emergency. Leave the valve wrench in position when using fuel cylinders that require a wrench for opening.
- Open the valve on the oxygen tank fully to safely engage the upper seat in the valve.
- Never stand in front of or look directly at a regulator while opening a cylinder valve.
- If you experience a flashback or a fire somewhere along the lines, immediately shut off the oxygen. Then shut off the acetylene and allow the equipment to cool.
- Always use an approved friction or electronic striker to light the flame. Using matches to light the torch brings fingers too close to the tip, and cigarette lighters can explode in the hand or in a pocket. (Think 1/8 the energy of a stick of dynamite!) See **Figure 28-24**.
- Never use acetylene at a pressure exceeding 15 psi. Refer to the torch manufacturer's recommendations for the correct operating pressures for the metal thickness and tip size being used.

- Keep the flame and heat away from the cylinder and hoses.
- Never leave a burning torch unattended.
- Do not weld or cut on containers if there is any possibility that they may have held flammable materials.
- Handle hot metal with pliers or tongs. Do not leave hot metal unattended— unsuspecting persons may touch it and be burned.

Setting up the Equipment

Proper setup of the equipment is the first step to safe operation of the oxyfuel torch. When assembling a torch set, have a plan. Setup should be completed once started. Leaving a torch assembly partially connected is a recipe for trouble. Since all of the connections are made with soft brass fittings, be careful not to cross-thread or overtighten the fittings.

Follow these steps to properly set up an oxyfuel torch assembly:

1. Ensure that the cylinders are secured properly with chains or clamps. Remove the protective caps. If the caps are stuck, they can be tapped lightly with a small, non-sparking (brass, wood, rubber) hammer to free the rusted threads.
2. Quickly purge the valves and attach the pressure regulators, using only finger pressure to tighten the connectors until they are snug. Then, perform final tightening using a torch wrench.
3. If using separate flashback arrestors or reverse flow check valves, install them at this time.
4. Next, install the hoses, red for fuel gas and green for oxygen. Tighten connections with the torch wrench and purge each line separately.
5. Connect the torch body to the hoses, securing each connection with the torch wrench.
6. Assemble the remaining attachments to the torch body. On most torches, these connections need be only hand-tightened. Connect attachments so the tip of the torch faces the same way as the torch control valves. This configuration keeps the valves in a position less likely to be accidentally adjusted by contact with the operator's clothes or the heavy cuff of a welding glove.
7. At this point, check for leaks using an approved testing solution. Retighten any connection that produces bubbles. See **Figure 28-25**. Do not test for leaks using a flame.

MCarper/Shutterstock.com

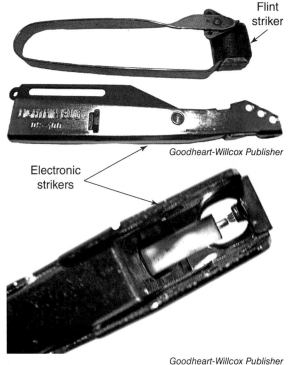

Goodheart-Willcox Publisher

Figure 28-24. Never use a butane lighter or a match to light a torch. Use a flint or electronic striker designed for the task.

Goodheart-Willcox Publisher

Figure 28-25. Check for leaks each time part of the torch system is connected or reconnected by applying a non-petroleum–based soapy solution to torch connections. The solution bubbles when a leak is present.

Lighting the Torch

Before lighting the torch, be sure that you and the work area are ready. The area should be well-ventilated, free from flammable materials, and cleared of hazards that could cause someone to trip. The material that is to be cut, welded, heated, or brazed should be well-supported and marked. Any additional supplies or tools should be handy and conveniently arranged.

Practice the following steps for lighting the torch until they become comfortable and can be completed in quick succession.

1. First check the pressure regulators to ensure that they are set to zero pressure. This setting reduces the chance that the introduction of high-pressure gas from the cylinder will damage the internal parts of the regulator.
2. Open the cylinder valves slowly while standing to the side. Turn the oxygen cylinder handwheel counterclockwise until it stops to ensure that the valve is completely seated in the open position. Open the fuel gas cylinder. Acetylene cylinders should be open no more than three-quarters of a turn, while propane cylinders should be fully open.
3. Adjust the pressure regulators according to the torch manufacturer's recommendations that match the thickness of the base metal and the tip size of the torch. The necessary tip selection charts are available in printed form and online.
4. Crack the torch fuel valve open just enough to allow gas to pass through while pointing the tip of the torch away from your body, **Figure 28-26A**.
5. Quickly ignite the fuel at the tip with a friction or electronic striker, **Figure 28-26B** and **C**.
6. Increase the acetylene flow by adjusting the torch acetylene valve until the flame just separates from the end of the tip. Then, reduce the acetylene flow until the flame just starts to produce black smoke around its edges. Finally, increase acetylene flow just enough to get rid of the black smoke.
7. Slowly open the torch oxygen control valve. If a welding or heating tip is installed, the control valve is on the torch body. If a cutting torch is installed, the control valve is located on the cutting attachment, and the torch oxygen valve is fully opened. Add just enough oxygen to the flame to create an intermediate flame between the outer flame and the inner blue cone. This third flame is called an excess acetylene *feather*.
8. Adding just a little more oxygen, shorten the excess acetylene feather until it merges with the inner blue cone. This results in a **neutral flame**, which has about the right characteristics for most welding and cutting purposes. Once the torch is lit, keep complete control of the torch—do not set down a lit torch.

Safety Note

Perform Steps 4 through 8 rapidly in order to reduce the release of excess fuel gas into the work area.

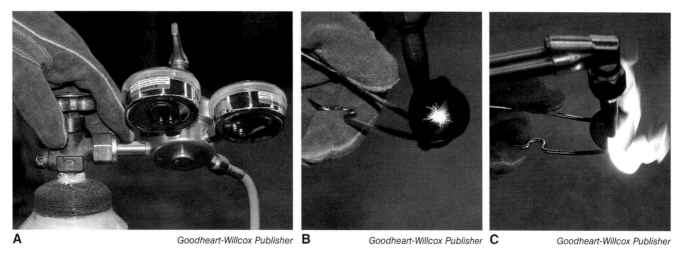

Figure 28-26. Lighting the torch. A—Open the cylinder valves. B—Use a striker to generate a spark. C—The spark ignites the fuel gas.

Types of Flame

With an oxyacetylene torch, the flame does all of the work. The torch should never touch the base metal or the filler material. The only functions of the torch are to create the correct gas mixture, which produces the correct flame, and direct the flame to the work.

There are three distinct types of oxyacetylene flames—the neutral flame, *carburizing flame*, and *oxidizing flame*. The flame type is determined by the ratio of oxygen to acetylene in the mixture produced by the torch. See **Figure 28-27**.

Neutral Flame

A neutral flame is produced when the ratio of oxygen to acetylene emitted from the torch is almost exactly one-to-one. When this quantity of oxygen combines with the oxygen available in the atmosphere, there is enough to completely burn or oxidize all of the acetylene coming from the tip. The flame is called *neutral* because it does not chemically react with the

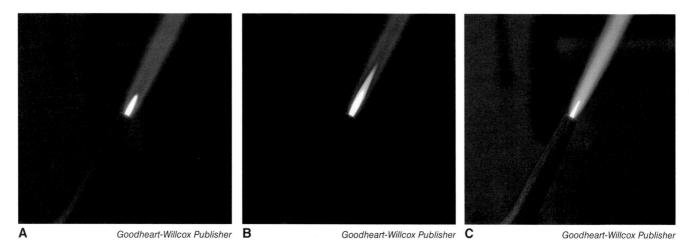

Figure 28-27. Types of oxyacetylene flames. A—Neutral. B—Carburizing. C—Oxidizing.

metal being welded. The consumption of the oxygen from the torch and the atmosphere prevents oxidization of the weld metal. By burning all of the fuel gas, the flame does not contribute to an increase in the carbon content of the welded metal. A neutral flame is required for most torch operations.

Carburizing Flame

A carburizing flame results when the proportion of acetylene to oxygen is higher than that required to produce a neutral flame. A carburizing flame can be identified by the presence of a distinctive feather of white-hot, unburned carbon particles appearing around and beyond the inner cone(s). The amount of acetylene excess can be determined by comparing the length of the acetylene feather to the length of the inner cone of the flame. For example, in a 2 × 2 excess acetylene flame, the feather is twice as long as the inner cone.

A carburizing flame tends to cause an increase in the carbon content of the weld metal when steel is welded. Some operations, such as hardfacing, require a substantial excess of acetylene. Hardfacing is a process in which a hard filler material is applied to cover and protect the surface of machine parts subject to extreme wear.

Oxidizing Flame

An oxidizing flame is created by an oxygen-rich fuel mixture at the tip of the torch. The excess oxygen combines with or burns some of the metal being welded. An oxidizing flame is easily recognizable by the sharp point created at the tip of the flame's shortened inner cone. It can also be distinguished by a distinct hissing noise that can be heard from quite a distance. Generally, in procedures that require a slightly oxidizing flame, it is more important that the fuel mixture *not* contain excess acetylene than it is for the mixture to have excess oxygen. A slightly oxidizing flame is used for procedures such as welding cast iron and brazing steel.

Torch Shut-Down Procedure

Using a safe and proper shut-down procedure is as important as being able to light the torch. The time to learn how to cut off the torch is prior to lighting it. All torch operators should be able to complete the shut-down procedure as one smooth series of quick tasks. To extinguish the flame, follow this procedure:

1. Quickly close the torch fuel gas valve.
2. Close the oxygen control valve.
3. Close both cylinder valves.
4. Purge the system by opening both torch valves to release all pressure from the torch, hoses, and regulators.
5. Check all of the gauges on the regulators; each should be at *0*.
6. Back out the pressure-adjusting screws of both regulators until there is no tension on them.
7. Close all torch valves.

Always follow the manufacturer's recommendations. Some manufacturers recommend closing the oxygen control valve first.

Using a Cutting Torch

A two-piece torch with a cutting attachment can be the most versatile tool in the shop. A cutting torch easily cuts most commonly used structural steels. It is also handy for removing damaged or stuck fasteners. See **Figure 28-28**. Even just the preheat flames of cutting torches provide plenty of heat for many common needs, such as bending or shaping small- to moderate-size parts.

Goodheart-Willcox Publisher

Figure 28-28. Use a torch to heat rusty connectors that are seized. The connectors will expand and loosen the connection so they can be removed without damage.

Burning Steel

As you begin to use the cutting torch, keep in mind how it actually works. The preheat flames bring the steel to kindling temperature, and then the stream of pure oxygen is released through the cutting orifice of the tip. The stream of oxygen ignites the steel and promotes the oxidation (burning) process, creating a narrow kerf through the base metal. The heat for the cutting part of the process is generated by the burning of the steel.

Steel is an alloy, or mixture, that consists primarily of iron and less than 1% carbon. When the steel is burned, where does it go? Iron combines with oxygen to form iron oxide, the most familiar form of which is rust. However, as the cutting process proceeds, iron oxide is carried from the kerf in the form of a waste material known as *dross*. The dross stream is a shower of sparks that cools to a dark gray material with particles ranging in size from dust to 1/2" or larger hollow globules. See **Figure 28-29**.

Goodheart-Willcox Publisher

Figure 28-29. A steady stream of dross flowing from the bottom of the kerf indicates that the cutting process is progressing correctly.

Cutting Table

When you are planning to cut with a torch, be sure that the hot stream of glowing dross has somewhere to go and does not create a fire hazard. Cut on top of a dedicated cutting table whenever possible. A *cutting table* is a steel working structure with a grated top set at a convenient working height for using a cutting torch. The grated top is typically constructed from replaceable steel slats and set over a nonflammable deflector that contains the dross stream to a safe area, away from the feet of operators and flammable objects. Cutting tables may include removable trays or similar methods of simplifying waste removal. See **Figure 28-30**.

Safety Note

Do not blow dross or direct flame on concrete surfaces. The heat could cause moisture that is trapped in the pores of the concrete to expand rapidly, blasting chips of concrete outward with dangerous force.

Goodheart-Willcox Publisher

Figure 28-30. A cutting table supports the work and provides a stable surface. It also collects the molten dross and small red-hot drop-off waste pieces.

Steps in the Cutting Process

Follow these steps to use an oxyfuel cutting torch:

1. Select the appropriate tip for the job. Make sure the torch tip is clean. If necessary, use tip cleaning tools to remove carbon from the orifice without enlarging or altering the shape of the orifice. See **Figure 28-31**. Purge the torch before use to remove any waste from the cleaning process.
2. Clean the base metal. While the oxyfuel torch is capable of working in the presence of a certain level of surface contamination, a clean surface almost always yields better results than attempting to cut through oxides, mill scale, and other contaminants. It may be necessary to grind or brush both the top and bottom surfaces of the material to clean it for cutting.

Safety Note

Be sure to wear the proper PPE when cutting. Always wear OSHA-approved (Z87.1) shaded goggles or glasses. Also wear a full face shield, leather welding gloves, long sleeves, and covered leather shoes. A leather apron or welding jacket is strongly recommended.

Goodheart-Willcox Publisher   *Goodheart-Willcox Publisher*

Figure 28-31. Torch tip cleaners include sets of file-like cleaning rods in sizes that exactly match the diameters of various torch orifices. The cleaning rods are made from a material that is softer than most torch tips to avoid damaging the tips.

3. Mark your cutting path on the surface with soapstone or another appropriate marking device. Remember to allow for the loss of material from the width of the kerf. Cut to the waste side of the mark, just like cutting lumber with a saw. See **Figure 28-32**.
4. Position your workpiece so that the cut will be easily accessible, and so that any pieces that drop off can do so without causing damage or injury.
5. Adjust the pressure regulators according to the manufacturer's recommendations for the metal thickness and tip size.
6. Light the torch and adjust it to a neutral flame.
7. Brace against the cutting table or other stable surface to stabilize both hands holding the torch. When holding the torch, rest your nondominant hand against the stable surface and use it to support the weight of the torch. With your dominant hand, grasp the torch handle to provide directional control and operate the cutting lever. See **Figure 28-33**.
8. Hold the torch flame with inner cones approximately 1/8″ above the steel. Make sure that the entire preheat flame is directed at the metal to reduce the preheat time.

Goodheart-Willcox Publisher

Figure 28-32. Mark steel to be cut with soapstone or an argent pencil. Cut on the waste side of the line to allow for the width of the kerf.

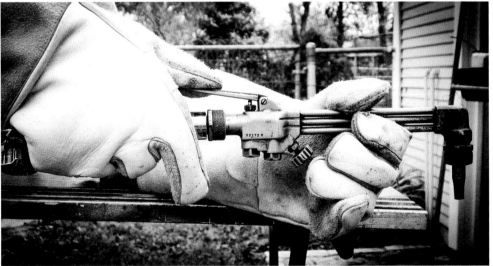

Goodheart-Willcox Publisher

Figure 28-33. When holding a cutting torch, rest your nondominant hand against a stable surface and use it to support the weight of the torch. Hold the torch handle in your dominant hand to guide the torch and operate the cutting lever.

Figure 28-34. As steel is heated, it changes color from red to orange and finally to a bright orange with slight movement in the surface. At this point, the steel is at kindling temperature.

9. Maintain the preheat until the base metal is bright orange and starting to glisten, indicating that kindling temperature has been achieved. See **Figure 28-34**.
10. With the torch held perpendicular to the surface of the material, press the cutting lever all the way down. Hold the lever in this position throughout the cut.
11. With the torch still held perpendicular to the material, move the torch at a steady speed to maintain the cut. The shower of dross exiting the kerf should determine the travel speed. It should remain constant throughout the cut.
12. When the cut is complete, release the cutting lever and close the torch control valves.

It is important to recognize the characteristics of a successful torch cut. First, the pieces should separate without hammering or prying. A good cut should follow the initial mark and have sharp, clean edges. Only a minimal amount of dross should be left adhering to the base metal. You should be able to easily chip away any remaining dross with a chipping hammer. The face of the cut should be free from deep gouges or high ridges—a good cut does not require grinding. See **Figure 28-35**. Time spent practicing with the torch will save grinding time.

Welding with Oxyfuel

Although oxyfuel welding has been performed for more than 100 years, its commercial use has waned in recent years. It is still useful in remote locations where electric power is unavailable. However, oxyfuel welding requires a great deal of manual manipulation by the welder. This makes it more time-consuming than most other welding methods, but an ideal choice

Figure 28-35. A properly cut section of steel (left) has clean edges with minimum adhering dross and shallow, parallel, slightly curved lines on the faces of the kerf. Poorly executed cuts (right) may have excess slag and jagged edges combined with cut lines that lack uniformity.

Hands-On Agriculture

Practicing Oxyfuel Welding

For this activity, you will need the following:
- 2″ × 4″ coupons of 1/8″ steel plate
- 3/32″ mild steel filler rods (For practice purposes, metal coat hanger wire works.)
- An oxyacetylene torch setup with a welding tip
- Shade #5 eye protection
- Leather welding gloves
- Pliers or tongs and a wire brush
- A welding table with a flat surface (Fire brick makes an excellent surface for this exercise.)
- Soapstone or a similar metal marker for identifying your completed coupon.

Follow this procedure:
1. Plan the path of your weld. Position the material and yourself where, as the welding process continues, your body will naturally shift toward a more comfortable position rather than toward a position that is uncomfortable or difficult to reach. Following this rule leads to smoother motion, resulting in higher-quality welds.
2. Set the regulator pressures according to the manufacturer's recommendations (usually a 1:1 ratio of acetylene to oxygen). Light the torch and adjust it so the middle blue feather and inner whitish-blue cone merge into a single region, indicating a neutral flame.
3. Apply the flame to the parts to begin heating. Use the region of the flame near the tip of the blue cone. Do not touch the metal with the inner cone of the flame.
4. The metal begins to glow. Continue heating both parts being welded until a small pool of molten metal appears near the edge of each of the parts. You must get molten pools on *both* parts simultaneously to create the weld. This may require adding more heat to one side than the other and takes some practice.
5. After the molten pools have formed on both sides of the weld, use the flame to gently stir the two pools together to form the weld. See **Figure A**. A small drop of filler metal from the filler rod may be required to begin this union. Introduce the filler rod into the edge of the flame with your free hand, keeping the tip just above and to the edge of the pool. Do not allow the filler rod to directly contact the base metal, the inner cone, or the torch. Instead, direct the droplets into the molten pool as the rod melts. This also takes practice.
6. After the two pools have joined, slowly move the flame along the weld line, lengthening the pool using metal from both parts. A gentle, semicircular, swirling motion helps mix the molten metal from both sides as the pool is lengthened. As you move the pool along the weld line, the bead (deposit of filler metal from a welding pass) solidifies as the heat is removed. This process is highly dependent on the materials and geometries of the parts being welded. Practice is the only way to develop better control.
7. Continue this process until the entire bead is complete.
8. Once you are done, turn off the flame by quickly closing both torch control valves.

Goodheart-Willcox Publisher

Figure A. Two pools, one in each of the parts to be welded, flow into a common molten pool of metal at the beginning of the weld joint

Safety Note

When oxyfuel welding, wear dark (shade 3–5) goggles or glasses to protect your eyes from the light of the flame and make the weld area more clearly visible. Wear a good pair of leather welding gloves. Make sure to wear a long-sleeved shirt made of all-natural fibers. Wearing a leather welding jacket provides added protection against burns.

for those wishing to develop a high level of skill in welding metals. Because the fusion process occurs slowly enough for an operator to get a clear picture of the process as it happens, many welding instructors choose to start novice welders with an oxyacetylene torch and some steel plate.

Mild steel, or steel having a low carbon content, is fairly easily welded with an oxyacetylene torch. With the proper fit-up and joint preparation, mild steel can be welded with or without the addition of a filler metal. A neutral flame is used for the process. Because burning acetylene uses atmospheric oxygen to reach total combustion, the pocket of gas around the flame and the weld pool contains little or no free oxygen to bond with the molten steel. No flux is required for the welding process—any impurities are floated to the surface of the weld pool.

Using 1/8″ thick or #10 gauge steel plate is ideal for practice because the material is lightweight, relatively inexpensive, readily available, and presents enough of a challenge to encourage the development of skill. A welder who is competent on thin materials can easily adapt to thicker metals. Welding an edge or outside corner joint without using a filler rod is a good weld to start with. See **Figure 28-36**. T-joints or butt welds with filler metal are a logical next progression. T-joints and butt welds are described in Chapter 29, *Shielded Metal Arc Welding*.

Goodheart-Willcox Publisher

Figure 28-36. An edge or corner joint can be welded without the addition of filler material.

Braze Welding and Brazing

The *braze welding* process heats and melts the filler material at a temperature above 800°F (427°C) but below the melting temperature of the base materials. It does not depend on capillary action for the distribution of the joining material as does *brazing*, which uses similar filler metals and fluxes. Both processes use a brass or bronze filler material that is an alloy consisting of roughly 60% copper and 40% zinc with small amounts of tin, iron, manganese and silicon. Brazing is similar to soldering except it is performed at temperatures higher than 800°F (427°C).

The two base metals in a braze-welded joint do not truly fuse. However, there is a very narrow zone, which can be seen only under a microscope, in which the atoms of the base metal and brazing alloy have mixed. The resulting bond is stronger than a soldered joint, which relies on the surface tension and tensile strength of the solder material. Like soldering, brazing and braze welding can be used to bond dissimilar metals. See **Figure 28-37**.

STEM Connection

Capillary Action

Capillary action is the result of molecules that have adhesion properties (dissimilar molecules attract each other and thus "stick" together) as well as cohesion properties (similar molecules stick together more than they break apart). These forces work together to cause capillary action that pulls liquid around the soldered or brazed joint. The metal tries to stick together in its liquid phase, but it also sticks to the side of adjoining metal. Because it is both sticking together and sticking to the metal, it is pulled into the joint and fills all the space.

Braze welding is often a better choice than fusion welding for bonding metals if heat input could damage the part. Cast iron parts are particularly susceptible to cracking due to heat from welding processes, but cast iron can be successfully braze-welded. Properly braze-welded steel and cast iron parts exhibit joint strength equal to, or even superior to, the original strength of the base metal. The ductility (ability to deform under stress) of bronze filler metal makes braze-welded joints successful in repairs of many parts subject to stresses of vibration. At temperatures over 400°F (204°C), braze-welded joints begin to lose integrity. For this reason, parts bonded by this method should be used only in applications at temperatures below this level.

A flux must be used with any brass or bronze filler material. Application of flux encourages complete and positive tinning. The flux reduces fuming of the zinc and works like a detergent in the molten filler, causing impurities to float to the surface. The flux may be precoated on the filler rod or applied to the rod by dipping the heated end of the filler into a can of powdered flux. See **Figure 28-38**.

When braze welding, select a welding tip that is at least one size smaller than would be used for fusion welding material of the same thickness. Use a slightly oxidizing flame to prevent carbon contamination of the joint. Start with a relatively soft neutral flame, less than the maximum size achievable with that tip. Then adjust the acetylene flow back just enough to cause a barely visible shortening of the flame's inner cone. Heat the base metal by moving the flame back and forth over an area larger than the joint. Heat the area until it just begins to glow. Using a precoated filler rod is recommended for

Yotsatorn Laonalongit/Shutterstock.com

Figure 28-37. Carbide cutting tips are often braze-welded to high-carbon tools to add strength and durability to saw blades, drill bits, and machining tools.

Goodheart-Willcox Publisher

Figure 28-38. Bronze filler rods.

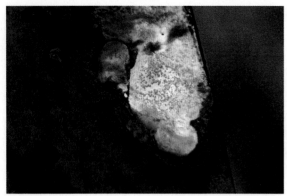

Figure 28-39. As metal is heated to the proper temperature, the bronze filler flows in a thin layer, tinning the surface of the base metal.

Figure 28-40. The underside of a joint that has undergone destructive testing. The joint was bent back on itself, showing the attachment of the root of the bead to the base metal.

beginners and preferred by most who have tried them. Otherwise, while heating the metal, heat the end of the filler rod and dip it in the flux to get it well-coated.

When the base metal begins to glow, concentrate the flame on the rod until a small portion drips off onto the hot base metal. If the metal has been heated to the proper temperature, the melted bronze will flow evenly across the surface of the metal, tinning it. See **Figure 28-39**. If it sits on the surface in a ball, the base metal is still not hot enough. If the bronze spreads, bubbles, and forms small droplets, the temperature is too high.

Once tinning action has begun, continue adding filler material and allow it to form a pool, or puddle. "Walk" the puddle along the joint to form a bead. Be cautious to not melt rod into the puddle faster than the base metal ahead of the puddle has had a chance to tin properly. Overheating the base metal or the puddle causes the bead to be too liquid and uncontrollable. Correct technique results in a uniform ripple effect on the surface of the bead. Properly executed braze joints are very strong. See **Figure 28-40**.

To develop skill controlling the heat and the filler rod, start practicing braze welding by laying a series of beads on the surface of a piece of thin steel plate. Then progress to butt and T-joints on 1/8" to 1/4" steel plate. Remember that parts must be absolutely clean throughout the area where the joint will be braze-welded.

Using a Multiflame Heating Tip

Oxyacetylene torches have been used for shaping and bending tasks, including delicate ornamental ironwork as well as straightening large structural steel parts. For making ornamental ironwork and other metal sculpture work, an oxyfuel torch is one of the artisan's most powerful tools. The ability to direct heat directly toward a specific area of the part allows precise bending as well as the ability to cut, braze, and weld.

Often, structural frame members such as angle iron, channels, and *T*s can be reshaped for special purposes by using the torch for both cutting and heating. A common example is the construction of a square or rectangular frame using angle iron. See **Figure 28-41**. The most efficient method of constructing such a frame is to use a single section of angle iron, notch it with a cutting torch at the corner marks, then use the torch to heat at the corners to form smooth, properly positioned corner bends. Where the cut edges come together, the frame is then welded. Selection of the correct tip size results in minimal use of fuel gas during the job. A larger tip builds up the required

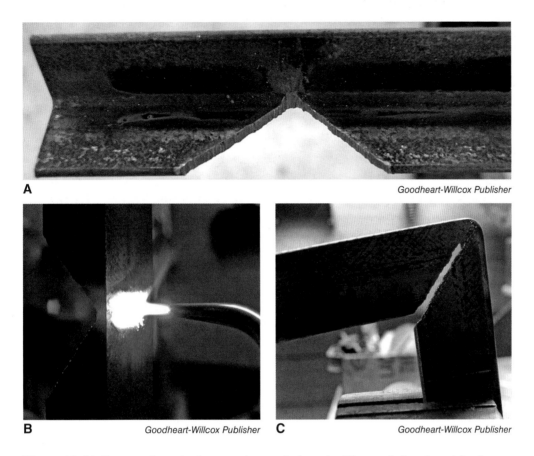

Figure 28-41. Construction of a frame using angle iron. A—The angle iron is notched. B—The angle iron is heated for bending. C—The angle iron is bent to form a 90° corner.

heat more quickly than a tip that is too small. For this example, a multiflame torch that burns 80 ft³ of acetylene per hour is more economical than a head that burns 20 ft³ of acetylene per hour.

What about the rule that acetylene should not be withdrawn from a cylinder at an hourly rate greater than 1/7th of full cylinder capacity? According to the rule, 80 cfh is a higher withdrawal rate than can be safely drawn from a single cylinder. How is this accomplished safely? The key is that continuous operation is not required. If the acetylene withdrawal rate is limited to 5–10 minutes at a time, the high withdrawal is restricted to short periods not to exceed 10 minutes, and such demands are spaced to a maximum of once every 30 minutes, then a large capacity tip can be used with a single cylinder.

Multiflame attachments should always be operated at pressures and delivery rates recommended by the torch manufacturers. Attempting to use a lower working pressure than recommended will result in *starving*, or choking, the torch. This is likely to cause overheating of the attachment and may result in a flashback. If the distinctive hissing sound of a flashback is heard, a flame is burning inside the nozzle of the heating head. Immediately turn off the oxygen control valve on the torch handle. Then, turn off the fuel valve and allow the torch to cool down before relighting. If a flashback reoccurs, have the torch and regulators checked by a qualified technician before reusing them.

CHAPTER 28
Review and Assessment

Summary

- An oxyfuel torch mixes a compressed fuel gas with oxygen to create a concentrated flame that can be directed and controlled for the purposes of heating, cutting, or bonding metals.
- Acetylene is preferred as a fuel gas for welding because it burns at higher temperatures than other options.
- A basic acetylene gas torch setup consists of a pressurized cylinder of acetylene, a pressurized cylinder of oxygen, pressure regulators for both tanks, gas hoses, the torch handle where fuel gas and oxygen are mixed, and the torch head attachment.
- Acetylene and propane cylinders should be kept in an upright position at all times.
- An adjustable gas pressure regulator controls the pressure of gas flowing toward the torch.
- Oxyfuel torches mix oxygen and the fuel gases inside the torch. Torch assemblies vary according to the intended use.
- A flashback is the recession of a flame into the tip or all of the way back into the mixing chamber of the oxyfuel gas torch. Flashback arrestors contain a mesh filter of stainless steel that is designed to stop the progress of flashback traveling through gas passages.
- When using oxyfuel equipment, take precautions to keep yourself safe from injury and ensure that your coworkers remain protected. Safety considerations include proper clothing and PPE, proper storage of oxyfuel, and correct use of oxyfuel equipment and torches.
- Oxyacetylene flame types are the neutral flame, carburizing flame, and oxidizing flame. The flame type depends on the ratio of oxygen to acetylene in the mixture produced by the torch. A neutral flame is required for most torch operations.
- The preheat flames of a cutting torch bring the steel to kindling temperature. A stream of pure oxygen released through the cutting orifice of the tip ignites the steel and promotes the oxidation process, creating a narrow kerf through the base metal. The heat for the cutting part of the process is generated by the burning of the steel.
- Oxyfuel welding requires a great deal of manual manipulation by the welder.
- Mild steel is fairly easily welded with an oxyacetylene torch, with or without the addition of a filler metal.
- Braze welding and brazing heats and melts the filler material at a temperature above 800°F (427°C) but below the melting temperature of the base materials.
- Brazing is similar to soldering except it is performed at temperatures higher than 800°F (427°C).

Words to Know

Match the key terms from the chapter to the correct definition.

A. acetylene
B. carburizing flame
C. cutting tip
D. dross
E. ethyl mercaptan
F. feather
G. ferrule
H. flashback
I. hose barbs
J. liquefied petroleum gas (LPG)
K. manifold system
L. neutral flame
M. oxidizing flame

1. Characterized by a sharply pointed inner cone and a distinct hissing noise.
2. Two or more fuel cylinders connected together to operate a large multiflame tip.
3. Attached to the head of the torch and secured with a retaining nut.
4. Tapered, ringed nipples used on the ends of flexible fuel gas and oxygen supply lines.
5. Brass collar used to crimp fittings on the ends of hoses.
6. Distinctive scent marker added to propane and other fuel gases.
7. Results when the ratio of oxygen to acetylene emitted from the torch is almost exactly proportional.
8. C_2H_2—the most popular fuel gas for torches.
9. Flame between the outer flame and the inner blue cone.
10. Mixture of propane and other fuels stored in liquid form.
11. Waste material from the kerf produced by oxyfuel cutting.
12. Hazardous situation in which the flame moves to the tip or into the mixing chamber of the torch.
13. An excess of fuel gas results in carbon particles appearing by the inner cone.

Know and Understand

Answer the following questions using the information provided in this chapter.

1. When mild steel reaches its _____ temperature, it can be cut with a concentrated stream of pure oxygen.
2. Acetylene is made by submerging calcium carbide in _____.
3. A(n) _____ should be in place to protect the soft valve whenever a compressed gas cylinder is not in use.
4. A standard size compressed oxygen cylinder contains 250 cubic feet of pure oxygen and is compressed at a pressure of _____ psi.
5. Which hose grade is designed for use with propane and other fuel gases?
6. The most common type of oxyfuel torch is the _____ torch.
7. Cutting tips designed for acetylene are made from solid pieces of _____.

8. Large multiflame tips, also called _____, may require acetylene to be withdrawn from a cylinder at a rate that is greater than 1/7th of the cylinder capacity per hour.
9. If the torch flame disappears and a shrill hissing sound is heard, the torch should be quickly shut down because this is an indication of a(n) _____.
10. Compressed gas cylinders should be transported, stored, and used in a secured, _____ position.
11. *True or False?* Petroleum-based lubricants are safe to use on oxyfuel equipment.
12. What should be used to handle hot metal?
13. How can a carburizing flame be recognized?
14. How can an oxidizing flame be recognized?
15. After a successful torch cut, a minimal amount of _____ should remain adhering to the base metal.
16. During oxyfuel welding, the welder should *not* allow the filler rod to directly contact the base metal, the _____ of the flame, or the torch.
17. Why must flux be used with brass or bronze filler material in the brazing process?
18. What would happen if a welder used a lower working pressure than recommended when operating a multiflame attachment?

STEM and Academic Activities

1. **Science.** The oxidation of acetylene is represented by the following formula:
   $2C_2H_2 + 5O_2 = 4CO_2 + 2H_2O + heat$
   What is the chemical formula for burning propane?
2. **Technology.** Compare the preheat flame of a cutting torch to the flame of a welding tip. Explain how each flame is specifically suited to the task for which it is designed.
3. **Engineering.** Draw a sketch of the torch cutting process. Label your drawing.
4. **Math.** Fit a torch tip cleaner to the orifice of a used welding tip and then use a micrometer to measure the diameter of the tip cleaner. How close to the manufacturer's specification (found online) is your measurement?
5. **Language Arts.** Take notes on a topic in this chapter, such as gases used in oxyfuel welding, cylinders, safety, torch shut-down procedures, or another topic. Using your notes, explain the topic to another student. Then reverse roles.
6. **Social Science.** Liquefied petroleum gas is used with oxyfuel torches. List several other agricultural and residential applications of LPG.

Thinking Critically

1. Research pressure regulator settings needed to cut 3/8" steel plate for two or more cutting torches from different manufacturers. Why are the settings different for torches of the same size? Why is it important to follow the recommendations that are specific to your torch?
2. What are the similarities and differences between cutting steel with propane and cutting steel with acetylene? How do the differences affect safety?

CHAPTER 29
Shielded Metal Arc Welding

Chapter Outcomes

After studying this chapter, you will be able to:
- Explain the shielded metal arc welding (SMAW) process.
- Properly use SMAW equipment.
- Describe various joint types.
- Describe AWS welding position classifications.
- Explain how to prepare the surfaces of pieces to be welded.
- Control welding variables to produce a good weld.
- Identify weld discontinuities.
- Take appropriate safety precautions when SMAW welding.
- Weld a practice pad using SMAW.
- Explain how to control distortion in a weld.
- Recall the special requirements for welding cast iron.
- Explain the certification process for professional welders.

Words to Know

arc force
arc length
auto-darkening lenses
beveling
butt joint
constant-current machine
corner joint
coupon
crack
destructive weld testing
discontinuities
duty cycle
edge joint
electrode
electrode holder
fast-fill electrode
fast-freeze electrode
ferrous
fillet weld
fill-freeze electrode
flux
groove weld
heat-affected zone
inclusions
incomplete fusion
incomplete joint penetration
inverter
keyhole
lap joint
nondestructive testing
out-of-position welding
overlap
polarity
porosity
radiographic testing
rod oven
root bead
shielded metal arc welding (SMAW)
slag
spatter
stringer bead
tack weld
tie-in
T-joint
transformer welder
travel speed
ultrasonic testing
unacceptable weld profile
undercutting
weaving bead
weave pattern
weld specifications
welding current
welding defects
welding leads
welding procedure specification (WPS)

Before You Read

Arrange a study session to read the chapter with a classmate. After you read each section independently, stop and tell each other what you think the main points are in the section. Continue with each section until you finish the chapter.

Lone Wolf Photography/Shutterstock.com

While studying this chapter, look for the activity icon to:
- **Practice** vocabulary terms with e-flash cards and matching activities.
- **Expand** learning with interactive activities.
- **Reinforce** what you learn by completing the end-of-chapter questions.

www.g-wlearning.com/agriculture

The *shielded metal arc welding (SMAW)* process, also known as *stick welding*, involves using a consumable *electrode*, or metal rod, coated in a flux to form the welded bead. Electricity from a welding machine creates an electric arc between the electrode and the base metals. *Welding current* is the flow of electricity that results in the welding arc. The amount of heat generated by the arc is directly proportional to the amount of current produced by the power source. When the welder establishes the arc, the base metal and the electrode melt to form the weld pool, **Figure 29-1**. As the molten pool cools, it resolidifies to form a bonded joint. The *flux* coating on the electrode chemically reacts to the heat of the arc, using up the oxygen in the vicinity of the weld to create an oxygen-free shield of gas around the molten pool. Residue from the flux forms a layer of slag that protects the weld from atmospheric contamination as it cools. See **Figure 29-2**.

Goodheart-Willcox Publisher

Figure 29-1. Welders use electric energy from a welding machine to form an electric arc between the electrode and the base metal.

Goodheart-Willcox Publisher

Figure 29-2. The SMAW process.

Development of Electric Arc Welding

The art of blacksmithing was developed during the Middle Ages. Blacksmiths, through forging processes, formed and welded items of iron by hammering red-hot pieces together until they bonded, **Figure 29-3**. Developed near the close of the nineteenth century, carbon arc welding (CAW) was the first arc welding process. In this process, an arc is created between a copper-clad carbon electrode and the base metal. The intense heat of the arc melts the surfaces of the base metal to be joined. A separate filler rod, which carries no electrical current, is often added to the weld, similar to the way filler metal is added in the oxyacetylene welding process.

The first US patent for an arc welding process using a metal electrode was awarded in 1890 to C. L. Coffin of Detroit, who used an uncoated, electrically conductive wire that melted to become part of the weld. The first coated metal electrodes, introduced in about 1900, had a thin coating of clay or lime, providing a more stable arc. During the 1920s, electrode coatings were developed that not only stabilized the arc but also improved the physical characteristics of the completed weld joint.

The shielded metal arc welding process is best suited to bonding *ferrous* (iron-containing) metals, including mild steel, carbon steel, stainless steel, and cast iron. Aluminum and other alloys can be welded using the SMAW process, but better options are usually available. A big advantage of the SMAW process is its adaptability. Switching electrodes and making simple output adjustments to the power source allows the welder to perform a variety of welding tasks.

Equipment

The setup for SMAW consists of the welding machine, electrode and workpiece welding leads, and the consumable electrode.

Welding Machines

There are several common types of arc welding machines or power sources, **Figure 29-4**. All machines for SMAW are *constant-current machines*, which means they provide an even supply of amperage to the electrode.

Valeriy Lebedev/Shutterstock.com

Figure 29-3. Blacksmiths were some of the first welders.

SAE Connection
Shielded Metal Arc Welding

Shielded Metal Arc Welding is generally more economical than other metal joinery methods and does not require a lot of space. If you have an interest in this process, you may be able to purchase the equipment needed to build metal projects using this technique for a reasonable price. Many backyard BBQ grills, small utility trailers, and artistic creations have been constructed by students using SMAW.

As you study this chapter, discuss with your agriculture teacher the potential for beginning a supervised agricultural experience (SAE) program that uses arc welding equipment to create projects. Estimate the cost of the equipment and other items needed to create your welding projects and then calculate the potential profit that can be generated by selling the items you create.

A welding-related SAE may lead to a profitable career as you progress in skill and obtain your welding certification. Remember to document your investment of time and money to allow you

Paul Broadbent/Shutterstock.com

to accurately complete a proficiency application. Take full advantage of the opportunities available in class and as an FFA member to advance in this area. Experienced welders make a good living and have the option of working at various companies, for themselves, or a combination of both.

Figure 29-4. SMAW power sources are available in many different designs.

Transformer Welders

A *transformer welder* is the oldest type. It directs the high voltage supplied by the power company through a primary coil wrapped around an iron core to create a magnetic field. A secondary coil is passed through the magnetic field, inducing a stepped-down or lower voltage. The actual welding current is produced by the secondary coil. The output from a transformer is AC power. Welding machines that incorporate a bridge rectifier in combination with the transformer can produce DC welding current output. A *rectifier* is an electrical device that converts AC into DC by causing the current to flow through the rectifier in one direction only. Transformer welders are heavy machines. They are available in many configurations ranging from the expensive, high-amperage, multipurpose units used in industry to simple, low-amperage, AC machines sold to home welders.

Inverters

A second type of welding machine uses an electronically controlled *inverter* power source to produce a usable welding current. An inverter passes single-phase or three-phase AC power through a bridge rectifier to create a DC current. The DC current is electronically broken into very high-frequency AC current, which is then passed through a small, efficient transformer to produce low-voltage, high-current AC power useful for welding. An additional bridge rectifier and an inductor are used to create smooth DC welding output. Inverter welders are lightweight and produce a smooth welding arc. Inverter power supplies provide AC and DC welding options.

Both transformers and inverters are most often powered by common AC electricity. Portable welders are available with their own generators or alternators, which are powered by internal combustion engines. These portable welders are commonly used in remote locations, such as construction sites or pipelines. They typically provide power for welding current as well as for other operations.

Duty Cycle

Welding machines are classified by their maximum amperage output and their *duty cycle*. A duty cycle is a rating that indicates how long a welding machine can be used at a given output current without damaging it. Duty cycle is based on a 10-minute time period while the machine is switched on. A welding machine with a 50% duty cycle can produce a welding current at its maximum rated output current for 5 out of every 10 minutes. It must be left running to cool down for the remaining five minutes. A welding machine is likely to overheat if the duty cycle is exceeded. Some machines have an automatic cutout, or switching device, to prevent damage to the power unit. At lower current settings, the duty cycle typically increases and the power source may be used for longer periods of time.

When selecting a welding power source, choose a unit that can meet most welding needs at the middle of its output range. For example, if most of your welding will be done at the 100 A to 120 A level, a machine rated at 200 A to 250 A maximum output will supply adequate power at a reasonable duty cycle.

Welding Leads

The cables that connect the welding power source to the work are known as *welding leads*. Welding leads must be able to carry large amounts of current while remaining flexible enough to be manipulated by the welder. Since electrons travel on the surfaces of a conductor, a multiple-strand conductor can carry more current than a single-strand conductor of the same size. Welding leads consist of a twisted copper wire core composed of many tiny strands bound into multiple bundles and covered by a heavy rubber or neoprene insulating sheath. See **Figure 29-5**. The leads are connected to the power source directly with crimp-on lugs or quick connectors, **Figure 29-6**.

Goodheart-Willcox Publisher

Figure 29-5. Welding cable is composed of many small strands of wire to carry large amounts of current and to remain flexible.

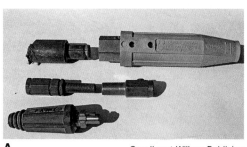
A   *Goodheart-Willcox Publisher*

B   *Goodheart-Willcox Publisher*

Figure 29-6. Welding lead connectors are available in assorted styles. A—Assorted quick connectors. B—Lug connector.

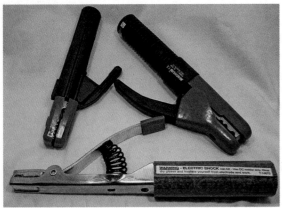

Figure 29-7. Electrode holders.

The electrode lead carries current from the machine to the *electrode holder*, an insulated clamp that holds the welding electrode. The electrode holder is sometimes referred to as a *stinger*, **Figure 29-7**. A heavy-duty connector engages a large surface area of the conductor to attach the electrode holder to the welding cable. Most common electrode holders are designed to accept the electrode at multiple angles between two spring-loaded contacts made from a copper alloy.

The workpiece lead is the lead from the work to the machine. It is terminated with a workpiece connector designed to make a firm electrical connection to the workpiece. Commonly, this lead and clamp are incorrectly referred to as the "ground." It is not a direct connection to ground. ANSI Z49.1, "Safety in Welding, Cutting and Allied Processes," states that the workpiece or the table that it rests on should be connected to ground. Two common workpiece connector types are spring-loaded models made from cast bronze alloy and those formed from light sheet steel with alloy inserts used at contact points. See **Figure 29-8**.

Safety Note

A poor workpiece connection can affect weld quality and create a safety hazard. An effective connection of the power supply to the workpiece ensures that adequate current is directed to the weld and that a shock hazard is not created by stray current. Connection devices should be regularly checked to ensure that they are in good condition. Repair or replace damaged or worn components.

Electrodes

Electrodes, or welding rods, are classified by size and by type. The size of an electrode is determined by its diameter measured at the wire core. Sizes range from 1/16" to over 3/8" in diameter. Electrodes with larger diameters require more current and therefore produce more heat than smaller electrodes of the same type. Electrode types are classified according to the makeup of their flux covering.

The insulating flux on the electrode protects the chemical integrity of the weld. The heat of the arc causes parts of the flux to burn. The oxygen in the immediate vicinity of the molten weld pool is used in the chemical reaction of the burning, or oxidation, process. This creation of a gaseous shield leaves little or no oxygen to bond with and contaminate the molten metal.

Figure 29-8. Different types of workpiece connections.

Other parts of the flux are mixed into the weld pool. Some of these parts work within the molten pool, like soap in water, to float out impurities and bind them to the waste that forms on top of the bead. This waste is called *slag*. The slag covering prevents the welded bead from cooling too quickly, which would result in brittleness. Slag is easily removed from a properly executed weld after cooling, **Figure 29-9**.

Some electrode types have iron powder and alloying elements included in the flux to improve the physical properties of the metal in the completed weld. Most steel electrodes have the same mild steel filler wire in their core but differ in the composition of the flux.

The heat of the electric arc causes the filler metal to melt faster than the flux can burn. This results in a cupped formation of the flux around the arc as it extends from the end of the electrode, **Figure 29-10**. The insulating cup has a stabilizing effect on the arc and helps to keep the arc directed at the work.

Electrode Classifications

Electrodes are classified under a numbering system developed by the American Welding Society (AWS). In accordance with AWS Specification A5.1, codes are stamped on all covered electrodes or their packaging to indicate the electrode type. See **Figure 29-11**.

Goodheart-Willcox Publisher

Figure 29-9. Slag easily separates from a cooling bead due to a difference in rates of contraction between the bead metal and the slag.

Each code begins with an *E* for electrode. The first two digits of a 4-digit number and the first three digits of a 5-digit number indicate minimum tensile strength of the completed weld in ksi (thousands of psi). The next-to-last digit indicates the welding position for which the electrode was designed. (Welding positions are explained later in this chapter.) A *1* (EXX1X) indicates an all-position electrode. An EXX2X electrode indicates horizontal and flat positions only, an EXX3X electrode is flat position only, and an EXX4X electrode can be used in the flat, overhead, horizontal, vertical-down positions. The last digit, together with the next-to-last digit, signifies the type of covering and current to be used.

An E6010 electrode, for example, has 60,000 psi tensile strength, may be used to weld in all positions, and works best with DC electrode positive current to provide deep penetration and a thin slag covering. An E7024 has 70,000 psi tensile strength and can only be used in the flat and horizontal positions. This electrode type works with any polarity to produce a soft arc, producing light penetration welds with a high deposition rate of filler material.

Goodheart-Willcox Publisher

Figure 29-10. Filler metal melts faster than the flux on the electrode, creating a cupped effect at the electrode tip. The slag formation on the core must be chipped away to restart an arc.

Figure 29-11. Electrode types are identified by a standard coding system.

Electrode Categories

AWS places SMAW electrodes into four broad categories. The classifications are fast-fill, fill-freeze, fast-freeze, and low-hydrogen.

Fast-Fill Electrodes

A *fast-fill electrode* has a heavy iron powder flux. This type of flux increases the deposition rate and produces a thick, easily removed slag that promotes the formation of a very smooth bead. Fast-fill electrodes are often called *cover rods* in reference to the smoothness of their beads. The arc length of electrodes with thick iron powder fluxes appears short because it is mostly concealed within the cup formed at the tip. An E7024 is an example of this type of electrode.

Fill-Freeze Electrodes

A *fill-freeze electrode* produces a medium arc and has a moderate deposition rate. Fill-freeze electrodes produce beads with uniform ripple patterns on their surfaces. Fill-freeze electrodes are particularly well suited for use on sheet metal because warpage is minimized. An E6013 is an example of this type of electrode.

Fast-Freeze Electrodes

A *fast-freeze electrode* provides deep penetration and only moderate amounts of fill. The thin flux of fast-freeze electrodes burns away at a comparatively rapid rate. The result is a shallow cup at the electrode tip that leaves most of the arc exposed. The highly exposed arc has the appearance of being more aggressive than the arc produced by other classes of electrodes. The beads they produce have a light slag that may adhere to the weld. Fast-freeze electrodes are preferred for out-of-position welding. E6010 and E6011 are common examples of fast-freeze electrodes.

Low-Hydrogen Electrodes

Low-hydrogen electrodes are all fill-freeze as well. They produce a medium penetrating arc and have a smooth, moderately high deposition rate. Special formulations in the flux reduce the formation of hydrogen compounds in the weld that can lead to cracking. After being exposed to the atmosphere for nine hours, low-hydrogen electrodes pick up enough moisture to counteract the low-hydrogen characteristic. The moisture absorbed into the coating can be transferred into the weld metal in the form of hydrogen. The hydrogen can lead to under-bead cracking, sometimes called *hydrogen-induced cracking*. The moisture does not affect the welding performance of the electrode.

The E7018 is the most commonly used low-hydrogen electrode. In welds that do not require low-hydrogen qualities, the E7018 electrodes could be used. An E6010 electrode normally deposits as much as 40 ml/100 g of hydrogen in weld metal.

Open packages of low-hydrogen electrodes should be stored in a *rod oven* to prevent the absorption of excess moisture and preserve the low-hydrogen properties, **Figure 29-12**.

Goodheart-Willcox Publisher

Figure 29-12. Rod oven for dry storage of electrodes.

Joint Types

The type of welding joint is determined by the physical relationship of the two pieces of metal to each other. Virtually all welded joints can be classified as butt, T-, corner, edge, or lap joints. Each requires different welding techniques and application of heat, **Figure 29-13**.

Butt Joint

A *butt joint* is welded between two pieces of metal positioned approximately in the same plane, beginning with a gap between the two parallel edges. The welded bead is referred to as a *groove weld* because it passes completely through the gap, or groove, between the two parallel edges. In many applications, the butt joint must be welded from one side

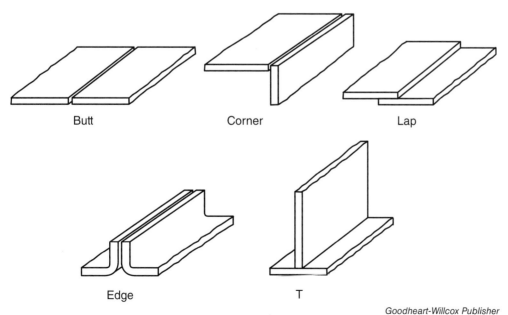

Goodheart-Willcox Publisher

Figure 29-13. Basic joint types for welding.

Figure 29-14. A fillet weld. Note the concave form of the bead.

only and must penetrate through the entire depth of the joint. Sometimes a backing plate, or narrow strip of material, is used in combination with a wider initial gap. This enables the welder to more easily penetrate the plates when welding from the bottom of the groove up to the surface.

T-Joint

A *T-joint* is made between two metal parts located approximately at right angles to each other in the form of a *T*. A gap may or may not be present at the start of a T weld. The welded bead is positioned at the intersection of the two parts. This is referred to as a *fillet weld*. See **Figure 29-14**. Like butt joints, T-joints sometimes are only accessible to weld from a single side.

Corner Joint

A *corner joint* is created between two members located perpendicular or at an angle to each other. Welded containers such as toolboxes, fuel tanks, and loader buckets usually have corner joints. The inside of a corner joint, if accessible, can be welded with a fillet weld similar to that used on a T-joint.

Welding the outside of a corner joint requires precise control of the heat input and the deposition rate of the electrode. On thin metals, care must be taken to avoid the overapplication of heat, which could lead to melting holes in the base metal. A skillfully welded outside corner is rounded slightly from the contour of the weld and requires little effort to clean the welded bead.

Edge Joint

An *edge joint* is formed when the edges of two or more parallel (or almost parallel) members are welded along one of the flush edges. Precise

STEM Connection

Heat and Temperature

Heat and temperature are related but are separate concepts. A material can contain a large amount of heat energy but have a low temperature. A material with high temperature can contain little heat. An acetylene flame burns at a very high temperature, but it must remain in contact with the metal long enough to transfer adequate heat for bending, welding, or cutting.

Heat is energy that is transferred by the movement of molecules in a substance. The transfer of energy results from a temperature difference—the energy moves from a high-temperature material to a low-temperature material. Temperature is a measurement of the level of kinetic energy (energy resulting from motion) in a material.

control of heat input and the amount of fill deposition is critical to a successful bonding. By its nature, an edge joint may only be welded from a single side.

Lap Joint

A *lap joint* is a bond between two overlapping parts. Lap joints are often welded along the edge of one piece where it contacts the surface of the second part. If the pieces need to be welded at a location that is not on an edge, they can be welded together through a hole created in one piece. This is called a *plug weld*. Whether the joint is welded along an outside edge or the edge of a hole, the resulting weld is a fillet weld.

Welding Positions

The welding position is determined by the location of the electrode in relation to the welded joint. The American Welding Society recognizes four basic welding positions—flat, horizontal, vertical, and overhead. See **Figure 29-15**. The AWS designates welding positions with numbers and letters. The numbers represent the positions—1 for flat, 2 for horizontal, 3 for vertical, and 4 for overhead. The letters *F* and *G* stand for fillet welds and groove welds, respectively. Welding flat is known as *position welding*, and all other orientations are known as *out-of-position welding*.

In industrial applications, weld engineers analyze the materials and joint characteristics to determine *weld specifications*. These specifications include the information a welder needs to perform a specific job, including how to position the weld. For many agricultural welding jobs, the welder must decide how to make the weld. Selecting the correct welding position can greatly affect the quality of the work. The following sections briefly describe each position.

Flat Position

In the flat position, the electrode is kept more or less in a vertical alignment and directed down toward the weld pool. AWS designations for welding in the flat position are 1G (groove weld in plate) and 1F (fillet weld with the joint down).

Flat welding is the easiest position for the beginning welder to master. When welding in this position, the welder has the benefit of the force of gravity. It is usually preferable, even for experienced welders, to weld in the flat position, and the work is repositioned to allow this if possible. The flat position is best for welding beads that require the addition of large amounts of fill. In these instances, a fast-fill electrode is used.

Horizontal Position

In the horizontal position, the electrode is held in a mostly horizontal position, and the weld progresses from side to side in a horizontal weld. In a horizontal groove weld (2G), the two parts are aligned approximately in

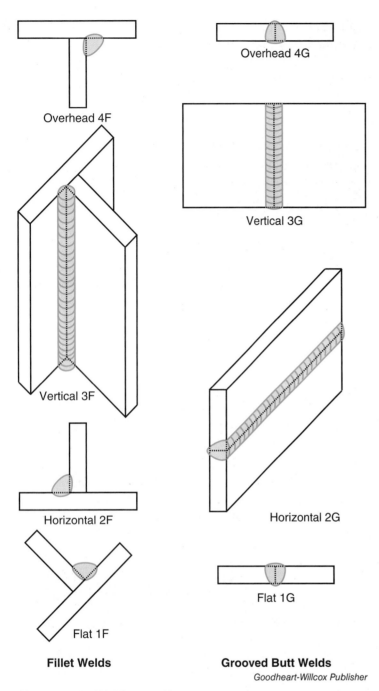

Figure 29-15. Welding positions.

the same vertical plane, and a butt weld is placed between them along a horizontal line. A horizontal fillet weld (2F) is placed in the joint between two parts, one located in a vertical plane and fitted to the surface of the second part lying in a horizontal plane.

Vertical Position

Vertical welds (3G and 3F) are welded similar to horizontal welds, except that the welding progresses from top to bottom (downhill welding, sometimes referred to as *vertical-down*) or from bottom to top (uphill welding, sometimes referred to as *vertical-up*). Because gravity affects the molten pool of vertical welds, a lower amperage setting than that used when welding flat or horizontally is helpful to control the weld pool. Best welding results are achieved by maintaining a short arc, moving the electrode at a uniform speed, and feeding the electrode at a constant speed as it melts.

Overhead Position

Welding in the overhead position (4G and 4F) is like welding flat, except that it is upside-down. Due to the effect of gravity, fast-freeze electrodes must be used. The weld pool must resolidify quickly to reduce dripping. A close fit of parts to be welded, with minimal gaps, reduces the heat required for welding. The width of the molten pool is kept small, and the amount of fill must be limited on each pass. Multiple passes may be needed on thicker material.

The term *overhead* can be misleading. It refers to the position of the electrode, which is pointing up at a downward facing joint, **Figure 29-16**. It does not necessarily mean the joint is over the welder's head. In fact, when welding overhead, welders should make every attempt to position their bodies out of the direct path of the resulting shower of hot slag and molten metal.

Pipe Welding Positions

Welding pipes that are fixed in position requires a combination of different welding positions in a single joint. For example, when joining two

Wes Hancock, Caldwell High School   *Lisa F. Young/Shutterstock.com*

Figure 29-16. Overhead welding refers to the position of the electrode, not the location of the welder. Both welders shown are welding in the overhead position.

horizontal pipes fixed in position end to end, the weld must be made all around the circumference of the joint. AWS designates this as a 5G weld. A 5F weld is formed when a horizontal pipe is connected to a flat surface or flange in a vertical plane. If the pipe is nearer a 45° angle than horizontal, the AWS designations are 6G and 6F. Skilled pipe welders are some of the highest paid welders. See **Figure 29-17**.

Surface Preparation

Surface preparation includes cleaning and fitting the pieces to be welded. Under ideal conditions, surfaces to be welded should be clean from all contaminants and oxides. Surface contaminants have an insulating effect that results in the need for higher amperage. Contaminants on the surface of welded materials can cause weak points or *inclusions*.

shinobi/Shutterstock.com

Figure 29-17. Pipeline welders work mostly outdoors and are well paid.

Inclusions are foreign materials, such as slag, trapped within the finished weld. Some electrodes, such as E6010 and E6011, have better cleaning properties than others. However, clean and properly fitted surfaces always yield better results for the welder.

Cleaning prior to welding may be as simple as brushing with a wire brush, **Figure 29-18**. Cleaning reclaimed steel may entail removal of rust, oil, grease, dirt, and paint residues. New steel may also have oily residues or mill scale from the manufacturing process. When cleaning with a wire brush is not enough, edges need to be dressed with a grinder prior to welding.

Welded parts should be closely fitted with a small but uniform gap between them. Thin materials can be welded with virtually no gap. Thicker metals require some type of edge preparation to enable adequate weld

Figure 29-18. Use a handheld or powered wire brush to clean steel before and after welding beads.

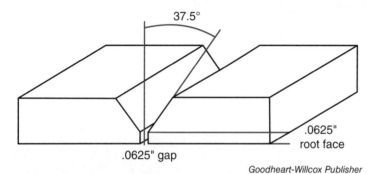

Figure 29-19. Bevels and a slight gap at the weld joint increase the surface area of the bond and help to ensure penetration through the base metal.

penetration. When parts are accessible for welding from both sides, weld a bead from the back side (backing bead). Then, penetrate and re-fuse the weld from the front with the next bead. Setting a gap of up to 1/8″ may be enough in many cases. With parts that are 5/16″ or thicker, some *beveling* may be required on one or both parts, **Figure 29-19**. Beveling removes part of a square edge to create a sloping edge.

The maximum weld width for butt, fillet, and multipass welds should be slightly wider than the depth of the weld. A width-to-depth ratio of 1.25:1 to 1.5:1 is recommended for these welds to reduce the possibility of cracking due to internal stress in the bead.

Oxyfuel or plasma cutting can be used for beveling edges, **Figure 29-20**. All loose scale and dross resulting from cutting must be removed by chipping or grinding prior to assembly and welding. It is not necessary to remove the thin oxide layer formed on cooling. This oxide is allowed to remain on the edges to help prevent rusting if the parts are to be stored for some time before welding. Although grinding may be used for beveling, in most instances it is inefficient.

Figure 29-20. A torch can be used to cut a bevel.

Controlling Welding Parameters

The key to creating a good weld is to maintain control of the welding process. *Parameters* are limits for how something should be done. Welders control welding parameters that include current, electrode selection, arc length, electrode angle, electrode motion, and travel speed.

Welding Current and Polarity

Welding current is expressed in amperes (amps, or A). The amperage required to weld is determined by the electrode diameter, the thickness of the pieces to be welded, and the position of the welding. Amperage is the most important setting on a SMAW welding machine. The amount of amperage controls the heat input at the weld joint, which affects the weld penetration, the deposition of filler material, and the conditions of the weld pool. **Figure 29-21**. Most welding machines designed for SMAW automatically control the voltage. The voltage in a welding circuit determines the ability of the arc to form between the electrode and the base metal.

Refer to the welding machine manufacturer's recommendations when setting the amperage. Manufacturers of welding equipment supply these charts in several different ways. Most often, amperage charts are attached to the power source for easy reference. They are also available in printed form and on company websites. Free smartphone apps can be used to determine the correct amperage setting. Search for welding apps using the key word *welding*.

Amperage charts provide a range of suggested amperages for a given set of conditions. Set the amperage on the machine within the recommended range. You can weld on scrap pieces to fine-tune the amperage setting. Each machine is a little different, so making minor adjustments when switching machines is common. Generally, a smaller electrode and lower amperage is needed to weld a small piece than would be needed for a larger piece of the same thickness. Thin metals require less current than thick metals, and a small electrode requires less amperage than a large one.

Many SMAW machines can be switched to different welding polarity settings. On some machines, polarity is switched by simply turning a dial or pushing a button. See **Figure 29-22**. On other machines, the welding leads must be moved to different terminals to change polarity.

Polarity is the direction of flow of electric current. Electric circuits have negative and positive poles. Direct current (DC) flows in only one direction, resulting in static polarity. Alternating current (AC) flows in one direction half the time and in the opposite direction the other half the time. AC current changes polarity 120 times per second with the 60-hertz current supplied

Goodheart-Willcox Publisher

Figure 29-21. Amperage is the most important setting on a SMAW welding machine. Adjustments vary greatly on different machines.

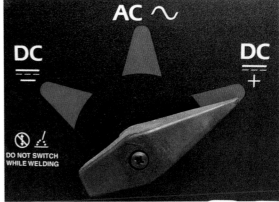

Goodheart-Willcox Publisher

Figure 29-22. The smoothest welding characteristics are obtained by using the correct polarity for the chosen electrode.

by power companies in the United States. Modern SMAW welding power supplies provide constant current (CC) that can be set to either AC or DC, depending on the choice of electrode. DC welding currents are designated as direct current electrode negative (DCEN) or direct current electrode positive (DCEP). The smoothest welding characteristics are obtained by using the correct polarity for the chosen electrode.

In a welding circuit, current flows from the negative terminal of the welding machine, through the arc, and back to the positive terminal of the welding machine. The useful effect of this is that more of the heat follows the direction of current flow—about 30% of the heat is distributed to the positive side of the circuit and the other 70% is released on the negative side. Under most circumstances, DCEP results in deeper penetration. DCEN results in faster melting of the electrode and yields a faster deposition rate. Different chemicals in the flux covering generate exceptions to this rule for some electrodes.

Electrode Selection

Choosing the proper electrode type and size for the job greatly affects the quality of the finished joint. There are hundreds of electrodes to choose from. For most agricultural needs, however, a collection of a half dozen is adequate. In fact, for most agricultural repair and construction needs, a stock of E6011s and E7018s is sufficient.

The first step in choosing an electrode is to determine the composition of the base metal. The electrode type must be matched to the base metal composition to ensure a strong weld. Begin by considering the source of the metal. Assess the appearance of metals from questionable sources.

If the base metal is magnetic, it is likely a carbon steel or alloy steel. Standard electrodes, such as E6011 or E6013 with a mild steel core, will work with these base metals. If the base metal has a coarse and grainy internal surface, it is likely a cast metal requiring a specialty electrode. If the base metal is not magnetic, the material could be stainless steel, manganese steel, or a nonferrous alloy such as titanium. A chisel bites into softer metals, such as mild steel or aluminum, and bounces off or chips against harder metals, such as high-carbon steel, chrome-molybdenum alloy, or cast iron. A spark test with a grinder may give some indication as to the type of steel. More flare in the sparks indicates a higher carbon content in steel, **Figure 29-23**.

Nagy-Bagoly Arpad/Shutterstock.com

Figure 29-23. Spark-testing steel to determine its relative carbon content.

High-carbon steels require low-hydrogen electrodes. These electrodes prevent the formation of oxides in the weld and limit the formation of hydrogen compounds in the steel that lead to cracking in the completed bond. Low-hydrogen electrodes are recommended for steels in which cracking can be a problem due to low alloy, high carbon, or high sulfur content and for high-deposition, out-of-position welds on thick plate. Many types of low-hydrogen electrodes, such as EXX16, EXX18, and EXX28, are available for use in SMAW. Low-hydrogen electrodes are the most widely used SMAW welding rods for power generation, general fabrication, shipbuilding, and pipeline welding.

An E7018 electrode is useful for welding in all positions, with the exception of vertical-down. An E7018 electrode has a 25% to 40% iron powder content that facilitates a smooth, quiet arc, generating low spatter. The E7018 produces medium penetration and high-fill deposition. This electrode produces a moderately heavy slag that is easy to remove.

Arc Length

Arc length is the distance from the electrode to the base metal. An arc with correct length and amperage produces a sharp, crackling sound like that of frying bacon. Correct arc length is related to electrode diameter. Maintaining the proper arc length is essential for developing and maintaining the oxygen-free shield around the molten weld pool. If the arc is too long, the protective envelope collapses, **Figure 29-24**. A long arc causes turbulence in the weld pool, resulting in splashing and dispersion of *spatter*, or metal droplets, around the area of the bead. See **Figure 29-25**.

Examine the weld bead to determine if the arc length is correct. Arc length for 1/16″ and 3/32″ diameter electrodes should be about 1/16″. Arc length for 1/8″ and 5/32″ electrodes should be about 1/8″.

Goodheart-Willcox Publisher

Figure 29-24. Excessive arc length results in spatter and collapse of the O_2-free envelope.

Electrode Angle

Under most conditions, the electrode is held nearly perpendicular to the work, although tilting it ahead 10° to 15° in the direction of travel (referred to as a *drag angle*) increases visibility of the weld pool. For welding in the flat, horizontal, and overhead positions, use a drag welding technique. When uphill welding, tilt the top of the electrode 10° to 15° down, away from the direction of travel. This is referred to as a *push angle*. See **Figure 29-26**.

The energy of the arc, combined with the directional cup formed by the flux at the electrode tip, creates arc force. *Arc force* is the force generated by the heat from the electric arc and the burning of the flux. When properly directed, the arc force aids in penetration of the base metal. If the electrode angle is allowed to drop too much, the arc force is directed over the molten weld pool at an angle that distorts the shape and disturbs the calmness of the molten pool. The result is a long arc and

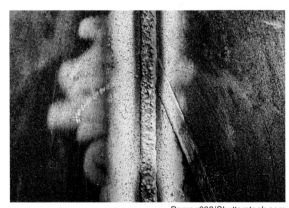

Roman023/Shutterstock.com

Figure 29-25. Spatter is the result of holding an arc length that is too long.

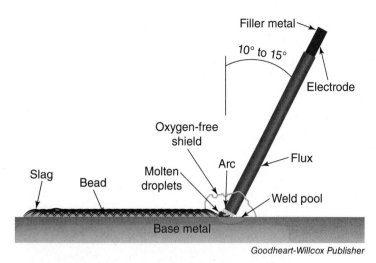

Figure 29-26. Tilt the electrode 10° to 15° from perpendicular to improve visibility of the pool.

the distribution of excess spatter around a welded bead that lacks penetration. The increased turbulence of a long arc created by too much electrode angle also destabilizes the O_2-free shield. Too much electrode angle results in a pointed effect in the ripple of the finished bead. This is evident during welding because the shape of the molten weld pool will also be pointed.

A common mistake made by novice welders is dropping the electrode angle to get a better look at the weld. It is important to learn to reposition your head and body in order to see the weld instead of modifying the electrode angle, **Figure 29-27**.

Electrode Motion

Welders develop their own individual techniques for manipulating the electrode. Practicing helps you to develop your own style, and studying the techniques of good welders will improve your own.

Two types of beads are used in arc welding: stringer beads and weave beads. *Stringer beads* are straight, narrow beads made with minimal side-to-side movement. *Weave beads* are wider weld beads that are formed using a *weave pattern* (repeated electrode motion). On material 1/4″ and thinner, weaving the electrode is typically not necessary and creates a wider bead than necessary. A single stringer bead is often satisfactory for narrow groove weld joints.

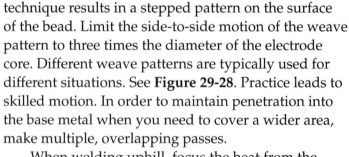

Figure 29-27. This unacceptable bead pattern was caused by too much electrode angle.

To create a wider bead on thicker material, manipulate the electrode from side to side in a continuous series of partially overlapping semicircles. This technique results in a stepped pattern on the surface of the bead. Limit the side-to-side motion of the weave pattern to three times the diameter of the electrode core. Different weave patterns are typically used for different situations. See **Figure 29-28**. Practice leads to skilled motion. In order to maintain penetration into the base metal when you need to cover a wider area, make multiple, overlapping passes.

When welding uphill, focus the heat from the electrode on the sides of the joint. Move across the middle of the joint slowly so the weld pool follows, pausing slightly at the sides to ensure a smooth *tie-in*, or blending, to the sidewall. If the layers of the finished bead look like fish scales, forward movement was too fast.

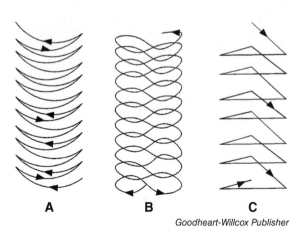

Figure 29-28. Weave patterns. A—Crescent, or 1/2 moon. B—Figure-eight. C—Zigzag.

When making a fillet weld, rock the back of the electrode slightly from side to side in order to direct the heat of the arc uniformly into each side of the joint. Welding vertical-up with a fast-freeze electrode requires more of a whipping pattern than a weaving motion. The arc is rapidly moved from the pool to the unmelted base metal and back in an effort to prevent excessive heating. On thinner material, a stringer bead may be adequate.

Travel Speed

Travel speed is the rate of forward motion during welding. The condition of the weld pool is the major indicator of correct travel speed. Welding is like riding a bike—you do not watch the bicycle, but rather focus your attention on the road ahead. When welding, focus on the shape and stability of the molten weld pool. It should be calm and oval-shaped, **Figure 29-29**, extending from the not-yet-welded base metal up into the edge of the solidifying bead. Keep the arc in the leading one-third of the weld pool. Correct travel speed results in a smooth, consistent ripple pattern on top of the bead. This appearance is often described as the *stacked dimes* effect; however, it should actually be smoother than a true stack of dimes. See **Figure 29-30**.

Goodheart-Willcox Publisher

Figure 29-29. A calm, properly shaped weld pool.

Traveling too slowly allows heat to be directed into the pool, rather than into the base metal. This produces a wide, convex bead with shallow penetration and leads to poor fusion. The result is a "cold" weld that appears to be simply sitting on the surface of the material.

Excessively rapid travel speed also decreases penetration. Moving too fast creates a narrower, possibly high-crowned bead and often results in undercutting. **Undercutting**, or poor tie-in, means that the outside edge of the welded bead is concave or recessed. See **Figure 29-31**. Traveling too fast can create a thin or undersized bead.

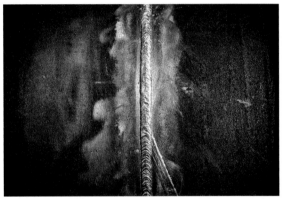

Roman023/Shutterstock.com

Figure 29-30. A bead with proper surface character has a uniform width and a uniform, unbroken pattern of semicircular waves across the top.

Anatomy of a Weld—Weld Quality

There are a number of reasons to inspect a completed weld. The most important reason is to determine if its quality is sufficient for its intended application. Welds are examined to check their size and to look for discontinuities. The size of a weld often correlates directly to strength and ultimate performance. The optimum size of a weld is determined by the thickness of the base metal—thicker materials require larger beads. Too much weld material, on the other hand,

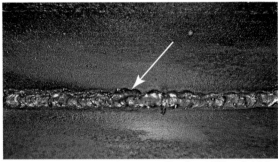

Goodheart-Willcox Publisher

Figure 29-31. A bead exhibiting severe undercutting.

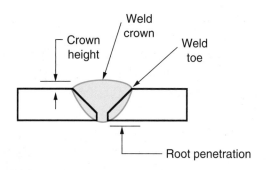

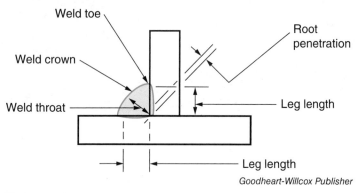

Figure 29-32. Weld terminology.

creates internal stresses that pull against the sides of the weld, resulting in weak spots. Welds that lack adequate size may not stand up to stresses encountered during service. Oversize welds produce stress concentrations and contribute to the likelihood of distortion in welded components.

Weld Terms

It is important to understand the terminology used to describe welded joints in order to effectively communicate about the quality of welds. The *crown* of a bead is its visible upper surface. The *root* of the bead is the lower or innermost portion. Root penetration is essential to the bonding quality of a welded bead. The *legs* of a bead are the extensions of the bead past the center of the joint, and the *toes* are where the edges of the bead end. The toes should blend smoothly into the base metal. See **Figure 29-32**.

Discontinuities

A key to success in welding is consistency. **Discontinuities** are breaks in the consistency of a welded bead. Some are visible, while others are inside the metal. Some discontinuities are severe enough to be considered a fault, while others are just inconsistencies.

When discontinuities are of an unacceptable size or in an unacceptable location, they are called **welding defects**. Defects reduce strength or produce stress concentrations within the welded bond. AWS welding codes and standards are used to determine the number of acceptable and unacceptable weld discontinuities for welding that is to be commercially inspected. Many times in agricultural settings, there are no prescribed codes to follow. It is important for welders to understand the basics of inspecting welds for quality and to be able to evaluate welds on equipment and structures before putting them into service.

Inspect welds for these common discontinuities:
- **Porosity** is a cavity-type discontinuity formed by gas trapped in the molten weld. The gas may form bubbles or pockets as the weld solidifies. The main causes of porosity are dirty base material, moisture contamination, or incorrect welding techniques. To control porosity, use clean base materials, properly store welding consumables, maintain welding equipment, use recommended procedures, and weld in suitable environmental conditions. See **Figure 29-33**.

- *Incomplete fusion* is a weld discontinuity in which blending does not occur between the weld metal and joint faces or adjoining weld beads. This reduced fusion results when the base material or previously deposited weld metal is not melted during the welding process. Incomplete fusion can occur at any location within the weld joint and occurs in both fillet and groove welds. It often is found on one leg of a fillet weld. Incomplete fusion is caused by an incorrect welding angle that unevenly distributes heat between the two sides of a joint. It also may be caused by oxides or other foreign material on the surface of the base material. Proper cleaning, correct welding techniques, and adequate amperage go a long way toward achieving excellent fusion.

- *Incomplete joint penetration* is the failure of the filler metal or base metal to fill the weld root completely, **Figure 29-34**. Common causes are a bad groove weld design or a fit-up that is unsuitable for the welding conditions. Incomplete joint penetration can occur if the root face dimensions are too large, the root opening is too small, or the included angle of a groove weld is too narrow. An incorrect electrode angle or holding the arc too long also results in insufficient penetration of the base metal.

- *Unacceptable weld profiles*, or poorly shaped beads, occur when too much or too little filler metal is added to a weld or when the heat is directed unevenly toward one side of the weld joint. These discontinuities include improper contour, undercut, and overlap. Many result in stress concentrations within the weld. Groove welds should be slightly convex, rising a little above the surface of the metal. Fillet welds should generally be slightly concave, blending smoothly into each leg of the joint. *Undercut* is a groove melted into the base metal adjacent the weld toe, or weld root, and left unfilled by weld metal. The usual cause of undercut is excessive travel speed. See **Figure 29-35**. *Overlap* is a protrusion and notch in the weld metal beyond the weld toe, or weld root, which results in a concentration of internal stresses. Overlap is caused by incorrect welding technique or insufficient current.

Leonid Eremeychuk/Shutterstock.com

Figure 29-33. The uniform porosity of this weld is likely a result of poor technique.

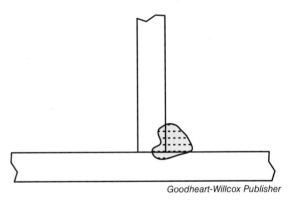

Goodheart-Willcox Publisher

Figure 29-34. A fillet weld with incomplete joint penetration results from failure to incorporate an adequate amount of the base metal into the bead.

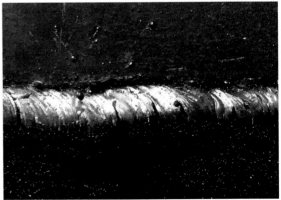

Goodheart-Willcox Publisher

Figure 29-35. This weld has an undercut on the top edge and exhibits excessive weld spatter.

- *Cracks* are severe, unacceptable discontinuities that are sharp at the edges, concentrating points of stress. See **Figure 29-36**. The damaging effect of a crack is greater than that of most other discontinuities because cracks have a tendency to spread, thus multiplying their damage. Cracking is often caused by stress concentration near discontinuities in welds and base metal and near sharp edges in the design of the structure. Hydrogen absorption can contribute to crack formation in some metals. Cracks can be *hot* or *cold*. Hot cracks form at high temperatures during the solidification of the weld. Cold cracks develop after the weld cools somewhat as a result of stresses. Cracks in the *heat-affected zone*, the area around the weld, most often occur in base metals that have a high enough carbon content that they can be hardened, **Figure 29-37**. Cracking of the weld material is usually a result of stresses external to the welded joint and compounded by poor welding technique. Cracks must be completely removed by grinding or gouging, and the void must be filled with sound weld metal. Weld cracking is best avoided by selecting the correct electrodes and amperage as well as using approved techniques that take into consideration heat input and the expansion properties of the base metal.

Goodheart-Willcox Publisher

Figure 29-36. Cracks are unacceptable weld discontinuities. A—A crack formed due to inclusion. B—Under-bead cracking. C—Weld cracking due to overwelding, poor technique, and improper electrode selection has resulted in a complete failure of this joint over time.

Visual Inspection

Visual inspection, if performed correctly, is an easy, inexpensive, convenient, and effective method of evaluation for many welding applications. Welders should be able to identify most of the significant welding discontinuities during visual inspection. They should also be able to evaluate the severity of those discontinuities to decide whether to accept or reject the welding job prior to placing equipment in service.

Destructive Weld Testing

Destructive weld testing involves destruction of the welded joint in order to establish weld integrity or performance. Destructive tests include sectioning, bending, or breaking the welded component in order to evaluate various mechanical or physical properties of the welded metal. See **Figure 29-38**. Examples of these tests are the guided bend test, tensile strength test, fracture test, and weld separation test. These tests are most often done during welder performance qualification or certification testing. A welding inspector is often required to conduct, supervise, and evaluate these tests.

Nondestructive Weld Testing

Nondestructive testing evaluates the integrity of welds without causing damage. With these methods, a welding inspector is able to examine the internal structure of the weld to establish the weld's integrity without destroying the welded component. Several of these tests are described in the following sections.

Surface Crack Detection

A welding inspector can test for surface cracks that are not visible using simple visual inspection. Testing methods, such as liquid penetrant dye and magnetic particle inspection, are used to indicate the presence and location of invisible cracks. Penetrating dyes, sometimes viewed under special lighting, flow into cracked surfaces and leave a visible residue at the crack's location. See **Figure 29-39**. The sharp edges of a crack, when subjected to a magnetic field, cause a sprinkling of iron particles to stand up along the line of the cracked surface. These testing methods can be used without destroying the weld. However, if cracks are found, the weld is unacceptable.

Radiographic Weld Inspection

Radiographic testing (X-ray testing) makes use of X-rays to penetrate through a weld onto a photographic film, resulting in an image of the welded joint's internal structure. See **Figure 29-40**. The amount of energy absorbed by the object depends on its thickness and density. Energy not absorbed by the metal causes exposure of the X-ray film. These areas are dark when the film is developed, while areas of the film exposed to less energy remain lighter. Discontinuities, such as porosity or cracks, appear as dark outlines on the film. Inclusions of low density, such as slag, also appear as dark spots. All discontinuities are detected by viewing shape and variation in density of the processed film.

Ultrasonic Weld Inspection

In *ultrasonic testing*, a stream of ultrasonic energy is directed into the weld to be tested. These ultrasonic waves travel through the weld with insignificant loss, except when they are intercepted and reflected by the sharp edges of a discontinuity. This system uses a transducer, which produces a high-frequency vibration. When the pulse of ultrasonic waves strikes a discontinuity in the test piece, it is reflected back

Goodheart-Willcox Publisher

Figure 29-37. The change in color around this bead shows the heat-affected zone.

Goodheart-Willcox Publisher

Figure 29-38. Penetration of a fillet weld can be destructively tested by exerting pressure back toward the welded side.

Goodheart-Willcox Publisher

Figure 29-39. Dye penetrant testing illuminates surface discontinuities of a weld.

shinobi/Shutterstock.com

Figure 29-40. Radiographic testing on a pipeline that will be buried.

Goodheart-Willcox Publisher

Figure 29-41. In ultrasonic weld testing, discontinuities are indicated by high or low peaks in the wave pattern of an oscilloscope.

to its point of origin, the transducer. The transducer then serves as a receiver for the reflected energy. The returned echoes from the discontinuities and the echo of the rear surface of the test piece are displayed on the screen of an oscilloscope, **Figure 29-41**.

Arc Welding Safety

Welders must always be aware of and take precautions to avoid the multiple hazards associated with arc welding. These hazards include the following:

- Electrical shock
- Electromagnetic fields (EMF)
- Burns from hot metal
- Burns from intense light
- Burns from steam
- Fire in the surrounding area
- Flying particles
- Fumes and gases
- Explosions from expanding gases
- Tripping on equipment or waste
- Excessive noise
- Moving parts

The preceding list includes most of the hazards associated with arc welding, but specific situations may create other hazards. The best protection against hazards is prevention. Many hazards can be avoided by following safety rules and using good judgment or common sense.

The Welding Environment

Planning procedures and safely arranging the work environment should be first considerations in avoiding hazards. Whenever possible, weld only in areas that are well-ventilated, dry, and free from clutter. Flammable materials should be stored well away from areas where hot metalwork is being performed.

A solid connection of the workpiece lead to the workpiece greatly reduces the likelihood of electric shock. Electricity always follows the path of least resistance to ground—make sure that path is not you. Inadequate ground connections also have the tendency to produce unwanted arcing. This arcing can cause sparks at the ground connection.

Other people in the vicinity of welding should be protected from the ultraviolet and infrared rays emitted by the welding arc. Many schools and businesses use translucent welding curtains for this purpose, **Figure 29-42**.

Position the welding machine close to the work to reduce tripping hazards caused by welding leads lying across floors. Pick up loose electrodes and electrode stubs left after welding to prevent slipping hazards on hard concrete floors.

When required to weld on objects close to a concrete floor, use a shield. The pores of a concrete floor contain water that can quickly turn to steam, resulting in small explosions in the surface of the concrete. Welding on containers should be avoided, since the expansion of gases within the container can result in violent reactions.

Good ventilation is critical for welder safety. Dedicated ventilation systems are excellent for removing contaminants from the air, but they can be noisy. See **Figure 29-43**. Working in open areas or opening large bay doors to increase ventilation when welding is a quieter alternative.

Personal Protective Equipment

Proper personal protective equipment (PPE) is essential. The PPE worn should be appropriate for all hazards that may be present.

A welding helmet that protects the entire face is required. Handheld observer's shields are for supervision purposes only and offer inadequate protection for welding. Welding helmets should have a minimum of a #10 shade or filter lens and a clear protective plastic cover lens in front, **Figure 29-44**. The shades available for arc welding filter lenses range in protection from #9 to #14. A #9 lens is for limited use in special situations only, and a #14 shaded lens is so dark that its use may be detrimental to the quality of welds

Goodheart-Willcox Publisher

Figure 29-42. Welding curtains protect bystanders from the hazardous light emitted by the welding process.

Goodheart-Willcox Publisher

Figure 29-43. An exhaust apparatus for welding should be placed as near as practical to the actual welding arc.

Goodheart-Willcox Publisher

Figure 29-44. Many styles of welding helmets are available to meet the different needs and personal preferences of welders.

produced by beginning welders. Many varieties and qualities of filter lenses are available. They may be made of plastic or glass, reflective or nonreflective.

Auto-darkening lenses are triggered to darken by the welding flash. A sensor detects bright light and causes a liquid crystal display (LCD) to polarize and distort the light to limit the amount of light passing through the lens. These lenses are popular with many welders and greatly improve visibility when a weld is begun. However, welding for extended periods of time warrants the use of lenses that offer more protection to reduce eye fatigue.

OSHA-approved (Z87.1) goggles or glasses must be worn at all times, including under the welding helmet. As welded beads cool, particles of hot slag often pop off and are propelled some distance due to the forces created by the contracting steel and slag. When the welding helmet is raised, eye protection must already be in place.

moomsabuy/Shutterstock.com

Figure 29-45. Gauntlet-type gloves for SMAW are made of insulated thick leather to provide protection from the intense heat of the welding arc.

Gauntlet-type, heavily insulated leather welding gloves protect the hands from the heat and radiant energy of the welding arc. See **Figure 29-45**. Covered leather shoes protect the feet and ankles from burns and injury from sharp or heavy objects. Synthetic soles with prominent tread are more resistant to slipping and provide some protection from electric shock.

Heavy cotton, long-sleeve shirts and pants are resistant to burning and protect the body from ultraviolet and infrared rays. Shirts should have covered pockets or no pockets, since open pockets can catch hot metal spatter. Pants should not have rolled or frayed cuffs. Do not wear clothing made of synthetic, petroleum-based fibers such as rayon or nylon while using a torch. Fire-retardant clothing is available for workers facing high levels of exposure. Leather aprons, leather sleeves, welding chaps, or a welding jacket may be necessary, especially when performing operations that require out-of-position welding.

Practice Welding

The formation of strong, uniform beads is the goal of every welder. Practice is required to attain the skill needed to correctly strike the arc and weld uniform beads.

Striking an Arc

A successful welder is able to strike an arc and maintain it without having to think about the process. The most common ways of starting an arc are by tapping the electrode to the grounded base metal or by scratching the electrode, like a match, across the steel.

Scratching is the easiest way to begin. Sweep the tip of the electrode against the metal as if striking a match. Once the electrode makes contact with the steel, the arc starts. The tricky part is to locate where the weld should begin and move the arc quickly to that spot. It takes practice to be consistent. The main reason for scratching is to allow a little more time to establish the arc length before the tip sticks to the steel. However, scratching tends to leave objectionable marks on the base metal near the weld. Under some conditions, the marking creates enough of a heat-affected zone to be considered a weld defect.

Tapping the electrode is a better method for starting the arc in the correct location, but it is more challenging to keep the electrode from sticking while still maintaining an arc. Tapping is poking the exact starting point of the weld with the tip of the electrode and then lifting the electrode back to correct arc length. Once the arc strikes, it is natural to pull the electrode away to prevent it from sticking. Novice welders tend to pull away too far and lose the arc. Sticking is caused by the filler metal melting to the steel before the electrode is pulled away to the correct arc length.

When the electrode does stick, quickly snap the electrode holder backward from the direction of travel to free the electrode. If this does not work, it is necessary to break the circuit by releasing the electrode from the holder. See **Figure 29-46**. Do not switch off the machine with an electrode stuck. Doing so causes arcing within the switch and results in premature part failure.

Goodheart-Willcox Publisher

Figure 29-46. When an electrode sticks, heat accumulates quickly. Safely break the electrode from the surface with pliers or tongs.

Practice striking an arc on scrap material before trying to weld beads. The motion becomes more natural with repetition. Start with clean metal that is free of major contaminants to make the process easier. Ensure that the workpiece lead is connected directly to the workpiece to provide a more positive connection and improve arcing characteristics. Begin with the scratching technique to get the feel of the arc, but switch to the tapping technique before developing the habit of using the scratching technique. Practice makes perfect; however, practicing poor techniques leads to imperfect habits.

Welding a Bead

Once the arc is struck, it must be held at a consistent distance from the base metal to form the molten pool. The arc is moved along at a pace that allows the pool to penetrate the base metal as it accepts the addition of filler metal from the melting electrode. Blending, or tie-in, is encouraged by a slight crescent-shaped weaving or whipping of the electrode. This pattern of motion should be kept to one and one-half to two times the diameter of the electrode and should not extend the arc length. Allowing the electrode movement to exceed these limits invites collapse of the O_2-free shield, leaving the molten pool exposed to the formation of oxides and nitrides in the bead.

Safety Note

When releasing the electrode from the holder, do not grab it with your hand. This can result in a bad burn, even if you are wearing welding gloves. Use pliers to twist off the stuck electrode. Leave your face shield down and break the rod free while your eyes continue to be protected—there will be a bright flash when it comes loose. Others should stand away from the area just behind the welder's elbow so as not to be struck by the motion as the stuck electrode is being freed.

Welding Beads on Flat Plate

Welding a series of beads on small rectangular flat plate, called a *coupon*, is the starting point for most beginning welders. See **Figure 29-47**. This procedure, typically called *welding a practice pad*, emulates a welding process used to build up worn parts with filler metal. See **Figure 29-48**. It is crucial that each successive weld completely blend into the last, leaving no holes or voids in the finished surface. Sometimes a harder or more durable surface than the original base metal is created.

Plates for practice pads should generally be 3/16″ to 3/8″ thick, rectangular, with no dimension longer than about 6″ or shorter than 3″. A typical size would be 1/4″ thick and 3″ wide by 4″ long. Working with thin materials teaches beginning welders to be conscious of controlling the heat input. The thin materials also allow for the use of smaller-diameter electrodes that are less expensive and produce less wear on the equipment. Welders who develop skills on thin materials can easily transition to thicker materials.

A correctly welded practice pad should have uniform beads that exhibit a relatively flat surface on top. There should be a minimum of high or low spots—each bead should smoothly transition into the next. The plate itself should display the development of a warped curve toward the welded surface, **Figure 29-49**. This distortion, or deflection, is evidence of adequate penetration and results from the contracting of the layer of beads as the metal cools.

Garsya/Shutterstock.com

Figure 29-47. The surface of a correctly welded practice pad should be uniform and relatively flat.

Goodheart-Willcox Publisher

Figure 29-48. The surface of this chisel plow point has been built up with welds and finished with hard-surfacing electrodes to improve wear resistance.

Welding a Practice Pad

The following is a procedure for creating a practice pad. See **Figure 29-50**.

1. Obtain a steel coupon from your instructor or cut one with a cutting torch.
2. Using a scrap piece of steel, set up the machine according to manufacturer's specifications and run a test bead. Make adjustments to fine-tune the settings. Remember that each machine is a little different.
3. Strike the arc with a tapping motion.

Safety Note
Before striking the arc, call out the word *cover* to alert anyone nearby to avoid looking at the arc without adequate eye protection.

4. Weld the 3″ × 4″ × 1/4″ steel coupon to the 10″ long 3/4″ or 1/2″ pipe handle to make the part safer to handle and provide an excellent location for attaching the workpiece connection. Flatten the end of the pipe with a heavy hammer on the anvil where it connects the coupon to simplify this step.

5. Properly attach the workpiece connection directly to the handle. Attaching the workpiece to the table is an indirect connection method that leads to poor arc quality.

6. Position the coupon on the welding table. Right-handed welders are usually most comfortable working from left to right, and left-handed welders are the opposite, **Figure 29-51**. Plan the welding process so that movement travels from a less comfortable position to a more comfortable position. Leaning on stationary objects for support further reduces unwanted movement.

7. Use a 3/32″ E6011 electrode to weld a single-pass bead along one edge of the coupon in the flat (1G) position, **Figure 29-52**.

8. Leave the welding machine running. Remember that the duty cycle requires a cool-down time to protect the internal components of the power source.

9. Thoroughly clean the bead with a chipping hammer and a wire brush, **Figure 29-53**. Lightly tap at the edge of the bead—avoid hammering hard enough to leave marks on the surface of the bead. Do not cool the coupon at this point. Rapid cooling deteriorates bead quality.

10. Weld a second bead between the previous bead and the plate. There should be a 25% to 30% overlap on the previous bead, **Figure 29-54**. Alternate the direction of travel with each successive bead to yield a smoother surface.

11. Thoroughly clean the bead and continue welding and cleaning until the plate is covered. It is very important to clean welds thoroughly between each pass.

12. Weld a second layer of beads perpendicular to and on top of the first layer, covering one-half of the coupon.

13. Clean the welds thoroughly. At this point it is acceptable to cool the pad prior to handing it to an instructor.

Goodheart-Willcox Publisher

Figure 29-49. Distortion of a welding pad caused by the contraction of cooling beads.

Goodheart-Willcox Publisher

Figure 29-50. Overlapping layers of beads will cover the practice pad.

Goodheart-Willcox Publisher

Figure 29-51. Grasp the handle of the electrode holder with the dominant hand. Use the other hand to provide support.

Goodheart-Willcox Publisher

Figure 29-52. The first pass weld on a practice pad is positioned very close to the long edge of the plate.

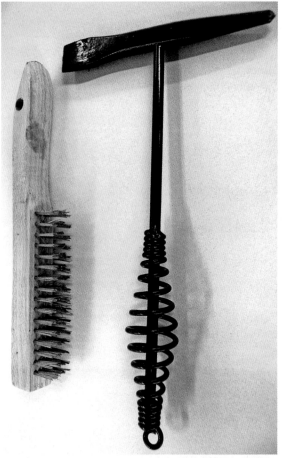

Figure 29-53. Use a wire brush and chipping hammer to clean each bead before welding the next.

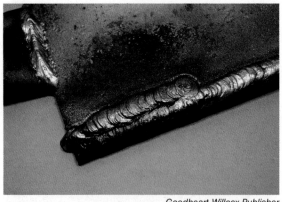

Figure 29-54. The second bead, and each successive bead thereafter, is placed between the previous bead and the plate.

Safety Note

Be aware that steam rising from dipping hot metal into water can cause severe skin burns. Position your hand, arm, face, and other body parts to the side of the rising steam.

14. Clean your work area and return all equipment to its proper place.

Fillet Welding in the Flat and Horizontal Positions

Two pieces fitted together in a more or less perpendicular arrangement are joined by a fillet weld. Performing a fillet weld in the 1F position is much like welding beads on flat plate, since the weld is placed in a 90° V-shaped groove. As in flat plate welding, you must ensure that the molten pool blends to each side, being careful to obtain full penetration, and leaving a uniform surface on the finished bead.

Shifting the parts to an upside-down T and welding the joint on one or both sides is welding in the horizontal (2F) position. See **Figure 29-55**. A sideways T-joint welded from the top is also a 2F weld. Because of the effect of gravity in this position, more of the weld tends to be placed on the lower plate than on the vertical plate. Also, the balance of heat should be considered. Heat rises, and it takes less heat to melt the edge of a plate compared to the flat side of a plate. If the two plates are of different thicknesses, the balance of heat is different. The goal is to melt both plates evenly at the weld joint, watching the edges of the pool and manipulating the arc to control the application of heat.

Fillet welds on thicker materials require multiple passes to complete a strong joint. Begin with the first weld directly in the joint between the plates; this is the *root bead*. The weld root needs to penetrate deeply into the joint and must be cleaned thoroughly prior to adding successive welds. Weld the next bead in a multiple-bead fillet joint into the joint between one side of the base metal and the root bead. Successive beads are layered to build a flat surface until the other plate is again welded into the joint. The thicker the plate, the more layers of bead are required. If the joint can only be accessed from a single side for welding, or if the plate is over 1/4" thick, the piece that comes to an edge in the joint may require beveling to allow for adequate penetration of the weld. Materials 3/8" and over in thickness generally require beveling on one or both sides prior to welding.

The contour of the completed bead can be slightly concave, dome-shaped, or slightly convex. A bead with a convex profile has a little more material in the joint but is better able to handle variable stresses. In either case, the toes (edges) of the bead should blend smoothly to the base metal without causing a reduction in the thickness of the material.

Fillet Weld Practice

Welding on 1 1/2″ × 4″ × 1/4″ to 1 1/2″ × 6″ × 1/4″ coupons is great practice for learning to perform fillet welds. The following procedure can be used for fillet weld practice.

Goodheart-Willcox Publisher

Figure 29-55. Welding in the 2F position.

1. Obtain two 1 1/2″ × 4″ × 1/4″ steel coupons from your instructor or cut them using the cutting torch. Grind the edge of one of the pieces straight and smooth using a bench grinder or disc grinder.

Safety Note

When using a grinder, observe the power tool safety rules described in Chapter 8 and these safety precautions:
- Wear OSHA-approved eye protection.
- Operate grinders and sanders in well-ventilated areas.
- Allow the tool to reach operating speed before contacting the work material.
- For disc grinders, apply only light pressure and let the tool do the work. Keep the tool in constant motion.
- For bench grinders, apply even pressure and move the work material continuously across the face of the stone/sander. Use the tool rest, and keep it adjusted to within 1/8″ from the grinding stone.

2. Using a scrap piece of steel, set the machine parameters according to manufacturer's specifications. Run a test bead, and then make any adjustments to fine-tune the settings.
3. Strike the arc. Remember to call out the word *cover* before striking the arc to alert anyone nearby to avoid looking at the arc without adequate protection.
4. Tack weld the two coupons lengthwise and perpendicular to each other by placing a small weld on each end of the joint. *Tack welds* are very short, temporary beads used to hold parts in place while welding. When the permanent weld is performed, the tack welds are remelted and included in the final bead. It may be necessary to realign the plates after placing the first tack due to contraction of the cooling tack weld.

5. Weld a 10″ long, 3/4″ or 1/2″ pipe handle to the tacked coupons to make the part safer to handle and provide an excellent location for attaching the workpiece connection.
6. Position the coupons on the welding table for comfort during welding. For practice in the 1F position, place the coupons so that the V formed by the joint is facing up. For practice in the 2F position, position the coupons so that the top of the T is upside-down on the table.
7. Use a 3/32″ or 1/8″ E6011 electrode to weld a single-pass bead along the joint of the coupons. This is the root bead.
8. Leave the welding machine running. The duty cycle requires a cool-down time to protect the internal components of the power source.
9. Thoroughly clean the root bead with a chipping hammer and a wire brush. Do not cool the coupon at this point because rapid cooling deteriorates bead quality.
10. Evaluate the root bead quality by answering these questions:
    - Is the surface uniformly textured?
    - Do the edges blend smoothly and evenly to each side of the joint?
    - Are there discontinuities? How many? How severe?
    - Is the penetration acceptable?
11. Weld a second bead on the other side of the joint to back up the root bead.
12. Weld another bead between the root bead and the plate. There should be about a 60% overlap on the previous bead. Repeat this step on the opposite side of the coupon.
13. Thoroughly clean the bead and repeat on the other side of the joint. Alternating the direction of travel from the previous bead yields a smoother finished surface. Repeat this step on the opposite side of the coupon.
14. Clean the welds thoroughly. If desired, continue filling the joint until it is full. Doing so affords you more practice and helps you improve your technique. A variation of this exercise, the "hammer" project, uses four coupons arranged in a cross pattern to create four fillet joints for practice welding. See **Figure 29-56**.
15. Clean your work area and return all equipment to its proper place.

Goodheart-Willcox Publisher

Figure 29-56. The "hammer" project is a version of practicing fillet welds that allows for a great deal of practice using a minimum of materials.

Groove Welds on Flat Plate

A groove weld is performed between two plates, usually of approximately equivalent thickness, positioned in a butt joint. On plates of 1/4″ and thinner, the edges may be square with no further preparation other than cleaning. On thicker plates, some beveling is usually required to ensure adequate bead penetration, and multiple passes may be required to completely bond the joint.

On open groove welds, the two plates are fused together at the bottom with a flat bead of weld metal. No backing plate is used. E6010 and E6011 SMAW electrodes are commonly used on mild steel due to their fast-freeze characteristics. To achieve complete penetration, the electrode is pushed down through and pulled along the gap. On the back side of the completed weld, the root bead should blend smoothly to both edges and exhibit a uniform surface free from discontinuities. See **Figure 29-57**.

During welding, a *keyhole*, or rounded enlargement, appears in the opening just ahead of the pool. See **Figure 29-58**. Before the keyhole expands beyond control, briefly whip the electrode slightly upward and ahead of the weld to allow the base metal to cool slightly. The keyhole size should remain the same as the bead at the back of the pool solidifies. Then, quickly whip the electrode back above the molten pool to allow another drop of filler metal to fall from the electrode. The entire whipping motion must occur very quickly. The rate of motion is determined by the level of heat observed in the weld. At the beginning of the weld, less time out of the pool is required. As heat builds up with the progression of the weld, the electrode should be moving in and out of the molten pool at a very rapid pace due to the high heat now in the base metal.

The ability to maintain the keyhole ahead of a uniform pool depends on understanding and controlling the heat of the weld. Success requires the welder to risk the application of too much heat—the common tendency is not staying in the pool long enough to achieve penetration. Developing skill at open groove welding requires practice and may involve burning through a few plates to find the correct balance between too much and too little heat.

Goodheart-Willcox Publisher

Figure 29-57. Penetration of the root bead on the back side of a groove weld is important to the strength of the welded bond.

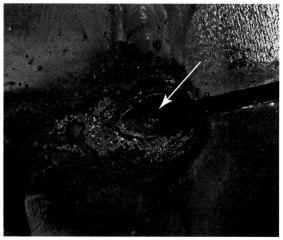

Goodheart-Willcox Publisher

Figure 29-58. In order to ensure adequate penetration, the welder should observe a keyhole effect just ahead of the weld pool.

Groove Weld Practice

Using 1 1/2" by 4" coupons of 1/4" plate is an economical way to practice performing groove welds. The following procedure outlines a simple method of practice.

1. Obtain two 1 1/2" × 4" × 1/4" steel coupons from your instructor or cut them using the cutting torch. For practice with 1/8" E6011 electrodes, grind the adjoining edges of the pieces straight and smooth with a bench or disc grinder. For practice with 3/32" E6011 electrodes, the adjoining edges of the pieces should be cut and ground at a 60° bevel, leaving a 3/32" root face.

2. Using a scrap piece of steel, set up the machine parameters according to manufacturer's specifications and run a test bead. Make any adjustments to fine-tune the settings.
3. Strike the arc. Remember to call out the word *cover* before striking the arc to alert anyone nearby to avoid looking at the arc without adequate protection.
4. Tack weld the two coupons lengthwise and parallel to each other leaving a gap no wider than 3/32", **Figure 29-59**. It may be necessary to realign the plates after placing the first tack due to contraction of the cooling tack weld.
5. Welding a 10" long 3/4" or 1/2" pipe handle to the tacked coupons will make the part safer to handle and provide an excellent location for attaching the ground clamp.
6. Position the coupons on the welding table so that the joint is facing up and the back side of the joint does not contact the top of the table. The coupons may have to be bridged across pieces of scrap to achieve this position. Contact with the table on the back side of the joint invites the introduction of contaminants into the back of the weld.
7. Use a 3/32" or 1/8" E6011 electrode to weld a single-pass bead along the joint of the coupons. This is the root bead.
8. Leave the welding machine running. The duty cycle requires a cool-down time to protect the internal components of the power source.
9. Thoroughly clean the root bead using a chipping hammer and a wire brush. Although it may be tempting to cool the coupon at this point, this is a bad habit. Rapid cooling deteriorates bead quality.
10. Evaluate the root bead quality by answering these questions:
    - Is the surface uniformly textured?
    - Is the top of the root bead slightly concave, and do the edges blend smoothly and evenly to each side of the joint?
    - Is the penetration complete?
    - Are there discontinuities? How many? How severe?
11. Weld a second bead on top of and penetrating into the root bead. This cover bead should blend smoothly to both edges of the joint.
12. Clean the welds thoroughly. If desired, continue filling the joint until it is full. Doing so affords you more practice and helps you improve your technique.
13. Clean your work area and return all equipment to its proper place.

Goodheart-Willcox Publisher

Figure 29-59. Tack weld for groove weld practice.

Vertical Fillet and Butt Welds

When welding in the vertical position (3G or 3F), keep in mind the effect of gravity on the molten pool as well as the effect of heat rising through the base metal. The trick to vertical welding is preventing the molten pool from dripping out of the joint. Vertical welding can be performed by traveling up or down. Uphill welding provides better penetration than downill welding, but the downhill technique can be used on thinner plate because less penetration is required. Whether running vertical-up or vertical-down welds, the welder should always begin by setting the amperage in the lower end of an electrode's amperage rating as published by the manufacturer.

Vertical uphill welding is a layering operation in which a welded base is built from the bottom up. Each new layer acts as a base on which the following layer is built. Each solidified section of the bead should be one and one-half to two times the diameter of the electrode.

Using a 1/8″ electrode, the weld pool should be about 1/4″ across. The weld pool should freeze in time to support the next weld. With an E7018 electrode, control the heat by weaving the electrode from side to side, mostly focusing on the sides of the joint. This motion allows the bead below the weld pool to cool just enough to form a supporting platform for the next section of the bead. Hold the electrode with the back tilted slightly down from perpendicular. Maintain the correct arc length to ensure penetration, **Figure 29-60**. The slag should drip away from the molten weld pool to ensure a strong weld that is free from slag inclusions.

Vertical uphill welding can also be performed with fast-freeze electrodes such as an E6011. When this electrode is used, the AWS recommends a whipping technique in which the electrode is positioned rapidly in and out of the weld pool. Using an E6011 electrode requires a stacking technique, rather than the weaving performed with an E7018. The bead will resemble a stack of dimes, with each "dime" supporting the next. Refer again to Figure 29-30. To stack a vertical-up weld, keep the electrode in the root of the material. The technique requires the welder to maintain the arc while whipping the electrode up, away from the weld pool. As the weld pool freezes, the electrode is returned to the molten pool to build another level of the weld. Whipping with an E6011 requires a higher level of skill than weaving with an E7018.

Safety Note

Due to the effects of gravity, the risk of dripping molten metal is increased when 3G and 3F welds are performed. When welding in these positions, stand to the side and wear full protective gear. Check the surrounding area for potential fire hazards and have a properly charged fire extinguisher nearby.

Wes Hancock, Caldwell High School

Figure 29-60. Welding in the vertical position. Note the position of the electrode.

Controlling Distortion

The biggest factor influencing the amount of distortion produced by a weld is the amount and distribution of heat input. The faster a weld is made, the less heat is absorbed into the base metal. Simply speeding up the welding process can greatly reduce the introduction of heat. Using a higher amperage setting to achieve a faster travel speed may actually result in a lower overall heat input. Welding a joint using intermittent short beads, rather than welding the entire joint with a single bead, is another method of controlling distortion.

The order and placement of intermittent welds can also affect distortion in a weld joint. The welding of a joint should usually begin at the center and progress toward each free end. If attached at one end, begin welding at the point of attachment and proceed toward the loose end. Use only the minimum number of passes required to complete a weld. Overwelding introduces excess heat to the metal and creates unnecessary points of stress. The following common procedures are used to minimize distortion:

- Preheat parts before welding. Preheating counteracts movement from distortion by creating a larger area of expansion and contraction than just the weld joint.
- Use clamps, jigs, or other fixtures to hold parts solidly in place while welding and cooling, thereby minimizing distortion.
- Position parts to be welded slightly out of the desired position. The distortion is planned to result in contraction that moves the pieces into final position.

Welding Cast Iron

Cast iron contains much more carbon and silicon than steel, making it less ductile and more challenging to weld. However, successful cast iron repair welds are possible and can lead to extreme savings on investments in equipment. Several different types of cast iron are commonly used in the construction of equipment. Each type reacts a little differently when welded. An important factor for success in welding cast iron is the age and condition of the casting.

Before welding cast iron, use cleaning solvents to clean the surface to be welded and the area surrounding it. Make sure the surface is free of defects and surface contaminants. Used parts are likely to be dirty and impregnated with oil or grease. Contaminants missed by solvents and other cleaners can often be burned out with an oxidizing oxyacetylene flame, followed by wire brushing or grinding. New cast parts may be contaminated with foreign material left from the casting process. This foreign material must be removed prior to welding.

To prepare a crack for welding, first drill a hole at each end of the crack to prevent it from spreading. Then, grind the crack out to the bottom. Be sure to remove any residue left by the grinder.

In most cases, preheating is required when cast iron is welded. Preheating reduces the heat input requirements and relieves internal stresses. Using a cast-iron filler metal requires a much higher preheat because the filler metal has low ductility and is susceptible to cracking. An alternative to preheating is to use a nickel-based electrode, which produces a welded bead that yields during cooling and absorbs stresses that could cause cracking in or near the weld.

Often a heat treatment needs to be applied to welded cast iron parts after welding. This can be done with a multi-flame heating attachment on an oxyfuel torch or applied in a furnace. The purpose is to relieve stresses and slow the cooling process. Welded casting can also be cooled slowly after welding by covering it with or burying it in insulating material, such as lime or vermiculite. Cooling parts should be left undisturbed until they reach atmospheric temperature.

Peening the weld with a ball-peen hammer immediately after welding expands the weld metal to counteract the effects of contraction. Longer sections of cast iron can be welded by making short welds with soft nickel electrodes followed by peening while the initial cooling takes place.

Welding Certification

Welding certification is written documentation that a welder has produced welds according to certain standards. The welder must perform the test under supervision and following prescribed conditions. Certified welders are considered desirable by potential employers and can often command higher salaries than uncertified coworkers. Ability to certify also indicates a willingness to commit to a goal. However, not all welders need to be certified. Many agricultural jobs require only occasional welding.

Different certification tests are administered for variances in type of material, thickness of material, welding position, intended use, and other factors. However, if written correctly, a single test may qualify a welder for several positions and thicknesses. For example, persons certified in vertical welding are often considered certified to weld in the flat and horizontal positions as well.

Certification tests are almost always based on a welding code and follow an approved *welding procedure specification (WPS)*, a document that outlines the parameters or conditions of the test. The completed weld must meet visual inspection criteria as well as pass destructive and nondestructive testing.

Companies that need certified welders often hire third-party companies to administer certification tests. If a certified welder changes jobs, even within the same company, a new certification is usually necessary because the company is legally responsible for the quality of welded joints. Certification by this method is generally not portable, or transferable, from one company to the next. However, previous certification is a competitive advantage when applying for jobs.

Certification testing is available at AWS-accredited test facilities. The resulting certificate is slightly more portable than a test performed for a specific company.

Many schools offer training programs that lead to certification, and most offer some type of certificate of completion. This is not welding certification; however, it indicates that a person is prepared to be tested. A welding certification test does not necessarily involve a difficult weld joint; however, most welding tests are challenging. If you know what the test will be ahead of time, practice is in order.

CHAPTER 29 Review and Assessment

Summary

- In the SMAW process, a consumable electrode coated in a flux is used to form the welded bead.
- All machines for SMAW are constant-current machines that provide an even supply of amperage to the electrode.
- Welding machines are classified by their maximum amperage output and their duty cycle. A duty cycle is a rating that indicates how long a welding machine can be used at a given output current without damaging it.
- Electrodes are classified by size and by type. The size of an electrode is determined by its diameter measured at the wire core. Electrode types are classified by the makeup of their flux covering.
- Four broad categories of electrodes are fast-fill, fill-freeze, fast-freeze, and low-hydrogen.
- Welded joints can be classified as butt, T-, corner, edge, or lap joints. Each requires different welding techniques and application of heat.
- The American Welding Society designates welding positions with numbers and letters. The numbers represent the positions—1 for flat, 2 for horizontal, 3 for vertical, and 4 for overhead. Fillet and groove welds are represented by the letters *F* and *G*.
- Welding variables controlled by the welder include current, arc length, electrode angle, electrode manipulation, travel speed, type of surface preparation, and selection of electrodes.
- The electrode type must be matched to the base metal composition to ensure a strong weld.
- Common discontinuities in welds include porosity, incomplete fusion, incomplete joint penetration, undercut, overlap, and cracks.
- Weld quality is determined using many different methods. These include visual inspection, destructive testing methods, nondestructive methods such as liquid penetrant dye and magnetic particle inspection, radiographic testing, and ultrasonic testing.
- Arc welding safety involves planning procedures, safely arranging the work environment, and always wearing the appropriate clothing and personal protective equipment (PPE).
- To form strong, uniform beads, the arc must be held at a consistent distance from the base metal to form the molten pool. The arc is moved along at a pace that allows the pool to penetrate the base metal as it accepts the addition of filler metal from the melting electrode.
- The amount and distribution of heat input strongly influences the amount of distortion produced by a weld.
- Cast iron contains much more carbon and silicon than steel, making it less ductile and more challenging to weld.
- Welding certification is written documentation that a welder has produced welds according to prescribed standards.

Words to Know

Match the key terms from the chapter to the correct definition.

A. arc force
B. arc length
C. coupon
D. discontinuities
E. distortion
F. duty cycle
G. fillet weld
H. flux
I. groove weld
J. inclusion
K. keyhole
L. out-of-position welding
M. overlap
N. porosity
O. root bead
P. slag
Q. spatter
R. stringer bead
S. tack weld
T. tie-in

1. Used to stabilize the welding arc and to reduce weld contamination.
2. Protrusion and notch in the weld metal beyond the weld toe, resulting in a concentration of internal stresses.
3. Percentage of time that a machine may be used under load while power is switched on.
4. Type of weld produced by welding two parallel flat edges butted together.
5. A weld between two perpendicular parts forming a *T*.
6. Welding material in any orientation other than 1G or 1F.
7. The distance from the electrode tip to the base metal.
8. Metal droplets that land and solidify in locations around, but not in, the welded bead.
9. Force created by the electric welding arc that can have a physical effect on the calmness of the molten puddle.
10. A bead created by moving the electrode along the weld without any side-to-side weaving motion.
11. Blending of a new weld with the base metal or with a previous bead.
12. Waste material that contaminates and weakens a weld by blending into small pockets of the cooling weld material.
13. Deviations from the consistency of a completed weld bead.
14. Bubbles or pockets formed into a welded bead by trapped gas.
15. A small piece of base metal used for practice welding or welder testing purposes.
16. Deflection or warpage of the welded part due to differential cooling of welded beads and welded parts.
17. Initial bead applied to a joint that will include multiple layers of welded beads.
18. Very short bead used to hold parts in position for further welding of joints.
19. Roundish opening formed just in front of the molten puddle in a joint being welded.
20. Waste material that serves to blanket a new weld to slow the cooling rate.

Know and Understand

Answer the following questions using the information provided in this chapter.

1. The SMAW process is best suited for bonding _____ metals.
2. The flux coating on an SMAW electrode chemically reacts to the heat of the arc, creating a(n) _____-free shield of gas around the molten weld pool.
3. The size of an electrode is determined by its _____, and the electrode type is determined by the flux coating.
4. An E7018 electrode will produce a finished weld with a tensile strength of _____ psi.
5. What are the five main classifications of welded joints?
6. What are the four basic welding positions recognized by the AWS?
7. What three main factors determine the amperage required to weld?
8. The deposition of spatter around a completed weld is an indication of a(n) _____ arc length.
9. A pointed effect in the ripple of the finished bead results from too much electrode _____.
10. If welding travel speed is too fast or too slow, the result will likely be a bead that lacks _____.
11. When welding material over 5/16" in thickness, it is advisable to _____ the joint edges to ensure complete penetration.
12. _____ electrodes should be chosen to weld on steels with a high carbon content.
13. Weld discontinuities that are at an unacceptable level are considered weld _____.
14. Name two nondestructive test methods used on welds.
15. Why should welders wear shirts with no pockets or covered pockets?
16. Why is tapping a better method of striking an arc than scratching?
17. Drilling holes at both ends of a crack and _____ the weld as it cools are measures used to reduce cracking when cast iron is welded.
18. Welders can obtain welding _____ by passing tests based on welding code in which they produce welds under prescribed conditions.

STEM and Academic Activities

1. **Science.** Working with a partner, coil the leads of welding machine in small loops on the floor while welding. As one person welds, have the other person sprinkle iron filings over and around the coiled leads. What happens to the iron filings? What causes this? What are the implications to safety and health? Be sure to store the equipment properly and clean up the iron filings.
2. **Technology.** List several factors you would consider when choosing a new SMAW welding machine to purchase for farm repairs.

3. **Engineering.** Practice preparing two 4″ × 4″ welding coupons of 3/8″ mild steel plate for a test groove weld. Bevel each edge at 37.5° and leave a 1/16″ root face. Why are these measurements and consistency so important in preparing coupons for testing?
4. **Math.** Use a micrometer or a caliper to measure several different types of electrodes at the filler metal and at the flux covering. How uniform is the thickness of the flux? What percentage, by volume, is the flux coating compared to the filler metal for different types of electrodes?
5. **Social Science.** How can you use the Internet and social media to improve your welding skills? Explore different welding-related websites and social media platforms. Report on the three you found most helpful and explain the reasons for your choices.
6. **Language Arts.** The acronym *CLAMS* helps beginning welders remember important parameters of welding that they control. Look up this acronym on the Internet. In small groups, discuss what each letter stands for. Each group member should choose one of the parameters and explain to the others how controlling it contributes to a quality weld.

Thinking Critically

1. You and a partner are asked to construct a new slatted top for an oxyfuel cutting table to be used in the lab. Describe your design, including materials, joint types, welding positions, and electrode selection.
2. When selecting an electrode for a job, one of the considerations is the electrode size (diameter). In which welding situations is a smaller diameter electrode preferable? In which welding situations is a larger diameter electrode preferable?

CHAPTER 30
Gas Metal and Flux Cored Arc Welding

Chapter Outcomes

After studying this chapter, you will be able to:
- Describe the gas metal arc welding (GMAW) process.
- Describe the flux cored arc welding (FCAW) process.
- Explain the purpose of GMAW/FCAW equipment components and settings.
- Select the appropriate electrode and shielding gas for a welding job.
- Change an electrode wire.
- Describe the four primary metal transfer modes.
- Control welding variables to produce a good weld.
- Take appropriate safety precautions when GMAW or FCAW welding.
- Weld a practice pad using GMAW.

Words to Know

automatic welding
backhand welding
backing plate
contact tip
diffuser
double groove joint
flowmeter
flux cored arc welding (FCAW)
flux cored electrodes
forehand welding
gas metal arc welding (GMAW)
globular transfer
inductance
machine welding
metal cored electrodes
metal inert gas (MIG) welding
MIG welder's pliers
open root weld
positioner
postflow timer setting
preflow timer setting
pulsed arc machine
push-pull welding gun
root face
semiautomatic welding
shielding gas nozzle
shielding gas solenoid valve
short-arc
short-circuit metal transfer
single groove joint
spool gun
spray transfer
stickout
surfacing
travel angle
welding gun
wire feeder
wire speed
work angle

Before You Read

Skim the Know and Understand questions at the end of the chapter first. Use them to help you focus on the most important concepts as you read the chapter.

 While studying this chapter, look for the activity icon to:

- **Practice** vocabulary terms with e-flash cards and matching activities.
- **Expand** learning with interactive activities.
- **Reinforce** what you learn by completing the end-of-chapter questions.

www.g-wlearning.com/agriculture

Lone Wolf Photography/Shutterstock.com

Gas metal arc welding (GMAW), also known as *wire welding*, is a welding process in which a consumable, continuously supplied electrode surrounded by a protective envelope of a shielding gas is used to form the welded bead. GMAW is the most commonly used welding process in industry. Because the shielding is supplied as a gas, GMAW is performed mostly indoors where air movement can be controlled.

Flux cored arc welding (FCAW) is similar to GMAW but may or may not involve a supplied shielding gas. FCAW is performed with a hollow wire electrode filled with powdered flux or a mixture of powdered metal alloying agents.

Like SMAW, GMAW and FCAW are electric arc welding processes. The arc is maintained between a continuous, consumable wire electrode and the base metal. A shielding gas supplied from a compressed gas cylinder shields the arc from atmospheric contaminants. See **Figure 30-1**.

Development of Gas Metal Arc Welding

Gas metal arc welding was developed in the late 1940s as a faster way to weld nonferrous metals such as aluminum. Because inert gases, such as helium and argon, were used as shielding gases, the process was originally known as *metal inert gas (MIG) welding*. The term *MIG* is still often used to describe the process; however, this term is nonstandard and the American Welding Society discourages its use.

Advances in power sources, wire feeders, and electrode and shielding gas composition have driven the usefulness and popularity of GMAW. Initially, the high cost of inert gases made the use of this process on steel cost-prohibitive, but the development of cheaper methods of collecting and concentrating gases to be used as flux changed the economics of GMAW. Some modern shielding gases are slightly reactive and have a useful effect on the arc characteristics during the welding process.

GMAW performed by hand is sometimes referred to as *semiautomatic welding* because the equipment controls the feeding of the wire electrode. The welder manipulates (moves) the *welding gun* (welding torch). In *machine welding*, a welding gun is connected to a controlling arm not physically operated by a welder's hand. See **Figure 30-2**. The welder operating this

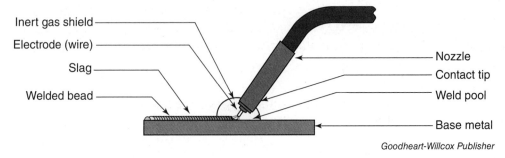

Goodheart-Willcox Publisher

Figure 30-1. Gas metal arc welding.

Praphan Jampala/Shutterstock.com

Figure 30-2. Machine-controlled GMAW welding and robotic welding reduce the element of human error in the welding process.

machine constantly sets and adjusts controls that position the machine-held welding gun. *Automatic welding*, the process used in robotic welding setups, employs sophisticated equipment to weld without constant adjusting of controls by a welder. In this process, automatic sensing devices trigger corrections to welding gun alignment in relation to a weld joint.

Advantages and Disadvantages of GMAW

GMAW offers the advantage of higher productivity. The process is fast and time is saved because rods do not have to be constantly changed. The weld does not require repeated cleaning, and cleanup can be done quickly. GMAW is an easy process to learn. A variety of metals (stainless, mild steel, and aluminum) can be welded in any position with the correct setup. Welding on thin materials is simplified with the GMAW process. Because the electrode is a continuously fed wire, the welder is able to concentrate on technique.

A disadvantage of GMAW is that, in most instances, it must be performed in a controlled environment out of the elements. More sophisticated and expensive equipment is required than for SMAW. Switching to a different material or joint may require complex changes to the equipment compared to switching a stick electrode and adjusting machine settings. The welding gun is bulky and may inhibit visibility. Inexperienced welders may produce a weld with a high-quality surface appearance but that lacks penetration and sufficient fusion.

AgEd Connection: Welding Opportunities

Gas metal arc welding (GMAW) and flux cored arc welding processes (FCAW) are commonly used in the manufacturing industry. A skilled worker in this area can easily find a productive, well-paying job. If the study of this chapter interests you and you enjoy the practice necessary to become proficient, then consider investigating potential careers and research to determine what necessary training you will need after high school. Some businesses will provide the training necessary for employment; others will require technical training from post-secondary institutions.

If you have access to GMAW/FMAW equipment and a location to work, you may even design and build your own projects for sale. For example, you could use your creativity and welding knowledge to build unique outdoor décor. People pay well for interesting metal garden sculptures.

The development of skills necessary to safely and efficiently use GMAW/FCAW equipment while working in your agricultural mechanics laboratory during the study of this chapter could be the foundation for a great metalworking profession.

Goodheart-Willcox Publisher

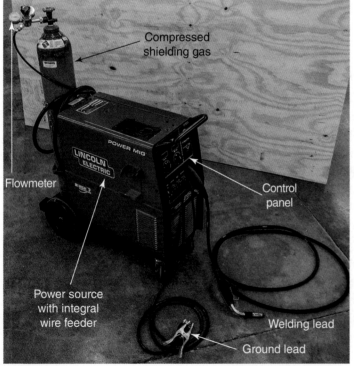
Goodheart-Willcox Publisher

Figure 30-3. A typical GMAW equipment setup.

Equipment

The equipment necessary for GMAW includes a power supply, a wire feeder, a welding lead, a welding gun, and a shielding gas supply. See **Figure 30-3**. There are several common types of GMAW machines, or power sources. The welding machine supplies electrical power for welding. The *wire feeder* contains the supply of wire and delivers it through a lead to the welding gun, where the electrode wire is directed to the joint. The shielding gas protects the weld pool from atmospheric contamination.

GMAW and FCAW Power Sources

A constant-voltage DC power source is most commonly used with GMAW and FCAW, but constant-current and AC power is used in some systems. Like machines for SMAW, GMAW power sources can be constructed around either a transformer or an inverter. Most power sources operate on an input of 240 V current, but some small machines designed for occasional use operate on 120 V. Many modern power sources have an integral wire feeder, but some have separate wire feeders that can be positioned well away from the main power source.

Basic GMAW machines have only simple adjustments for wire speed and open circuit voltage. The shielding gas flow rate is adjusted externally through a *flowmeter* or flow gauge at the cylinder. As the sophistication and cost of power sources increases, so do the number of adjustments available to set different welding parameters. See **Figure 30-4**. Two other common adjustments are *preflow* and *postflow* timer settings. These settings allow shielding gas to envelop the weld area prior to and after the arc to help prevent atmospheric contamination. Preflow shielding gas purges the weld area of atmospheric gases, including oxygen. Postflow shielding gas briefly maintains a protective shield on the weld area as it begins to cool and serves to cool down welding torch components. Newer GMAW power sources are programmable, automatically making adjustments according to weld parameters such as metal type and thickness, electrode size, and type of shielding gas.

Wire Feed Speed

Setting the wire feed speed regulates how fast the electrode is fed to the welding arc. **Wire speed** is calculated in inches per minute (ipm) and serves to regulate the welding amperage. The amperage output in GMAW varies depending on the wire speed and the amount of electrode extending beyond the welding gun, or *stickout*. Increasing the feed rate of the electrode wire increases its time in proximity to the base metal. Increasing the proximity time allows more amperage to pass through the wire and results in increased heat available at the weld pool. On machines with integral feeders, the wire speed adjustment is located on the main control panel. Independent wire feeders have a separate control.

Voltage

The voltage setting on most GMAW welders is selected based on electrode size, base metal thickness, and type of shielding gas. The setting remains relatively constant during the welding process. Typical GMAW

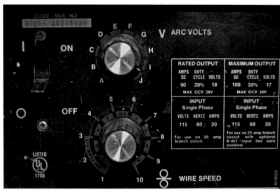
Goodheart-Willcox Publisher

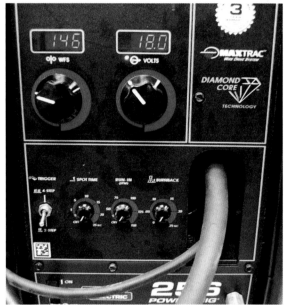
Goodheart-Willcox Publisher

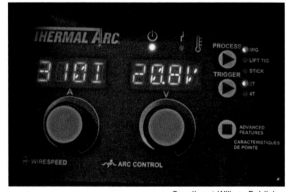
Goodheart-Willcox Publisher

Figure 30-4. GMAW power sources vary greatly in their ability to adjust welding parameters.

power sources can produce a range of output voltages from 0 V to 50 V. This range of voltages is usually adjusted through a switch located on the front of the machine. The welder initially sets the open circuit voltage, and then after the weld begins, machines with a voltmeter display the actual load voltage (the voltage at the arc). Load voltage establishes the working arc length occurring at the weld pool. Fine adjustments to the voltage can be made while welding to achieve optimum arc characteristics.

Inductance

Some machines have an *inductance* setting. Inductance, sometimes referred to as *pinch effect*, is the squeezing forces exerted on the electrode wire at the weld due to the amount of current passing over its surface. These forces affect the formation of molten drops of filler metal into the weld. In general, increasing the inductance increases the fluidity of the molten pool. Some increase often results in a smoother bead, but too much inductance can make the weld pool difficult to control.

Pulsed Arc

Early machines for wire welding processes were *short-arc* only—the electrode had to touch or short-circuit into the base metal in order to build up enough heat to melt off. As a result, the arc is extinguished and reestablished many times a second. In the newer *pulsed arc machines*, the working current is first run through a pulsing unit that rapidly cycles between periods of higher voltage and lower voltage, and the arc is maintained throughout the welding process. A smoother, more uniform weld joint is the result. The ratio of time on high voltage compared to low voltage can be adjusted to modify the heat input and weld deposition rate. Pulsed arc is most effective on machine-controlled or robotic welders that reduce human error.

Wire Feeders

The purpose of a wire feeder is to constantly supply the welder with fresh electrode. In the GMAW process, the electrode moves from a storage spool through the welding lead to the weld pool. Wire feeders may be included in the power source, or they may be independent and mounted externally. Wire feeders operate by engaging the wire between two hardened steel, grooved rollers that pull the electrode from the spool and push it through the lead and welding gun. See **Figure 30-5**. These rollers are easily interchangeable and must be configured to match the size and type of electrode being used.

Aluminum and flux cored electrodes are soft and susceptible to deformation. These electrodes must be paired with the correct size and texture grooves on the drive wheels. Follow the machine manufacturer's recommendations for set up and pressure settings.

Goodheart-Willcox Publisher

Figure 30-5. GMAW wire feeder rollers exert pressure on the electrode and push it through the liner of the welding lead and through the welding gun.

Most wire feeders have an electric motor that runs on 120 V and uses a potentiometer to control the speed.

A *push-pull welding gun* is equipped with a set of auxiliary rollers that pull the wire through the lead. Push-pull setups are especially good for aluminum electrodes because the softness of aluminum makes it problematic to push. See **Figure 30-6**. *Spool guns* are remote wire feeders built into the welding torch. The electrode is fed from small spools of wire mounted directly to the gun. See **Figure 30-7**. The drive rollers for these units are usually powered by small 12 V DC motors. The power needed to drive the spool gun is diverted from the main power supply through a transformer within the welding machine. Spool guns eliminate the need to force the wire electrode through a long lead. Feed rate is usually adjusted at the gun.

Welding Leads

The workpiece lead for GMAW is similar to that used for SMAW. Having a secure ground connection is essential to a quality bead and operator safety. Workpiece connections and cables should be rated to adequately handle the maximum welding currents delivered by the machine. See **Figure 30-8**. A welding gun, **Figure 30-9**, is used instead of the electrode holder used in SMAW, and the welding cable is more complex.

Wes Hancock, Caldwell High School

Figure 30-6. A push/pull welding setup feeding soft aluminum electrode wire. A set of small rollers in the welding gun pull the electrode at the same time it is being pushed by the rollers in the machine wire feeder.

Goodheart-Willcox Publisher

Figure 30-7. Using a spool gun eliminates the need to force electrode wire through long lengths of cable.

Goodheart-Willcox Publisher

Figure 30-8. The composition, construction, and quality of workpiece connections vary. The three shown, from left to right, can carry increasing amounts of welding amperage.

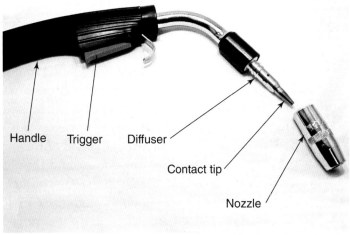

Goodheart-Willcox Publisher

Figure 30-9. Parts of a GMAW welding gun.

A GMAW welding lead is constructed with multiple components to perform several tasks:

- A conductor carries the current necessary for the welding arc.
- A coiled steel liner, much like a long spring, forms a pathway for the wire electrode to be pushed through.
- An airtight tube delivers clean, uncontaminated shielding gas to the tip of the welding gun. On many machines, this shielding gas also serves to cool the welding gun.
- On heavy-duty machines, two water tubes carry cooling water to and away from the welding gun.
- At least two small-gauge wires connect the microswitch at the gun to control solenoids in the power source, wire feeder, and shielding gas supply.
- An insulating cover binds the components into a single cable.
- The entire lead is often encased in a secondary protective sheath to guard it from external damage.

Handle leads with care. Unroll only enough lead to reach the welding location to reduce contact with dirty floors and minimize damage. Extra length equates to more wear and tear and requires more effort from wire feeders. If a 12′ lead can reach all foreseeable welding operations, a 25′ lead is excessive. Store leads in large loops rather than winding them tightly. Tightly winding the lead can cause the liner to exert undue friction on the electrode wire, resulting in erratic feed rates. See **Figure 30-10**.

Goodheart-Willcox Publisher

Figure 30-10. Store welding leads off the floor in large loops.

Welding Gun

The welding gun, or welding torch, functions as the electrode holder. It also directs the shielding gas to the area around the weld, brings power to the wire, and controls the on/off functions of the wire feeder and the shielding gas solenoid valve, which is described in the next section.

The welding torch housing is a plastic or composite encasement that insulates the power conductors from the welder's hand. Most welding guns have a short, sometimes positional, insulated metal tube (neck) through which the liner or electrode conduit extends and terminates. A brass *diffuser* attaches to the end of this tube. The diffuser has holes that distribute the shielding gas around the electrode. The diffuser also serves as the attachment for the copper alloy *contact tip*. Welding current is transferred to the electrode through the contact tip. Contact tips are easily replaceable and must be matched to the electrode size. They have a bore size just a little larger than the wire electrode diameter.

The *shielding gas nozzle* fits around the base of the diffuser and directs the flow of shielding gas around the contact tip and electrode. See **Figure 30-11**. Keep nozzles clean to ensure an uninterrupted, smooth flow of shielding gas. For the FCAW process, the nozzle can be removed to provide better visibility.

Each welding gun has a microswitch to trigger solenoids in the power source, wire feeder, and shielding gas supply. Pressing the trigger energizes the electrode wire, turns on the motor drive of the wire feeder(s), and opens the solenoid valve to release shielding gas through the welding lead and torch.

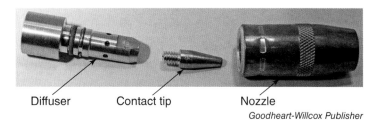

Figure 30-11. Consumable parts of a welding gun.

Compressed Gas Cylinders and Connections

Most shielding gases are stored, sold, and used from compressed gas cylinders similar to the oxygen cylinders used with an oxyfuel torch. The cylinders, which vary in size, are all equipped with a brass control valve. The fittings for shielding gases are different than those for fuel gas or oxygen in order to prevent potentially dangerous combinations.

Delivery of the shielding gas to the arc is crucial to maintaining quality in the finished weld. A regulator measures cylinder pressure and controls the flow rate of shielding gas from a single gas cylinder or from a manifold system of several cylinders connected together. Regulators indicate the flow rate with a flow gauge or with a flowmeter. See **Figure 30-12**. The rate of gas flow is measured in either cubic feet per hour (cfh) or liters per minute (L/min). From the regulator, the shielding gas passes through a flexible hose to a *shielding gas solenoid valve*, an electrically operated valve located within the power source or the wire feeder. This solenoid valve switches the gas flow to the welding gun on and off.

Consumables

Welding consumables are items that are used up and replaced. Consumables for GMAW include electrodes and shielding gas.

A

B

Figure 30-12. Flowmeters and flow gauges are used to manage the volume of shielding gas supplied to the GMAW welding arc. A—Flowmeter. B—Flow gauge.

Electrodes for GMAW

The electrode for GMAW is supplied in the form of a solid wire. Electrodes may be bare wire or coated with a very thin layer of a copper alloy. This alloy coating controls oxidation in storage and improves contact with the tip within the welding torch.

GMAW electrodes are sold on various spool sizes ranging from one-pound rolls intended for use in spool guns to huge spools handled by forklifts or other mechanical means for use on automated welding systems. See **Figure 30-13**. Like SMAW electrodes, size is determined by the diameter of the welding wire. Common GMAW electrode sizes for manual operation include 0.023", 0.030", 0.035", 0.045", and 0.052". Larger sizes are used with semiautomatic machine-driven and automated systems. The size of the electrode depends more on the machine type, voltage, and wire-feed options than on the thickness of the base metal.

If the correct shielding gas is used, the filler material combined with the base metal should result in a completed weld joint that matches the chemical and physical properties of the base metal. When choosing an electrode, consider the base metal, shielding gas, joint design, and type of service the finished part will be subjected to.

Filler metals for steel typically include alloying elements to improve the weld quality. Silicon is the element most commonly added to the electrode wire. In concentrations of up to 1%, silicon helps remove oxygen from the weld pool and results in a slight increase in weld strength. The carbon content of electrode wire is usually kept to a low level to provide strength without causing an appreciable increase in weld porosity. Small amounts of aluminum, manganese, titanium, and zirconium may also be used in the alloy formulation of GMAW electrode wires.

Both silicon and manganese have a significant cleaning effect. Using electrodes with higher amounts of these alloying elements can improve weld quality on steel with rust, oil, or mill scale contamination. Silicon and manganese also tend to increase the fluidity of the weld pool. For out-of-position welding, titanium or zirconium alloying elements may be a better choice.

The GMAW process can be used on carbon steels that would require low-hydrogen electrodes in the SMAW process. Because there is no flux to absorb moisture that could be introduced into the weld, the GMAW process produces a low-hydrogen deposit. The absence of a powdered flux also results in a finished weld without slag. Only small amounts of contaminants are floated to the surface in shiny flakes of silicon that must be removed prior to painting.

Goodheart-Willcox Publisher

Figure 30-13. This machine accepts 10 lb or 2 lb spools of electrode. Larger GMAW machines use electrodes packaged in larger quantities.

Like SMAW electrodes, GMAW electrodes are labeled according to a uniform coding system. See **Figure 30-14**. A spool of ER 70S-6 is a common mild steel electrode.

The *ER* indicates *electrode*. The next two or three digits, *70* in this case, represents the strength of the weld. This electrode yields a minimum of 70,000 pounds of tensile strength per square inch of weld in mild steel. The *S* indicates a solid wire requiring a shielding gas. The last character of the code, in this case *6*, represents materials added to the wire to improve weld quality.

Changing the Electrode Wire

Changing the electrode wire on most self-contained GMAW machines is a simple process that every welder should master. Before beginning, have a spool of electrode wire and a pair of **MIG welder's pliers**, or welpers, on hand. These pliers, **Figure 30-15**, have multiple functions designed around the GMAW welding torch. You may also need a new contact tip, shield gas nozzle, or diffuser when you change the electrode wire.

Begin by unplugging the power source. Remove and inspect the shielding gas nozzle. If the machine has been in use, let it cool down before handling the nozzle, tip, and diffuser. Clean nozzles that are lightly contaminated with weld spatter using the squared, tapered jaws of the MIG welder's pliers. Use the pliers as a reamer to loosen the debris. Be sure that the contact tip, nozzle, and diffuser are in good condition and are compatible with each other and the electrode diameter. The pliers also have recesses for gripping and turning these parts.

Goodheart-Willcox Publisher

Figure 30-14. Spool of ER 70S-6 GMAW electrode.

Safety Note
When changing an electrode wire, be sure to wear safety goggles or safety glasses.

- Long-nosed jaws for gripping wire
- Squared edges on jaws for cleaning the nozzles
- Small round jaw opening for gripping contact tips
- Hammer bosses for lightly tapping stuck torch parts
- Offset cutting edges for trimming wire to correct stickout
- Large round jaw opening for gripping hot shielding nozzles

Goodheart-Willcox Publisher

Figure 30-15. MIG welder's pliers can be used to pull and cut wire, remove and install tips or nozzles, and remove spatter from nozzles.

Figure 30-16. Remove loose dust and debris from the liner using compressed air and a blower nozzle

Next, open the wire access door. Release the tension arm on the wire feed rollers, allowing the two rollers to separate. If the wire has just run out, there will be a length of electrode wire remaining in the welding lead. Grasp the end of the remaining wire and pull it completely out of the liner. It may be necessary to remove the contact tip to accomplish this task. The wire can be pulled from either end of the lead. Be careful when the end of the wire is released that it does not recoil toward anyone's face. Use compressed air and a blower nozzle to remove loose dust and debris from the liner. See **Figure 30-16**. If you are changing electrode size, it may be necessary at this point to reverse or replace the rollers to match the wire type.

Release the retainer mechanism that holds the spool on the spindle and remove the spool, paying close attention to the orientation of the wire. Some machines pull the wire from the bottom of the spool, while others pull the wire from the top. They will not function properly if the wire orientation is reversed. See **Figure 30-17**. If unsure about this, check the operator's manual. Set the new spool of wire in place using the correct orientation. Be careful not to let go of the free end of the electrode—it will uncoil rapidly, creating a mess. Reattach the spool retainer mechanism.

Figure 30-17. Wire feeders are offered in top feeder and bottom feeder configurations.

Pull and straighten about 6″ on the end of the wire and insert it through the wire guide, over the bottom roller, and into the end of the liner. Swing the top roller into place, ensuring that the wire is properly gripped in the grooves on the roller. Engage the tension arm. Only light pressure should be exerted on the electrode wire. Refer to the operator's manual for exact settings.

Plug in and switch on the machine. With the lead extended in a straight line, depress the trigger to feed the electrode through the liner. On some machines, feeding through the welding gun is smoother with the contact tip removed until the end of the wire emerges. Replace the contact tip and the shielding gas nozzle. Use the MIG welder's pliers to clip the electrode to about 3/8″, the correct initial stickout for most applications. See **Figure 30-18**. Finally, close and latch the access door to the wire feeder.

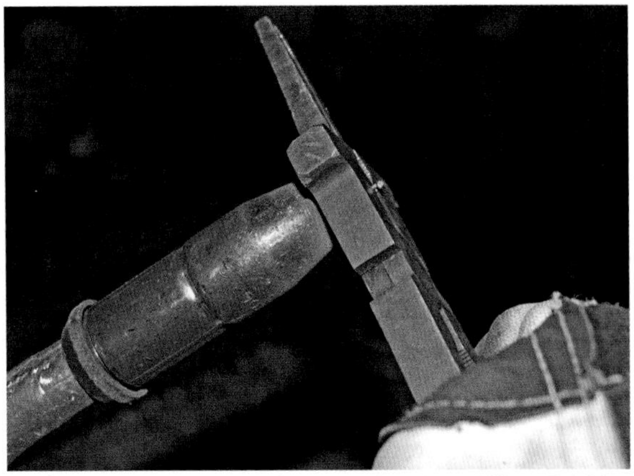

Shielding Gases

A gaseous shield protects the arc and molten weld metal from atmospheric contamination. Using an external shielding gas can also increase the smoothness of the arc and the finished bead. The type of shielding gas can be easily changed for welding on different types of material. Changing the welding process, in some cases, also requires switching to a different shielding gas.

Goodheart-Willcox Publisher

Figure 30-18. Clip the electrode to about 3/8″ by aligning the open face of the MIG welder's pliers with the end of the shielding gas nozzle.

Shielding gases can be inert, intentionally reactive, or a mixture of the two. Shielding gas typically makes up less than 5% of the total cost of welding; however, the correct choice of gas is critical to welding success. The most common shielding gases used for GMAW and FCAW are argon, carbon dioxide (CO_2), and oxygen. Helium is occasionally used in certain blends for welding aluminum and other nonferrous alloys. Argon and CO_2 can be used in pure form or mixed together or with other gases in different combinations to form shielding gas blends. Blended gases are labeled by their percentage, such as 75% argon/25% CO_2. See **Figure 30-19**. Blends are commonly referred to by the percentages only, such as 75/25. This can be confusing unless it is known which two gases the percentages are referring to.

Each shielding gas has different properties that determine its effects on the welding arc and bead formation. CO_2 shielding gas provides very deep weld penetration but also generates more turbulence in the weld pool. The turbulence results in a higher peak to the bead with greater material loss due to spatter. High levels of CO_2 tend to produce a convex bead contour. Argon blended shielding gases provide a more stable weld pool that results in well-shaped beads with minimal spatter. A high level of argon in the shielding gas mixture tends to produce beads with a narrow profile of penetration and a flatter surface. See **Figure 30-20**. Small levels of oxygen (2%–5%) in the shielding gas increase

Goodheart-Willcox Publisher

Figure 30-19. Labels on the shoulder of each cylinder indicate the contents and associated hazards.

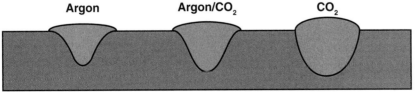

Goodheart-Willcox Publisher

Figure 30-20. The type of shielding gas affects the penetration of the weld.

Figure 30-21. These students are welding outdoors in an area that is shielded to protect the shielding gas envelopes surrounding their welds from the effects of wind.

the fluidity of the weld pool and have a significant cleaning action.

The flow rate of the shielding gas should be set high enough to blanket the area around the weld with a stable protective envelope. The shielding gas flow is set and measured in cfh. Gas flow rates and working pressures for GMAW are very low—3 to 8 psi typically achieves an adequate flow rate.

Manufacturers of electrodes and welding equipment offer recommendations for shielding gas selection and flow rates. Recommended flow rates range from 10 to 65 cfh; more is not necessarily better. Too much flow can result in turbulence at the weld that introduces atmospheric contaminants to the weld rather than producing a productive barrier. Flow rate recommendations typically trend toward the high side to account for errors in technique and occasional drafts. Welding outside of a protected indoor environment requires extra shielding. Attempting to compensate for drafts with excessive gas flow is usually ineffective. See **Figure 30-21**.

Changing Shielding Gas Cylinders

Handle cylinders containing compressed shielding gases carefully. A completely filled cylinder may contain pressure exceeding 2200 psi, and commonly used cylinders weigh between 65 and 108 pounds. When in use, compressed gas cylinders must be secured by chains, straps, or a similar device to prevent accidental tipping. See **Figure 30-22**. When not in use, regulators should be removed and the cylinders stored in a manner to secure them from falling. Their brass valves should be protected with the steel caps screwed into place.

To move a compressed gas cylinder on concrete floors, first ensure that the protective cap is in place and secure. Then tilt the cylinder slightly and roll it across the floor along its bottom rim. Roll the cylinder with a slow and deliberate motion, maintaining the tilt to keep the weight balanced on the bottom rim. See **Figure 30-23**. Move electrical power cords, welding leads, and air supply hoses out of the path to avoid rolling over them. It is a good idea to practice this skill the first time with an empty cylinder.

Figure 30-22. Many GMAW welding rigs are mounted on carts or running gears that provide a stable location for securing shielding gas cylinders.

Methods of Metal Transfer

The four primary methods of metal transfer in GMAW are short-circuit mode, globular transfer mode, spray transfer mode, and pulsed arc mode. Each has distinct characteristics, advantages, and drawbacks. Each transfer method requires different consumables and machine settings. Not every power source is capable of all methods of transfer. As a general rule, the short-circuit mode of welding requires lower welding current settings than spray-arc methods. The metal transfer mode is determined by the type of material being welded, the power source, and the wire feeder.

Short-Circuit Metal Transfer

In the *short-circuit metal transfer* mode, a continuously fed solid or metal-cored wire electrode melts and is deposited through repeated short-circuits of the welding current between the electrode and the base metal. The entire metal transfer occurs when the electrode is in physical contact with the weld pool, causing the short circuit. Each short circuit transfers a single molten droplet from the electrode. At each contact, the current in the electrode rises and the magnetic force surrounding the end of the electrode increases. This electromagnetic force squeezes, or pinches off, the molten droplet. See **Figure 30-24**. Short-circuits can cycle up to 200 times each second.

Short-circuit GMAW is a relatively low heat input mode of metal transfer, making it ideal for sheet metal. The short-circuit method is typically used on thin materials up to just under 1/4″. Pure CO_2 and

Wes Hancock, Caldwell High School

Figure 30-23. Rolling compressed gas cylinders along the bottom edge is a safe method for moving cylinders short distances.

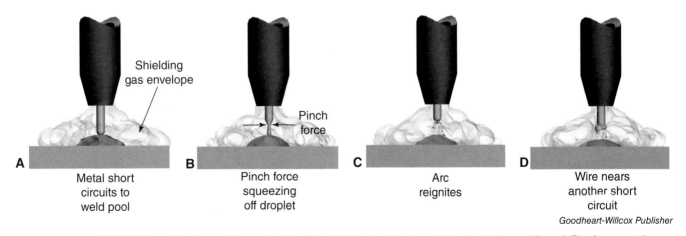

Goodheart-Willcox Publisher

Figure 30-24. GMAW short circuit metal transfer. A—The electrode wire short-circuits to the weld pool. B—A magnetic force pinches off a droplet of molten electrode metal. C—The welding arc reignites. D—The process repeats itself.

argon/CO_2 blends are commonly used as shielding gas for short-circuit transfer. Electrode diameters generally range from .023″ to .045″. When performed correctly, welding in short-circuit mode emits a sound similar to that of frying bacon.

Globular Transfer Mode

Globular transfer is the state of transfer between short-circuiting and spray arc transfer. Before the wire short-circuits, large droplets of metal are melted off the end of the electrode wire and deposited in the weld pool. In this mode, the drop diameter is greater than the electrode diameter. See **Figure 30-25**. Globular transfer of filler metal results when voltage and wire feed speed are set higher than the optimum for short-circuit transfer. Welding in this mode of transfer can be erratic, creating undesirable amounts of spatter and a less desirable weld appearance than spray arc transfer.

Globular transfer is limited to flat position welding of horizontal fillet welds on metal that is 1/8″ or thicker. Globular transfer can take place with any shielding gas when the current is relatively low in relation to wire speed and voltage. For most purposes, this transfer method is undesirable due to lack of fusion, insufficient penetration, and excessive filler metal deposition.

Spray Transfer Mode

Spray transfer maintains a fine arc column and sends a stream of small molten droplets through the arc to the weld. Typical droplet diameter is equal to or less than the electrode diameter. The metal spray is axially directed, which means the droplets are expelled from the end of the electrode following a path similar to water leaving the end of a garden hose. See **Figure 30-26**. The metal transfer rate may be as high as several hundred droplets per second, and the wire speed may be as much as 800″/min. Spray transfer requires a shielding gas with a high percentage of argon and relatively high voltage settings. Welding in the spray transfer mode is a fast procedure—the travel speed must keep the arc moving to prevent melting through the base metal. This transfer mode produces a distinct humming sound.

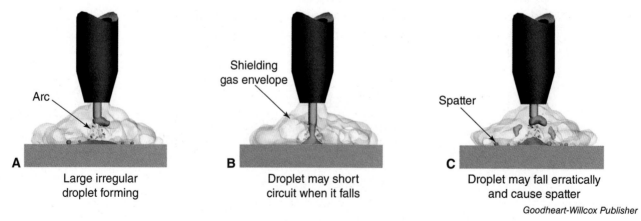

Goodheart-Willcox Publisher

Figure 30-25. GMAW globular metal transfer. A—Large irregular droplets form. B—Droplets may short-circuit when they fall. C—Droplets fall erratically and may cause splatter.

Only power sources capable of higher output can deliver enough energy to operate in spray transfer mode. Using the spray transfer mode offers high deposition rates, good fusion, and a deep, fingerlike penetration. Very little spatter is deposited around the weld, and the weld pool has a highly fluid character.

The spray transfer mode should be limited to welding in the flat and horizontal fillet positions on metal greater than or equal to 1/8" thick. Clean metal and proper fit-up are important. The fluid character of the weld pool makes filling open roots, or large gaps, problematic.

Pulsed Arc Mode

Pulsed arc transfer is a variation of the spray transfer mode. Only power supplies engineered to deliver a pulsed welding current are capable of welding in this mode. The power supply shifts the current between a high spray transfer current and a low background current. The low level of current is below the level at which metal transfer occurs, while the high level is well into the spray arc transfer. Filler metal is transferred to the weld only during the periods of high current. Ideally, one droplet is transferred during each high-current pulse. Arc stability is good, similar to that of spray transfer without pulsing.

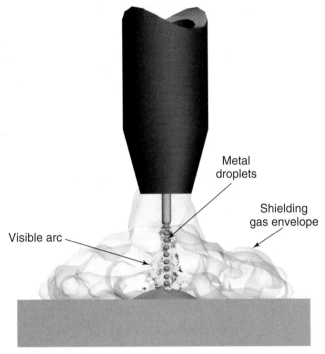

Goodheart-Willcox Publisher

Figure 30-26. Spray transfer requires higher amperage than globular transfer. The higher amperage breaks the electrode into many tiny droplets that are deposited into the weld without disturbing the arc.

The periods of low background current maintain the arc and reduce the average current. The lower average current makes welding thinner gauge materials and out-of-position welding on thicker materials possible. The frequency of the cycle is the number of times the shift from high to background current occurs per second. As the wire speed is increased, the cycles per second increase proportionally. The same high-percentage argon shielding gas mixtures used for spray transfer are also used for pulsed arc mode.

Flux Cored Arc Welding (FCAW)

Introduced in the 1950s, flux cored arc welding (FCAW) is closely related to GMAW. Both use similar power supplies, continuous wire feeds, and welding torches. In the FCAW process, a hollow, tubular, mild steel electrode filled with flux is fed through the center of the welding gun. A gas shield supplied by the welding gun protects the weld pool from oxidation. FCAW that uses a shielding gas as well as a flux cored electrode is referred to as *gas-shielded FCAW* (FCAW-G), or informally as *dual shield welding*.

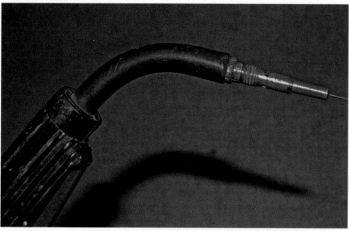

Figure 30-27. The shield gas nozzle has been removed from this GMAW torch for better visibility with a self-shielded FCAW electrode.

However, FCAW may also be performed without a shielding gas. This is referred to as *self-shielded FCAW*, or FCAW-S. See **Figure 30-27**.

FCAW also has similarities to SMAW. Self-shielded electrodes create the gaseous shield at the arc through the decomposition of the flux. The powdered flux inside the electrode forms a slag covering that protects the weld from the atmosphere. The physical barrier provided by the flux and slag allows the welding procedure to be more easily adapted to outdoor conditions.

FCAW is used in all welding positions. It yields high deposition rates, rapid travel speeds, and high-quality welds with excellent weld appearance. Dual shield welding is particularly well suited to structural steels. It is often the choice for welding thicker sections of carbon steel, stainless steel, and low-alloy steels used in the construction industry. Most nonferrous metals, including aluminum, cannot successfully be welded using FCAW.

FCAW Electrodes

Flux cored electrodes are tubular, usually formed from a mild steel alloy, and filled with powdered flux material similar to the coating found on SMAW electrodes. See **Figure 30-28**. Some electrodes are formulated to be used with a shielding gas, while others are self-shielding. Self-shielding electrodes create the O_2-free shield around the weld from the disintegration and combustion of the flux core.

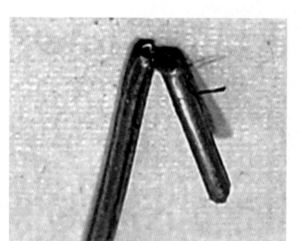

Figure 30-28. An FCAW electrode is composed of a hollow tube filled with flux material.

FCAW electrodes are formed from flat metal strips using a series of rollers. In the middle of the forming process, the strip is formed in a *U* shape that is filled with flux and alloying materials before the roll is completed. Allowing the electrode to kink or coil too tightly can cause separation of the longitudinal seam and deformation of the round profile of the hollow wire. *Metal cored electrodes* are filled with minimal or no flux. These electrodes contain powdered mixtures of metals engineered to improve the properties of the alloy in the finished weld.

Using cored electrodes results in high deposition rates with relatively small-diameter electrodes. Increasing arc voltage beyond recommendations increases weld pool fluidity, flattens the weld bead, and increases spatter. Higher voltages also tend to reduce penetration and may cause loss of alloying elements.

FCAW electrodes can be particularly well suited to hardfacing, in which a tough, wear-resistant filter metal is deposited on the surface. Special hardfacing electrodes are used to apply the filler to the entire surface or in a waffle-like, basket weave pattern.

Welding Using GMAW

For successful gas metal arc welding, the welder must gain and maintain control over a number of different parameters. Prior to starting the arc, the welder selects the machine settings, prepares the surface, chooses the welding process, selects consumables, and sometimes determines the welding position.

Machine Settings

Setting the appropriate wire speed and voltage is essential to performing quality welds. On modern machines, digital readouts indicate wire speeds and voltage outputs. See **Figure 30-29**. Wire speed charts are supplied with new machines and posted on manufacturer's websites. These charts can also be found in free welding apps that can be downloaded to smartphones and other devices.

Older machines have only these two settings, which are adjusted by turning two dials. Sometimes the markings around the dials do not indicate the actual voltage or wire speed, but instead have a graduated series of numbers or letters. To set up these machines, use the specific manufacturer's recommendations.

Goodheart-Willcox Publisher

Figure 30-29. Digital readouts indicating wire feed speed and voltage output.

Without a chart, wire speed can be estimated in the following manner:
1. Convert the measurement of the base metal thickness into a decimal by dividing the numerator by the denominator. For example, if the steel is 3/16" thick, divide 3 by 16 to get 0.1875".
2. Each 1000th of an inch of base metal requires one ampere to weld. Therefore, multiply the decimal by 1000. For this example, 0.1875" thick metal requires 187.5 amps to weld.
3. Multiply the required amperage of your weld by the burn rate for the electrode size selected to calculate the wire feed speed. For this example, choose 0.035" wire from the burn rates chart in **Figure 30-30**. Using this example, multiply 187.5 amps by the burn rate of 1.6" to calculate a feed speed of 300 ipm for 3/16" steel using 0.035" wire.

GMAW Burn Rates	
Wire Size	Burn Rate
0.023"	3.5"/amp
0.030"	2.0"/amp
0.035"	1.6"/amp
0.045"	1.0"/amp

Goodheart-Willcox Publisher

Figure 30-30. GMAW burn rates.

If the markings on the machine do not accurately indicate the wire speed, it can easily be calculated at any setting. To manually calculate wire speed, turn the welder on and pull the trigger on the welding torch for exactly six seconds. Measure the length of the wire run during this time and multiply by ten to determine the current wire speed at this setting.

To set the wire speed on most modern power sources, start with the machine manufacturer's recommendations. On many machines, a chart supplying the needed information is attached to the machine. To use the chart, the welder must know the thickness of the base metal, the welding wire diameter, and the shielding gas composition. Read the recommended setting from the chart and set both the wire speed and the voltage with the controls. Programmable machines are even easier—just scroll through the digital menu and select the correct process.

Surface Preparation

Preparing the surface to be welded is crucial to the quality of GMAW welds. Surface preparation includes the removal of contaminants that would be detrimental to the completed weld, as well as shaping the contours of the edges to be welded. Joints are typically cleaned by chemical and/or mechanical processes. The joint area must be cleaned a sufficient distance from the weld so contaminants are physically kept away from the molten pool. Often this is approximately a one-inch-wide path. In addition to the joint area, the place where the workpiece lead is connected to the part must also be cleaned to ensure a solid electrical connection.

Cleaning

Field equipment to be repaired should be precleaned of dirt and organic matter with a steam cleaner, pressure washer, or high-pressure water or air. Mechanical brushing or scraping may also be necessary. Perform the precleaning outdoors if possible to avoid extra cleanup. Heavy grease and oil deposits may require the use of chemical degreasers and proper collection and disposal of the waste.

All metals, whether parts to be repaired or new material, have surface corrosion that must be cleaned. Failure to remove surface contaminants often leads to inclusions in the finished bead. Paints, other finishes, and rust must be mechanically removed from repair parts to be welded. Mill scale and

Thinking Green

Waste solvents and petroleum products, such as grease and oil, are harmful if released into the environment. Do not throw them in the trash or pour them on the ground or down a drain. They should be collected in leakproof containers. These waste containers can be taken to collection centers for proper disposal. Many counties and municipalities sponsor regular collection days for disposal of hazardous wastes. FFA chapters can coordinate with government agencies to publicize collection dates and locations.

oxides must be removed from new metal prior to welding. See **Figure 30-31**.

Files, scrapers, and wire brushes with steel bristles are used by hand to clean surfaces for welding. Disc grinders equipped with grinding discs, sanding attachments, carbide burrs, or steel brushes are faster and usually more effective than hand methods of cleaning.

When cleaning aluminum or stainless steel for welding, always use wire brushes with stainless steel bristles that are kept for cleaning only that specific material. Using regular steel brushes or mixing uses with a stainless brush cross-contaminates stainless steel and aluminum.

Goodheart-Willcox Publisher

Figure 30-31. The mill scale and rust have been brushed away from the edge of this steel plate to leave a clean metal edge for welding.

When using a power tool to prepare surfaces for welding, be careful not to remove too much material—thinner sections are weaker sections. Also, remove material so that any striations (small scratches left by the grinder) all run longitudinally along the part. Striations running across the part act like a perforated line on a piece of paper, creating a fault line that encourages separation.

Joint Preparation

The same basic joints found in SMAW welding are used in GMAW and FCAW welding. When the completed bead needs to penetrate completely through the joint, beveling of the edges is often necessary to facilitate the welding process. Bevel groove joints are common because they increase the surface penetration at the weld compared to a square joint configuration. Groove angles may be cut with a saw, a torch, a grinder, or a combination of methods. In industry, machines guide torches and other tools to cut precise angles in preparation for welding. When doing repair work, welders are often required to achieve the same result with handheld tools.

A bevel groove may be a *single groove joint*, beveled on one side only, or a *double groove joint*, beveled on both sides of the joint. A groove is typically beveled to a 60° or 75° angle, which translates to a 30° or 37.5° angle removed from each side of the joint. Bevels may extend all of the way through the plate or stop short, leaving a narrow square section known as the *root face*. See **Figure 30-32**. Too little angle

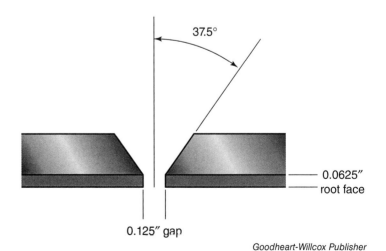

Goodheart-Willcox Publisher

Figure 30-32. Properly configured edges at the weld joint improve penetration, increase the surface area of the bond, and provide space for filler material.

Figure 30-33. The edge of this plate is beveled to improve weld penetration, increase the effective surface area of the bond for greater strength, and create space for the filler material.

does not provide enough clearance for adequate penetration, while too much angle requires more welding passes to fill the joint. Bevel grooves may be set up with a *backing plate* that supports the root weld as it is formed or as *open root welds* with no backing support.

Fillet joints can also be beveled on the part where the edge is welded to increase penetration. See **Figure 30-33**. When beveled joints are welded, correct fit-up is crucial. A little gap between parts is beneficial, but too much gap can make quality welds impractical.

Choosing a Welding Process

Understanding the modes of transfer and the capabilities of the available equipment helps you select the best welding method for a particular job. Most welding jobs can be accomplished satisfactorily by using more than one method. Choosing a particular process may offer advantages in heat input, rate of deposition, speed, clean up, and welding conditions. Study the options offered by the available equipment and consumables and select them according to your needs.

Consumable Selection

Selecting the mode of transfer narrows down the selection of consumables. The electrode wire must be compatible with the composition of the base metal and sized according to the thickness of the material. The size of the electrode also influences the amperage and thus the heat of the weld. Depending on the desired welding process, select a solid, metal cored, or flux cored electrode.

Shielding gases are selected to affect the penetration profile of the weld, the surface character of the bead, and the stability of the arc. Specific modes of transfer require minimum levels of argon or other shielding gases in the mixture. Welding with self-shielding flux cored electrodes requires no shielding gas.

Some joint profiles require specific setups on the welding torch. Long, tapered contact tips or modified shielding gas nozzles may be needed to allow for better access in confined spaces.

Welding Position

Welding in the flat position is almost always preferable to out-of-position welding. Horizontal is preferred next, followed by vertical-up, vertical-down, and overhead. When practical, position projects to enable flat welding, and reposition projects for additional welds. Many types of clamping tools are available for holding parts in position. See **Figure 30-34**.

Commercial fabrication facilities use special equipment called *positioners* for aligning parts to be welded in the optimum positions. When complete repositioning is impractical, sometimes a partial repositioning can reduce the need to weld in the vertical or overhead positions.

Safety Note

Use caution and common sense when moving and welding on large objects. Lift with the proper equipment, and support projects to prevent accidental shifting while welding or cleaning welds. Never work under unsupported or inadequately supported loads.

Parameters Managed during Welding

The following sections discuss the parameters controlled by the welder during the welding process. These include stickout, work angle, travel angle, electrode manipulation, and travel speed.

Stickout

Stickout is measured from the end of the contact tip to the free end of the electrode. The proper wire stickout for most solid wire GMAW applications is about 3/8". See **Figure 30-35**. Increasing the stickout slightly increases the resistance and lowers the heat input. Shortening the stickout reduces resistance, allowing for a greater heat input at the weld. Understanding this concept and making slight adjustments to stickout length can help compensate for poor fit-up often experienced during repairs. Maintaining 3/8" to 1/2" stickout length works for most applications. Most MIG welder's pliers are set up to cut the wire at an ideal stickout length when the open face of the cutter is placed against the nozzle while clipping the wire.

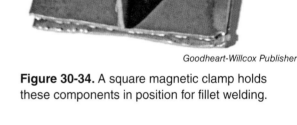

Goodheart-Willcox Publisher

Figure 30-34. A square magnetic clamp holds these components in position for fillet welding.

Work Angles and Travel Angles

The *work angle* is the angle between the welding gun and electrode and the center of the weld joint. In most cases, the work angle should be 90° to the surface of the material, or pointed directly into the center of the joint. See **Figure 30-36**. Maintaining the correct work angle ensures that both edges of the bead blend evenly to each side of the weld joint. When welding materials of different thicknesses, shift the work angle slightly to concentrate more heat to the thicker of the two parts.

Figure 30-35. Practice visualizing what 3/8″ looks like to help you maintain the correct stickout while welding.

Figure 30-36. The correct work angle is perpendicular to the weld joint.

The *travel angle* is the angle between the electrode and the direction of weld progression. See **Figure 30-37**. In SMAW, the electrode is almost always pulled or dragged, focusing the arc and the heat toward the back of the weld pool and the forming bead. In GMAW, the electrode can be either pushed or pulled. There is some debate among experts as to which technique is best—both methods work.

The pushing welding technique, also known as *forehand welding*, involves pushing the gun ahead of the weld pool. The forehand technique often produces shallower penetration and a wider, flatter bead because the arc force tends to be directed away from the weld pool.

In *backhand welding*, the dragging or pulling technique, the welding gun is pointed at the back of the weld pool and is pulled away from the forming bead. Welding backhand typically yields deeper penetration and a narrower bead with more deposition. The key is to tilt the electrode no more than 10° to 15° from perpendicular whether pushing or pulling. Too much angle extends the stickout and creates arc blow across the weld pool. The closer the torch is held to perpendicular, the deeper the penetration.

When welding thin materials, the forehand technique allows faster travel and creates less penetration into the base metal. Excessive penetration in thin metals melts a hole through the material. The backhand technique yields more penetration and better fusion when welding thicker steel.

An old welder's saying is, "If it leaves a slag, you drag," indicating use of the drag technique for SMAW and FCAW. Either technique is appropriate for GMAW, but welding forehand usually provides better visibility around the shielding gas nozzle. Whether welding forehand or backhand, concentrate on feeding electrode to the leading edge of the weld pool. Failure to maintain a consistent feed to the leading edge results in gaps in the completed bead. The resulting insufficient fill is susceptible to forces that can cause cracks and weakness in the weld.

Manipulation of the Electrode

Maintaining a uniform weave pattern is just as important with GMAW as with SMAW. However, the motion is simpler because the welder does not have to continually feed the electrode. Feed is controlled by setting the wire

 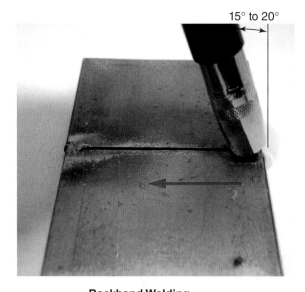

Forehand Welding   **Backhand Welding**

Goodheart-Willcox Publisher

Figure 30-37. Maintain the correct travel angle of 15° to 20° whether using the forehand or backhand technique.

speed in the beginning and maintaining a correct stickout. The welder can then concentrate more closely on weaving to maintain smooth tie-in on both sides of the joint while controlling the contour of the deposition of the bead. There are many different variations of weave patterns. Patterns that loop over the top of the weld pool should be avoided because these types of movement lead to bubbles or inclusions in the finished weld. A slight crescent-shaped weaving or whipping of the electrode, kept to less than 3/8″ in width, is fairly standard, but there is always more than one welding technique that works. Select a pattern that is comfortable and practice it until you become skilled.

Travel Speed

GMAW is much faster than SMAW. Maintain a travel speed that produces the desired conditions in the molten weld pool. The pool should be consistently shaped, showing a slight oval that is longitudinally oriented in the line of travel. It should be of proper size in relation to the thickness of the base metal. Move the welding torch at a pace that maintains these conditions. On thin metals, also consider the heat-affected zone—signs of too much heat indicate the need to travel faster.

Safety

The hazards associated with SMAW also apply to GMAW. In addition, the self-feeding aspect of the GMAW process involves extra moving parts that the welder should take into consideration. The best protection against hazards is always prevention. Most hazards can be avoided through understanding the welding process, following safety rules, and exercising good judgment.

Personal Protective Equipment

Proper personal protective equipment (PPE) is essential, although it is not a substitute for good planning, common sense, and safety engineering. When selecting PPE, consider the specific hazards that may be present. Follow these guidelines for PPE:

- Wear OSHA-approved (Z87.1) goggles or glasses at all times, including under the welding helmet. See **Figure 30-38**. As welded beads cool, particles of hot slag or silicon chips often pop off and are propelled some distance. Eye protection must already be in place when the welding helmet is raised. Others in the vicinity of welding should be protected from the ultraviolet and infrared rays emitted by the welding arc. Welding curtains placed around the work area serve this purpose.

Wes Hancock, Caldwell High School

Figure 30-38. Always wear safety glasses under the welding helmet to continue eye protection when the hood is raised.

- A welding helmet that protects the entire face is required. Do not use handheld observer's shields. Welding helmets should have at minimum of a #10 shade or filter lens and a clear protective plastic cover lens in front. Consider a darker shade if you are welding for extended periods or experience eye fatigue after welding. Inspect welding lenses on a regular basis. Look for cracks and proper fit in the helmet. Ensure that gaskets are in place and there is no leakage. When using an auto-darkening lens, verify that it is functioning correctly prior to use.
- Gauntlet type, heavy, insulated leather welding gloves protect the hands and forearms.
- Protective footwear should completely cover the foot with leather and be constructed with synthetic soles having significant tread.
- Heavy cotton shirts should have no pockets or covered pockets. Long-sleeved shirts and pants are required. Avoid rolled or frayed cuffs that could catch flying sparks.
- For out-of-position welding, leather aprons, leather sleeves, welding chaps, or a welding jacket are recommended because of the greater risk of burns.

The Welding Environment

GMAW is generally performed in controlled welding shop environments. FCAW-S, on the other hand, is well suited to field work. Working outdoors presents additional challenges, such as uneven working surfaces, heat from the sun, and insects.

Choose to weld in dry, clutter-free, and well-ventilated areas. Avoid out-of-position welding whenever possible and safely arrange the work environment to avoid hazards. Store flammable materials well away from locations in which hot metal work is being performed. Ventilation systems designed for welding remove contaminants from the air; however, they can be noisy and may create drafts that are detrimental to the GMAW process. Position ventilation systems and orient the work to minimize drafts while still removing smoke and other contaminants. When possible, weld in draft-free open areas or open large bay doors to increase the ventilation as well as reduce the noise level.

The electrically energized components of GMAW equipment present a risk of electric shock. Be sure to shut off the welding machine power before making internal adjustments. Keep welding equipment and cables dry. Regularly check cables for damaged insulation. Always solidly ground the equipment and the workpiece. Ensure that all electrical connections are tight. Avoid standing on a damp floor, and keep clothing and gloves dry, changing damp items when needed, to reduce the possibility of electric shock.

Uncoil only enough workpiece lead to do the job. Avoid having coils of lead in the work area to reduce the formation of electromagnetic fields. The welding gun lead, however, should be fully extended to minimize the coiling and binding of the electrode as it passes through the liner. See **Figure 30-39**. Position both leads as much as possible away from walking paths. Keep the welding leads clean and avoid dragging them roughly across sharp edges.

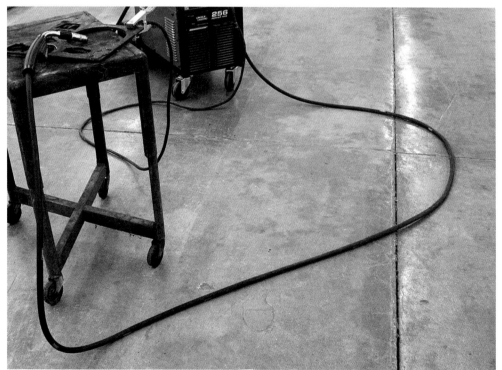

Goodheart-Willcox Publisher

Figure 30-39. Forming a large loop with the GMAW welding lead minimizes binding of the electrode in the liner.

Heat Hazards

Hot metal looks like cold metal! Assume that any loose piece of scrap is hot and handle hot coupons with tongs or pliers, not gloves. Welding a handle to practice coupons is a safety measure—a handle provides a safe place to pick up the hot metal and is a good place for the grounding connection.

Avoid the habit of cooling welds in water because this can produce undesirable brittleness in welded steel. For some practice exercises, however, cooling the coupons can save time, allowing for more practice. Be very careful when cooling hot metal in water—the rising steam produced is capable of causing severe burns. Always wear heavy gloves and position your body away from the path of rising steam.

The shielding gas nozzle and contact tip get hot enough to burn through heavy gloves while welding. Use MIG welder's pliers for changing tips and cleaning the nozzle.

As with other hot metalworking processes, use a sheet metal shield to prevent welding directly next to concrete. The shield protects against pavement damage and minimizes the risk of injury from flying bits of concrete.

Use extreme caution when welding on containers. Expansion of gases within the container can cause violent reactions. Avoid using steel containers of questionable origin. Containers such as 55 gallon and 35 gallon drums are often recycled and contain residues that are benign at normal temperatures but degrade into harmful chemicals when exposed to heat. Even seemingly safe food-grade containers have been engineered with coatings to contain leaks. These containers are not designed to be safe for building common backyard barbecue devices.

Moving Parts

The moving parts of a GMAW wire feeder require the welder to take precautions. Turn off the power when changing electrode wire to prevent fingers or clothes from being caught in the feed mechanism. On machines that enclose the wire feeder inside a protective housing, close the cover before beginning to weld. When feeding a new electrode wire through the liner and welding gun, protect your eyes. Avoid looking directly into the contact tip—the new wire will come through that hole very rapidly.

Practice Welding

Welders must practice to become skilled. Welding a series of beads on small rectangular coupon, or practice pad, is a good starting point for GMAW practice.

Welding coupons is similar to *surfacing*, a welding process used to build up worn parts with filler metal. Surfacing can also be used to create a harder or more durable surface than that of the original base metal.

Welding Beads on Flat Plate

Plates for practice pads should generally be 3/16″ to 3/8″ thick, rectangular in shape, with no dimension longer than 6″ or shorter than 3″. A typical size would be 1/4″ thick and 4″ wide by 4″ long. A correctly welded practice pad has uniform beads exhibiting a relatively flat surface on top. The width of each overlapping bead should be from 5/16″ to 3/8″. Each bead should be a minimum of 1/8″ high with smooth transition into the next. The completed plate should show pronounced distortion toward the welded surface as evidence of adequate penetration.

Welding a Practice Pad

The following is a procedure for creating a practice pad.

1. Cut a 4″ × 4″ × 1/4″ steel coupon using the cutting torch.
2. Using a scrap piece of steel, set up the machine parameters according to manufacturer's specifications for 0.035 solid electrode wire and 75% argon/25% CO_2 shielding gas. Run a test bead, and then make any adjustments to fine-tune the settings. Remember that each machine is a little different.
3. Strike the arc.
4. Weld the 4″ × 4″ × 1/4″ steel coupon to a 10″ long 3/4″ or 1/2″ pipe handle. See **Figure 30-40**.
5. Properly ground the work by attaching the workpiece connection directly to the handle.
6. Position the coupon on the welding table for comfort during welding.
7. Weld a single pass bead 1/4″ from the edge down the edge of the coupon in the flat position. Keep the bead dimensions to 5/16″ wide and less than 1/8″ high.
8. Leave the welding machine running, respecting the duty cycle.
9. Visually inspect the bead and allow the coupon to air-cool a bit.
10. Alternate the direction of travel, and weld a second bead between the previous bead and the plate. Allow a 25% to 30% overlap on the previous bead.
11. Continue repeating steps 9 and 10 until the entire surface of the coupon is covered.
12. Weld a second layer of beads perpendicular to and on top of the first layer, covering one-half of the coupon.
13. Clean the coupon by brushing the welds thoroughly. Be cautious of steam rising from dipping hot metal into water.

Safety Note
Before striking the arc, call out "Cover!" to alert anyone nearby to avoid looking at the arc without adequate eye protection.

Goodheart-Willcox Publisher

Figure 30-40. Begin a practice pad by attaching a handle that will serve as a place to secure the ground clamp as well as a safe way to move the hot metal.

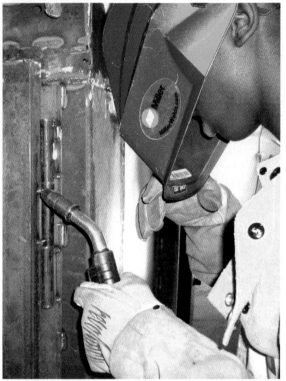

Figure 30-41. This student is welding in the 3F position attaching pipe hinges to a flat surface using the GMAW process.

Figure 30-42. Weld enough beads to completely fill the valleys formed by the angles on all four sides of the fillet welding exercise.

Safety Note
Be aware that steam rising from dipping hot metal into water can cause severe skin burns. Always wear heavy gloves and position your body away from the path of rising steam.

14. Clean your work area and return all equipment to its proper place.

Fillet Welding in Flat, Horizontal, and Vertical Positions

Performing a fillet weld in the flat (1F) position is the process of placing the weld in a 90° V-shaped groove. The welder must ensure that the molten weld pool blends to each side, full penetration is achieved, and the finished bead has a uniform surface.

Shifting the parts 45° creates an upside-down T. Welding this joint on one or both sides is welding in the horizontal (2F) position. A sideways-oriented T-joint welded from the top is also the 2F position. Fillet joints may also be welded using GMAW in the vertical (3F) position. See **Figure 30-41**. Vertical welding can be performed uphill or downhill. Uphill generally is preferred in structural applications for its tendency toward deeper penetration.

Fillet Weld Practice

Welding on 1 1/2″ × 4″ coupons of angle iron is great practice for learning to perform fillet welds. See **Figure 30-42**. The following is a procedure for fillet weld practice.

1. Cut two 1 1/2″ × 4″ steel angle sections with a hydraulic shear or a metal-cutting saw.

Safety Note
Be sure to get your instructor's approval and follow all of the safety precautions required for using metal-cutting equipment.

2. Using a scrap piece of steel, set up the machine parameters according to manufacturer's specifications for 0.035 solid electrode wire and 75% argon/25% CO_2 shielding gas. Run a test bead, and then make any necessary adjustments to fine-tune the settings.

3. Strike the arc.
4. Tack weld the two angle sections lengthwise at the vertex (highest point) of their respective angles to form a cross of four 90° angles when observed from the end of the piece. See **Figure 30-43**.
5. Weld a 10″ long 1/4″ or 1/2″ pipe handle to the tacked angles.
6. Position the coupon for comfort during welding.
7. Weld a single-pass root bead along the joint of the angles in the 1F position.
8. Leave the welding machine running. The duty cycle requires a cool-down time to protect the internal components of the power source.
9. Evaluate the root bead quality by answering these questions:
   - Is the surface uniformly textured?
   - Do the edges blend smoothly and evenly to each side of the joint?
   - Are there discontinuities? How many? How severe?
   - Is the penetration acceptable?
10. Position the coupon for welding in the 2F position and weld a second bead on the other side of the joint to back up the root bead.
11. Weld a filler bead between the first root bead and one side of the angle in the 1F position. There should be about a 50% overlap on the previous bead.
12. Repeat on the opposite side of the coupon in the 2F position.
13. Repeat steps 11 and 12 until each side of the coupon is smoothly filled level across with beads.
14. In the 3F position, fill the other two sides of the coupon. Weld one side using the vertical uphill technique. Use the vertical downhill technique to complete the other side.
15. Clean the welds thoroughly, cool the metal, and submit the coupon for inspection and grading.
16. Clean your work area and return all equipment to its proper place.

> **Safety Note**
> Before striking the arc, call out "Cover!" to alert anyone nearby to avoid looking at the arc without adequate eye protection.

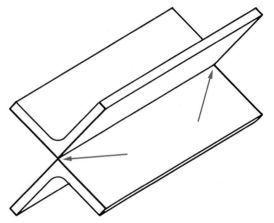

Goodheart-Willcox Publisher

Figure 30-43. Tack weld where shown by the arrows.

GMAW Certification Practice

The AWS D1.1 3G GMAW Welding Certification qualifies a welder to weld in the 1G, 2G, and 3G positions, as well as the 1F, 2F, and 3F positions, without having to take a separate test for 1G and 2G certification. Successfully performing the certification test on 3/8″ thick material certifies a welder as qualified to weld material from 1/8″ to 3/4″ thick. Obtaining this welding certification is a worthy goal for aspiring welders because it can aid in gaining employment.

Welder certification tests are performed according to specific welding parameters listed in a document known as a *welding procedure specification*

(WPS). According to the American Welding Society, the following technical information should be included in the WPS for the AWS D1.1 3G GMAW Welding Certification test:

- Welding Process—GMAW
- Progression—Vertical up with at least one restart
- Joint Type—Butt (at least 6" long)
- Backing Strip—Yes
- Transfer Mode—Spray
- Current—DCEN
- Base Metal—ASTM A36
- Backing Material—ASTM A36
- Root Opening—1/4" to 5/16"
- Groove Angle—45°
- Thickness and Groove—.375"
- Shielding Gas Composition—Argon > 90%, CO_2 < 10%, O_2 < 5%
- Flow Rate—15 to 35 cfh
- Gas Cup Size—1/2" to 5/8"
- Technique—Stringer or weave bead
- Multipass—Yes
- Cleaning between Passes—Mechanical or wire brush
- Preheat Temperature—Min. 60°
- Amps—150 to 175 (achieved by setting wire speed and controlling stickout)
- Volts—24 to 26

The GMAW welding machine for the certification test needs to be set up according to these procedure specifications. Begin with a wire speed that will keep the amperage at the lower end of the range. The following steps outline the welding procedure:

1. Prepare the 3/8" thick plate by using a disc grinder or sander to remove traces of mill scale or rust from the weld area and back at least 1".

2. Cut and grind a uniform 45° groove bevel extending completely across the edge of both plates.

3. Attach the two coupons by tack welding them to a backing plate, leaving a 1/4" root opening between the edges.

4. Position the piece vertically for welding—do not move it from this position until completed.

5. Weld the root pass with a tight side-to-side weave, resembling an upside down *V*. Pause slightly on the sides until the feathered edges create a crater, then fill the crater and move on. The pattern should trace the leading edge of the weld pool.

6. Inspect the root. According to standards, the maximum root surface concavity can only be 1/16", and the melt-through can be no more than 1/8". If the root weld meets these standards, grind down the center of the weld to make it flat. Between-pass cleaning is allowed. This makes space for the first filler pass to deposit fill while generating enough heat to blend smoothly to both sides.

7. Allow the plate to air-cool to the touch, and then use the same settings and technique as for the root pass to weld a second bead.

8. Clean the weld as before and allow the coupon to cool.

9. Weld the second fill pass, pausing slightly (about one second) at each side. This second fill bead will be wider than the first two beads. It should be slightly below flush with the outside of the groove and flat across the top of the bead.
10. Clean and cool the weld again.
11. Before welding the final bead, adjust the machine settings to compensate for the added thickness of the coupon at the weld joint. A faster wire speed feed rate increases the amperage, deposition rate, and travel speed.
12. Standards limit the reinforcement, or height, of the final bead to no more than 1/8″ above the surface of the base metal. To meet this standard, weave the electrode from side to side, hesitating about two seconds on each side. Keep the overall width of the finished bead under 1″.
13. Clean and inspect the finished weld for discontinuities.

A weld completed for the actual AWS test is subjected to a guided bend test by a certified weld inspector. The backing plate is removed, and both surfaces of the weld are ground back to surface level. The plate is then cut into strips that are bent through the weld joint, one with the root up and the other with the final bead up. See **Figure 30-44**. The bent specimens are then inspected for discontinuities. Mastering this welding procedure is challenging and requires practice.

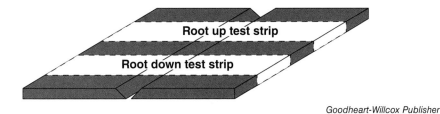

Goodheart-Willcox Publisher

Figure 30-44. A destructive guided bend test may be the culmination of a welder certification test.

CHAPTER 30
Review and Assessment

Summary

- In the gas metal arc welding process, a consumable, continuously supplied electrode surrounded by a protective envelope of a shielding gas is used to form the welded bead.
- In the flux cored arc welding process, a hollow wire electrode filled with powdered flux or a mixture of powdered metal alloying agents is used. The process may or may not involve a supplied shielding gas.
- The equipment necessary for gas metal arc welding includes a power supply, a wire feeder, a welding lead, a welding gun, and a shielding gas supply.
- A GMAW electrode moves from a storage spool through the welding lead to the weld pool. Wire feeders may be integral to a power source or externally mounted.
- GMAW electrodes may be bare wire or coated with an alloy coating that controls oxidation in storage and improves contact with the tip within the welding torch.
- The shielding gases most commonly used for GMAW and FCAW are argon, CO_2, and oxygen. Argon and CO_2 can be used pure form or mixed together or with other gases to form shielding gas blends.
- Metal transfer modes in GMAW include short-circuit mode, globular transfer mode, spray transfer mode, and pulsed arc mode. The mode used is determined by the type of base material, the power source, and the wire feeder.
- Flux cored electrodes are filled with powdered flux material similar to the coating found on SMAW electrodes. Some are formulated for use with a shielding gas, while others are self-shielding.
- Setting the appropriate wire speed and voltage is essential to performing quality welds.
- Shielding gases are selected to affect the penetration profile of the weld, the surface character of the bead, and the stability of the arc. Specific modes of transfer require minimum levels of argon or other shielding gases in the mixture.
- The electrode wire selected for a job must be compatible with composition of the base metal and sized according to the thickness of the material.
- During welding, the welder controls several parameters, including stickout, work angle, travel angle, manipulation, and travel speed.
- The best protection against hazards is prevention. Arranging the welding environment for safety, good ventilation, and appropriate PPE are important steps in avoiding injuries.

Words to Know

Match the key terms from the chapter to the correct definition.

A. contact tip
B. diffuser
C. inductance
D. open root weld
E. postflow timer setting
F. preflow timer setting
G. pulsed arc machine
H. push-pull welding gun
I. root face
J. spool gun
K. stickout
L. travel angle
M. wire feeder
N. wire speed
O. work angle

1. Unrolls the spool of electrode to provide a constant supply of filler material through a lead to the weld gun.
2. Allows the weld area to be covered with shielding gas prior to the arc being struck.
3. Typically calculated in inches per minute, this machine setting regulates the welding amperage.
4. Protects the cooling weld and cools torch components.
5. Should be maintained at about 3/8″ for most solid wire GMAW applications.
6. Machine setting found on some power sources that controls the fluidity of the weld pool.
7. Technology found in some power sources that supplies power to the weld in repeating cycles of higher voltage and lower voltage, resulting in increased uniformity.
8. Uses a set of auxiliary rollers to assist the main wire feeder in supplying electrode to the weld.
9. GMAW welding torch that contains a small supply of welding wire electrode.
10. Distributes shielding gas around the electrode.
11. Transfers welding current to the electrode.
12. A bevel groove joint that is set up without the use of a backing plate.
13. Angle between the welding gun and electrode and the center of the weld joint.
14. Angle between the welding gun and electrode and the direction of weld progression.
15. Section of an edge that has been left square while the remainder of the edge has been beveled for welding.

Know and Understand

Answer the following questions using the information provided in this chapter.

1. In the GMAW process, the arc is shielded from atmospheric contaminants by a(n) _____ gas supplied from a compressed gas cylinder.
2. Describe the type of electrode used for FCAW welding.
3. Most commonly, a constant-_____ power source with DC current is used for GMAW welding.
4. Setting the wire feed speed determines the inches per minute of electrode supplied at the weld and regulates the welding _____.

5. The contact tip serves as the electrical connection to the electrode and is attached to the _____ in the welding torch.
6. Dual-shield welding is accomplished by using GMAW equipment, a compressed shielding gas, and a(n) _____ electrode.
7. Using CO_2 as a shielding gas provides very deep weld penetration but generates more _____ in the weld pool.
8. What special tool is used by welders when changing an electrode wire?
9. Why is the short-circuit method of metal transfer ideal for thin materials such as sheet metal?
10. Surface contaminants must be removed from metal before it is welded in order to prevent _____ in the finished bead.
11. The size of an electrode wire for a particular welding job is selected according to the _____ of the base metal.
12. Whenever possible, work should be positioned so the welder can weld in the _____ position.
13. What effect does increasing the stickout have on resistance and heat input?
14. Why is the forehand welding technique preferred for welding thin materials?
15. The GMAW welding gun lead should be positioned to avoid tight coils to reduce the chances of the electrode binding as it passes through the _____.
16. How should a welder pick up a loose piece of scrap?
17. On a practice pad, each bead should overlap the previous bead by _____%.

STEM and Academic Activities

1. **Science.** Prepare a chart that compares five shielding gases and shielding gas blends used in gas metal arc welding. List the properties of each gas or blend, its effect on the welding arc, and its effect on penetration and bead formation.
2. **Technology.** Practice loading a new roll of wire electrode into the GMAW wire feeder. Develop a flyer that outlines the steps required to replace the wire electrode and display it near the welding area.
3. **Engineering.** Study the way pressure is regulated between the rollers on a wire feeder. Explain why the tension should be less for aluminum wire than for steel electrode wire. How would the use of flux cored wire affect the required tension and why?
4. **Math.** Use the method described in this chapter to calculate the actual wire speed of a GMAW wire feeder. Compare the calculated speed to the speed indicated by the machine. Are the two values the same? Repeat the procedure for a different wire speed setting on the machine, and compare the results to the first test.
5. **Social Science.** Search the Internet for welding and fabrication businesses that serve the agricultural industry in your area. Choose a company website and make a list of the services provided by that company.
6. **Language Arts.** Download one of the Safety & Health Fact Sheets from the American Welding Society website. Choose a sheet that contains information pertaining to GMAW/FCAW welding. Give a short presentation to the class on the contents of the sheet.

Thinking Critically

1. GMAW is faster and more efficient than SMAW in many applications. Practice welding the same type of joint using both processes and time your progress. How much time can be saved using GMAW rather than SMAW? What would be the daily savings per welder on a job?
2. Why is it possible with GMAW to produce beads with great surface characteristics and poor penetration? Explain how this problem can be avoided.

CHAPTER 31
Gas Tungsten Arc Welding

Chapter Outcomes

After studying this chapter, you will be able to:
- Describe the gas tungsten arc welding (GTAW) process.
- Identify and use GTAW equipment.
- Select the appropriate tungsten electrode and shielding gas for a welding job.
- Take appropriate safety precautions when GTAW welding.
- Properly clean and prepare a joint for welding.
- Control welding variables to produce a good weld.
- Practice GTAW welding on various base metals.

Words to Know

autogenous weld	E3® tungsten electrode	lanthanated tungsten electrodes	thoriated tungsten electrodes
ceriated tungsten electrodes	gas lenses	lift start	truncated cone tip
collet	gas tungsten arc welding (GTAW)	pointed cone	tungsten extension
collet body	hemispherical	scratch start	walking the cup
diffuser	high-frequency start	shield gas nozzle	

Before You Read

Skim the chapter by reading the first sentence of each paragraph. Use this information to create an outline for the chapter before you read it.

While studying this chapter, look for the activity icon to:

- **Practice** vocabulary terms with e-flash cards and matching activities.
- **Expand** learning with interactive activities.
- **Reinforce** what you learn by completing the end-of-chapter questions.

www.g-wlearning.com/agriculture

Lone Wolf Photography/Shutterstock.com

This chapter will provide you with a solid understanding of the gas tungsten arc welding process. **Gas tungsten arc welding (GTAW)** is also known as *tungsten inert gas (TIG)*, or *heliarc welding*. GTAW employs a tungsten electrode to create the arc for welding. The tungsten electrode does not melt and become incorporated into the weld joint. See **Figure 31-1**. An inert shielding gas, such as argon or helium, protects the weld from atmospheric contamination. Most of the time, a filler metal is used with the GTAW process, although some joints may be welded without the addition of extra material. GTAW is most often used to weld thin sections of stainless steel and nonferrous metals such as aluminum. It can also be used to weld mild steels and carbon steels.

Gas tungsten arc welding offers the welder more control over the weld than other arc welding processes, allowing for stronger, higher-quality welds. The tradeoff is that GTAW is more complex and challenging to master, and it is significantly slower than most other welding techniques. Similar to welding with an oxyfuel torch, manual GTAW requires the welder to use both hands. One hand controls the movement of the torch, and the other manually feeds a filler metal into the weld area. See **Figure 31-2**.

Photo Love/Shutterstock.com

Figure 31-1. Tungsten electrodes do not melt and become part of the weld joint.

Development of GTAW

As early as the 1930s, compressed inert gases were used as shielding gas with DC power sources to weld magnesium for the aircraft industry. By 1941, the process had evolved to the use of negatively charged tungsten electrodes in a helium shielding gas. *Heliarc welding*, as it was first called, worked well on steel and stainless steel but was still unsuitable for aluminum and other nonferrous materials. The development of AC power sources allowed TIG to become the chosen method for many welds on aluminum and other nonferrous parts. Shielding gas mixtures

Corepics VOF/Shutterstock.com

Figure 31-2. Manual gas tungsten arc welding requires the welder to use both hands.

858 Agricultural Mechanics and Technology Systems

containing small amounts of hydrogen or nitrogen were developed for a few special welding conditions. Although not commonly used, these gases are reactive rather than inert, so the name of the process was changed from TIG to GTAW.

Equipment for GTAW

Equipment for GTAW welding includes the power source, welding torch, and gas cylinders and connections. Consumables used in the GTAW process include filler metals, shielding gases, and electrodes.

GTAW Power Sources

The same type of power supply used for SMAW is used for GTAW. In fact, most machines designed for GTAW are also capable of SMAW. See **Figure 31-3**. A constant-current welding power supply produces a consistent amperage and heat output independent of the arc distance and voltage adjustment. GTAW power sources have controls for adjusting the polarity, amperage, and shielding gas flow settings. On some machines, arc frequency and pulsing can be adjusted.

Typically, direct current electrode negative (DCEN) polarity is used for welding steel, nickel alloys, and titanium. Alternating current is used for welding aluminum or magnesium because the constant switching of direction of the electron flow produces a cleaning action. This cleaning action helps to remove the surface oxides that form almost immediately on these metals when they are exposed to the atmosphere.

Two features found on GTAW machines that are not used on SMAW welders are high-frequency start and remote control. *High-frequency start* generates a very high-frequency arc that can jump long distances between conductors. This feature allows the welder to initiate the arc without touching the tungsten electrode to the base metal, reducing the chances of contamination of the weld and the electrode. Often there is an option to use high frequency continuously or just for starting the arc.

Remote controls are located in a foot pedal or a thumb switch attached to the torch. See **Figure 31-4**. Using the remote control, a welder can switch the power on and off, as well as vary the amperage as needed to manage the heat input at the weld pool. Traditional remote control units rely on switch wires connected to the power supply. Switch wires should be located away from foot traffic areas to prevent damage and reduce tripping hazards. Newer GTAW remote controls feature wireless technology to eliminate the need for tethered switch wires.

Goodheart-Willcox Publisher

Figure 31-3. This GTAW machine is also capable of SMAW welding. Note the settings for *STICK* and *TIG*.

Goodheart-Willcox Publisher

Figure 31-4. A GTAW foot pedal remote control. A foot pedal can be linked to the power source with wires and a pin connector or through a wireless connector.

GTAW Welding Torch

The GTAW welding torch is set up similar to a GMAW welding torch, except that the tungsten electrode is mounted in place of an automatically fed electrode wire. Most GTAW torches have interchangeable parts, making them capable of multiple configurations for a variety of welding conditions. Air-cooled torches are available in capacities up to about 200 A and are relatively small, lightweight, and simple. Water-cooled torches, with capacities of up to 600 A, are more complex and require the use of a water-cooling unit that is much like the radiator on an automobile. The welding torch is connected to the power source by a welding lead containing a conductor cable, a hose for the shielding gas, and, depending on the configuration, small control wires and the water supply and return lines.

Inside the torch, the metal parts are made of hard copper or brass alloys to effectively conduct electricity. See **Figure 31-5**. The tungsten electrode is held firmly in the center of the torch with a holding device called a *collet*, which must be properly sized to match the diameter of the tungsten electrode. When the end cap on the rear of the torch is tightened, the collet is forced into the *collet body* end cap, causing the collet to firmly grip the electrode. See **Figure 31-6**. Ports around the base of the collet body emit a constant flow of shielding gas into the ceramic *shielding gas nozzle*, or cup, surrounding the tip of the electrode. The main body of the torch is made of heat-resistant plastic, providing insulation from heat and electricity.

The parts of a GTAW torch are interchangeable. In order to change electrode size, the collet and collet body must also be changed to match the electrode diameter. Special collet bodies with wire mesh outlets called *gas lenses* are often used to reduce turbulence in the flow of shielding gas around the weld pool. A gas lens requires a larger shielding gas nozzle. See **Figure 31-7**. For working in extremely close quarters, short end caps can be combined with short tungsten electrodes to decrease clearance requirements. Under normal operating conditions, the metal parts of a GTAW torch last for a long time.

Ceramic nozzles can be durable, but they are brittle and subject to damage from minor jolts. Nozzles made of fused quartz, a glass-like substance, provide better visibility when a gas lens is used.

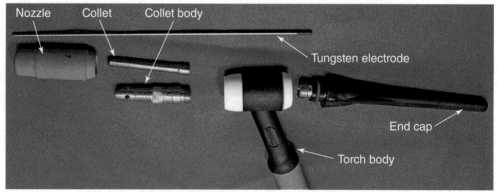

Goodheart-Willcox Publisher

Figure 31-5. Parts of a GTAW torch.

Welders can change the shielding gas nozzle size based on the desired size of the weld, electrode size, or accessibility to the joint. Nozzle sizes represent the inside diameter of the nozzle and are identified by a standard coding system based on graduations of 1/16″. For example, a #6 nozzle is 6/16″, and a #7 nozzle has a 7/16″ inside diameter. See **Figure 31-8**. The recommended nozzle size for welding with 1/16″ tungsten ranges from about #4 to #6. Welding with 3/32″ and 1/8″ electrodes requires nozzles ranging in size from #6 to #8.

Compressed Gas Cylinders and Connections

Shielding gases for GTAW are stored, sold, and used from compressed gas cylinders. Cylinders vary in size, but all are equipped with brass control valves. Shielding gas cylinders must be refilled by qualified persons using approved procedures. The fittings for shielding gases are different from those used for fuel gas or oxygen in order to prevent potentially dangerous combinations.

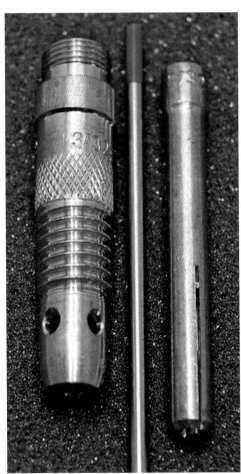

Goodheart-Willcox Publisher

Figure 31-6. The collet and collet body are matched to fit the exact diameter of the tungsten electrode.

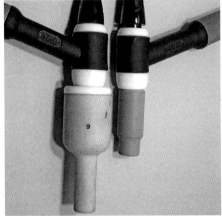

Goodheart-Willcox Publisher

Figure 31-7. A gas lens setup requires a shielding gas nozzle with a larger base. Both the gas lens (left) and standard nozzle (right) shown are size #6.

Goodheart-Willcox Publisher

Figure 31-8. The numbers stamped on nozzles indicate their size. What are the sizes (in fractions of an inch) of the nozzles shown?

Figure 31-9. Read the flowmeter at the top of the floating ball. This flowmeter is set to 18 cfh.

Figure 31-10. A pencil-type GTAW torch with a shielding gas flow valve included in the torch handle.

Figure 31-11. The identification markings stamped on these rods indicate the composition of the filler material.

Delivery of the shielding gas to the arc is crucial to maintaining quality in the finished weld. A gauge measures cylinder pressure and a regulator controls the delivery of shielding gas to the torch. A flow gauge or a flowmeter is used to measure the rate of gas flow in cubic feet per hour (cfh). Flowmeters are preferred for GTAW because they are more precise than flow gauges. See **Figure 31-9**. Typical flow rates range from 5 cfh to 25 cfh. Generally, the flow rate should be just enough to protect the weld. Too much flow creates turbulence that can mix atmospheric air into the shielding area.

From the flowmeter, the shielding gas passes through a flexible hose to an electrically operated solenoid valve in the power source. This solenoid valve switches the gas flow to the welding torch on and off. Simplified GTAW setups bypass the power source to a manually operated valve at the torch, **Figure 31-10**.

Consumables for GTAW

Consumables for GTAW include filler metals, shielding gas, and tungsten electrodes. Learning how to choose the correct consumables is an important step to producing high-quality welded joints. Consumables are matched to the composition and thickness of the base metal for each welding job.

Filler Metals

Filler metals for GTAW are usually bare metal rods engineered to match the composition of the metal being welded. The filler metals are classified and labeled similar to electrodes used for GMAW. GTAW filler metals in the form of cut-to-length rods 24″ to 36″ long are available in diameters ranging from 1/16″ to 1/4″. Often the designation for the rod is stamped into a flat area at the end of the rod. See **Figure 31-11**. The copper-coated mild steel filler metal rods used for oxyfuel welding are not suitable for use in the GTAW process. An ER70S-6 is a common GTAW filler for steel, and ER4043 is a common filler for aluminum. Many compositions are available to allow welders to closely match the filler material to base metals. The closer the match, the better the chance for a flawless weld joint. For example, if a welder is welding 308 stainless steel, then selection of ER308 filler metal is appropriate.

Shielding Gas

The recommended shielding gas for most GTAW applications is argon or helium. Both of these are inert gases. Argon is used more commonly because it performs with a lower arc voltage; produces a smoother, quieter arc; and provides easier arc starting. Argon is the most readily available shielding gas, and the most economical. Using helium as a shielding gas results in deeper penetration and faster travel speeds, making it suitable for welding thicker materials. Because of its lower density and tendency to rise in the atmosphere, helium requires higher flow rates to maintain an adequate shield around the weld area. Argon-helium mixtures are blended to balance the characteristics of the two. Hydrogen and nitrogen are added in small amounts to shielding gas mixtures for special-purpose welding.

Tungsten Electrodes

Tungsten electrodes are often called *nonconsumable* because they are not intended to become incorporated into the bead. In normal use, however, tungsten electrodes are eroded from the extreme heat. They are in fact a true consumable. Electrodes typically are sold in 7″ lengths and range in size from .01″ to 1/4″ in diameter. The sizes most commonly used in manually operated torches are 3/32″ and 1/8″. See **Figure 31-12**.

Several types of tungsten electrodes with different welding properties are available. A color-coding system identifies the electrodes:

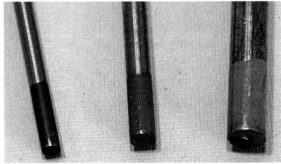

Goodheart-Willcox Publisher

Figure 31-12. Tungsten electrodes are available in a variety of sizes and types.

- Pure tungsten electrodes have a green band. These electrodes are the least expensive, highest consumption electrodes, but they provide good arc stability when used with balanced AC current. This is an advantage when welding aluminum and magnesium.
- Two percent *thoriated tungsten electrodes* have a red band. These electrodes contain approximately 2% thorium, which results in easier starting and allows a higher current capacity. Thoriated electrodes are commonly used for welding ferrous metals.

STEM Connection

Laminar Flow

Think of the screen or diffuser on a common water faucet. The screen distributes the water to a laminar flow that reduces splashing. If the screen is removed from the faucet, the flow is turbulent. A diffuser on a handheld hair dryer works in a similar fashion, smoothing the flow of heated air to prevent violently blasting the hair with a concentrated stream of hot air. The laminar flow of shielding gas prevents oxygen and other atmospheric contaminants from entering the weld area. What other metal-working processes depend on a laminar flow of gas to produce a favorable result?

Safety Note

Thorium is slightly radioactive. Follow all safety guidelines in the safety data sheet provided with new thoriated electrodes. Thorium dioxide, a stabilizer used in thoriated electrodes, is hazardous when its dust or fumes are inhaled. Normal handling of thoriated electrodes does not pose significant risk of external radiation exposure.

- *Ceriated tungsten electrodes* have an orange band. These electrodes contain about 2% cerium and are commonly used with DC on thin sheet metal work. They are also used for welding small, delicate parts.
- *Lanthanated tungsten electrodes* have a gold band. These electrodes contain approximately 1.5% lanthanum and have excellent arc starting, a low consumption rate, and good arc stability. Lanthanated tungsten electrodes perform equally well on AC or DC currents and can be used to weld steels as well as aluminum. The addition of 1.5% lanthanum to a pure tungsten electrode of a given size increases the maximum current-carrying capacity by approximately 50%.
- *E3® tungsten electrodes* have a purple band. These electrodes contain approximately 1.5% lanthanum oxide, .08% zirconium dioxide, and .08% yttrium oxide. This combination allows these electrodes to be used for mild steel, stainless steel, and aluminum welding. E3® electrodes are engineered to use AC or DC welding currents and are marketed as a safer alternative to thoriated electrodes. Some testing indicates that they offer increased stability and longer tip life than thoriated electrodes.

To maximize weld efficiency, refer to the machine manufacturer's charts to determine the smallest diameter tungsten possible for the selected amperage. Tungsten electrodes 1/16″ in diameter can be operated between 50 A and 150 A; 3/32″ electrodes are designed for current loads of 65 A to 250 A; and 1/8″ tungsten electrodes require an amperage in the 135 A to 400 A range. Make sure that the tungsten electrode size matches the collet and collet body size.

Tungsten Electrode Preparation

After selecting the appropriate electrode, the next step is to properly configure the tip of the electrode. Tungsten electrode shape affects the arc characteristics, which in turn directly affect the bead size and profile, the weld penetration, and the durability of the electrode tip. The three tip configurations are hemispherical, truncated cone, and pointed. See **Figure 31-13**.

Welding aluminum with GTAW requires AC current and a *hemispherical tip*, or balled tip, on a pure tungsten electrode. A hemispherical tip distributes the arc over a wider area than a sharp pointed electrode, providing a wider path around the weld for the AC current to clean contamination. The center of the hemisphere concentrates part of the arc to weld the joint. To shape the tip, strike an arc using DCEP current on a thick piece of copper or very clean steel. The heat from the arc causes a rounded point to form. The ball should be no larger than the diameter of the electrode. If a smaller hemisphere is desired, use a smaller electrode or taper the electrode before forming the ball.

A *truncated cone tip* is a tapered cone with a truncated (flat) tip. This tip configuration is used for welding steel and stainless steels with DC welding processes. Under ideal conditions, a dedicated grinding tool with diamond

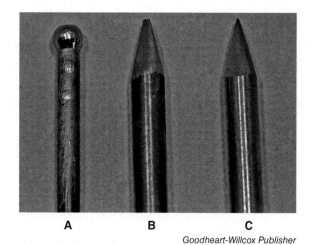

Goodheart-Willcox Publisher

Figure 31-13. Electrode tip configurations. A—Hemispherical. B—Truncated cone. C—Pointed.

or borazon stones is used to grind the electrode tip. It is best to use a grinder or sander with a fine grit that is dedicated only to sharpening tungsten to avoid contaminating the tip. Hold the electrode against the abrasive surface in a manner that leaves only longitudinal striations. See **Figure 31-14**. Use a tool rest and rotate the electrode between the fingers while grinding to keep the point centered. Leave a flat tip about 1/4 of the diameter of the electrode. The length of the tapered section of the tip should be between 2 and 2 1/2 times the diameter of the electrode. The blunt tip reduces the chance of a sharpened tip popping off into the weld pool.

For welding very thin sections of metal, less than 1/16", only at low amperages, grind the tip to a sharp point to further concentrate the arc.

Figure 31-14. Electrodes must be ground using the proper technique. A—Grinding around the circumference of the tip leaves ringed striations that lead to an erratic arc. B—Grind parallel to the length of the electrode to leave only longitudinal striations.

GTAW Safety

Welding with the GTAW process does not produce the metal spatter common with other types of arc welding. GTAW does generate intense heat and large amounts of light. Due to the reduced amount of smoke around the GTAW arc, up to twice the amount of infrared and UV rays are released compared to SMAW. Skin exposed to these rays is subject to damage similar to severe sunburn. See **Figure 31-15**. Cotton shirts alone should not be relied on for protection. Cotton cloth provides insufficient protection and can deteriorate from exposure to the infrared and UV rays produced by the GTAW process. Wear dark clothing to reduce reflection of light into the back side of the helmet.

Figure 31-15. Exposed skin is subject to damage similar to sunburn.

Personal Protective Equipment (PPE)

Proper personal protective equipment (PPE) is essential, although it is not a substitute for good planning, common sense, and safety engineering. When selecting PPE, consider the specific hazards that may be present. Follow these guidelines for PPE:

- OSHA-approved (Z87.1) goggles or glasses must be worn at all times, including under the welding helmet. When the helmet is raised, eye protection must already be in place.
- A welding helmet that fits on the head and protects the entire face is required. Welding helmets for GTAW should have at minimum a #10 shade or filter lens and a clear protective plastic cover lens in front. A darker shade is recommended if visibility can be maintained. Inspect welding lenses on a regular basis for damage and proper fit in

> **Safety Note**
> When shaping thoriated tungsten electrodes, avoid inhaling the slightly radioactive dust. Wear breathing protection rated for fine dust particles. Collect the dust for proper disposal according to the safety data sheet.

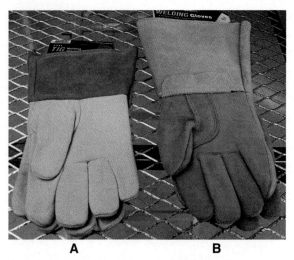

Figure 31-16. A thinner glove is desired for GTAW welding to allow the required dexterity. A—Glove for GTAW welding. B—Typical glove used for other welding processes.

Goodheart-Willcox Publisher

the helmet. Gaskets should be in place, permitting no leakage. When using an autodarkening lens, verify that it is functioning correctly prior to use and that it is set to a dark enough filter setting.

- Gauntlet-type, heavy insulated leather welding gloves protect the hands and forearms. A better grade of glove than is typically used in other arc welding procedures is recommended for GTAW. See **Figure 31-16**. Gloves must provide protection from the heat and light while remaining supple enough for the welder to precisely manipulate the torch and filler electrode at the same time.
- Protective footwear should completely cover the foot with leather and should have synthetic soles with significant tread.
- Heavy cotton shirts should have no pockets, or the pockets should be covered. Undershirts should be worn as an extra protection from the extreme light. Long-sleeved shirts and pants are required. Fire-resistant clothing is recommended.
- Leather aprons, leather sleeves, welding chaps, or a welding jacket are recommended.

Other Safety Precautions

Other precautions for GTAW welding include:
- Ensure that all electrical, shielding gas, and water connections are secure prior to use.
- To reduce the chance of accidental arcing or shock, detach the SMAW electrode holder from the dual or multiprocess power source before GTAW welding.
- Extend leads only enough to reach the work in order to reduce tripping hazards and to protect leads from accidental damage.
- Adequate ventilation should always be provided. Inert gases can displace breathing air in enclosed spaces. Ozone and fumes from vaporized metals produced during welding can increase to harmful levels in poorly ventilated areas.
- Welding curtains should be used to protect others in the vicinity from the ultraviolet and infrared rays emitted by the welding arc.
- Never touch the filler rod to the electrode. You become part of the path to ground and you will receive an electric shock.

Welding with GTAW

GTAW differs from other arc welding methods in the way the filler metal is introduced to the weld. Welding can also occur without the addition of filler metal—an *autogenous weld* consists only of the base metal. If a filler metal is used, it must be fed into the weld pool with the free hand while the other hand manipulates the torch. Most manual GTAW welders use the dominant hand to handle the torch because it offers the most control.

Torch Setup

Use the following guidelines when setting up the GTAW welding torch prior to and during the welding process:

- Keep the torch and work leads as short as possible. Move the power source as close as possible to the work.
- Check and tighten all electrical and gas connections. Place the workpiece connection as close as possible to the weld joint to reduce resistance.
- The inside diameter, or size, of the shielding gas nozzle should be at least three times the diameter of the tungsten electrode to allow for adequate shielding gas flow. (For example, if using a 1/8″ electrode, the nozzle should be a minimum of 3/8″ in diameter, or a #6.)
- The *tungsten extension*, or the distance the tungsten extends beyond the tip of the nozzle, should be no longer than the inside diameter of the nozzle. See **Figure 31-17**.
- Refer to the machine manufacturer's recommendations to set the amperage according to the thickness of the base metal

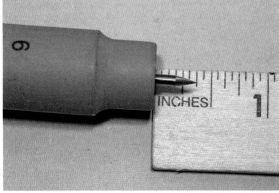

Goodheart-Willcox Publisher

Figure 31-17. A #6 standard nozzle has an opening of 3/8″, limiting the maximum allowable tungsten extension to 3/8″.

SAE Connection

Agricultural Mechanics Design and Fabrication

Gas tungsten arc welding involves equipment and training that is typically used in specialized situations, such as the construction of items from stainless steel. If you are interested in welding using the GTAW process, ask your teacher how you can locally gain experience in this area. Skill in gas tungsten arc welding will provide unique opportunities for employment that could lead to increased pay and demand for your skill. Remember to record the hours invested and the skills learned to accurately complete a proficiency application for a supervised agricultural experience (SAE) program in Agricultural Mechanics Design and Fabrication or Agricultural Mechanics Repair and Maintenance.

senlektomyum/Shutterstock.com

and the electrode diameter. Apps are available for smart devices that consolidate these recommendations into an easy-to-use format.

Joint Preparation

Joint preparation is critical for successful welding in the GTAW process. All joints must be cleaned with a grinder or clean sander. Aluminum is also cleaned with chemicals. Welded surfaces must be free from rust, oxidation, mill scale, paint, oil, or any other contaminants on the bare metal. GTAW potentially produces the highest-quality welds, but quality depends on starting with a clean joint.

When using the GTAW process, the welder must fit joints closely. The uniformity of the final bead depends on the welder being able to form the weld pool with a consistent motion. Gaps between pieces that are too large or vary in width along the weld joint cause the welder to modify welding technique, resulting in variations in the weld that can lead to discontinuities.

GTAW Techniques

As with other methods of arc welding, the welder selects the parameters (machine settings, shielding gas, joint preparation, electrode and filler metal selection) prior to welding. See **Figure 31-18**. During welding, the welder must maintain simultaneous control of the following in order to create a quality weld:

- **Arc length.** The arc length is the distance from the tip of the tungsten electrode to the base metal. For most applications, the arc length should be held under 1/4″. Many times, however, the tungsten electrode should be as close as 1/16″ to the weld pool. Under no circumstances should the tungsten electrode be allowed to come in contact with the molten base metal.

- **Work angle.** The work angle is the angle of the torch to the base metal in a plane that is perpendicular to the direction of travel. On groove welds, the work angle should be 90°, pointing equally to both sides of the joint. For fillet welds, the torch should be pointed directly at the joint, applying equal heat to both sides of the weld joint. A technique known as *walking the cup* allows the welder to use the edge of the shield gas nozzle to guide the path of the torch. This method makes it easier to maintain a uniform arc length and working angle. When walking the cup, keep the rim of the nozzle in contact with the metal to provide stability.

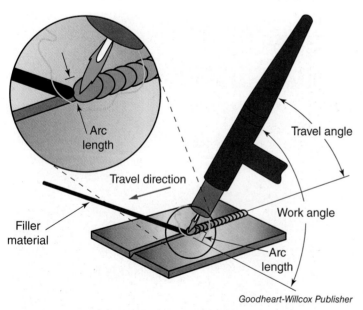

Goodheart-Willcox Publisher

Figure 31-18. The welder controls the arc length, work angle, travel angle and speed, amperage, and the feeding of the filler material during welding.

- **Travel angle.** The travel angle is the lean, or tilt, of the torch in relation to the direction of travel. Holding the torch perfectly perpendicular would result in the most uniform weld, but visibility would be poor. GTAW is forgiving of a variable travel angle as long as other variables are maintained. Tilting the torch about 25° to 30° and pushing the arc in the direction of travel is typically effective.
- **Amperage.** The amperage at the power source is set based on the metal thickness. Thicker base metals dissipate heat very quickly. Thin metals are just the opposite, requiring lower maximum amperages. GTAW allows the welder to fine-tune the amperage while welding using a foot pedal or thumb control on the welding torch. As the weld pool size starts to increase, backing off on the foot pedal decreases the amperage and heat input. Depressing the pedal slightly more when the pool size starts to diminish adds amperage.
- **Travel speed.** The rate of travel for GTAW is slower than that of other arc welding processes. Travel speed is determined by the amount of penetration and deposition. The molten pool should equally penetrate each side of the joint and should remain consistent in size and shape throughout the welding process. The amount of filler metal in the completed bead should not be excessive—usually about one-third the thickness of the base metal is allowed to extend above the top of a groove weld. The leg of a fillet weld should be equal to the thickness of the parts welded.
- **Feeding the filler material.** The key to manually feeding filler metal is to melt the base metal and dip the rod into the weld area. Incomplete fusion is the likely result of dipping the rod into unmelted base metal. With the free hand, hold the filler metal rod at a 10° to 15° angle to the base metal ahead of the weld pool. Keep the tip of the filler within the envelope of the gaseous shield, but never allow it to touch the tungsten electrode. Touch the filler rod to the front edge of the molten pool. The heat from sufficient amperage draws molten filler metal into the approaching body of the weld pool. A repetitive movement that results in a consistent drop-by-drop addition to the weld pool results in a high-quality bead. The foot pedal can be used to pulse the amperage, allowing the melting and solidification process to proceed without the danger of melting through. Pulsing power sources can be set to pulse automatically.

Starting a GTAW Arc

There are three ways to initiate the GTAW arc—scratch start, lift start, and high-frequency start. Not all power sources can perform all three methods.

Scratch Start

To *scratch start*, scratch the electrode on the metal and then lift it while the remote control pedal is depressed. Scratch starting is the least desirable method of starting the arc because of a high probability of electrode contamination.

Lift Start

When using a machine with a *lift start* option, the welder touches the electrode to the base metal for one to two seconds, and then an arc is established as the electrode is withdrawn into welding position. Normal welding voltage is not present before the tungsten electrode touches the base metal. A low-voltage signal indicates contact between electrode and work. The solid-state output current contactor energizes after the electrode is signaled. This reduces the chance that the electrode will become contaminated. To perform lift starting, touch the electrode and the rim of the nozzle to the base metal at the same time, and then roll back the electrode using the nozzle as a pivot to establish the arc.

High-Frequency Start

In the *high-frequency start* procedure, a high-frequency voltage initiates the arc without the welder touching the electrode to the work. Increasing the frequency, or the rate at which the AC electricity changes direction, also increases the tendency for arcing. See **Figure 31-19**. To use high-frequency start, position the electrode within about 1″ of the work and ramp up the current using the remote control until an arc is established.

Photo Love/Shutterstock.com

Figure 31-19. High-frequency AC allows for no-touch starting of the weld.

GTAW Practice

GTAW practice should begin with simple exercises, gradually moving to exercises that require increasingly higher levels of skill. Traditionally, welders were trained to perform oxyfuel welding before learning to weld with GTAW because the body motions are similar. The first GTAW exercises should be autogenous in order to allow the welder to gain skill and confidence with the torch before requiring the second hand to control the filler rod.

Mild steel is good for beginning practice because it is less expensive and relatively forgiving when it comes to heat input. Stainless steels react to welding similarly to mild steel. The weld pool in many grades of stainless is calmer and a bit more fluid than in mild steel. Although stainless steel is more sensitive to heat input, the resulting beads tend to have a pleasing appearance.

Aluminum, which is highly sensitive to heat, has a much smaller window of workable temperatures for welding. Color changes as aluminum is heated are much more subtle than with other metals. Inexperienced welders working on aluminum have trouble maintaining a workable temperature for welding, tending to transition quickly from not enough heat to too much heat. Aluminum must be clean and free from all contaminants and oxides for successful welding.

Perform the following exercises on mild steel first. Repeat the exercise with stainless steel, and finally, with aluminum to gradually increase skill level.

Chasing Puddles

Chasing a series of puddles (molten weld pools) on thin plate is a good way to develop the hand-eye coordination needed for gas tungsten arc welding. In this exercise, the welder uses the welding torch to melt and walk a puddle in a thin mild steel plate without the addition of a filler metal. By not using filler metal, the welder can concentrate on manipulating the torch. The result should be a very concave bead that penetrates the lower surface without melting through. Uniformity and heat control are the goals.

1. Begin by thoroughly cleaning a thin, 14 gauge to 16 gauge sheet metal plate about 3″ by 4″ in size.
2. Adjust the machine settings according to the manufacturer's recommendations for the base metal thickness.
3. Secure the workpiece connection and place the coupon in the flat position.
4. Start the arc and begin a puddle about 1/4″ from one edge.
5. Maintain a uniform arc length, work angle, travel angle, travel speed, and puddle size as you walk the puddle slowly in a straight line parallel to the edge. The result should be a uniform concave bead. See **Figure 31-20**.
6. Repeat this exercise until uniformity is achieved. For additional practice, repeat this exercise using filler metal.

Autogenous Edge Joint

In this exercise, you will perform autogenous welds on edge or corner joints using thin steel coupons. This type of weld, without the use of filler, takes full advantage of the GTAW process to produce a clean bond on the edge of two plates that requires little or no surface treatment after the weld is completed.

1. Cut coupons from thin plate (10 gauge or thinner, which is .102″ thick), 1″ to 1 1/2″ wide, and about 4″ long.
2. Clean the coupons prior to welding.
3. Position two coupons in a 90° angle with the vertex pointing up and the workpiece connection attached.

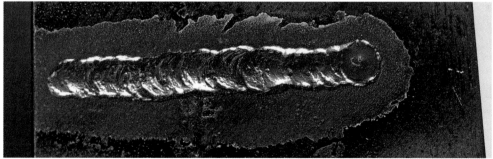

Goodheart-Willcox Publisher

Figure 31-20. A bead created by chasing a puddle (weld pool).

4. Adjust the machine settings according to the manufacturer's recommendations for the base metal thickness.
5. Tack weld the two pieces on both ends of the joint.
6. Start a weld pool on one end of the joint by working the heat from one side to the other. Walk the puddle along the joint to form a uniform bead from the melted edges of the two coupons. The root of the bead should penetrate slightly into the inside of the joint. See **Figure 31-21**.

Open Root on 1/8″ Plate

Complete this exercise in two beads, both welded from one side of the plate only. The first pass is autogenous. The second is made using 3/32″ filler material.

1. Thoroughly clean two mild steel 1/8″ coupons, 1 1/2″ wide and 4″ long.
2. Adjust the machine settings according to the manufacturer's recommendations for the base metal thickness.
3. Set up the coupons with the square edges butted tightly at the joint.
4. Attach the workpiece connection.
5. Tack weld each end.
6. The first pass should be similar to chasing a puddle (weld pool) on flat plate. Be sure that the weld pool is equally distributed to each plate and penetrates to the lower side of the joint. At the end of the root pass, a slightly convex bead should be formed on the back side of the joint and a concave bead should be formed on the topside of the joint. See **Figure 31-22**.
7. Finish the coupon with a second filler pass using a 3/32″ filler rod to build up a bead on the top. The bead should have a slightly convex profile that blends smoothly to both sides and rises no more than 3/32″ above the surface of the base metal.

Goodheart-Willcox Publisher

Figure 31-21. An autogenous weld on an edge joint.

Figure 31-22. A properly penetrated root bead on the back side of a weld.

2G Fillet Weld on 10 Gauge Plate

Most fillet welds made with the GTAW process require the use of a filler metal. Enough material must be supplied to ensure that the edges of the bead blend smoothly without reducing the thickness of the base metal. Use a 1/16" or 3/32" filler material when working with 10 gauge material. The coupons used for practicing edge welds can be used for this exercise as well.

1. Position the two coupons at a 90° angle to form an upside down *T*.
2. Attach the workpiece connection.
3. Adjust the machine settings according to the manufacturer's recommendations for the base metal thickness.
4. Tack weld the two pieces on both ends of the joint.
5. Start a weld pool on one end of the joint. By working the heat from one plate to the other, walk the puddle along the joint to form a uniform bead.
6. Feed the filler metal into the leading edge of the molten weld pool. The goal is to maintain enough fill to complete a bead with a slightly concave profile that blends evenly to both sides of the joint. See **Figure 31-23**.
7. For additional practice, weld a second bead on the other side of the joint.

Figure 31-23. A fillet weld blended evenly to both sides of the joint.

CHAPTER 31
Review and Assessment

Summary

- In the GTAW process, a tungsten electrode creates the arc for welding but does not melt and become incorporated into the weld joint. An inert shielding gas protects the weld from atmospheric contamination. Most of the time a filler metal is used, but some joints may be welded without the addition of extra material.
- GTAW is most often used to weld thin sections of stainless steel and nonferrous metals, although it can also be used to weld mild steels and carbon steels.
- GTAW offers the welder more control over the weld than other arc welding processes, but is more challenging to master and is slower than most other welding techniques.
- DCEN polarity is used for welding steel, nickel alloys, and titanium. AC is used for welding aluminum or magnesium because it produces a cleaning action.
- Most GTAW torches have interchangeable parts, making them adaptable for a variety of welding conditions. Torches may be air-cooled or water-cooled.
- Filler metals for GTAW are usually bare metal rods engineered to match the composition of the metal being welded.
- The recommended shielding gas for most GTAW applications is argon or helium.
- Several types of tungsten electrodes with different welding properties are available. The electrodes are identified by a standard color-coding system.
- Tungsten electrode shape affects arc characteristics, which in turn affect the bead size and profile, the weld penetration, and the durability of the electrode tip. The three tip configurations are hemispherical, truncated cone, and pointed.
- GTAW generates intense heat and large amounts of infrared and UV rays. Exposed skin is subject to damage similar to severe sunburn. Proper clothing and PPE is critical to safety.
- To avoid electric shock, never touch the filler rod to the electrode.
- When practicing GTAW, begin with autogenous exercises to gain skill and confidence with the torch before using the second hand to control the filler rod.
- Practice GTAW on mild steel first; then move to stainless steel, and finally, to aluminum to gradually increase skill level.

Words to Know

Match the key terms from the chapter to the correct definition.

A. autogenous welding
B. ceriated tungsten electrodes
C. collet
D. gas lens
E. hemispherical tip
F. high-frequency start
G. lanthanated tungsten electrodes
H. scratch start
I. shielding gas nozzle
J. thoriated tungsten electrodes
K. truncated cone tip
L. tungsten extension
M. walking the cup

1. Allows the welding arc to jump from the electrode to the base metal without contact.
2. The part within a GTAW torch that holds the electrode in place.
3. Ceramic cup that surrounds the electrode tip.
4. Collet body containing a stainless steel mesh that reduces shielding gas turbulence.
5. Marked with a red band; dust from sharpening is slightly radioactive.
6. Marked with an orange band; good for thin sheet metal.
7. Marked with a gold band; increases current capacity about 50% compared to pure tungsten.
8. Electrode tip shaped like a ball.
9. Tapered conical tip that is flat on the end.
10. The distance the GTAW electrode extends past the tip of the nozzle.
11. Welding without the addition of filler material.
12. Torch method in which the edge of the nozzle remains in contact with the base metal to help maintain a consistent arc length.
13. Requires contact between the base metal and the electrode; prone to electrode and weld contamination.

Know and Understand

Answer the following questions using the information provided in this chapter.

1. In the GTAW process, the welder typically uses one hand to manipulate the torch and the other to feed _____ into the weld pool.
2. The GTAW process is most often used to weld what types of metals?
3. A foot pedal or thumb switch can be used by the welder to vary the _____.
4. Why is alternating current used for welding aluminum?
5. As the collet is pressed into the collet body, it forms a tight grip to hold the _____ in the torch.

6. List three reasons why argon is most commonly used as a shielding gas for GTAW.
7. Tungsten electrodes coded with a gold band are _____ electrodes.
   A. thoriated
   B. ceriated
   C. pure tungsten
   D. lanthanated
8. Why should the flow rate for GTAW be just enough to protect the weld?
9. A(n) _____ cone tip is used for welding stainless with DC welding processes.
10. Although a minimum shade of #_____ is required for welding helmets worn for gas metal arc welding, a darker shade should be used if the visibility is adequate.
11. To reduce the chance of electric shock, the _____ electrode holder should be disconnected from dual or multiprocess machines prior to GTAW welding.
12. Inert gases can _____ breathing air in enclosed spaces.
13. List three types of contaminants that must be cleaned from bare metal before it is welded.
14. The tilt of the torch in relation to the direction of travel is called the _____.
    A. travel speed
    B. travel angle
    C. work angle
    D. arc length
15. The inside diameter of the shielding gas nozzle should be at least _____ times the diameter of the tungsten electrode to allow for adequate shielding gas flow.
16. The tip of the filler metal should be kept within the gaseous shield at all times while welding, but never allowed to make contact with the _____.
17. Describe how an arc is started using the lift start method.
18. Compared to other metals, _____ has more subtle color changes when subjected to welding temperatures and a narrow window of working temperatures.

STEM and Academic Activities

1. **Science.** With your instructor's permission, prepare the ends of two tungsten electrodes with truncated cones for welding steel. Grind the first electrode correctly with longitudinal striations and grind the second electrode improperly with radial (rays coming from a common center) striations. Strike a low amperage arc with each electrode. What are the differences between the two arcs? Formulate a hypothesis as to why each arc reacts the way it does. Be prepared to explain your answer in class.
2. **Technology.** Compare the starting processes of lift start versus high-frequency start. Which process is likely to increase the useful life of the tungsten electrodes? Be prepared to explain your answer in class.

3. **Engineering.** Study the internal parts of a GTAW torch. Sketch the torch with colored pencils, showing the path of the flux gas as it flows through the torch. Draw in additional passages that are found on a water-cooled torch. Explain how the water is kept safely separated from the electricity.
4. **Math.** Locate a metal container. With a micrometer, measure the thickness of the metal. Use a chart or online app to calculate the correct settings and consumables required to weld this material.
5. **Language Arts.** Prepare flash cards for the different types of GTAW electrodes. On the front, draw a color image of the electrode. On the back, write the name of the electrode and a description of its properties and uses. With a partner, use the cards to quiz each other on the electrode types.
6. **Social Science.** Prepare a time line that shows the major developments in welding from the 1800s to the present. Include all the types of welding that you have studied in this textbook.

Thinking Critically

1. Explain the difference between consumables and nonconsumables for GTAW. Can certain torch parts be considered consumable? Why or why not?
2. It is possible to weld steels of dissimilar content. What would be the likely result of such a joint? What effects could result from the addition of a filler metal?

CHAPTER 32 Plasma Cutting

Chapter Outcomes

After studying this chapter, you will be able to:
- Describe the plasma cutting process.
- Identify metals best suited for plasma cutting.
- Explain the functions of plasma cutting equipment components.
- Take appropriate safety precautions when plasma cutting.
- Use proper techniques to produce a clean cut with a plasma torch.
- Describe CNC plasma cutting equipment and processes.

Words to Know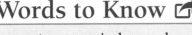

computer numerical control (CNC)	drag shield	pilot arc	standoff
	.dxf file format	plasma	standoff guide
diffuser	nesting	sacrificial part	tool path

Before You Read

Arrange a study session to read the chapter aloud with a classmate. Take turns reading each section. Stop at the end of each section to discuss what you think its main points are. Take notes of your study session to share with the class.

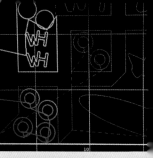

Lone Wolf Photography/Shutterstock.com

While studying this chapter, look for the activity icon to:

- **Practice** vocabulary terms with e-flash cards and matching activities.
- **Expand** learning with interactive activities.
- **Reinforce** what you learn by completing the end-of-chapter questions.

www.g-wlearning.com/agriculture

Plasma is a gas heated to an extremely high temperature and ionized to the point that it becomes electrically conductive. Plasma is often described as the fourth state of matter. The difference between states of matter is their energy levels. For example, the first three energy levels of water are ice, water, and steam. Adding energy to ice in the form of heat melts the ice and forms water. Additional heat vaporizes the water into steam. When even more energy is added to steam, it becomes ionized and electrically conductive, converting to plasma.

Lightning is an example of naturally occurring plasma. Ionized gases in the atmosphere conduct electricity from one place to another. In the plasma cutting process, plasma containing ionized gases and metal fumes transfers an electric arc to the work, where the intense heat of the arc melts the metal. The pressurized jet of plasma gas blows the molten metal away, creating a cut.

Plasma Cutting

The plasma cutting process transfers energy through the electrically conductive gas from the plasma cutting torch to the base metal. The plasma jet melts a highly concentrated area of the base metal. See **Figure 32-1**. The arc forces its way through the work, removing the molten metal. As in other electric arc processes, the base metal must be securely connected to the power source with a workpiece connection. See **Figure 32-2**. Plasma cutting is limited to cutting electrically conductive metals.

Plasma cutting is well suited for cutting most metals in intricate patterns and straight lines. Cutting beveled edges in preparation for welding joints is also easily completed with plasma. Remember that the oxyfuel cutting process generates heat for the cut from the burning, or oxidation, of the metal. In the plasma cutting process, heat from the electric arc that is transferred through the plasma melts the base metal.

Advantages of plasma torches over oxyfuel torches include a smaller heat-affected zone around the cut; clean, narrow kerfs; and easier clean-up after the cut. Plasma cutting is used to cut nonferrous metals, such as aluminum and stainless steel, which cannot be cut with the oxyfuel process. Because cutting with a plasma torch relies on conductivity rather than oxidation, metals with thin layers of rust or paint can be cut effectively. In some materials, cuts almost as clean as those made with a saw are possible. Thin sheet metal properly cut with a plasma torch has a smoother edge than the knife-sharp edges produced by traditional shears or snips.

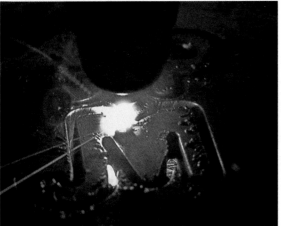

Goodheart-Willcox Publisher

Figure 32-1. A plasma jet and arc being used to create initials by melting and removing a highly concentrated area of the base metal.

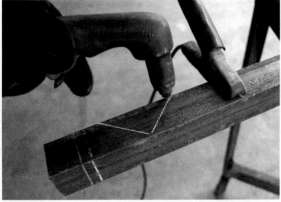

Goodheart-Willcox Publisher

Figure 32-2. The base metal must be connected to the power source using a secure workpiece connection.

Plasma Cutting Equipment

Plasma cutting systems include a constant-current DC power source, a compressed gas supply, an arc-starting circuit, and a torch. See **Figure 32-3**. Power sources vary in size. Large stationary machines are used in industrial shops, and smaller portable units the size of a small suitcase are taken from job to job. Most power sources operate with a primary output voltage in the range of 120 V to 400 V DC. They have an adjustable amperage setting that can be tailored to the thickness of the material and the desired speed of the cut. The primary circuit provides the energy to maintain the cut. A secondary high-frequency AC circuit of 5,000 V to 10,000 V provides the starting arc. This voltage creates a high-intensity arc inside the torch to excite the electrons of the gas, initiating the production of plasma.

Plasma cutting torches, like power sources, may be large, bulky units for automated machines or small, manually operated units. The torch is the framework that holds a consumable nozzle and electrode. In addition, the torch provides a method of cooling these parts. See **Figure 32-4**.

Goodheart-Willcox Publisher

Figure 32-3. Common plasma torch setup.

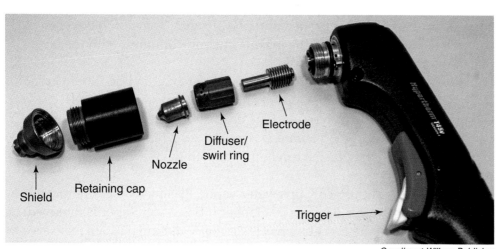

Goodheart-Willcox Publisher

Figure 32-4. Parts of a plasma torch.

In general, heavy-duty automated torches need to be water-cooled, while handheld medium- and light-duty torches are adequately cooled by flowing gases.

Pulling the trigger on the plasma cutting torch energizes the electrode, and a valve opens to allow gas flow to begin. The gas is released through a *diffuser* (swirl ring) that causes it to swirl and pass through the high-frequency current, which ionizes it to form plasma. Energy in the electrode from the primary current flows across the plasma to the nozzle to form a *pilot arc*. The pilot arc initiates the plasma process—it is not the cutting arc. The pilot arc and plasma are forced through the cone-like nozzle opening and brought into contact with the base metal. At this point, the energy of the arc transfers from the nozzle to the base metal; this is the cutting arc. The flowing plasma jet and the heat of the arc then penetrate and cut through the base metal.

Plasma torch tips are often protected by shields of a copper alloy. Shields can be configured to leave a lot or just a little of the nozzle exposed. If the nozzle contacts the base metal or is contaminated with molten dross, the energy flow can be diverted to secondary arcing between the nozzle and the electrode, resulting in premature wear of both. Shielding reduces this diversion and promotes longer consumable life. A properly configured *drag shield* allows the torch to be dragged along a line to trace a design while preventing the nozzle from touching the base metal. See **Figure 32-5**.

Goodheart-Willcox Publisher

Figure 32-5. The torch on the left is equipped with a drag shield that protects the nozzle from accidental contact with the base metal. The torch on the right has an exposed nozzle that must be carefully held at a continuous gap, or standoff, from the base metal.

Compressed Gas Sources for Plasma Cutting

Compressed air supplied directly from a shop compressor is the gas most often used for plasma cutting. The compressed air can also be supplied from a cylinder when using a compressor is inconvenient. Ultraportable plasma cutting units have self-contained air compressors. These units are limited to cutting very thin metal. Atmospheric air is composed of about 78% nitrogen and 21% oxygen. Some of the oxygen in compressed air is reactive in the plasma cut and combines with the base metal to form oxides in the dross.

Pure oxygen can be used to produce cuts of excellent quality in carbon steel, but it results in rough edges in aluminum and stainless steels. Compressed pure nitrogen can also be used and is preferred for precision cuts in stainless steel. The volume (in cfm) and the pressure (in psi) of any gas used with the plasma torch must match the machine's specific requirements.

Atmospheric air also contains water vapor. Plasma torches operate more efficiently and consumable parts last longer if this water vapor is removed, or at least reduced. Many machines have externally mounted filters that serve to trap the water from the air. In humid environments, the trapped water must be expelled periodically and the filter elements replaced. See **Figure 32-6**.

Plasma Cutting Safety

A number of hazards are associated with plasma cutting. The plasma cutting process produces a stream of hot metallic waste. Depending on the material being cut, a plume of smoke, ash, and fumes is released into the air. Adequate ventilation is essential to safely removing these airborne hazards from the work area. Plasma cutting also generates intense heat and significant amounts of bright light. The rush of gases past the edge of the material being cut generates noise levels above 85 decibels. The hot stream of plasma can penetrate most conductive materials, including the human body. There is also a risk of electric shock because water is introduced into an electrical environment. A good understanding of the plasma cutting process and common sense are needed to stay safe.

Personal Protective Equipment (PPE)

Wearing the appropriate personal protective equipment when plasma arc cutting is critical to safety. Essential PPE for plasma cutting does not have to offer the same protection associated with other electric arc processes. However, plasma cutting presents specific hazards that require appropriate PPE. The following PPE should be worn when plasma cutting:

- OSHA-approved (Z87.1) goggles or glasses must be worn at all times.

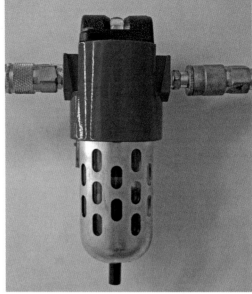

Goodheart-Willcox Publisher

Figure 32-6. A filter or water trap incorporated into the compressed air supply line removes foreign particles and excess moisture to improve cutting characteristics and extend the life of torch components.

AgEd Connection — Plasma Cutting

The plasma cutting process produces very neat, precise cuts and can be used to create intricate designs in thin metals. Coupled with computer-aided design, the plasma cutting process can replicate parts quickly and easily. Ask your agriculture teacher for information about local businesses that utilize this process and could potentially provide experience in this area.

You may decide to invest in your own equipment and develop a fabrication and design business to meet a demand in your area. Research the potential demand for jobs involving plasma cutting in your community and ways you can develop your skills while working in the agricultural mechanics laboratory at your school.

Vereshchagin Dmitry/Shutterstock.com

Shaded Lens Requirements

Torch Amperage	Minimum Shade Number
Fewer than 300A	#9
300A–400A	#12
400A–800A	#14

Goodheart-Willcox Publisher

Figure 32-7. Shaded lenses must be worn when using a plasma torch. The shade number required depends on the amperage of the torch.

- Shaded lenses are necessary. The actual shade required depends on the output amperage of the plasma torch. See **Figure 32-7**.
- Gauntlet-type, heavy, insulated leather welding gloves protect the hands and forearms from heat and intense light. Higher-quality gloves provide protection from the heat and light while remaining supple enough to precisely handle the torch. Gloves must be kept dry.
- Protective footwear should completely cover the foot with leather and be constructed with synthetic soles having significant tread.
- Heavy, long-sleeve cotton shirts with no pockets or covered pockets are required. Fire-resistant clothing is recommended.
- Leather aprons, leather sleeves, welding chaps, or a welding jacket are not usually necessary but may be worn as extra protection.
- Hearing protection (earplugs or earmuffs) should be worn during lengthy operation of plasma arc equipment.

Other Safety Precautions

- Ensure that all electrical, compressed air, and water connections are secure prior to use.
- To reduce chances of accidental shock, always check that the workpiece connection is in good condition and makes secure contact with the workpiece.

STEM Connection

Hexavalent Chromium

Chromium is often added to metal alloys to increase their hardness and resistance to corrosion. It is a component of stainless steel as well as several nonferrous metals. Chromium-containing alloys pose a particular threat to welders because the welding process converts chromium to a chemical state called *hexavalent chromium*, or Cr(VI). This form of chromium is present in the fumes released during the welding process. It is considered by OSHA and NIOSH to be both toxic and extremely carcinogenic. Inhalation of Cr(VI) is known to cause several types of cancer, including lung and sinus cancer. Even topical, or skin, contact can cause skin irritation, skin ulcers, and dermatitis. Appropriate PPE and excellent ventilation are therefore critical for welders who work with chromium-containing alloys.

- Extend leads only enough to reach the work to reduce tripping hazards and protect leads from accidental damage.
- Adequate ventilation should always be provided. Smoke and fumes from vaporized metals produced during cutting can reach harmful levels in poorly ventilated areas. Increased ventilation is required when materials with coatings such as paint or galvanizing are cut.
- Never use a plasma torch to cut a container that contains or may have previously contained a flammable material.
- Although the heat of a plasma torch is very concentrated, parts can be extremely hot when first cut. Handle parts with pliers or tongs.

Using a Plasma Torch

A plasma torch, unless it is used to cut a bevel, should be held perpendicular to the surface being cut. The torch can be pushed or pulled. If the torch is allowed to tilt, it should be operated with the arc trailing rather than leading.

The best way to cut a straight line is to follow a straightedge as a guide. See **Figure 32-8**. The narrow heat-affected zone of a plasma torch makes it possible to use a metal, wood, or even cardboard guide. The resulting cut can be as clean and straight as one cut with a shear. The best cuts are achieved by maintaining a *standoff*, the distance from the torch tip to the base metal, of about 1/8" to 3/16". The closer the standoff is, the narrower the kerf. Some torches are equipped with drag shields that help to maintain this distance when thin materials are cut.

A *standoff guide* is a shield-like apparatus that clips onto the end of the torch and is electrically isolated from the nozzle and electrode. The standoff guide can be dragged along the work surface while the torch tip is held at a consistent standoff distance. Standoff guides are accessories mounted on torches that are not equipped with drag shields. While they function in a similar manner, standoff guides are capable of cutting thicker materials.

Simple standoff guides are formed from a single piece. More complex units have adjustable guide wheels that reduce friction and can be adjusted with an offset (each wheel at a different height) for cutting beveled edges. See **Figure 32-9**.

Goodheart-Willcox Publisher

Figure 32-8. A plasma torch equipped with a drag shield can be used with a straightedge to cut precisely straight lines in thin materials. This student is splitting a section of 3/4" pipe.

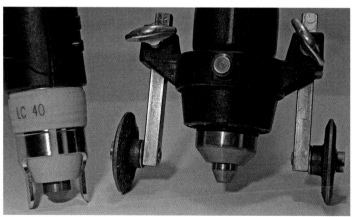

Goodheart-Willcox Publisher

Figure 32-9. Simple and wheeled standoff guides can be used to cut along a straightedge or to follow a pattern in thick materials.

Torches equipped with standoff guides can be used in conjunction with a straightedge to cut long, straight lines. A section of angle iron can be easily clamped in place, and its rigidity makes it a sturdy guide for the torch. Intricate shapes can be cut freehand or with the aid of templates that guide the motion of the torch.

Travel speed greatly affects the quality of the cut and the amount of dross accumulated on the lower edges of the kerf. When operated at the correct speed, the draglines left on the edges of the kerf should curve back at about a 15° angle. There should be a minimum of adhering dross. See **Figure 32-10**. A travel speed that is too slow results in a wide kerf and excess dross, while traveling too fast can result in loss of penetration and an incomplete cut. Optimal travel speed varies according to the thickness and type of metal and may also be affected by amperage and air pressure settings. Cutting smaller parts tends to build up more heat than long cuts, requiring adjustments to the travel speed. Machine manufacturers' cutting speed charts provide good starting points for most jobs. Making practice cuts allows machine operators to tweak settings for the best quality performance.

Goodheart-Willcox Publisher

Figure 32-10. Properly performed plasma cutting leaves drag lines on the edges of the kerf. The drag lines curve back at about a 15° angle with a minimum of adhering dross.

A plasma torch can be used to cut holes into metal. To pierce a hole, position the torch at a 45° angle to the work and press the trigger to start the cutting arc. The arc will begin gouging a hole in the surface of the base metal. Quickly roll the torch tip to a 90° angle, and the plasma jet will pierce through the base metal. While maintaining the arc and the correct standoff, complete the cut to form a hole. Initiating the cut at an angle reduces the chances of dross being blown back into the nozzle tip.

Correctly performed plasma cuts have very little dross; however, many cuts are less than perfect. Cleaning dross is a simple process. On exposed edges, release adhering dross by tapping the edge lightly with a ball peen or other metalworking hammer, **Figure 32-11**. Direct gentle hammer blows at the point where the dross is attached to the metal to pop it off. Any dross that remains after light tapping can be broken off with a cold chisel or end. See **Figure 32-12**. An electric grinder works, but these simpler methods are actually more efficient.

Normal use of a plasma torch creates predictable wear on the consumable parts of the torch. The electrode tip eventually deteriorates, shortening the electrode. The orifice in the nozzle wears, creating an increasingly larger opening. See **Figure 32-13**. The results of this wear are a larger kerf and lower-quality cuts, leading to eventual nonfunctioning of the torch. Occasional touching of the nozzle to the workpiece also causes elongation of the orifice, leading to a wandering or "off-target" plasma jet. Handheld torches tend to require parts replacement more often than

> **Safety Note**
> Wear proper eye protection when removing dross.

Figure 32-11. Adhering dross can be easily removed from most plasma cuts by striking the edge of the cut lightly with a ball peen hammer.

Figure 32-12. An alternate method of removing adhering dross from plasma-cut parts is to use end nippers to pinch it gently from the corner of the cut.

automated torches due to more inconsistency in positioning the working torch. Torch electrodes and nozzles are easy to replace, but replacement electrodes and nozzles must be compatible with the torch and with each other. Replacement parts vary by manufacturer and are sold according to the amperage ranges for which they are engineered. Some are available in matched sets to reduce confusion. See **Figure 32-14**.

Safety Note
Torch parts are very hot just after making a cut. Be sure to turn off the power and allow the torch to cool before servicing a plasma torch.

CNC Plasma Cutting

Computer numerical control (CNC) is a system in which a computer controls the positions of machines or tools using numerical values. CNC can operate machines such as milling machines, lathes, routers, engravers, torches, and plasma cutters. See **Figure 32-15**. An external computer is

Figure 32-13. Damage from normal wear on the tip of these plasma torch electrodes has decreased cutting efficiency.

Figure 32-14. Plasma torch electrodes and nozzles must be matched to fit the torch and to be compatible with each other and the amperage range.

Goodheart-Willcox Publisher

Figure 32-15. A CNC plasma table cuts parts quickly and accurately, eliminating the error associated with using handheld torches.

programmed to run the machines with a series of codes. *M* codes control the tooling, and *G* and *F* codes control the movement. In the case of a plasma cutter, the *M* codes turn the torch on and off, and *G* and *F* codes control the torch movement on an X,Y, Z grid. The operator can manually input codes or use a computer program to direct the code-driven torch's movements within the grid.

The plasma cutting process is extremely precise. The human hand manipulating the torch is responsible for the imperfections in manually cut plasma parts. CNC results in smoother shapes and faster cutting than would be possible with manual or bandsaw cutting. Designs can include intricate curves and sharp corners. See **Figure 32-16**. In CNC cutting, shapes are arranged on the computer screen and cut automatically. The operator does not touch the material except to load raw materials and to remove parts and waste. CNC technology enables the efficient creation of machine parts that are decorative as well as structurally correct. See **Figure 32-17**.

Most CNC cutting begins with shapes drawn with computer-aided drafting (CAD) software. Files drawn in CAD programs should be saved or exported to a *.dxf file format* that allows exchange of information to other computer programs. CAD programs associated with CNC equipment use the

.dxf file to generate a *tool path*. The tool path is the "map" for the torch to follow when cutting the part. This path can be any direction within the limits of the X, Y grid. Tool paths follow the lines of a drawing in a logical order. Interior cuts are usually made first and small waste pieces are dropped. Then, the cuts around the perimeter are made. These cuts ultimately separate the desired parts from the stock material. After the appropriate tool paths are included in the .dxf file, the file should be saved again. The .dxf file is then imported into a computer-aided machining (CAM) program that directs the CNC.

To account for the width of the kerf, an option within the program software allows the operator to position the tool path either directly on the drawing lines, to the inside of lines (female cuts), or to the outside of lines (male cuts). To produce a small part, select a male cut. To cut a hole into a larger plate, select the female cut. Cutting directly on the line divides the width of the kerf equally to both sides of the drawing line.

CAM programs translate the tool path in *G code* and saves it in another file format, such as .fgc, that is more readily used by the operating system of the particular CNC drive mechanism. G code is an organized series of signals sent to direct the movements of the torch drive mechanism within the X, Y grid. Within the torch control module

Goodheart-Willcox Publisher

Figure 32-16. Identical parts are easily repeated using a plasma table to make cuts.

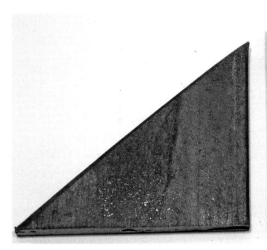

Goodheart-Willcox Publisher

Figure 32-17. CNC technology was used to make the decorative gusset on the right. A gusset is a bracket for strengthening angles in a structure.

signal generator, *G* codes are combined with *F* codes. *F* codes determine the feed rate of the torch and are relative to the thickness of the material. The *F* codes are manually entered or automatically set by the CAM software. The *Z* direction of the grid is the vertical positioner for the torch head, which determines the standoff distance of the plasma torch. Modern automated torch drivers electronically sense the *Z* position of the stock material and set standard standoff distances according to thickness parameters preprogrammed into the software.

Each manufacturer of automated torch drivers uses slightly different software and procedures. Be sure to study the operating instructions for the particular equipment that you use. Many online instructional videos are available, and most manufacturers offer technical support.

Single parts can be cut individually, or multiple parts can be cut during one combined torch operation from a single sheet of material. Multiple parts are positioned in the cut file to be arranged on the metal plate for maximum efficiency of motion and minimum waste of material. This process, called *nesting*, is shown in **Figure 32-18**. Operators can use the computer rendition of the material to nest the parts or allow the CAM program to nest parts automatically.

To make a cut with the CNC, position the sheet metal on the cutting bed and locate a zero (origin) point where the entire part or parts will be inside the dimensions of the sheet. Check the standoff distance for accuracy. Be sure that the ventilation system is on, and push the start button.

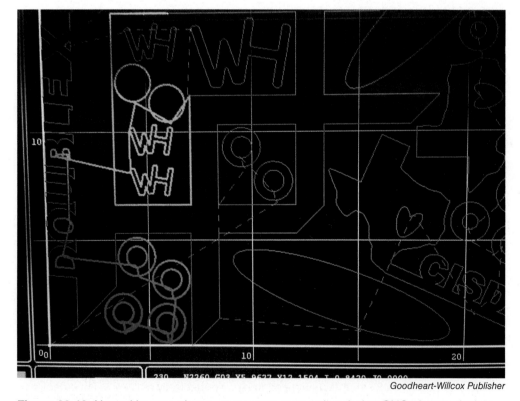

Goodheart-Willcox Publisher

Figure 32-18. Nested images shown on a computer monitor during CNC plasma cutting.

After the part is cut, jog or move the cutting head clear of the part. Lift the part(s) from the table with pliers or tongs. Some parts may need light tapping with a hammer to loosen them from the sheet. This is often an indication that the feed rate should be slightly increased. Dross accumulations should be minimal, but any dross can be removed with the same methods used when manually cutting parts.

The cutting bed is generally constructed in the form of a grate that has easily replaceable sacrificial steel bars. A *sacrificial part* is one that is meant to be damaged during regular functioning of the machine. The steel slats that make up the cutting bed of the plasma table are damaged during normal use. See **Figure 32-19**. The slats must be periodically cleared of large debris that affect the level surface, or replaced if they are damaged. A surface that is no longer flat does not allow the material to be positioned with enough stability to ensure precise, clean cuts. Periodic cleaning of the top edges of the slats results in a more stable cutting surface and better-quality cuts.

> **Safety Note**
> Avoid looking at the arc without adequate shading.

Goodheart-Willcox Publisher

Figure 32-19. Even light use of the automated plasma torch can damage the cutting bed. The steel slats that make up the cutting bed must be periodically cleared of large debris or replaced when damaged.

The water in the pan that contains the grate absorbs some of the sound and much of the smoke and fumes created by the torch. The water is usually treated with chemicals to reduce oxidation of the machine parts and increase the cleansing action, lowering the release of air contaminants. Water must be added to maintain a level just below the cutting action. Draining the water and cleaning sediments from the pan on a regular basis is also necessary.

Safety Note

Waste from cleaning the cutting bed may contain heavy metals and other chemical components potentially harmful to the environment. Always allow the contaminants to settle out and responsibly dispose of the residue. Check with your ordinances and rules.

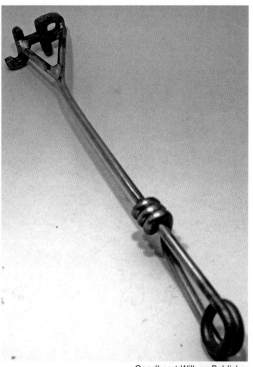

Goodheart-Willcox Publisher

Figure 32-20. Miniature branding iron.

CNC Plasma Cutting Practice

Design your own brand using a CAD program to fit in a 2″ × 2.75″ space, the size of a standard business card. The lines of the brand should be between 1/4″ and 5/16″ wide.

1. Draw the shape from the back side; this is the way the brand will leave an imprint. Draw in traditional 2D style at a scale of 1:1.
2. Save the drawing as a .dxf file.
3. Use the software program to create a tool path that can be uploaded into the driver program for the CNC torch.
4. Upload the tool path for your brand and allow the torch to complete a trial run without a cutting arc.
5. If everything appears to be on the correct path, cut the piece from 1/4″ plate.
6. After the cut piece has cooled, clean the excess dross and attach your brand to a handle formed from 1/4″ round stock at least 14″ long. See **Figure 32-20**.

CHAPTER 32 Review and Assessment

Summary

- The plasma cutting process transfers energy through the electrically conductive ionized gas from the plasma cutting torch to the base metal.
- Plasma cutting is used to cut ferrous and nonferrous metals, such as aluminum and stainless steel, which cannot be cut with the oxyfuel process.
- Plasma cutting systems include a constant-current DC power supply, a compressed gas supply, an arc starting circuit, and a torch.
- A plasma torch holds a consumable nozzle and electrode and provides a method of cooling these parts.
- Compressed air supplied directly from a shop compressor is the gas most often used for plasma cutting. Compressed gas from a cylinder can also be used.
- Plasma cutting hazards include the hot stream of plasma; hot metallic waste; smoke, ash, and fumes; intense heat; bright light; noise; and risk of electric shock. It is important to follow all safety rules and wear proper personal protective equipment (PPE).
- A plasma torch, unless it is used to cut a bevel, should be held perpendicular to the surface being cut.
- Maintaining correct standoff and travel speed results in a clean cut with a minimum of adhering dross.
- In CNC plasma cutting, an external computer is programmed to run the machines using a series of codes.
- Most CNC cutting begins with shapes drawn with CAD software. A CAM program translates the tool path to a file format understood by the operating system of a particular CNC drive mechanism.

Words to Know

Match the key terms from the chapter to the correct definition.

A. diffuser
B. drag shield
C. .dxf file format
D. nesting
E. pilot arc
F. plasma
G. sacrificial part
H. standoff
I. standoff guide
J. tool path

1. Electrically conductive, superheated, ionized gas.
2. A device, also called a swirl ring, that directs pressurized gas through the high-frequency current.
3. Initial current flow from the electrode across the ionized gas to the nozzle.
4. A device that isolates the torch nozzle from contact with the base metal, allowing the torch to be drawn directly along a line to mark a cut in thin metal.
5. The distance between the plasma torch nozzle and the surface of the base metal.
6. Used to save and transfer CNC information between different computer systems.
7. Serves as a guide, providing directions for a computer-controlled torch.
8. Arrangement of multiple parts into a single cut file for maximum machine efficiency.
9. Accessory used to maintain a gap between the exposed torch nozzle and the base metal for cutting thicker materials.
10. A part of a machine that is meant to be damaged during the normal functioning of the machine.

Know and Understand

Answer the following questions using the information provided in this chapter.

1. Adding energy to a gas can ionize it to create _____.
2. Plasma cutting can be performed only on electrically _____ metals.
3. Plasma cutting is faster than oxyfuel cutting and offers the advantage of a smaller _____-affected zone around the cut.
4. List the four main components of a plasma cutting system.
5. A person operating a 35A plasma torch should wear shade #_____ lenses.
6. Heavy-duty plasma torches are cooled by water, while handheld units are cooled by flowing _____.
7. The pilot arc in a plasma torch is formed between the _____ and the _____.
8. Cutting coated materials with a plasma torch requires increased levels of _____ for safety.
9. A standoff guide allows a plasma torch operator to use a(n) _____ to assist in cutting long straight lines.

10. What is the recommended method for removing excess dross adhering to the kerf of a plasma cut?
11. Shapes for CNC plasma cutting are drawn with _____ software.
12. CAD software that works with CNC equipment generates a(n) _____ that is converted into G code, which directs the movements of the automated plasma torch.
13. When multiple parts are to be cut from a single steel plate, a process called _____ is used to maximize efficient use of time and materials.

STEM and Academic Activities

1. **Science.** Investigate the effects of plasma acting in nature. Prepare an illustrated report on the formation of fulgurites.
2. **Technology.** Explain how computer file formats such as .dxf or .pdf that are readable across multiple programs are important for the development of technology.
3. **Engineering.** Study the parts of a plasma torch. Develop a flowchart depicting the formation and use of the plasma stream for cutting metal.
4. **Math.** Use a CAD program to design a miniature branding iron that fits within the dimensions of a standard business card.
5. **Language Arts.** Research the word *plasma* and be prepared to answer the following questions in class. What does this term mean in the ancient Greek language? When and by whom was the term *plasma* first used to refer to ionized gases? Why was this term chosen?
6. **Social Science.** List three specific applications of plasma cutting on a farm or ranch.

Thinking Critically

1. The steel slats that make up the cutting bed of a plasma table are sacrificial parts. What is another example of a sacrificial part used in agricultural mechanics?
2. Compare and contrast oxyfuel welding and plasma cutting. Write a short report that discusses the benefits and drawbacks of each.

CHAPTER 33
Internal Combustion Engines

Chapter Outcomes

After studying this chapter, you will be able to:
- Discuss uses, history, and operation theory of the internal combustion engine.
- List ways engines are classified and discuss the operation for two-stroke and four-stroke engines.
- Describe the mechanical components of an engine.
- Discuss the workings of fuel, ignition, lubrication, and cooling systems.
- Describe safety hazards and safety procedures for working with engines.

Words to Know

camshaft	cylinder	jet	stroke
connecting rod	engine	piston	venturi
crankshaft	flywheel	safety data sheet	
cylinder head	ignition coil	scavenging	

Before You Read

Read the chapter title and tell a classmate what you have experienced or already know about the topic. Write a paragraph describing what you would like to learn about the topic. After reading the chapter, share two things you have learned with a classmate.

While studying this chapter, look for the activity icon to:

- **Practice** vocabulary terms with e-flash cards and matching activities.
- **Expand** learning with interactive activities.
- **Reinforce** what you learn by completing the end-of-chapter questions.

www.g-wlearning.com/agriculture

Elena Elisseeva/Shutterstock.com

Many pieces of agricultural equipment, as well as vehicles used for agriculture, are engine driven. As an agricultural worker, you will need to use and maintain engine-driven equipment and vehicles. Understanding how engines work and possible causes of problems with engines will be helpful.

Internal Combustion Engines

An *engine* is a device that changes the chemical energy in fuel to mechanical energy. In an internal combustion engine, this process takes place within the engine itself. See **Figure 33-1**. Internal combustion engines allow many agricultural tasks to be completed faster and more efficiently than by other means. In fact, the modern agricultural business would not exist without the engines that do the work that not so long ago was done by horses and men.

History of the Internal Combustion Engine

Where and when the internal combustion engine was invented is still debated. The Dutch physicist, Christian Huygens, began experimenting with an idea for an internal combustion engine in the late 1600s, but a

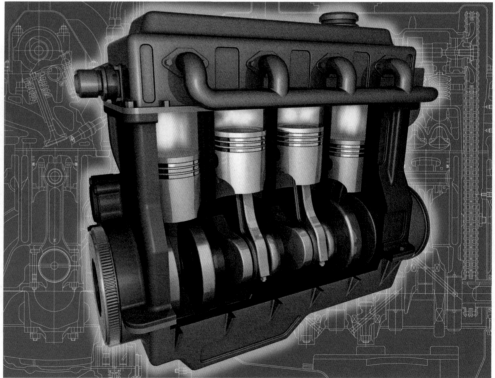

Nikonaft/Shutterstock.com

Figure 33-1. The conversion of chemical energy in fuel to mechanical energy occurs inside the engine.

workable engine was not developed until the 1800s. In 1826, a patent was issued to American inventor Samuel Morey for a four-stroke internal combustion engine. However, it was not until the late 1870s that a working model of this engine, which operated on a mixture of coal gas and air, was built by German engineer Nikolaus Otto. Coal gas is not gasoline. Instead, it is a gaseous mixture—mainly hydrogen, methane, and carbon monoxide—made by the destructive distillation (heating in the absence of air) of bituminous coal. Coal gas was used to fuel street lamps in the 1800s. This early coal gas-powered prototype has led many to credit Otto as the father of the four-stroke engine. However, in 1885, Gottlieb Daimler built a gasoline-fueled internal combustion engine that is widely recognized as the forerunner of the modern gasoline engine. As a result, he is credited with being the key figure in the final development of the four-stroke engine that we know today.

Theory of Engine Operation

In its most basic form, an engine consists of a cylinder, piston(s), connecting rod, and crankshaft. Understanding the function of these parts will help you understand the combustion process.

- *Cylinder*—the bore in which a piston travels.
- *Piston*—an engine part that slides back and forth in a cylinder and creates reciprocating (up and down) motion during the combustion process.
- *Connecting rod*—an engine part that links a piston to the crankshaft.
- *Crankshaft*—an engine part that converts reciprocating motion created by the engine pistons to rotating motion.

If an internal combustion engine is to start and run properly, a specific sequence of coordinated events must occur. This sequence involves the following steps:

1. Intake—the cylinder must be filled with a flammable mixture of air and fuel.
2. Compression—the mixture of air and fuel must be compressed into a smaller space and then ignited by a spark.
3. Power—once the air and fuel mixture is ignited, the force of the expanding, burning mixture can be used for power production.
4. Exhaust—the burned mixture must be removed from the cylinder.

This engine operating sequence must be repeated over and over in precisely the same order.

If gasoline is used for fuel, it must be mixed with the proper amount of air. The correct proportion of air to fuel is maintained by the fuel system (discussed later in this chapter). In a gasoline engine, an electric spark is used to ignite the fuel. A diesel engine does not use a spark plug to ignite the fuel. In a diesel engine, the air charge is compressed in the cylinder. Diesel fuel is injected into the cylinder under high pressure near the end of the compression stroke. The diesel fuel vaporizes and ignites as it comes into contact with the super heated, compressed air.

AgEd Connection: Engine-Related Research

Opportunities

Many people are unaware of the science behind internal combustion and the physics that make up the basic operating principals of internal combustion engines. As you study this chapter, think about the potential areas of research that could be conducted in relation to engines, engine components, and small engine theory. From the physics behind the conversion of linear motion to rotary motion to the variety of lubricants, fuels, and fuel additives used in internal combustion engines, there are many potential areas of research available.

Ask your agriculture teacher how you can get started and identify the necessary resources that are available to conduct your research. Identify a question to answer and begin planning the details for setup. You may be able to conduct the research in your school laboratory, at home, or even with a local business. There is an excitement and a satisfaction that comes with answering difficult questions through research. Internal combustion engines provide many potential research projects.

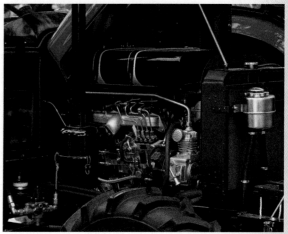

Baloncici/Shutterstock.com

Engine Classifications

Internal combustion engines can be classified in a variety of ways. One of the most common engine classifications is by the number of piston strokes (up or down movements) required to complete one engine operating sequence. A *stroke* refers to the full travel of the piston in the cylinder, in either direction. In the past century, there have been many engine designs. Some of them have incorporated two, four, and even six strokes per cycle. However, due to many practical factors, it is the four-stroke engine that is now used in most power applications. Two-stroke engines are commonly used to power small hand-held equipment, such as chain saws and string trimmers.

Engine classifications can be based on the number of cylinders; engine size (in cubic inches, centimeters, or liters), horsepower rating, and the type of fuel used. Engines are also classified based on crankshaft orientation and cylinder arrangement. The crankshaft is a steel shaft that transfers the force created by the pistons into a force that moves the gears, wheels, or other parts in a rotating motion. If the crankshaft is in the horizontal position, the engine is commonly called a horizontal shaft engine. Conversely, if the crankshaft is oriented in the vertical position, then the engine is referred to as a vertical shaft engine. If multiple cylinders are arranged in a straight line, the engine is referred to as an inline engine. In a V-type engine, the cylinders are arranged in two banks that form a V shape and are offset from each other. For example, a V6 contains two banks with three cylinders each; a V8 has four cylinders per bank. See **Figure 33-2** for a visual representation of cylinder arrangement.

Four-Stroke Cycle Design

As previously mentioned, four piston strokes are needed to complete one operating cycle in a four-stroke engine. A stroke refers to the actions that occur as a piston moves from the top of the cylinder to the bottom or from the bottom of the cylinder and back to the top. In a four-stoke engine, each stroke corresponds to one of the four events in engine operating sequence—intake, compression, power, and exhaust. These events must occur in sequential order for the engine to perform properly.

During the intake stroke, the first of four strokes, the intake valve opens, the exhaust valve closes, and the piston travels downward in the cylinder bore until it reaches bottom dead center (BDC). As the piston moves downward, a vacuum or suction is created to aid in drawing air and fuel into the cylinder. When the piston reaches the bottom of the cylinder, the first stroke is completed and both valves will be closed to trap the mix of fuel

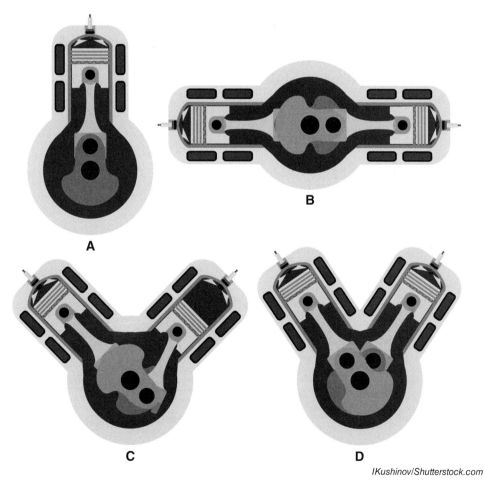

IKushinov/Shutterstock.com

Figure 33-2. Engines can be classified based on their crankshaft orientation and cylinder arrangement. Note that different V-type engines can have angles between their cylinders. A—Inline arrangement. B—Horizontally opposed, or boxer. C—90° V configuration. D—60s4 V configuration.

and air. See **Figure 33-3**. At this precise moment the piston starts traveling upward, and the compression stroke, the second of four strokes, begins.

During the compression stroke both valves are closed. The air and fuel that entered the cylinder during the first stroke are being compressed into a small space between the combustion chamber of the cylinder head and the top of the piston. This is why the second stroke is called the compression stroke because the air and fuel mixture is literally being compressed into a small space. Both valves remain closed and the compressed air and fuel have no means of escaping, **Figure 33-4**.

After the completion of the compression stroke and when the piston is near top dead center (TDC) in the cylinder, the air and fuel mixture is ignited, initiating the power stroke. The compressed air and fuel mixture burns; it does not explode. During the third stroke, the power stroke, the piston will be pushed down by the expanding gases created by the burning of the air and fuel mixture. Both valves remain closed until the piston reaches bottom dead center. The action caused by the power stroke imparts rotary motion to the crankshaft. See **Figure 33-5**.

The fourth and final engine stroke is the exhaust stroke, which completes the four-stroke cycle. During the fourth stroke, the exhaust valve opens and the piston pushes the burned air and fuel out of the open valve as it moves up to the top of the cylinder. Once the piston reaches its highest point of travel (TDC) all combusted materials and residual air has been forcefully expelled out of the engine. The fourth stroke is over. At this moment one full cycle has been completed. The piston traveled up and down twice during the four strokes of this one full cycle, **Figure 33-6**.

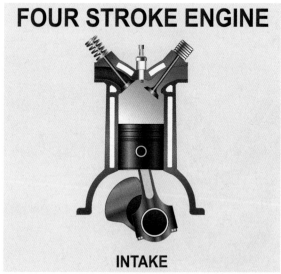

Yugu-Design/Shutterstock.com

Figure 33-3. During the intake stroke, the first of four strokes, the intake valve opens, the exhaust valve closes, and the piston travels downward in the cylinder until it reaches bottom dead center.

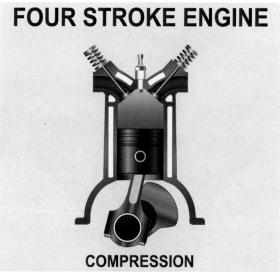

Yugu-Design/Shutterstock.com

Figure 33-4. During the compression stroke, both valves are closed. The air and fuel that entered the cylinder during the first stroke, the intake stroke, are being compressed into a small space between the combustion chamber of the cylinder head and the top of the piston.

Without hesitation, another four-stroke cycle begins. The exhaust valve closes, the intake valve opens, and the piston travels down the cylinder to draw in a fresh dose of fuel and air. This four-stroke process repeats over and over as the engine runs.

During one four-stroke cycle, the piston moves from top dead center (TDC) to bottom dead center (BDC) twice and from BDC to TDC twice. Therefore, during one cycle there are four strokes with two complete rotations of the crankshaft. The *camshaft*, which is a lobed shaft used to open and close the valves, operates at one-half of the rotation of the crankshaft. For every rotation of the crankshaft, the camshaft rotates one-half of a revolution. In other words, it takes two full rotations of the crankshaft to rotate the camshaft one full rotation.

A *flywheel* is a device used to store rotational energy. It is bolted directly to the crankshaft. The rotation of the crankshaft matches the rotation of the flywheel. A flywheel is heavy. When it is spinning during engine operation, the flywheel stores a great amount of energy in the form of inertia. It releases this energy to keep the crankshaft spinning smoothly and at a consistent speed between power strokes. See **Figure 33-7**.

Sharomka/Shutterstock.com

Figure 33-7. The flywheel uses kinetic energy in the form of inertia to reduce engine vibration by smoothing out the power stroke as the air and fuel mixtures in each cylinder ignites.

Yugu-Design/Shutterstock.com

Figure 33-5. During the third stroke, the power stroke, the piston will be pushed down by the expanding gases created by the burning of the air and fuel mixture. Both valves remain closed until the piston reaches bottom dead center.

Yugu-Design/Shutterstock.com

Figure 33-6. During the fourth and final stroke, the exhaust valve opens and the piston pushes the burned air and fuel out of the open valve as it moves up to the top of the cylinder.

Two-Stroke Cycle Design

An understanding of what occurs during the four-stroke cycle will aid in your understanding of the two-stroke cycle. Air and fuel are brought into the combustion chamber and are compressed. Once compressed, the mixture is ignited. Burning gases expand and move the piston, which initiates reciprocating motion. Finally, the burnt mixture is expelled through the exhaust. Once completed, the process is repeated.

When comparing a four-stroke to a two-stroke engine cycle, you will notice that all of the same events occur in exactly the same order when converting the chemical energy of the fuel into mechanical energy. Except, they occur within a shorter time in relationship to crankshaft revolutions in a two-stroke engine cycle. The two-stroke engine completes the cycle in one revolution of the crankshaft whereas the four-stroke engine needs two revolutions of the crankshaft to complete a cycle. See **Figure 33-8**.

The two-stroke engine is simpler in its design than a four-stroke engine. It typically is made of lighter materials, has fewer moving parts, and can operate at higher speeds than a four-stroke engine. Therefore, it can be used on portable equipment where less weight and higher speeds are needed for specific applications. In a two-stroke engine, oil is mixed with the fuel to lubricate the engine. However, in a four-stroke engine, lubrication of moving parts is accomplished through an oil reservoir.

The two strokes of a two-stroke engine are commonly referred to as the compression stroke and the combustion stroke. See **Figure 33-9**. The compression stroke of the two-stroke engine combines the intake and compression strokes of a four-stroke engine. Similarly, the combustion stroke of a two-stroke engine combines the power and exhaust strokes of the four-stroke cycle. As the piston travels from BDC to TDC during the compression stroke, a mixture of air, fuel, and oil is drawn from the intake manifold into the crankcase, where it lubricates moving parts. Once the piston reaches TDC, the compressed air and fuel mixture in the combustion chamber is ignited. This drives the

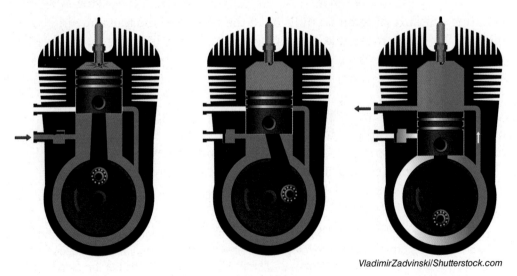

VladimirZadvinski/Shutterstock.com

Figure 33-8. The two-stroke engine completes one cycle in one revolution of the crankshaft, whereas the four-stroke engine needs two revolutions of the crankshaft.

piston toward BDC. As the piston moves downward during the power stroke, a new charge of air, fuel, and oil are drawn into the cylinder. As the air, fuel, and oil mixture rushes into the cylinder, it pushes out exhaust gases, completing the combustion stroke.

A negative consequence of two-stroke engine design is higher emissions. One factor that increases emissions in a two-stroke engine is the fact that the air-fuel mixture also contains oil, which does not burn as quickly or completely as the air and fuel. Another factor that increases emissions in a two-stroke engine is the fact that a small portion of the air, fuel, and oil mixture escapes with the exhaust gases during the combustion stroke. A two-stroke engine normally does not efficiently combust fuel as efficiently as its four-stroke counterpart, but it is able to produce equivalent power with fewer strokes. Due to their power-to-weight advantage, the two-stroke design is often used for handheld power equipment.

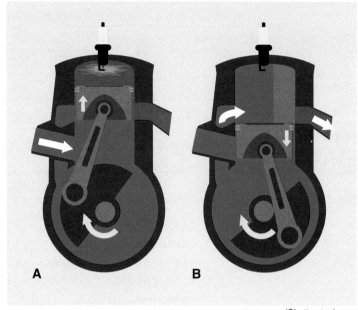

gameanna/Shutterstock.com

Figure 33-9. Two stroke operating cycle. A—The compression stroke of the two-stroke engine combines the intake and compression strokes of a four-stroke engine. B—The combustion stroke of a two-stroke engine combines the power and exhaust strokes of the four-stroke cycle.

A two-stroke engine does not use the same kind of valves that a four-stroke engine uses. A four-stroke engine uses poppet, or mushroom, valves actuated by the camshaft. The valve systems used in two-stroke engines are simpler. The most common type of valve used in two-stroke engines is the reed valve. Reed valves are a type of check valve, which restrict the flow of fluids to a single direction, opening and closing under changing pressure on each side of the valve face. Reed valves control the fuel and air mixture admitted to the cylinder through piston movement. As the piston rises in the cylinder, a vacuum is created in the crankcase beneath the piston. The resulting pressure differential pulls open the valve and the fuel and air mixture flows into the crankcase. As the piston descends, it raises crankcase pressure. This causes the valve to close, trapping and pressurizing the fuel mix for transfer to the combustion chamber. Simplicity, low cost, and reduced rotational mass are the engineering advantages of reed valves.

Through piston movement, reed valves are opened or closed to allow the mixture of air, fuel, and oil in and to exhaust burnt gases out. In two-stroke engines, this process is called *scavenging*. A fundamental part of an internal combustion engine cycle is the supply of fresh air and removal of exhaust gases. This is the gas exchange process. Scavenging is the removal of exhaust gases by forcing in fresh air.

Operators should understand the characteristics of two-stroke and four-stroke engines so they can make an informed decision when purchasing new equipment or selecting a replacement engine for existing equipment. See **Figure 33-10**. It is also important to understand that (due to its higher

	Two-Stroke Engine	Four-Stroke Engine
Material in construction	Low in weight	High in weight
Engine speed (RPM)	High	Low
Efficiency (overall)	Low	High
Engine emissions	High	Low
Operating position	More positions	Fewer positions
Fuel system	Mixed fuel and oil	Separate fuel and oil reservoirs

Goodheart-Willcox Publisher

Figure 33-10. Comparison between designs of small gas engines.

emissions levels) manufacturers are abandoning the two-stroke engine in favor of the four-stroke engine. There are many two-stroke engines still in use, but the long term trend overwhelmingly favors adoption of the four-stroke engine in most applications.

Engine Components

An internal combustion engine is made of many specific parts, which are all important for proper engine operation. Understanding how these parts function and how they are constructed is important for using engines effectively. The major components of an internal combustion engine are shown in **Figure 33-11**.

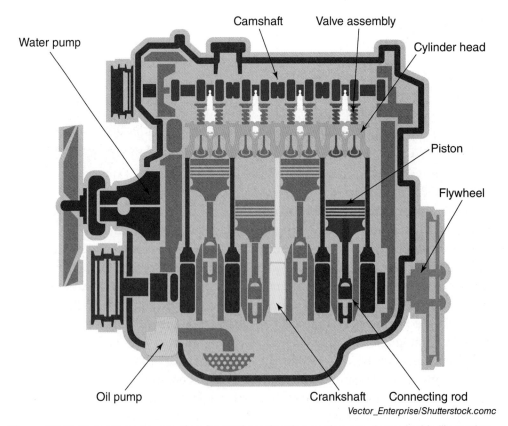

Vector_Enterprise/Shutterstock.comc

Figure 33-11. Major components of an internal combustion engine are shown in this illustration.

Engine Block

The main structural component of the engine is the engine block. See **Figure 33-12**. It provides support for all the other components and houses some internal parts of the engine. Normally, it is cast in one piece and is made of cast iron or aluminum. Many four-stroke engines have a space or opening in the engine block called the cylinder. As described earlier, the cylinder is a round opening or tube where the piston travels from BDC to TDC.

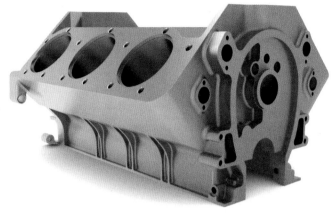

Cherezoff/Shutterstock.com

Figure 33-12. The main structural component of the engine is the cylinder block. It is cast in one piece and is made of cast iron or aluminum.

In some heavy duty applications, cylinder sleeves can be inserted into the engine block. This is referred to as a sleeved engine block. An advantage of a sleeved engine block is that the entire sleeve can be driven out of the block and be replaced if it does not work properly or is damaged. To repair a cylinder that is bored or machined into an engine block, it must be further enlarged to a specific dimension in order to house replacement parts.

The lower part of the engine block houses the crankshaft, and is referred to as the crankcase. The crankcase provides support for the crankshaft and the camshaft (in engines that do not have overhead cam configurations). On small engines, it also serves as the oil reservoir for lubrication of moving engine parts. On larger engines, an oil pan is bolted to the bottom of the crankcase and serves as the oil reservoir.

On top of the engine block sits the *cylinder head*. The cylinder head seals the combustion chamber. On gasoline engines, it also houses the spark plug, which is used to ignite compressed air and fuel mixture. For overhead valve (OHV) configuration engines, the cylinder head has intake and exhaust ports as well as housing for the valves and related engine parts. Cylinder heads are usually cast in a one-piece design and can be made of aluminum or cast iron alloys.

Pistons and Connecting Rods

The piston is a cylindrical engine part that fits into the cylinder and moves from TDC to BDC and back again (reciprocating motion) as the engine runs. The piston is connected to the crankshaft by a connecting rod. The piston, with the aid of piston rings, seals the piston to the walls of the cylinder to make a tight fit. The fit is so tight that the gas produced during combustion cannot escape past the piston rings. Many small gas engine pistons have an indentation or mark on them that indicates its correct orientation in the cylinder. The tight-fitting piston rings usually consist of two compression rings and one set of oil rings. There are many different styles of piston rings. You should consult the engine manufacturer for replacement ring specifications.

The piston is attached to the connecting rod by a wrist pin. A wrist pin is a short length of tubular steel that slides into a hole in the side of the piston,

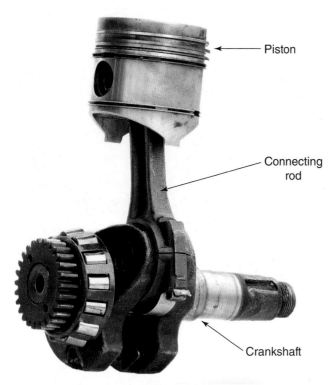

Sanit Fuangnakhon/Shutterstock.com

Figure 33-13. Connecting rods connects pistons to the crankshaft, enabling the reciprocating (up and down) motion of the piston to be converted to rotary (turning) motion by the crankshaft.

through a bearing in the small end of the connecting rod, and into a hole on the other side of the piston. Circular clips hold the wrist pin in place. Although it is small, the wrist pin is not only an important link, but it is also the most stressed part of the engine. Stress from cylinder pressure and inertia load expose the wrist pin to crushing force many times per second.

As stated earlier, the piston is attached to the connecting rod. The other end of the connecting rod, the large end, connects to the crankshaft. The connecting rod connects the piston(s) to an offset journal on the crankshaft, enabling the reciprocating (back-and-forth) motion of the piston to be converted to rotating (turning) motion of the crankshaft. See **Figure 33-13**. The connecting rod is secured to the crankshaft at its large end by a connecting rod cap that uses high strength bolts. A torque wrench is used to fasten these high strength bolts to ensure a secure connection that can withstand high heat and extreme stress.

Crankshaft

The crankshaft is securely held by main caps at the bottom of the engine block and is attached to the piston(s) through connecting rod(s). The spinning crankshaft converts the reciprocating movement of the pistons into rotating motion and transfers it to other parts, such gears and pulleys.

Attached on the other end of the crankshaft is the flywheel. The flywheel can be made out of cast iron, alloy, or aluminum. Depending on the type or style of engine, the flywheel can also house magnets that help generate electricity for the ignition system and electrical accessories. The flywheel may also have teeth on the outside of the hub to mesh with an electrical starter.

Valve Train and Timing

Two-stroke engines use simple reed valves or rotary valves. In a four-stroke engine, the valve train consists of all the parts associated with opening and closing the intake and exhaust valves. The camshaft is connected to the crankshaft either through gears or by a chain or belt. The camshaft has lobes for each valve used in the engine. The lobes are oblong (egg shaped), and they will open and close the valve as they rotate during engine operation. The shape, or grind, of the lobe dictates how long the valve is opened (duration) and how high the valve is opened (lift). The camshaft lobe regulates lift and duration of valve timing. See **Figure 33-14**. Riding on top of the camshaft lobe(s) are lifters or tappets. The lifters transfer the upward or downward motion caused by the cam lobe to the

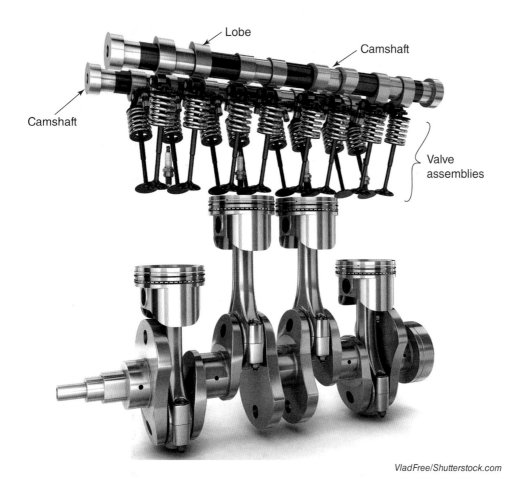

Figure 33-14. One or more camshafts regulate the opening and closing of the intake and exhaust valves. There is a camshaft lobe for each valve used in the engine.

valve either directly or through a push rod and rocker arm assembly. Constant spring tension keeps the valves closed until the valve train sends them a mechanical signal (the cam pushing on the lifter) to open.

If the valves are located in the engine block, the design is commonly called an L-head design or flat head. If the valves are located in the cylinder head above the engine block and the camshaft is located in the engine block, the arrangement is called an overhead valve (OHV) design. If the valves and the camshaft are both located in the cylinder head, the design is called an overhead cam (OHC) design.

Valves rest on valve seats, which are precision manufactured to form an air-tight seal. See **Figure 33-15**. When the valves are closed, the air and fuel being compressed in the combustion chamber cannot leak

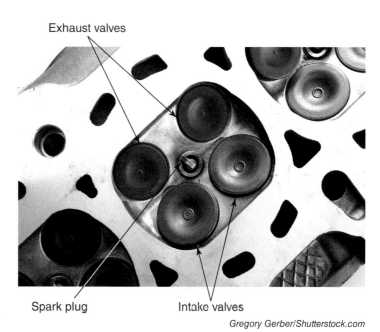

Figure 33-15. The bottom of an cylinder head is shown here. The recesses in the head form the top of the combustion chamber. Intake valves are typically slightly larger than the exhaust valves.

out. Valve timing is accomplished through the rotary motion of the camshaft. As the lobes turn and fully open the valves (when the oblong portion of the lobe reaches its highest point), the air and fuel mixture is drawn into the combustion chamber or burnt gas is expelled.

Related Systems

Several related systems are needed for an internal combustion engine to function properly in a vehicle or piece of equipment. These systems include the fuel system, ignition system, lubrication system, and cooling system.

Fuel System

The purpose of a fuel system is to deliver the proper amount of atomized fuel to the engine. There are three main types of fuel systems in use: carburetor, indirect fuel injection, and direct fuel injection. Although these systems perform the same basic function, they use different methods to meter (measure) the proper amount of fuel for combustion.

Regardless of the metering system used, all fuel systems have a fuel reservoir, commonly called a fuel tank. In some fuel systems on small engines, gravity alone causes fuel to flow from the tank to the metering device. Other engines use a fuel pump to transfer the fuel from the fuel tank to the metering device. Fuel pumps can be electrically or mechanically powered. Electrical fuel pumps use an electric motor to pump the fuel. Mechanical pumps use the rotating motion of the crankshaft or camshaft to drive a pivoted lever to pump fuel to the engine. The actuated lever moves up and down, flexing a diaphragm, which pumps fuel into the engine.

A carburetor is a device that mixes fuel and air together in the proper ratio for efficient combustion. During engine operation, a combination of atmospheric pressure and the downward stroke of the piston draw the air and fuel mixture into the carburetor through the open intake valve and into the combustion chamber. Carburetors are available in many sizes and types. Although different, they all have the same goal: to mix air and fuel at the correct ratio for efficient combustion.

All carburetors have a throttle valve, and most carburetors have an additional choke valve. The throttle valve regulates airflow through the carburetor throat, and the choke valve regulates the amount of air allowed into the carburetor. Most carburetors have floats that regulate how much fuel is held in reserve by the carburetor. If the fuel line cannot supply a sufficient amount of fuel during hard acceleration or cornering, the carburetor still contains enough fuel to supply the engine for smooth operation.

Most carburetors have jets that precisely meter the amount of fuel delivered into the engine. A carburetor *jet* is a tiny hole in the *venturi*, which is a narrow point in the carburetor throat. This narrowing in the venturi increases the speed of airflow, creating a low pressure area. This is called the venturi effect. It is into this low-pressure, high-velocity airstream that the jet squirts fuel for efficient atomization. Atomization is important because too much fuel in a combustion chamber can create a condition in which there is not enough air to burn all of the fuel, causing inefficient engine power output. Having not enough fuel is called a lean condition, which is a cause of a poorly running engine. The throttle valve provides control of the amount of fuel delivered by varying airflow. When the throttle valve is fully open, more air flows through the venturi, and all carburetor jets are open. When the throttle valve is partially closed, less air flows through the carburetor, and one or more of the jets are blocked.

An indirect fuel injection system uses a fuel injector to spray a specific amount of fuel into the intake airflow before it enters the combustion chamber. A fuel injector is essentially a solenoid-operated valve. In indirect fuel injection systems, a low-pressure fuel pump draws fuel from the fuel tank and pressurizes the fuel line to the fuel injectors. Sensors in the engine monitor engine operating conditions, such as engine speed, crankshaft position, and intake airflow, and send that information to a control unit. The control unit, in turn, sends an electrical signal to the fuel injector, causing it to open. When the fuel injector opens, pressurized fuel is sprayed into the intake air. The control unit keeps the injector open long enough to inject the proper amount of fuel for the current operating conditions.

A direct fuel injection system is very similar to indirect fuel injection system, but sprays fuel directly into the combustion chamber rather than into the intake airstream. Diesel engines all have a direct fuel injection system. In recent years, a greater number of gasoline engines have also been designed with direct fuel injection.

In a direct injection fuel system, the fuel is sprayed into the combustion chamber near the end of the compression stroke. The control systems for direct fuel injection systems are very similar to the control systems for indirect fuel injection systems, but the fuel delivery components are very different. In order for the fuel to spray into the cylinder in a fine atomized mist, it must be squirted in at a much higher pressure than the pressure of the compressed air in the combustion chamber. To achieve this degree of pressurization, a direct fuel injection system uses a second, high-pressure, fuel pump called an injection pump. The low-pressure fuel pump draws fuel from the fuel tank and delivers it to the injection pump. The injection pump highly pressurizes the fuel line to the injector. When it receives the signal from the control unit, the injector opens and sprays this highly pressurized fuel directly into the combustion chamber, where it ignites instantly.

Ignition System

For an engine to operate as designed, an electric spark must be created and delivered to the combustion chamber at the appropriate time. There are two main classifications of ignition systems: magneto ignition and battery ignition. In a magneto ignition system, the flywheel holds magnets which work in combination with an ignition coil. An *ignition coil* is a device that transforms low voltage current into high voltage current that can cause a spark to ignite the fuel and air mixture in an engine. As the magnets approach the coil, they induce current in the ignition coil's primary windings, which builds a strong magnetic field. As the magnets move away from the ignition coil, the magnetic field in the primary windings collapses, generating a strong current in the secondary windings. The high-voltage current that is generated causes the spark plug to fire.

A battery ignition system works in a similar way. The battery produces current that flows through the ignition coil's primary winding, building a magnetic field. In old engines, a mechanical switch, known as contact points, would open and close the electrical circuit through the primary windings based on the position of the camshaft. In newer engines, an electronic control module opens and closes the circuit through primary windings. When the primary circuit is opened, current stops flowing, and the magnetic field collapses. This induces high voltage current in the secondary winding.

Because batteries eventually drain, engines with battery ignition systems must also be equipped with a charging system. The charging system uses a magneto or an alternator to generate electricity to charge the battery. The magneto or alternator produces alternating current, which is then converted into direct current and used to charge the battery. A properly working charging system can charge the battery faster than it is drained by powering the ignition system and accessories. This keeps the battery fully charged.

Lubrication System

Keeping moving parts of an engine lubricated is very important for good performance and long engine life. Friction causes engine parts to wear and reduces engine performance. The rubbing of metal parts against each results in friction. Heat caused by the friction will soften the part and accelerate wear. Heat and friction dramatically reduce the life of an engine. To lessen the effects of friction, the proper lubricants (oils) must be used.

Lubricants accomplish many tasks. They reduce friction between moving parts and allow smooth movement. They clean metal parts, absorb and remove heat, and keep foreign particles that could cause engine damage from binding together. Always refer to manufacturer's manuals to find lubricant specifications and recommendations for machines and equipment.

In four-stroke engines, an oil sump (located in the engine crankcase) stores lubricating oil. In small engines, the connecting rod has a protrusion on the large end that dips into the oil in the sump and slings it onto the internal engine parts. Since the crankcase is small, this oil spatter is able to lubricate all the moving parts of the engine. This is referred to as splash lubrication.

Splash lubrication is not appropriate for larger engines because the oil would not be able to reach all of the parts in need of lubrication. The resistance of multiple dippers moving through the oil would also result in an unacceptable power loss. In larger engines, an oil pump is used to pump oil from the sump to various parts of the engine. Oil passages, called galleries, deliver the oil directly to bearings and other wear surfaces. In two-stroke engines, oil is added to the fuel and is dispersed throughout the interior of the engine during operation.

Cooling System

Engines are either air cooled or liquid cooled. The cooling system must be maintained to ensure that overheating does not occur and damage the engine. On air-cooled engines, debris and other foreign materials must be kept away from the cooling fins and air ducts. These types of engines are cooled by airflow removing heat from the engine. See **Figure 33-16**. Air-cooled engines often incorporate a shroud (metal enclosure) into their design. A fan directs airflow under the shroud and past the engine cooling fins to remove excess heat.

For engines that use liquid-based cooling systems, the liquid coolant mixture is routed through the engine to remove heat from around cylinder heads and cylinders. See **Figure 33-17**. A water pump circulates the water around the cylinders and to a heat exchanger, such as a radiator. The radiator allows airflow to remove the captured heat from the coolant mixture. After shedding the excess heat, the liquid coolant mixture is pumped back into the engine to repeat the process.

Water alone can be used as a coolant; however, using pure water as a coolant has two disadvantages. First, water can freeze and expand, resulting in engine damage. Second, water can only absorb and transport a limited amount of heat. However, when water is combined with the chemicals found in antifreeze and pressurized, it will freeze at a lower temperature, allow more heat to be absorbed and dissipated, tolerate higher engine temperatures, and lubricate the water pump.

Safety Note
Always ensure that coolant provide freeze protection below the lowest outdoor temperature expected. Never open a cooling system (radiator cap) under pressure when it is overheated. Doing so may cause burns or other injuries.

Safety

Safety is always an important factor when working in any agricultural setting. When working with engines and related chemicals or substances, you should identify possible hazards and conditions and protect against them. Liquids or chemicals that could be caustic should be handled with care, and appropriate safety attire should be worn at all times. Be aware of the dangers of working with heavy parts, which may be sharp and have protruding pieces that could cause injuries.

Piotr Krzeslak/Shutterstock.com

Figure 33-16. Air flowing past the cooling fins is the only way an air-cooled engine stays within its operating temperature range. A coating of oily dirt will insulate the cooling fins and lead to overheating. This image illustrates why the cooling fins must be kept clean and free of debris.

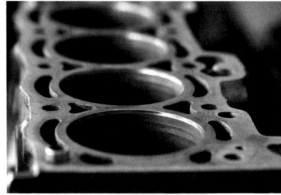

Anupong Nantha/Shutterstock.com

Figure 33-17. The cavities around the cylinder in this engine block are called water jackets. The cylinder head and cylinder head gasket have matching openings. These water jackets allow liquid coolant to flow through the engine to remove excess heat.

Sirastock/Shutterstock.com

Figure 33-18. Proper attire and eyewear are important for safety in work settings. Safety glasses, gloves, hearing protection, and proper clothing should always be worn.

Keep your work area clean and organized, work in a well-ventilated area, and use tools properly. Clean floors in the work area daily to remove debris and liquids or chemicals. If liquid or chemical spills occur, use appropriate materials and methods to clean the spill. Ensure that the floor is dry and safe for you and other workers.

Safety glasses or goggles and proper clothing should always be worn when working in hazardous areas or conditions. See **Figure 33-18**. Safety glasses should always be worn when using power tools, hand tools, compressed air, and fitting together engine parts. Goggles or full face shields should be used when working with liquids/chemicals. Proper shoes should be worn to protect against toe and foot injuries. Also, hearing protection should be worn when equipment is operated to reduce damage from high noise levels. Loss of hearing may not occur instantly, but the effects over time may result in hearing loss later in life.

Air quality is often an overlooked safety issue. Engines produce carbon monoxide (CO) gas, a colorless and odorless gas emitted as the exhaust of engines. Persons inhaling small amounts of CO can have headaches and feel drowsy. If large amounts of CO are taken in by the body, death can occur. Any enclosed areas where engines are running must be well ventilated. Many systems can be used to remove harmful gases in the work area. Check with manufacturers to determine the best solution for the area where you are working.

Besides hazardous gases produced from engines, chemicals used with engine maintenance or operation can release toxic fumes that are flammable. All flammable liquids should be disposed, stored, and used properly to limit the danger associated with their use.

The Occupational Safety and Health Administration (OSHA) is a governmental agency that is part of the U.S. Department of Labor, created to "assure safe and healthful condition by setting and enforcing standards" (OSHA, 2013). Safety hazards in the work area should immediately be addressed. It is the responsibility of all individuals in the work area to handle hazardous materials as outlined in *Safety Data Sheets* (SDS). Safety Data Sheets are documents required by OSHA and provided by chemical manufacturers, distributors, or importers. They are a great source of information about hazards, including environmental hazards, and provide comprehensive information about a substance or mixture for use in workplace chemical management.

CHAPTER 33 Review and Assessment

Summary

- Internal combustion engines allow many agricultural tasks to be completed faster and more efficiently than by other means.
- In 1885, Gottlieb Daimler built a gasoline-fueled internal combustion engine that is widely recognized as the forerunner of the modern gasoline engine.
- A sequence of coordinated events must occur for an internal combustion engine to operate: intake, compression, combustion, and exhaust.
- Engines may be classified by the number of strokes in the operating sequence, number of cylinders, engine size, horsepower rating, type of fuel used, crankshaft orientation, and cylinder arrangement.
- Four piston strokes are needed to complete one operating cycle in a four-stroke engine. These strokes are called the intake, compression, power, and exhaust strokes.
- The two-stroke engine is simpler in its design than a four-stroke engine. This type of engine completes all four operating events in only two strokes, called the compression stroke and the combustion stroke.
- The main structural component of the engine is the cylinder block. It provides support for all the other components and houses some internal parts of the engine.
- A piston fits into the cylinder and moves up and down as the engine runs. A connecting rod attaches the piston to the crankshaft.
- The spinning crankshaft converts the reciprocating motion of the pistons into rotating motion and transfers it to other parts, such as gears and pulleys.
- In a four-stroke engine, the valve train consists of all the parts associated with opening and closing the intake and exhaust valves. Valve timing is accomplished through rotating motion of the camshaft.
- Fuel systems can be categorized as carburetor, indirect fuel injection, and direct injection. All fuel systems have a fuel tank, and some have one or two fuel pumps.
- An ignition coil transforms low voltage current into high voltage current that can cause a spark to ignite the fuel and air mixture in an engine.
- Keeping moving parts of an engine lubricated is very important for good performance and long engine life. Lubricants reduce friction between moving parts, clean metal parts, and absorb and remove heat.
- Engines are either air cooled or liquid cooled. The cooling system helps ensure that overheating does not occur, damaging the engine.
- People using or repairing engines should be aware of safety precautions, follow safety procedures, and wear the appropriate clothing and safety equipment.

Words to Know

Match the key terms from the chapter to the correct definition.

A. camshaft
B. connecting rod
C. crankshaft
D. cylinder head
E. cylinder
F. engine
G. flywheel
H. ignition coil
I. jet
J. piston
K. safety data sheet
L. scavenging
M. stroke
N. venturi

1. A device that transforms low voltage current into high voltage current that can cause a spark to ignite the fuel and air mixture in an engine.
2. An engine part that converts reciprocating motion created by the engine pistons to rotary motion that moves gears or other parts in a device or vehicle.
3. The bore in which a piston travels.
4. A document that provides information about hazards and provides comprehensive information about a substance or mixture for use in workplace chemical management.
5. A narrow point in the carburetor throat.
6. A tiny hole in the narrow part of a carburetor throat.
7. An engine part that slides back and forth in a cylinder and creates reciprocating (up and down) motion during the combustion process.
8. The removal of exhaust gases from an engine by forcing in fresh air.
9. A device that changes the chemical energy in fuel to mechanical energy.
10. A lobed shaft, used to open and close the valves, that operates at one-half of the rotation of the crankshaft.
11. The part of an engine that seals the combustion chamber. On gasoline engines it also houses the spark plug.
12. An engine part that links a piston to the crankshaft.
13. A device used to store rotational energy.
14. The full travel in either direction of a piston in the cylinder.

Know and Understand

Answer the following questions using the information provided in this chapter.

1. Who is credited with inventing a gasoline-fueled engine that was the forerunner of the modern gasoline engine?
   A. Rudolf Diesel.
   B. Gottlieb Daimler.
   C. Nikolaus Otto.
   D. Christian Huygens.

2. What parts are included in an engine in its most basic form?
3. If an internal combustion engine is to start and run properly, what sequence of coordinated events must occur?
4. What are some characteristics by which engines may be classified?
5. Describe the four strokes of the four-stroke operating cycle.
6. How does the process of a four-stroke engine cycle compare to that of a two-stroke engine cycle?
7. Where is the crankcase located, and what is its function?
8. How is valve timing accomplished in a four-stroke engine?
9. What are the three types of fuel systems?
10. *True or False?* The purpose of a carburetor is to mix fuel and air together in the proper ratio for efficient combustion.
11. Why is proper atomization of the air and fuel mixture important?
12. What are the two main types of ignition system and how are they different?
13. What are some tasks that lubricants accomplish in an engine?
14. What are the two types of engine cooling systems, and what is their purpose?
15. When should safety glasses or goggles be worn when using or maintaining engines?

STEM Connections and Academic Activities

1. **Science.** Compare the qualities of different lubricating oils used in engines.
2. **Technology.** Research career options in the small engine repair field.
3. **Engineering.** Describe the heat value of fuels used in engines.
4. **Math.** Calculate the distance a piston travels in a minute in a four-stroke engine when it is operating at 1750 revolutions per minute. Assume the piston has a stroke of 3″.
5. **Social Science.** Research the impact that small engines have had on society, in regards to the ability to accomplish work.
6. **Language Arts.** Write a short essay about the role of engines in society.

Thinking Critically

1. The Daimler engine, the first gasoline powered four-stroke engine, was developed in 1885. Today, there are hundreds of millions of engines in operation throughout the world. Discuss the positives and negatives of this statement with the class.
2. You and a partner are given an engine that does not operate. What could cause the engine not to operate? Use the theory of operation to formulate your choices.
3. Describe the differences between the two-stroke and four-stroke engine designs.
4. Why are two-stroke engines being phased out?

CHAPTER 34
Small Engine Performance, Maintenance, and Troubleshooting

Chapter Outcomes

After studying this chapter, you will be able to:
- Explain how engine performance is measured.
- List factors affecting an engine's suitability for a particular application.
- Demonstrate proper engine maintenance procedures.
- Describe the role that service information plays in the service and maintenance of an engine.
- Indentify the three basic things that an engine needs in order to run.
- List the sequence of checks used to systematically troubleshoot an engine.

Words to Know

bore	displacement	horsepower	torque
compression ratio	force	power	work

Before You Read

Review the chapter headings and use them to create an outline for taking notes during reading and class discussion. Under each heading list any Words to Know. Write two questions you expect the chapter to answer.

While studying this chapter, look for the activity icon to:

- **Practice** vocabulary terms with e-flash cards and matching activities.
- **Expand** learning with interactive activities.
- **Reinforce** what you learn by completing the end-of-chapter questions.

www.g-wlearning.com/agriculture

Elena Elisseeva/Shutterstock.com

Small engines are used for many purposes in agricultural businesses. As an agricultural worker, you may need to use engines in your work. Understanding engine performance will help you select an appropriate engine for a particular job and be aware of problems that may cause an engine to perform poorly. Understanding and preforming routine maintenance on small engines will improve performance and increase engine life. Knowing and following troubleshooting procedures is also important for keeping engines working properly.

Measuring Engine Performance

Engine performance is not a single attribute, but rather a group of measurements used to determine an engine's suitability for different tasks or applications. An understanding of basic terminology is needed in order to understand power and performance of engines. Remember that engines convert heat from burning fuels into reciprocating motion of the piston. This motion is then changed into rotating motion at the crankshaft. The performance of an engine can be defined as the amount of work an engine can do in a given period of time and the efficiency with which the engine does that work.

The engine has a cylinder, in which the piston moves from one end to the other, **Figure 34-1**. The diameter of this cylinder is called the engine *bore*. The measurement of the piston's movement from one end of the cylinder to the other end (travel from top dead center to bottom dead center) is called the stroke. Top dead center (TDC) is the highest point of upward travel of the piston within the cylinder. Bottom dead center (BDC) is the lowest point of piston travel within the cylinder. See **Figure 34-2**. Because the crankshaft rotates 180° during each engine stroke, the length of the stroke equals two times the offset between the centerlines of the rod journal and main journal of the crankshaft. More offset results in a longer stroke and less offset results in a shorter stroke. If two engines have equal diameter pistons, the engine with the longer stroke will be able to bring more fuel and air into the cylinder per stroke. If the combustion chambers of the two engines are of equal volume, the engine with the greater stroke should produce more power.

Figure 34-1. In many small engines, the cylinder block and crankcase are integrated in a single casting.

Goodheart-Willcox Publisher

Engine Displacement

Engine displacement is a common way to indicate engine size. *Displacement* refers to the volume of air and fuel the engine can draw into its cylinder(s) during an intake stroke. If the engine has one cylinder, displacement can be defined as the volume displaced by the piston through its movement from TDC to BDC. See **Figure 34-3**. Therefore, it can be calculated as the volume increase in the space above the piston as the piston travels from its highest point of travel (TDC) to its lowest point of travel (BDC).

To calculate the displacement of a single cylinder engine, the following formula could be used:

$(\pi/4) \times D^2 \times L$ = engine displacement

Where

D = cylinder diameter
L = length of stroke

However, to simplify the calculation, the formula can be shortened as follows:

$0.7854 \times D^2 \times L$ = engine displacement.

As an example, imagine a single-cylinder engine with a bore of 3.00″ and a stroke of 3.50″. The calculation would be $0.7854 \times 3.00^2 \times 3.50$. Therefore, the displacement of the engine would be 24.74 cubic inches displaced (CID).

For engines with more than one cylinder, the solution found for a single cylinder would be multiplied by the number of cylinders in the engine:

$0.7854 \times D^2 \times L \times N$ = engine displacement

where

D = cylinder diameter
L = length of stroke
N = number of cylinders in the engine

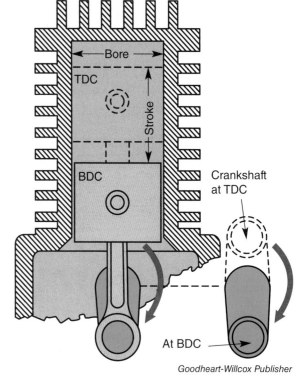

Figure 34-2. Engine bore is the diameter of the piston, and stroke is the distance the piston travels during 180° of crankshaft rotation.

As an example, if the engine described in the previous example had three cylinders instead of one, the engine displacement would be $0.7854 \times 3.00^2 \times 3.50 \times 3$, which equals 74.22 CID.

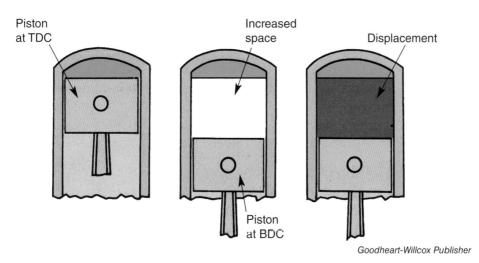

Figure 34-3. Displacement of an engine equals the difference in volume above the piston when the piston is at bottom dead center and when it is at top dead center.

Compression Ratio

Now that you have a basic understanding of engine displacement, you can start to understand compression ratio. *Compression ratio* is the relationship of the total volume of empty space above the piston (in the cylinder and combustion chamber) when the piston is at BDC compared to the volume when the piston is at TDC. See **Figure 34-4**. In other words, the compression ratio equals the volume above the piston when the piston is at BDC *divided by* the volume above the piston when the piston is at TDC. If the volume is 10 cubic inches when the piston is at BDC and 1 cubic inch when the piston is at TDC, the compression ratio of the engine is 10 to 1 (10:1). If the volume with the piston at BDC is 11 cubic inches and the volume with the piston at TDC is 1.25 cubic inches, the compression ratio would be 8.8 to 1 (11 ÷ 1.25). Most small gasoline engines have a compression ratio between 6.5 to 1 and 9.5 to 1.

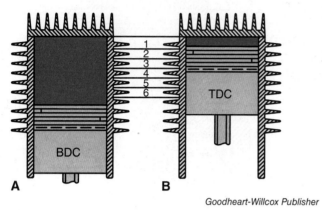

Goodheart-Willcox Publisher

Figure 34-4. Compression ratio is calculated by dividing the volume above the piston when the piston is at bottom dead center by the volume above the piston when it is at top dead center.

Force, Work, and Power

Understanding force, work, and power are necessary to understanding performance in engines. *Force* is an influence (such as a push or pull) that tends to change the motion (speed or direction) of an object or cause stress in a stationary object. For example, the Earth's gravity exerts a force on objects and people, pulling them to the ground. The force that gravity exerts on an object determines its weight. If the object is in the air, the force of gravity acts on the object, moving it toward the ground. However, if the object is resting on the ground, the ground is exerting an equal and opposite force on the object. The condition of equal and opposite forces being in balance is known as equilibrium.

For example, imagine you are standing on the ground and jump up onto a platform. To jump, you must exert force. An action in which a force applied to an object causes a displacement of the object is called *work*. When you land on the platform, equilibrium is reestablished. You become stationary once again, but work has been performed.

In order to calculate the work performed, multiply the force exerted on an object by the distance the object travels:

Work = Force × Distance

Normally, distance is measured in feet and force in pounds. Work is measured in units called foot pounds. A foot pound is the energy needed to raise 1 pound a distance of 1 foot. For example, if an object that weighs 100 pounds is lifted 4 feet, it would take 400 foot pounds to accomplish this work. It takes 100 pounds of force to overcome gravity and lift the object, and that force is applied through a distance of 4 feet.

Power is a measure of the amount of work that can be performed in a given amount of time. Whether the object was lifted in 1 second, 1 minute, or 1 hour, the same amount of work occurred. Power describes the rate at which that work is performed. For example, imagine two employees are stocking boxes of floor tile at a building supply store. The tile comes in a case that contains four individual 50 pound boxes. The first employee can lift and place the four 50 pound boxes on a 4 foot high shelf in 8 seconds. The second employee is a competitive weightlifter, and he can lift the entire case and place it on the shelf in 2 seconds. Although both employees have performed the same amount of work, 800 foot pounds, the first employee has worked at a rate of 100 foot pounds per second and the second employee has worked at a rate of 400 foot pounds per second.

Horsepower and Torque

Most commonly, the unit of horsepower is used when discussing power in engines. *Horsepower* is a unit of power equal to 550 foot pounds per second. Before engines were invented, work was done with the aid of animals, especially horses. Thus, comparisons between the amount of work that a horse could perform and the amount of work that an engine could do in the same amount of time allowed people to understand the engine's capability. Early inventor and engineer James Watt measured the amount of work that a horse could perform in a given time and designated that rate of work as a horsepower (HP). Through experimentation, he found that horses could perform approximately 33,000 foot pounds per minute, or 550 foot pounds per second. He designated that value as one horsepower. To determine the horsepower equivalent of an amount of work performed during a given time, divide the amount of work performed by the amount of time it required. Next, divide that amount by 550 foot pounds per second:

$$(W \div T) \div 550 \text{ foot pounds per second} = \text{horsepower}$$

Where

$$W = \text{work performed}$$
$$T = \text{time required}$$

As an example, remember the weightlifter who lifted a 200 pound case of tile 4 feet in 2 seconds. He performed work at a rate of 400 foot pounds per second. Now, divide that work amount by 550 foot pounds per second to determine the equivalent horsepower:

$$400 \text{ foot pounds per second} \div 550 \text{ foot pounds per second} = 0.73 \text{ HP}$$

Torque is defined as rotating or twisting force. Torque is equal to a force applied across a distance. The force applied is measured in pounds and the distance is measured in feet. Therefore, most measurements of torque are made in standardized units called pound feet. Although we commonly think of force as linear, either pulling or pushing on an object, in engines there is also rotary motion. In calculating work, the force applied must result in motion for work to be accomplished. The same is *not* true when calculating torque. The force applied results in torque, whether the object is moved or not.

Engine Selection

To determine an engine's suitability for any given application, a number of performance and design criteria must be considered. For example, if an engine on a pump is replaced, the engine must have sufficient torque to drive the pump at the pressures required. It also must have the proper horsepower rating to allow the pump to deliver the desired flow rate. Also, the shaft must be oriented in proper direction, either horizontal or vertical. The engine must be the proper size to fit in the space available, and the cooling system must be adequate for the operating conditions. For example, the coolant in a liquid-cooled engine is able to transfer heat more effectively than air. Therefore, a liquid-cooled engine is a better choice for a very hot environment. Another important design consideration is an engine speed control. Certain engines, such as those in a generator, must have a governor system that operates the engine at a specific speed. Others require throttle controls that allow the operator to control the engine speed. See **Figure 34-5**.

The list of things that must be considered when determining an engine's suitability varies by use. However, these common factors should be considered when replacing an engine or selecting an engine for a new project:

- Horsepower rating.
- Torque rating.
- Fuel efficiency.
- Crankshaft diameter, length, and orientation.
- Physical engine size.
- Fuel system (type of carburetor or fuel injection system and fuel pump).
- Engine speed controls (throttle and governor).
- Bearing type.
- Lubrication system type (pressurized or splash).
- Cooling system type (liquid cooled or air cooled).

yanami/Shutterstock.com

Figure 34-5. Consider the criteria that a chainsaw engine must meet. It must be lightweight and compact with a high power-to-weight ratio. The fuel system must be capable of delivering fuel when the engine is held in any position. Likewise, the method of lubrication must be effective when the engine is operated in any position.

Engine Maintenance

Proper maintenance is crucial for the efficient operation and longevity of an engine. Many malfunctions and breakages are caused by a lack of proper maintenance. Maintenance procedures are tasks that must be performed to sustain proper engine performance. They might also be thought of as preventive procedures because they minimize engine wear.

Maintenance schedules and tasks are based on typical use. If engines are operated under extreme conditions or subjected to heavy use, maintenance should be performed more often, and additional maintenance tasks may be required. Many manufacturers list separate maintenance schedules for regular use and for heavy use of engines. Heavy use is normally defined as operation under abnormal conditions, such as extreme heat, dust, heavy load, or extended periods of operation. Any use that is beyond the identified normal use parameters is consideded heavy use. As an example, if a small engine's normal use is based on 20 hours of use per week and the engine is used for 30 hours per week, that use would be categorized as heavy use.

Engines should be kept clean and free of dirt and debris. If dirt and debris surround the engine or engine block, excessive heat can build up in the engine. On air-cooled engines, dirt or debris can form an insulating layer. This layer can prevent heat from transferring from the surface of the engine to the air circulating around the engine. On liquid-cooled engines, the radiator should be kept clean and free of debris. Dirt and debris can block airflow and prevent the transfer of heat from the coolant.

Safety Note

An engine remains hot for a period of time after being operated. The engine's exhaust components get especially hot. To prevent burns, allow the engine to cool before performing maintenance. Some maintenance, such as changing the oil, should be performed with the engine warm, but not hot.

Lubrication System

Engine oil in most four-stroke engines is held in the engine's crankcase. Before each use, the engine should be checked to make sure that it has the proper oil level. Different engines require different methods of checking the oil level. The oil level in some engines is checked with a dip stick. Some engines have a filler plug with a check stick, and some engines have only a simple filler plug, **Figure 34-6**. Dip sticks usually have marks indicating the maximum and minimum acceptable oil levels. The oil level must be kept between these marks. If the engine has a filler plug with a check stick attached, the check stick may have a mark for the correct oil level. The check stick may instead have maximum and minimum marks like those on a dipstick. If the

A
Charles Knowles/Shutterstock.com

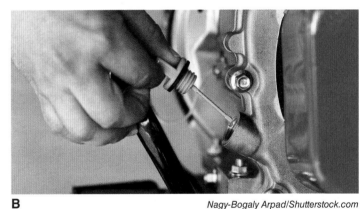
B
Nagy-Bogaly Arpad/Shutterstock.com

C
Goodheart-Willcox Publisher

Figure 34-6. Different engines have different provisions for checking the oil level. A—On this lawn mower, oil level is checked with a dipstick accessible from the top of the engine. B—This engine has a plug with a check stick that is accessed directly at the crankcase. C—This engine has a simple plug at the crankcase. Oil level should be right to the bottom of the plug.

engine has only a filler plug, the oil level should be right at the bottom of the plug bore (do not overfill). See the engine manufacturers' service information for proper methods of checking the oil level and the correct fill levels.

Oil has three main functions in an engine. It cleans, lubricates, and cools surfaces in the engine. The interior of the engine will develop deposits of carbon and other contaminants during normal engine operation due to the combustion of fuel. The engine oil has detergents that clean the interior of the engine. Further, oils hold contaminants in suspension (trapped in the oil) and away from surfaces. Engine oils form a barrier film between moving parts, which helps to prevent the parts from rubbing together. Oil also provides a slick surface on parts that reduces the effects of friction if the parts do rub together.

Because oil traps dirt, debris, and foreign materials and reduces the damaging effect they would otherwise have, it can look dirty even when it can still do its job in the engine. It is not true that oil used in today's engines is necessarily bad if it has a dark color. The reason oil darkens is because it is holding foreign materials. Oil can become dark after only a short time of use. Some of the contaminants contained in the oil could be a lighter color, and the oil could potentially be contaminated without any visible signs. For these reasons, you should change the oil at the recommended times outlined by the engine manufacturer rather than based on the color of the oil.

Manufacturers recommend different grades (viscosities) of oils for different levels of use and temperature ranges. The higher the grade, the more viscous, or thick, the oil will be. **Figure 34-7** illustrates major manufacturers' recommended lubricant grades for four cycle engines. Always follow the manufacturer's specific oil recommendation for your application and the temperature range in which the engine will be used. Single-grade oils (such as SAE 30) perform as 30 weight oil at any temperature. Multigrade oils act as lightweight oil in cold conditions and heavyweight oil in warm temperatures. For example, a 10W30 oil acts like a 10 weight oil in the winter and as a 30 weight oil in the summer.

Fuel and Air System

The fuel system meters the amount of fuel (chemical energy) and air delivered to the combustion chamber. It includes a fuel storage space, referred to as a fuel tank. The fuel tank should be kept free of debris, which could block fuel lines or passages in the carburetor. Fuel should be drained at the end of the operating season, if possible. Fuel breaks down over time. It can cause clogging and corrosion problems if allowed to remain in the fuel system. Fuel stabilizers, which are readily available, reduce the deterioration of fuel, allowing it to be stored for longer periods of time. If the engine will not be used for extended periods between uses, it is a good idea to use fuel treated with a fuel stabilizer. In addition, keep the fuel tank nearly full. This will minimize the amount of water that gets into the fuel due to condensation. Always follow the manufacturer's recommendations regarding the use of a fuel stabilizer.

With the increasing use of ethanol in fuel blends, it is imperative that the engine manufacturer's fuel standards be closely followed. Most small engines are *not* designed to operate on fuel with an ethanol content greater than 10 percent.

Four-Cycle Crankcase Lubrication (Viscosity-Grade) Recommendations of Manufacturers

Manufacturer	Above 40°F	Above 32°F	Below 5°F	Below 0°F	Below −10°F
Briggs and Stratton	SAE 30 or 10W-30		5W-20 or 10W		
Kohler		SAE 30	SAE 10W		5W or 5W-20
Tecumseh		SAE 30		10W-30	

Goodheart-Willcox Publisher

Figures 34-7. Manufacturers' oil viscosity recommendations vary depending on the ambient air temperature.

Using a fuel that contains more than 10 percent ethanol can result in poor performance, deterioration of engine parts, and corrosion. Fortunately, most states require fuel suppliers to identify the maximum amount of ethanol that their fuel contains on the pump. Fuel containing up to 10 percent ethanol is often referred to as E10. If the fuel contains up to 15 percent ethanol, it is often labeled as E15. E85 fuel contains up to 85 percent ethanol.

When mixing fuel and oil for two-stroke engines, make sure that the fuel and oil are mixed at the proper ratio based on the manufacturer's recommendation. This will ensure that the engine performs well and has adequate lubrication to avoid premature wear. Too little oil in the mixture causes excessive engine wear. Too much oil causes oil fouling and poor performance.

The carburetor, which regulates fuel and air, should be kept clean and free of debris. Fuel filters should be replaced at the recommended intervals. Air filters must also be inspected and serviced periodically. Some paper filter elements, **Figure 34-8**, can be cleaned by tapping them on a hard surface. This will knock dust and dirt from the filter element without damaging it. If the filter is dirty to the point that light will not shine through the element, it must be replaced. Do not use compressed air to blow dirt out of a clogged paper air filter element. The air pressure can enlarge the pores in the filter, allowing particles to be drawn into the carburetor when the filter is put back into service.

Some air filters have foam elements that use oil to trap foreign material from entering the carburetor. See **Figure 34-9**. To clean a foam element, it is washed with soap and water and then rinsed thoroughly to remove oil and debris. Once dry, the foam element is lightly coated with oil and reinstalled.

Warren Price Photography/Shutterstock.com

Figure 34-8. A pleated paper filter, such as the one shown here, can be lightly cleaned by tapping it with the pleats down on a hard surface. If it is very dirty, it must be replaced.

Goodheart-Willcox Publisher

Figure 34-9. Foam-type air filters can be cleaned by washing them in a solution of mild detergent and water.

Cooling System

As discussed earlier, many small engines use directed airflow around the outside of the engine to aid in cooling. Any dirt, debris, oil, or other buildup on the external surfaces of the engine will act as a layer of insulation, preventing efficient heat transfer from the engine surfaces to the surrounding air, **Figure 34-10**. The external areas of the engine must be kept clean so the airflow can cool the engine adequately. Also, the engine flywheel and grass screen must be kept free of debris to aid in cooling and proper function.

If the engine has a liquid-cooling system, the parts of that cooling system must be properly maintained to provide adequate engine cooling. The liquid, referred to as coolant, flows through water jackets in the engine where it collects heat. The coolant then flows to the radiator where it transfers that heat to air flowing around the radiator coolant tubes. The radiator must be kept free of debris. Any debris on the surface of the radiator can block airflow and reduce the transfer of heat between the coolant tubes and air. Also, the cooling fan must be periodically inspected to ensure it is undamaged and functioning properly.

As is commonly known, water corrodes many metals. Therefore, other chemicals must be added to the water in a cooling system to inhibit corrosion. These chemicals, which are referred to as antifreeze, also increase the cooling efficiency of the system and prevent the coolant from freezing. The coolant must have the proper ratio of antifreeze to water. Too much antifreeze in the coolant will degrade system performance. Too little antifreeze will fail to provide adequate protection against freezing and corrosion. The coolant's ability to provide the needed protection can be checked with a hydrometer, which measures the coolant's capacity to provide cooling and resist freezing based on the ratio of antifreeze to water.

Goodheart-Willcox Publisher

Figure 34-10. Engines should be kept free of dirt and debris. The buildup of grass clippings and other debris on this engine would act as thermal insulation, causing the engine to quickly overheat.

Service Information

In order for an engine to perform properly and provide the longest possible service life, it must be assembled and maintained as specified by the manufacturer. Some parts, such as spark plugs, must be replaced periodically. Those replacement parts must be adjusted to the manufacturer's specification for the exact engine being serviced. In the example of the spark plug, the correct replacement plug must be selected, its electrode gap must be set to the manufacturer's specification, and it must be tightened to the correct torque. Those specifications are provided in the manufacturer's printed or online service literature.

Similarly, if an engine is disassembled for repair, the internal engine parts must be measured for wear. The manufacturer's service literature lists the maximum allowable wear for all critical engine parts. Any internal parts that are worn beyond the manufacturer's specification must be replaced. If a carburetor is disassembled for service, the manufacturer's specified idle speed must be reset, as should the mixture screw setting(s) on adjustable carburetors.

Often, the manufacturer's service literature will list specifications for more than one engine. The literature also typically lists specifications for different variations of an engine design. For example, if an engine is available with either a carburetor or a fuel injection system, the specifications for each may be found in the same service literature. It is very important that you are able to properly identify the engine you are working on.

The first step in properly identifying an engine is to determine the manufacturer of the engine. Usually, the manufacturer's name is printed or stamped on one or more prominent places on the engine. The engine model may also be printed on the engine somewhere. Often, the manufacturer is printed on the blower shroud and a model number is stamped in one or more places on the engine. Record any numbers or codes printed on the engine. Most small engines have model numbers, serial numbers, and other codes that can be used to correctly identify the engine. See **Figure 34-11**.

When servicing an engine, use only the manufacturer's specifications for that particular model of engine. Relying on generic or general specifications can lead to misdiagnosis of problems, unnecessary repairs, and improper adjustments. Although there are no universally applicable specifications, there are some basic operating standards that all engines must meet in order to function properly:

- Clean, fresh fuel must be available to the carburetor.
- Air must be able to be drawn into the carburetor for atomization of the fuel.
- Spark at the spark plug must be strong and precisely timed.
- High-quality oil should be used in engines and kept at the proper level.
- The engine housing should be clear of debris and foreign materials to aid in cooling.

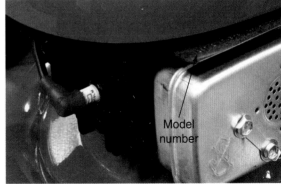

Goodheart-Willcox Publisher

Figure 34-11. The model number of this Briggs and Stratton engine is stamped on a flange just above the muffler.

Engine Troubleshooting

Engine troubleshooting is the systematic diagnosis of engine problems. Because of the complexity of engines, many actions occur very rapidly. Even minor variations in the fit or condition of parts or the sequence or timing of engine events can cause performance issues. By following the sequential steps presented in this chapter, you can correctly diagnose and resolve the majority of failures. Remember that there are many types of engines and procedures vary. Consult the manufacturer for specific steps and procedures to follow. Remember that every engine needs fuel, air, and spark to run. The fuel and air must be mixed in the right ratio, properly atomized, and adequately compressed in the combustion chamber. The ignition spark must be properly timed to ignite the compressed air and fuel mixture at the beginning of the power stroke. The engine relies on the energy released during the power stroke to drive the piston through the remaining strokes of the engine operating cycle.

Troubleshooting engines can be discouraging or exciting. If you understand how all the systems are supposed to work and perform, deciding what is wrong can be easy. It is important that you follow logical steps when trying to determine why something is not working correctly. If you do not, you can easily misdiagnose a problem and make several failed attempts at correcting the problem before stumbling onto the correct diagnosis.

Fundamental troubleshooting steps should be followed in the following sequential order:

1. Make sure the engine is getting fuel.
2. Make sure the engine is getting air.
3. Check for spark.
4. Check the compression.

> **Safety Note**
>
> Before doing any repair, service, or diagnosis of a small engine, always disconnect the spark plug wire and ground it to the engine. This will prevent accidental sparks or starting of the engine. The spark plug wire can be reconnected as needed during the diagnostic process, but it should be immediately disconnected and grounded when no longer needed.

These steps are performed in a logical progression to systematically eliminate potential causes and isolate the true source of the engine malfunction.

Checking the Fuel Supply

If the engine will not start, the first step is always to ensure that a supply of clean, fresh fuel is available for the engine. Begin by making sure there is clean, fresh fuel in the tank. Then, place a metal can or similar metal container under the fuel supply hose where it connects to the carburetor. Next, carefully disconnect the fuel supply line at the carburetor. See **Figure 34-12**. On gravity fed systems, fuel should flow out of the hose and into the catch basin. If no fuel comes out of the hose, there is a blockage somewhere in the fuel supply line. The

Goodheart-Willcox Publisher

Figure 34-12. To check the fuel supply, place a container under the fuel line, and then remove the fuel line at the carburetor. If no fuel flows out, work backward toward the tank, disconnecting the fuel line at each junction until fuel flows freely.

> ## SAE Connection
> ### Small Engine Troubleshooting and Repair
>
> Small engines are an everyday part of American life. From weed trimmers and lawn mowers to recreational vehicles, they are found everywhere. We take for granted that they will start and run when we need them, and when they malfunction, we generally look for a repair person. Are you curious about how small engines work? Do you think you would enjoy troubleshooting and repairing them? If you answered yes to these questions, you may want to consider a supervised agricultural experience (SAE) program that focuses on small engine troubleshooting and repair. All you need to begin is a place to work and a few simple tools. You might start by fixing old mowers that do not run and then reselling them for a profit. As your skill increases, you will probably have no shortage of customers asking you to perform maintenance and repair tasks on their small engine–operated equipment.
>
> Discuss the potential of starting a small engine repair business with your agriculture teacher or find a local repair shop looking for part-time help. Either way, you will gain knowledge and skills that will be useful as a future homeowner and in the completion of agricultural mechanics proficiency applications.

problem could be a clogged fuel filter or screen, a closed or clogged fuel shut-off valve (petcock), and even a deteriorated and clogged fuel line. Trace the fuel supply back to the tank, disconnecting the fuel line at each junction until fuel flows from the hose. By noting where fuel flow is restored, you can determine the component that is causing the blockage.

If the engine has a fuel pump, the process is essentially the same. However, you will need to have an assistant crank the engine after the fuel line is disconnected at each junction. In a pressurized system, fuel will only be pumped when the engine is turning over. If the engine is equipped with a vacuum-operated fuel pump, ensure that vacuum line is delivering vacuum to the fuel pump.

> **Safety Note**
> Keep all sparks and open flames away from the work area when work is being performed on the fuel system.

Checking the Air Supply

Once you verify that fuel is being delivered to the carburetor, you should check the condition of the air filter. A clogged air filter can prevent the engine from getting enough oxygen for combustion to occur. If the air filter is relatively clean, the next step in the troubleshooting process is to determine if there is spark at the spark plug.

Checking the Spark

A spark testing tool can be used to check for spark. Begin by attaching the spark tester between the spark plug wire and the top of the spark plug, **Figure 34-13**. The engine is then turned over, as if trying to start. If a spark appears in the tester as the engine is turned over, the ignition system is working properly. If no

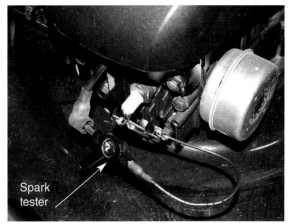

Goodheart-Willcox Publisher

Figure 34-13. Placing the spark tester between the spark plug wire and the tip of the spark plug will test the operation of the entire ignition system, including the spark plug.

Goodheart-Willcox Publisher

Figure 34-14. If no spark is seen in the previous test, install the spark tester between the spark plug wire and a good ground, such as a cooling fin. If a spark is visible when the engine turns over, the problem is with the spark plug.

spark appears, the system can be furthered diagnosed by attaching the spark tester between the spark plug wire and a good ground, usually a cooling fin on the cylinder head. See **Figure 34-14**. Again, the engine is turned over. A spark should be visible within the spark testing tool. If a spark appears with the tester attached in this configuration, the spark plug is the source of the problem. Properly gap and install a new plug, and then try to start the engine.

If no spark appears in the spark tester when it is connected between the plug wire and a ground, the problem is somewhere else in the ignition system. Follow these steps to check for other causes of the problem:

- Check the air gap between the flywheel and ignition assembly.
- Ensure that the flywheel key is present and not sheared.
- Ensure that there are no safety interlocks, such as operator presence switches or low oil shutoff switches, or kill switches interfering with ignition system operation.
- Check for ignition module failure.

Most engines have safety interlock switches that prevent the engine ignition system from generating spark unless certain operational conditions are met. One common safety interlock switch is a low-oil switch, which disables the ignition system if the engine's oil level is low. When diagnosing an engine's ignition system, refer to the manufacturer's service literature to determine what safety interlocks are installed and how to test them.

If the fuel and air are getting to the carburetor and the engine is producing a strong, properly timed spark, the problem preventing the engine from starting is either a mechanical problem or a carburetor problem.

Checking the Compression

Mechanical problems within the engine can allow the air fuel mixture to leak out of the cylinder rather than being compressed as the piston moves upward during the compression stroke. As a result, the low-pressure air and fuel mixture left in the combustion chamber at the end of the compression stroke will not release enough energy to maintain proper engine operation when it is ignited.

Engine compression can be checked using a compression gauge, compression tester, or bump test. A compression gauge is essentially a pressure gauge that is installed in the spark plug opening in the cylinder head. See **Figure 34-15**. It measures the maximum pressure achieved during the compression stroke as the engine is turned over. Because cylinder pressure is directly related to compression, comparing the measured cylinder pressure to the manufacturer's specification will reveal whether the cylinder has adequate compression.

Many factors affect compression readings. Manufacturers' compression specifications are usually given for an engine that is at normal operating temperature. A cold engine will have a lower compression reading. Also, if the engine is equipped with an automatic compression release device for easy starting, the engine will need to be rotated backward (opposite normal rotation) in order to get an accurate measurement.

Another method of checking compression is to use a differential compression tester to measure cylinder leakage. To use this device, the piston is positioned at top dead center on the compression stroke and then locked in place so it cannot move. The tester is connected to a supply of compressed air at one end, and the hose at the other end is threaded into the spark plug hole. Regulated compressed air is pumped into the cylinder. One pressure gauge on the tester measures the pressure of the compressed air. A second gauge

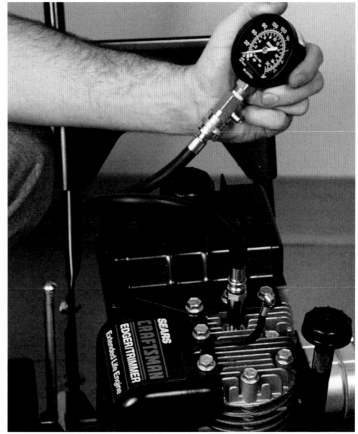

Goodheart-Willcox Publisher

Figure 34-15. A compression gauge is installed in the spark plug hole and used to measure the pressure developed in the cylinder as the engine is turned over.

on the tester measures the air pressure at the cylinder. The pressure drop between the two gauges indicates the amount of leakage taking place inside the cylinder. The technician can listen for the sound of escaping air at various points on the engine to determine where the leaks are occurring. If escaping air is heard at the exhaust, the exhaust valve is bad. If escaping air is heard at the intake, the intake valve is bad. If escaping air is heard in the crankcase, the piston rings or cylinder wall are worn or damaged.

If a compression gauge or differential compression tester is not available, you can perform a simple bump test to evaluate compression. For this test, give the engine flywheel a hard spin in the direction opposite normal rotation. The piston should rebound sharply during the compression stroke. If the flywheel does not rebound after being spun, the engine does not have enough compression to run.

The following are common causes of compression loss:
- Loose spark plug or missing compression washer.
- Loose cylinder head bolts.
- Blown head gasket.
- Bad valve seat or face clearance.

- Insufficient lifter/tappet clearance.
- Warped cylinder head.
- Worn cylinder walls.
- Worn or broken piston rings.

Checking the Carburetor

Perform a quick check of the throttle and choke controls to ensure that they move the throttle and choke valve plates into the proper positions. Then, make sure the governor linkages are not binding.

If no problems were found with the engine throttle, choke, and governor controls, remove the carburetor from the engine. Follow the manufacturer's instructions to completely disassemble, clean, and inspect the carburetor. See **Figure 34-16**. Replace any worn or damaged parts as well as any gaskets and diaphragms. Inspect the float and needle valve closely, and replace them if there is any sign of damage, corrosion, or wear. Carefully follow the manufacturer's instructions to reassemble and adjust the carburetor.

Goodheart-Willcox Publisher

Figure 34-16. This typical float-type carburetor has been disassembled for inspection and cleaning. Carburetors contain many small parts; work carefully and keep track of the way things are disassembled.

CHAPTER 34 Review and Assessment

Summary

- Engine performance refers to an engine's suitability for different tasks or applications.
- Engine displacement is a common way to indicate engine size. Displacement refers to the volume of air and fuel the engine can draw into its cylinder(s) during an intake stroke.
- Compression ratio compares the volume of space above the piston at its lowest point of travel to the volume above the piston when it is at its highest point.
- Force refers to the energy used to push or pull on an object. Force that results in motion is called work. Power is the rate at which work is performed. Horsepower is a unit of power equal to 550 foot pounds per second. Torque is a twisting or rotating motion—the angular equivalent of linear force.
- Some factors that should be considered when selecting an engine include the horsepower rating, torque rating, fuel efficiency, crankshaft orientation, and engine size.
- Proper maintenance is crucial for the efficient operation and longevity of an engine. Maintenance schedules and tasks are based on typical use.
- The lubrication system, fuel system, and cooling system must be maintained and used according to manufacturer's specifications for proper engine operation.
- A small engine may be air cooled or liquid cooled. Both types of cooling systems must be properly maintained.
- To ensure proper operation, engines must be serviced and maintained in accordance with the manufacturer's service information. Service information includes important specifications for a given engine model. These specifications typically include wear tolerances, torque values, and gap settings.
- Engine troubleshooting is the systematic diagnosis of engine problems. Troubleshooting steps include checking the fuel, air supply, spark, compression, and carburetor.

Words to Know ↗

Match the key terms from the chapter to the correct definition.

A. bore
B. compression ratio
C. displacement
D. force
E. horsepower
F. power
G. torque
H. work

1. The diameter of a hole, such as the distance measured across the cylinder in an engine.
2. An action in which a force applied to an object causes a displacement of the object.
3. An influence or energy that tends to change the motion of an object or cause stress in a stationary object.

4. The relationship of the total volume of space above the piston when the piston is at bottom dead center compared to the total volume of space above the piston when the piston is at top dead center.
5. A measure of the amount of work that can be performed in a given amount of time.
6. The volume of air and fuel the engine can draw into its cylinder(s) during an intake stroke.
7. A twisting or rotating force.
8. A unit of power equal to 550 foot pounds per second (or the power of an engine expressed in these terms).

Know and Understand

Answer the following questions using the information provided in this chapter.

1. What is a piston stroke and how does it relate to engine performance?
2. What formula can be used to calculate the displacement of a single cylinder engine?
3. Most small gasoline engines have a compression ratio within what range?
4. How is the amount of work performed calculated?
5. How is the horsepower equivalent of an amount of work performed in a given time determined?
6. What are some factors that should be considered when replacing an engine or selecting an engine for a new project?
7. What are some problems that may occur if an engines is not kept clean and free of dirt and debris?
8. What are the main functions oil performs in an engine?
9. How does temperature affect the action of multigrade oils?
10. What is the purpose of a fuel stabilizer?
11. Most small engines can be damaged by using fuel with ethanol content greater than what percent?
12. Why are chemicals added to water that is used in an engine cooling system?
13. Why is it important that you properly identify the engine (manufacturer and model number) before you begin working on the engine?
14. What are some basic operating standards that all engines must meet in order to function properly?
15. What are the fundamental steps that should be followed when troubleshooting engine problems?
16. What problem can a clogged air filter cause?
17. What are safety interlock switches and how do these switches relate to engine maintenance?
18. If an engine is equipped with an automatic compression release device for easy starting, what must be done to get an accurate compression reading?
19. What are three ways in which engine compression can be checked?
20. What are common causes of engine compression loss?

STEM Connections and Academic Activities

1. **Science.** Devise and conduct experiments to determine the effects that various contaminants, including water, fuel, and carbon, have on oil. Devise a means of measuring the outcomes so you can test the repeatability of your results.
2. **Technology.** Develop a troubleshooting guide for small engines. The guide should be sequential. If a step does not reveal the cause of the problem, the user should be directed to the next step in the troubleshooting process.
3. **Engineering.** Build a cylinder and piston out of paper or cardboard. Also, build two combustion chambers (with different sizes) that will fit on top of the cylinder you built. Calculate the compression ratio of each design (cylinder and combustion chamber combination) and record the ratios found. For each design, imagine that the compressed air and fuel mixture is prematurely ignited 85 percent of the way through the compression stroke. What would the effect be, and how would the effect differ for the two designs?
4. **Math.** A portable generator with a four-stroke engine running at 1,800 RPM burns 1 gallon of gas in an hour and five minutes. Determine the average amount of fuel, in ounces, that the engine burns during each combustion event.
5. **Social Science.** Describe the impact lawn mowers have had on the aesthetic values of communities in the United States.
6. **Language Arts.** Find a manufacturer's manual for an engine. Use the information in the manual to write a condensed inspection and maintenance guide for that engine. Use your own words to paraphrase the appropriate information from the manual.

Thinking Critically

1. You have bought your first home that has a small yard. You will buy your first push mower. What are some factors you should consider regarding the engine options when purchasing this mower?
2. As an agriculturalist, you have many pieces of equipment that have small gasoline engines. What are some steps you should undertake to minimize the annual expenses required to keep them operating properly?
3. You are replacing an engine on a piece of equipment that you use often in your work. Describe the equipment, and determine the optimum engine horsepower and torque of the replacement engine.
4. The small gasoline engine you used when you lived in the coastal region does not perform well when you moved to a higher elevation in the mountains. What could cause this problem?
5. What are some maintenance procedures you should perform before storing an engine for the winter?

CHAPTER 35
Drive Trains and Power Systems

Chapter Outcomes
After studying this chapter, you will be able to:
- Name the components of a power train.
- Understand the differences among transmission types.
- Know how a power train transmits power to output sources.
- Discuss the ways power can be harnessed for applications other than driving wheels.
- Know how to calculate torque and RPM.

Words to Know 📱

automatic transmission	final drive	limited-slip differential	pulley
belt	gear	output shaft	sprocket
clutch	gear ratio	power take-off (PTO)	transmission
differential	hydraulics	power train	
driveline	hydrostatic		

Before You Read
As you read the chapter, put sticky notes next to the sections where you have questions. Write your questions on the sticky notes. Discuss the questions with your classmates or teacher.

While studying this chapter, look for the activity icon to:

- **Practice** vocabulary terms with e-flash cards and matching activities.
- **Expand** learning with interactive activities.
- **Reinforce** what you learn by completing the end-of-chapter questions.

www.g-wlearning.com/agriculture

Elena Elisseeva/Shutterstock.com

There are many ways for power to be transferred in most power systems. The power may be transferred by friction or gears, or through hydraulic fluids. Most clutches use friction to transmit power. A belt drive is an example of a friction system commonly used to transmit power. A manual transmission is an example of a gear-to-gear system for transmitting power. Fluid power is used in applications such as hydrostatic systems to turn a final output drive. No matter how power is transferred (friction, gears, or fluid), the power train transfers power and motion and converts it to usable force or motion for final output.

Power Trains

Power trains transmit power to output sources such as drive wheels and power take-offs (PTOs). The power train:
- Connects and disconnects the power source.
- Determines speed: torque ratios.
- Equalizes drive wheel power.
- Connects and disconnects power to the PTO.
- Allows reverse travel.

A typical power train includes a clutch, transmission, differential, and final drive. The clutch connects or disconnects power to the transmission and to the final drive. The transmission gives operators the ability to select directional travel (forward or reverse) and operational speed through the transmission's gear ratios. The final drive is the part of a power transmission system between the drive shaft and the differential. It directs power to the drive wheels and provides a final gear ratio. Additionally, a power take-off (PTO) can transfer power to attached or towed equipment.

Clutches

In agricultural applications, a clutch is the most common means of coupling and decoupling an engine or motor from a transmission. A *clutch* is a friction disc assembly that engages and disengages the engine flywheel to transfer power through gear sets. Power from the engine is transferred through the engaged clutch to the input shaft of the transmission. The input shaft is the shaft that carries the driving gear, which applies the power.

Think of the clutch as a switch. In most cases, clutches are either mechanical or hydraulic. Clutches operate by depressing or releasing a foot pedal located in the cab of the equipment. On mechanical clutches, the pedal is connected to linkages, cables, and springs that engage or disengage the clutch from the power source. A hydraulic clutch uses fluid power to aid in its engagement or disengagement. A clutch can also be either a dry clutch or a wet clutch. Both the wet and dry clutches operate in essentially the same way, but the wet clutch is immersed in a fluid to aid in cooling.

The main components of a typical clutch system are depicted in **Figure 35-1**. The flywheel is bolted to a flange on the engine crankshaft. The pressure plate is bolted to the flywheel. The friction disc is positioned between the flywheel and pressure plate and is splined to the input shaft of the transmission. When the clutch pedal is released, spring pressure from springs in the pressure plate keeps all three components in contact with each other. The friction material on the friction disc grips the flywheel and pressure plate, creating a strong mechanical connection between the engine and transmission. When the friction disc and flywheel are connected by friction, the power train is transmitting the turning force of the power source to the final drive.

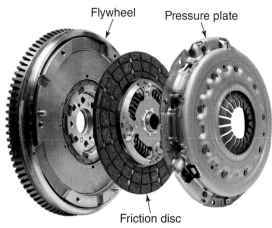

Figure 35-1. The main components of a clutch assembly are the pressure plate, friction disc, and flywheel. When the clutch is engaged, the friction disc turns with the flywheel and pressure plate. When the clutch is disengaged, the friction disc is disconnected from flywheel and pressure plate.

The clutch assembly also includes a clutch fork that is moved by the operator by engaging a pedal. This action applies pressure to the pressure plate through the throwout bearing. When the pedal is pressed, the pressure from the throwout bearing causes the pressure plate to retract away from the friction disc, and the friction between the flywheel and clutch is released. When the pedal is released, springs inside the pressure plate push the pressure plate, friction disc, and flywheel back into contact with each other. The friction connection between the clutch and flywheel is restored, and the rotating motion of the power source is transmitted through the system to the final drive.

Transmissions

The *transmission* consists of many gears that allow the power source to propel the final drive wheels in multiple forward and reverse speeds. A *gear* is a disc- or cone-shaped object with a series of teeth cut into its edge. These teeth mate with the teeth of another gear or part that is in contact with the first gear. The size and number of teeth on the meshing gears determines the relative speeds obtained.

A transmission has different arrangements of gears of varying sizes. The exact path of power through the transmission can be changed by the operator, which allows the operator to select appropriate speeds for a specific application. If, for example, more torque is needed at the rear drive wheels, the operator can change gear ratios in the transmission to have the power source turn the drive wheels at a slower speed. Different gear selections will increase or decrease output shaft speed, increase or decrease torque, or even change the direction of travel.

Gear drives are engineered to transmit power efficiently. When different gear sizes (outside diameter) are chosen, adjustments to speed and torque are the result. For example, if a small driver gear (attached more directly to the power source) is used with a larger driven gear, more torque is produced but with less output speed (RPM). If, however, a large driver gear is used with a small driven gear, then more speed (RPM) is produced at the final gear, but with less torque.

The difference between the two gears is called the *gear ratio*. A gear ratio is the number of teeth on the drive gear compared to the number of teeth on the driven gear (pinion gear). For example, if you have a driven gear that has 20 teeth (or 20″) and a driven gear that is 10 teeth (or 10″) then the gear ratio would be 2:1. If the sizes or teeth were reversed, the gear ratio would be 1:2. Output speed (RPM) is also proportionally increased or decreased as a result of the gear ratio selection.

Consider the gear combination shown in **Figure 35-2**. The large gear on the right has 40 teeth, and the small gear on the left has 16 teeth. If the large gear is used to drive the small gear, the gear ratio would be 2.5:1. The small gear would make 2.5 revolutions for every 1 revolution of the large gear, but it would have only 40% of the torque of the large gear. If the small gear is used to drive the large gear, the gear ratio would be 1:2.5. It would take 2.5 revolutions of the small gear to produce 1 revolution of the large gear. However, the torque of the large gear would be 2.5 times greater.

It is also worth noting that two gears in direct contact with each other rotate in opposite directions. Therefore, if there are two gears in a power transmission path, the gears rotate in opposite directions, **Figure 35-3A**. If there are three gears, the gears on the end rotate in the same direction, but the gear in the middle rotates in the opposite direction, **Figure 35-3B**. If there are four gears in the power transmission path, the two outer gears rotate in opposite directions, and the two inner gears rotate in opposite directions, **Figure 35-3C**. This pattern of alternating rotations holds true regardless of the number of gears in the path, and it is this principle that gives transmissions the ability to change vehicle direction in addition to speed and torque.

Fouad A. Saad/Shutterstock.com

Figure 35-2. The large gear on the right has 40 teeth, and the small gear on the left has 16 teeth. Depending on which gear is driving the other, the gear ratio will be either 2.5:1 or 1:2.5.

Automatic and manual transmissions share the same basic purposes. They connect and disconnect the engine from the drive wheels, multiply engine torque as required by load and vehicle speed, and offer a way to efficiently reverse the direction of the power flow. Three basic types of transmissions are used in agricultural mechanics:

- Automatic.
- Hydrostatic.
- Mechanical.

Most people are familiar with the concept of automatic transmissions, because they are the overwhelming choice for automobiles. An *automatic transmission* is one in which gear ratio changes are activated without direct operator input. Based on the demand for power and speed, a pressurized hydraulic fluid causes valves to move that redirect the hydraulic fluid to

Chapter 35 Drive Trains and Power Systems 943

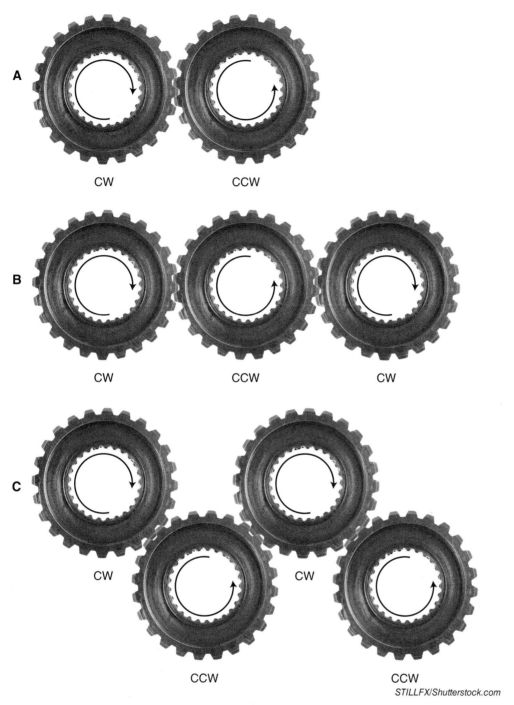

Figure 35-3. Two gears that are meshed with each other will always spin in opposite directions. A—In two-gear systems, both gears rotate in opposite directions. B—In three-gear systems, the two outer gears spin in the same direction, and the center gear spins in the opposite direction. C—This alternating pattern of rotation is consistent, no matter how many gears are in the path, or the point of contact between gear pairs.

initiate gear ratio selection. Most automatic transmissions using this type of design are referred to as *hydrodynamic-drive transmissions*. In a hydrodynamic-drive transmission, fluid is continuously flowing, whether or not it is driving the vehicle.

Hydrostatic is defined as the ability to transfer power through fluids at rest or under pressure. In a hydrostatic transmission, mechanical power is converted to hydraulic power through a pump/motor combination that modulates fluid pressure. Because the fluid is not compressible, energy transfer is accomplished through the pump/motor in this closed system. Speed is controlled through fluid flow rates. The direction of travel (forward and reverse) is manipulated using internal valves to direct the flow of the fluid. In a hydrostatic drive transmission, fluid flows only when it is driving the vehicle. The pump and motor are contained in a single housing.

The hydraulic assist transmission also uses fluid pressure to change gear ratios. This type of transmission uses fluid to shift the power train's gears without interrupting power flow. The clutch, friction discs, and plates are located internally in this type of transmission. These internal parts control fluid power to shift gears, which determines the required speed or torque.

The mechanical transmission, also known as a *standard transmission*, is a manual type of transmission and has been used for power and speed selection for longer than any other type of transmission. A hand-operated selector lever is used to mesh different gears, allowing the operator's choice of gear ratios to be transmitted through the transmission. Each gear ratio produces a different speed and torque output. A clutch is used to decouple the power source from the transmission so that the gear selection can occur while the power source is running. Once the gear selection is made, the clutch is manually reengaged and power is transmitted through the mechanical transmission toward the final drive.

Each transmission type has advantages and disadvantages, depending on specific applications. Therefore, machine designers and operators choose a transmission type that is appropriate for the kind of work the machine is intended to do. Although it is the oldest design, the mechanical transmission has proven to be very reliable and remains in widespread use. Manual transmissions are the most efficient in terms of power transfer. However, some energy is lost through this design. Because it uses gears, there is minimum loss of energy through power conversion. Where does it lose efficiency? If you thought about the friction clutch system, you would be correct.

The hydraulic assist transmission was designed to be more reliable than an automatic, and for certain applications it was supposed to be more economical. Unfortunately, in many instances, it has not proven to be as versatile as operators need it to be.

The automatic has allowed operators to transmit power more easily, but its ability to handle the larger horsepower power plants typically found in agricultural applications has increased its cost. Automatic transmissions are the least efficient type of transmission. The use of multiple clutches is the source of their inherent inefficiency. More clutches equals more places to dissipate power, but in some applications the ease of use still gives the automatic transmission the edge.

Hydrostatic transmissions, such as the drive systems in some garden tractors, have advantages and disadvantages as well. However, they lack the ability to coast, and they are often more expensive than alternative transmission options.

Driveline

A *driveline* connects the transmission to the final drive. There are many types of drivelines, but most often it is just a straight shaft or metal tube. Connectors at both ends of the driveline link the transmission to the differential. It is important to understand that the transmission and the driveline are on different planes; they are not directly in line with each other. Also, the driveline moves due to suspension travel and therefore requires a connector flexible enough to move up or down while rotating. The elegant engineering solution is the universal joint or U-joint. See **Figure 35-4**. The U-joint is in the form of an X with four equidistant points. These points have caps with internal bearings. The caps allow the U-joint to pivot and move, even while rotating. This design allows for dissimilar mating, so the connection to the driveline does not have to be perfectly centered. In addition, the drive shaft usually has a splined section, either in the propeller tube or on one of the yokes. This splined section allows the drive shaft to slide forward and backward with suspension travel.

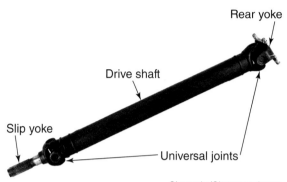

Figure 35-4. This drive shaft assembly is made up of two yokes, two universal joints, and a drive shaft. A splined transmission yoke, called a *slip yoke*, allows the end of the assembly to slide into and out of the transmission as needed.

Final Drives

Final drives are the mechanism(s) transmitting power to the drive wheels. The *differential* transmits power from the transmission to the final drive. It is an assembly of gears used to provide power to the drive axles and allow them to rotate at different speeds as necessary.

Connecting the differential to the transmission can be accomplished with or without a driveline. The driveline creates distance between the two parts of the drive train.

One of the most important characteristics of the differential is that it allows the drive wheels to turn at different speeds, **Figure 35-5**. This is very important because if an operator turns the steering wheel to make a turn, and the drive wheels turn at exactly the same

Figure 35-5. In a differential assembly, the drive gear rotates the ring gear. When the vehicle is moving straight ahead, the pinion gears revolve without rotating on their shafts. This causes the end gears to rotate at the same rate as the ring gear. When one of the drive axles encounters resistance, such as in cornering, that drive axle slows down. The pinion gears begin rotating on their shafts to allow the axles to spin at different speeds while continuing to be driven by the ring gear.

rotation and speed, the outer drive wheel, which needs to travel a further distance, will skid and place extreme stress on the axle and drive gears.

To solve this problem, the differential uses a ring gear and pinion drive gear. The pinion drive gear receives power from the transmission and transmits it to the ring gear. Two pinion (spider) gears are attached to the ring gear. Power is transmitted to the axle end gears through these spider gears. When the car is moving straight ahead, the spider gears revolve with the ring gear, but do not rotate on their shafts. When the axles must turn at different speeds, such as during cornering or if a wheel is stuck, the spider gears can rotate on their shafts. This drives one axle at the same speed as the ring gear, but allows the other axle to turn at a different speed, or even remain stationary, without damaging the differential.

This type of differential is referred to as a *standard* or *open differential*. Although it solves the mismatched wheel speed problem, it is not without faults. The main fault is demonstrated in loss of traction situations, such as mud or snow, in which one of the wheels begins to slip. The slipping wheel will get all of the power, while the contacting wheel does not get enough power and remains stationary. This is the opposite of what is desired.

When slippage of the drive wheel occurs, a mechanical device called a ***limited-slip differential***, or locking differential, prevents excessive power from being devoted to one wheel. Instead, it keeps both wheels in powered rotation.

The differential is normally where the final gear reduction of the power train takes place. This gear reduction is accomplished through the ratio of the pinion and ring gears in the differential. Wheel and tire size affect the final drive gear ratio. However, equipment manufacturers take this into account and specify tire and wheel size based on the intended application.

AgEd Connection Drive Train Service and Repair

Agricultural equipment is made up of many components. Drive trains take the power generated by an engine or motor and convert it into useful motion that performs work. Drive train components and others are potentially a part of the agricultural technology and mechanical systems career development event (CDE). As you study this chapter, ask your teacher about the CDE and how you could be a part of the team.

Also, your supervised agricultural experience (SAE) program may include the use of tractors or power take-off (PTO) driven equipment that would make the material in this chapter useful to know. As you complete your proficiency applications for your SAE, remember to keep up with the new skills that you learn in agriculture class and record them in your application. Repair shops often have employees who specialize in certain components like transmissions. If this chapter is of interest to you, a career in drive train repair would be worth exploring.

Dejan Dundjerski/Shutterstock.com

Power Systems

The power system may be used for many applications other than driving wheels/tires. In agricultural practices, operators can use the power generated by the equipment to accomplish tasks such as running a wood chipper, generator, mixer, tiller, or any number of agricultural implements. The power take-off is an extension of the rotational power of an engine's crankshaft. It allows an engine to transmit its power to an application and is the most commonly used agricultural power system. It transmits this power through a splined output shaft to a corresponding input shaft on the application end. Belts and chains can also be used to transmit power to the drive wheels or to accessory attachments. Again, an output shaft from the power source provides the rotation to drive the belts or chains.

Another power system used in agricultural applications is hydraulic power. Fluid power is generated by a pump driven by the power source. Pressurized fluid from the pump is routed to hydraulic actuators, such as hydraulic motors or cylinders. These actuators provide the power to operate the equipment.

Belts and Chains

A belt drive may also be a part of the power train, especially in low-power applications. A belt drive is a system of two or more pulleys connected by belts. This is called a *friction connection* because friction keeps the belt from slipping on the pulley. The friction between the belt and the pulley surface allows the transfer of forces from one pulley to another pulley, **Figure 35-6**. A *belt* is a flexible loop designed to connect two or more pulleys to transfer rotational power. A *pulley* is a device connected to the power source. It is one of the most simple, yet useful, machines ever devised. It allows the direction of a force to be changed. When used in multiple sets, pulleys create a mechanical advantage.

Subsequent pulleys are located on output devices or in an idler. The idler is a pulley that is used when there is no need for power to be transferred. In many applications, idler or tensioner pulleys are used to aid in the friction connection. Tensioner pulleys help reduce friction loss in the pulley system.

Often in heavier load applications, multiple belts and pulleys are used per shaft to reduce slippage in the friction connection. Belts can be made from rubber (most common), synthetics, leather, or other flexible materials that have good tensile strength. Connections must be in proper alignment and have the correct tension for efficient transfer of power. Horsepower and torque in belt systems can be calculated as shown in **Figure 35-7**, assuming 100% efficiency of the friction connection.

Grandpa/Shutterstock.com

Figure 35-6. This belt drive system consists of two pulleys with multi-groove sheaves and three parallel belts. The use of multiple belts reduces the risk of slippage under heavy load.

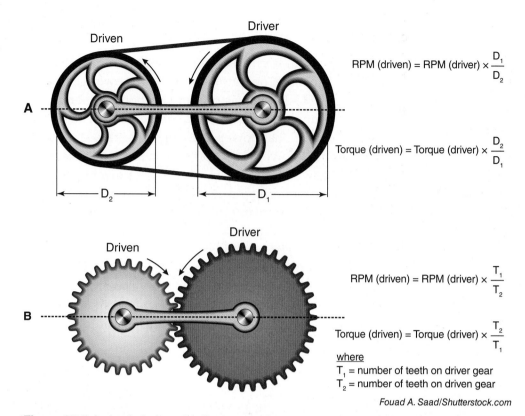

Figure 35-7. A simple belt-and-pulley system is compared to a gear system. A—The ratio between pulley diameters determines the torque and speed. Both the driver and driven pulleys rotate in the same direction. B—In a gear system, the number of teeth on the driver and driven gears determines the speed and torque. The gears rotate in opposite directions.

There will always be a driver pulley, which is the pulley connected to the power source. The other pulleys are the driven pulleys. If the driver pulley is larger than a driven pulley, speed of the driven pulley increases and torque decreases. Conversely, if the driver pulley is smaller than the driven pulley, speed of the driven pulley decreases and torque increases. To calculate revolutions per minute (RPM), the RPM (driven) = RPM (driver) × (driver diameter ÷ driven diameter). To calculate torque, the formula is Torque (driven) = Torque (driver) × (driven diameter ÷ (driver diameter).

Chains can also be used to connect a power source to an output. Chain drive systems are more efficient than belt drive systems because they have little or no slip. This advantage can also be considered a disadvantage, because if something impairs the movement of the output device, breakage of the drive system or connected parts could occur. For example, a forage cutting machine that operated on a belt drive system could experience drive system slippage in heavy crop conditions. The machine would not cut the crop evenly because of drive belt-to-pulley friction loss. If the forage cutting machine were chain-driven, then theoretically, it would cut the crop evenly in heavy conditions because there would be no slipping belt to disrupt the cutting action. Unfortunately, if the cutting knives encountered a stump or a stone, the chain would not slip, and damage to the machine could occur, resulting in expensive repairs.

Chain drives use profiled wheels that are similar to gears, but are called *sprockets*. A **sprocket** is a toothed wheel with evenly spaced teeth located around its perimeter. A sprocket does not mesh directly with another sprocket like a gear does. The difference between sprockets and pulleys is that sprockets have teeth and pulleys are smooth. Sprockets are connected from the output to the power source by a chain, **Figure 35-8**. Bicycle sprockets connected by a chain are a familiar example. The chain's links are designed to allow the sprocket to mesh with the chain. Chains have many interconnected links that can be added or subtracted to change the chain's length.

Chain drives are used in various agricultural applications due to their long service life and minimal maintenance requirements. They are best used in applications in which no slip is expected and positive connections between the power source and output devices are desired.

enterphoto/Shutterstock.com

Figure 35-8. A simple chain drive system has one sprocket fitted to the power source and another fitted to an output shaft. Speed and torque multiplication are calculated as they are for gear systems, by comparing the number of teeth on the driver sprocket and driven sprocket. However, the sprockets rotate in the same direction, like the pulleys in a belt drive system, rather than in opposite directions, like adjacent gears in a gear train.

Power Take-Off (PTO)

A *power take-off (PTO)* is a gear-driven shaft that normally extends past the differential on the rear of equipment. It is driven by the power source and has a splined output shaft that turns with great torque at either 540 or 1000 RPM. The *output shaft* provides a connection point for an implement to be attached.

Most agricultural implements and attachments that need an external power source to operate will couple to this power supply through a PTO drive shaft, **Figure 35-9**. The PTO is an excellent way to transfer power from the source supply to an added implement. PTOs are used in many applications, and those working around them must be extremely cautious because of the inherent potential for *extreme bodily harm*. A typical shaft connected to the PTO can rotate nine times per second. Any contact between the operator and the PTO shaft is a potential for disaster.

Safety Note

Tractor accidents have been identified as the leading cause of deaths and disabling injuries on farms (National Coalition for Agricultural Safety and Health, 1988). Injuries that happen when working with machinery include entanglements in belts, chains, gears, power take-offs at the tractor and along the PTO drive, and crop gathering and moving mechanisms.

It is imperative that operators make sure that safety shields are in place and in good working order before making connections between the power source

PTO drive shaft   *smerka/Shutterstock.com*

Figure 35-9. A PTO drive shaft connects the PTO output shaft to the implement. The drive shaft is splined so it can slide back and forth on the PTO output shaft as the implement travels over uneven terrain. A universal joint at the implement end of the drive shaft allows the shaft to change positions as the tractor and implement turn.

Safety Note

Most incidents involving PTO stubs are caused by clothing caught by an engaged but unguarded PTO stub. Make it a specific point to keep all components of PTO systems shielded and guarded.

and implement or attachment. The power source must be turned off before connecting anything to the PTO.

Hydrostatic Drives

A hydrostatic drive uses oil pressure and flow to transmit power to the drive wheels. *Hydraulics* is the use of a liquid to perform work, such as extending a cylinder to lift a heavy load. A hydrostatic drive is a hydraulic system that consists of a pump, motors, valves, and oil lines, which replace the traditional clutch and transmission. The engine's mechanical power is converted to hydraulic power by a hydraulic pump. The hydraulic motors then take the hydraulic power and convert it to mechanical power to drive the wheels.

A hydrostatic drive is not the prime mover in the way an engine powers an automobile. Rather, it is a drive unit in which a pump provides fluid to drive a motor. Yet, hydrostatic drives do depend on an engine for power. The advantage is that power is transmitted efficiently and the operator can easily adjust and accurately control speed, torque, horsepower, and the direction of wheel rotation.

Another advantage is that hydrostatic drive systems do not need electrical connections but instead rely on control valves to direct fluid power and direction of flow. Fluid pressure produces power in a hydrostatic drive system. The direction of the flow dictates the direction of travel. Also, the rate of fluid flow affects the speed of travel. Therefore, if quicker directional changes are needed, especially through different drive wheels, a hydrostatic drive can be an appropriate choice. In addition, attachments and implements can access and use the source of fluid power as their own power source.

A hydrostatic drive is coupled to the power source through an input shaft. In most applications, a hydraulic pump is mounted to the crankshaft. The pump in turn creates fluid flow or fluid power. This fluid power is then used to power motors or cylinders for a specific application. The power can then be manipulated externally and individually. A good example is a skid steer. It has a power source (engine) that is coupled to a hydraulic pump to create fluid flow which is then diverted to drive motors at each wheel.

The operator can independently control the direction of wheel rotation direction through control valves between the pump and drive wheels. As a result, the individual drive wheels can rotate at different speeds, and even in opposite directions. This provides good maneuverability and allows the drive system to double as the steering system for applications such as skid steers and windrowers.

The fact that the drive motors can be placed remotely, at each drive wheel, means that this type of system can provide great clearance. For this reason, hydrostatic drives are commonly used in sprayers. See **Figure 35-10**.

Stockr/Shutterstock.com

Figure 35-10. This sprayer is equipped with hydrostatic drive motors at each wheel, eliminating the need for a transmission, driveshaft, differential, and axles. As a result, the machine has great ground clearance, allowing it travel down rows of crops without damaging them.

CHAPTER 35 Review and Assessment

Summary

- Power trains allow operators to direct power to the drive wheels.
- In the transmission and final drive units, gear ratios can be manipulated in order to achieve torque or power needs.
- Meshed gears, belts, and pulleys, as well as gears with chain drives, can transfer power and can increase or decrease power and torque to the end piece of equipment used.
- Hydraulic fluid can provide power to drive units. A hydrostatic drive is an example.
- All of the different units that transfer power to the drive units allow operators to select final output ratios and direction of travel to the drive wheels.

Words to Know ↗

Match the key terms from the chapter to the correct definition.

A. automatic transmission	E. gear	I. power take-off (PTO)
B. belt	F. gear ratio	J. sprocket
C. clutch	G. hydrostatic drive	K. transmission
D. final drive	H. limited-slip differential	

1. Gives operators the ability to select directional travel (forward or reverse) and operational speed through gear ratios.
2. Directs power to the drive wheels and provides a final gear ratio.
3. Can transfer power to attached or towed equipment.
4. Type of drive that uses oil pressure and flow to transmit power to the drive wheels.
5. A friction disc assembly that engages and disengages the engine flywheel to transfer power to the transmission.
6. A disc- or cone-shaped object with teeth that engage the teeth on another.
7. A comparison of the number of teeth between two gears.
8. The type of transmission in which gear ratio changes are activated without direct operator input.
9. A toothed wheel used to drive a chain.
10. A flexible loop designed to connect two or more pulleys to transfer rotational power.
11. When slippage of the drive wheel occurs, this mechanical device keeps both wheels in powered rotation.

Know and Understand

Answer the following questions using the information provided in this chapter.

1. Three ways power can be transferred are ____, ____, and ____.
2. *True or False?* Often, in heavier-load applications, multiple belts and pulleys are used per shaft location to reduce slippage in the friction connection.
3. *True or False? Slip* is a term used to describe the loss of friction between pulley and belt.
4. A(n) ____ is used to press down on the belt and keep it tight against the pulley.
5. The ____ connects or disconnects power to the transmission and/or the final drive.
6. The ____ gives operators the ability to select directional travel (forward or reverse) and operational speed through the transmission's gear ratios.
7. The ____ directs power to the drive wheels and provides a final gear ratio.
8. The ____ can transfer power to attached or towed equipment.
9. A clutch can be either ____ or ____.
10. A(n) ____ clutch uses fluid power to aid in its engagement or disengagement.
11. The difference between two gears is called a(n) ____.
12. A(n) ____ connects the transmission to the final drive.
13. The ____ allows the drive wheels to turn at different speeds during cornering.
14. *True or False?* A special kind of locking differential called a *limited-slip differential* keeps both drive wheels in powered rotation to greatly enhance traction.
15. *True or False?* Wheel and tire size have an effect on the final drive gear ratio.
16. A(n) ____ and sprockets can be used to transmit power to the drive wheels or the accessory attachments.
17. What kind of connection do belts and pulleys make?
18. The ____ pulley is connected to the power source.
19. A(n) ____ is a toothed wheel that does not mesh directly with another gear.
20. What is the main difference between a sprocket and a pulley?

STEM and Academic Activities

1. **Science.** Research Bernoulli's principle. How does it relate to hydrostatic power trains used in agricultural applications?
2. **Technology.** Conduct research to find out more about continuously variable transmissions. Write a report explaining how these transmissions work and how they are or could be used in agricultural applications.
3. **Engineering.** Explain the mechanical differences between a standard or open differential and a limited-slip differential.
4. **Math.** What is the gear ratio if the drive gear has 20 teeth and the driven gear has 48 teeth?

5. **Social Science.** Write a short essay outlining the development of agricultural power take-off systems and explaining how these systems have impacted the agriculture industry.
6. **Language Arts.** Draw a poster showing the formulas for calculating horsepower and torque. Give a presentation to the class on how to make these calculations correctly.

Thinking Critically

1. In your opinion, which is better for agricultural work: a manual transmission or a hydrostatic transmission? Write a paragraph explaining your reasoning.
2. Think about the advantages of a hydrostatic drive system for agricultural applications. Do you think a hydrostatic drive system provides enough advantages to justify replacing an existing dry-clutch, manually-shifted drive system? Write a paragraph justifying your answer.

CHAPTER 36
Machinery Maintenance and Management

Chapter Outcomes

After studying this chapter, you will be able to:
- Schedule regular maintenance and maintain service records on machinery.
- Identify mechanical systems and regular service needs.
- Perform preventive maintenance.
- Troubleshoot basic mechanical systems.
- Calculate a precise measure of slippage.
- Explain the difference between fixed costs and variable costs.

Words to Know

anti-friction bearing	fixed costs	potential energy	thermal energy
ballasting	friction bearing	pressure energy	total energy
clearance	hydrometer	preventive maintenance	variable costs
electrical energy	kinetic energy	rolling resistance	viscosity
energy	mechanical energy		

Before you Read

The summary at the end of the chapter highlights the most important concepts. Read the chapter and write a summary of it in your own words. Then, compare your summary to the summary in the text.

While studying this chapter, look for the activity icon to:

- **Practice** vocabulary terms with e-flash cards and matching activities.
- **Expand** learning with interactive activities.
- **Reinforce** what you learn by completing the end-of-chapter questions.

www.g-wlearning.com/agriculture

Elena Elisseeva/Shutterstock.com

Repairing equipment is more cost-effective than buying new equipment, so preventive maintenance and timely repair will go a long way in reducing the total costs of running a business or operation. Before attempting to work on any machinery, ensure that you thoroughly understand all safety guidelines. Follow and strictly adhere to safety guidelines when performing maintenance tasks. Wear proper attire and use safety equipment at all times. Follow all manufacturers' guidelines while performing maintenance or service.

The engine and equipment should be cool and clean when serviced. Equipment cleanliness allows users and technicians to diagnose leaks, wear, and possible damages for timely repairs. Following these simple steps throughout your career will result in reduced operating costs and equipment failures. Most of the parts discussed in this chapter are wearable parts, meaning that they are specifically designed to be replaced instead of replacing more costly items. For example, there is a wearable surface on the connecting rod(s) of most engines. It is much cheaper to replace the connecting rod bearing than the connecting rod.

Manuals, Schedules, and Service Records

Mechanical systems require proper maintenance to perform as designed. Preventive maintenance is performed to prevent premature equipment failure. Without the proper knowledge about how to care for equipment and when maintenance should be done, mechanical system failures will occur. Equipment failure can be a result of improper lubrication, incorrect parts matching, or failure to check equipment regularly for signs of wear, overheating, or overloading.

SAE Connection

Equipment Maintenance

The life of equipment can be extended with regular and proper maintenance performed to manufacturer's recommendations. Both residential and agricultural equipment benefit from properly performed periodic maintenance.

If you have a supervised agricultural experience (SAE) program that utilizes equipment, be sure to document the skills gained and hours spent performing maintenance activities. You may be required to change engine oil in a mower for a landscape maintenance SAE or lubricate the picker head on a cotton picker in a fiber/oil crop SAE. No matter the equipment, there are maintenance tasks to perform. Pay attention to the details discussed in this chapter and recognize the important benefits of taking care of your equipment.

Fotokostic/Shutterstock.com

Systems Inspection and Maintenance

Inspection of mechanical systems should be done routinely and should be based on manufacturers' specifications. Inspection intervals also depend on conditions. If the equipment's operating environment is especially severe, maintenance intervals will be shortened. Once again, when working with any equipment or mechanical systems, safety should always be a top priority!

Air Intake and Exhaust System

For an engine to run properly, air and fuel must be adequately supplied to the combustion chamber. Both the fuel and the air are filtered before they enter the combustion chamber.

The air should be clean and free from foreign particles or dirt. Air may appear to be clean, but in reality it is filled with many contaminants; everything from tiny particles of dust and dirt, to ash, soot, and pollen. The clearances inside an engine are tight. Any airborne contaminants that make it into the engine can have an abrasive effect on bearings, cylinder walls, and moving parts.

Dirt contaminates the oil, and the abrasive dirt particles rapidly load up and clog the oil filter to a point where it will bypass dirty oil to circulate freely through the engine. All oil filters use an internal bypass valve. This bypass is designed to prevent a situation in which the filter gets so clogged that it can no longer supply a sufficient volume of oil to the engine. The idea behind a bypass valve is that a supply of dirty oil is better than no oil. As the bypassed oil gets dirtier and dirtier, the abrasive effect on the engine's internal moving parts is amplified, causing accelerated wear and very soon (as few as 20,000 miles on an automobile) premature engine failure. Changing the oil is a temporary fix because the lack of effective air filtration will quickly load the new oil with a fresh dose of engine-killing grit. Therefore, the very first line of defense for clean engine oil and long engine life is to start with an efficient air filter and a leak-free air cleaner system.

To ensure that the air brought into the engine is as clean as possible, an air cleaner is used to filter all incoming air. Most air cleaners used in agricultural applications, whether they are small-displacement or high-horsepower engines, use either a paper or foam air cleaner element.

A paper element air cleaner has a replaceable paper cartridge that filters the incoming air. The paper element can be round, flat, or most any shape. All that matters is that any air entering the engine must first pass through the paper filter element. Therefore, a secure and leak-free fit is absolutely critical.

Inspect the air cleaner filter as a matter of routine engine maintenance. Paper elements can be lightly cleaned by tapping them on a hard surface to dislodge trapped debris. Compressed air should not be used to clean a paper element because the pressurized air can enlarge the pores in the filter, allowing larger particles to enter the system. You should not clean paper elements with liquid cleaner because doing so will damage the filter's ability to remove particulate matter.

When the paper element becomes heavily soiled, it should be replaced. A paper filter element is inexpensive. This is one place you should not compromise. The potential damage to the engine is way out of proportion to the cost of a paper filter.

Many agricultural equipment air cleaning systems use dual-element air cleaners. Dual-element air cleaners have a paper element as the primary filter, with a secondary filter made of foam. In dusty and dirty conditions, this type of filter system is common. The outer filter (foam) traps the large particles before they can enter and clog up the primary paper filter, which would quickly reduce the efficiency of the entire air filtration system. Also, foam elements are easy to clean when cleaning intervals are shortened because of extra-filthy conditions. Secondary foam elements are cleaned with a solution of mild soap and water, squeezed out, and then thoroughly air-dried.

Another type of air cleaning system is an oil foam air cleaner. This type of air cleaner is used on older small gas engines. A common procedure for cleaning an oil foam filter is:

1. Remove the air cleaner assembly.
2. Disassemble the air cleaner and remove the oil foam filter.
3. Wash the oil foam filter in warm water with mild soap.
4. Dry the foam filter completely.
5. Using a small amount of oil, squeeze the foam filter to thoroughly spread oil throughout.
6. Ensure that only a light coat of oil is present by squeezing the filter and removing excess oil.
7. Clean the air filter box.
8. Reassemble the air cleaner and reattach it to the engine.

Never use gasoline or other flammable materials to clean air filtration elements. The use of a flammable material could result in combustion of the filter materials or even an explosion. Also, never leave too much oil on a foam element. Oil could cause engine damage if it collects in the combustion chamber. And finally, always ensure that the air cleaner assembly is fitted properly and tightly so that unfiltered air cannot enter the engine.

Lubrication System

Engine wear is reduced by minimizing friction between moving parts. Lubricants are an intermediary substance used between the parts to reduce the friction between them. Not all lubricants are suitable for use with all friction-producing parts. Some parts have larger or smaller clearances between them and require specific types of lubricants based on the viscosity needed for maintaining the clearances. The *clearance* is the manufacturer's specified distance between two wearable parts. *Viscosity* is oil's resistance to flow. Lubricants with higher viscosity are thicker and flow less freely than lubricants with lower viscosity.

Viscosity is the single most critical physical property of oil because it affects both the rate of wear between moving parts and fuel efficiency. To gain a better understanding of viscosity, compare water with maple syrup. Water is a low-viscosity fluid; maple syrup is a high-viscosity fluid. With oil, like syrup, as you increase temperature, viscosity is reduced, so it flows more easily.

There are three main types of lubricants. Lubricants may be liquid (oil), semisolid (grease), or solid (graphite), **Figure 36-1**. Oil is a common lubricant because it can easily flow into the tiny clearances between friction-producing parts and just as easily flow away to remove heat. In some instances, lubricants such as oil are instrumental in the removal of heat from a localized area. Grease, being highly viscous, creates a lubricating film. Unlike liquid oil, grease does not flow away, but leaves a residue on parts to aid in lubrication. Grease can also act as a barrier to the entry of foreign particles that might increase wear.

Graphite, or dry lubricant, is best suited for extreme conditions. Its ability to provide lubrication at extreme temperatures and excessive loads makes it advantageous in specific heavy-duty applications.

Bearings

In agricultural equipment, bearings are commonly used because they reduce wear between moving parts and are easy to replace when they wear out. Wearable bearings help protect major parts and are much less expensive to replace in the event of failure. For example, it is much less expensive to replace an axle bearing than it is to replace an axle. Bearings can also aid in radial, axial, or both types of load reduction, extending the service life of shafts, **Figure 36-2**.

A — maxuser/Shutterstock.com B — Tossapol/Shutterstock.com C — Goodheart-Willcox Publisher

Figure 36-1. Lubricant types. A—Motor oil is an example of a liquid lubricant. B—Grease is an example of a semisolid lubricant. C—Graphite is an example of a solid lubricant.

960 Agricultural Mechanics and Technology Systems

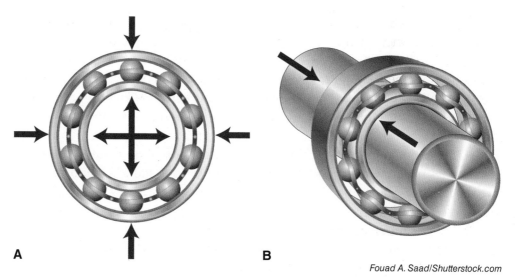

Figure 36-2. Bearing load from inner to outer races. A—Radial loads are applied perpendicular to the shaft. B—Axial loads are applied parallel to the shaft.

TraXXXe/Shutterstock.com

Figure 36-3. Friction bearings have no moving parts. They are designed to wear out before the part in which they are installed.

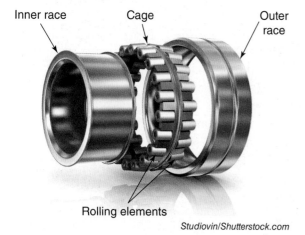

Studiovin/Shutterstock.com

Figure 36-4. Anti-friction bearings are an assembly of rolling elements and inner and outer races. The assembly typically includes a cage that holds the rolling elements in position relative to each other as they roll along the races.

The two most common types of bearings are friction and anti-friction bearings. A *friction bearing*, also called a bushing, is a replaceable surface used between moving parts. Without a friction bearing, the major parts would rub directly against each other. A good example is a connecting rod attached to a crankshaft. Over time, the part (either the rod or the crankshaft) would wear out and require replacement. When a friction bearing is placed between the moving parts, it is the friction bearing that wears, not the moving parts. Instead of replacing the expensive rod or crankshaft, the inexpensive worn-out bearing is replaced. Friction bearings are normally made from bronze, aluminum, or Babbitt materials. The bearing on the connecting rod (where it connects to the crankshaft) of a small engine is most commonly a Babbitt bearing. See **Figure 36-3**.

An *anti-friction bearing* provides a low-friction surface between two parts that could generate high friction if they came in direct contact. Common anti-friction bearings include roller and ball bearings. An example of a situation in which an anti-friction bearing would be appropriate is an axle shaft that connects a drive wheel, which in turn supports the weight of a large piece of equipment. The housing that holds the axle would create excessive friction because of the rotational friction of the axle turning the drive wheel against the housing. By using a lubricated anti-friction bearing, the point where the axle housing rubs against the axle generates much less friction and therefore creates significantly less heat and wear. See **Figure 36-4**.

Copyright Goodheart-Willcox Co., Inc.

An anti-friction bearing has an inner race that is mated to the shaft of the moving part. Once again, a good example of this would be an axle bearing. On the outside of the race, rolling elements, such as ball bearings, are used to allow movement of the part and also to reduce friction buildup. A cage is used to confine the rollers, and an outer race provides a second surface for mating the two friction-producing parts.

Seals

Oil seals are used to prevent lubricants from leaking out where a shaft passes through a housing. Often called grease, fluid, or dirt seals, they close up the spaces between stationary and moving components in mechanical equipment to prevent lubricant escape. They protect against leakage and aid in the service life of the part by keeping foreign materials from entering into the area being protected, especially in severe environments. The oil seal lip must be flexible but also be firm enough to provide a sealing surface compatible with and conforming to the moving part.

Cooling System

The two major types of engine cooling systems used in agriculture equipment are air cooling and liquid cooling. Air-cooled engines use air to remove the excess heat produced by the engine. Liquid-cooled engines use a liquid coolant to remove heat from the system. Remember that engines create heat, which produces work. Too much or not enough heat can affect engine efficiency. If the engine runs too cool, it will not convert fuel into work as efficiently as it could. Too much heat will cause the engine to operate poorly and will eventually result in premature engine failure.

An engine is designed to operate in accordance with the manufacturer's specifications for specific applications. Users or operators should make every effort to ensure that the air-cooled engine can shed heat as it was designed to do. Clear away any debris, dirt, or oily gunk from the cooling fins because buildup acts as an insulator and will cause the engine to retain excessive amounts of heat.

Air-cooled engines rely on airflow around the engine to provide cooling. The heat produced by the engine must be removed by airflow to allow the engine to operate efficiently, in an optimum operating temperature range.

As engine sizes increase, so too does the amount of heat they produce. Liquid-cooled engines were developed as a better way to remove heat from the engine. Liquid-cooled engines rely on a circulating liquid coolant to remove heat. The liquid coolant travels through passages cast into the engine. Those passages, and the rest of the cooling system, are specifically designed to maintain an optimum operating temperature. Engine operating temperature is the temperature the engine coolant reaches under normal running conditions.

As the coolant is pumped through the engine, it absorbs heat and carries it away to a radiator. The radiator transfers heat from the coolant to the outside air. After the radiator removes heat, the cooled liquid coolant is pumped back into the engine, where it once again absorbs engine heat. This process is repeated continuously as the engine runs.

The thermostat is the brain of the engine coolant system. A thermostat is a temperature-sensitive valve that maintains engine operating temperature in the proper range by controlling the flow of coolant through the engine, **Figure 36-5**.

Liquid-cooled systems are pressurized to raise the boiling point of the coolant. Although pure water can be used as the liquid in a cooling system, it rarely is. Most cooling systems call for a 50/50 mix of coolant and water. Coolants contain an additive package that imparts desirable qualities to the mix. Pure water will cause rust. The by-products of corrosion can quickly clog up the narrow radiator passages and also send abrasive grit into the water pump bearings. Coolants contain an additive to fight galvanic corrosion. Also, the coolant contains a lubricant additive to help the water pump last longer.

The main advantage of adding coolant to the system is that it extends the range of the engine's operating temperature. A 1:1 ratio of water to coolant will protect an engine from freezing to –34°F. This is important because water expands when it freezes. If the water in an engine were to freeze, it could literally crack the engine block. On the other side of the temperature range, the 1:1 coolant to water mix protects the engine from overheating to 265°F.

Coolants should not be mixed in proportions other than those recommended by the manufacturer. Most coolants come with a mixing chart printed on the container that recommends the correct ratio for the geographic area where the equipment is to be used or stored. To prepare the cooling system, first determine the lowest temperature the engine or equipment will be exposed to. Then, prepare the appropriate mixture and fill the system to its capacity.

For example, suppose you wanted an engine to be prepared for a temperature of –32°F. Referring to the chart in **Figure 36-6**, you would mix 50% water to 50% coolant. If, on the other hand, your machine will be operating in Montana where winter temperatures can dip below –32°F,

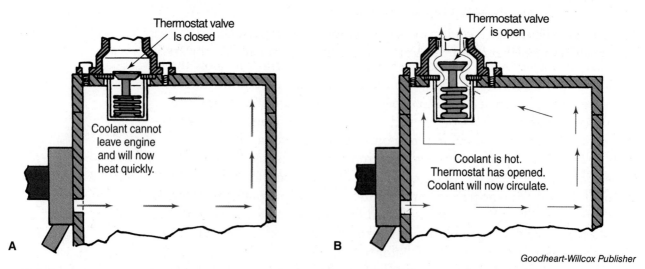

Figure 36-5. Thermostat operation. A—While the engine is warming up, the thermostat remains closed. This allows the coolant in the engine water jackets to heat up. B—When the engine reaches its normal operating temperature, the thermostat opens, allowing hot coolant to circulate to the radiator as it is replaced by cold coolant from the water pump.

Freeze Protection Limit (°F)	Antifreeze Concentration	Water Concentration
32	0%	100%
25	10%	90%
20	16%	84%
15	21%	79%
10	25%	75%
5	29%	71%
0	33%	67%
−10	39%	61%
−20	44%	56%
−30	48%	52%
−34	50%	50%
−50	56%	44%
−60	59%	41%
−70	64%	36%
−80	70%	30%

Goodheart-Willcox Publisher

Figure 36-6. Typical coolant mixing chart. To provide adequate lubrication and anticorrosion properties, the coolant mixture should contain at least 50% antifreeze.

you would prepare a 70% coolant-to-water mix and protect the engine to temperatures as low as −80°F. Always check with your coolant's manufacturer for specific mixing ratios and temperatures.

Coolant breaks down and does not hold its properties over time, so review the coolant manufacturer's suggestions for change intervals. A *hydrometer* can be used to determine the integrity of the mixture present in a system. The hydrometer is a tool that measures the specific gravity and strength of a coolant.

Fuel System

The fuel systems used in most engines are similar. They all contain fuel storage (tank), a fuel line, fuel filter(s), and a device to meter and distribute fuel to the engine. Many spark ignition (gasoline) engines use a fuel pump and a carburetor or fuel injection system to deliver fuel to the combustion chamber, **Figure 36-7**. Compression ignition engines (diesel) use an injection pump and fuel injectors. The fuel pump (spark ignition) and injector pump (compression ignition) both provide pressurized fuel to the carburetor/fuel injectors for precise distribution into the combustion chamber.

In diesel engines, the injector pump and injector meter the fuel for each cylinder. Injectors serve as valves allowing fuel flow but also restricting it when fuel is not needed.

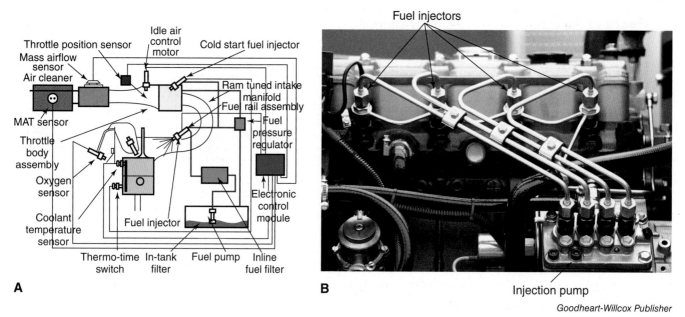

Figure 36-7. Fuel injection systems. A—In a typical port fuel injection system for a gasoline engine, a single fuel pump is used. The fuel injectors spray fuel into the intake manifold. B—In a typical diesel fuel injection system, a high-pressure injector pump is used in addition to a low-pressure fuel delivery pump. The injectors spray fuel directly into the combustion chamber.

Fuel systems must be properly maintained. Having clean, fresh fuel is imperative for the optimum performance of any engine. It is important that routine maintenance be followed on fuel filters to ensure that foreign materials do not get into the fuel metering components and enter the engine. Many fuel filters are located in fuel lines and can easily be replaced or removed. Follow manufacturer's recommendations for service, replacement intervals, and maintenance procedures for all fuel system components.

Electrical System

Most agricultural equipment and machinery use direct current (DC) voltage. DC voltage is stored in batteries for use as needed. Batteries are classified based on their voltage and amperage. Voltage is a measure of how much electrical force is pushing the electrons. Amperage indicates the rate of electron flow. When selecting batteries, you are most concerned with the voltage of the battery and the cold cranking amperage (CCA).

Battery voltage must be compatible with the system. For example, if you have a 12-volt system, you must use a 12-volt battery. You can then select the CCA based on your needs. If you have a diesel compression ignition engine, you will want a higher CCA. If you have a lower compression gasoline engine, a lower CCA is needed to start the engine.

Batteries

A 12-volt battery, such as an automobile battery, is an electrochemical device that can produce and store direct current (DC). The battery is normally a plastic case that has lead plates with a sulfuric acid and water (electrolyte) solution filling the case. Positive (+) and negative/ground (–) terminals are used to direct the voltage. The electrolyte solution and lead plates result in an excess of electrons (−) on one side of the battery and a shortage of electrons (+) on the other. If given a path, the surplus of electrons on the negative side of the battery will flow to positive side until both sides reach equilibrium. This is what gives the battery its ability to produce current when connected to an electrical circuit.

Some batteries have vents or filler plugs where water and sulfuric acid can be added. Most modern batteries are sealed and are maintenance-free. For older batteries and those that need continuous recharging, maintaining the correct water levels is imperative. The water levels should be checked daily and should cover the lead plates completely. Do not overfill the battery. A hydrometer can be used to check the battery's specific gravity level to indicate whether the battery's electrolyte is weak or strong. A low specific gravity indicates a low-charged battery or one that can no longer take and hold a charge. A battery that cannot take and hold a charge should be replaced.

Batteries can be connected in series or parallel for specific applications. Connecting batteries in series increases the voltage output. For example, if two 12 V batteries are connected in series, the voltage is doubled to equal 24 V. If the two 12 V batteries are connected in parallel, the voltage is not changed and still equals 12 V, **Figure 36-8**.

The battery does not supply electricity to the engine once it is running. After the engine has started, it makes its own electricity.

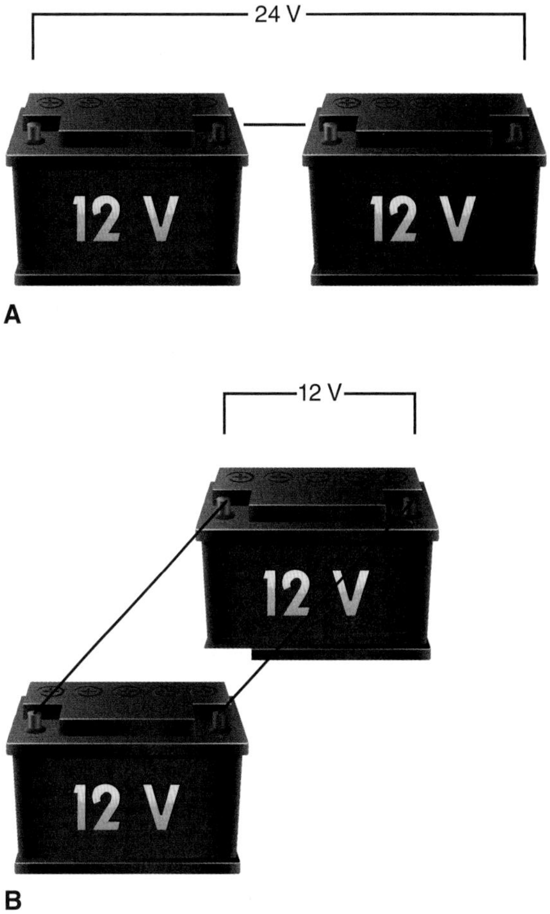

Figure 36-8. The way 12 V batteries are connected affects their combined voltage. A—If the batteries are connected in series, their voltage is doubled. B—If the batteries are connected in parallel, their voltage is unchanged, but they are able to operate longer without becoming drained.

Starting/Ignition Systems

A century ago, engines were started with a hand crank. Today, one of the most common uses for batteries is to start the engine. Engines with electric start systems use battery voltage to spin an electric motor (starter motor) which engages the teeth on the flywheel to rotate the crankshaft and move the pistons in the cylinder block.

The parts of a starting system include an ignition switch, starter motor, starter solenoid, neutral safety switch (optional), and the battery. The ignition switch allows the operator to start or shut down the engine. The starter solenoid acts as a relay—it uses a relatively low-amperage switch to open and close a much higher-amperage circuit. When the key is turned to the "on" position, the starter solenoid allows the starter motor to rotate and engage the flywheel to "crank" the engine.

A neutral safety switch ensures operator safety by requiring the transmission to be in neutral or park before the starter can crank the engine.

The ignition system works in conjunction with the starting system. Current travels from the coil to the distributor cap. The distributor cap then "distributes" current to each cylinder through plug wires and spark plugs. See **Figure 36-9**.

For compression ignition engines (diesel), the starting system is very similar to spark ignition engines. Compression ignition engines normally need more amperage for starting but do not need more voltage. Therefore, multiple batteries are used and connected in a parallel circuit. A compression ignition engine does not need an ignition system because it operates without spark plugs. Ignition of the air/fuel mixture is accomplished when the cylinder's fuel injector sprays fuel into the superheated, compressed air charge at the end of the compression stroke.

Machinery Management

Managing machinery and equipment effectively is important to the success of any production agricultural business. For the business to be successful, costs must be maintained at a low enough level that a profit can be made. Careful attention to preventive maintenance, the ability to troubleshoot mechanical systems, and an understanding of how to set up and use the equipment most efficiently are all part of managing machinery. Knowing the fixed and variable costs of each machine also helps the agricultural producer to control costs and manage the equipment effectively.

Preventive Maintenance

Proper maintenance prevents part failure and extends the life expectancy of wearable surfaces. For example, maintaining the correct tire pressure can prolong tire life. Tires should be inflated to the recommended pressure and checked regularly to ensure that the proper pressure is maintained. Recommended inflation pressures will vary, depending on implement type, axle load, and tire type. It is critical to complete proper maintenance procedures when recommended. It is foolish to invest a small fortune in agricultural equipment and then be stingy on relatively inexpensive maintenance.

Maintenance that involves routine or scheduled inspections is referred to as *preventive maintenance*. This type of maintenance is intended to prevent potentially catastrophic and expensive equipment failure by finding and dealing with small problems before they become big ones. Preventive maintenance also includes minor equipment adjustments or repairs intended to keep the equipment working as designed. Preventive maintenance should be done in a logical fashion and according to the manufacturer's recommendations.

Whenever working with any equipment, it is imperative that the power source be shut down and in the off position. Any movable parts should be locked out (made immovable), and tagged out. The lockout/tagout (LOTO) statement informs personnel that the machine parts are in the off position, meaning that they cannot move and cause injury. Nobody has permission to disturb the equipment while others are performing inspections or maintenance on the equipment. Always refer to manufacturer's guidelines for performing maintenance and how to properly perform a LOTO.

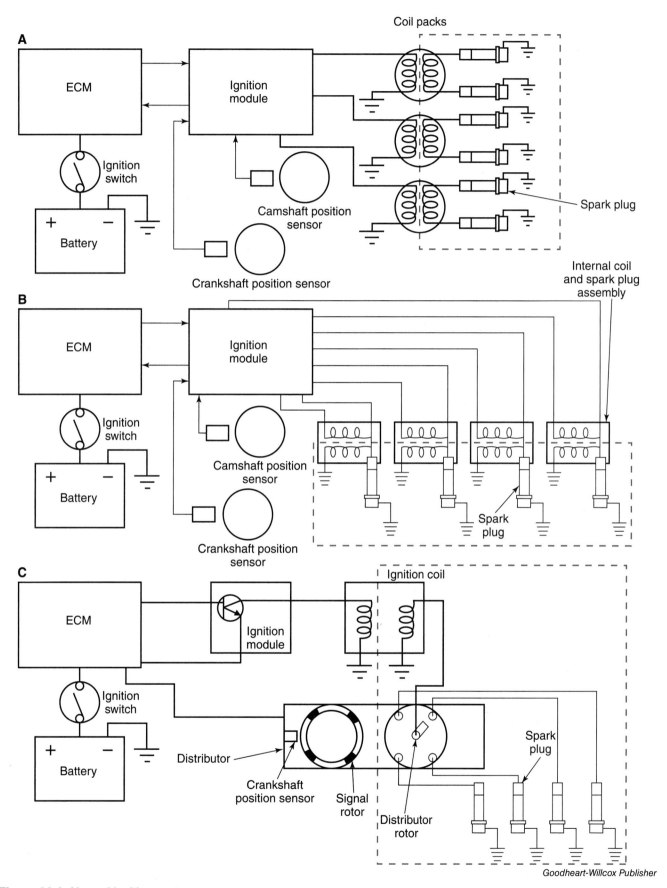

Figure 36-9. Normal ignition system.

Troubleshooting Mechanical Systems

As with routine inspections, troubleshooting requires personnel to engage in a linear progression of processes. Mechanical systems normally fail because of excessive wear due to friction and overloading. Even when the best preventive maintenance is performed, at some point in the life of a part, it will fail. When inevitable equipment failure does occur, it is important to determine the cause of the failure in a logical way.

Most equipment failures do not happen without the operator first hearing unusual sounds or feeling excessive or even violent vibration. Often, excessive heat can be felt or smelled during a diagnosis. This is why it is always a good idea to have a conversation with the operator before beginning a troubleshooting analysis. Afterward, a visual analysis of the parts involved in the failure may lead to an early diagnosis of what might have caused the problem.

Power and Performance of Equipment

For equipment to do its work, *energy* is needed to power the equipment. Energy is usable power that exists in one of two states: potential or kinetic, **Figure 36-10**. *Kinetic energy* is energy that is in motion. Examples of kinetic energy used in agricultural applications include the rotation of the crankshaft in engines, fluids flowing in pipe, hydraulic cylinder actuation, or belt and pulley rotation. The product of kinetic energy normally is mechanical energy.

Potential energy is energy in its stored form. Envision a dam that stores water to power hydroelectric drives for electrical energy. Once the water is released, it turns and powers the drives to produce electricity—this is kinetic energy. When the water flow is stopped, the energy that the water can produce is stored until it is needed—this is potential energy. The flow of water is what creates its ability to do work or produce kinetic energy. Remember that energy is neither created nor destroyed, it only changes form.

Thermal energy is energy based on particle or molecular density. With the addition of heat, molecules will move faster, and when heat is reduced or removed, the molecular movement will slow down or stop. Thermal energy is the ability of an energy source to perform work through heat. Kinetic energy is produced by the motion of the excited molecules. The more efficiently the energy is produced, the more efficient the conversion to power and performance will be. Many agricultural applications use thermal efficiency as a rating of how efficiently energy is converted to power.

Electrical energy is one of the more common types of energy used by agricultural equipment. *Electrical energy* is the flow of electrons through a conductor. For example,

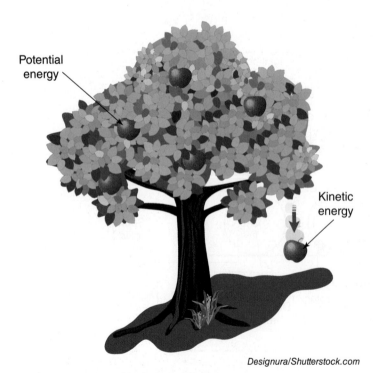

Designura/Shutterstock.com

Figure 36-10. Example of potential and kinetic energy.

an auger has an electric motor that turns a shaft with a helical screw blade that carries the seed or grain from one location to another. Electrical energy allows the motor to do the work needed to move the seed or grain.

Mechanical energy is energy produced and transferred using pulleys and belts, or by gears. When the rotational motion is transferred to the gear or pulley, mechanical energy is produced and transferred to the other gear(s) or pulley(s). See Chapter 35, *Drive Trains and Power Systems*, for calculations of power in belt and gear systems.

When force is applied to a constrained fluid, the result is called *pressure energy*. Fluid power is based on the principle of pressure energy because fluids are incompressible. Therefore, when force is exerted on fluids, work can be performed. When pressure is exerted on the fluid, flow is created. When the flow meets resistance (piping), pressure energy is created.

More than one form of energy can be used in a system, and we can use a fluid power system as an example. When an electric motor (utilizing electrical energy) turns a hydraulic pump, gears create fluid flow (mechanical energy). Pressure energy is then created when the fluid flows through a restriction. When multiple forms of energy are used to perform a given task, the sum of the forms of energy is called *total energy*.

Tractor Weight and Ballasting

Machinery power loss in tractors can be traced to wheel slippage and rolling resistance. Wheel slippage occurs when the tires do not have traction. It is caused by a combination of soil conditions and tractor/equipment weight. Too much weight increases the amount of rolling resistance. Not enough weight creates too much slippage of the tires.

Slightly broken tire tread imprints in the soil indicate good traction and just the right amount of slippage. Why do we not want total traction? There are a couple of reasons. First, if no wheel slippage occurs, then 100% of the grip of the tire would be transferred to the moving parts. This would impart more stress on the driveline and result in premature part failure.

Another reason relates to *rolling resistance*, which is the resistance created by the imprint/traction of a tire in soil and its efficient use of power to roll up and out of the tire imprint made in the soil. Even with proper air pressure in the tires, the tire must roll slightly up and out of its imprint, thus increasing the need for power and resulting in less efficiency. Having the proper amount of rolling resistance and the proper amount of wheel slippage provides a more efficient use of power by the equipment.

In some cases, we may want a more precise measurement of slippage than a visual inspection of tire contact with soil can provide. The following procedure can be used in these cases:

1. Mark a rear tire at the center of its contact with the soil (no equipment attached).
2. Mark the soil corresponding to the tire mark.
3. Move the tractor forward for 10 tire revolutions.
4. Mark the ground where the tire completed its 10 revolutions.

5. Attach the equipment.
6. Return the tractor so the center of the marked wheel is once again aligned with the starting mark.
7. Using the same gearing and RPM, drive the tractor forward until the center of the marked wheel is once again aligned with the finish mark in the soil. Have an assistant count the number of revolutions of the tire as this is done.

To calculate percent of slippage, you will subtract the unloaded number of turns (UT) (step #3) from the loaded number of turns (LT) (step #7). Divide the result by the loaded number of turns (LT) and multiply this number by 100. For example, if it took 13 tire rotations when loaded, the percent slippage would be about 23% wheel slippage. It is recommended that wheel slippage be around 15%. This example indicates too much slippage and should be corrected, **Figure 36-11**.

Ballasting is a common method of correcting wheel slippage. It involves changing the weight distribution of the tractor by adding or subtracting weight where needed. Adding weight creates more soil compaction and more load on the power train of the tractor. There are many ways to ballast a tractor. Liquid or powder can be added to tires to increase weight. Also, the addition of wheel weights on rear wheels, or the use of dual rear wheels, can add ballast. Adjustments of weight of the implement attached to the hitch can add rear wheel weight. Additionally, front-end weighting can be used to modify tractor weight. All of these methods increase load on the tires to increase traction and lessen wheel slippage, but all of these solutions add more weight. This added weight must be compensated by the drive train and power source. Furthermore, tire damage can occur through overweighting.

Because many types of tractor drive systems are used today, it is important to understand how weight affects tractor performance. Tractors should have a given weight based on PTO horsepower performance. Also, the optimum weight distribution per axle based on drive type should be taken into consideration. See **Figure 36-12** for acceptable weights and percentages.

Slippage	Weight Recommendation
<15%	Remove weights
15%	Proper weighting
>15%	Add weights

Goodheart-Willcox Publisher

Figure 36-11. Percent wheel slippage and recommendations.

Tractor Weight per HP Drive Type/Equipment Type (weight distribution ratio)						
Speed (mph)	2WD/Towed (25:75)	FWD/Towed (40:60)	4WD/Towed (55:45)	2WD/Mounted (35:65)	FWD/Mounted (45:55)	4WD/Mounted (60:40)
4	160	140	140	160	140	140
5	135	115	115	135	115	115
6	110	95	95	110	95	95

Goodheart-Willcox Publisher

Figure 36-12. Tractor weight by PTO horsepower and weight distribution percentages.

Machinery Costs

Owners of agricultural equipment are concerned with keeping machinery costs as low as possible. Based on use, the amount and type of land that is worked, and management practices, the cost of machinery ownership is a reality that must be reckoned with. All of the associated costs must be understood and planned for to realize maximum efficiency and profitability. The two types of machinery costs are *fixed costs* (those that do not change) and *variable costs* (those that do change).

Fixed Costs

Fixed costs include the total costs of the actual equipment. The fixed costs over the life of a machine include depreciation, interest, taxes, housing, and insurance. Most equipment and machines have a known number of years of expected productivity. The machine will be worth progressively less over time as it is used. This is called depreciation. The item still may have value after several years, but time and use continue to reduce its real value.

Agricultural equipment is expensive; few can afford to pay cash for it. Therefore, most agricultural machinery is purchased on credit. When credit is used to purchase high-value items, a down payment is made to allow producers to purchase the item and gain immediate use. The remaining cost of the equipment is paid for over time. The lender charges interest on the unpaid amount. Interest rates are based on various factors, such as the current cost of money, the borrower's credit score, and the confidence the lender has in the borrower's ability to pay back the loan. Normally, the lender charges interest on a yearly basis that is based on the amount financed.

Many states consider agricultural machinery to be personal property. Thus, personal property taxes are assessed based on the value of the machinery. State and local laws govern this expenditure. Insurance allows owners to protect their investment in case of damage. Insurance rates are normally based on the value of the machine. Housing or storing the machinery and equipment is another fixed cost. Because bad weather adversely affects equipment and machines, appropriate housing, such as the construction of a large steel building, might be considered a necessary cost to protect the owner's investment.

Variable Costs

Variable costs are those costs associated with operating the machinery. This can include labor to operate, fuel and lubricants, and repairs or maintenance. Labor is a major variable cost based on the use of the item. The time allocated and costs associated with labor vary. Labor costs are normalized for specific applications, but some tasks may take more time than allocated. The time it takes to accomplish tasks adds other costs as well. For example, if the task takes longer to finish, the amount of fuel and lubricants used will increase and so will overall variable costs.

Proper maintenance of equipment and machinery can reduce variable costs, but as the equipment ages, increases in repairs and maintenance become an additional cost to be considered. Most agricultural equipment is used in harsh conditions, and it is inevitable that it will break and, of course, repairs will be needed. Also, when equipment is not in use, repair and maintenance may be needed during this downtime to ensure that it operates as it should when it is used again. Owners of agricultural equipment are concerned with keeping machinery costs as low as possible. Based on use, the amount and type of land that is worked, and management practices, the cost of machinery ownership is a reality that must be reckoned with. All of the associated costs must be understood and planned for to realize maximum efficiency and profitability.

Storing Engine-Powered Equipment

Several precautions should be taken when storing an engine-powered equipment for an extended period, such as when storing a tractor for the winter season. Proper storage procedures will help ensure long engine life and easy starting at the beginning of the following season.

Begin by thoroughly cleaning the engine. Layers of dirt and debris tend to hold moisture, encouraging the formation of rust and corrosion.

Next, add a fuel stabilizer to the fuel to prevent carburetor varnish from forming or partial plugging of the carburetor jets. After adding the fuel stabilizer, run the engine for approximately 5 minutes to distribute the stabilizer throughout the fuel system. Then, shut off the engine and close the fuel shut-off valve

While the engine is still warm, change the oil and filter (if so equipped). Then, lubricate all grease fittings, and service or replace the air cleaner. On implements with liquid-cooled engines, make sure the coolant is strong enough to protect against freezing at the lowest possible temperatures. If necessary, drain and refill the cooling systems.

Next, clean the battery and remove it from the equipment. Check the battery's electrolyte level and charge it fully. Then store the battery in a cool dry place.

Wash the equipment with a soap-and-water solution and dry it thoroughly. Then, inspect the equipment carefully and repair or replace damaged parts. To prevent rust, apply grease or paint to bare metal.

Finally, store the equipment in a dry location. If it must be stored outside, place a waterproof cover over it. If necessary, place blocks or chocks under the equipment's tires to keep it from rolling.

> **Safety Note**
>
> After operating any engine-powered equipment, move it to an open area, shut off the engine, and allow it to cool completely. Hot engine parts can cause serious burns and may ignite flammable materials near the equipment.

CHAPTER 36 Review and Assessment

Chapter Summary

- Equipment failure can be a result of improper lubrication, incorrect parts matching, or failure to check equipment regularly for signs of wear, overheating, or overloading.
- Inspection of mechanical systems should be done routinely and should be based on manufacturer's specifications.
- To ensure that the air brought into an engine is as clean as possible, an air cleaner is used to filter incoming air.
- The three main types of lubricants are liquid oil, semisolid grease, and solid lubricants, such as graphite.
- The two major types of engine cooling systems used in agriculture equipment are air cooling and liquid cooling.
- Maintenance that involves routine or scheduled inspections is referred to as preventive maintenance.
- Energy generally exists in one of two states: potential or kinetic.
- Machinery power loss in tractors can be traced to wheel slippage and rolling resistance.
- The cost of owning and operating agricultural machinery is a factor that every agricultural producer needs to investigate and plan for.

Words to Know

Match the key terms from the chapter to the correct definition.

A. anti-friction bearing
B. clearance
C. electrical energy
D. energy
E. fixed costs
F. friction bearing
G. hydrometer
H. kinetic energy
I. mechanical energy
J. potential energy
K. pressure energy
L. preventive maintenance
M. rolling resistance
N. thermal energy
O. total energy
P. variable costs
Q. viscosity

1. An assembly, commonly made up of roller or ball bearings and races, that is used between two parts that generate high friction.
2. The costs associated with operating machinery, including labor, fuel and lubricants, and repairs or maintenance.
3. Cannot be created or destroyed, only changed from one form to another.
4. The cost of the equipment that can be calculated at the time of purchase.
5. The ability of an energy source to perform work through heat.

6. A replaceable surface used between moving parts; also called a bushing.
7. The kind of energy created when force is applied to a constrained fluid.
8. Energy that is in motion.
9. Manufacturer's specified distance between two wearable parts.
10. An oil's resistance to flow.
11. Energy in its stored form.
12. Maintenance that involves routine or scheduled inspections.
13. The type of energy produced and transferred by pulleys and belts or by gears.
14. The resistance created by the imprint/traction of a tire in soil and its efficient use of power to roll up and out of the imprint made in the soil.
15. Tool used to determine the integrity of the coolant mixture present in a liquid cooling system.
16. The flow of electrons through a conductor.
17. The sum of several forms of energy that are used to perform a task.

Know and Understand

Answer the following questions using the information provided in this chapter.

1. What is the difference between a friction bearing and a non-friction bearing?
2. Why is oil viscosity important to an engine?
3. Give an example of kinetic energy and an example of potential energy.
4. Explain why preventive maintenance is considered a variable cost.
5. Describe a "total energy" scenario in an agricultural application. Use at least three types of energy in your description.
6. Wheel _____ occurs when tires do not have traction.
7. Explain rolling resistance.
8. A hydrometer is a tool that is used to measure the _____.
    A. specific gravity and strength of a coolant
    B. airflow around an air-cooled engine
    C. viscosity of engine oil
    D. efficiency of an air cleaner filter element
9. Explain the difference between variable costs and fixed costs.
10. Explain why clean engine oil and long engine life depend on an efficient air filter and a leak-free air cleaner system.

STEM and Academic Activities

1. **Science.** Fluid power is becoming increasingly common in heavy-duty machinery. What are the advantages and disadvantages of fluid power?
2. **Technology.** Oil is the key to engine maintenance. In fact, the longevity of farm equipment depends on checking on and replacing oil at regular intervals. What kinds of oil are typically used in agricultural equipment? What about synthetic oil? Write a paragraph on the newest developments and trends in lubricating oil for agricultural applications.
3. **Engineering.** Describe the different kinds of bearings a technician should expect to find when performing routine maintenance on a tractor. What would you do to make sure these different bearings last a long time?
4. **Math.** Select two other students and work as a team to create a small poster that teaches others how to use the formula to calculate the percentage of slippage for the rear wheels of a farm tractor.
5. **Social Science.** Air filtration is very important to the longevity of an engine. Get together with another student. Use the Internet to do research on air filtration for engines and then compile a comprehensive step-by-step list of what will happen if an air filter is never changed.
6. **Language Arts.** Write a report discussing fixed costs versus variable costs and how they are relevant to successfully running an agricultural operation.

Thinking Critically

1. Imagine that an air filter has not been properly replaced and an air space between the paper replacement filter and the air cleaner box is allowing unfiltered air into the engine. What will this stream of unfiltered air eventually do to the engine?
2. Why is preventive maintenance important and how does it relate to the total cost of machinery operation?

CHAPTER 37
Hydraulic and Pneumatic Power

Chapter Outcomes

After studying this chapter, you will be able to:
- Understand basic fluid power principles and systems.
- Understand the multiplication of force in a fluid power system.
- Identify the components of a hydraulic system.
- Describe the operation of gear, rotary, and piston fluid pumps.
- Briefly explain how system controls operate a fluid power system.
- Summarize the characteristics of hydraulic pipe, tubing, and hoses.
- Understand pneumatic power principles.
- Explain the basic operation of an air compressor.
- Identify the components of a pneumatic power system.
- Describe the maintenance required on pneumatic systems.

Words to Know ↗

air compressor	head	ports	rod
cap	hydraulic system	pressure	schedule number
centrifugal force	hydraulic cylinder	pump	seal
fluid power	piston	quick-disconnect couplings	vane pump
force	piston pump		
gear pump	pneumatic system		

Before You Read

Before you read the chapter, interview someone in the workforce (your supervisor, a parent, relative, or friend). Ask the person why it is important to know about the chapter topic and how this topic affects the workplace. Take notes during the interview. As you read the chapter, look for the items from your notes that are discussed in the chapter.

While studying this chapter, look for the activity icon to:

- **Practice** vocabulary terms with e-flash cards and matching activities.
- **Expand** learning with interactive activities.
- **Reinforce** what you learn by completing the end-of-chapter questions.

www.g-wlearning.com/agriculture

Elena Elisseeva/Shutterstock.com

Throughout history, many inventions have allowed individuals to accomplish tasks more easily and efficiently. Engines allowed agriculturalists to accomplish more work in a shorter time period. No longer were they limited by the strength and endurance of their available workforce and animals. The use of fluid power in agricultural applications greatly increased productivity and expanded the abilities of workers. *Fluid power* refers to using fluids (liquids or gases) to complete work. If liquids are used to perform work, it is classified as a *hydraulic system*. If gases are used to perform work, it is classified as a *pneumatic system*. Both types of systems are commonly referred to as *power transfer systems*.

Fluid Power Advantages and Disadvantages

Different situations and work requirements create the need for certain system attributes. Fluid power systems are advantageous for many reasons:
- Multiplication of force.
- Multiplication of torque.
- Fewer moving parts to accomplish tasks than a mechanical system.
- Variation of speed control.
- Accurate control of speed in selected applications.
- Adaptability to various designs and size applications.
- Minimal damage to work (actuator) mechanisms under load stress.
- Storage of source/power materials for use (liquids and gases).

Although they are versatile, fluid power systems have some disadvantages. Based on the application of the system, these disadvantages must be considered:
- Fluid leakage and spills, which can create hazardous work environments.
- Component wear via fluid contamination by foreign materials.
- Fluids and gases under extreme pressure, which pose a constant safety threat.
- Costs associated with fluid power system components.
- Noise levels of operating equipment.

Fluid Power Fundamentals

Fluid power systems transfer power from a source (prime mover) to accomplish work (actuator). The power source drives a hydraulic or pneumatic *pump*, which creates fluid flow. As the fluid flows through the fluid conductors of the system, it meets resistance in the form of a load on an actuator or a restricting orifice. Pressure develops in the system due to flow restrictions. Within the system, the area of a moving part in the actuator is where the force created by hydraulic pressure is applied to perform work.

To envision the difference between the two types of fluid power systems (pneumatic and hydraulic), picture a room filled with air. If one wall began sliding toward the opposite wall and there was no way for the air to escape, the air would be compressed or compacted. Air and other gases can be compressed, but there is a limit to how tightly they can be compressed. However, if the same room was filled with hydraulic oil and the sliding wall was activated, the oil would not be compressed. All liquids are nearly incompressible. In addition to explaining this characteristic of liquids and gases, this example describes the most important concept of hydrostatic (fluids under pressure) fluid power systems.

Pascal's Law

Blaise Pascal was a 17th-century French physicist and mathematician who developed the basic principle of hydrostatic fluid power systems. Pascal's law states that when pressure is applied to a confined liquid, it results in an equal pressure transmitted in all directions throughout the system. Pascal's law can be stated in terms of force, area, and pressure. *Force* is a pushing or pulling action applied to an object. *Pressure* is a force applied to a specific area. The size of the area on which the pressure acts determines the output force. The formula can be arranged to calculate each of the variables.

Force = F (measured in pounds or Newtons)

Pressure = P (measured in pounds per square inch, bar, or kilopascals)

Area = A (measured in square inches or square millimeters)

Pressure = Force ÷ Area   P = F ÷ A

Force = Pressure × Area   F = P × A

Area = Force ÷ Pressure   A = F ÷ P

Multiplication of Force in a Fluid Power System

One of the greatest advantages of a fluid power system is force multiplication. In a simple closed hydraulic system, a small input piston is plumbed to a large output piston. According to Pascal's law, if a force is applied to the input piston, the force will be fully transmitted to the output piston and all parts of the system. The developed pressure acts on the output piston's larger area, which multiplies the output force.

To understand how this works, imagine a basic system built out of two syringes connected with a piece of hose. One syringe has a plunger with a 2 in^2 surface area and the other syringe has a plunger with a 12 in^2 surface area. If 10 pounds of force is applied to the 2 in^2 plunger surface area, a fluid pressure of 5 psi (10 lb ÷ 2 in^2 = 5 psi) develops in the system. When the 5 psi is applied to the 12 in^2 plunger surface area, the force is multiplied to 60 lb (F = 5 psi × 12 in^2). See **Figure 37-1** for a diagram of these principles.

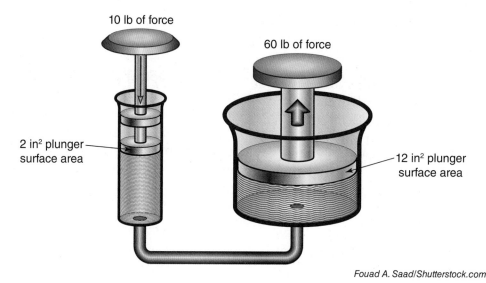

Figure 37-1. An illustration showing multiplication of force in a simple hydraulic system. This effect occurs based on Pascal's law.

Bernoulli's Principle

Daniel Bernoulli, an 18th-century Swiss physicist, demonstrated an important concept of hydrodynamic (fluids in motion) fluid power systems. Bernoulli's principle explains that in a system with a constant fluid flow rate, if the diameter of the fluid conductor (pipe, hose, or tubing) changes, fluid velocity and pressure will change proportionally in opposite directions. When pressure increases at a larger diameter section of conductor, fluid velocity proportionally decreases. When pressure decreases at a smaller diameter section of conductor, fluid velocity proportionally increases. See **Figure 37-2**.

SAE Connection
Pneumatic and Hydraulic Equipment

Pneumatic and hydraulic equipment is used in a variety of agricultural applications. Pneumatic tools, such as sanders and wrenches, are often lighter and more durable than their electric counterparts. Hydraulic equipment, such as tractor-mounted front loaders and backhoes, perform tasks like digging trenches and loading large amounts of livestock feeds quickly and efficiently. Imagine how difficult and time-consuming it would be to load a ton of silage with only a shovel. Understanding how to use, maintain, and repair pneumatic and hydraulic equipment will be a huge benefit to you if you use this type of equipment in your supervised agricultural experience (SAE) program. Be sure to document any new skills you learn in your SAE, such as operating hydraulic or pneumatic equipment, when completing your proficiency application.

System Design Factors

Many different actuators are used in fluid power systems. Gears, pulleys and belts, shafts, and other mechanical devices are designed to work in sync with fluid power systems to complete the job.

Although fluid power systems can be used in numerous and versatile applications, many system design considerations must be taken into account:
- Accuracy of movement.
- Speed of movement.
- Pressure and force needed.
- Cost.

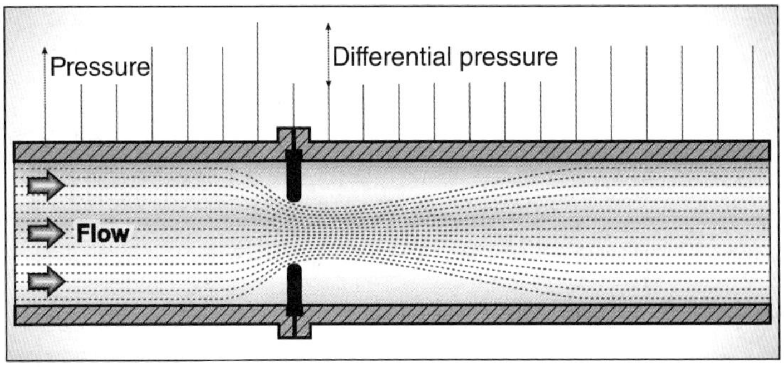

Fouad A. Saad/Shutterstock.com

Figure 37-2. Bernoulli's principle is demonstrated with a venturi. With a constant fluid flow rate in a system, if the diameter of the conductor decreases, the system pressure decreases and the fluid velocity increases in proportion to each other.

The decision to use a hydraulic or pneumatic system depends on the requirements of the specific application. Remember that the compressibility of gases and liquids is quite different. Hydraulic oil can be compressed only slightly; the air or gas in a pneumatic system can be compressed much more. When transmitting force, hydraulic oil behaves like pushing on a solid piece of wood. Air and gas behave like pushing on a spring. Therefore, hydraulic systems produce steadier, more accurate actuator movement. Pneumatic systems are not as suitable in applications that require high accuracy. In contrast, pneumatic applications are more easily adapted to applications in which speed of movement is imperative. Additionally, although hydraulic systems lack speed of movement, they are much more adept when high pressure is used to create more power.

Due to the cost of components and the complexity of many modern fluid power systems (especially hydraulics), the systems and machines are initially very expensive. However, their versatility and expected long service life with proper maintenance offset the initial investment.

Hydraulic Components

Hydraulic power systems use many components, **Figure 37-3**, in order to accomplish tasks, including:
- Power source—converts energy (electrical or chemical) into rotary motion.
- Coupler—connects the power source to the system pump.
- Pump—creates fluid movement through the system.
- Fluid conductors—house and safely carry fluid throughout the entire system.
- Actuators—convert fluid flow to motion and system pressure to force in order to perform work.
- Reservoir—holds and cools the system fluid.
- Directional control valve—guides pump flow to the appropriate actuator and back to the reservoir.

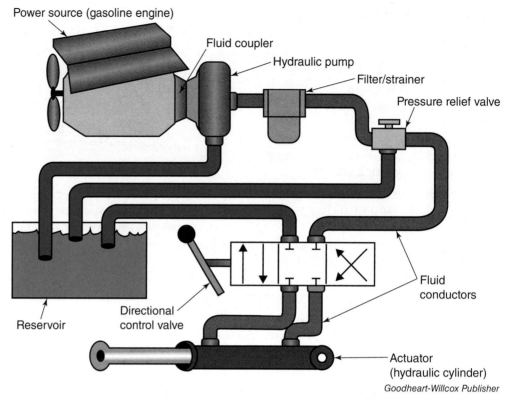

Figure 37-3. The components of a basic hydraulic system. Note that the schematic symbol for a directional control valve is used in this illustration.

- Pressure relief valve—ensures safe operation of a high-pressure system by releasing excess pressure in certain situations.
- Filter/strainer—aids in keeping the system fluid free of contaminants.

Hydraulic Power Sources

The power source of hydraulic systems can be either an electric motor or a gasoline or diesel engine. System designers make application-specific choices to determine the suitable source of power. If electric power is readily available, the system will use an electric motor. If electric power is not available, a gas or diesel engine is used as a power source due to its ease of portability. Whichever method is used, the motor or engine powers the pump to create fluid flow. Flow ensures that the system operates properly and can accomplish work tasks. The power source must have a connection to the hydraulic pump.

Fluid Coupler

Fluid can be used to transfer mechanical power from the power source to the hydraulic pump. The power source drives an impeller that is connected to its drive shaft. A turbine connects to the driven shaft of the hydraulic pump. The turbine and impeller look like bowls with blades around the inside. As the impeller blades spin, centrifugal force throws fluid outward into the turbine blades, causing the turbine to rotate. *Centrifugal force* is a force that causes a rotating object to move outward as it spins. Changes in the power source's speed increase or decrease the impeller speed, which produces faster or slower turbine speeds, respectively. The amount of power transferred to the pump allows the pump to create the needed pump flow to operate the actuators. A common example of a fluid coupler is a torque converter in an automobile with an automatic transmission.

Pumps

Hydraulic pumps are normally classified by their displacement, pumping motion, or fluid delivery characteristics. Displacement classification is based on the volume of fluid moved from the pump inlet to outlet in a single pump cycle. The two major categories of pumps are positive-displacement and non-positive-displacement pumps.

A fluid coupling demonstrates how a non-positive-displacement hydraulic pump operates: fluid enters at the inlet of an impeller and is thrown toward the pump outlet by centrifugal force. Non-positive-displacement pumps are constructed to less strict machining tolerances. This can create fluid flow variations at the pump outlet. Non-positive-displacement pumps are used in low-pressure applications.

Positive-displacement hydraulic pumps create a constant flow output and are the most common type of pump. In order to create fluid flow, positive-displacement pumps use a change in volume from the pump inlet to the pump outlet to move fluid through the pump. An increase in volume is created during the intake portion of pump operation. A decrease in volume then expels the fluid out of the pump.

Pumps in fluid power systems can be further classified as rotary or reciprocating. If the pump uses a circular path to create fluid flow, it is a rotary pump. Rotary pumps use a process similar to the blowers commonly seen on heating and air conditioning units. An inlet allows the fluid to flow into an enclosed rotating wheel that creates flow toward the outlet of the pump.

If an up-and-down motion is produced by a piston, the pump is classified as a reciprocating pump. This type of pump operates very similarly to a two-stroke engine; when the piston is at its lowest travel, fluid is brought into the cylinder. As the piston moves toward its highest point of travel, it applies force and creates fluid flow in the direction of piston travel.

Gear Pumps

One of the most common types of positive-displacement pumps is the *gear pump*. These pumps operate with rotary motion to create fluid flow. Gears are located in a housing that has both an inlet port and an outlet port. The gears are normally meshed together, with one driven by the source of rotary power. The second gear is the driven gear. As the gears rotate, the displaced volume created by the gear teeth coming out of mesh near the inlet port causes atmospheric pressure to push fluid into the pump. As the gears rotate further, they carry the fluid until the teeth mesh again. This forces the fluid to be discharged through the outlet port, **Figure 37-4**.

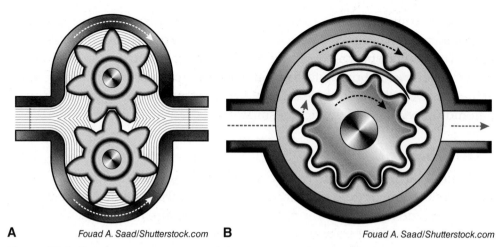

Figure 37-4. Gear pumps use two rotating gears to produce fluid flow at the pump outlet. A—An external gear pump. B—An internal gear pump design.

Vane Pumps

Another type of positive-displacement pump is the *vane pump*, **Figure 37-5**. This type of pump uses vanes (fins) to create fluid flow through volume displacement. Much like gear pumps, the rotary motion of the vanes moves the fluid through the pump. The parts of a vane pump include:

- Housing.
- Cam ring.
- Slotted rotor.
- Vanes.
- Inlet and outlet ports.

The vanes are rotated inside the cam ring via the slotted rotor. Because the vanes are movable, centrifugal force extends them out to touch the cam ring. This creates a seal between the vane and cam ring. As the vanes move in an offset rotary motion, small chambers open between the vanes at the pump inlet. Fluid enters the small chambers and moves through the pump. The fluid is forced out of the pump as the vanes retract near the pump outlet and the chamber spaces get smaller. Vane pumps can be used in high-pressure situations with high fluid-flow properties.

Piston Pumps

Piston pumps are positive-displacement pumps that can be found in many hydraulic systems. Not to be confused with a hydraulic cylinder, the piston pump has a housing that supports the piston and a motion conversion mechanism. The conversion mechanism changes the rotary motion of the power

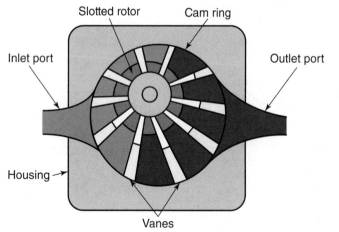

Figure 37-5. Vane pump operation.

source's drive shaft to the usable motion needed by the reciprocating piston. As the piston moves downward in its cylinder, fluid is drawn in. The piston then starts upward in the cylinder and forces the fluid out the pump outlet.

Different piston pump designs can have multiple pistons or pistons that are set parallel to the pump's drive shaft. Piston pumps are very efficient in their design and can be used in very high-pressure applications.

System Controls

Fluid flow control in a hydraulic system is accomplished through the use of various valve types. A pressure relief valve installed in many systems directs flow back to the reservoir when system pressure exceeds the valve's safety pressure setting. Basic check valves allow fluid to flow in one direction only, preventing flow in the opposite direction. These types of valves are part of the system but cannot be directly controlled by the machine operator.

A directional control valve is the most commonly used control valve in fluid power systems. To create movement of the actuator(s), it uses a spool valve to control and limit fluid flow through the system. The spool valve has a single inlet and multiple outlets that are separated by a spool in the valve body. Depending on the configuration, the operator uses a lever with multiple positions to incrementally change the flow route and limit flow through the directional control valve. Fluid flows into the valve and is then either held or is fully or partially directed toward the actuator to be used. See **Figure 37-6**. Depending on the hydraulic system design, the fluid could also bypass through the control valve and back to the reservoir until the operator redirects flow to an actuator.

Most hydraulic systems operate with either an open-center or a closed-center directional control valve design. The open-center system design uses continuous fluid flow through the directional control valve (neutral position) back to the reservoir until the directional control valve is moved to direct flow to an actuator. The closed-center system uses a directional control valve that blocks and holds fluid flow and pressure when it is not positioned to divert flow to an actuator (neutral position). The closed-center design requires a hydraulic pump that can vary its output flow based on system demand.

Hydraulic Fluid Conductors

The fluid flow created by the pumping mechanism needs a fluid conductor—a path to reach the actuator or work mechanism. Hydraulic pipe, tubing, and hoses are used for this purpose. Hydraulic system designers consider the strength of the conductor, its ability to withstand shock from pressure loads and reversals, and its ability to limit resistance to fluid flow when choosing a fluid conductor for a system.

Fluid conductors must exceed maximum system pressure ratings. SAE-approved hydraulic hoses and tubing are manufactured with a safety rating of four, which means that the pressure at which the hose or tube will fail is four times greater than its rated operating pressure. Conductors must also be

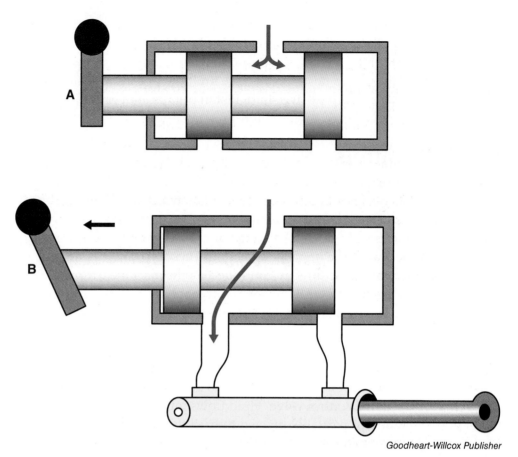

Figure 37-6. A—When the directional control valve in this closed-center system is moved to the neutral position, fluid flow is blocked at both valve outlets. B—When the directional control valve is moved to extend the cylinder, the spool valve slides in the valve body to open the outlet that leads to the actuator's extend port but blocks the second valve outlet. To retract the cylinder, the spool is shifted in the opposite direction so that the actuator's retract port is open to fluid flow, but the spool blocks the first valve outlet.

able to handle pressure shocks created by some loads on the system. Using conductors with the correct operating pressure rating can minimize system pressure shocks.

Due to the physics of fluids moving through conductors at high pressures and velocities, fluids and conductors can create resistance. System pressure will be dramatically lowered if conductors create added resistance. Resistance to fluid flow in hydraulic conductors can be caused by:

- Conductor cross-sectional area.
- Conductor bends or fittings.
- Turbulence in the fluid stream from excessive fluid velocity or sharp fluid conductor turns in the system's design.
- Fluid viscosity (measure of the thickness of a fluid).
- Inhibition of fluid flow in conductors, pumps, and actuators.

Each of the three types of fluid conductors exhibits different characteristics that can possibly reduce or increase these causes of resistance within a hydraulic system.

Pipe

Pipe is a rigid fluid conductor typically made from mild steel. Pipe ends and fittings are threaded to facilitate connections within the system. However, it is important to limit the number of pipe seams due to the possibility of leakage at the threaded connections. Pipe size is listed in nominal sizes that do not reflect the true inside or outside diameter of the pipe. A pipe with a 2″ nominal size might have an outside diameter of 2.375″ but an inside diameter of 2.067″ (schedule 40), 1.939″ (schedule 80 HD), or 1.689″ (schedule 160).

Pipe is available in different wall thicknesses, which are identified by established *schedule numbers*, **Figure 37-7**. In the 2″ nominal size pipe example, a higher schedule number means a thicker wall and decreased inside pipe diameter. As the wall thickness increases, the pipe is able to handle higher operating pressures. Schedule numbers can range from 10 to 160, but most hydraulic fluid power systems use 40, 80, or 160. Schedule 40 is normally used in low-pressure pipes, and schedules 80 or 160 are used in high-pressure pipes.

Tubing

Tubing, like pipe, is a metal conductor used in fluid power systems. It can be bent and formed at different angles to fit around components or into tight spaces. Galvanized or copper tubing is not used because the properties of these two metals can cause chemical reactions with the hydraulic fluid, resulting in damage to the system. As with piping, tubing has varying inside diameter measurements based on different wall thicknesses. The outside dimension, however, is the actual outside diameter. As wall thickness increases, the tube is able to safely house higher hydraulic operating pressures.

Hose

Hose is a flexible fluid conductor commonly made of synthetic rubber. If system components require flexible connections or when severe vibration is present in the system, flexible hoses are most adaptable. A hose is typically constructed with three layers: an inner layer that is flexible and made of an oil-resistant material, a middle layer for reinforcement, and an outer

Nominal Size	Outside Diameter	Inside Diameter		
		Schedule 40	Schedule 80	Schedule 160
1/4″	0.540″	0.364″	0.302″	—
3/8″	0.675″	0.493″	0.423″	—
1/2″	0.840″	0.622″	0.546″	0.466″
3/4″	1.050″	0.824″	0.742″	0.618″
1″	1.315″	1.049″	0.957″	0.815″
1 1/2″	1.900″	1.610″	1.500″	1.338″

Goodheart-Willcox Publisher

Figure 37-7. Nominal sizes of pipe.

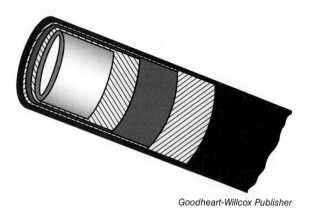

Figure 37-8. Hydraulic hose is manufactured with a minimum of three distinct layers.

protection layer that resists abrasion and shields against extreme environmental conditions, **Figure 37-8**. Sometimes additional layers between the three layers help reinforce and bind them together.

Fittings

Fittings are a main component in fluid power systems. Fittings allow the ends of conductors and system components to be mated together in a secure and leak-free manner. Due to the use of component attachments and conductors of different sizes, fittings may also serve as adapters to properly mate components. Fittings called *quick-disconnect couplings* are used to frequently or rapidly connect or disconnect components and tools from a fluid power system. These fittings are usually operated by hand, can be locked in place, and have check valves to prevent fluid and pressure loss.

Cylinders

Hydraulic cylinders convert fluid flow into linear motion or force. This force or movement is used to actuate the load. A basic hydraulic cylinder has many different parts, **Figure 37-9**:

- Cap.
- Piston.
- Rod.
- Ports.
- Head.
- Seals.

The cylinder is a tube-shaped container that contains a movable piston and rod. A *cap* is used to seal the end of the cylinder. The *piston* is attached to the rod and is used to provide a movable seal inside the cylinder. System pressure created by fluid flow and the load on the cylinder acts on the surface area of the piston. The *rod* is connected to the piston and transfers the generated force in a linear motion. *Ports* located on the cylinder direct fluid flow into or out of the cylinder. The *head* of the cylinder encloses the cylinder tube end where the rod exits and enters. Located on the head of the cylinder and creating an area where fluid cannot escape is the *seal*. Seals are used to prevent internal or external leakage of fluid.

Recall that fluid power systems create fluid flow. As the flow enters the cylinder, force is generated on the piston located

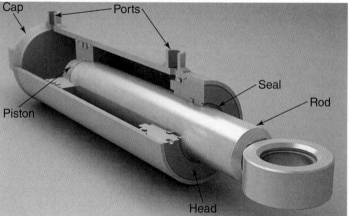

Figure 37-9. Study the parts of a hydraulic cylinder.

inside the cylinder. The output force of the cylinder and cylinder speed are determined by controlling the system pressure and the fluid flow rate, respectively. A larger piston surface area creates more output force, but a shorter piston stroke, than a smaller piston surface area given similar fluid flow. Therefore, controlling the rate of fluid flow and pressure will result in different force and speed applied to the piston and rod.

Motors

Hydraulic motors operate in a similar manner to hydraulic cylinders. The difference is that instead of producing linear motion from fluid power, motors produce rotary motion. They are commonly called *rotary actuators*. Hydraulic motor design closely resembles positive-displacement hydraulic pump design. Fluid power drives the motor to constantly turn an output shaft that delivers rotary actuator output force.

Reservoirs and Fluids

A hydraulic reservoir stores fluid for the system to use and also acts as a storage place for returning fluid in the system. Generally, the reservoir has two to three times the capacity of the pump. Pump capacity is rated in gallons per minute, or gpm. As fluid is moved through the system, the reservoir also aids in cooling, but cooling in systems that produce too much heat should also incorporate an oil cooler.

Hydraulic fluids must be kept clean. Contamination control in a hydraulic system, especially large, complex machines, is a very important process that can prevent major system and component problems. Filters and strainers are sometimes used to ensure that contaminants are removed from the system fluid. Strainers are usually installed in-line and use an element containing fine mesh screens to remove larger contaminants. Filters are typically round and contain many layers of porous material to trap small contaminants. It is essential that strainers and filters remove contaminants to ensure safe operation of the system and to reduce component wear. Always ensure that selected fluids for hydraulic applications meet the manufacturer's specifications.

Pneumatic Power Fundamentals

Pneumatic systems use pressurized air as an energy source to perform operations. Recall the compressibility characteristics of gases described earlier in this chapter. Compared to a hydraulic system, transmitting force in a pneumatic system is less precise, and actuator movement can be jerky. A pneumatic system, however, can produce faster, more responsive actuator movement. For applications requiring low operating pressures and quick movement, a pneumatic system is commonly the more economical option.

Air for a pneumatic system is provided by a compressor that takes air from outside the unit and compresses it to increase the air's pressure. An electric motor or, on some larger or remote applications, a gasoline or diesel engine is used to power the compressor. The electric motor or

990 Agricultural Mechanics and Technology Systems

AlexLMX/Shutterstock.com

Figure 37-10. A typical air compressor.

internal combustion engine creates rotary motion that drives the input shaft of the air compressor. The *air compressor* draws external air through a filter and into the unit, where the air is compressed inside the unit's chamber or cylinder, **Figure 37-10**.

Similar to hydraulic pumps, compressors are broadly classified as positive-displacement or non-positive-displacement units. Positive-displacement compressors compress air in a chamber that can be mechanically reduced in volume to increase the pressure of the air. Non-positive-displacement compressors use vanes or impellers that rotate at high speeds to compress air through the velocity of air movement.

Boyle's Law

Robert Boyle, a 17th-century scientist, developed a law about the behavior of gases that is the foundation of pneumatic systems. Boyle's law states that in a closed system with a constant temperature, the pressure of a gas is inversely proportional to its volume. For example, in a pneumatic system, when an air compressor reduces the volume of air to 1/3 its original volume, the air's pressure becomes three times its original pressure. See **Figure 37-11**.

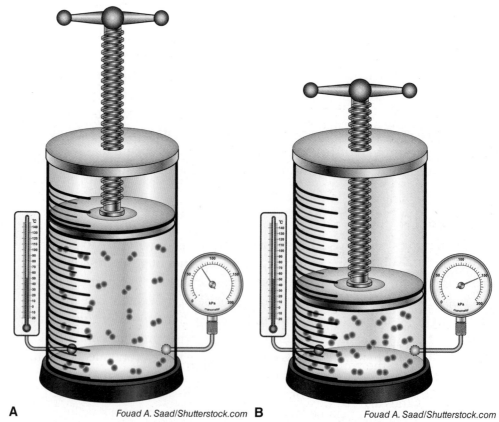

A *Fouad A. Saad/Shutterstock.com* B *Fouad A. Saad/Shutterstock.com*

Figure 37-11. An illustration of Boyle's law. A—A given volume of gas at 50°C and a pressure of 75 kPa. B—When the volume is reduced by half, the pressure doubles to 150 kPa.

Copyright Goodheart-Willcox Co., Inc.

Pneumatic Components

Most pneumatic systems have a power source that drives the air compressor. The air compressor builds air pressure to be used by actuators. The compressed air is delivered to the actuator(s) through an air line (conductor). Air lines can be made of flexible or non-flexible materials. The most suitable supply pathway of compressed air for the actuator and job site is dictated by the application. Many of the components of pneumatic systems are similar to their hydraulic counterparts, with a few exceptions.

Compressor Design Types and Operation

Numerous air compressors with different designs are available. Each design, however, works on the same principle of decreasing the volume of the space containing the air in order to increase its pressure.

Reciprocating-Piston

One of the most common designs is the reciprocating-piston compressor. This design uses a piston in a cylinder to pull in atmospheric air and compress it. Various designs may use one piston and cylinder or multiple pistons, each with a separate cylinder. A single-acting compressor operates much like an internal-combustion engine. A crankshaft turns in a circular motion, driving the piston up or down via the connecting rod. Air is sucked into the cylinder by the vacuum created on the piston's downstroke and is compressed when the piston decreases the volume of the cylinder on the upward stroke. Many units use valves that regulate airflow and pressurization.

Double-acting compressors use more than one piston and cylinder to increase air pressure. The difference between this design and a single-acting compressor is that air intake and compression occur simultaneously; one piston is drawing air into its cylinder as the other piston is compressing the air in its cylinder. It is an efficient design that is capable of generating higher pressures than single-acting compressors. The air that is compressed in the first cylinder passes into the second cylinder, where it is further compressed, further increasing the air pressure.

Rotary Vane

Another air compressor design is the rotary vane compressor, which uses vanes to create air pressure. The vanes of the pump, extending off the rotary pump shaft, are rotated or spun. As in fluid power pumps, the shaft is offset, which allows the rotor to seal the sides of the enclosed air chamber at specific times. This design draws atmospheric air into the chamber as the volume of the chamber increases, and compresses and expels the air when the vanes seal to the wall and decrease the volume of the chamber. The rotary motion of the vane pump creates centrifugal force, which aids in sealing the vanes against the chamber wall because the vanes are not fixed. This allows for a lower tolerance requirement in production and helps reduce component wear.

Rotary Screw

Rotary screw compressors use two meshing helical screws that rotate to form air chambers to compress air. The power source drives one screw, which meshes with the driven screw to move air in a linear fashion through the compressor. At the pump inlet, the screws create a larger volume space for the air. As the air is captured in the spaces between the meshed screws, it is compressed due to the smaller volume of airspace at the pump outlet. These positive-displacement pump designs provide high, constant volume with low maintenance.

Pressure Regulator

Many pneumatic systems have pressure regulators that can be adjusted by the operator for specific actuators. A pressure regulator allows a more precise operation of actuators in the system. Pneumatic systems commonly produce more pressure than actuators use. The pressure regulator adjusts the pressure so that only the required pressure is applied to the actuator. Conversely, with prolonged use, actuators may use more pneumatic energy than is stored, so regulators help provide a more consistent source of energy. See **Figure 37-12**.

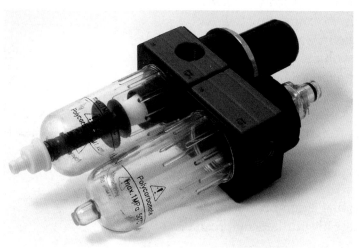

BEEE/Shutterstock.com
Figure 37-12. A pneumatic pressure regulator.

Lubricators

Air-line lubricators introduce friction-reducing oil into the pneumatic system to improve performance and service life of actuators and pneumatic tools. Lubricators must be adjusted properly, because too much lubrication can damage tools and actuators and lessen their performance. Manufacturers are now designing pneumatic components that are permanently lubricated to remove the cost and possible hazards of lubricators.

Water Separators

Air contains other elements in addition to oxygen, including water. A moisture separator is located in the air pathway to actuators and is normally found in air storage containers on the air compressor itself. The separator traps and removes accumulated moisture in the air before it can damage pneumatic tools and actuators.

Pneumatic Conductors

Rigid or flexible conductors are used to move pressurized air to the actuator or tool. Flexible conductors are most commonly used, especially when tools are used in various places around the air compressor. Rigid-supply conductors are normally used when multiple outlets of pressurized air are needed in an area where a large, central air compressor is used. Numerous outlets can be piped off the supply line of the pressurized rigid conductor, allowing multiple operators to use pneumatic power from the system.

Most pneumatic tools and actuators are connected to pneumatic conductors with either fixed connections or quick-disconnect connections. Fixed connections allow for tight, leak-free connections when installed correctly. Quick-disconnect couplings also provide leak-free connections, but allow operators to swiftly change implements without the aid of tools when needed. The female end of the quick-disconnect connections has a sliding collar that must be pulled back when pressed onto the male end and then slid forward to create the locking connection with the system.

Safety Note

Exercise caution when using flexible conductors. Due to the high pressure some conductors carry, they could burst or whip around erratically, causing injury to workers and damage to equipment.

System Controls

Pneumatic system controls resemble hydraulic system controls. Pressure relief valves and check valves operate in the same manner. Directional control valves in pneumatic systems are also similar, although they are constructed of lighter material due to the lower system pressures. Directional control valves use a spool valve inside the valve body that can be adjusted by the operator to block or limit airflow to actuators, just as they do in a hydraulic system.

Pneumatic System Maintenance

Pneumatic system maintenance should be practiced daily. Air compressors should be checked to ensure that they are in good working order and meet the manufacturer's minimum performance standards. Many air compressors use oil for the moving parts, which aids in its operation and prolongs the life of the compressor.

Operators should drain water that is collected in storage containers on air compressors. Draining the water ensures that less moisture will enter the pneumatic system. Water or moisture can damage tools and actuators. If moisture separators and drains are located in the lines of the pneumatic system, they should be drained to reduce moisture in the system. Furthermore, all system connections should be checked daily to ensure that no leaks are present. Pneumatic component damage, system leaks, and other system problems must be repaired, if any are found, before the system is put back into operation.

CHAPTER 37 Review and Assessment

Summary

- Fluid power systems use liquids (hydraulic) or gases (pneumatic) to perform work.
- Pascal's law defines the relationships between force, pressure, and a specific area.
- Bernoulli's principle explains that in a system with a constant fluid flow rate, if the diameter of the fluid conductor changes, fluid velocity and pressure change proportionally in opposite directions.
- Hydraulic power systems include a power source, coupler, pump, fluid conductors, actuators, a reservoir, directional control valves, pressure relief valves, and filters and strainers.
- The three most common types of hydraulic pumps are gear pumps, vane pumps, and piston pumps.
- Compared to a hydraulic system, transmitting force in a pneumatic system is less precise, and actuator movement can be jerky.
- Components of a pneumatic power system include an air compressor, pressure regulator, lubricators, water separators, pneumatic conductors, and system controls.
- Pneumatic system maintenance should be practiced daily.

Words to Know ↗

Match the key terms from the chapter to the correct definition.

A. air compressor	E. hydraulic system	H. rod
B. centrifugal force	F. piston pump	I. seal
C. fluid power	G. pneumatic system	J. vane pump
D. gear pump		

1. A fluid power system that uses gases to perform work.
2. A positive-displacement pump that uses a piston to create an up-and-down motion to create fluid flow.
3. A positive-displacement pump that uses rotary motion through gears to create fluid flow.
4. A positive-displacement pump that uses fins to create fluid flow through volume displacement.
5. The part of a hydraulic cylinder that is connected to the piston and transfers the generated force in a linear motion.
6. A device that draws in external air through a filter and compresses it for use as an energy source to perform work.
7. A device located on the head of the cylinder that creates an area where fluid cannot escape; used to prevent internal or external leakage of fluid.

8. Power obtained by using fluids (liquids or gases) to complete work.
9. A fluid power system that uses liquids to perform work.
10. A force that causes a rotating object to move outward as it spins.

Know and Understand

Answer the following questions using the information provided in this chapter.

1. What are the two major types of fluid power systems, and how are they different?
2. *True or False?* Pascal's law can be applied in fluid power systems to multiply output force.
3. Name the components of a hydraulic power system.
4. In a(n) _____ pump, a conversion mechanism changes the rotary motion of the power source's drive shaft to reciprocating motion.
5. What types of valves are used in the system control of a fluid power system?
6. Schedule _____ pipes are normally used in low-pressure hydraulic system applications.
   A. 10
   B. 40
   C. 80
   D. 160
7. What component provides the air for a pneumatic system?
8. Boyle's law states that in a closed system with a constant temperature, the pressure of a gas is inversely proportional to its _____.
9. Name three types of air compressors that are commonly used in pneumatic power systems.
10. Briefly describe the maintenance required on pneumatic systems.

STEM and Academic Activities

1. **Science.** Investigate Charles's law of volumes. Write a short report explaining how this law pertains to fluid power technology.
2. **Technology.** Create a poster illustrating how directional control valves are used in a fluid power system.
3. **Engineering.** Conduct research to find a fluid power application that uses a reciprocating-piston air compressor. Study the design and explain why the engineer chose that type, rather than a rotary air compressor.
4. **Math.** Consider a simple hydraulic system that has an input piston with a surface area of 6 in^2 and an output piston with a surface area of 20 in^2. If 24 pounds of force is applied to the input piston, how much force is exerted by the output piston?
5. **Social Science.** Describe three applications for fluid power on a farm or ranch. For each, state whether the system is hydraulic or pneumatic and why.
6. **Language Arts.** Write a report describing safety precautions that should be taken when working with fluid power systems.

Thinking Critically

1. Consider a food processing plant that prepares and packages peanut butter. What type of fluid power system might be used to transfer the peanut butter into its retail packaging (jars)? Why?
2. Can you think of a fluid power application that might contain both pneumatic and hydraulic system components? Why might such a system be needed?

Reference Section

Units of Measure

U.S. CUSTOMARY	METRIC
LENGTH 12 inches = 1 foot 36 inches = 1 yard 3 feet = 1 yard 5,280 feet = 1 mile 16.5 feet = 1 rod 320 rods = 1 mile 6 feet = 1 fathom	1 kilometer = 1000 meters 1 hectometer = 100 meters 1 dekameter = 10 meters 1 meter = 1 meter 1 decimeter = 0.1 meter 1 centimeter = 0.01 meter 1 millimeter = 0.001 meter
WEIGHT 27.34 grains = 1 dram 438 grains = 1 ounce 16 drams = 1 ounce 16 ounces = 1 pound 2000 pounds = 1 short ton 2240 pounds = 1 long ton 25 pounds = 1 quarter 4 quarters = 1 cwt	1 tonne = 1,000,000 grams 1 kilogram = 1000 grams 1 hectogram = 100 grams 1 dekagram = 10 grams 1 gram = 1 gram 1 decigram = 0.1 gram 1 centigram = 0.01 gram 1 milligram = 0.001 gram
VOLUME 8 ounces = 1 cup 16 ounces = 1 pint 32 ounces = 1 quart 2 cups = 1 pint 2 pints = 1 quart 4 quarts = 1 gallon 8 pints = 1 gallon	1 hectoliter = 100 liters 1 dekaliter = 10 liters 1 liter = 1 liter 1 deciliter = 0.1 liter 1 centiliter = 0.01 liter 1 milliliter = 0.001 liter 1000 milliliters = 1 liter
AREA 144 sq. inches = 1 sq. foot 9 sq. feet = 1 sq. yard 43,560 sq. ft. = 160 sq. rods 160 sq. rods = 1 acre 640 acres = 1 sq. mile	100 sq. millimeters = 1 sq. centimeter 100 sq. centimeters = 1 sq. decimeter 100 sq. decimeters = 1 sq. meter 10,000 sq. meters = 1 hectare

FAHRENHEIT		CELSIUS
32° F	Water freezes	0° C
68° F	Reasonable room temperature	20° C
98.6° F	Normal body temperature	37° C
173° F	Alcohol boils	78.34° C
212° F	Water boils	100° C

Goodheart-Willcox Publisher

Millimeter-Inch Equivalents

Inches		Milli-meters	Inches		Milli-meters
Fractions	Decimals		Fractions	Decimals	
	.00394	.1	15/32	.46875	11.9063
	.00787	.2		.47244	12.00
	.01181	.3	31/64	.484375	12.3031
1/64	.015625	.3969	1/2	.5000	12.70
	.01575	.4		.51181	13.00
	.01969	.5	33/64	.515625	13.0969
	.02362	.6	17/32	.53125	13.4938
	.02756	.7	35/64	.546875	13.8907
1/32	.03125	.7938		.55118	14.00
	.0315	.8	9/16	.5625	14.2875
	.03543	.9	37/64	.578125	14.6844
	.03937	1.00		.59055	15.00
3/64	.046875	1.1906	19/32	.59375	15.0813
1/16	.0625	1.5875	39/64	.609375	15.4782
5/64	.078125	1.9844	5/8	.625	15.875
	.07874	2.00		.62992	16.00
3/32	.09375	2.3813	41/64	.640625	16.2719
7/64	.109375	2.7781	21/32	.65625	16.6688
	.11811	3.00		.66929	17.00
1/8	.125	3.175	43/64	.671875	17.0657
9/64	.140625	3.5719	11/16	.6875	17.4625
5/32	.15625	3.9688	45/64	.703125	17.8594
	.15748	4.00		.70866	18.00
11/64	.171875	4.3656	23/32	.71875	18.2563
3/16	.1875	4.7625	47/64	.734375	18.6532
	.19685	5.00		.74803	19.00
13/64	.203125	5.1594	3/4	.7500	19.05
7/32	.21875	5.5563	49/64	.765625	19.4469
15/64	.234375	5.9531	25/32	.78125	19.8438
	.23622	6.00		.7874	20.00
1/4	.2500	6.35	51/64	.796875	20.2407
17/64	.265625	6.7469	13/16	.8125	20.6375
	.27559	7.00		.82677	21.00
9/32	.28125	7.1438	53/64	.828125	21.0344
19/64	.296875	7.5406	27/32	.84375	21.4313
5/16	.3125	7.9375	55/64	.859375	21.8282
	.31496	8.00		.86614	22.00
21/64	.328125	8.3344	7/8	.875	22.225
11/32	.34375	8.7313	57/64	.890625	22.6219
	.35433	9.00		.90551	23.00
23/64	.359375	9.1281	29/32	.90625	23.0188
3/8	.375	9.525	59/64	.921875	23.4157
25/64	.390625	9.9219	15/16	.9375	23.8125
	.3937	10.00		.94488	24.00
13/32	.40625	10.3188	61/64	.953125	24.2094
27/64	.421875	10.7156	31/32	.96875	24.6063
	.43307	11.00		.98425	25.00
7/16	.4375	11.1125	63/64	.984375	25.0032
29/64	.453125	11.5094	1	1.0000	25.4000

Goodheart-Willcox Publisher

METRIC PREFIXES, EXPONENTS, AND SYMBOLS

DECIMAL FORM	EXPONENT OR POWER	PREFIX	SYMBOL	MEANING
1 000 000 000 000 000 000	$= 10^{18}$	exa	E	quintillion
1 000 000 000 000 000	$= 10^{15}$	peta	P	quadrillion
1 000 000 000 000	$= 10^{12}$	tera	T	trillion
1 000 000 000	$= 10^{9}$	giga	G	billion
1 000 000	$= 10^{6}$	mega	M	million
1 000	$= 10^{3}$	kilo	k	thousand
100	$= 10^{2}$	hecto	h	hundred
10	$= 10^{1}$	deka	da	ten
1				base unit
0.1	$= 10^{-1}$	deci	d	tenth
0.01	$= 10^{-2}$	centi	c	hundredth
0.001	$= 10^{-3}$	milli	m	thousandths
0.000 001	$= 10^{-6}$	micro	μ	millionth
0.000 000 001	$= 10^{-9}$	nano	n	billionth
0.000 000 000 001	$= 10^{-12}$	pico	p	trillionth
0.000 000 000 000 001	$= 10^{-15}$	femto	f	quadrillionth
0.000 000 000 000 000 001	$= 10^{-18}$	atto	a	quintillionth

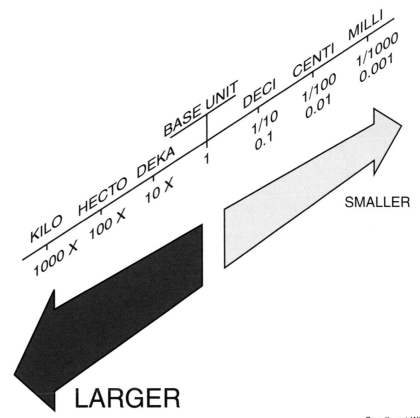

Goodheart-Willcox Publisher

Conversion Table: US Customary to SI Metric

When You Know:	Multiply By:		To Find:
	Very accurate	Approximate	
Length			
inches	* 25.4		millimeters
inches	* 2.54		centimeters
feet	* 0.3048		meters
feet	* 30.48		centimeters
yards	* 0.9144	0.9	meters
miles	* 1.609344	1.6	kilometers
Weight			
grains	15.43236	15.4	grams
ounces	* 28.349523125	28.0	grams
ounces	* 0.028349523125	0.028	kilograms
pounds	* 0.45359237	0.45	kilograms
short ton	* 0.90718474	0.9	tonnes
Volume			
teaspoons		5.0	milliliters
tablespoons		15.0	milliliters
fluid ounces	29.57353	30.0	milliliters
cups		0.24	liters
pints	* 0.473176473	0.47	liters
quarts	* 0.946352946	0.95	liters
gallons	* 3.785411784	3.8	liters
cubic inches	* 0.016387064	0.02	liters
cubic feet	* 0.028316846592	0.03	cubic meters
cubic yards	* 0.764554857984	0.76	cubic meters
Area			
square inches	* 6.4516	6.5	square centimeters
square feet	* 0.09290304	0.09	square meters
square yards	* 0.83612736	0.8	square meters
square miles		2.6	square kilometers
acres	* 0.40468564224	0.4	hectares
Temperature			
Fahrenheit	*5/9 (after subtracting 32)		Celsius

* = Exact

Goodheart-Willcox Publisher

Conversion Table: SI Metric to US Customary

When You Know:	Multiply By:		To Find:
	Very accurate	Approximate	
Length			
millimeters	0.0393701	0.04	inches
centimeters	0.3937008	0.4	inches
meters	3.280840	3.3	feet
meters	1.093613	1.1	yards
kilometers	0.621371	0.6	miles
Weight			
grains	0.00228571	0.0023	ounces
grams	0.03527396	0.035	ounces
kilograms	2.204623	2.2	pounds
tonnes	1.1023113	1.1	short tons
Volume			
milliliters		0.2	teaspoons
milliliters	0.06667	0.067	tablespoons
milliliters	0.03381402	0.03	fluid ounces
liters	61.02374	61.024	cubic inches
liters	2.113376	2.1	pints
liters	1.056688	1.06	quarts
liters	0.26417205	0.26	gallons
liters	0.03531467	0.035	cubic feet
cubic meters	61023.74	61023.7	cubic inches
cubic meters	35.31467	35.0	cubic feet
cubic meters	1.3079506	1.3	cubic yards
cubic meters	264.17205	264.0	gallons
Area			
square centimeters	0.1550003	0.16	square inches
square centimeters	0.00107639	0.001	square feet
square meters	10.76391	10.8	square feet
square meters	1.195990	1.2	square yards
square kilometers		0.4	square miles
hectares	2.471054	2.5	acres
Temperature			
Celsius	*9/5 (then add 32)		Fahrenheit

* = Exact

Goodheart-Willcox Publisher

Concrete Mixtures for Various Applications

Construction Application	Water/Cement Ratio Gal Per Bag	Consistency (amount of slump)	Maximum Size of Aggregate	Appropriate Cement Content Bags Per Yd	Probable 28th Day Strength (psi)
Footings	7	4″ to 6″	1½″	5.0	2800
8″ basement wall moderate groundwater	7	4″ to 6″	1½″	5.0	2800
8″ basement wall severe groundwater	6	3″ to 5″	1½″	5.8	3500
10″ basement wall moderate groundwater	7	4″ to 6″	2″	4.7	2800
10″ basement wall severe groundwater	6	3″ to 5″	2″	5.5	3500
Basement floor 4″ thickness	6	2″ to 4″	1″	6.2	3500
Floor slab on grade	6	2″ to 4″	1″	6.2	3500
Stairs and steps	6	1″ to 4″	1″	6.2	3500
Topping over concrete floor	5	1″ to 2″	⅜″	8.0	4350
Sidewalks, patios, driveways, porches	6	2″ to 4″	1″	6.2	3500

Goodheart-Willcox Publisher

Mortar Types for Classes of Construction

ASTM Mortar Type Designation	Construction Suitability
M	Masonry subjected to high compressive loads, severe frost action, or high lateral loads from earth pressures, hurricane winds, or earthquakes. Structures below grade, manholes, and catch basins.
S	Structures requiring high flexural bond strength, but subject only to normal compressive loads.
N	General use in above grade masonry. Residential basement construction, interior walls and partitions. Concrete masonry veneers applied to frame construction.
O	Nonloadbearing walls and partitions. Solid load bearing masonry of allowable compressive strength not exceeding 100 psi.
K	Interior nonloadbearing partitions where low compressive and bond strengths are permitted by building codes.

Goodheart-Willcox Publisher

Mortar Proportions by Volume

Mortar Needed for Concrete Masonry Units	
Nominal Height and Length of Units in Inches	Cubic Feet of Mortar per 100 Sq Ft
8 × 16	6.0
8 × 12	7.0
5 × 12	8.5
4 × 16	9.5
2¼ × 8	14.0
4 × 8	12.0
5 × 8	11.0
2 × 12	15.0
2 × 16	15.0

Goodheart-Willcox Publisher

Wood Characteristics	Specific Gravity 16 per ft³	Weight (at 12% moisture content)	Working Properties — 1. Excellent 2. Good 3. Average 4. Fair 5. Poor					
Species			Planing	Drilling	Sanding	Turning	Gluing	Nail and Screw Holding (includes split resistance for nailing)
Alder, red	0.41	28	2					
Ash	0.49	34	3	2	2	2	2	3
Banak	0.44	30	1	3	3	3	3	2
Basswood	0.37	24	2	1	1	4	1	1
Beech	0.64	45	1	1	2	3	1	2
Birch	0.62	43	2	2	2	1	2	2
Butternut	0.38	27	1	2	2	2	2	4
Cedar, Aromatic Red	0.37	26	4	1	1	1	2	4
Cherry	0.50	35	1	3	2	3	2	5
Chestnut	0.43	30	1	1	1	1	2	3
Cottonwood	0.40	28	1	1	3	2	1	2
Cypress, bald	0.46	32	2	1	2	2	1	1
Ebony	1.22	63	4	2	2	2	1	3
Elm, American	0.63	44	3	4	4	3	4	2
Fir, Douglas	0.48	34	3	3	2	3	2	3
Gum, red	0.52	35	3	2	2	4	2	3
Hackberry	0.53	37	2	3	3	2	2	2
Hickory	0.72	50	3	2	2	2	2	3
Lauan (Philippine Mahogany)	0.49	35	2	3	2	3	3	4
Limba	0.49	35	1	2	1	2	2	2
Mahoganies, genuine	0.49	35	2	1	1	2	2	3
Mahogany, African	0.49	35	2	2	1	2	2	2
Maple, hard	0.63	38	3	2	1	2	2	2
Maple, soft	0.54	44	3	2	2	2	4	5
Oak, red or white	0.62	43	3	2	2	2	3	4
Paldao	0.59	44	3	2	2	3	3	3
Pecan	0.66	46	3	3	2	3	3	3
Pine, Ponderosa	0.39	28	1	3	2	3	3	4
Pine, sugar	0.35	26	2	1	2	2	2	2
Pine, yellow	0.40	28	2	1	2	2	2	2
Primavera	0.40	30	2	2	3	2	2	3
Redwood	0.40	28	1	2	1	1	2	2
Rosewood	0.75	50	3	1	1	2	1	3
Santos Rosewood (Pau Ferro)			4	4	4	4	4	3
Sapele	0.54	40	2	2	2	2	2	2
Sassafras	0.46	32	3	2	2	3	2	4
Satinwood	0.83	67	3	4	3	4	3	3
Spruce	0.40	28	2	2	3	3	1	3
Sycamore	0.49	34	3	3	3	3	3	2
Teak	0.62	43	4	4	4	4	4	3
Tulip, American (Yellow Poplar)	0.42	30	2	2	2	2	1	1
Walnut, American	0.55	38	2	2	2	3	3	3
Willow	0.39	26	2	3	3	2	1	1
Zebrawood	0.62	48	3	3	3	4	2	4

(Continued)
Goodheart-Willcox Publisher

	1. Excellent	2. Good	3. Average	4. Fair	5. Poor		
	Physical Properties						
Species	Bending Difficult (1) to Easy (5)	Hardness Hard (1) to Soft (5)	Compression Strength Weak (1) to Strong (5)	Shock Resistance Low (1) to High (5)	Stiffness Limber (1) to Stiff (5)	Availability L (Lumber) V (Veneer)	Cost Expensive (1) to Inexpensive (2)
Alder, red							
Ash	3	2	4	1	3	L	5
Banak	4	2	4	4	4	LV	3
Basswood	4	5	3	2	3	L	3
Beech	5	5	2	2	1	L	5
Birch	4	2	4	5	4	LV	4
Butternut	2	2	4	4	4	LV	3
Cedar, Aromatic Red	2	4	2	4	2	LV	2
Cherry	3	3	2	1	1	LV	3
Chestnut	2	3	4	4	4	LV	2
Cottonwood	4	4	2	3	2	L	2
Cypress, bald	2	4	1	2	2	LV	5
Ebony	2	4	3	4	3	L	3
Elm, American	1	1	4	5	5	L	1
Fir, Douglas	5	2	3	4	4	LV	4
Gum, red	3	4	2	4	4	LV	4
Hackberry	3	2	3	4	4	LV	3
Hickory	3	2	3	4	3	LV	4
Lauan (Philippine Mahogany)	4	4	5	5	4	LV	3
Limba	3	3	3	4	3	LV	3
Mahoganies, genuine	2	3	3	4	3	L	3
Mahogany, African	3	3	3	4	3	LV	2
Maple, hard	3	3	3	4	3	LV	2
Maple, soft	2	1	5	5	5	LV	3
Oak, red or white	4	2	5	4	2	LV	3
Paldao	1	4	4	4	4	LV	3
Pecan	2	3	4	4	4	LV	2
Pine, Ponderosa	4	4	4	4	4	LV	3
Pine, sugar	3	4	2	2	3	L	4
Pine, yellow	5	4	2	1	2	L	3
Primavera	3	3	4	4	4	L	4
Redwood	3	3	3	4	3	LV	3
Rosewood	3	3	2	4	4	L	3
Santos Rosewood (Pau Ferro)	2	1	5	4	4	V	1
Sapele	3	3	5	4	4	V	2
Sassafras	2	3	2	4	4	L	3
Satinwood	4	3	4	4	4	V	1
Spruce	4	4	2	2	3	L	4
Sycamore	3	3	3	4	3	L	3
Teak	4	2	4	4	4	LV	1
Tulip, American (Yellow Poplar)	4	4	3	2	3	LV	5
Walnut, American	2	3	4	4	4	LV	2
Willow	4	5	1	4	2	L	5
Zebrawood	4	1	4	4	4	V	1

(Continued)
Goodheart-Willcox Publisher

Finishing Selected Wood Species

Species	Wood Type			Recommended Finishes				
	Softwood	Hardwood		Filler	Stain	Build-Up Topcoat	Penetrating Oil	Paint
		Open Pores	Closed Pores	(R = Required for flat surface)	(O = Optional)			
Alder, red			●		O	■	■	
Ash		●		R	O	■	■	
Banak	●				O	■	■	
Basswood			●					■
Beech			■		O			
Birch			●		O			
Butternut		●		R	O	■	■	
Cedar, Aromatic Red	●					■	■	
Cherry			●		O	■	■	
Chestnut		●		R	O	■	■	
Cottonwood			●					■
Cypress, bald	●					■		■
Ebony			●					
Elm, American		●		R	O	■	■	
Fir, Douglas	●					■	■	■
Gum, red			●		O	■	■	
Hackberry		●		R	O	■	■	
Hickory		●			O	■	■	
Lauan (Philippine Mahogany)		●		R	O	■	■	
Limba		●		R	O	■	■	
Mahoganies, genuine		●		R	O	■	■	
Mahogany, African		●		R	O	■	●	
Maple, hard			●		O	■	●	
Maple, soft			●		O	■		
Oak, red or white		●		R	O	■	■	
Paldao		●		R	O	■	■	
Pecan			●		O	■	■	
Pine, Ponderosa	●				O	■	■	■
Pine, sugar	●				O	■	■	
Pine, yellow	●				O	■	■	
Primavera		●		R	O	■	■	
Redwood	●				O	■	■	
Rosewood			●			■		
Santos Rosewood (Pau Ferro)			●		O	■	■	
Sapele		●		R	O	■	■	
Sassafras		●		R	O	■	■	
Satinwood			●		O	■	■	
Spruce	●							■
Sycamore			●		O	■		
Teak			●					
Tulip, American (Yellow Poplar)			●		O	■		■
Walnut, American		●		R	O	■	■	
Willow			●		O	■	■	
Zebrawood			●			■		

Goodheart-Willcox Publisher

Plumbing Symbols

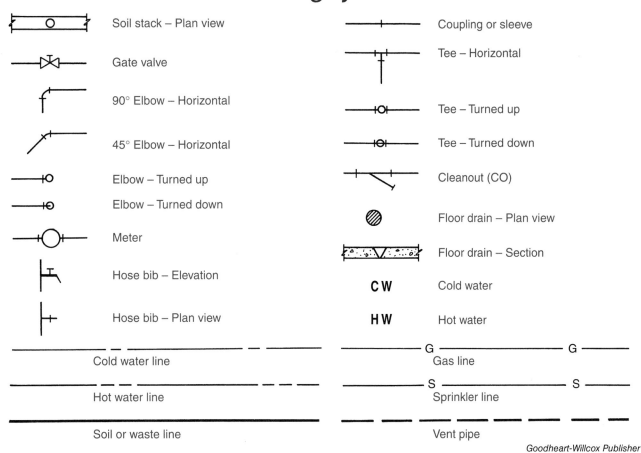

Electrical Symbols

Symbol	Description	Symbol	Description	Symbol	Description
☉	Ceiling fixture	⊖	Single receptacle outlet	$	Single-pole switch
⊙	Recessed fixture	⊖	Duplex receptacle outlet	$_2$	Double-pole switch
⊙D	Drop cord fixture	⊕	Triplex receptacle outlet	$_3$	Three-way switch
⊙F	Fan hanger fixture	⊕	Quadruplex receptacle outlet	$_4$	Four-way switch
⊙J	Junction box	⊖	Split-wired duplex receptacle outlet	$_{WP}$	Weatherproof switch
	Fluorescent fixture	△	Special-purpose single receptacle outlet	$_L$	Low-voltage switch
◀	Telephone	⊖	240-volt receptacle outlet	▫	Push button
◁	Intercom	⊖$_{WP}$	Weatherproof duplex receptacle outlet	CH	Chimes
⊙S	Ceiling fixture with pull switch	⊖$_S$	Duplex receptacle outlet with switch	TV	Television antenna outlet
T	Thermostat	⊖$_{GFCI}$	GFCI receptacle outlet	$_D$	Dimmer switch
⊙ A, B, C, Etc.	Special fixture	⊖ A, B, C, Etc.	Special duplex receptacle outlet	$_{A, B, C, Etc.}$	Special switch
	Flush-mounted panel box				

Goodheart-Willcox Publisher

Formulas

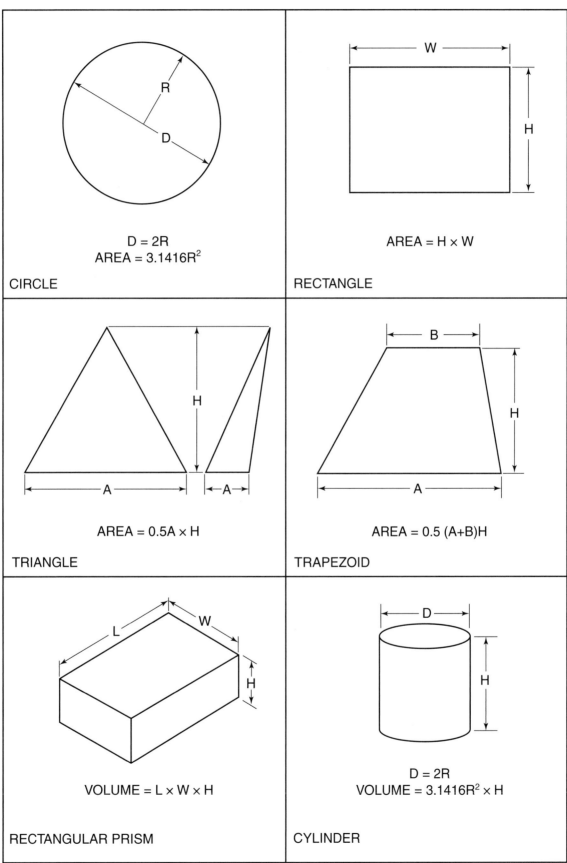

Formulas

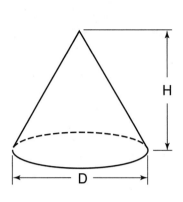

$D = 2R$
$$\text{VOLUME} = \frac{3.1416 R^2 \times H}{3}$$

CONE

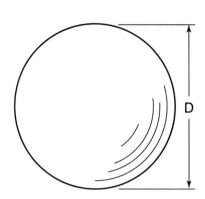

$D = 2R$
$$\text{VOLUME} = \frac{4 \times 3.1416 R^3}{3}$$

SPHERE

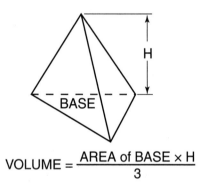

$$\text{VOLUME} = \frac{\text{AREA of BASE} \times H}{3}$$

TRIANGULAR PYRAMID

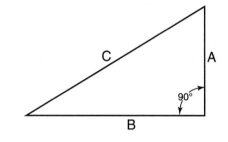

$A = \sqrt{C^2 - B^2}$
$B = \sqrt{C^2 - A^2}$
$C = \sqrt{A^2 + B^2}$

PYTHAGOREAN THEOREM

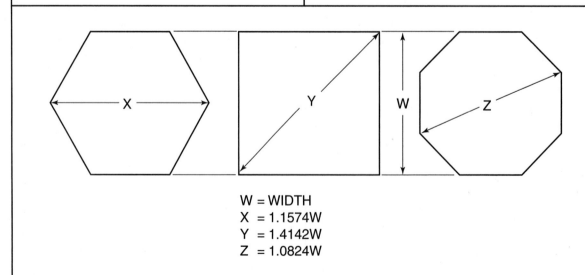

W = WIDTH
X = 1.1574W
Y = 1.4142W
Z = 1.0824W

Goodheart-Willcox Publisher

Copyright Goodheart-Willcox Co., Inc.

Common Shapes of Metals

Shapes		Length	How Measured	How Purchased
	Sheet less than 1/4″ thick	Up to 144″	Thickness × width, widths to 72″	Weight, foot, or piece
	Plate more than 1/4″ thick	Up to 20′	Thickness × width	Weight, foot, or piece
	Band	Up to 20′	Thickness × width	Weight or piece
	Rod	12′ to 20′	Diameter	Weight, foot, or piece
	Square	12′ to 20′	Width	Weight, foot, or piece
	Flats	Hot rolled 20′ to 22′ Cold finished	Thickness × width	Weight, foot, or piece
	Hexagon	12′ to 20′	Distance across flats	Weight, foot, or piece
	Octagon	12′ to 20′	Distance across flats	Weight, foot, or piece
	Angle	Up to 40′	Leg length × leg length × thickness of legs	Weight, foot, or piece
	Channel	Up to 60′	Depth × web thickness × flange width	Weight, foot, or piece
	I-beam	Up to 60′	Height × web thickness × flange width	Weight, foot, or piece

Goodheart-Willcox Publisher

Color Codes for Marking Steels

S.A.E. Number	Color Code	S.A.E. Number	Color Code	S.A.E. Number	Color Code	S.A.E. Number	Color Code
	Carbon steels	2115	Red and bronze	T1340	Orange and green	3450	Black and bronze
1010	White	2315	Red and blue	T1345	Orange and red	4820	Green and purple
1015	White	2320	Red and blue	T1350	Orange and red		**Chromium steels**
X1015	White	2330	Red and white		**Nickel-chromium steels**	5120	Black
1020	Brown	2335	Red and white	3115	Blue and black	5140	Black and white
X1020	Brown	2340	Red and green	3120	Blue and black	5150	Black and white
1025	Red	2345	Red and green	3125	Pink	52100	Black and brown
X1025	Red	2350	Red and aluminum	3130	Blue and green		**Chromium-vanadium steels**
1030	Blue	2515	Red and black	3135	Blue and green		
1035	Blue		**Molybdenum steels**	3140	Blue and white	6115	White and brown
1040	Green	4130	Green and white	X3140	Blue and white	6120	White and brown
X1040	Green	X4130	Green and bronze	3145	Blue and white	6125	White and aluminum
1045	Orange	4135	Green and yellow	3150	Blue and brown	6130	White and yellow
X1045	Orange	4140	Green and brown	3215	Blue and purple	6135	White and yellow
1050	Bronze	4150	Green and brown	3220	Blue and purple	6140	White and bronze
1095	Aluminum	4340	Green and aluminum	3230	Blue and purple	6145	White and orange
	Free cutting steels	4345	Green and aluminum	3240	Blue and aluminum	6150	White and orange
1112	Yellow	4615	Green and black	3245	Blue and aluminum	6195	White and purple
X1112	Yellow	4620	Green and black	3250	Blue and bronze		**Tungsten steels**
1120	Yellow and brown	4640	Green and pink	3312	Orange and black	71360	Brown and orange
X1314	Yellow and blue	4815	Green and purple	3325	Orange and black	71660	Brown and bronze
X1315	Yellow and red	X1340	Yellow and black	3335	Blue and orange	7260	Brown and aluminum
X1335	Yellow and black		**Manganese steels**	3340	Blue and orange		**Silicon-manganese steels**
	Nickel steels	T1330	Orange and green	3415	Blue and pink		
2015	Red and brown	T1335	Orange and green	3435	Orange and aluminum	9255	Bronze and aluminum
						9260	Bronze and aluminum

Goodheart-Willcox Publisher

Metal Sheet Materials Chart

Material (Sheet less than 1/4″ thick)	How Measured	How Purchased	Characteristics
Copper	Gage number (Brown & Sharpe and Amer. Std.)	24″ × 96″ sheet or 12″ or 18″ by lineal feet on roll	Pure metal
Brass	Gage number (Brown & Sharpe and Amer. Std.)	24″ × 76″ sheet or 12″ or 18″ by lineal feet on roll	Alloy of copper and zinc
Aluminum	Decimal	24″ × 72″ sheet or 12″ or 18″ by lineal feet on roll	Available as commercially pure metal or alloyed for strength, hardness, and ductility
Galvanized steel	Gage number (Amer. Std.)	24″ × 96″ sheet	Mild steel sheet with zinc plating, also available with zinc coating that is part of sheet
Black annealed steel sheet	Gage number (Amer. Std.)	24″ × 96″ sheet	Mild steel with oxide coating, hot-rolled
Cold-rolled steel sheet	Gage number (Amer. Std.)	24″ × 96″ sheet	Oxide removed and cold-rolled to final thickness
Tin plate	Gage number (Amer. Std.)	20″ × 28″ sheet 56 or 112 to pkg.	Mild steel with tin coating
Nickel silver	Gage number (Brown & Sharpe)	6″ or 12″ wide by lineal feet on roll	Copper 50%, zinc 30%, nickel 20%
Expanded	Gage number (Amer. Std.)	36″ × 96″ sheet	Metal is pierced and expanded (stretched) to diamond shape; also available rolled to thickness after it has been expanded
Perforated	Gage number (Amer. Std.)	30″ × 36″ sheet 36″ × 48″ sheet	Design is cut in sheet; many designs available

Goodheart-Willcox Publisher

Number and Letter Size Drills Conversion Chart

Drill No. or Letter	Inch	mm	Drill No. or Letter	Inch	mm	Drill No. or Letter	Inch	mm	Drill No. or Letter	Inch	mm	Drill No. or Letter	Inch	mm	Drill No. or Letter	Inch	mm
	.001	0.0254					.101	2.5654	7	.201	5.1054	N	.301	7.6454		.401	10.1854
	.002	0.0508	38	.1015			.102	2.5908		.202	5.1308		.302	7.6708		.402	10.2108
	.003	0.0762					.103	2.6162		.203	5.1562		.303	7.6962		.403	10.2362
	.004	0.1016	37			13/64	.104	2.6416		.2031	5.1594		.304	7.7216	Y	.404	10.2616
	.005	0.1270					.105	2.6670	6	.204	5.1816		.305	7.7470		.405	10.2870
	.006	0.1524	36		1065	5	.106	2.6924		.205	5.2070		.306	7.7724		.406	10.3124
	.007	0.1778					.107	2.7178		.206	5.2324		.307	7.7978	13/32	.4062	10.3187
	.008	0.2032					.108	2.7432		.207	5.2578		.308	7.8232		.407	10.3378
	.009	0.2286					.109	2.7686		.208	5.2832		.309	7.8486		.408	10.3632
	.010	0.2540			7/64	4	.1094	2.7781		.209	5.3086		.310	7.8740		.409	10.3886
	.011	0.2794	35				.110	2.7940		.210	5.3340		.311	7.8994		.410	10.4140
	.012	0.3048	34				.111	2.8194		.211	5.3594		.312	7.9248		.411	10.4394
	.013	0.3302					.112	2.8448		.212	5.3848	5/16	.3125	7.9375		.412	10.4648
80	.0135		33				.113	2.8702	3	.213	5.4102		.313	7.9502	Z	.413	10.4902
79	.0145			.015	0.3810		.114	2.8956		.214	5.4356		.314	7.9756		.414	10.5156
	1/64	0.3969		.0156			.115	2.9210		.215	5.4610		.3150	8.0000		.415	10.5410
78	.016	0.4064	32				.116	2.9464		.216	5.4864		.315	8.0010		.416	10.5664
	.017	0.4318					.117	2.9718		.217	5.5118	O	.316	8.0264		.417	10.5918
77	.018	0.4572					.118	2.9972		.218	5.5372		.317	8.0518		.418	10.6172
	.019	0.4826				7/32	.1181	3.0000		.2187	5.5562		.318	8.0772		.419	10.6426
76	.020	0.5080					.119	3.0226		.219	5.5626		.319	8.1026		.420	10.6680
75	.021	0.5334	31				.120	3.0480		.220	5.5880		.320	8.1280		.421	10.6934
	.022	0.5588					.121	3.0734	2	.221	5.6134		.321	8.1534	27/64	.4219	10.7156
74	.0225			.023	0.5842		.122	3.0988		.222	5.6388		.322	8.1788		.422	10.7188
73	.024	0.6096					.123	3.1242		.223	5.6642	P	.323	8.2042		.423	10.7442
72	.025	0.6350					.124	3.1496		.224	5.6896		.324	8.2296		.424	10.7696
71	.026	0.6604			1/8		.125	3.1750		.225	5.7150		.325	8.2550		.425	10.7950
	.027	0.6858					.126	3.2004		.226	5.7404		.326	8.2804		.426	10.8204
70	.028	0.7112					.127	3.2258		.227	5.7658		.327	8.3058		.427	10.8458
	.029	0.7366				1	.128	3.2512		.228	5.7912		.328	8.3312		.428	10.8712
69	.0292			.030	0.7620	30	.1285			.229	5.8166	21/64	.3281	8.3344		.429	10.8966
68	.031	0.7874					.129	3.2766		.230	5.8410		.329	8.3566		.430	10.9220
	1/32			.0312	0.7937		.130	3.3020		.231	5.8674		.330	8.3820		.431	10.9474
67	.032	0.8128					.131	3.3274		.232	5.8928		.331	8.4074		.432	10.9728
66	.033	0.8382					.132	3.3528		.233	5.9182	Q	.332	8.4328		.433	10.9982
	.034	0.8636					.133	3.3782	A	.234	5.9436		.333	8.4582		.4331	11.0000
65	.035	0.8890					.134	3.4036					.334	8.4836		.434	11.0236
64	.036	0.9144	29			15/64	.135	3.4290		.2344	5.9531		.335	8.5090		.435	11.0490
63	.037	0.9398					.136	3.4544		.235	5.9690		.336	8.5344		.436	11.0744
62	.038	0.9652					.137	3.4798		.236	5.9944		.337	8.5598		.437	11.0998
61	.039	0.9906					.138	3.5052		.2362	6.0000		.338	8.5852	7/16	.4375	11.1125
	.0394	1.0000	28	.1405			.139	3.5306		.237	6.0198		.339	8.6106		.438	11.1252
60	.040	1.0160			9/64		.140	3.5560	B	.238	6.0452	R	.340	8.6360		.439	11.1506
59	.041	1.0414					.1406	3.5519		.239	6.0706		.341	8.6614		.440	11.1760
58	.042	1.0668					.141	3.5814		.240	6.0960		.342	8.6868		.441	11.2014
57	.043	1.0922					.142	3.6068		.241	6.1214		.343	8.7122		.442	11.2268
	.044	1.1176	27			C	.143	3.6322		.242	6.1468	11/32	.3437	8.7312		.443	11.2522
	.045	1.1430					.144	3.6576		.243	6.1722		.344	8.7376		.444	11.2776
56	.0465			.046	1.1684		.145	3.6830		.244	6.1976		.345	8.7630		.445	11.3030
	3/64			.0469	1.1906	26				.146	3.7084	D	.245	6.2230		.346	8.7884		.446	11.3284
	.047	1.1938					.147	3.7338		.246	6.2484		.347	8.8138		.447	11.3538
	.048	1.2192	25	.1495			.148	3.7592		.247	6.2738		.348	8.8392		.448	11.3792
	.049	1.2446					.149	3.7846		.248	6.2992	S	.349	8.8646		.449	11.4046
	.050	1.2700					.150	3.8100		.249	6.3246		.350	8.8900		.450	11.4300
	.051	1.2954	24				.151	3.8354	8	1/4	.250	6.3500		.351	8.9154		.451	11.4554
55	.052	1.3208					.152	3.8608		.251	6.3754		.352	8.9408		.452	11.4808
	.053	1.3462	23				.153	3.8862		.252	6.4008		.353	8.9662		.453	11.5062
	.054	1.3716					.154	3.9116		.253	6.4262		.354	8.9916	29/64	.4531	11.5094
54	.055	1.3970					.155	3.9370		.254	6.4516		.3543	9.0000		.454	11.5316
	.056	1.4224			5/32		.156	3.9624		.255	6.4770		.355	9.0170		.455	11.5570
	.057	1.4478	22				.1562	3.9687		.256	6.5024		.356	9.0424		.456	11.5824
	.058	1.4732					.157	3.9878	F	.257	6.5278		.357	9.0678		.457	11.6078
	.059	1.4986					.1575	4.0000		.258	6.5532		.358	9.0932		.458	11.6332
53	.0595			.060	1.5240	21			.158	4.0132		.259	6.5786	T	.359	9.1186		.459	11.6586
	.061	1.5494					.159	4.0386		.260	6.6040		.3594	9.1281		.460	11.6840
	.062	1.5748	20			G	.160	4.0640		.261	6.6294	23/64				.461	11.7094
	1/16			.0625	1.5875		.161	4.0894		.262	6.6548		.360	9.1440		.462	11.7348
52	.0635			.063	1.6002		.162	4.1148		.263	6.6802		.361	9.1694		.463	11.7602
	.064	1.6256					.163	4.1402		.264	6.7056		.362	9.1948		.464	11.7856
	.065	1.6510					.164	4.1656		.265	6.7310		.363	9.2202		.465	11.8110
	.066	1.6764	19			17/64	.165	4.1910		.2656	6.7469		.364	9.2456		.466	11.8364
51	.067	1.7018				H	.166	4.2164		.266	6.7564		.365	9.2710		.467	11.8618
	.068	1.7272					.167	4.2418		.267	6.7818		.366	9.2964		.468	11.8872
	.069	1.7526					.168	4.2672		.268	6.8072		.367	9.3218			
50	.070	1.7780	18	.1635			.169	4.2926		.269	6.8326	U	.368	9.3472	15/32	.4687	11.9062
	.071	1.8034					.170	4.3180		.270	6.8580		.369	9.3726		.469	11.9126
	.072	1.8288			11/64		.171	4.3434		.271	6.8834		.370	9.3980		.470	11.9380
49	.073	1.8542					.1719	4.3656		.272	6.9088		.371	9.4234		.471	11.9634
	.074	1.8796	17				.172	4.3688		.273	6.9342		.372	9.4488		.472	11.9888
	.075	1.9050					.173	4.3942		.274	6.9596		.373	9.4742		.4724	12.0000
48	.076	1.9304					.174	4.4196		.275	6.9850		.374	9.4996		.473	12.0142
	.077	1.9558					.175	4.4450		.2756	7.0000	3/8	.375	9.5250		.474	12.0396
47	.0785			.078	1.9812	16			.176	4.4704		.276	7.0104		.376	9.5504		.475	12.0650
	5/64			.0781	1.9844		.177	4.4958		.277	7.0358	V	.377	9.5758		.476	12.0904
	.0787	2.0000					.178	4.5212		.278	7.0612		.378	9.6012		.477	12.1158
	.079	2.0066	15				.179	4.5466		.279	7.0866		.379	9.6266		.478	12.1412
	.080	2.0320					.180	4.5720		.280	7.1120		.380	9.6520		.479	12.1666
46	.081	2.0574	14			K	.181	4.5974		.281	7.1374		.381	9.6774		.480	12.1920
45	.082	2.0828					.182	4.6228	9/32	.2812	7.1437		.382	9.7028		.481	12.2174
	.083	2.1082					.183	4.6482		.282	7.1628		.383	9.7282		.482	12.2428
	.084	2.1336	13				.184	4.6736		.283	7.1882		.384	9.7536		.483	12.2682
	.085	2.1590					.185	4.6990		.284	7.2136		.385	9.7790		.484	12.2936
44	.086	2.1844					.186	4.7244		.285	7.2390	W	.386	9.8044	31/64	.4844	12.3031
	.087	2.2098			3/16		.187	4.7498		.286	7.2644		.387	9.8298		.485	12.3190
	.088	2.2352					.1875	4.7625		.287	7.2898		.388	9.8552		.486	12.3444
43	.089	2.2606	12				.188	4.7752		.288	7.3152		.389	9.8806		.487	12.3698
	.090	2.2860					.189	4.8006		.289	7.3406		.390	9.9060		.488	12.3952
	.091	2.3114				L	.190	4.8260		.290	7.3660	25/64	.3906	9.9219		.489	12.4206
	.092	2.3368	11				.191	4.8514		.291	7.3914		.391	9.9314		.490	12.4460
42	.0935			.093	2.3622		.192	4.8768		.292	7.4168		.392	9.9568		.491	12.4714
	3/32			.0937	2.3812	10	.1935			.193	4.9022		.293	7.4422		.393	9.9822		.492	12.4968
	.094	2.3876					.194	4.9278		.294	7.4676		.3937	10.0000		.493	12.5222
	.095	2.4130	9			M	.195	4.9530		.295	7.4930		.394	10.0076		.494	12.5476
							.196	4.9784		.296	7.5184		.395	10.0330		.495	12.5730
41	.096	2.4384					.1969	5.0000	19/64	.2969	7.5406		.396	10.0584		.496	12.5984
	.097	2.4638					.197	5.0038		.297	7.5438	X	.397	10.0838		.497	12.6238
40	.098	2.4892					.198	5.0292		.298	7.5692		.398	10.1092		.498	12.6492
39	.0995			.099	2.5146		.199	5.0546		.299	7.5946		.399	10.1346	1/2	.499	12.6746
	.100	2.5400				8	.200	5.0800		.300	7.6200		.400	10.1600		.500	12.7000

Goodheart-Willcox Publisher

Copyright Goodheart-Willcox Co., Inc.

Feeds and Speeds for HSS Drills, Reamers, and Taps

Material	Brinell	Drills			Reamers		Taps (sfm)			
		(sfm)	Point	Feed	(sfm)	Feed	Threads per Inch			
							3–7 1/2	8–15	16–24	25–up
Aluminum	99–101	200–250	118°	M	150–160	M	50	100	150	200
Aluminum bronze	170–187	60	118°	M	40–45	M	12	25	45	60
Bakelite	...	80	60-90°	M	50–60	M	50	100	150	200
Brass	192–202	200–250	118°	H	150–160	H	50	100	150	200
Bronze, common	166–183	200–250	118°	H	150–160	H	40	80	100	150
Bronze, phosphor, 1/2 hard	187–202	175–180	118°	M	130–140	M	25	40	50	80
Bronze, phosphor, soft	149–163	200–250	118°	H	150–160	H	40	80	100	150
Cast iron, soft	126	140–150	90°	H	100–110	H	30	60	90	140
Cast iron, medium soft	196	80–110	118°	M	50–65	M	25	40	50	80
Cast iron, hard	293–302	45–50	118°	L	67–75	L	10	20	30	40
Cast iron, chilled*	402	15	150°	L	8–10	L	5	5	10	10
Cast steel	286–302	40–50*	118°	L	70–75	L	20	30	40	50
Celluloid	...	100	90°	M	75–80	M	50	100	150	200
Copper	80–85	70	100°	L	45–55	L	40	80	100	150
Drop forgings (steel)	170–196	60	118°	M	40–45	M	12	25	45	60
Duralumin	90–104	200	118°	M	150–160	M	50	100	150	200
Everdur	179–207	60	118°	L	40–45	L	20	30	40	50
Machinery steel	170–196	110	118°	H	67–75	H	35	50	60	85
Magnet steel, soft	241–302	35–40	118°	M	20–25	M	20	40	50	75
Magnet steel, hard*	321–512	15	150°	L	10	L	5	10	15	25
Manganese steel, 7% – 13%	187–217	15	150°	L	10	L	15	20	25	30
Manganese copper, 30% Mn.*	134	15	150°	L	10–12	L	...	...	...	...
Malleable iron	112–126	85–90	118°	H	...	H	20	30	40	50
Mild steel, .20 –.30 C	170–202	110–120	118°	H	75–85	H	40	55	70	90
Molybdenum steel	196–235	55	125°	M	35–45	M	20	30	35	45
Monel metal	149–170	50	118°	M	35–38	M	8	10	15	20
Nickel, pure*	187–202	75	118°	L	40	L	25	40	50	80
Nickel steel, 3 1/2%	196–241	60	118°	L	40–45	L	8	10	15	20
Rubber, hard	...	100	60-90°	L	70–80	L	50	100	150	200
Screw stock, C.R.	170–196	110	118°	H	75	H	20	30	40	50
Spring steel	402	20	150°	L	12–15	L	10	10	15	15
Stainless steel	146–149	50	118°	M	30	M	8	10	15	20
Stainless steel, C.R.*	460–477	20	118°	L	15	L	8	10	15	20
Steel, .40 to .50 C	170–196	80	118°	M	8–10	M	20	30	40	50
Tool, S.A.E., and forging steel	149	75	118°	H	35–40	H	25	35	45	55
Tool, S.A.E., and forging steel	241	50	125°	M	12	M	15	15	25	25
Tool, S.A.E., and forging steel*	402	15	150°	L	10	L	8	10	15	20
Zinc alloy	112–126	200–250	118°	M	150–175	M	50	100	150	200

*Use specially constructed heavy-duty drills.
Note: Carbon steel tools should be run at speeds 40% to 50% of those recommended for high speed steel.
Spiral point taps may be run at speeds 15% to 20% faster than regular taps.

Goodheart-Willcox Publisher

Tap Drill Sizes for Fractional Size Threads (75% Depth of Thread)

Tap Size	Threads per Inch	Diameter Hole (in.)	Drill	Tap Size	Threads per Inch	Diameter Hole (in.)	Drill
1/16	72	.049	3/64	1/2	20	.451	29/64
1/16	64	.047	3/64	1/2	13	.425	27/64
1/16	60	.046	56	1/2	12	.419	27/64
5/64	72	.065	52	9/16	27	.526	17/32
5/64	64	.063	1/16	9/16	18	.508	33/64
5/64	60	.062	1/16	9/16	12	.481	31/64
5/64	56	.061	53	5/8	27	.589	19/32
3/32	60	.077	5/64	5/8	18	.571	37/64
3/32	56	.076	48	5/8	12	.544	35/64
3/32	50	.074	49	5/8	11	.536	17/32
3/32	48	.073	49	11/16	16	.627	5/8
7/64	56	.092	42	11/16	11	.599	19/32
7/64	50	.090	43	3/4	27	.714	123/32
7/64	48	.089	43	3/4	16	.689	11/16
1/8	48	.105	36	3/4	12	.669	43/64
1/8	40	.101	38	3/4	10	.653	21/32
1/8	36	.098	40	13/16	12	.731	47/64
1/8	32	.095	3/32	13/16	10	.715	23/32
9/64	40	.116	32	7/8	27	.839	27/32
9/64	36	.114	33	7/8	18	.821	53/64
9/64	32	.110	35	7/8	14	.805	13/16
5/32	40	.132	30	7/8	12	.794	51/64
5/32	36	.129	30	7/8	9	.767	49/64
5/32	32	.126	1/8	15/16	12	.856	55/64
11/64	36	.145	27	15/16	9	.829	53/64
11/64	32	.141	9/64	1	27	.964	31/32
3/16	36	.161	20	1	14	.930	15/16
3/16	32	.157	22	1	12	.919	59/64
3/16	40	.155	23	1	8	.878	3/8
3/16	24	.147	26	1 1/16	8	.941	15/16
13/64	32	.173	17	1 1/8	12	1.044	1 3/64
13/64	30	.171	11/64	1 1/8	7	.986	63/64
13/64	24	.163	20	1 3/16	7	1.048	1 3/64
7/32	32	.188	12	1 1/4	12	1.169	1 11/64
7/32	28	.184	13	1 1/4	7	1.111	1 7/64
7/32	24	.178	16	1 5/16	7	1.173	1 11/64
15/64	32	.204	6	1 3/8	12	1.294	1 19/64
15/64	28	.200	8	1 3/8	6	1.213	1 7/32
15/64	24	.194	10	1 1/2	12	1.419	1 27/64
1/4	32	.220	7/32	1 1/2	6	1.338	1 11/32
1/4	28	.215	3	1 5/8	5 1/2	1.448	1 29/64
1/4	27	.214	3	1 3/4	5	1.555	1 9/16
1/4	24	.209	4	1 5/8	5	1.680	1 11/16
1/4	20	.201	7	2	4 1/2	1.783	1 25/32
5/16	32	.282	9/32	2 1/8	4 1/2	1.909	1 29/32
5/16	27	.276	J	2 1/4	4 1/2	2.034	2 1/32
5/16	24	.272	I	2 3/8	4	2.131	2 1/8
5/16	20	.264	17/64	2 1/2	4	2.256	2 1/4
5/16	18	.258	F	2 5/8	4	2.381	2 3/8
3/8	27	.339	R	2 3/4	4	2.506	2 1/2
3/8	24	.334	Q	2 7/8	3 1/2	2.597	2 19/32
3/8	20	.326	21/64	3	3 1/2	2.722	2 23/32
3/8	16	.314	5/16	3 1/8	3 1/2	2.847	2 27/32
7/16	27	.401	Y	3 1/4	3 1/2	2.972	2 31/32
7/16	24	.397	X	3 3/8	3 1/4	3.075	3 1/16
7/16	20	.389	25/64	3 1/2	3 1/4	3.200	3 3/16
7/16	14	.368	U	3 5/8	3 1/4	3.325	3 5/16
1/2	27	.464	15/32	3 3/4	3	3.425	3 7/16
1/2	24	.460	29/64	4	3	3.675	3 11/16

Goodheart-Willcox Publisher

Metric Tap Drill Size

Metric Tap Size	Recommended Metric Drill				Closest Recommended Inch Drill			
	Drill Size (mm)	Inch Equivalent	Probable Hole Size (in.)	Probable Percent of Thread	Drill Size (mm)	Inch Equivalent	Probable Hole Size (in.)	Probable Percent of Thread
M1.6 × .35	1.25	.0492	.0507	69	—	—	—	—
M1.8 × .35	1.45	.0571	.0586	69	—	—	—	—
M2 × .4	1.60	.0630	.0647	69	#52	.0635	.0652	66
M2.2 × .45	1.75	.0689	.0706	70	—	—	—	—
M2.5 × .45	2.05	.0807	.0826	69	#46	.0810	.0829	67
*M3 × .5	2.50	.0984	.1007	68	#40	.0980	.1003	70
M3.5 × .6	2.90	.1142	.1168	68	#33	.1130	.1156	72
*M4 × .7	3.30	.1299	.1328	69	#30	.1285	.1314	73
M4.5 × .75	3.70	.1457	.1489	74	#26	.1470	.1502	70
*M5 × .8	4.20	.1654	.1686	69	#19	.1660	.1692	68
*M6 × 1	5.00	.1968	.2006	70	#9	.1960	.1998	71
M7 × 1	6.00	.2362	.2400	70	15/64	.2344	.2382	73
*M8 × 1.25	6.70	.2638	.2679	74	17/64	.2656	.2697	71
M8 × 1	7.00	.2756	.2797	69	J	.2770	.2811	66
*M10 × 1.5	8.50	.3346	.3390	71	Q	.3320	.3364	75
M10 × 1.25	8.70	.3425	.3471	73	11/32	.3438	.3483	71
*M12 × 1.75	10.20	.4016	.4063	74	Y	.4040	.4087	71
M12 × 1.25	10.80	.4252	.4299	67	27/64	.4219	.4266	72
M14 × 2	12.00	.4724	.4772	72	15/32	.4688	.4736	76
M14 × 1.5	12.50	.4921	.4969	71	—	—	—	—
*M16 × 2	14.00	.5512	.5561	72	35/64	.5469	.5518	76
M16 × 1.5	14.50	.5709	.5758	71	—	—	—	—
M18 × 2.5	15.50	.6102	.6152	73	39/64	.6094	.6144	74
M18 × 1.5	16.50	.6496	.6546	70	—	—	—	—
*M20 × 2.5	17.50	.6890	.6942	73	11/16	.6875	.6925	74
M20 × 1.5	18.50	.7283	.7335	70	—	—	—	—
M22 × 2.5	19.50	.7677	.7729	73	49/64	.7656	.7708	75
M22 × 1.5	20.50	.8071	.8123	70	—	—	—	—
*M24 × 3	21.00	.8268	.8327	73	53/64	.8281	.8340	72
M24 × 2	22.00	.8661	.8720	71	—	—	—	—
M27 × 3	24.00	.9449	.9511	73	15/16	.9375	.9435	78
M27 × 2	25.00	.9843	.9913	70	63/64	.9844	.9914	70
*M30 × 3.5	26.50	1.0433						
M30 × 2	28.00	1.1024						
M33 × 3.5	29.50	1.1614						
M33 × 2	31.00	1.2205	\} Reaming Recommended to the Drill Size Shown					
M36 × 4	32.00	1.2598						
M36 × 3	33.00	1.2992						
M39 × 4	35.00	1.3780						

Goodheart-Willcox Publisher

Cutting Fluids for Various Metals

Aluminum and its Alloys	Kerosene, kerosene and lard oil, soluble oil
Plastics	Dry
Brass, Soft	Dry, soluble oil, kerosene and lard oil
Bronze, High Tensile	Soluble oil, lard oil, mineral oil, dry
Cast Iron	Dry, air jet, soluble oil
Copper	Soluble oil, dry, mineral lard oil, kerosene
Magnesium	Low viscosity neutral oils
Malleable Iron	Dry, soda water
Monel Metal	Lard oil, soluble oil
Slate	Dry
Steel, Forging	Soluble oil, sulfurized oil, mineral lard oil
Steel, Manganese	Soluble oil, sulfurized oil, mineral lard oil
Steel, Soft	Soluble oil, mineral lard oil, sulfurized oil, lard oil
Steel, Stainless	Sulfurized mineral oil, soluble oil
Steel, Tool	Soluble oil, mineral lard oil, sulfurized oil
Wrought Iron	Soluble oil, mineral lard oil, sulfurized oil

Goodheart-Willcox Publisher

Sheet Metal and Wire Gage Designation

Gage Number	American or Brown & Sharpe's A.W.G. or B. & S.	Birmingham or Stubs Wire B.W.G.	Washburn & Moen or American S.W.G.	United States Standard	Manufacturers' Standard for Sheet Steel	Gage Number
0000000	—	—	.4900	.500	—	0000000
000000	.5800	—	.4615	.469	—	000000
00000	.5165	—	.4305	.438	—	00000
0000	.4600	.454	.3938	.406	—	0000
000	.4096	.425	.3625	.375	—	000
00	.3648	.380	.3310	.344	—	00
0	.3249	.340	.3065	.312	—	0
1	.2893	.300	.2830	.281	—	1
2	.2576	.284	.2625	.266	—	2
3	.2294	.259	.2437	.250	.2391	3
4	.2043	.238	.2253	.234	.2242	4
5	.1819	.220	.2070	.219	.2092	5
6	.1620	.203	.1920	.203	.1943	6
7	.1443	.180	.1770	.188	.1793	7
8	.1285	.165	.1620	.172	.1644	8
9	.1144	.148	.1483	.156	.1495	9
10	.1019	.134	.1350	.141	.1345	10
11	.0907	.120	.1205	.125	.1196	11
12	.0808	.109	.1055	.109	.1046	12
13	.0720	.095	.0915	.0938	.0897	13
14	.0642	.083	.0800	.0781	.0747	14
15	.0571	.072	.0720	.0703	.0673	15
16	.0508	.065	.0625	.0625	.0598	16
17	.0453	.058	.0540	.0562	.0538	17
18	.0403	.049	.0475	.0500	.0478	18
19	.0359	.042	.0410	.0438	.0418	19
20	.0320	.035	.0348	.0375	.0359	20
21	.0285	.032	.0317	.0344	.0329	21
22	.0253	.028	.0286	.0312	.0299	22
23	.0226	.025	.0258	.0281	.0269	23
24	.0201	.022	.0230	.0250	.0239	24
25	.0179	.020	.0204	.0162	.0209	25
26	.0159	.018	.0181	.0188	.0179	26
27	.0142	.016	.0173	.0172	.0164	27
28	.0126	.014	.0219	.0156	.0149	28
29	.0113	.013	.0150	.0141	.0135	29
30	.0100	.012	.0140	.0125	.0120	30
31	.0089	.010	.0132	.0109	.0105	31
32	.0080	.009	.0128	.0102	.0097	32
33	.0071	.008	.0118	.00938	.0090	33
34	.0063	.007	.0104	.00859	.0082	34
35	.0056	.005	.0095	.00781	.0075	35
36	.0050	.004	.0090	.00703	.0067	36
37	.0045	—	.0085	.00624	.0064	37
38	.0040	—	.0080	.00625	.0060	38
39	.0035	—	.0075	—	—	39
40	.0031	—	.0070	—	—	40
41	.0028	—	.0066	—	—	41
42	.0025	—	.0062	—	—	42
43	.0022	—	.0060	—	—	43
44	.0020	—	.0058	—	—	44
45	.0018	—	.0055	—	—	45
46	.0016	—	.0052	—	—	46
47	.0014	—	.0050	—	—	47
48	.0012	—	.0048	—	—	48

Precision Sheet Metal Setback Chart

90° Bend Radius / Material Thickness

Bend Radius	.016	.020	.025	.032	.040	.051	.064	.072	.078	.081	.091	.102	.125	.129	.156	.162	.187	.250
1/32	.034	.039	.046	.05	.065	.081	.102	.113	.121	.125	.139							
3/64	.041	.046	.053	.062	.072	.090	.108	.119	.127	.131	.145							
1/16	.048	.053	.059	.068	.079	.093	.110	.122	.134	.138	.152							
5/64	.054	.060	.066	.075	.086	.100	.117	.127	.138	.144	.158							
3/32	.061	.066	.073	.082	.092	.107	.124	.134	.142	.146	.160							
7/64	.068	.073	.080	.08	.099	.113	.130	.141	.148	.153	.167	.181						
1/8	.075	.080	.086	.095	.106	.120	.137	.147	.155	.159	.172	.186	.216	.221				
9/64	.081	.087	.093	.102	.113	.127	.144	.154	.162	.166	.179	.193	.223	.228	.263			
5/32	.088	.093	.100	.109	.119	.134	.150	.161	.169	.173	.186	.200	.230	.235	.270	.278		
11/64	.095	.100	.107	.116	.126	.140	.157	.168	.175	.179	.192	.207	.236	.242	.277	.284	.317	
3/16	.102	.107	.113	.122	.133	.147	.164	.174	.182	.186	.199	.213	.243	.248	.283	.291	.324	.405
13/64	.108	.114	.120	.129	.140	.154	.171	.181	.189	.193	.206	.220	.250	.255	.290	.298	.330	.412
7/32	.115	.120	.127	.136	.146	.161	.177	.188	.196	.199	.212	.227	.257	.262	.297	.305	.337	.419
15/64	.122	.127	.134	.143	.153	.167	.184	.195	.202	.206	.219	.233	.263	.269	.304	.311	.344	.426
1/4	.129	.134	.140	.149	.160	.174	.191	.201	.209	.213	.226	.240	.270	.275	.310	.318	.351	.432
17/64	.135	.141	.147	.156	.166	.181	.198	.208	.216	.220	.233	.247	.277	.282	.317	.325	.357	.439
9/32	.142	.147	.154	.163	.173	.187	.204	.215	.223	.226	.239	.254	.284	.289	.324	.332	.364	.446

Developed Length = X + Y − Z
Z = Setback Allowance from the chart.

Grade Marking for Bolts

Bolt Head Marking	SAE = Society of Automotive Engineers ASTM = American Society for Testing and Materials	Bolt Material	Minimum Tensile Strength in Pounds per Square Inch (psi)
No Marks	SAE Grade 1 SAE Grade 2 Indeterminate quality	Low-carbon steel Low-carbon steel	65,000 psi
2 Marks	SAE Grade 3	Medium-carbon steel, cold worked	110,000 psi
3 Marks	SAE Grade 5 ASTM – A 325 Common commercial quality	Medium-carbon steel, quenched and tempered	120,000 psi
Letters BB	ASTM – A 354	Low-alloy steel or medium-carbon steel, quenched and tempered	105,000 psi
Letters BC	ASTM – A 354	Low-alloy steel or medium-carbon steel, quenched and tempered	125,000 psi
4 Marks	SAE Grade 6 Better commercial quality	Medium-carbon steel, quenched and tempered	140,000 psi
5 Marks	SAE Grade 7	Medium-carbon alloy steel, quenched and tempered, roll-threaded after heat treatment	133,000 psi
6 Marks	SAE Grade 8 ASTM – A 345 Best commercial quality	Medium-carbon alloy steel, quenched and tempered	150,000 psi

Goodheart-Willcox Publisher

Bolt Torquing Chart

Metric Standard						SAE Standard/Foot Pounds					
Grade of Bolt	5D	.8G	10K	12K		Grade of Bolt	SAE 1 & 2	SAE 5	SAE 6	SAE 8	
Min. Ten. Strength	71,160 P.S.I.	113,800 P.S.I.	142,200 P.S.I.	170,679 P.S.I.		Min. Ten. Strength	64,000 P.S.I.	105,000 P.S.I.	133,000 P.S.I.	150,000 P.S.I.	
Markings on Head	5D	.8G	10K	12K	Size of Socket or Wrench Opening	Markings on Head	⬢	⬢	⬢	⬢	Size of Socket or Wrench Opening
Metric		Foot Pounds				Metric	U.S. Standard	Foot Pounds			U.S. Standard
Bolt Dia.	U.S. Dec. Equiv.				Bolt Head	Bolt Dia.					Bolt Head / Nut
6mm	.2362	5	6	8	10	10mm	1/4	5	7	10	10.5	3/8	7/16
8mm	.3150	10	16	22	27	14mm	5/16	9	14	19	22	1/2	9/16
10mm	.3937	19	31	40	49	17mm	3/8	15	25	34	37	9/16	5/8
12mm	.4720	34	54	70	86	19mm	7/16	24	40	55	60	5/8	3/4
14mm	.5512	55	89	117	137	22mm	1/2	37	60	85	92	3/4	13/16
16mm	.6299	83	132	175	208	24mm	9/16	53	88	120	132	7/8	7/8
18mm	.7090	111	182	236	283	27mm	5/8	74	120	167	180	15/16	1
22mm	.8661	182	284	394	464	32mm	3/4	120	200	280	296	1 1/8	1 1/8

Goodheart-Willcox Publisher

Glossary

4-H. A youth development program of the nation's US Department of Agriculture's Cooperative Extension System. The program is available in every county and parish in the United States. (2)

A

Abrams cone. A container 12″ high with an 8″ diameter bottom (widest part) and 4″ diameter top used in testing concrete. Also called *slump cone*. (14)

abrasives. 1. Hard substances used to wear away another material. 2. Small, sharp-edged particles bound to certain consumable power tool attachments that are used for the removal of material. (27, 8)

ABS. Acrylonitrile butadiene styrene, a type of plastic pipe used only for DWV pipes in plumbing systems because of its low resistance to heat. (24)

accelerant. Anything that increases the temperature of the flame in an oxyfuel torch and increases the combustion rate. (28)

accuracy. The ability to record the same measurement every time for a specific quantity, distance, or weight. (6)

acetylene. A hydrocarbon gas (C_2H_2), produced by submerging calcium carbide in water, which burns at higher temperatures than other fuel gases. (28)

adhesive. A substance (generally a liquid) that forms a physical and/or chemical bond with materials and causes them to adhere. (9)

adjustable wrench. A wrench with a stationary jaw and a movable jaw that allows the mouth to be adjusted to any size within its opening range. (7)

admixture. Something added to concrete (such as coloring agents, retarders, or accelerators) during the mixing or pouring phase to change the characteristics of the mixture. (14)

aesthetics. The look or style of something, such as a building or landscape. (10)

aggregate. Sand, gravel, or other similar substances added to the mixture of Portland cement to create concrete or masonry units. (14)

agricultural mechanics. The design, construction, maintenance, repair, management and use of agricultural technology and mechanical systems. (1)

agriculture. The science and practice of sustainably producing food, fiber, and fuel to meet the needs of a global population. (1)

agriscience. The field of science devoted to improving agriculture. (1)

air compressor. A device that draws in external air through a filter and compresses it for use as an energy source to perform work. (37)

air quality. A measure of the degree to which air is free of pollution or other harmful substances. (17)

alloy. A base metal mixed with another element. (9)

all-terrain vehicle (ATV). A small motor vehicle that travels on low-pressure tires, with a seat that is straddled by the operator and is steered using handlebars. (5)

alternating current (AC). Electric current in which the power source constantly alternates, or switches, polarity from positive to negative and back. (20)

aluminum. A lightweight metal widely used in agricultural applications because of its relatively low cost, light weight, and strength. (9)

American Society for Testing and Materials (ASTM). An international society that tests materials and develops standards for their use. (15)

American Society of Mechanical Engineers (ASME). An organization that provides educational programs, training, codes and standards, and other information related to engineering. (10)

American Wire Gauge (AWG). A standard set of wire conductor sizes (diameters) used in the United States. (21)

ampere (amp). The unit of measurement in which electric current is measured. One ampere is approximately 6,240,000,000,000,000,000 electrons passing through a given point on a conductor per second. (20)

anaerobic digesters. Microorganisms that do not require oxygen to live and can break down crop residue and animal manure to produce a biogas that can be used to generate electricity or heat, while reducing odors and greenhouse gas emissions. (4)

anchor bolt. A specially designed fastener used to hold wood framing to concrete masonry units. (15)

annealing. The process of heating a metal and letting it cool slowly, reducing hardness and making the material tougher and more flexible. (24, 27)

anti-friction bearing. An assembly of roller or ball bearings and races that provides a low-friction surface between two parts that could generate high friction if they came in direct contact. (36)

anvil. Steel or cast iron device used to back up or support the work when shaping metal parts. (7)

1023

Copyright Goodheart-Willcox Co., Inc.

appearance lumber. Wood used for making projects that appear free from defects. (11)

apprenticeship. A combination of on-the-job training and formal classroom education in the skilled trades, generally sponsored by a trades group. (2)

aquaculture. A system of growing and harvesting animals that normally grow or live in water. (18)

aquaponics. A combination of aquaculture and hydroponics in which plants and aquatic animals live and grow in the same system. (18)

aquatic plant. A plant that can only grow in water or in a water-saturated environment. (18)

aquifer. An underground source of water. (25)

arc blow. The force of motion generated by the heat from the electric arc and the burning of the flux. When properly directed, the arc blow aids in penetration of the base metal. (29)

arc construction. A type of structure that uses rafters or trusses which are curved from one side of the building to the other side. (16)

architect. A professional who designs buildings and landscapes. (10)

arc length. The distance from the welding electrode to the base metal. (29)

argent pencil. A silver pencil that uses a heat-resistant substance for marking metals for welding. (6)

armored cable. Cable that has a flexible metallic sheathing that allows for extra protection. It is often used in commercial construction. (21)

array. Multiple photovoltaic modules wired together to produce a larger amount of electricity. (22)

asset. Property, equipment, and other items owned by a person or company. (3)

audit. Official inspection of an individual's or organization's accounts. (3)

auto-darkening lenses. Lenses in welding helmets that are triggered to darken by the welding flash. (29)

autogenous weld. A weld made without using a filler metal, consisting only of the base metal. (31)

automatic transmission. A transmission in which gear ratio changes are activated without direct operators input. (35)

automatic watering system. A watering system that applies water automatically according to moisture or a specific time interval. (18)

automatic welding. Robotic welding in which the equipment uses automatic sensing devices to trigger corrections, allowing welds without constant adjusting of controls by a welder. (30)

automation. Automatic control of the operation or function of equipment, processes, or systems by mechanical or electronic devices. (4)

autonomous. Not requiring human intervention for direction or control. (4)

aviation snips. Cutting pliers used to cut sheet metal. This tool may have a straight jaw or a jaw angled to the left or the right. (7)

B

backflow preventer. An irrigation system device used to ensure that water pumped by the irrigation system is not returned to the groundwater in the well, or to the main irrigation line. (25)

backhand welding. Welding technique in which the welding gun is pointed at the back of the weld pool. (30)

backing plate. A metal plate used below the root opening to help control joint penetration and support the root weld as it is formed. (30)

backsaw. A precision tool for cutting very straight lines. This tool is a crosscut saw with small teeth and a stiffening rib on the top of the rectangular blade. (7)

backsight. A rod measurement taken at a known elevation point and used in differential leveling. (13)

ball peen hammer. A tool used for shaping metal and driving punches and chisels. The head has a face that is cylindrical and a face that is hemispherical. (7)

ballasting. Changing the weight distribution of a tractor by adding or subtracting weight where needed to correct wheel slippage. (36)

balloon framing. A building method that uses studs that are long enough to extend through two stories instead of just one. (16)

band saw. A saw that uses a continuous belt-like metal blade that is driven around at least two pulleys to cut material. (8)

barbed wire. Most common type of wire used in agriculture. A noticeable characteristic of this type of wire is the barbs placed certain distances apart along the wire. (19)

basin irrigation. A common form of surface irrigation in which the crop is flooded for a brief period of time. Small berms along the borders of the field keep the water from running out of the field. (25)

batter boards. Stakes and horizontal cross pieces used as a temporary framework when marking corners and lines of a foundation. (13)

battered foundation. A base for a building that has a flexible pole-like structure (piles) driven or drilled into underlying bedrock or deep into the soil. (14)

battery. A storage device for an electric charge. (20)

battery charger. A device that uses standard 120-volt AC current to fully charge a battery in 1 to 16 hours. (8)

bearings. Friction-reducing machine parts that support the drive shaft within the frame or motor housing, usually consisting of inner and outer races that serve as tracks for a set of rollers made of very hard steel alloys. (23)

bed of mortar. The bonding agent (mortar) that is placed under the first course of masonry units to support and bond the units to the supporting surface. (15)

belt. A flexible loop designed to connect two or more pulleys to transfer rotational power. (35)

benchmark. A point with a known or expected elevation from which all site elevation measurements are referenced. (13)

beveling. Removing part of a square edge to create a sloping edge that provides better weld penetration. (29)

bill of materials. A complete list of all parts and supplies used in a project, such as wood, metal, hardware, fasteners, and adhesive. (10, 27)

bimetal blade. Reciprocating saw blade that combines the toughness of carbon steel with the hardness of HSS. (8)

biofuels. Fuels, produced from natural and agricultural products, that are considered renewable resources because they can be replaced naturally. (4)

biologging. The use of wireless sensor networks to record multiple animal health variables. (4)

biosecurity. The use of procedures specifically to protect humans and animals against disease or harmful biological agents. (4)

biotechnology. The use of living organisms to make or modify products that serve a useful purpose. (4)

biscuit jointer. A tool used to make crescent shaped cuts in two pieces of wood that are to be joined together. Also called *plate jointer*. (12)

bit. The biting or cutting edge or part of a tool. (8)

black pipe. A type of steel pipe that gets its color from the carbon formed when the iron is manufactured into pipe. It is often used in pressurized gas supply lines. (24)

bladder. A soft bag that fills with water or air and expands. (24)

blade connector. Flat male/female connectors, often crimp-on, that are used to connect wires. (22)

blade set. The difference between the overall width of a saw blade's teeth and the thickness of the blade material. (8)

bleed water. Water that rises to the surface of concrete during the curing process soon after placement and floating are complete. (14)

blind cut. A cut that does not pass completely through the wood to the other side. (11)

blind rivet. A nonthreaded fastener used in applications in which only one side of the joint is accessible. (27)

blueprints. Two-dimensional plans that include all dimensions, measurements, and construction details for a building. (16)

board foot. Lumber that is 12″ long by 12″ wide by 1″ thick or some combination that equals this size. (10)

bolt. A threaded fastener, usually with a flattened tip and straight (not tapered) shank, used with a nut to secure materials. (9)

bond. To join items or substances together securely by means of an adhesive, heat, or pressure. (15)

border irrigation. A form of surface irrigation in which water is pumped along the border of an evenly sloping field. Gravity and the slope of the land are used to disperse the water from various locations. (25)

bore. The diameter of a hole, such as the distance measured across the cylinder in an engine. (34)

bottle jack. Vertical cylinder with a piston or ram that is driven upward by the application of hydraulic force. (8)

brace and bit. An offset frame, two handles, and a chuck for holding a bit. The top handle is round and flat, which allows the user to apply adequate pressure on the drill bit.

brad nail. A very thin wire nail, with almost no head, that is used to attach thin trim materials. (9)

brad nailer. Smaller versions of nail guns that drive very small fasteners for fine trim work. (8)

braided wire. Wire made of a combination of polyester fibers with small-gauge smooth wire interwoven. It is used for high visibility in electric fences. (19)

brass. A metal composed of mainly copper and zinc. (9)

braze welding. A process that heats and melts filler material at a temperature above 800°F (427°C) but below the melting temperature of the base materials. It does not depend on capillary action for the distribution of the joining material. (28)

brazing. A process similar to braze welding, but depends on capillary action for the distribution of the joining material. (28)

brick. A molded rectangular block of clay that has been baked in an oven to increase the strength and durability of the material. (15)

brick chisel. A type of chisel to scribe a brick and then cut it to the appropriate length. Also called *brick set*. (15)

brick hammer. A tool used to drive nails, chip stone, or break brick or masonry units. (15)

brick set. A type of chisel to scribe a brick and then cut it to the appropriate length. Also called *brick chisel*. (15)

bridging. A brace of wood or steel installed in framing to provide extra support and to distribute loads more evenly. (16)

bronze. A metal composed of mainly copper and tin. (9)

broom finish. A texture created with a stiff bristle broom that is pushed across the floated concrete slab, leaving behind tiny grooves and rough areas that provide traction. (14)

brushes. The carbon elements that rub across the commutator to transfer electrical energy to the rotor winding in some electric motors. (23)

budget. An estimate of income and expenditures for a set period of time. (3)

build-up roofing. A type of asphalt roofing that is made from alternating layers of asphalt and roofing tar. (17)

bullet connector. Round male/female connectors used to connect wires to wires, providing a somewhat temporary connection that allows for simple repairs. (22)

bull float. A tool with a long handle attached to a large, flat surface that is moved over the top of fresh concrete to completely settle aggregate, level the surface, and smooth high and low spots. Also called *Darby level*. (14)

burr. A bit of leftover metal remaining after a pipe is cut. (24)

bushings. Friction-reducing machine parts that support the drive shaft within the frame or motor housing, consisting of a single-piece sleeve made of a material with a lower coefficient of friction than the drive shaft or housing. (23)

business plan. Document that helps provide directions for the future of a business, containing formal statements of business goals, the reasons they are attainable, and the plans for reaching them. (3)

butt joint. A joint welded between two pieces of metal positioned approximately in the same plane, with a gap between the two parallel edges. (29)

C

cable ripper. A tool that cuts the outside sheath of a cable for removal while minimizing damage to the insulation on the conductors inside. (21)

CAD software. A program that allows the user to design architectural, engineering, and construction projects on a computer. (16)

caliper. An adjustable measuring tool used to measure inside and outside dimensions of small parts. (6)

camshaft. a lobed shaft used to open and close the valves, operates at one-half of the rotation of the crankshaft. (33)

cap. The part of a hydraulic cylinder seals the end of the cylinder. (37)

capacitor-start motor. A type of motor that has an electrolytic capacitor mounted on the outside of the motor housing, which causes the voltage in the circuit to lag behind the current by a predetermined degree of the phase to increase electromagnetic force. (23)

cap screw. A threaded fastener with a finished appearance used in precision applications. Similar to a *machine screw*. (27)

carburizing flame. The type of flame that results when the proportion of acetylene to oxygen is higher than that required to produce a neutral flame. (28)

career development event (CDE). A competitive event sponsored by the National FFA Organization in which students learn, practice, and demonstrate skills. (2)

carpenter's level. A rectangular frame with one or more vials embedded along its length. The center vial is aligned parallel with the long edge of the tool for checking levelness. (6)

cast iron. An iron alloy high in carbon that is hard, brittle, resistant to wear, and difficult to crush. It may be classified as white cast iron, malleable iron, or gray iron. (9)

cattle guard. A device that is installed at ground level and functions as a gate because it is difficult for cattle to cross, although it can easily be crossed by vehicles or tractors. (19)

ceiling joist. A framing component made of boards that connects each side of a building where a rafter attaches to a wall. (16)

center-pivot sprinkler system. Agricultural irrigation system in which the sprinkler system pivots and moves in a circular pattern across the field. (25)

center punch. A punch that not only marks the location, but also creates a depression that guides and stabilizes a drill bit as it starts the hole. (6, 7)

centrifugal force. A force that causes a rotating object to move outward as it spins. (37)

ceriated tungsten electrode. A tungsten electrode that contains approximately 2% cerium. It is used with DC on thin sheet metal and for welding small, delicate parts. Marked with an orange band. (31)

certificate. Evidence that a student has specialized agricultural mechanics skills, evaluated through some type of testing. (2)

chaining pins. Metal stakes used during surveying to incrementally mark taping measurements on the ground. (13)

chain-link fence. A type of fence commonly used as a barrier in kennels and yards. Wire is manufactured and woven into a chainlike pattern. (19)

chamfered. A beveled edge, which is included on some tools to reduce the damaging effects from repeated impact. (7)

charge memory effect. Adverse result from charging certain battery types before they are completely discharged, which causes only a portion of the battery to accept charge. (8)

chisel. A hand tool made of hardened steel used to scribe (mark) and cut bricks, blocks, or stones. (15)

chlorinated polyvinyl chloride (CPVC). A type of plastic pipe that is similar to PVC pipe but is more flexible. (24)

chuck. The part of a drill that holds the bit in place. (8)

chuck key. Geared tool used to tighten and loosen the hold on the bit. (8)

circuit. A closed, logical path followed by electricity. (20)

circuit breaker. A device that breaks a circuit if the current in the circuit rises to an unsafe level. Unlike fuses, circuit breakers can be reset and reused. (20)

circular saw. Portable power saw with a round blade. (8)

claw hammer. A hammer designed for driving nails. It has a slightly convex face designed to contact the nail during the downward swing without marking the wood. (7)

clearance. Manufacturer's specified distance between two wearable parts. (36)

clinker. A stone-like, unburnable lump left after the heating process in the manufacture of Portland cement. (14)

closure block. The last block or brick laid in the course (complete row of blocks or bricks). (15)

cloud server. A network of remote computer servers hosted on the Internet to store, manage, and process data. (4)

clutch. A friction disc assembly that engages and disengages the engine flywheel to transfer power through gear sets. (35)

cold working. The process of forming, bending, or hammering metals well below their melting point to improve their strength and hardness. (27)

collet. The holding device in a GTAW torch that holds the tungsten electrode firmly in the center of the torch. (31)

collet body. The part of a GTAW torch that receives the collet when the end cap on the rear of the torch is tightened, holding the electrode firmly. (31)

combination square. A tool that performs many different measuring and layout tasks. The standard form incorporates a rigid steel ruler and a standard head—a sliding frame that includes a 90° and a 45° fence. A spirit level and a hidden scribe for marking materials are included as part of the standard head. (6)

combination wrench. A tool that consists of an open-end wrench on one end and a box-end, or completely closed, wrench on the other. (7)

combustible liquid. Any liquid having a flash point at or above 100°F (37.8°C), but below 200°F (93.3°C), except mixtures having components with flash points of 200°F (93.3°C) or higher that make up 99% or more of the total volume of the mixture. (5)

commutator. Thin strips of conductors aligned side-by-side around the diameter of the drive shaft and connected to the windings of the stator to provide an electrical connection between the windings and brushes. (23)

composite material. A substance made of a variety of material in a way that makes the finished product different from the original members of the mixture. (9)

compression ratio. The relationship of the total volume of empty space above the piston (in the cylinder and combustion chamber) when the piston is at bottom dead center compared to the volume when the piston is at top dead center. (34)

compression strength. The ability to withstand heavy weights or forces that reduce size. (10)

computer-aided design (CAD). The use of computers and software to design, draw, and model parts, objects, or structures. (10)

computer numerical control (CNC). A system in which a computer controls the positions of machines or tools using numerical values, or codes, programmed into a computer. (32)

concave joint. A type of mortar joint that is made with a finishing tool to leave an indentation that curves inward in the mortar. (15)

concrete. A composite material made of Portland cement, aggregate (coarse gravel or fine sand), and water. (9)

concrete masonry unit (CMU). A hollow or solid block that is made from Portland cement, aggregates, admixtures, and water and is used as a building product. (15)

conduction. The transfer of heat by contact between a warm object and a cool object. (17)

conductor. Any material that readily supports electron flow. (20)

conduit. Rigid tubing made of galvanized metal or plastic tubing that provides a raceway to protect electrical conductors. (21)

conduit bender. A tool consisting of a long handle and semicircular mandrel used to bend conduit. (27)

connecting rod. An engine part that links a piston to the crankshaft. (33)

constant-current machine. A welding machine that provides an even, or constant, supply of amperage to the electrode. (29)

construction procedure. A step-by-step plan for making subassemblies and the final product. (12)

contact tip. The part of the welding gun through which welding current is transferred to the electrode. (30)

continuing education training. Training in which an employee continues to study and gain new skills in a chosen profession. (2)

continuity. Refers to a circuit that is complete and allows electricity to flow. (20)

continuous hinge. A fastener with two plates with alternating pins that allow one plate to fold back upon the other. Also called *piano hinge*. (9)

control joint. A groove cut into a slab that allows the concrete to crack at this point so that it does not crack in a more random, unsightly pattern. Also called *expansion joint*. (14)

controlled environment agriculture (CEA). A growing area for plants or animals in which the growing environment is controlled to enhance growth or production. (18)

convection. The transfer of heat through a gas or liquid. (17)

cooling system. A system installed in a greenhouse to help lower the temperature in the growing area. (18)

coping saw. A saw with a lightweight, U-shaped metal frame. Thin, narrow blades are used for cutting shapes in sheet materials. (7)

copper. A ductile metal that has excellent heat and electrical conductivity. (9)

cordless power tool. Battery-operated tool that commonly uses between 9.6 volts–36 volts DC power. (8)

corner joint. A joint between two members located perpendicular or at an angle to each other. (29)

corner lead. A group of bricks or concrete blocks laid several units long and rows high to establish an outside corner of a structure. (15)

corner pole. A masonry tool with markings used to indicate the top of each course of bricks or blocks. Also called *story pole*. (15)

corner post. A post that starts or ends a fence and provides extra strength when a fence material is installed. (19)

corrosive. A chemical that causes visible destruction of or irreversible alterations in living tissue by chemical action at the site of contact. (5)

corrugated fastener. A wavy, thin piece of metal, about 1/2″ to 3″ long, that is driven into a wood joint (usually a miter or butt joint) that pulls each piece toward the joint. (9)

coupling. A device used in direct-drive systems to absorb shock from starting and stopping and vibration from slight misalignment between the motor and equipment. (23)

coupon. A small, rectangular piece of flat plate used as a practice pad to practice welding techniques. (29)

course. A complete or continuous row of bricks or blocks laid in a wall, structure, or project. (15)

covering. A material that separates the inside environment of a greenhouse from the outside environment. (18)

cover letter. A letter sent with a résumé that introduces a job candidate and refers the reader to the accompanying résumé. (2)

crack. A severe, unacceptable discontinuity that is sharp at the edges, concentrating points of stress. (29)

crankshaft. An engine part that converts reciprocating motion created by the engine pistons to rotating motion. (33)

crawlspace. The area below a building that allows access underneath a structure. (17)

cribbing. A method of using stacked timbers in a crisscross pattern to support a heavy load. (8)

cross peen hammer. Metalworking hammer that has a rounded horizontal wedge opposing the flat hammer face, and the body hammer, which is used for repairing dents in auto bodies and other sheet metal work. (7)

crosscut saw. A saw with pointed knifelike teeth designed for cutting lumber across the grain. (7)

crosscutting. Cutting across the grain of lumber. (8)

cross-linked polyethylene (PEX). A type of plastic pipe that can be used to supply both hot and cold water in a plumbing system. (24)

cut list. A detailed list of the exact lengths of each material required for the project. (27)

cuts and fills. Amounts of soil to be removed from or added to the landscape or build site. (13)

cutting attachment. The part of an oxyfuel torch that performs specific jobs, such as welding or cutting various ferrous metals, and contains valves for controlling the oxygen. (28)

cutting lever. An on/off valve that controls the oxygen supply in a cutting attachment. (28)

cutting list. A list of exact dimensions and cutting instructions for each piece of material in a project. (10)

cutting table. A steel working structure with a grated top set at a convenient working height for using a cutting torch. (28)

cutting tips. Replaceable tips attached to the head of an oxyfuel torch that direct the gases and flame to the work area. (28)

cylinder cap. A metal cap that fits over the valves of a gas cylinder and screws to the top of the cylinders to protect the valves when not in use. (28)

cylinder head. The part of an engine that seals the combustion chamber. On gasoline engines it also houses the spark plug. (33)

cylinder. 1. A pressurized container in which oxygen and fuel gases are stored. 2. The bore in which a piston travels. (28, 33)

D

Darby level. A tool with a long handle attached to a large, flat surface that is moved over the top of fresh concrete to completely settle aggregate, level the surface, and smooth high and low spots. Also called *bull float*. (14)

dead-blow hammer. A hammer designed to deliver a single blow while minimizing the rebound of the hammer head. The head is hollow and filled with sand or steel shot. (7)

dead load. The weight of materials used to build a structure. (16)

decibel (dB). The unit of measure for noise. (5)

Department of Labor (DOL). US federal government department that administers and enforces rules and regulations to ensure workers' rights to fair, safe, and health working conditions. (5)

depth gauge. A sliding tail, found on some calipers, that extends into a recess to measure depth. (6)

destructive weld testing. Weld testing that destroys the welded joint, including sectioning, bending, or breaking the component in order to evaluate various mechanical or physical properties of the welded metal. (29)

detention pond. A pond used to capture and filter excess irrigation and rainwater. It serves as a flood control mechanism and may dry out completely. Also called *dry pond*. (25)

dew point. The temperature at which the moisture in the air begins to condense or form droplets that are too large to remain suspended in the air. (26)

diagonal pliers. A tool designed to cut light-gauge wire. Two cutting edges pinch together to cut the wire. (7)

die. A tool used to cut threads on the outside of round stock. (7)

die grinder. Pneumatic tool used for finish work on small parts and for shaping surfaces with intricate contours that fit closely with other parts. (8)

dielectric gel. A nonconductive, silicone-based substance designed to seal out moisture and prevent corrosion on electrical connectors. (21)

diesel exhaust fluid (DEF). A reducing agent for selective catalytic reduction that consists of automotive-grade urea (synthetic ammonia and CO_2). (4)

differential. An assembly of gears used to provide power to the drive axles and allow them to rotate at different speeds as necessary. (35)

diffuser. 1. A consumable part of a welding gun that has holes to distribute the shielding gas around the electrode and serves as the attachment for the contact tip. 2. The ring in a plasma cutting torch that causes the gas to swirl as it passes through the high-frequency current within the torch. Also called *swirl ring*. (30, 32)

dimensional lumber. Wood that is cut to a standardized width and depth specified in inches. (9)

diode. An electric gate that allows current to pass in one direction but not the other. (22)

direct current (DC). Electricity that flows in only one direction. (20)

discontinuities. Breaks in the consistency of a welded bead. (29)

displacement. The volume of air and fuel the engine can draw into its cylinder(s) during an intake stroke. (34)

distortion. Deflection or warpage of a welded part due to different cooling rates of welded beads and welded parts. (29)

distribution panel. An enclosure that protects the circuit breakers and main electrical connections. Also called *breaker box*. (21)

door jamb. The component of a door frame that provides a resting location for a door when closed. (17)

door sill. The door frame component located on the very bottom of a door frame. (17)

double groove joint. A bevel groove that is beveled on both sides of the joint. (30)

double-insulated. Tools that have plastic housing parts wherever a user normally touches the machine. The electric current is isolated from any exposed metal parts. (8)

drafting plans. Technical drawings that illustrate construction details and measurements for parts. Also called *blueprints*. (12)

drag shield. A copper alloy shield used to reduce diversion of energy flow that causes arcing between the nozzle and the electrode when the nozzle contacts the base metal. The shield allows the torch to be dragged along a line to trace a design. (32)

drain-waste-vent (DWV) system. The part of a plumbing system that removes water, waste, and gases from the system and maintains atmospheric pressure within the system. (24)

drawing. An image, figure, or sketch used to develop and manufacture a part or project. (27)

dressed dimensions. The final measurements of a piece of lumber after it has been kiln dried, milled, and planed. (11)

drill press. A large, variable-speed power drill mounted to a vertical stabilizing shaft with an adjustable table for securing workpieces. (8)

drip irrigation. An irrigation system in which water drips out of an emitter at a slow pace that ensures that the plants are able to take up the water. Also called *trickle irrigation*. (25)

drip watering system. A type of watering system that applies water through water lines at the soil level. (18)

drive ratio. The ratio by which torque is increased and speed is decreased, or vice versa, between two connected pulleys. (23)

driveline. The component that connects the transmission to the final drive. (35)

drive shaft. A solid shaft that rotates in the center of a motor, transferring mechanical power from the motor to other equipment. (23)

drone. An aircraft that is controlled remotely and does not have a human pilot on board. (4)

dross. Iron oxide waste material from the kerf produced by oxyfuel cutting. (28)

dry-bulb temperature. The temperature of air measured with a standard thermometer. (26)

ductility. 1. A property of metal that allows it to be drawn or stretched into thin flexible wires. 2. The property of metal that allows it to be permanently deformed (reshaped) by stretching (tensile force) without breaking. (27, 9)

dumpy level. 1. A surveyor's optical instrument used in agriculture for planning new construction sites, laying out fence lines, and determining drainage slopes. This tool can only sight at horizontal angles. 2. An optical tool that is used along with a target rod to measure the elevation rise and fall of a piece of land. (6, 13)

duty cycle. 1. A rating that indicates how long a welding machine can be used at a given output current without damaging it. 2. The amount of time an electric motor can safely operate in relation to the time it needs to rest or cool down. (29, 22)

.dxf file format. A file format that allows the exchange of information among CAD files and other file formats. It is used in CNC equipment to create a tool path. (32)

E

E3® tungsten electrode. A tungsten electrode that contains approximately 1.5% lanthanum oxide, .08% zirconium dioxide, and .08% yttrium oxide. It is used for mild steel, stainless steel, and aluminum welding. (31)

ears. The two flanges (ridges) on each end of a stretcher block where the motor is placed just prior to the unit being laid. (15)

earthen structure. A structure that is made from natural components, such as soil, sod, and topography. (18)

edge joint. A joint formed when the edges of two or more parallel, or almost parallel, members are welded along one of the flush edges. (29)

edger. A tool used to settle aggregate, provide a seal in the concrete, and give a smooth finish to the edges of a concrete slab. (14)

effort. The magnitude and direction of a force being applied to a machine. (1)

electrical energy. The flow of electrons through a conductor. (36)

electrical tape. Pressure-sensitive tape used as a standard insulating material for electrical connections. (22)

electric current. Electricity that flows along a path, or circuit. (20)

electric fence. A fence that uses a small electric shock to discourage animals from breaking through the fence. (19)

electric fence charger. An electric device that supplies the voltage necessary to make electric fences operate. (19)

electricity. A type of energy that consists of charged particles, such as electrons. (20)

electrocution. Death from electrical shock. (5)

electrode. In SMAW, a consumable metal rod used to combine with the base metal, forming the weld pool. Also called *welding rod*. (29)

electrode holder. An insulated clamp that holds the welding electrode. (29)

electrolytic capacitor. A capacitor in which the anode electrode is made of a metal such as aluminum or tantalum, and an electrolyte performs as the cathode electrode. It is used in capacitor-start motors to make the voltage lag behind the current by a predetermined degree of the phase. (23)

electromagnet. A magnet in which the magnetic field is produced by an electric current. (23)

electronic distance measuring (EDM) device. A piece of equipment that has circuit boards that calculate distance between two points through algorithms based on the speed of light. (13)

EMT conduit. A galvanized steel conduit made of thin-walled electrical metallic tubing that is thick enough to accept threaded fittings. (21)

encoder. A complex controller used to control servo motors. (23)

end area method. A method used to calculate cuts and fills in which you calculate the area of each end of a given space, take an average of the two areas, and then multiply by the length between the ends to find the volume. (13)

end splice. A type of splice in which wires pointing in the same direction are securely twisted together and then insulated and secured to prevent loosening when the wires are moved. (22)

energy. Usable power. (36)

energy efficient. A descriptive term for something that has low use of natural resources or electricity to heat, cool, use appliances, or operate equipment. (17)

engine. A device that changes the chemical energy in fuel to mechanical energy. (33)

engine control unit (ECU). A computer used on vehicles to help make internal combustion engines more efficient by monitoring and controlling various engine functions. (4)

engineer. A professional who designs solutions to problems with creativity and specialized knowledge of science and math. (10)

engineered lumber. A wide variety of wood products made from scrap from other wood manufacturing processes. (9)

enterprise. Another name for a business. (3)

enthalpy. The total heat energy in the air expressed in Btu per pound of air. (26)

entrepreneurship SAE. An SAE that involves supervised activities conducted by students who are owners or managers for profit. (3)

environment. The surroundings or conditions that affect animals, plants, or objects. (26)

environmental control. Managing conditions, such as temperature, humidity, and light, inside a structure or area. (17)

environmental load. A weight or stress caused by a meteorological event (wind, rain, snow, or earthquake) or a condition of the surrounding area, such as dirt pressing against a foundation. (16)

ergonomic. Comfortable and unlikely to cause stress or injury to the human body. (27)

erosion. The washing away of soil caused by excess water or wind. (25)

ethyl mercaptan. A chemical added to propane and LPG gases as a safety measure because these gases are colorless and odorless. The ethyl mercaptan emits a distinctive, disagreeable odor that helps with leak detection. (28)

eutectic. An alloy that melts and solidifies at the same temperature. (27)

evaporative cooling. A process in which water vapor is added to the air and lowers the dry-bulb temperature without altering the amount of heat in the air. (26)

evapotranspiration. The combination of water loss from large bodies of water (evaporation) and water lost through plant growth processes (transpiration). (25)

exhaust gas recirculation (EGR). The process of drawing cooled exhaust gases into the air-fuel mixture entering an engine's cylinders to reduce combustion temperatures, which results in lower nitrous oxide emissions. (4)

exothermic. A reaction or process that releases heat. (14)

expansion joint. A groove cut into a slab that allows the concrete to crack at this point so that it does not crack in a more random, unsightly pattern. Also called *control joint*. (14)

experiential learning. The process of creating understanding by doing something. (3)

exploratory SAE. A foundational SAE project that allows students to explore agriculture course topics and career pathways. (3)

explosive. A chemical that causes a sudden, almost instantaneous release of pressure, gas, and heat when subjected to sudden shock, pressure, or high temperature. (5)

extractor. An instrument that can be used to remove a damaged fastener. (7)

F

Fair Labor Standards Act (FLSA). Provisions to ensure safe working conditions for youths. (5)

Glossary

fascia. Building trim made from vinyl, metal, or wood that is installed on the vertical edge of the overhang on a building. (17)

fast-fill electrode. A type of electrode with a heavy iron powder flux. The flux increases the deposition rate and produces a thick, easily removed slag that promotes the formation of a very smooth bead. Also called *cover rod*. (29)

fast-freeze electrode. A type of electrode that contains a thin flux that burns away rapidly, providing deep penetration and a moderate amount of fill. It is preferred for out-of-position welding. (29)

feather. An intermediate flame, between the outer flame and the inner blue cone, caused by excess oxygen in the fuel mix. (28)

feeler gauge. Gauge designed to fit into spaces to determine the clearance between two surfaces. (6)

fencing material. A fence component that provides a barrier or boundary between the posts of a fence. (19)

ferrous. Iron-containing. (29)

ferrous metal. An iron-based metal, such as cast iron or steel. (9)

ferrule. A brass collar that fits over the hose barbs on a hose connector and is crimped on to secure the connector to the hose. (28)

file. A hand tool with raised areas or teeth used to cut or smooth material, such as wood, metal, or plastic. (11)

fill dirt. Soil that is used to fill low areas or to create higher levels in an area of land. (13)

fillet weld. A weld in which the weld bead is positioned at the intersection of two parts. (29)

fill-freeze electrode. A type of electrode that provides a moderate deposition rate and produces beads with uniform ripple patterns. These electrodes are used on sheet metal because they minimize warpage. (29)

filtration. The removal of anything in a water or air supply that would negatively affect plant health. (18)

final drive. The part of a power transmission system between the drive shaft and the differential. (35)

finish nail. A nail similar in design to brad nails except the wire is somewhat thicker and the head is bigger and often concave so that a nail set can easily drive it below the surface of the work. (9)

finish. A product, such as paint, stain, or sealer, used to seal wood to protect it from damage, rot, and other hazards or to add beauty to the project. (12)

finishing. The process of covering the framing structure with a variety of materials, such as metal, wood, or brick. (16)

firing. The process of heating a brick in an oven to over 2,000°F. (15)

first aid. Emergency care provided for injury before professional medical treatment is available. (5)

fitting. A small part used to join, turn, or otherwise change a plumbing system. (24)

fixed costs. Total costs of the equipment that will not change over the lifetime of the equipment. (36)

fixture. A permanent part of a plumbing system that uses water or drains the system, such as hose bibs and showerheads. (24)

flammable liquid. Any liquid having a flash point below 100°F (37.8°C), *except* mixtures having components with flash points of 100°F (37.8°C) or higher that make up 99% or more of the total volume of the mixture. (5)

flash point. The lowest temperature at which a liquid gives off a sufficient concentration of vapor to form an ignitable mixture in the air above the liquid. (5)

flashback. The recession of a flame into the tip or all of the way back into the mixing chamber of an oxyfuel gas torch. (28)

flashback arrestor. A device containing a stainless steel mesh filter designed to stop the progress of flashback traveling through gas passages. It may be built into torch fittings, regulator fittings, or both. (28)

float. A tool used to smooth the surface of concrete as part of the finishing process. (14)

floor. The lower, supporting surface of a room or building. (16)

floor jack. A device that uses a horizontally mounted bottle jack that pushes against a lever to lift the saddle arm and platform vertically. (8)

floor scraper. A tool used to knock down bumps and lumps of excess concrete or other imperfections left on the top of concrete as part of the finishing process. (14)

floor squeegee. A tool used to remove excess water from a concrete surface and to spread sealant onto a concrete slab. (14)

flowmeter. Device on a gas cylinder that allows the shielding gas flow rate to be adjusted. (30)

flow-through aquaculture system. An aquaculture system that has a constant flow of fresh water through the system. (18)

fluid power. Power obtained by using fluids (liquids or gases) to complete work. (37)

flutes. The spiral hollows formed in drill bits to assist in the removal of waste material or the cutting edges of router bits. (8)

flux. 1. A cleaner used to clean the pieces being soldered and to exclude air and other impurities from the soldering metal. 2. A coating on an electrode that reacts chemically in the heat of the arc, using up the oxygen in the vicinity of the weld to create an oxygen-free shield of gas around the molten weld pool. (27, 29)

flux cored arc welding (FCAW). A welding process in which a hollow wire electrode filled with powdered flux or a mixture of powdered metal alloying agents is used to form the welded bead. A shielding gas may or may not be used. (30)

flux cored electrodes. Tubular FCAW electrodes filled with powdered flux material and alloying materials. (30)

flux paste. A slightly abrasive cleaner used to remove oxidation before a joint is soldered. (24)

flywheel. A device used to store rotational energy. (33)

folding rule. A measuring device that is 6′ in length, folds for easy storage, and is the standard measuring tool used for laying bricks or blocks. Also called *mason's rule* or *ruler*. (15)

footing. The underground support for a post used to stabilize the post and hold it firmly in place. (19)

force. 1. A pushing or pulling action applied to an object. 2. An influence or energy that tends to change the motion (speed or direction) of an object or cause stress in a stationary object. 3. Any action that causes motion or a change of motion. (37, 34, 1)

forced induction. The delivery of compressed air to the intake of an internal combustion engine, making higher power possible with smaller engines and improving fuel economy. (4)

forehand welding. Welding technique in which the gun is pushed ahead of the weld pool. (30)

foresight. A rod measurement taken to determine the elevation at an unknown point and used in differential leveling. (13)

forge. A furnace used to heat metal to be forged. (27)

forging. A forming process involving the use of force and pressure to shape metal. (27)

form. A mold or frame used to hold fresh concrete until it sets enough to hold the desired shape. (14)

formed-in-place gasket. A gasket formed by dispensing a bead of sealing adhesive onto the contact surfaces of two parts. (27)

framing. 1. The basic structural components of a building or the process of assembling these components. 2. The main support structure for a greenhouse. (16, 18)

framing square. A tool used to position parts or for marking materials. It has a longer leg (blade) and a shorter leg (tongue). (6)

friction bearing. A replaceable surface used between moving parts. Also called *bushing*. (36)

fulcrum. The pivot point on which a lever rests or is supported. (1)

furrow irrigation. A form of surface irrigation in which water is pumped into furrows between the rows of the crop. Gravity and the slope of the land disperse the water from various locations. (25)

fuse. A device that contains a thin wire that is inserted in series with a circuit to protect the circuit from electrical surge. If a surge occurs, the thin wire melts, breaking the circuit. (20)

G

galvanized. Metal coated in zinc to protect against corrosion. (27)

galvanized pipe. Zinc-coated steel pipe. (24)

gap. An opening in a fence spanned by one or two strands of wire attached to a post on both sides of the opening. (19)

gas lens. Wire mesh outlet on a collet body that reduces turbulence in the flow of shielding gas around the weld pool. (31)

gas metal arc welding (GMAW). A welding process in which a consumable, continuously supplied electrode surrounded by a protective envelope of a shielding gas is used to form the welded bead. Also called *wire welding* and *MIG welding*. (30)

gas tungsten arc welding (GTAW). A type of arc welding in which a tungsten electrode is used to create the arc. Also called *tungsten inert gas (TIG) welding* or *heliarc welding*. (31)

gate. An opening in a fence that provides a point of entry. Gates are usually hung on gate posts using hinges. (19)

gate post. A post that supports the full weight and function of a gate. (19)

gauge. Measuring tool used to determine precise part clearances and dimensions, material thickness, or enclosed system pressures. (6)

gear. A disc- or cone-shaped object with a series of teeth cut into its edge. These teeth mate with and engage the teeth of another gear or part that is not concentric with the first gear. (35)

gear pump. A positive-displacement pump that uses rotary motion through gears to create fluid flow. (37)

gear ratio. The number of teeth on the drive gear compared to the number of teeth on the driven (pinion) gear. (35)

genetic engineering. The scientific process of making changes to the genetic code of plants or animals to produce a desired trait. (4)

geographic information system (GIS). Data management software used to map, model, query, and analyze large quantities of data within a single database according to a reference location. (4)

girder. Boards installed to provide an attachment point for other materials or to provide support for the structure. (16)

global positioning system (GPS). A navigation system made up of a network of satellites in orbit around the earth that transmit signals used by GPS receivers to pinpoint their exact geographic location. (4)

global positioning system/ geographic information system (GPS/GIS). Software and equipment that use advanced technology and linked satellites to measure distance. (13)

globular transfer. A method of metal transfer in which droplets of metal larger than the electrode diameter are melted off the end of the electrode wire and deposited in the weld pool. (30)

gouge. A chisel with a U-shaped cutting edge that produces a concave cut into the material. (7)

grade. The slope, pitch, or incline of a surface or line. (13)

grading. Changing the slope of the land, especially to control water drainage. (25)

grain. The pattern the cells of a tree make because of the growth rings. (9)

green brick. A freshly molded block of clay that has not yet been heat treated in a kiln. (15)

greenhouse. An agricultural structure used to start and maintain plants in specialized growing environments. (18)

grinder. Machine that uses high-speed rotation and an abrasive to smooth and shape metal.

groove-joint pliers. A tool with a long slot at the fulcrum, or joint, which allows for multiple working positions and for gripping objects within a large range of sizes. (7)

groove weld. A weld bead that passes completely through the gap between the two edges to be welded. (29)

ground fault circuit interrupter (GFCI). 1. A device that monitors the electric current flow along the hot and neutral wires of a circuit and interrupts electricity to the circuit if there is a difference between these current values. 2. A very fast circuit breaker built into convenience outlets to protect humans from electric shock. (8, 20)

grounded. Term that refers to a power tool that uses 120-volt AC power and has three separate conductors included in the power cord. (8)

grounding rod. A copper-clad steel rod that is 8′ to 10′ long and 5/8″ in diameter and is driven into stable earth as a ground. (21)

grout. A fluid cement mixture used to fill gaps or voids between wythes or other areas in masonry construction. (15)

growing environment. The characteristics of an area where plants are grown, including all factors affecting plant health, such as temperature, humidity, and moisture. (18)

growing medium. A substance or material that contains the proper nutrients in the proper amounts to start and grow plants. (18)

growth cycle. Growth of a plant from a beginning point through some end point, such as from seed germination to transplanting. (18)

gullet. The space between saw teeth that allows for waste removal. (8)

gutter-connected greenhouse. Consists of two or more single greenhouse structures that share common walls. (18)

H

hacksaw. A saw that cuts material using replaceable blades held under tension across an adjustable metal frame. (7)

hammer drill. A tool that resembles a conventional power drill but has a stronger driving force. It is useful for drilling holes in hard materials such as concrete, masonry, and rock. (8)

hand plane. A tool used for shaping wood by removing small shavings when the plane is moved over the wood. (11)

hands-on learning. Performing experiments or building projects, as well as applying technical skills in FFA and SAE activities. (3)

handy box. An enclosure that is press-formed from steel to have rounded, smooth corners. It is used to enclose wire connections and electrical devices. (21)

hardware cloth. A type of wire fabric that has very small openings between wires and is suitable for making small animal cages. (19)

hardwired. Wired directly to the building circuit without an easy means of disconnecting it from the circuit. Often used for large machinery that is too big to be moved easily or for machinery that must remain in a specific location for safety reasons. (21)

hardwood. Lumber cut from an angiosperm (a tree whose seeds are inside an ovule or fruit). (9)

head. The part of a hydraulic cylinder that encloses the cylinder tube end where the rod exits and enters. (37)

header brick. A brick that is laid flat with the short end exposed. (15)

headhouse. An area of a greenhouse that is not used for growing plants. These areas are typically used for instruction, storage, potting, or other tasks. (18)

heat. A form of energy that causes atoms to move. (26)

heat transfer. The movement of heat from warm areas into cool areas. (17)

heat-affected zone. The area around the weld that is heated during the welding process. (29)

heat-treating. The controlled heating and cooling of metals to obtain certain desirable changes to their physical properties. (27)

hemispherical tip. An electrode tip that is ball-shaped to provide a wider path around the weld for the AC current to clean contamination, while the center of the hemisphere concentrates part of the arc to weld the joint. (31)

hertz (Hz). Cycles per second. (20)

hex key. An L-shaped tool used to turn set screws. The tool is formed from a single piece of hexagonal bar stock. (7)

high-frequency start. A feature on GTAW machines that generates a very high-frequency arc that can jump long distances, allowing the welder to initiate the arc without touching the electrode to the base metal. (31)

high-tensile wire. Wire that is made to have high tensile strength and not break easily. (19)

hinge. Two flat plates held together by a pin that allows a lid or door to be attached but not stationary. (9)

hollow ground blades. Blades in which the tips of the teeth are thicker than the body of the blade, resulting in the removal of less material and reduced waste. (8)

homeostasis. The maintenance of a nearly constant internal temperature despite external changes. (26)

homoeothermic. Maintaining a nearly constant core body temperature regardless of the surrounding environment. (26)

hone. To grind or shape to a razor-sharp edge. (11)

hook tooth configuration. A saw blade type that uses an aggressive 10° positive rake angle to make fast cuts. (8)

horsepower. A unit of measurement that equals the force needed to lift 550 lb a distance of 1 foot in 1 second. (1, 34)

hose barbs. Tapered, ringed nipples on hose connectors that are inserted into the ends of a hose and secured by crimped-on collars. (28)

hose traveler system. A type of irrigation system that operates by pulling a cart equipped with a high-pressure irrigation nozzle through a field. As the cart is pulled through the field, it distributes the water to irrigate the crops. (25)

hot bars. Two metal strips in a breaker box that distribute electric power to the individual breakers. 120 V breakers connect to one hot bar; 240 V breakers connect to both hot bars. (21)

hot work. Any process that can be a source of ignition when flammable material is present. (5)

humidistat. A control device that senses the relative humidity of the air. (26)

humidity. The amount of water vapor in the atmosphere. (26)

hydration. An exothermic chemical reaction in which a substance combines with water. (14)

hydraulic cement. A type of cement that sets and hardens under water, such as Portland cement. (15)

hydraulic cylinder. A tube-shaped container that contains a piston and rod. It converts fluid flow into linear motion or force. (37)

hydraulic pipe bender. A hydraulic-powered machine used to bend heavier gauge pipe and tubing. (27)

hydraulic press. Sturdily framed machine that uses a hydraulic cylinder to apply pressure to objects for a variety of purposes. (8)

hydraulics. Use of a liquid to perform work, such as extending a cylinder to lift a heavy load. (35)

hydraulic system. A fluid power system that uses liquids to perform work. (37)

hydrogen fuel cell. A device that uses hydrogen gas to produce the electricity to power an electric motor. (4)

hydrometer. Tool used to determine the integrity of the coolant mixture present in a liquid-cooling system by measuring its specific gravity and strength. (36)

hydroponics. A system for growing plants in a water solution. (18)

hydrostatic. Having the ability to transfer power through fluids at rest or under pressure. (35)

hyperthermy. A condition is which body temperature is above the norm. (26)

hypothermy. A condition is which body temperature is below the norm. (26)

I

I-beam. A laminated structural member made with oriented strand board (OSB) as the center component and dimensional lumber as the top and bottom pieces. (9)

ignition coil. A device that transforms low voltage current into high voltage current that can cause a spark to ignite the fuel and air mixture in an engine. (33)

impact driver. A tool that combines the torque of a power drill with an impact mechanism much like a hammer drill.

impact sockets. Tools that look similar to sockets attached to ratchet handles but are made of much stronger steel to withstand the hammering effect of an impact driver. (8)

improvement activities. Learning activities that improve the value or appearance of the place of employment, home, school or community; the efficiency of an enterprise or business; or the living conditions of the family. (3)

inclined plane. A flat, sloping surface or ramp. (1)

inclusions. Foreign materials, such as slag, that are trapped within the finished weld and weaken the weld. (29)

incomplete fusion. A weld discontinuity in which blending does not occur between the weld metal and joint faces or adjoining weld beads. (29)

incomplete joint penetration. Failure of the filler metal or base metal to fill the weld root completely. (29)

inductance. The squeezing forces exerted on an electrode wire at the weld due to the amount of current passing over its surface. Also called *pinch effect*. (30)

induction. The process of moving a conductor through a magnetic field to generate current in the conductor. (23)

in-line fuse. A fuse placed individually in a circuit that is dedicated to protect a single circuit. (22)

in-line splice. A type of splice in which wires that point toward each other are joined together and are then insulated and sometimes soldered or crimped to hold securely. (22)

insulated concrete form (ICF). A frame made with insulating materials on each side into which concrete is poured and which remains in place as part of the wall. (14)

insulation. A material that slows the movement of heat from warm areas to cool areas. (17)

insulation classes. Classes of insulation based on thermal capacity using code letters A, B, F, and H, in which a higher code letter means a higher thermal capacity. (23)

insulator. 1. A fence component that prevents electrical flow in electric fences and provides an attachment point for fence material. 2. Any material that resists the free movement of electrons from one atom to another. (19, 20)

internal combustion engine. An engine that burns fuel within a cylinder, creating a force to drive a piston. (1)

internship. A placement experience, lasting from 3 to 12 months, in which a student works at a paid or unpaid job to learn more about an aspect of the business. (3)

inverse-time breaker. A type of breaker that trips more quickly with higher overcurrents than with lower overcurrents. (23)

inverter. A power source that passes single-phase or three-phase AC power through a bridge rectifier to create a DC current. (29)

irrigation. The application of water to crops, landscaping plants, or turf to aid in plant growth. (25)

irrigation controller. A device that controls the timing and watering locations of an irrigation system. (25)

irrigation zone. A group of sprinklers that are turned on to water a specific area. (25)

isometric drawing. A method for depicting three-dimensional objects in two dimensions by rotating the object 30° from the front view and tilting it forward, showing all three dimensions of an object in a single image. (10)

J

jack stand. A stand with adjustable height, used to support a vehicle that has been raised by a jack. (8)

jam nut. A fastener torqued into place against another nut to hold it in place. (27)

jet. A tiny hole in the narrow point in a carburetor throat (called a *venturi*). (33)

jig. A pattern used to accurately guide the formation of a part. (27)

jig saw. A handheld saw that uses a reciprocating blade action. It is used to make precise, often curved cuts in wood, metal, plastics, and other building materials. (8)

job application. A form supplied by an employer to collect information about a prospective employee. (2)

jointer. 1. A stationary tool that uses rotating blades to straighten and finish the edges of lumber for applications where boards are fitted side by side. 2. A tool used to finish the surface of a mortar joint between bricks or concrete blocks. (8, 15)

joist. A structural member laid horizontally from rim to rim on a floor or ceiling that carries the load. (9)

journaling. Writing about one's own experiences. (3)

junction box. An enclosure designed to protect connections between electrical wires. (21)

K

kerf. The physical slot made by a saw or other tool that cuts by removing a thin section of waste material. (7, 11)

keyhole. A rounded enlargement that appears in the opening just ahead of the weld pool. (29)

keyhole saw. A saw with thin, tapering steel blades, used to make inside cuts on covering materials such as paneling and plywood. (7)

kickback. Event that can occur when a power tool sticks or binds momentarily in the work. The interruption can cause the tool to suddenly jump backward out of the work and toward the operator. (8)

kiln. An oven or furnace that is used to heat green bricks to a very high temperature. (15)

kindling temperature. The temperature at which steel readily reacts with oxygen—approximately 1600°F (871°C). (28)

kinetic energy. Energy that is in motion. (36)

knee board. A rectangular or round platform on which the user kneels to distribute body weight while moving over a concrete surface during the finishing process. (14)

L

laminated dimensional lumber (LDL). Lumber produced much like plywood, except all the veneer panels have their grain running in the same direction. Also called *laminated veneer lumber (LVL)*. (9)

laminated lumber. Building materials formed by adding adhesives and then compressing layers of wood in specific orientations. (9)

laminated structural member. An engineered piece used in floor joist and ceiling joist applications. (9)

laminated veneer lumber (LVL). Lumber produced much like plywood, except all the veneer panels have their grain running in the same direction. Also called *laminated dimensional lumber (LDL)*. (9)

lanthanated tungsten electrode. A tungsten electrode that contains approximately 1.5% lanthanum. It is used with AC or DC and provides excellent arc starting, low consumption rate, good arc stability, and a 50% higher current-carrying capacity than pure tungsten. Marked with a gold band. (31)

lap joint. A joint formed between two overlapping parts. (29)

laser level. A level that incorporates a laser beam instead of the scope for increased speed, improved accuracy, and reduced labor requirements. (6)

latent heat. Energy that changes a substance's physical phase from one state to another (liquid, gas, or solid). (26)

lateral line. A secondary irrigation line that carries water from the main line to the sprinkler head or emitter. These lines are only under pressure when the irrigation system is in use. (25)

law of conservation of energy. The principle that states that energy cannot be created or destroyed. (1)

leach field. A series of perforated pipes placed in covered, gravel-lined trenches downstream from a septic tank to allow the liquid from the septic tank to percolate through the soil. (24)

lead. A toxic, very dense, heavy metal used in anchoring materials, in batteries, and as shielding material for radiation. (9)

level. 1. A tool used to determine whether a surface is exactly horizontal or vertical. 2. Exactly horizontal (parallel) to the gravitational pull of the Earth. (15, 6)

lever. A rigid bar free to turn on a fulcrum. It can be used to lift or move loads when force is applied to the opposite end. (1)

lift start. A method of initiating a GTAW arc by touching the electrode and the rim of the nozzle to the base metal at the same time, and then rolling back the electrode using the nozzle as a pivot to establish the arc. (31)

limited-slip differential. A mechanical device that prevents excessive power from being devoted to one wheel. Instead, it keeps both wheels in powered rotation. Also called *locking differential*. (35)

limit switch. A switch that is activated or triggered when a limit is reached, such as when a grain bin becomes full. (22)

line holder. A device used to hold a mason's line in place when positioned on the bricks or blocks. (15)

lineman splice. A type of splice in which the wires are twisted together in such a way that the electrical connection tightens as the conductors pull against each other. (22)

lineman's pliers. A heavy-duty tool combining flat jaws and a side-cutting, pinch-type wire cutter. (7)

liquefied petroleum gas (LPG). A mixture of propane and similar gases, such as butane and propylene. (28)

liquid layer. The layer in a septic tank that consists of liquids such as water, urine, and liquid waste products. (24)

lithium-ion (Li-ion) battery. A power source for cordless power tools that delivers uniform speed and torque down to about 20% of a full charge. (8)

live load. The stress caused by or the weight of everything placed on, in, or temporarily attached to a structure, such as machinery, people, furniture, and animals. (16)

livestock. Animals in an agricultural setting. (19)

load. 1. A stress or weight that is supported by a structure or member. 2. The part of a circuit that does the work, such as a light or a heating element. (16, 20)

load-bearing member. A component of a framing system that must support an amount of weight that is important for the strength of a building. (16)

locking pliers. Pliers that offer strong clamping strength combined with a jaw adjustment screw and quick release feature. (7)

lodging. The result of soft metal particles building up as deposits in the surface of an abrasive attachment. (8)

long-term goal. A goal that typically takes several years to accomplish. (3)

loom. A plastic covering used as an added layer of protection for wires and bundles of wire. (22)

M

machine. Any mechanical device designed to make work easier by changing the direction of the force applied, transferring a force from one location to another, increasing the magnitude of a force, or changing the distance a force moves. (1)

machine bolt. Threaded fastener used in nonprecision applications and often furnished with a matching nut. (27)

machine guard. Part attached to a machine to protect the user from moving parts and flying particles. (8)

machine screw. A threaded fastener 1/4″ or less in diameter. Similar to a *cap screw*. (27)

machine welding. Welding in which the welding gun is connected to a controlling arm that is not physically operated by a welder's hand. The welder sets and adjusts the controls to position the welding gun correctly. (30)

main line. A large-diameter pipe that carries water to the lateral lines of an irrigation system. (25)

malleability. The property of metal that allows it to be hammered or pressed (compressive force) into shape without breaking. (9, 27)

mallet. An impact tool with a head made of wood, rubber, or rawhide. This tool is often used to drive wood chisels or leatherworking tools. (7)

mandrel. A tool used as a form or guide for bending metal parts. (27)

manifold system. A system of two or more cylinders connected together so that gas can be drawn from multiple cylinders at once. (28)

manometer. An instrument used to measure pressure in air, gas, or liquid or to determine the difference between two pressures. (26)

margin trowel. A trowel with a rectangular front edge and a V-shaped head that is used to hold mortar or fill small areas. (15)

mason's level. A level that is built like a carpenter's level but is usually four feet or longer in length for leveling and plumbing longer surfaces. (6)

mason's line. A strong nylon cord used to show the layout of a building or to keep a course (row) of bricks or concrete blocks in a straight line. (15)

mason's rule. A measuring device that is 6′ in length, folds for easy storage, and is the standard measuring tool used for laying bricks or blocks. Also called *folding rule*. (15)

masonry nail. A nail with a hardened, extra thick shank that is often grooved to give it increased holding power. This type of nail is used for joining wood to masonry or concrete. (9)

materials list. A record showing all items needed in the construction of a project, such as wood, hardware, and paint. (12, 19)

mechanical advantage. The ratio of the resistance to the effort applied by a machine. (1)

mechanical energy. A type of energy produced and transferred by pulleys and belts or by gears. (36)

mechanical refrigeration. A process that uses the principles of various scientific gas laws to absorb heat into a refrigerant to produce a low temperature. (26)

metal cored electrodes. FCAW electrodes that contain minimal or no flux. Instead, they contain powdered mixtures of metals engineered to improve the properties of the alloy in the finished weld. (30)

metal inert gas (MIG) welding. A nonstandard term for *gas metal arc welding*. (30)

micrometer. A device used to take very precise measurements. (6)

MIG welder's pliers. Pliers that have multiple functions designed around the GMAW welding torch. (30)

mild steel. A low carbon steel that is hard, easily welded, and is often formed into structural components because it holds its shape well. (9, 28)

milling. Cutting a piece of lumber to a certain size. (11)

miter box. A piece of equipment used to guide a backsaw when cutting precise angles in trim work. (7)

miter gauge. A device that can be used to set the position of a board so that it is cut at a particular angle. (11)

miter saw. A circular saw mounted on a pivot that is attached to a stabilizing frame or table. (8)

modern braced framing. A building method in which all main structural components are assembled before subflooring is installed. (16)

moisture. This term is used to describe the amount of moisture present, especially the amount of water that can be used by plants. (18)

Morse taper. A tapered shank that holds a drill bit entirely by the friction between the shank and the tapered socket. (8)

mortar. A cement paste consisting of Portland cement, sand, and water mixed in a ratio that is plastic when wet and rock hard when cured. (9)

mortar board. A device made of plywood or polymer that is used to hold mortar when laying bricks or concrete blocks. (15)

mortar box. A container used to hold ingredients that are mixed to create mortar. (15)

mortar stand. A device, made of metal or stacked concrete blocks, used to hold and position a mortar board at the correct height when laying bricks or blocks. (15)

motor housing. The frame that encloses and supports the components of a motor. (23)

multiflame heating tip. Welding attachment with a bulbous tip that has a ring or multiple rings of orifices, each of which supplies a separate flame. Also called *rosebud*. (28)

multimeter. A device that can measure voltage, current, or resistance in a circuit. (20)

multiple-strand insulated conductor. Wiring made up of several strands of wire twisted together into a single conductor and covered with insulation. (21)

N

nail. A small metal spike that can be driven into wood or other materials to hold materials together. (9)

nail gun. A pneumatic tool that drives a sharp spike into wood and similar materials with a single power stroke. (8)

National Electric Code® (NEC). The definitive guide in the United States for proper electrical installation and inspections to prevent people and physical property from damage by electrical hazards. Published by the National Fire Protection Association (NFPA). (10, 21)

National FFA Organization. An organization for students interested in agriculture and leadership. (2)

National Institute for Occupational Safety and Health (NIOSH). A division of the CDC that works to advance worker health and safety by tracking work-related hazards and developing solutions. (5)

needle-nose pliers. A tool with long, narrow jaws that are engineered to grip very small objects and work in tight places. (7)

networking. Getting to know people in various professions through self-introduction or introductions through friends and associates and informing others that you are looking for employment. (2)

neutral bar. A neutral conductor from the power supplier that is connected to every neutral from all of the branch circuits in a breaker box so the neutral is continuous and common to all circuits. (21)

neutral flame. The type of flame produced when the ratio of oxygen to acetylene emitted from the torch is almost exactly one-to-one. (28)

nickel-cadmium (NiCad) battery. The oldest type of rechargeable battery used in cordless power tools. Easily deteriorates when not charged and discharged properly. (8)

nickel-metal-hydride (NiMH) battery. A rechargeable battery that is very sensitive to improper charging methods and storage conditions. (8)

nominal dimensions. The measurements of a piece of wood before kiln drying, milling, and planing. (11)

nondestructive testing. A type of weld testing in which the weld remains intact (is not destroyed). (29)

nonferrous metal. Metal that does not contain iron. (9)

nonmetallic sheathed cable (NMC). Two or more insulated conductors bundled in a nonmetallic sheath. (21)

nonstress graded. Wood used as construction material. Also called *common lumber*. (11)

normally closed N/C switch. A momentary switch that completes a circuit except when the switch is physically triggered. When triggered, the switch opens the circuit. (22)

normally open N/O switch. A momentary switch that completes a circuit only when the switch is physically triggered. (22)

nut. A tubular, threaded piece of metal whose threads must match the threads of the bolt with which it is used so they will fit together securely. (9)

nut driver. A tool that resembles a screwdriver with a socket wrench attached to the tip. (7)

nylon. A relatively old form of thermoplastic that is lightweight and heat tolerant. It may be combined with other materials and used to make replacement parts. (9)

O

Occupational Outlook Handbook (OOH). A handbook detailing information and predicted demand for most careers in the United States. (2)

Occupational Safety and Health Administration (OSHA). A division of the DOL that has the authority to set and enforce health and safety standards for workers in most private industries. (5)

ohm. The unit of measurement for resistance in a circuit, represented by the omega symbol (Ω). (20)

Ohm's law. A set of formulas that define the relationships among voltage, amperage, resistance, and power. (20)

opening. A free space in a wall or other building component that allows passage of people, animals, or objects or where another component, such as a door or window, can be installed. (16)

open root weld. A bevel groove joint set up without the use of a backing plate. (30)

oriented strand board (OSB). A laminated wood product made from very thin rectangular wood strands and resin that is compressed into cross-oriented layers. (9)

orthographic drawing. A two-dimensional image that shows a three-dimensional object from different views, usually the front, side, and top. (10)

oscillating multi-tools. Portable power tools designed with a mechanism that rapidly vibrates a blade or other attachment back and forth in a narrow arc. (8)

out-of-position welding. Welding performed in any welding position other than the flat position. (29)

output shaft. A shaft that provides a connection point for an implement to be attached. (35)

overall rise. The measurement of how many inches or feet of rise are included in a roof from eave to peak. (17)

overburden. Soil, rock, or material that is not wanted on land or a building site. (13)

overhang. The part of the roof that extends beyond the edge of the outside wall and along the roof line. (17)

overlap. A weld discontinuity that consists of a protrusion and notch in the weld metal beyond the weld toe, or weld root, which results in a concentration of internal stresses. (29)

oxidize. To combine with oxygen to form corrosion as when iron oxidizes and forms rust. (9)

oxidizer. A chemical other than a blasting agent or explosive that initiates or promotes combustion in other materials, thereby causing fire either of itself or through the release of oxygen or other gases. (5)

oxidizing flame. The type of flame created by an oxygen-rich fuel mixture at the tip of an oxyfuel torch. The flame is characterized by a sharply pointed inner cone and a distinct hissing noise. (28)

oxyfuel cutting. The process of using oxygen and a fuel gas to burn a cut into a ferrous material. (28)

oxyfuel welding. The process of joining metals through the introduction of heat generated by burning gases. (28)

P

paddock. A smaller portion of land that is divided from a larger property and used as a living or grazing area for livestock. (19)

parallel circuit. A circuit in which each load is wired separately so that the failure of one load does not have an adverse effect on the other loads. (21)

pattern bond. A design or arrangement formed by masonry units and mortar joints on the face of a masonry wall. (15)

perimeter fence. A fence that is installed all the way around an area where livestock are grown. (19)

personal interest inventory. List of questions to assess one's own natural interests and provide guidance in developing a career. (3)

personal protective equipment (PPE). Safety devices or clothing used or worn to protect an individual against the health or safety hazards that may be present in the work environment. (5, 15)

pest. An unwanted insect or animal in a growing environment. (18)

Phillips head screwdriver. Common screwdriver type that has four tapered flutes that are rounded by design. (7)

photosynthesis. The process by which plants create food from carbon dioxide and water. (18)

photovoltaic cell. A device that converts light directly into DC electricity at the atomic level. (20, 22)

photovoltaic module. A group of solar cells mounted in a frame and electrically connected to each other. (22)

photovoltaic system. A solar electric energy system that converts light energy (photons) into electricity to provide power. (4)

piano hinge. A fastener with two plates with alternating pins that allow one plate to fold back upon the other. Also called *continuous hinge*. (9)

pier footing. A foundation component made of concrete poured in a box shape that is thick enough to spread the weight from a post or structural member into the surrounding soil. (14)

pigtail. A short piece of wire tapped into another and leading to a device in the same enclosure. (21)

pilot arc. The arc formed when energy in the electrode from the primary current flows across the plasma to the nozzle. The arc initiates the plasma process. (32)

pipe roller. A machine tool used to create long curves in metal parts. (27)

pipe wrench. An adjustable wrench designed to grip and hold or turn pipes of various diameters. (7)

piston. 1. An engine part that slides back and forth in a cylinder and creates reciprocating (up-and-down) motion during the combustion process. 2. The part of a hydraulic cylinder that provides a movable seal inside the cylinder. (33, 37)

piston pump. A positive-displacement pump that uses a piston to create an up-and-down motion to create fluid flow. (37)

placement SAE. An SAE that involves placement of a student as an employee or intern for a production agriculture enterprise, agribusiness, or agency. (3)

plain-sawn lumber. Lumber cut in slabs starting with one edge of the log and continuing until all the log is cut into boards with a variety of grain patterns. (9)

planer. Tool used to shave layers of wood from lumber to smooth surfaces, remove saw marks, or adjust the thickness. (8)

planing. Shaving and smoothing the edges of wood. (11)

plank and beam framing. A building method used to create large open areas and to show off exposed wood components, such as support beams, for their visual appeal. (16)

plant disease. An infection caused by a harmful organism that can negatively affect plant health and can cause plant death. (18)

plasma. A gas heated to an extremely high temperature and ionized to the point that it becomes electrically conductive. (32)

plate. A relatively thin metal piece designed for general strengthening of a project made in Tee, ELL, or straight shapes with pre-drilled holes. (9)

platform framing. A building method in which each level or story of a building is built as a separate unit. Also called *western framing*. (16)

plumb. 1. A term used to describe an element, such as a wall, that is straight vertically and exactly perpendicular to another component, such as the floor. 2. Aligned perfectly perpendicular to the pull of gravity. (16, 6)

plumb bob. A measuring tool that consists of a pointed weight that is tied to the end of a string. When the string is hung from the opposite end, gravity stretches the string in a vertical, plumb line. (6)

plywood. A manufactured wood product made of very thin wood layers (veneers) and laminated under high pressure. (9)

pneumatic hammer. An air-powered tool used to cut metal. Also called *air hammer* or *air chisel*. (27)

pneumatic system. A fluid power system that uses gases to perform work. (37)

polarity. 1. A condition of having a positive and negative charge or electrical state. 2. The direction of flow (positive or negative) of electric current. (20, 29)

pole frame construction. A type of structure in which poles or posts are used as the main structural components for walls. (16)

polyethylene. A type of plastic commonly used as a greenhouse covering material. (18)

polyethylene pipe. A soft, flexible irrigation system pipe (usually black) that can be used on top of the ground. (25)

polyvinyl chloride (PVC). A type of rigid plastic pipe that is often used in DWV systems. (24)

pond system. An aquaculture system in which the water supply is contained in one location that is created by digging and shaping the soil. (18)

porosity. A cavity-type discontinuity formed by gas trapped in the molten weld, resulting in bubbles or pockets as the weld solidifies. (29)

portable power drills. Handheld tools intended primarily for boring holes in wood, metals, and other common construction materials. (8)

portable power tools. Tools that are traditionally powered through an electric power cord plugged into standard 120-volt AC current. Larger tools may require 240-volt AC current. Some tools are powered by direct current (DC) power sources. (8)

portfolio. A record of skills, activities, training, and education. (2)

Portland cement. A substance that is made from limestone and clay and that hardens as it cures, even under water. (14)

ports. Holes located on the cylinder that direct fluid flow into or out of the cylinder. (37)

positioner. A device that aligns parts to be welded in the optimum positions. (30)

positive ground system. An older system in which a circuit for mobile vehicles and equipment is connected from positive to negative, so that the positive side of the circuit is connected to ground and the negative side is hot. (22)

post. 1. A long, sturdy piece of wood or metal that is set upright in the ground or a structure and often used in agricultural buildings as an attachment point for other structural components. 2. A fence component that supports and holds fencing materials. (16, 19)

post-and-rail fence. A type of fence that has posts for vertical components and boards, or rails, for horizontal components. (19)

post driver. A tool used to mechanically drive posts into the ground. Post drivers are either manually operated or tractor mounted. (19)

postflow timer setting. Setting on a GMAW machine that allows shielding gas to continue to flow after the arc is extinguished, maintaining a protective shield on the cooling weld area and also cooling the welding torch components. (30)

post frame and truss framing. A building method that results in large open areas inside a building and is the most common type of framing. (16)

potential energy. Stored energy. (36)

poultry wire. A type of wire commonly used with poultry and specifically chickens. (19)

power. The rate of doing work or the time it takes to perform a specific amount of work. (1, 34)

powered auger. A tool used to drill a hole in the ground quickly using a mechanical power source such as an engine, tractor, or skid-steer loader. (19)

power shear. Large, stationary, hydraulic- or electric-powered metal-cutting shears. (27)

power take-off shaft (PTO). 1. A metal rod attached to a power source (tractor) at one end, and an attachment at the other. 2. A gear-driven shaft that normally extends past the differential on the rear of equipment. It is driven by the power source (engine) and allows the power to be used to drive accessory attachments or other applications. (5, 35)

power train. The system that transmits power to output sources such as drive wheels and power take-offs (PTOs). Typically includes a clutch, transmission, differential, and final drive. (35)

power trowel. A piece of motorized equipment used to apply a smooth finish to a concrete surface. (14)

precision. The ability to be exact within fractions of a unit when taking measurements. (6)

precision agriculture. A systematic approach to site-specific agricultural management in which variations in crop growth are observed and recorded, and the data is used to improve productivity. (4)

prefabricated agriculture structure. An agricultural structure that is built at one location and transported to another location for installation and use. (18)

preflow timer setting. The setting on a GMAW machine that allows shielding gas to envelop the weld area before the arc in order to purge the weld area of atmospheric gases. (30)

prehung door. A door that is sold as a complete unit and already installed in a door frame when purchased. (17)

pressure. A force applied to a specific area. (37)

pressure energy. Energy created when force is applied to a constrained fluid. (36)

pressure regulator. A device that controls the pressure of gas flowing from the storage cylinder toward the oxyfuel torch. (28)

pressure-treated lumber. Wood that has been processed with chemical preservatives to be highly resistant to rot and insect damage. (9)

preventive maintenance. A type of maintenance that consists of routine or scheduled inspections to catch minor issues before they become major problems. (36)

prick punch. A tool that is ground to a sharp point and is primarily used to mark metals. (6)

primer. A liquid used to clean and prepare the surface of PVC or CPVC pipe before a solvent is used to join the pipe. (24)

production agriculture. Agriculture for the specific purpose of producing a marketable product, such as food or fuel. (1)

proficiency award. An honor given to FFA members who have demonstrated excellence in their SAEs. (3)

propagation. Plant cultivation and growth. (18)

propane. A fuel gas (C_3H_8) that produces very high heat but with an inner cone temperature that is cooler than that of acetylene. (28)

psychrometric chart. A graphical representation of the physical properties of air. (26)

psychrometrics. The study of science involving thermodynamic properties of air and water vapor. (26)

pulley. 1. A wheel that spins freely on an axle supported in a frame or is attached to the object to be moved. The wheel has a groove to receive a rope or belt. 2. A simple machine that can change the direction of a force. When used in multiple sets, pulleys create a mechanical advantage. (1, 35)

pulsed arc machine. A type of machine for wire welding processes in which the working current is run through a pulsing unit that rapidly cycles between periods of higher voltage and lower voltage, maintaining the arc throughout the welding process. (30)

pump. A device that provides the force needed to move or compress fluids or gases. (37)

punching. A metal shearing process that creates holes using a punch and die. (27)

purlin. A board that is part of roof framing, connects trusses or rafters, and functions as sheathing. (17)

push-pull welding gun. A welding gun equipped with a set of auxiliary rollers that pull the electrode wire through the lead. (30)

push stick. A piece of wood or plastic used to guide lumber through a saw. (11)

Q

quarter-sawn lumber. Lumber that is sawn from the log in such a way as to have all the growth ring (grain lines) perpendicular to the face of the board. (9)

quenching. The process of rapidly cooling a heated metal with a fluid to harden it. (27)

quick-disconnect couplings. Fittings used to frequently or rapidly connect or disconnect components and tools from a fluid power system or other system. (37)

quote. A document that shows a price given by a dealer or seller for an item to be purchased. (10)

R

raceway. A conduit for electrical conductors to protect them from damage prior to entering the main service panel. (21)

racking. The process of ending a brick or block course prior to the previous course, creating a step effect in the courses. Also called *racking back the lead*. (15)

racking back the lead. The process of ending a brick or block course prior to the previous course, creating a step effect in the courses. Also called *racking*. (15)

radial arm saw. A heavy-duty circular saw mounted to a rotating, tilting, yoke-like frame that slides below a horizontal support arm. Ideally suited for use as a cut-off saw. (8)

radiation. The transfer of heat through heat waves traveling from an emitting heat source to a person or object. (17)

radiographic testing. A type of weld testing that uses X-rays to penetrate through the weld to show an image of the internal structure of the joint. (29)

rake angle. The number of degrees off perpendicular to the back of the blade each tooth in a band saw blade is angled. (8)

raked joint. A type of mortar joint where a portion of the mortar is removed to create a joint recessed below the surface of the brick. (15)

rasp. 1. A cutting tool with rows of multiple triangular teeth that remove controlled amounts of material aggressively from a workpiece. 2. A hand tool that is similar to a file except that the cutting teeth are much coarser. (7, 11)

ratchet handle. A tool that provides leverage to rotate a socket to loosen or tighten fasteners. (7)

ratchet tensioner. A device installed in a wire fence that is used to increase fence tension using a wrench or ratchet. (19)

reactivity. The tendency to react or undergo chemical changes. (5)

rebar. A steel rod used for reinforcement in concrete. (14)

receptacle. An electrical device that provides a means of conveniently connecting to or disconnecting a load from a circuit. (21)

receptacle tester. A device that plugs into a convenience outlet to verify that circuits are wired properly. (20)

reciprocating saw. Handheld demolition tool that operates by mechanically moving a blade in a fast back-and-forth motion. (8)

recirculating system. An aquaculture system in which water is filtered and recirculated through the system. (18)

reclaimed water. Wastewater that is captured and treated. (25)

rectifier. A device that uses diodes to convert AC to DC electricity. (22)

reference. A person who is willing to speak or write well of a job applicant because of past work experience or other association with the candidate. (2)

registration marks. Points on a pattern that are used to align the pattern in the correct location when it is repeated or to align parts of a pattern that are printed on separate sheets of paper. (12)

relative humidity. The amount of water vapor present in the air. (17)

relay. An electromagnetic switch connected to a primary circuit and a secondary circuit, allowing a high-amperage secondary circuit to be controlled by the much lower-amperage primary circuit. (22)

relief cut. A cut made to release waste material and reduce the chances that the blade will bind. (11)

rendering. A precise drawing of an object or the act of creating a drawing as in a technical drawing. (10)

repulsion-start motor. A motor that uses a strong electromagnetic force to start and rotate heavy loads. (23)

research SAE. An SAE that consists of an original project that involves agriscience—the application of scientific principles to an agricultural problem. (3)

reservoir. A manmade or natural lake that is used for a water supply. (25)

resistance. 1. The force that resists the effort force in a machine. 2. Restriction of the flow of electricity, measured in ohms. (1, 20)

résumé. A summary of a job applicant's experience, education, and skills. (2)

retention pond. A pond used to capture and filter excess irrigation and rainwater. It holds a permanent pool of water. (25)

reverse flow check valve. A valve installed on torch hose connections to reduce the chance of backflow by forcing gas to flow in one direction only. (28)

rheostat. An adjustable resistor used to vary the resistance in an electric circuit. (22)

rigid conduit. A type of galvanized steel conduit that is similar to water pipe and is thick enough to be threaded to accept threaded fittings. (21)

rigid construction. A type of structure in which posts and rafters are assembled as a unit and are open the entire span of a building. (16)

rigid panels. Panels made of sturdy materials such as polycarbonate or acrylic that are stronger, lighter, and longer lasting than other greenhouse covering materials. (18)

rip fence. A piece of metal that can be attached to the table for a saw and used to guide a board as it is being cut. (11)

ripping. Cutting with the grain. (8)

ripsaw. A tool used to cut boards in a direction parallel to the grain of the wood. (7)

rivet. A metal pin with a head and a tail piece made from a soft piece of steel or other alloy. (9)

rod. The part of a hydraulic cylinder that is connected to the piston and transfers the generated force in a linear motion. (37)

rod oven. A heat- and humidity-controlled storage container that prevents low-hydrogen electrodes from absorbing excess moisture and preserves their low-hydrogen properties. (29)

rolling resistance. The resistance created by the imprint/traction of a tire in soil and its use of power to roll up and out of the soil. (36)

rollover protection system (ROPS). Device designed to protect operators from injuries caused by vehicle overturns or rollovers. (5)

Romex®. A type of wiring commonly used in residential applications. Types include nonmetallic sheathed cable (NMC) and underground feeder (UF) wiring. (21)

roof. The external covering on the top of a building that is exposed to the environment. (16)

roofing nail. A nail with a short shank and large head most often used to attach roofing paper, asphalt shingles, and roofing sheet metal. (9)

root bead. The first weld bead in a multiple-pass weld, placed directly in the joint between the plates. (29)

root face. The distance from the end of the bevel angle to the root of the joint. (30)

rotor. The rotating portion of an electric motor. (23)

router. 1. A portable woodworking tool consisting of a powerful electric motor, a rotary cutting tool, and an adjustable base plate to control cutting depth. 2. A power tool that can be used for cutting a decorative edge or design on wood. (8, 11)

rowlock brick. A brick that is laid on the face edge with the end of the brick visible in the wall face. (15)

rowlock stretcher brick. A brick that is laid on edge with the largest surface of the brick exposed in the wall face. Also called *shiner*. (15)

runoff. The draining away of water not absorbed by plants (typically crops). (25)

S

sacrificial part. A part that is meant to be damaged during regular functioning of a machine. (32)

safety data sheet. A document that provides information about hazards and provides comprehensive information about a substance or mixture for use in workplace chemical management. (5, 33)

safety roller. A tool used for stamping a pattern into concrete to create a textured finish. (14)

sailor brick. A brick laid on end with the largest surface of the brick exposed in the wall face. (15)

sander. A machine that uses abrasives to smooth rough wood surfaces and to remove small amounts of a wood surface. (8)

sash. Panels or bars in a window that hold the glass in the window. (17)

scaffolding. An elevated, temporary work platform. (5)

scale drawing. An image that uses a ratio (the scale) to reduce or enlarge the dimensions of an object for an accurate representation of the object. (10)

scavenging. The removal of exhaust gases from an engine by forcing in fresh air. (33)

schedule number. An established number that represents pipe of a specific internal diameter or wall thickness. (37)

scope. An optical device (telescope) on a surveying level through which the user looks to sight a level horizontal line. (13)

Scotchlok® connector. A wire connector that consists of a plastic frame and a slotted metal blade that is pressed into the conductors using a pair of pliers. (22)

scratch awl. A tool that looks similar to an ice pick. It leaves a precise, permanent mark that is unaffected by heat and will mark any material softer than itself. (6)

scratch start. A method of initiating a GTAW arc by scratching the electrode on the metal and then lifting it while pressing the remote control pedal. (31)

screed. A straight 2″ × 4″ piece of lumber used to level fresh concrete and settle the aggregate into the paste. (14)

screw. 1. A simple machine that consists of an inclined plane wrapped around a cylinder. 2. A fastener that has a head, a threaded shank, and a tip and is twisted into wood or other material. (1, 9)

screw terminal. 1. A connector that uses screw pressure to secure a wire or cable to another electrical device. 2. Wire connector in which the wire is wrapped clockwise around the screw and the screw is then tightened to form the connection. (21, 22)

scribe. Tool used to make scratch marks for cutting or laying out materials. (6)

scum. The top layer in a septic tank, which consists of very lightweight solids, fats, and other waste materials. (24)

seal. A device located on the head of the cylinder that creates an area where fluid cannot escape. It is used to prevent internal or external leakage of fluid. (37)

secondary agriculture. Practices performed by a farmer or on a farm as an incident to or in conjunction with farming operations. (5)

SEER rating. An abbreviation for seasonal energy efficiency ratio, which is a measure that describes the energy efficiency of electrical equipment or appliances. (17)

selective catalytic reduction (SCR). A technology in which a liquid reducing agent is injected into the exhaust stream of a diesel engine to break down nitrous oxide into harmless nitrogen and water. (4)

self-drilling screw. A screw configured with the tip end of the incline plane more open, allowing the threads to start a hole by cutting through the material. (9)

self-piercing screw. A screw with a long taper, an extremely sharp point, and threads that extend to the very tip. (9)

semiautomatic welding. Gas metal arc welding performed by hand. The welder manipulates the welding gun, but the equipment controls the feeding of the wire electrode. (30)

sensible cooling. Removing heat from the air while keeping a constant humidity ratio. (26)

sensible heat. Energy that changes the temperature of a substance with no change in physical state. (26)

sensible heating. Adding heat to air without changing its humidity ratio. (26)

sensor. A switch that is triggered by a means other than an operator's intentional manipulation of a device such as a handle or knob. (22)

septic system. A system designed to clean the wastewater from individual residences. It consists of a septic tank, a leach field, and a series of connecting pipes. (24)

series circuit. A circuit in which multiple loads are arranged so each load is a conductor necessary for the function of the entire circuit. If one load fails, the circuit is broken and none of the loads will function. (21)

service disconnect. A manually operated switch or other means of disconnecting the electric power to the entire system. (21)

service meter. A device that measures the amount of electric current for which the power company charges the customer. (21)

servo motor. A motor that has four to eight poles controlled by an encoder that allows it to rotate incrementally for controlled distances. (23)

sewage. Waste material, especially urine and feces. (24)

sewer gas. A toxic, noxious gas that results from the decomposition of sewage. (24)

shaver. A hand tool with a blade that is used to smooth wood by removing small shavings when drawn along the wood. (11)

sheathing. Material (wood or fibers) attached to framing to strengthen the structure and provide an attachment point for coverings, such as roofing or siding. (16)

shield. A part attached to a machine to protect the user from moving parts and flying particles. (8)

shielded metal arc welding (SMAW). A welding process that uses a consumable electrode coated in a flux and an electric arc to form the welded bead. (29)

shielding gas nozzle. A nozzle that fits around the base of the diffuser and directs the flow of shielding gas around the contact tip and electrode. (30, 31)

shielding gas solenoid valve. An electrically operated valve located within the GMAW welding machine or the wire feeder that switches the gas flow to the welding gun on and off. (30)

shiner brick. A brick that is laid on edge with the largest surface of the brick exposed in the wall face. Also called *rowlock stretcher*. (15)

shingle. A type of roof covering made from asphalt or wood that is installed in overlapping layers. (17)

short-arc. A type of machine for wire welding processes in which the electrode has to touch or short-circuit into the base metal in order to build up enough heat to melt off, so that the arc is extinguished and reestablished many times per second. (30)

short-circuit metal transfer. A method of metal transfer in which a continuously fed wire electrode melts and is deposited through repeated short-circuits of the welding current between the electrode and the base metal. (30)

shorted. Condition in which a conductor makes contact to ground prematurely or in an unintended location. (22)

short-term goal. Goal that can be accomplished in a short time, such as a day, a week, or even later in the same year. (3)

shrink tube. A heat-sensitive plastic sleeve that can be slid over a wire connection, then heated to shrink over the connection, insulating it and protecting it from moisture. (22)

siding. Material, such as brick, vinyl, or wood that is placed on the outside walls of a building. (17)

signal word. A word specified by ANSI to call attention to a specific hazard or safety message, such as **WARNING** or **DANGER**. (5)

sill. A horizontal board placed on the foundation to which vertical members, such as wall studs, are attached. Also called *sill plate*. (16)

sill plate. A horizontal board placed on the foundation to which vertical members, such as wall studs, are attached. Also called *sill*. (16)

single groove joint. A bevel groove that is beveled on one side only. (30)

single-phase power. Power supplied by a single AC signal traveling through a conductor. (20)

single-piece torch. Oxyfuel torch that consists of one piece. (28)

single-pole, double-throw (SPDT) switch. A switch that has two positions and one set of contacts. (20)

single-pole, single-throw (SPST) switch. A switch that has one common terminal and two opposing terminals that can connect to the common. (20)

single-strand insulated conductor. Wiring that consists of a single, solid conductor covered by insulation. (21)

site-specific management. Observing and measuring the differences in specific locations within fields and using this information to manage improvements according to individual needs. (4)

sketch. A hand drawing with little or no detail, numbers, or measurements. (12)

skilled trade. A job that requires a specific set of manual skills, such as electrical wiring, welding, construction, and masonry. (2)

SkillsUSA. A vocational student program associated with students studying skilled trades. (2)

skip tooth configuration. A saw blade type used for wood, plastics, and soft metals because the extended gullet allows more room for chips to clear from the kerf. (8)

slab-on-grade foundation. A single layer of concrete with thick edges that serve as footers and that typically rests on a bed of gravel. (14)

slag. Waste matter produced during the SMAW process. (29)

sledge hammer. A large hammer designed for heavy-duty work. It usually has an octagonal head with two identical, opposing flat faces. (7)

sliding T-bevel. An adjustable tool used to reproduce angles. (6)

slip. The difference between the full-load speed of an induction motor and its no-load speed. (23)

slip-joint pliers. Pliers that have an elongated, figure-eight-shaped hole at the fulcrum that allows the pliers to be shifted to two positions—one for gripping small objects and the other for gripping larger parts. (7)

slitting chisel. A type of cold chisel used to cut thin sheet metal. (27)

slope. The lay of the land that determines the direction runoff will flow. (25)

slotted-tip screwdriver. A screwdriver with a tip that fits screws with a single straight-walled slot that extends through the diameter of the head. Also called *standard screwdriver* or *flat-head screwdriver*. (7)

sludge. A mixture of solids and liquids that forms the bottom layer in a septic tank. (24)

slump. A measure of the consistency of fresh concrete as determined by performing a slump test. (14)

slump cone. A container 12″ high with an 8″ diameter bottom (widest part) and 4″ diameter top used in testing concrete. Also called *Abrams cone*. (14)

slump test. A method of checking the consistency of fresh concrete by seeing how much a sample of the concrete subsides (slumps down) when the surrounding container is removed. (14)

smooth wire. A type of fencing material that is made from mild steel and has no barbs, twists, or woven patterns. (19)

soapstone. An instrument that leaves a mark consisting mostly of talc. (6)

socket wrench. A tool composed of a short, steel tube that looks like a box-end wrench that fits into a ratchet handle to tighten nuts and bolts. (7)

soffit. Building trim placed underneath the roof structure in the space between the fascia board and the edge of the building. (17)

soft-faced hammer. A hammer designed to shift parts into position without causing material damage or producing sparks. The face is usually a replaceable insert made of soft metal. (7)

soft skills. A person's ability to function with other people. These skills include communication, language, personal habits, people management, and friendliness. (2)

softwood. Lumber cut from a gymnosperm (a tree that is generally evergreen and bears its seed without a fruit). (9)

soil structure. The way that individual soil particles stick together. (25)

soil texture. The size of the individual soil particles. (25)

solder. The filler material used in soldering, usually an alloy of tin and lead. (27)

soldering. A process used to join metal pieces by melting another alloy that has a lower melting point and allowing it to flow between the two metals to be joined. (9, 24, 27)

soldier brick. A brick that is laid on end with the narrow side exposed in the wall face. (15)

solenoid. An electromagnetic switching mechanism. (22)

solvent cement. An adhesive that bonds the plastic PVC or CPVC pipes and fittings together. (24)

span. The full extent of something from end to end, such as the complete distance across a building. (16)

spatter. Metal waste droplets around the area of the weld bead. (29)

SPDT (single-pole, double-throw) relay. A relay that consists of a coil and terminals (primary circuit), one common terminal, one N/C terminal, and one N/O terminal (secondary circuit). When the coil of the relay is not energized, the common terminal and the N/C terminal have continuity. When the coil is energized, the common terminal and the N/O terminal have continuity. (22)

specialized agricultural structure. Structures used for specific agricultural functions that are less common than some other areas of the agricultural industry. (18)

specific humidity. The moisture content of air as a measure of the actual amount of water held in the air in the form of vapor. (26)

Speed® Square. Versatile, triangular-shaped measuring tool that could be used as a saw guide for 45° and 90° cuts. (6)

spirit level. A tool that includes a liquid-filled glass or plastic vial that is used to check an object for levelness. (6)

splice. A connection from wire to wire when one section is not long enough. (21)

splicing. A fencing procedure in which the ends of two wires are overlapped approximately 8–12 inches and wrapped together. This procedure is used for repairing broken fences or installing new fence. (19)

split-phase motor. A motor powered by single-phase AC that divides the AC power between a starting and a running circuit. (23)

split-phase power. Power supplied using two hot conductors with phases offset by 180°. (20)

spool gun. A remote wire feeder built into a welding torch so that the electrode is fed from small spools of wire mounted directly to the gun. (30)

spray pattern. The area that one sprinkler head can cover. (25)

spray transfer. A method of metal transfer that maintains a fine arc column and sends a stream of small molten droplets through the arc to the weld. (30)

spreader. A tool or device used to distribute material, such as moving fresh concrete in a form. (14)

spread footing. A type of foundation that uses an inverted *T* shape to spread the weight of the structure over the soil or bedrock below. (14)

spring steel. A relatively mild steel that returns to its original shape after being bent. It is used for parts such as tension springs, compression springs, clock springs, and various types of wire. (9)

springback. The tendency of metals to spring back into position after bending. (27)

sprinkler head. A device attached to an irrigation line that dispenses water. (25)

sprinkler system. A type of watering system that applies water above plants and resembles rainfall. (18)

sprocket. A toothed wheel similar to a gear, with evenly spaced teeth around its perimeter, that does not mesh directly with another sprocket or gear. (35)

square. 1. Term for two objects set perpendicular, or 90°, to one another. 2. A term used to describe an element, such as a wall, that is straight horizontally and has exactly perpendicular sides (at a 90° angle to each other). (6, 16)

stainability. The ability of wood to take stain or paint. (11)

stainless steel. A type of metal with a chromium content high enough to make the steel corrosion resistant. It has low electrical conductivity, may or may not be magnetic, can be lustrous when polished, is expensive, and is difficult to repair. (9)

standards. Rules that describe methods for ensuring safety. (5)

standoff. The distance from the plasma torch tip to the base metal. (32)

standoff guide. A shield-like apparatus that clips onto the end of the torch and is electrically isolated from the nozzle and electrode, providing a consistent standoff distance. (32)

starving. Using a lower working pressure than recommended with a multiflame attachment, which chokes the torch, causing overheating of the attachment and possibly causing flashback. (28)

stationary power tools. Tools constructed with relatively heavy bases intended to stabilize the machine during operation. (8)

stator. The stationary portion of an electric motor. (23)

steel. A metal that is a mixture of mainly iron and carbon. (9)

step-down transformer. A device that increases current in a circuit in order to cause a proportional decrease in voltage. (20)

stepper motor. A motor that moves incrementally using electrical pulses released by an electronic controller. (23)

step-up transformer. A device that decreases current in a circuit in order to cause a proportional increase in voltage. (20)

stickout. The amount of electrode extending beyond the welding gun in GMAW welding. (30)

story pole. A masonry tool with markings used to indicate the top of each course of bricks or blocks. Also called *corner pole*. (15)

straightedge. An even piece of lumber (without curves or bends) or an aluminum or magnesium tool similar in shape to a 2″ × 4″ piece of lumber used to strike off concrete level at the top of a form. (14)

strap hinge. A hinge that is pinned in the center, but completely exposed when working. The strap is attached to the outside of the door and the other plate part is attached to a solid part of the door frame or box. This type of hinge allows a door or lid to lay flat back against a wall or side surface. (9)

stress graded. Wood used in applications where strength and stability are important. Also called *dimensional lumber*. (11)

stretcher block. A type of concrete block that is laid between two corner blocks and usually has ears (flanges) on both ends of the unit. (15)

stretcher brick. A brick that is laid flat with the long side exposed in the wall face. (15)

stringer bead. A weld bead that is straight and narrow. It is made with minimal side-to-side movement. (29)

string level. Small tubes containing a single vial that are suspended from a string stretched tautly between two points. Used for jobs such as building fences or laying bricks that require that points be level even when set long distances apart. Also called *line level*. (6)

strip gauge. Two parallel lines with a space between them molded into the plastic housing of a switch or receptacle. (21)

stroke. The full travel in either direction of a piston in the cylinder. (33)

stud. A vertical board or piece of metal that is the main structural component of a wall in stud wall construction. (16)

stud wall construction. A type of structure in which vertical boards or pieces of metal (studs) are used as the main structural component in walls. (16)

style. A distinctive manner or expression that is often related to the form, appearance, or character of a project. (10)

subassembly. Parts that are put together separately and used as a unit in a finished product. (12)

subfloor. A layer of material, such as plywood, placed over the floor joists. (16)

subsistence farming. Growing only enough food or other products for the use of the farmer and immediate family. (1)

sump pump. A small, portable pump that is used to pump excess water out of holes when repairing irrigation lines. (25)

supervised agricultural experience (SAE). Planned and supervised activity that allows a student to apply knowledge gained in the school-based agricultural education program. (3)

surface irrigation. The flooding of entire rows or fields of crops. The process uses gravity and the slope of the field to disperse the water. (25)

surfacing. A welding process used to build up worn parts with filler metal or to create a harder or more durable surface than that of the original base metal. (30)

survey. A procedure used to establish exact points, locations, boundaries, and property lines. (19)

surveyor. A professional who takes measurements to determine accurate points or places on land. (13)

swale. A type of ditch used in production agriculture that drains excess water (runoff) away from crops. (25)

swarf. Sharp-edged waste material created by the action of drill bits boring into metal. (8)

swell. Increases in soil volume due to moisture and air content. (13)

switch. Any device that is used to interrupt the flow of electrons in a circuit, stopping or redirecting the flow of electricity. (20)

T

table saw. A stationary or semi-portable version of a circular saw that is ideal for ripping lumber into narrower boards and for making cuts in plywood. (8)

tack weld. A very short, temporary weld bead used to hold parts in place while welding. (29)

tamper. A tool or material used to press down and settle substances, such as soil or concrete. (14)

tap. 1. A tool used to cut threads in a machined hole. 2. A type of wire-to-wire connection in which one wire is connected along the length of another to create a branch in another direction. (7, 21, 27)

tape measure. Extendable, flexible ruler that retracts into a durable case for storage. (6)

tapered punch. A punch used for aligning holes in two or more parts so the parts can be bolted or screwed together. It has a long, tapered shaft ending in a flat point. (7)

taping. A process used in land surveying where a metal tape marked with length measurements is used to determine the distance between two points. (13)

tap splice. A type of splice in which a conductor is joined to another conductor by connection to a wire in a circuit without breaking or cutting the original wire. (22)

target rod. A stick or pole marked with measurements and used with a leveling instrument to determine elevation. (13)

technical drawings. A set of plans that includes measurements and construction details needed to build a project. (12)

technology. The application of science to solve a practical problem. (4)

telematics. Technology that integrates wireless telecommunications with information technology to enable long-distance transmission of information in real time. (4)

temperature. The degree or amount of heat in something, such as the air, an object, or a substance. (17)

tempering. The process of reheating a metal following quenching for the purpose of relieving internal stress and decreasing brittleness. (27)

template. A pre-formatted document for a specific form, such as a résumé. (2)

tensile strength. The ability to withstand forces that pull apart or stretch something. (10)

terminal connection. A point at which wire conductors terminate and are connected to electrical devices. (21)

theodolite. A surveyor's optical instrument similar to a dumpy level, but is able to be tilted to sight at a specific vertical angle. (13)

thermal energy. The ability of an energy source to perform work through heat. (36)

thermal equilibrium. A balance of temperature between two bodies. (26)

thermal pane window. A window that contains two or more panes of glass to reduce heat transfer. (17)

thermal radiation. The transfer of heat by electromagnetic waves. (26)

thermodynamics. The branch of science dealing with the relationships between heat and other forms of energy. (26)

thermometer. An instrument used to measure temperature. (26)

thermoplastic. A plastic formed from a long-chain hydrocarbon that is relatively easy to recycle. This material is lightweight, heat and chemical resistant, noncorrosive, and inexpensive. (9)

thermostat. A device used to control a heating or cooling system. (26)

thickness gauge. A gauge made with graduated slots into which the material fits (go) or does not fit (no go) for measurement. (6)

thoriated tungsten electrode. A tungsten electrode that contains approximately 2% thorium, which allows easier starting and a higher current capacity. Marked with a red band. (31)

thread locking compound. A machinery adhesive used to prevent threaded fasteners from vibrating loose. (27)

threads. A ridge of metal that spirals on an incline from the tip to the head of a screw. (9)

thread series. A formal definition of the size and shape of a screw thread. The two most common are the Unified Thread Standard and ISO Metric thread series. (27)

thread tape. A nonsticky, stretchy white tape that is used to provide a tight fit when joining threaded galvanized pipe. (25)

three-phase motor. A motor in which the phases are synchronized so that each phase peaks in power every 60°, so that the magnetic field produced is rotating and no separate starting windings are needed. (23)

three-phase power. Power supplied using three hot conductors with phases offset by 120°. (20)

tie-in. Blending of the weld pool with the sidewalls of the joint. (29)

tile. A type of hollow clay masonry unit that is fired to harden it. (15)

tinning. The process of applying a thin coating of solder to a surface prior to the soldering process. (27)

T-joint. A joint made between two metal parts located approximately at right angles to each other in the form of a *T*. (29)

toggle switch. A switch that has only two positions: *on* and *off*. (22)

tolerance. The maximum allowable variation between measurement specifications and the actual dimensions of the final part or product. (6)

tool path. The "map" for a CNC-guided torch to follow when cutting parts. (32)

tool steel. A tough steel with a relatively high carbon content that is often machined into tools that can work other steels. (9)

torch body. The handle of an oxyfuel torch, which can be fitted with a variety of attachments for different jobs. (28)

torch wrench. A short wrench with a sharp grip that is made to fit all of the fittings associated with an oxyfuel torch outfit. (28)

torpedo level. A level that is typically nine to ten inches long and is often tapered in thickness on each end, handy for taking readings in confined spaces. (6)

torque. 1. A twisting force that, when applied to an object, causes the object to rotate. 2. A twisting or rotating force. (6, 34)

torque wrench. A tool used to apply a specific amount of torque to a nut or bolt. (6)

Torx®. Type of screws that have a six-pointed, star-shaped recess with rounded corners. (7)

total energy. The sum of many forms of energy used together to perform work. (36)

toughness. A property of metal that allows it to withstand tearing or breaking when bent or deformed. (27)

toxicity. The degree to which a poisonous substance or chemical can harm living beings. (5)

t-post. A steel fence post that is constructed with a cross-section in the shape of the letter *t*. (19)

transformer welder. The oldest type of welding machine, which produces welding current using a step-down transformer. (29)

transmission. The assembly or mechanism that transmits power from an engine to the wheels and other parts of a vehicle. It allows the final drive wheels to be propelled in multiple forward and reverse speeds. (35)

transplanting. Describes the process of moving a plant from one growing area to another. (18)

trap. A device that seals water into a vented pipe to stop sewer gas from entering a plumbing fixture. (24)

travel angle. The angle between the electrode and the direction of weld progression. (30)

travel speed. The rate of forward motion during welding. (29)

trim. Nonstructural building components installed in a structure to enhance appearance (such as baseboard or crown moulding) or help seal areas where water or air can penetrate a building. (17)

trowel. A hand tool used to work with concrete or mortar for masonry construction. (14)

truncated cone tip. An electrode tip that consists of a tapered cone with a flat tip, used to weld steel and stainless steels using DC welding processes. (31)

truss. A frame made of beams, bars, or rods and used in roof or floor construction. (16)

try square. An L-shaped tool, consisting of a steel blade and stock, which is used for checking perpendicular parts. (6)

tuck pointer. A type of trowel with a narrow blade used to fit into tight spaces, such as between brick or block joints. (15)

tungsten extension. The distance a tungsten electrode extends beyond the tip of the nozzle. (31)

turf rotor. A removable sprinkler head used on large grassy areas such as athletic fields and golf courses. (25)

turning point. A temporary elevation reference point used in differential and profile leveling. (13)

twisted shank nail. A nail with a slight twist to the shank. The nail turns as it is being driven into wood and forms a very tight bond. (9)

U

ultrasonic testing. A type of weld testing in which a stream of ultrasonic energy is directed into the weld to reveal discontinuities. (29)

unacceptable weld profile. A poorly shaped weld bead, including improper contour, undercut, and overlap. (29)

undercutting. Poor tie-in, so that the outside edge of the welded bead is concave or recessed. (29)

underground feeder (UF). A type of Romex® that is similar to NMC cable except that UF cables have a solid plastic core. It is used for buried cable and similar applications. (21)

underlayment. Material, such as tar paper, that is installed before final roofing materials to help prevent water from leaking into the building. (17)

unit of measurement. A specific quantity accepted as a standard of measurement. (6)

universal motor. A series-wound motor that can operate on single-phase AC or DC power. (23)

unmanned aerial vehicle (UAV). An aircraft that is controlled remotely and does not have a human pilot on board. (4)

utility lumber. Stress graded wood used for rough framing that may have distinct warping or twisting. (11)

V

valve. A fixture or fitting that allows water to leave the system, restricts the flow of fluids in the system, or diverts fluids to another part of the system. (24)

vane pump. A positive-displacement pump that uses vanes, or fins, to create fluid flow through volume displacement. (37)

vapor barrier. A material, such as thin plastic sheeting, used to retard the movement of moisture from one area to another. (14)

variable costs. The costs associated with operating machinery. This can include labor, fuel and lubricants, and repairs or maintenance. (36)

variable frequency drive. A type of motor controller that electronically manipulates AC frequency, resulting in a change in the speed of rotation in three-phase motors. (23)

variable-rate technology (VRT). A technology that allows farmers to vary the rate of crop inputs based on precision agriculture methods and site-specific needs. (4)

veneer. Very thin sheets of lumber, cut parallel to the length of the log, that are often used to form engineered lumber, such as plywood. (9)

ventilation. 1. The movement of fresh air into a building or controlled environment. 2. The process of exchanging air in a growing environment using vents and exhaust fans. (17, 18)

venturi. The narrow point in a carburetor throat. (33)

Vernier caliper. A caliper with a graduated main scale and graduated sliding bar, known as a Vernier scale, that indicates the precise width between the open jaws. (6)

viscosity. An oil's resistance to flow. (36)

vise. A tool that typically mounts to a bench and secures the workpiece in a tight grip while other tools are used. (7)

voltage. The potential energy between two points along an electrical conductor. (20)

voltage detector. A device that determines if voltage is present in a circuit. (20)

voltage drop. Decrease in voltage due to resistance. (20)

volume. The amount of space occupied by a substance or object. (26)

W

walking the cup. A technique in which the welder uses the edge of the shielding gas nozzle to guide the path of the torch. (31)

wall. A vertical component of a building that is used to separate inside and outside environments, divide spaces into smaller rooms, and support the weight of other building components. (16)

wall anchor. A fastener that provides support for a screw, allowing that screw to support more weight. This fastener is used in instances when the screw cannot be driven directly into a structural member. (9)

washer. A flat steel disk that has a hole in the center, used to keep a bolt from pulling through material. (9)

waste side. The edge of a cutting line that is opposite the project piece and nearest to the part of the board that will be discarded. (12)

water bed. A large container constructed to hold a few inches of water to allow growing trays to float and provide a growing area for plants. (18)

water cycle. The cyclic movement of water and precipitation in its various forms between the earth's oceans, rivers, lakes, and streams. (25)

water level. A basic measuring tool that functions based on Pascal's law and atmospheric pressure. (6)

water pressure. The force that water applies to its surroundings. (25)

water quality. The characteristics of the available water supply. Good-quality water for plant growth has no contaminants or particulates that would negatively affect plant health. (18)

water supply system. The part of a plumbing system that includes the water source and all of the pipes and fixtures that bring the water from the source to the final point of use. (24)

water system. The method and equipment used to deliver water to plants. (18)

watt. The standard unit of electric power—the rate at which work is done. (20)

weave bead. A relatively wide weld bead that is formed using a weave pattern created by moving from side to side in a continuous series of partially overlapping semicircles. (29)

weave pattern. A weld bead pattern created by repeated electrode motion. (29)

wedge. Two inclined planes placed back to back to form two inclined surfaces that meet at a thin edge at one end. (1)

welded wire fence. A type of fence made from heavier gauge wire for use with larger animals such as hogs. (19)

welding current. The flow of electricity that results in the welding arc. (29)

welding defects. Weld discontinuities that are of an unacceptable size or in an unacceptable location. (29)

welding gun. Another name for the welding torch used with GMAW. (30)

welding leads. The cables that connect the welding power source to the work. (29)

welding procedure specification (WPS). A document that outlines the parameters or conditions of a welding certification test. (29)

weld specifications. Specifications that include the information a welder needs to perform a specific job, including how to position the weld. (29)

well. A deep, manmade hole in the ground from which water can be pumped. (24)

western framing. A building method in which each level or story of a building is built as a separate unit. Also called *platform framing*. (16)

wet-bulb temperature. The temperature of air measured with a standard thermometer that takes into account the moisture content of the air. (26)

wheel and axle. A simple machine that consists of a wheel attached to an axle so that both turn together to form a continuous lever. (1)

window. An opening in a wall or roof that may be fitted with glass or another covering, allows light and air to pass through, and provides a view of the outside. (17)

window frame. The part of a window (sill and jams) that provides the structure for the sash to work properly and allows the window to be installed into a building. (17)

wing nut. Winged fasteners used on equipment that must be serviced or adjusted frequently. (27)

wire fabric. A fence material that can be made of a variety of wire sizes and designs. Wire fabric is manufactured in rolls or panels. (19)

wire feeder. A device that contains the supply of wire (electrode) and delivers it through a lead to the welding gun. (30)

wire nut. A plastic or ceramic cone with a tapered, spiral metal insert that grips the ends of wires to make a connection. (21)

wire speed. The rate at which the electrode in GMAW is fed to the welding arc, calculated in inches per minute (ipm). (30)

wire stripper. A tool that resembles a pair of pliers with a series of openings fitted to different gauges of wire. (21)

wiring harness. A bundle of wires bound together within the same insulating cover. (22)

wood chisel. A tool with a blade sharpened to a 25° to 30° angle on only one side to create a sharp edge. (7)

wood joint. A connection between two pieces of wood that is created using the wood itself rather than fasteners, such as nails or screws. (12)

work. 1. The result of a force that pushes or pulls an object through a specific distance. 2. An action in which a force applied to an object causes a displacement of the object. (1, 34)

work angle. The angle between the welding gun and electrode and the center of the weld joint. (30)

working pressure. The gas pressure at which oxyfuel welding or cutting is performed. (28)

woven wire. A type of fence material that has the largest openings of any of the wire fabric materials. (19)

wrap. A synthetic material installed over sheathing to help prevent air movement out of or into the building. (17)

wythe. A continuous vertical section of masonry that is one unit in thickness and constructed of a single line of masonry units. (15)

X

xeriscaping. The process of using water-efficient plants in the landscape. (25)

x-post fencing. A type of fencing that is installed above ground using posts arranged in an *X* shape. Horizontal rails are then placed between the posts, similar to post-and-rail fencing. (19)

Y

yield monitoring. The use of sensors to measure and record data about moisture and other conditions during harvest. (4)

Z

zoning code. Local rules for private property use that generally cover safety, land and building uses, and aesthetics. Also called *zoning law*. (10)

zoning law. Local rules for private property use that generally cover safety, land and building uses, and aesthetics. Also called *zoning code*. (10)

Index

.dxf file format, 888
4-H, 24

A

Abrams cone, 352
abrasive, 109, 215–217, 720
    cutting, 720–721
    friction, 216
    heat buildup, 216
    lodging, 216–217
abrasive cut-off saw, 721
abrasive cutting, 720–721
abrasive material, 720
ABS. See acrylonitrile butadiene styrene
AC. See alternating current
accelerant, 745
access, 509–510
AC circuit planning, 549–555
    armored cable, 551
    conduit, 551–552
    connector, 552–554
    enclosure, 552–554
    hardwiring, 554–555
    Romex® conductor, 550–551
    wire color, 549–550
    wire size, 549
accuracy, 121
acetylene, 744–745
acetylene cylinder, 748
acid-based flux, 734
acrylonitrile butadiene styrene, 236, 636
AC to DC conversion, 571
adapting plan, 273
adhesion, 316
adhesive, 244–246, 315, 732–733
    adhesion, 316
    machinery, 732
    structural, 732
    surface preparation, 732–733
adjustable wrench, 165–166
admixture, 353
aesthetic, 257
aesthetic fence, 497
aggregate, 350, 352
Agricultural Experience Tracker (AET), 59

agricultural fence, 495–496
    electric, 496
    post-and-rail, 496
    x-post, 496
agricultural machinery engineer, 586
agricultural mechanics, 4–5
agricultural mechanics careers, 22–23
agricultural mechanics hazards, 88–89
    NIOSH, 88–89
    ROPS, 89
agricultural mechanics job search, 34–36
agricultural mechanics mathematics, 10–11
agricultural trailer component, 587–590
agricultural trailer wiring, 586–593
agriculture, 4
    precision, 73–79
agriculture revolution, 5–9
    early mechanical invention, 6–9
agriscience, 4
agronomic irrigation, 659–661
agro-terrorism, 71
air chisel. See pneumatic hammer
air compressor, 990
air hammer. See pneumatic hammer
air intake system, 957–958
air pollution, 72–73
air quality, 453, 479, 686–688
    axial fans, 687, 688
    centrifugal fans, 687, 688
    fan design, 687–688
    fan operation, 687–688
air temperature, 479
Allen wrench. See hex key
alloy, 231
all-terrain vehicle. See ATV
alternating current, 530–532, 791
    ground, 530–531
    hertz, 530
    hot, 530–531
    neutral, 530–531
    phase, 531–532
    polarity, 530
alternative fuel, 66–67
    biofuel, 67

hydrogen fuel cell, 67
alternator, 533
aluminum, 234, 709
American National Standards Institute (ANSI), 92, 219
    safety standards, 92
American Society for Testing and Materials (ASTM), 388
American Society of Agricultural and Biological Engineers (ASABE), 93
American Society of Mechanical Engineers (ASME), 257
American Welding Society (AWS), 783, 820, 850
American Wire Gauge (AWG), 549
ampere (amp), 522, 791
anaerobic digester, 67–68
anchor bolt, 399
anchor bolt installation, 399
angle grinder, 210–211
annealing, 637, 738
ANSI. See American National Standards Institute
answering the right questions, 34
anti-friction bearing, 960
anvil, 161–162
    farrier, 161
appearance lumber, 282
apprenticeship, 22–23, 25
    US Department of Labor, 25
aquaculture, 482–485
aquaculture system, 482–485
    location, 484–485
    types, 483–484
aquaponic, 485
aquatic plant, 483
aquifer, 651
arc blow, 793
arc construction, 410
architect, 261
arc length, 793
arc welding safety, 800–802
areas with job opportunities, 36
argent pencil, 133
armored cable, 551
array, 572
ASABE. See American Society of Agricultural and Biological Engineers

ASME. *See* American Society of Mechanical Engineers
asphalt roofing, 442
assembling metal, 730–736
assembly drawing, 268
asset, 55
ASTM. *See* American Society for Testing and Materials
ATV, 111
audit, 59
auto-darkening lens, 802
autogenous weld, 866
auto level, 339
automatic punch, 134
automatic transmission, 942
automatic watering system, 472
automatic welding, 821
automation, 71
autonomous, 76–77
aviation snips, 157
award, 47
AWG. *See* American Wire Gauge
awning window, 450
AWS. *See* American Welding Society
AWS D1.1 3G GMAW Welding Certification, 849–851
AWS Specification A5.1, 783
axial fans, 687, 688
axle, 13

B

backflow preventer, 654
backhand welding, 842
backing plate, 840
backsaw, 146, 286
    miter box, 146
backsight, 341
ballasting, 970
balloon framing, 413–414
    sheathing, 414
ball-peen hammer, 159
band saw, 200–203
band saw blade, 202–203
    hook tooth configuration, 202
    rake angle, 202
    skip tooth configuration, 202
barbed wire, 502–503
bar clamp, 155
basin irrigation, 661
batter board, 334
battered foundation, 363–364
battery, 532, 964–965
battery charger, 182
battery-operated power tool, 182
BDC, 920–922. *See also* bottom dead center

beam, 415–416
bearing, 603, 959–961
bed of mortar, 393
belt, 616, 947–949
    drive ratio, 616
    friction connection, 947
    pulley, 947
    sprocket, 949
belt-drive system, 615–616
bench grinder, 211–212
benchmark, 340–341
bending metal, 722–723
    cold working, 723
    jig, 723
    mandrel, 723
    pipe and tubing, 723–724
Bernoulli's principle, 980
beveling, 790
bid package preparation, 273
bill of materials, 259–260, 710–711
    cutting list, 260
bimetal blade, 204
biofuel, 67
biosecurity, 71
biotechnology, 79–81
biscuit jointer, 315
bit, 186
black pipe, 640
bladder, 628
blade connector, 581
blade set, 198
bleed water, 359–360
blind cut, 289
blind rivet, 732
block corner lead, 395–399
block wall, 395–399
    anchor bolt installation, 399
    ears, 395
    finishing joint, 399
    laying courses of lead, 395–398
    stretcher, 395
    stretcher block, 398–399
blueprint, 408
BM. *See* benchmark
board foot, 261
body protection, 98
bolt, 241–242
bolt cutter, 719
bond, 378, 381–385
    brick types, 382
    concrete masonry unit, 384–385
    grout, 383
    mortar joint, 383–384
    pattern bond, 381–382
bonded armored cable, 551
border irrigation, 661
bore, 920
bottle jack, 194

bottom dead center, 903. *See also* BDC
bow, 284
box-end wrench, 164–165
box nail, 238–239
Boyle's law, 990
brace and bit, 147–148
brad nail, 239
brad nailer, 193–194
braided wire, 504
brass, 233
braze welding, 768–770
brazing, 768–770
brazing tip, 753–754
breaker box, 546–547
breathing protection, 97–98
    respirator, 97–98
    ventilation, 97
brick, 374
brick chisel. *See* brick set
brick corner lead, 393–395
brick hammer, 375–376
brick manufacture, 380–381
brick set, 378
brick siding, 438
bridging, 418–419
bronze, 233
broom, 171
broom finish, 361
brush, 171, 603
BS. *See* backsight
budget, 57, 60
budgeting, 258–261
    bill of materials, 259–260
    material cost, 260–261
    total project cost, 259
building code, 390
building construction, 454–458
building design, 406–408, 689–690
    computer aided design, 408
    load, 407–408
    material, 408
building load, 407–408
building material, 408
building site selection, 326–331
    land preparation, 330–331
    soil analysis, 326–328
    soil quality, 329
    soil testing, 329
build-up roofing, 442
bullet connector, 581
bull float, 359
burning gas, 745–746
burning steel, 763
burr, 636
bushing, 603
business plan, 57
butane, 745

butt hinge, 246
butt joint, 785–786

C

cabinetmaker, 294
cable ripper, 561
cable sheathing removal, 564
CAD. See computer-aided design
CAD software, 408
calculating material cost, 260–261
caliper, 123–124
call cover, 810
calling cover, 804, 807
CAM. See computer-aided machining
camshaft, 903
cap, 988
capacitor-start motor, 606
capillary action, 769
cap screw, 731
carbide bit, 188–189
carbon arc welding, 778
carbon monoxide gas, 914
carburetor, 910–911
carburetor jet, 910
carburizing flame, 761, 762
cardiopulmonary resuscitation. See CPR
career development event, 24, 47, 120, 683
carpenter's level, 125–126
carriage bolt, 242
casement window, 449
cast iron, 231, 706
cat's paw, 164
cattle guard, 509
CAW. See carbon arc welding
C-clamp, 154
CDE. See career development event
CEA. See controlled environment agriculture
ceiling framing, 426–427
ceiling joist, 427
Celsius, 683
center-pivot irrigation, 659
center-pivot sprinkler system, 660
center punch, 133–134, 162
centrifugal fans, 687, 688
centrifugal force, 982–983
ceriated tungsten electrode, 864
certificate, 24–25
certification, 24–25
CFC. See chlorofluorocarbon
CGA. See Compressed Gas Association
chain, 947–949

chain-drive system, 615
chaining pin, 337
chain-link fence, 503–504
chamfered, 162
charge memory effect, 182
charger, 505
check, 284
chemical safety, 106–107
chemical type, 106–107
chemistry of burning gas, 745–746
chicken wire. See poultry wire
chisel, 150, 297–298, 377–378
    brick set, 378
    cold, 150
    wood, 150
chlorinated polyvinyl chloride (CPVC), 635–636
chlorofluorocarbon, 72
choking. See starving
chop saw, 721
chuck, 190
chuck key, 190
circuit, 521–527
circuit breaker, 528–529
    electric, 521–527
    ground fault, 529
circuit protection, 527–530, 611–612
    circuit breaker, 528–529
    fuse, 528
    ground fault circuit interrupter, 529–530
circuit type, 548
circular saw, 196–199, 287–288
circular saw blade, 197–199
clamp, 154–156
    use, 714–715
clamping pliers, 157–158
classroom instruction, 45
claw hammer, 158–159
cleaning existing thread, 729
cleaning tool, 170–171
clearance, 958
clinker, 351
closure block, 398
cloud server, 74
clutch, 940–941
CMU. See concrete masonry unit
CNC. See computer numeric control
CNC plasma cutting, 887–892
CO. See carbon monoxide
cobalt bit, 188
cold chisel, 150, 717
    cutting metal, 717
cold saw, 717
cold working, 723
collet body, 860
color code, 103–104

Colorado River Storage Project, 648–649
colored marker, 132–133
combination square, 129–130
combination wrench, 164–165
combine, 7, 22
combustible liquid, 107
common agriculture metal, 706–709
common nail, 238
communication tool, 74
community development, 406
commutator, 603
competition, 24
composite material, 235–237
compressed gas, 107
Compressed Gas Association (CGA), 752
compressed natural gas, 745
compression fitting, 637–638
compression ratio, 922
compression ring, 637–638
compression strength, 262, 357–359
compressor operation, 991–992
compressor type, 991–992
computer-aided design, 256, 260, 270–271, 408, 720
    program, 270–271
computer-aided machining, 889
computer application, 74–75
computer numeric control (CNC), 726, 887
concave joint, 383
concrete, 235
    application, 350
    characteristics, 350
    compression strength, 357–359
    finishing, 359–362
    form, 364–366
    forming, 350–353
    tensile strength, 357–359
    testing, 359–360
concrete application, 350
concrete block, 384–385
concrete characteristic, 350
concrete equipment, 357
concrete finishing, 359–362
concrete form, 364–366
concrete forming, 350–353
concrete masonry unit, 384–385
concrete preparation, 353–357
concrete testing, 359–360
concrete tools, 357
    bull float, 359
    screed, 359
    spreader, 358
    straightedge, 359
    tamper, 358
conduction, 454

conductor, 520–521, 589
   ground, 531
   neutral, 531
   valence electron, 520
conduit, 545, 551–552
   EMT, 552
   rigid, 552
conduit bender, 723
connecting enclosure, 554
connecting rod, 899, 907–908
connecting wire, 559–563
   device installation, 562–563
   insulation stripping, 561–562
   terminal, 560
   wire-to-wire, 559–560
connector, 552–554
constant current (CC), 792
constant-current machine, 779
construction, 354
construction order, 263
construction procedure, 308
   plan, 310
construction procedure plan, subassembly, 310
construction standard, 273
contact tip, 826
continuing education, 25
continuity, 536–537
continuous hinge. *See* piano hinge
control joint, 360
controlled environment agriculture (CEA), 470
controlling torque, 169
convection, 454
cooling system, 473, 913, 961–963
   hydrometer, 963
coping saw, 145–146, 286–287
copper, 233, 636–639, 709
   soldering, 233, 637–638
copper pipe, 636–639
cordless power tool, 182
corner joint, 786
corner layout, 332–333
corner lead, 393
corner pole, 392
corner post, 502
corrosive, 107
corrugated fastener, 243–244
cotton gin, 6–7
coupling, 614
coupon, 804
course, 375
covering, 477
cover letter, 28, 30
cover rod, 784
CPR, 113
CPVC. *See* chlorinated polyvinyl chloride

crack, 798
crankshaft, 899, 908, 920
crawlspace, 459
Crescent® wrench. *See* adjustable wrench
cribbing, 195
crook, 284
crop storage, 693–694
crosscut saw, 145, 286
crosscutting, 197
cross-linked polyethylene (PEX), 634–635
cross-peen hammers, 160
crowbar, 163–164
crown, weld bead, 795
cup, 284
curing method, 367
cut list, 712
cut nail, 238
cuts and fills, 330
   calculation, 331
cutting, 313
cutting attachment, 754
cutting lever, 754
cutting list, 260
cutting metal, 713–721
   abrasive, 720–721
   cold chisel, 717
   hardness estimate, 713
   pneumatic hammer, 717–718
   securing your work, 713–715
cutting pliers, 157
cutting process, 764–766
cutting table, 763
cutting tip, 754–755
cutting tool, 144–152, 196–209
   band saw, 200–203
   chisel, 150
   circular saw, 196–199
   file, 151–152
   hand-powered drill, 147–150
   handsaw, 144–147
   jointer, 208–209
   miter saw, 200
   oscillating multi-tool, 206
   plane, 150–151, 208–209
   radial arm saw, 199–200
   rasp, 151–152
   reciprocating saw, 203–204
   router, 206–208
   safety, 196
   saw blade design, 198
   scroll saw, 205–206
   table saw, 199
cutting tool safety, 196
cutting torch use, 763–766
cutting wood, 285–294
   handsaw, 285–287

   power saw, 287–289
   stationary saw, 289–294
cylinder, 746, 899, 920, 988–989
cylinder cap, 747
cylinder head, 907

D

darby level. *See* bull float
dB. *See* decibel
DC circuit, 570, 572
DC circuit troubleshooting, 593–594
DCEP current, 864
DC motor, 604–605
DC power source, 570–573
dead-blow hammer, 160
dead load, 407–408
decibel (dB), 96
deck screw, 240
Deere, John, 6
DEF. *See* diesel exhaust fluid
delivery schedule, 390
Department of Labor (DOL), 90
depth gauge, 123
design consideration, 261–263
designing metal project, 709–712
destructive weld testing, 798
detail drawing, 268
detail sander, 215
detention pond, 653
device installation, 562–563
dew point, 681
diagonal pliers, 157
die, 149–150
die grinder, 212
dielectric gel, 560
diesel exhaust fluid (DEF), 72
diesel-powered tool, 184
differential, 945
differential leveling, 340–341
diffuser, 826, 882
dimension, 266–267, 282–284
   dressed, 284
   milling, 282
   nominal, 282
   planing, 282
dimensional lumber, 226–228
diode, 571
direct current (DC), 532, 791
direct current electrode negative (DCEN), 792, 859
direct current electrode positive (DCEP), 792
direct drive, 614
direct-drive mounting system, 614
direct fuel injection, 910–911
discontinuity, 796–798

disease control, 481
disk sander, 213
displacement, 920
distance measurement, 335–337
distribution panel, 546
document, 28–31
DOL. *See* Department of Labor
door, 451–452
doorjamb, 451
door sill, 451
double groove joint, 839
double hung window, 449
double-insulated, 185
double-pole single-throw switch, 558–559
dovetail saw, 286
downdraft table, 698
downhill welding, 788
drag angle, 793
drag shield, 882
drain-waste-vent system, 628
drawing, 308–309
    computer-aided drafting and design software, 710
    metalworking, 710
drawing principle, 264–265
drawing to scale, 265–266
dressed dimension, 284
dressing metal tools, 726–727
drill, 147–150
drill bit, 186–189
drill bit material, 188–189
drill bit size, 189
drilling, 729
drilling metal, 727–729
drilling tool, 186–194
    bit, 186
    brad nailer, 193–194
    drill bit, 186–189
    drill press, 191–192
    hammer drill, 190
    impact, 190–191
    nail gun, 193–194
    portable, 189–190
    safety, 192–193
drilling wood, 301
drill press, 191–192
drill safety, 192–193
drip irrigation, 659–660
drip watering system, 472
driveline, 945
driven sprocket, 615
drive ratio, 616
drive shaft, 603
drive sprocket, 615
drive train, 940–950
driving tool, 186–194
    bit, 186

    brad nailer, 193–194
    drill bit, 186–189
    drill press, 191–192
    hammer drill, 190
    impact, 190–191
    nail gun, 193–194
    portable, 189–190
    safety, 192–193
drones, 77–79. *See* unmanned aerial vehicles
dross, 763, 886
dry-bulb temperature, 680
dry pond. *See* detention pond
drywall screw, 241
dual shield welding, 835
ductile, 721
ductility, 231, 721
dumpy level, 128, 338–339
duplex receptacle, 555–556
Dust Bowl, 648
duty cycle, 574, 781
DWV system, 628–629
DWV system, plumbing trap, 629

E

E3® tungsten electrode, 864
early mechanical invention, 6–9
ears, 395
earthen structure, 486
earthquake fastener, 243
ECU. *See* engine control unit
edge joint, 786–787
edger, 360
EDM. *See* electronic distance measuring
education, 26–27
education continuation, 26
effort, 12
EGR. *See* exhaust gas recirculation
Egyptian irrigation, 648
electrical device, 555–559
electrical device enclosure box, 553–554
electrical energy, 968–969
electrical load, 573–576
electrical system, 522, 964–966
    battery, 964–965
    ignition, 965–966
    starting, 965–966
electrical tape, 583
electric arc welding development, 778–779
electric brake, 575
electric circuit, 521–527
    amp, 522
    load, 521, 523–526

    switch, 526–527
    watt, 522
electric current, 520
electric fence, 496
electric fence charger, 505
electric fence wire, 504
electric fencing material, 504–505
electrician, 538, 548
electricity, 520
electricity regulation, 544
electricity safety, 108, 535–538
electricity safety tool, 536–537
electric motor, 574–575, 605
    advantage, 600–601
    duty cycle, 574
    installation, 611–617
    maintenance, 617–620
    safety, 620
    selection, 610–611
    solenoid, 574–575
    type, 604–608
electric motor component, 603
electric motor installation, 611–617
    circuit protection, 611–612
    motor control, 612–613
    mounting, 613–616
    overload protection, 612
electric motor magnetism, 601–602
electric motor mounting, 613–616
electric power distribution, 534–535
electric power source, 533–534
electrocution, 108
electrode, 778, 782–785, 829
    category, 784–785
    classification, 783
    slag, 783
electrode angle, 793
electrode category, 784–785
electrode classification, 783
electrode holder, 782
electrolytic capacitor, 606
electromagnet, 575–576, 602
    electric brake, 575
electromagnetic solenoid, 575–576
electronic distance measuring (EDM), 336
electronic measurement, 337
employment, 27–34
EMT conduit, 552
enclosure, 552–554
encoder, 608
end area method, 331
end splice, 582
energy, 968–969
energy efficiency, 452–462
energy-efficient lighting, 70–71
energy management, 69–71
    NCAT, 70

engine, 898
   selection, 924
   theory of operation, 899–900
engine block, 906
engine classification, 900–906
engine component, 906–910
engine control unit (ECU), 72
engine displacement, 920–921
engineer, 257
engineered lumber, 228–230
engineering consideration, 262
engine maintenance, 924–928
engine performance, 920–923
engine-related research opportunity, SAE, 900
engine safety, 913–914
   carbon monoxide gas, 914
engine service information, 929
engine system, 910–913
engine troubleshooting, 930–934
enterprise, 52
enthalpy, 681
entrepreneurship, 53, 511
   SAE, 53
environment, 678
environmental control, 436, 452–453, 678–698
   Celsius, 683
   Fahrenheit, 683
   humidistat, 684
   Kelvin, 683
   manometer, 684
   thermocouple probe, 683
   thermometer, 683
   thermostat, 684
environmental instrument, 683–684
environmental load, 408
environmental system, 69–73
equipment maintenance, 956
equipment performance, 968–969
equipment power, 968–969
ergonomic, 714
erosion, 653
estimating hardness, method, 713
ethylene gas, 694
eutectic alloy, 734
evaporation, 650
evaporative cooling, 696–697
evapotranspiration, 650
excess concrete, 366–367
exhaust gas recirculation (EGR), 72
exhaust system, 697–698, 957–958
   fumes, 697–698
exothermic, 350–351
expansion joint, 360
experiential learning, 44
exploratory SAE, 49–50
explosive, 107

extension publication, 272
exterior wall, 446
extractor, 169–170
eye protection, 93–95

F

fabric, 425
face protection, 93–95
face shield, 95
Fahrenheit, 683
Fair Labor Standards Act (FLSA), 90–91
fan design, 687–688
fan operation, 687–688
farm irrigation, 648–671
farrier, 161
fascia, 444
fastener, 237–244, 314–315
   adhesive, 315
   biscuit jointer, 315
   bolt, 241–242
   corrugated, 243–244
   earthquake, 243
   hurricane, 243
   nail, 238–239, 314
   plate, 243
   rivet, 244
   screw, 240–241, 314–315
   wall anchor, 243
fast-fill electrode, 784
fast-freeze electrode, 784
FCAW, 820
FCAW electrode, 836–837
feather, 760
Federal Insecticide, Fungicide, and Rodenticide Act (FIFRA), 92
feeler gauge, 125
fence construction planning, 497–500
fence controller, 576
fence line, 498–499
fence tool, 511–512
   post driver, 512
   post hole digger, 511
fencing
   application, 494–495
   equipment, 511–512
   inspection, 512–513
   installation, 507–510
   maintenance, 512–513
   material, 500–507
   tool, 511–512
   type, 495–497
ferrous, 778–779
ferrous metal, 230–232, 706–707
   alloy, 231

   cast iron, 231, 706
   ductility, 231
   higher-carbon steel, 707
   luster, 231
   malleability, 231
   mild steel, 232
   spring steel, 232
   steel, 231–232, 706–707
   tool steel, 232
ferrule, 752
FFA. *See* National FFA Organization®
FFA emblem, 46
FFA level, 46
FFA motto, 46–47
FIFRA. *See* Federal Insecticide, Fungicide, and Rodenticide Act
file, 151–152, 295–296
fill dirt, 330
filler material, 733
filler metal, 862
fillet weld, 786
fillet weld practice, 848–849
fill-freeze electrode, 784
filtration, 473
final drive, 945–946
financial record, 60
financial risk, 50
finish, 308, 316–320
   hardware, 318–319
   paint, 318
   sealer, 318–319
   stain, 318
   type, 317
finishing, 360–362, 406
   broom finish, 361
   floor scraper, 362
   floor squeegee, 362
   knee board, 361–362
   power trowel, 360–361
   safety roller, 361
   trowel, 360–361
finishing concrete, 359–362
finishing joint, 399
finish nail, 239
finish type, 317
fire extinguisher, 104–105
firing, 381
first aid, 108, 113
fitting, 632, 988
fixed cost, 971
fixed window, 449
fixture, 627
flame type, 761–763
flammable liquid, 107
flared fitting, 638
flashback, 756
flashback arrestor, 756

flash point, 107
flat position, 787
flatwork curing, 366–367
flatwork form, 365–366
flatwork placement, 366–367
flexible pneumatic conductor safety, 993
float, 353
floor, 414–419
floor framing, 414–419
floor jack, 194
floor joist, 416–418
floor scraper, 362
floor squeegee, 362
flowmeter, 823
flow-through aquaculture system, 483
flow-through system, 483
FLSA. See Fair Labor Standards Act
fluid conductor, 981
fluid coupler, 981–983
fluid power, 978–981
   advantage, 978
   Bernoulli's principle, 980
   component, 981–989
   disadvantage, 978
   Pascal's law, 979
   pump, 978
   system design, 981
fluid reservoir, 981, 989
fluid system design, 981
flush hinge, 246
flute, 207
flux, 734, 778
flux-cored arc welding (FCAW), 820, 835–837
flux-cored electrode, 836
flux-cored solder, 734
flux paste, 637
flywheel, 903
folding rule, 376
follow-up letter, 34
food distribution, 71
food safety, 71
footing, 507
foot protection, 100
force, 12, 922, 979
forced induction, 72–73
forehand welding, 842
foresight, 341
forge, 722
forging, 722
form, 355
formed-in-place gasket, 732
foundation staking, 334
foundation system, 362–364
four-stroke cycle design, 901–903

framing, 406, 476–477
   greenhouse, 476–477
   type, 411–414
framing square, 128–129
framing type, 411–414
framing wall, 424–425
friction, 216
friction bearing, 960
friction connection, 947
FS. See foresight
fuel, alternative, 66–67
fuel system, 910–911, 963–964
   carburetor, 910–911
   direct fuel injection, 910–911
   indirect fuel injection, 910–911
fulcrum, 13
fume exhaust system, 697–698
function, 261
furrow irrigation, 661
fuse, 528, 579–580
   in-line, 580

G

galvanized, 731
galvanized pipe, 639–640, 655–656
   repairing, 667
galvanized steel, 655
gap, 505–507
gas metal arc welding (GMAW), 820, 837–843
   advantages and disadvantages, 821
   cleaning surface, 838–839
   consumable selection, 840
   electrode, 828–831
   electrode manipulation, 842–843
   joint preparation, 839–840
   machine setting, 837–838
   parameters managed during welding, 841–843
   stickout, 841
   surface preparation, 838–839
   travel speed, 843
   welding position, 840–841
gasoline-powered tool, 184
gas-shielded FCAW (FCAW-G), 835
gas tungsten arc welding (GTAW), 858–873
   compressed gas cylinders and connection, 861–862
   consumables, 862–865
   equipment, 859–862
   starting an arc, 869–870
gate, 505–507, 509–510
   cattle guard, 509
gate post, 509
gauge, 125

gear, 941
   gear ratio, 942
   hydraulic assist, 944
   hydrodynamic-drive, 943
   hydrostatic, 944
   standard, 944
   types, 942
gear-box system, 614
gear-driven rotary head, 658–659
gear pump, 983
gear ratio, 942
gene mapping, 80
general safety, 100–105
general tool safety, 143
generator, 533
genetically modified crop, 80–81
genetic engineering, 80
geographic information system (GIS), 75–76, 336
GFCI. See ground fault circuit interrupter
GFCI receptacle, 556
GHS. See Globally Harmonized System of Classification and Labeling of Chemicals
girder, 415–416, 437, 444–445
GIS. See geographic information system
glass, window, 450
Globally Harmonized System of Classification and Labeling of Chemicals (GHS), 106
global positioning system (GPS), 76–77, 336
globular transfer, 834
gloves, 99–100
GMAW. See gas metal arc welding
GMAW certification practice, 849–851
GMAW equipment, 822–827
GMAW safety, 843–846
GMO. See genetically modified crop
goals, 54–55
goggles, 94
good impression, 33–34
gouge, 150
government agency, 36
GPS. See global positioning system
grade, lumber, 280–282, 333
grade R, oxyfuel hose, 751
grade RM, oxyfuel hose, 751
grade T, oxyfuel hose, 751
grain, 226
grain dryer, 691–692
grain storage, 690–693
   ethylene gas, 694
graphite, 959
grease, 959

green brick, 381
greenhouse, 470–481
    controlled environment agriculture, 470
    covering, 477
    framing, 476–477
    growing environment, 470
    location, 471
    maintenance, 478
    material, 476–478
    orientation, 471
    plant health, 478–481
    repair, 478
    style, 473–476
    water availability, 471–473
    water quality, 471–473
greenhouse covering, 477–478
greenhouse frame, 476–477
greenhouse maintenance, 478
greenhouse material, 476–478
greenhouse repair, 478
greenhouse roof structure, 474
greenhouse style, 473–476
greenhouse technology selection, 473–474
green industry, 650
grinder, 209–213, 215–217
    abrasive, 109, 210, 215–217
    angle, 210–211
    bench, 211–212
    die, 212
    safety, 209–210
    stationary surface, 212–213
grinding metal, 724–725
gripping tool, 152–158
groove-joint pliers, 156–157
groove weld, 785, 808–809
ground, 530–531
ground conductor, 531
grounded, 185
ground fault, 529
ground fault circuit interrupter, 185, 529–530
grounding, 589–590
grounding rod, 546
groundwater, 651
groundwork, 355
grout, 383
growing environment, 470
growing medium, 480
growth cycle, 480
GTAW. *See* gas tungsten arc welding
GTAW development, 858–859
GTAW power source, 859
GTAW practice, 870–873
    autogenous edge joint, 871–872
    chasing puddle, 871
    open root on 1/8″ plate, 872

GTAW safety, 865–866
GTAW welding torch, 860–861
gullet, 197
gully erosion, 653
gutter-connected greenhouse, 475

H

hacksaw, 145
    cutting metal, 715–716
    procedure, 716
    teeth per inch, 715
halogen, 524
hammer, 375–376
hammer drill, 190
hand drawing, 269–270
hand drill, 148
handle, 248
hand plane, 151, 296
hand-powered drill, 147–150
handsaw, 144–147, 285–287
    backsaw, 146, 286
    coping, 145–146, 286–287
    crosscut, 145, 286
    dovetail, 286
    hacksaw, 145
    kerf, 145, 286
    keyhole, 146, 287
    ripsaw, 145, 286
    use, 146–147
hands-on learning, 44
hand tools, 101–102
    classification, 143–144
    cleaning, 170–171
    cutting, 144–152
    extractor, 169–170
    gripping, 152–158
    impact, 158–163
    leverage, 163–164
    safety, 101–102, 143
    torque, 164–169
hand tool care, 172–173
hand tool classification, 143–144
hand tool naming, 143–144
hand tool purchase, 171–172
hand tool storage, 172–173
handy box, 553
hard hat, 95–96
hardware, 246–248, 318–319
    handle, 248
    hinge, 246–248, 318–319
    knob, 319
    latch, 248
    lockset, 248
    pull, 319
hardware cloth, 503
hardwiring, 554–555

hardwood, 226
hazard, grain storage, 693
head, 988
header, 422–424
header brick, 382
headhouse, 477
head protection, 95–96
hearing protection, 96–97
heat, 769
heat-affected zone, 798
heat application, 525
heat buildup, 216
heat conduction, 680
heat convection, 680
heat for soldering, 734–735
heat transfer, 445–446, 679–680
    conduction, 680
    convection, 680
    thermal radiation, 680
heat-treated metal, 737–738
heat-treating, 737
heavy-duty receptacle, 556
heavy-duty single-piece torch, 752
heavy-duty torch, 752
heliarc welding, 858
hemispherical tip, 864
hertz (Hz), 530
hexavalent chromium, 884
hex bolt, 242
hex key, 167–169
higher-carbon steel, 707
high-frequency start, 859, 870
high-speed steel, 188, 198, 205
high-speed steel bit, 208
high-tensile wire, 503
hinge, 246–248, 318–319
    butt, 246
    flush, 246
    piano, 248
    strap, 247
hollow ground blade, 198
home development, 406
homeostasis, 685
homeothermic, 685
hone, 296
hook tooth configuration, 202
hopper window, 450
horizontal fillet weld (2F), 788
horizontal groove weld (2G), 787
horizontal position, 787–788
horsepower, 12, 923
hose, 987–988
hose barb, 752
hose reel system, 659–660
hose traveler system, 659–660
hot, 530–531
hot bar, 546
hot work, 104

housing livestock, 684–690
HSS. *See* high-speed steel
humidistat, 684
humidity, 680
hurricane fastener, 243
hydration, 351
hydraulic, 950
hydraulic actuator, 981
hydraulic assist transmission, 944
hydraulic cement, 388
hydraulic component, 981–989
    actuator, 981
    cylinder, 988–989
    directional control valve, 981
    filter, 982
    fluid conductor, 981, 985–988
    fluid coupler, 981–983
    fluid reservoir, 981, 989
    motor, 989
    power source, 981–982
    pressure relief valve, 982
    pump, 981, 983–985
    strainer, 982
    system control, 985
hydraulic cylinder, 988–989
hydraulic directional control valve, 981
hydraulic equipment, 980
hydraulic filter, 982
hydraulic fluid conductor, 985–988
hydraulic jack, 194–196
hydraulic pipe bender, 723
hydraulic power source, 981–982
hydraulic press, 194–196
hydraulic pressure relief valve, 982
hydraulic strainer, 982
hydraulic system, 978
hydrodynamic-drive transmission, 943
hydrogen fuel cell, 67
hydrogen-induced cracking, 784–785
hydrometer, 963
hydroponic, 481–482
hydroponic system, 481–482
hydrostatic, 944
hydrostatic drive, 950
hydrostatic transmission, 942
hyperthermy, 685
Hz. *See* hertz

I

I-beam, 230
ICF. *See* insulated concrete form
ignition coil, 911–912
ignition system, 911–912, 965–966
impact driver, 190–191
impact rotary head, 659

impact socket, 190–191
impact tool, 158–163
impact wrench, 191
improvement activity, 53–54
incandescent, 523
inclined plane, 14–15
inclusion, 789
incomplete fusion, 796
incomplete joint penetration, 796
indirect fuel injection, 910–911
inductance, 824
induction, 602
industrial machinery mechanic, 619
in-line fuse, 580
in-line splice, 582
insect-borne illness, 110
insulated concrete form, 365
insulation, 426, 436, 455–456
insulation class, 609
insulation requirement, 456–457
insulation stripping, 561–562
insulation type, 457–458
insulator, 508, 520–521
interest, 54–55
interior wall, 421–422, 446
internal combustion engine, 9, 898–914
    early mechanical invention, 9
    history, 898–900
    theory of operation, 899–900
    tractor, 9
Internet source, 272
internship, 52
inventory, 60
inverse-time breaker, 611
inverter, 780
irrigation, 648
    equipment, 654–656
    history, 648–650
    system design, 661–665
    system management, 665–668
    troubleshooting, 668–671
irrigation controller, 663–664
irrigation equipment, 654–656
irrigation history, 648–650
irrigation line, 655–656
    repair, 665–668
    thread tape, 667
irrigation repair technician, career, 668
irrigation symbol, 662
irrigation system, 656, 658–661
    agronomic irrigation, 659–661
    controller, 663–664
    design, 661–665
    design finalizing, 664
    installation safety, 664–665
    management, 665–668

    psi, 662
    sprinkler irrigation, 658–659
    symbol, 662
    troubleshooting, 668–671
    water pressure, 662
irrigation system design, 661–665
irrigation system management, 665–668
irrigation system troubleshooting, 668–671
    ground within irrigation zone soggy, 671
    irrigation controller will not turn off, 668–669
    sprinkler heads not covering intended area, 669–671
    sump pump, 670
irrigation zone, 663
isometric drawing, 264
isometric sketch, 264
ISO metric thread series. *See* metric thread series

J

jack stand, 195
jam nut, 731
jet, 910
jig, 723
jigsaw, 204–205, 289
    blind cut, 289
jigsaw blade, 205
job application, 30–31
job interview, 31–34
job opportunity, 36
job safety, 91
job skill, 23–27
jointer, 208–209, 375
jointer safety, 209
joint type, 785–787
journaling, 60
junction box, 553–554

K

Kelvin, 683
kerf, 145, 286, 886
keyhole, 809
keyhole saw, 146, 287
kickback, 186
kiln, 381
kindling temperature, 744
kinetic energy, 968
knee board, 361–362
knob, 319
knot, 285
knothole, 285

L

laboratory instruction, 45
lab safety, 103
ladder, 110–111
ladder safety, 110–111
lag bolt, 242
laminar flow, 863
laminated dimensional lumber (LDL), 229
laminated lumber, 228–230
laminated structural member, 230
laminated veneer lumber (LVL), 229
land preparation, 330–331
landscape irrigation, 648–671
lanthanated tungsten electrode, 864
lap joint, 787
laser level, 128, 340
latch, 248
latent heat, 681, 696
lateral line, 655
law of conservation of energy, 11
lay of the land, 652
layout, 311–313
LDL. See laminated dimensional lumber
leach field, 630
lead, 234
    poison, 734
    safety, 734
leadership, 47–48
learning activity, 24
leg protection, 100
legs, weld term, 795
level, 125–128, 377
    carpenter's, 125–126
    mason's, 125–126
    plumb, 125, 127, 377
    plumb bob, 127
    plumb line, 127
    spirit, 125
    string, 126
    surveying, 128
    torpedo, 126
leveling, 340–342
lever, 13
leverage tool, 163–164
lift start, 870
light-duty torch, 752
lighting, 523–525, 573–574
    halogen, 524
    incandescent, 523
Li-ion battery. See lithium-ion battery
llmlted-slip differential, 946
limit switch, 577
line holder, 378
lineman splice, 584
lineman's pliers, 156

liquefied petroleum gas (LPG), 746
liquid crystal display (LCD), 802
liquid layer, 630
liquid lubrication, 959
lithium-ion battery, 182
live. See hot
live load, 408
livestock, 494
livestock body temperature, 685–686
load, 407, 521, 523–526
    building, 407
    heat application, 525
    lighting, 523–525
    motion application, 525–526
load-bearing member, 407
load-bearing wall, 421–422
local adaptation, 257–258
location, 471, 484–485
locking pliers, 157–158
lockset, 248
lodging, 216–217
long-term goal, 54
loom, 584
low-hydrogen electrode, 784
LPG. See liquefied petroleum gas
LPG cylinder, 749
lubrication system, 912, 958–961
    bearing, 959–961
    clearance, 958
    graphite, 959
    grease, 959
    oil, 959
    seal, 961
    viscosity, 958
lubricator, 992
lumber defect, 284–285
luster, 231
LVL. See laminated veneer lumber

M

machine, 12–15
machine bolt, 731
machine guard, 185
machine mixing, 389
machinery adhesive, 732
machinery cost, 971–972
machinery management, 966–972
    cost, 971–972
    energy type, 968–969
    performance, 968–969
    power, 968–969
    preventive maintenance, 966
    tractors, 969–970
    troubleshooting, 968

machinery system maintenance inspection, 957–966
    air intake, 957–958
    cooling, 961–963
    electrical, 964–966
    exhaust, 957–958
    fuel, 963–964
    lubrication, 958–961
machine screw, 241, 730
machine welding, 820
machinist's vise, 152
magnetic clamp, 155–156
main line, 655
malleability, 231, 721
malleable, 721
mallet, 160
mandrel, 723
manifold system, 756
manometer, 684
manual, 956
manual shearing tool, 718–719
manure storage exhaust systems, 697
manure storage ventilation, 697
MAPP gas, 745
margin trowel, 375
marking metal, 712–713
marking tools, 131–134
    colored marker, 132–133
    pencil, 131–132
    punch, 133–134
    scratch awl, 133
    scribe, 133
    soapstone, 132
masonry building products, 379–381
    brick manufacture, 380–381
masonry equipment. See masonry tools
masonry nail, 239
masonry safety, 378–379
    PPE, masonry, 378–379
masonry tools, 374–378
masonry unit, ears, 395
masonry unit course height, 391–392
masonry unit laying, 390–399
masonry unit laying operation, 390–393
masonry unit layout on the footing, 393
masonry unit layout on the slab, 393
mason's level, 125–126
mason's line, 378
mason's line holder, 378
mason's rule, 376
material, 258, 353–357
    concrete equipment, 357

concrete tool, 357
    estimation, 355–357, 499–500
    site preparation, 353–355
    vapor barrier, 357
material cost, 260–261
material estimate, 355–357, 499–500
materials list, 308, 310, 500
material standard, 273
mathematics, 10–11
McCormick, Cyrus, 7
measurement taping, 336–337
measuring system, 120–121
measuring tool care, 135
measuring tools, 121–131, 376
    accuracy, 121
    caliper, 123–124
    care, 135
    folding rule, 376
    gauge, 125
    level, 125–128
    mason's rule, 376
    micrometers, 124
    precision, 121
    purchasing, 134–135
    ruler, 121
    square, 128–130
    tape measure, 122
    tolerance, 121
    torque wrench, 130–131
measuring tool storage, 135
mechanical advantage, 12–13
    effort, 12
    force, 12
    inclined plane, 14–15
    lever, 13
    machine, 12–15
    pulley, 13–14
    resistance, 12
    screw, 15
    simple machines, 12–15
    wedge, 15
    wheel and axle, 13
mechanical energy, 969
mechanical reaper, 7
    combine, 7
mechanical refrigeration, 694
mechanical system troubleshooting, 968
mechanical transmission, 942
mechanical ventilation systems, 698
    animal housing, 687
mechanic's vise, 152
medium-duty torch, 752
metal, 230–234
    sawing, 715–717
metal cored electrode, 836

metal inert gas (MIG) welding, 820
metal project planning, 709–712
metal roofing, 440–443
metal siding, 438
metal transfer method, 833–835
metalworking, 706–738
    careers, 746
metalworking hammer, 159–160
methane digesters. *See* anaerobic digesters
metric thread series, 730
micrometers, 124
MIG welder's pliers, 829
mild steel, 232, 768
mill, 725
milling, 282
milling machine, 725–726
    CNC, 726
miter box, 146
miter gauge, 290
miter saw, 200, 289–290
mixing concrete, 356–357
modern braced framing, 411–412
    subfloor, 411
moisture, 479
monomer, 236
morse taper, 191–192
mortar, 235, 385, 388–389
mortar board, 376–377
mortar box, 378
    mixing, 389
mortar joint, 383–384
mortar material, 388
mortar mixing, 389
mortar placement technique, 392
mortar property, 388
mortar spreading, 391
mortar stand, 376–377
mortar type, 388
    selection, 388–389
motion application, 525–526
motor, 989
motor control, 612–613
motor housing, 603
motor nameplate, 609–610
    insulation class, 609
    slip, 609
multiflame heating tip, 755–756, 770–771
    manifold system, 756
    starving, 771
    use, 770–771
multimeter, 536–537
multiple-strand insulated conductor, 549
multiple structures, 474–476
multipositional switch, 578

N

N/C switch. *See* normally closed switch
N/O switch. *See* normally open switch
nail, 238–239, 314
    box, 238–239
    types, 238–239
nail gun, 193–194
National Center for Appropriate Technology (NCAT), 70
National Electrical Code® (NEC), 257, 544
National FFA Organization®, 24, 27, 44, 45–48
    award, 47
    career development event, 47
    emblem, 46
    leadership, 47–48
    level, 46
    local program, 27
    motto, 46–47
National Fire Protection Association®, 257
National Institute for Occupational Safety and Health (NIOSH), 88–89, 92
NEC. *See* National Electric Code®
needle-nose pliers, 157
nesting, 890
networking, 35
neutral, 530–531
neutral bar, 547
neutral conductor, 531
neutral flame, 760–762
NFPA. *See* National Fire Protection Association®
nibbler safety, 720
NiCad battery. *See* nickel-cadmium battery
nickel-cadmium battery, 182
nickel-metal-hydride battery, 182
NiMH battery, 182. *See* nickel-metal-hydride battery
NIOSH. *See* National Institute for Occupational Safety and Health
NMC, 551. *See* nonmetallic sheathed cable
nominal dimension, 282
nonconsumable electrode, 863
nondestructive weld testing, 799
nonferrous metal, 233–234, 707–709
    aluminum, 234, 709
    brass, 233
    bronze, 233
    copper, 233, 709
    lead, 234

nonmetallic sheathed cable (NMC), 551
nonstress graded, 282
nonthreaded fastener, 731–732
   blind rivet, 732
   galvanized, 731
normally closed switch, 577
normally open switch, 577
nut, 241–242
nut driver, 168
nutrients, 479–480
nylon, 236

O

Occupational Outlook Handbook (OOH), 23–24
Occupational Safety and Health Act, 91–92
Occupational Safety and Health Administration (OSHA), 34, 91–92, 219, 535, 844, 914
ohm, 522
Ohm's law, 522–523
oil, 959
on/off switch, 577
OOH. *See Occupational Outlook Handbook*
open differential, 946
open-end wrench, 164–165
opening, 422
open root weld, 840
order of operation, 712
orientation, 471
oriented strand board (OSB), 230
orthographic drawing, 268
oscillating multi-tool, 206
OSHA. *See* Occupational Safety and Health Administration
outdoor equipment, machinery, 110–112
   ATV, 111
   ladder safety, 110–111
   scaffolding safety, 110–111
   tractor, 112
out-of-position welding, 787
output shaft, 949
overall rise, 439
overburden, 330
overburden removal, 331
overhang, 443
overhead, 788
overhead cam (OHC), 909
overhead position (4G/4F), 788
overhead valve (OHV), 907, 909
overlap, 797
overload protection, 612

oxidize, 233
oxidizer, 107
oxidizing flame, 761, 762
oxyfuel cutting, 744–745
oxyfuel equipment, 746–756, 758–759
   hose, 751–752
   regulator, 749–751
   safety device, 756
   setting up, 759
   torch, 752–756
oxyfuel hose, 751–752
oxyfuel torch, use, 758–759
   lighting, 760
oxyfuel torch safety, 756–759
oxyfuel welding, 744–745, 766–768
   kindling temperature, 744
   mild steel, 768
   practicing, 767
oxygen, 745
oxygen cylinder, 747

P

paddock, 499
paint, 318
painting area, 698
painting preparation, 317–318
parallel circuit, 548
parallel clamp, 154
parameter, 790
Pascal's law, 979
pattern bond, 381–382, 385
patterns, masonry, 381–385
PE. *See* polyethylene
pencil, 131–132
perimeter fence, 495
perishable storage, 693–694
   ethylene gas, 694
personal interest inventory, 54
personal protective equipment (PPE), 91, 93–100, 356, 378–379, 537–538, 665, 727
   breathing, 97–98
   care, 100
   clothing, 98–100
   eye, 93–95
   face, 93–95
   head, 95–96
   hearing, 96–97
   size, 100
pest, 481
pest control, 481
PEX, 655. *See* cross-linked polyethylene
PEX pipe, 656
   repair, 666–667
phase, 531–532

Phillips-head screwdriver, 168
photosynthesis, 477
photovoltaic cell, 534, 572
photovoltaic module, 572
photovoltaic system, 68
physical sciences, 11–15
physics, 11–15
piano hinge, 248
pier footing, 364
pigtail, 559
pilot arc, 882
pinch effect, 824
pin punch, 162–163
pin vise, 148
pipe, 987
pipe bending, 723–724
pipe fitting, 640–642, 656–658
pipe roller, 724
pipe valve, 640–642
pipe vise, 153
pipe welding position (5F), 788–789
pipe wrench, 166
piston, 899, 907–908, 920, 988
piston pump, 984–985
placement, SAE, 52–53
   enterprise, 52
   internship, 52
plain-sawn lumber, 227
plan, 57, 269, 272–273, 308–309
   adapting, 273
   budget, 57
   business, 57
   extension publication, 272
   Internet source, 272
   project book, 272
plane, 150–151
planer, 208–209
planing, 282
plank and beam framing, 414
planning, 311
planning process, 256–258
plan source, 272–273
plant disease, 481
plant health, 478–481
   air quality, 479
   air temperature, 479
   disease control, 481
   moisture, 479
   nutrients, 479–480
   pest control, 481
plasma, 880
plasma cutting, 880–892
   CNC, 887–892
   CNC practice, 892
   compressed gas source, 882
   equipment, 881–882
   process, 880–882
   safety, 883–885

plasma torch use, 885–887
plastic pipe, 634–636
plastic, types, 634–636
plate, 243
platform framing. See western framing
pliers, 156–158
plug weld, 787
plumb, 125, 127, 377, 420
plumb bob, 127
plumb line, 127
plumber's snake, 631
plumbing, 631
plumbing code, 626
plumbing pipe, 633–640
    copper, 636–639
    plastic, 634–636
    steel, 639–640
plumbing system, 626–630
    DWV, 628–629
    septic, 629
    water supply, 627–628
plumbing tools, 630–632
plumbing trap, 629
plywood, 229
pneumatic chisel. See pneumatic hammer
pneumatic components, 991–993
pneumatic equipment, 980
pneumatic hammer, 717–718
pneumatic power, 989–990
pneumatic power tools, 183–184
pneumatic system, 978
    components, 991–993
    maintenance, 993
polarity, 530, 791
pole frame construction, components, 409
pollution, 72–73
polyethylene (PE), 235, 477, 655
polyethylene pipe, 656, 667
polymer, 236
polymerization, 235
polystyrene, 235
polyvinyl chloride (PVC), 236, 635
polyvinyl chloride pipe, 655
    repair, 666
pond system, 483
pop-up sprinkler heads, 658
porosity, 796
port, 988
portable belt sander, 213–214
portable drill, 189–190
portfolio, 27–28
portland cement, 350–351
position welding, 787
positive ground system, 570
post, 409, 415–416, 500
    types, 500–502

post-and-rail fence, 496
post driver, 512
post hole digger, 511
postflow, 823
potential energy, 968
poultry wire, 503
power, 12, 923
powered auger, 511
power saw, 287–289
    cold saw, 717
    cutting metal, 716–717
power shear, 719–720
    safety, 720
power system, 940–950
    belt, 947–949
    chain, 947–949
    hydrostatic drive, 950
    power take-off, 949–950
    types, 947–950
power take-off, 940, 949–950
    output shaft, 949
power take-off (PTO), safety, 112, 949
power tool care, 219
power tool classifications, 180–184
Power Tool Institute, 219
power tools, 101–102
    care, 219
    cutting, 196–209
    drilling, 186–194
    driving, 186–194
    grinder, 209–213, 215–217
    hydraulic jack, 194–196
    hydraulic press, 194–196
    purchase, 217–219
    safety, 101–102, 184–186
    sander, 209–210, 213–217
    selection, 217–219
    storage, 219
power tool use, 185
power train, 940–946
    clutch, 940–941
    driveline, 945
    final drive, 945–946
    transmission, 941–945
power transfer system, 978
power trowel, 360–361
PPE. See personal protective equipment
precipitation rate, 662
precision, 121
precision agriculture, 73–81
prefabricated agricultural structures, 485–486
preflow, 823
prehung door, 451
preparation, 353–357

press fitting, 730
pressure, 979
pressure energy, 969
pressure per square inch (psi), 662
pressure regulator, 749, 992
pressure-treated lumber, 227–228
preventive maintenance, 966
prick punch, 133–134, 162
primer, 635
privacy fence, 496–497
production agriculture, 4
professional organization, 36
professional planner, 272
proficiency award, 47
profile leveling, 342
project assembly, 313–315
project book, 272
project finishing, 316–320
    finish types, 317
    hardware, 318–319
    painting, 317–318
    sealer, 318–319
    staining, 317–318
project part cutting, 313
project planning, 308–310, 389–390
    building code, 390
    construction procedure, 308, 310
    delivery schedule, 390
    drawing, 308–309
    finish, 308
    materials list, 308, 310
    plan, 308–309
    sketch, 308
    subassembly, 308, 310
    technical drawing, 308
propagation, 480
propane, 745–746
    liquefied petroleum gas, 746
propane cylinder, 749
proper tool selection and use, 144
propylene, 745
protective clothing, 98–100
pry bar, 163–164
psi. See pressure per square inch
psychrometrics, 680–682
    psychrometric chart, 681–682
PTO. See power take-off
pull, 319
pulley, 13–14, 616, 947
    drive ratio, 616
pulsed arc, 824
pulsed arc machine, 824
pulsed arc mode, 835
pump, 978, 981, 983–985
    gear, 983
    piston, 984–985
    vane, 984

punch, 133–134, 162–163
　automatic, 134
　center, 133–134, 162
　chamfered, 162
　pin, 162–163
　prick, 133–134, 162
　tapered, 163
punching, 718
purlin, 437, 444–445
push angle, 793
push drill, 148
push-pull welding gun, 825
push stick, 291
PVC, 655. *See also* polyvinyl chloride
PVC pipe, 655
　repair, 666

Q

quarter-sawn lumber, 226–227
quenching, 737
questions, 34
quick-disconnect coupling, 988
quote, 260–261

R

raceway, 545
racking, 393
racking back the lead, 393
radial arm saw, 199–200, 291–292
radiation, 455
radiographic weld inspection, 799
rafter, 426–427
rail fence installation, 509
rainwater harvesting, 654
rake angle, 202
raked joint, 384
random orbital sander, 214–215
rasp, 151–152, 295–296
ratchet handle, 165
ratchet tensioner, 513
reactivity, 106
readiness test, 360
rebar, 355
receptacle, 555
　duplex, 555–556
　GFCI, 556
　heavy-duty, 556
receptacle tester, 536
reciprocating-piston, 991
reciprocating saw, 203–204, 288
　bimetal blade, 204
recirculating system, 484
reclaimed water, 651
rectifier, 571, 780

references, 28–30
refrigeration, 694–695
　mechanical, 694
registration mark, 312
regulator, 749–751
　pressure, 749
　torch wrench, 751
　working pressure, 749
relative humidity, 453, 680
relay, 578–579
　SPDT, 578
relief cut, 283
rendering, 264
repulsion-start motor, 606–607
required education, 26–27
required lighting, 587–588
research, 50–52
research SAE, 50–52
reservoir, 650
resistance, 12, 522
resource, 55
respirator, 97–98
résumé, 28–29
retention pond, 653
reverse flow check valve, 756
rheostat, 578
rigid conduit, 552
rigid construction, 409–410
rigid panel, 477–478
rill erosion, 653
rip fence, 290
ripping, 197
ripsaw, 145, 286
　kerf, 286
rivet, 244
robotics, 81
rod, 988
rod oven, 785
rolling resistance, 969
rollover protection systems (ROPS), 89
Romex® conductor, 550–551, 561
roof, 409, 439–445
　types, 426
　underlayment, 439–440
roof framing, 426–427
roofing nail, 239
roof material, 440–443
roof overhang, 443–444
roof slope, 439
roof window, 450
root, weld term, 795
root bead, 806
root face, 839
ROPS. *See* rollover protection systems
rosebuds. *See* multiflame heating tip
rosin-based flux, 734

rotary saw, 299
rotary screw, 992
rotary vane, 991
rotor, 603
router, 206–208, 298–299
　router bit, 207–208
router table, 208
routing wire, 592–593
rowlock brick, 382
rowlock stretcher brick. *See* shiner brick
RPM, 941–942
ruler, 121
runoff, 652

S

sacrificial part, 891
SAE. *See* supervised agricultural experience
SAE plan development, 54–58
SAE, plasma cutting, 883
SAE program components, 53–54
SAE recordkeeping, 58–60
SAE selection, 56–57
SAE types, 49–53
safe practice, 537–538
safety, 86–114, 256–257
　ASME®, 257
　chemicals, 106–108
　drills, 192–193
　driving tools, 192–193
　electricity, 108–109, 535–538
　electric motors, 620
　electrocution, 108
　engineer, 257
　engines, 913–914
　first aid, 113
　food, 71
　general, 100–105
　grinders, 209–210
　hand tools, 101–102, 143
　hazards, 88–89
　insect-borne illnesses, 110
　in the lab, 103
　ladders, 110–111
　lead, 734
　masonry, 378–379
　NEC®, 257
　NFPA®, 257
　on the job, 91
　outdoor equipment and machinery, 110–113
　pneumatics, 993
　power tools, 101–102, 184–186
　PPE, 91, 93–100, 356, 378–379, 537–538, 665, 727
　regulation, 90–93

sanders, 209–210
scaffolding, 110–111
shearing metal, 720
signal words, 103–104
soldering, 737
   tools, 143
weather hazards, 109–110
welding, 843–846, 865–866
zoning code, 257
zoning law, 257
safety data sheet (SDS), 106, 914
safety device, 756
   flashback, 756
   flashback arrestor, 756
   reverse flow check valve, 756
safety glasses, 94
safety helmet. *See* hard hat
safety regulation, 90–93
safety roller, 361
sailor brick, 382
sander, 209–210, 213–217, 299–301
   abrasive, 109, 213, 215–217
   detail, 215
   disk, 213
   portable belt, 213–214
   random orbital, 214–215
   safety, 209–210
   sheet, 214–215
   stationary, 215
sash, 448
saw blade design, 198
sawing metal, 715–717
saw use, 146–147
scaffolding, 111
   safety, 110–111
scale drawing, 265–266
scavenging, 904
schedule, 956
schedule number, 987
school employment office, 36
scientific method, 50
scope, 338
Scotchlok® connector, 582
SCR. *See* selective catalytic reduction
scraper, 170–171
scratch awl, 133
scratch start, 869
screed, 359
screw, 15, 240–241, 314–315
   deck, 240
   drywall, 241
   machine, 241
   self-drilling, 241
   self-piercing, 241
   sheet metal, 240
   wood, 240
screwdriver, 167–168

nut driver, 168
   tips, 167–168
screw terminal, 546, 581
scribe, 130, 133
scroll saw, 205–206, 293–294
   blade, 206
scum, 630
SDS. *See* safety data sheet
SDS bit. *See* special direct system bit
seal, 961, 988
sealer, 318–319
seasonal energy efficiency ratio
      rating, 452
sectional drawing, 268
securing work, 713–715
security fence, 496–497
SEER rating. *See* seasonal energy
      efficiency ratio rating
selective catalytic reduction
      (SCR), 72
self-drilling screw, 241
self-piercing screw, 241
self-shielded FCAW (FCAW-S), 836
semiautomatic welding, 820
semisolid lubrication, 959
sensible cooling, 696
sensible heating, 681, 696
sensor, 578
sensor cable, 692
septic system, 629
series circuit, 548
service disconnect, 546
service entrance, 544–547
service entrance
      component, 545–546
service meter, 545
service record, 956
servo motor, 608
setting post, 507–508
sewage, 628
sewer gas, 628
shake, 285
shaping metal, 721–727
shaping wood, 295–301
shaver, 297
shearing metal, 718–720
sheathing, 414, 425, 436–437
sheet erosion, 653
sheet metal screw, 240
sheet sander, 214–215
shield, 185
shielded metal arc welding,
      778, 820, 859
   cast iron, 812–813
   distortion control, 811–812
   PPE, 801–802
   SAE, 779
   welding certification, 813

shielding gas, 831–832, 863
shielding gas cylinder,
      changing, 832
shielding gas nozzle, 827, 860
shielding gas solenoid, 827
shiner brick, 382
shingle, 442
shop safety, 103
short-arc, 824
short-circuit metal transfer, 833–834
shorted, 593
short-term goal, 54
shrink tube, 583
siding, 437–438
signal word, 103–104
sill, 415
sill plate. *See* sill
simple machines, 12–15
   inclined plane, 14–15
   lever, 13
   machine, 12–15
   pulley, 13–14
   screw, 15
   wedge, 15
   wheel and axle, 13
single groove joint, 839
single hung window, 449
single-phase power, 531
single-pole, double-throw relay. *See*
      SPDT relay
single-pole, double-throw
      switch, 527
single-pole, single-throw switch, 557
single-strand insulated
      conductor, 549
single structure, 474–476
site design, 332–334
site layout, 332–334
site preparation, 353–355
site-specific management, 75
site staking, 333–334
sketch, 308
sketching, 263–265
   computer-aided drafting and
         design software, 710
   drawing principle, 264–265
   isometric, 264
   metalworking, 710
skill documentation, 27–28
skilled trades, 22
SkillsUSA, 24
skip tooth configuration, 202
skylight window, 450
slab-on-grade foundation, 363
slag, 783
sledge hammer, 160
sliding T-bevel, 130
sliding window, 449

slip, 609
slip-joint pliers, 156
slitting chisel, 717
slope, 652
slotted-tip screwdriver, 167–168
sludge, 629
slump, 352–353
slump cone, 352
slump test, 352–353
small engines, 920–934
S.M.A.R.T. goal standard, 54–55
smartphone application, 74–75
SMAW. See shielded metal arc welding
SMAW equipment, 779–782
SMAW practice welding, 802–811
    fillet welding in flat and horizontal positions, 806–808
    fillet weld practice, 807–808
    groove weld practice, 809–810
    striking arcs, 802–803
    vertical fillet and butt weld, 811
    welding beads, 803–806
    welding beads on flat plate, 804
    welding practice pads, 804–806
SMAW surface preparation, 789–790
SMAW travel speed, 795
SMAW welding environment, 800–803
SMAW welding machine, 779–781
    duty cycle, 781
    inverter, 780
    rectifier, 780
    transformer, 780
SMAW welding parameters, 790–795
    arc length, 793
    current, 791–792
    electrode angle, 793
    electrode motion, 794–795
    electrode selection, 792–793
    polarity, 791–792
    travel speed, 795
smoothing wood, 295–301
    chisel, 297–298
    file, 295–296
    hand plane, 296
    rasp, 295–296
    rotary saw, 299
    router, 298–299
    sander, 299–301
    shaver, 297
smooth wire, 502
soapstone, 132
socket, 165
socket wrench, 165
    ratchet handle, 165

soffit, 444
soft-faced hammer, 160
soft skills, 37–38
softwood, 226
soil analysis, 326–328
soil bearing capacity, 354–355
soil conservation technician, 498
soil quality, 329
soil structure, 653
soil testing, 329
soil texture, 653
solar energy, 534, 572–573
    array, 572
    photovoltaic cell, 534, 572
    photovoltaic module, 572
solder, 733
soldered joint, completing, 736–737
soldering, 233, 582–583, 637–638, 733–737
    alloy, 733–734
    copper pipe, 637–638
    flux, 637, 734
    heat, 734–735
    joint completion, 736–737
    joint preparation, 735
    safety, 737
    tinning, 585, 735
soldering alloy, 733–734
soldering flux, 734
soldering tools, 735
soldier brick, 382
solenoid, 574–575
solenoid valve, 664
solid lubrication, 959
solvent cement, 635
span, 230, 409–410
spatter, 793
SPDT switch. See single pull, double throw switch
SPDT relay, 578
special direct system bit, 190
specialized agricultural structure, 470
specific humidity, 680
speed square, 129
spirit level, 125
splice, 559
splicing, 509
split, 285
split-phase motor, 605–606
split-phase power, 531
spool gun, 825
spray pattern, 658
spray transfer, 834
spray transfer mode, 834–835
spreader, 358
spread footing, 363
springback, 724

spring clamp, 155
spring steel, 232
sprinkler heads, 658–659
    sprinkler irrigation, 658–659
sprinkler system, 472
sprocket, 949
SPST switch. See single-pole, single-throw switch
square, 128–130, 420
stacked dimes effect, welding, 795
stain, 318
stainability, 280
staining preparation, 317–318
stainless steel, 232
standards, 91–92, 946
standard transmission, 944
standoff, 885
standoff guide, 885–886
star-drive. See Torx®
starting system, 965–966
starving, 771
static electricity, 520
stationary power tool, 181
stationary sander, 215
stationary saw, 289–294
stationary surface grinder, 212–213
stator, 603
steam-powered tractor, 8
steel, 231–232, 639–640, 706–707
steel moldboard plow, 6
steel pipe, 639–640
steel post, 501
steel square, 377
step-down transformer, 534
stepper motor, 608
step-up transformer, 534
stickout, 823
stick welding, 778
stinger, 782
story pole. See corner pole
straightedge, 359
strap hinge, 247
stress graded, 282
stretcher block, 382, 395
stringer, 418
stringer bead, 794
string level, 126
strip gauge, 561
stroke, 900
structural adhesive, 732
structural clay product, 380
structure type, 409–411
stud, 410
stud wall construction, 410–411
style, 262
subassembly, 308, 310
subfloor, 411, 419
subsistence farming, 4

sump pump, 670
supervised agricultural experience
 (SAE), 44, 48–54, 91, 406
 award, 47
 plan development, 54–58
 program component, 53–54
 recordkeeping, 58–60
 types, 49–53
supplemental heating, animal
 housing, 688–689
surface crack detection, 799
surface irrigation, 659, 661
survey, 497–498
surveying, 236, 268, 334–342
 application, 342–343
 distance measurement, 335–337
 equipment, 337–340
 leveling, 340–342
 surveyor, 332
surveying equipment, 337–340
surveying levels, 128
surveyor, 332
sustainable energy, 66–69
swale, 653
swarf, 186
swell, 330
switch, 526–527, 557–559, 576–580
 double-pole, single-throw,
  527, 558–559
 fuse, 579–580
 multipositional, 578
 on/off, 577
 relay, 578–579
 single-pole, single-throw, 527,
  557
system control, 985, 993

T

table saw, 199, 290–291
 miter gauge, 290
 push stick, 291
 rip fence, 290
tack weld, 807
tamper, 358
tap, 149–150, 559, 728
tape measure, 122
tapered punch, 163
taping, 336
tapping a hole in metal, 728–729
tap splice, 582
target rod, 338
TDC, 920–922
technical drawing, 265–271, 308
 types, 267–269
technology, 66
teeth per inch (TPI), 715

telematics, 77
temperature, 440
temperature probe, 683
tempering, 737
template, 29
temporary post, 502
tensile strength, 262, 357–359
terminal connection, 560
terrarium, 481
testing concrete, 359–360
T-foundation, 362–363
theodolite, 128
thermal energy, 968
thermal equilibrium, 679
thermal pane window, 450–451
thermal radiation, 680
thermocouple probe, 683
thermodynamics, 679–684
thermometer, 680, 683
thermoplastic, 235–236
thermoset plastic, 237
thermostat, 684
thickness gauge, 125
thoriated tungsten electrode, 863
thread, 240
threaded fastener, 730–731
threading metal part, 728–729
thread locking compound, 732
thread tape, 667
three-phase motor, 607–608
three-phase power, 531
tie-in, 794
tile, 380
timetable, 262
tinning, 585, 736
T-joint, 786
toes, weld term, 795
toggle switch, 577
tolerance, 121
tool evolution, 142–143
tool path, 889
tool safety, 143
tool selection, 144
tool steel, 232
tool use, 144
top dead center (TDC), 902, 903
torch, 752–756
torch body, 753
torch shut-down procedure,
 762–763
torch wrench, 751
torpedo level, 126
torque, 130, 923
 control, 169
 tool, 164–169
 torque wrench, 130–131
 Torx®, 168
torque control, 169

torque tool, 164–169
torque wrench, 130–131
Torx®, 168
total energy, 969
total project cost, 259
tough, 721
toughness, 722
TP. See turning point
TPI. See teeth per inch
t-post, 501
tractor, 9, 112
 accidents, 949
 ballasting, 969–970
 power take-off shaft, 112
 rolling resistance, 969
 safety, 949
 weight, 969–970
tractor accidents, 949
trade union, 36
trailer brake, 588
trailer wiring connector, 590–591
training, 26
transform welder, 780
transmission, 941–945
transpiration, 650
transplanting, 480
trap, 629
travel angle, 841–842
traveling gun systems. See hose
 traveler system
travel speed, 795, 886
trickle irrigation. See drip irrigation
trim, 443, 452
trowel, 360–361, 374–375
trowel loading, 391
truncated cone tip, 864
truss, 409, 426–427, 660
try square, 129
tube bending, 723–724
tubing, 987
tuck pointer, 375
tungsten electrode, 863
 preparation, 864–865
tungsten extension, 867
tungsten inert gas (TIG), 858
turf rotor, 659
turning point, 340
twist, 285
twisted shank nail, 239
two-stroke cycle design, 904–906

U

UAV. See unmanned aerial vehicle
UF. See underground feeder
UL. See Underwriters Laboratories
ultrasonic testing, 799–800

ultrasonic weld inspection, 799
unacceptable weld profile, 797
UNC thread series. *See* Unified National coarse thread series
undercut, 797
undercutting, 795
underground feeder, 551
underlayment, 439–440
Underwriters Laboratories, 92
UNF thread series. *See* Unified National fine thread series
Unified National coarse thread series, 730
Unified National fine thread series, 730
unified thread standard, 730
United States Department of Agriculture (USDA), 272
unit of measurement, 120–121
universal motor, 608
unmanned aerial vehicle (UAV), 77–79
uphill welding, 788
USDA. See United States Department of Agriculture
US Department of Labor, 25
US Department of Labor, apprenticeships, 25
utility lumber, 282

V

valence electron, 520
valve, 642
valve train and timing, 908–910
vane pump, 984
vapor barrier, 357
variable cost, 971–972
variable frequency drive, 613
variable-rate technology (VRT), 73–74
veneer, 229
ventilation, 97, 458–462, 473, 686–688
    axial fans, 687, 688
    centrifugal fans, 687, 688
    fan design, 687–688
    fan operation, 687–688
    manure storage, 697
    mechanical, 687
ventilation system, 697–698
venture, 910
vernier caliper, 123
vertical band saw, 292–293
vertical-down. *See* downhill welding
vertical position, 788
vertical-up. *See* uphill welding

vertical welds (3G/3F), 788
vinyl post, 501–502
vinyl siding, 438
viscosity, 958
vise, 152–154
Vise-grip® pliers. *See* locking pliers
vise maintenance, 153–154
volt, 522
voltage, 522, 823–824
voltage detector, 536
voltage drop, 535
volume, 355, 681
VRT. *See* variable-rate technology

W

wall, 409, 445–448
wall anchor, 243
wall framing, 419–425
wall material, 446–448
washer, 242
waste side, 313
water availability, 471–473
water bed, 472
water calculation, 662–663
water cannons. *See* hose traveler system
water conservation, 633, 652–654
    detention pond, 653
    erosion, 653
    rainwater harvesting, 654
    retention pond, 653
    runoff, 652
    xeriscaping, 654
water cycle, 650
water level, 135
water level construction, 135
water pollution, 73
water pressure, 662
water quality, 471–473
water separator, 992
water source, 650–651
water supply system, 627–628
water system, 472
watt, 522
Watt, James, 8
weather hazard, 109–110
weave bead, 794
weave pattern, 794
wedge, 15
weld anatomy, 795–800
welded wire fence, 503
welding area, 698
welding, careers, 822
welding, GMAW, 837–843
welding, oxyfuel, 766–768
welding beads on flat plate, 847–848
welding consumable, 827–832

welding current, 778, 791–792
welding defect, 796–798
welding discontinuity, 796–798
welding flat, 787
welding gun, 820, 826–827
welding helmet, 94–95
welding lead, 781–782, 825–826
    electrode holder, 782
    stinger, 782
welding positions, 787–789
welding practice, 846–851
welding practice pads, 847–848
welding polarity, 791–792
welding position, 787–789
welding procedure specification (WPS), 813, 849–850
welding shield, 94–95
welding tip, 753–754
welding with GTAW, 866–870
weld quality, 795–800
weld specification, 787
weld terms, 795
well, 627
western framing, 413
wet-bulb temperature, 681
wheel, 13
wheel and axle, 13
wheel dresser, 217
Whitney, Eli, 6–7
wind energy, 68–69, 533–534
window component, 448–449
window frame, 448–449
window glass, 450
window installation, 461
windows, 448–451
window sash, 448
window types, 449–450
wing nut, 731
wire, 502–503
wire color, 549–550
wire connection, 559–563, 580–586, 590–593
    lineman splice, 584
    protection, 583–584
    routing, 592–593
    soldering, 582–583
    splice soldering, 584–586
    trailer connector, 590–591
    type, 580–582
wire connection protection, 583–584
wire connection type, 580–582
wire fabric, 503–504
wire feeder, 824–825
wire feed speed, 823
wire fence installation, 508–509
wire fencing material, 502–504
wireless sensor network, 79
wire net, 504

wire nut, 559–560
wire size, 549
wire speed, 823
wire splice soldering, 584–586
wire stripper, 561
wire-to-wire connection, 559–560
    wire nut, 559–560
wire welding, 820
wiring, agricultural trailer, 586–593
wiring harnesses, 584
wood, 226–230
    characteristic, 280
    cutting, 285–294
    dimensional lumber, 226–228
    drilling, 301
    engineered lumber, 228–230
    grain, 226
    hardwood, 226
    selection, 280–285, 310–311
    shaping, 295–301
    smoothing, 295–301
    softwood, 226
    types, 280
wood characteristic, 280
wood chisel, 150
wood joint, 313–314
wood post, 500–501
wood roofing, 443
wood screw, 240
wood selection, 280–285, 310–311
    dimension, 282–284
    grade, 280–282
    lumber defect, 284–285
    stainability, 280
wood siding, 438
wood type, 280
woodworking, 284
woodworking bench vise, 153
work, 11, 922
work angle, 841–842
working pressure, 749
woven wire, 504
WPS. *See* welding procedure
        specification
wrap, 437
wrecking bar. *See* crowbar
wrench, 164–166
wythe, 383

X

xeriscaping, 654
x-post fence, 496

Y

yield map, 75
yield monitoring, 75

Z

zoning code, 256–257
zoning law, 257